Using LANDCADD

Using LANDCADD

Kent Gordon, Fullerton College

an International Thomson Publishing company I(T)P®

Albany • Bonn • Boston • Cincinnati • Detroit • London • Madrid
Melbourne • Mexico City • New York • Pacific Grove • Paris • San Francisco
Singapore • Tokyo • Toronto • Washington

NOTICE TO THE READER

Publisher does not warrant or guarantee any of the products described herein or perform any independent analysis in connection with any of the product information contained herein. Publisher does not assume, and expressly disclaims, any obligation to obtain and include information other than that provided to it by the manufacturer.

The reader is expressly warned to consider and adopt all safety precautions that might be indicated by the activities herein and to avoid all potential hazards. By following the instructions contained herein, the reader willingly assumes all risks in connection with such instructions.

The publisher makes no representation or warranties of any kind, including but not limited to, the warranties of fitness for particular purpose or merchantability, nor are any such representations implied with respect to the material set forth herein, and the publisher takes no responsibility with respect to such material. The publisher shall not be liable for any special, consequential, or exemplary damages resulting, in whole or part, from the readers' use of, or reliance upon, this material. Autodesk does not guarantee the performance of the software and Autodesk assumes no responsibility or liability for the performance of the software or for errors in this manual.

***AT&T WorldNet[SM] Service Offer Details:** Through an alliance with AT&T WorldNet Service, ITP can now offer you one of the most reliable Internet service providers around. When you purchase selected ITP products you will receive a CD-ROM which provides one free month* of unlimited access to AT&T WorldNet Service and the Internet. At the end of the free month, you may choose to continue the service for a low hourly or monthly fee. And to add even more value, ITP has customized AT&T WorldNet Service to include links specific to selected subjects and courses. This customization grants you immediate access to course-related Internet sites and resources that expedite your search for helpful, interesting and relevant links.

*Telephone access and other charges and taxes may apply. Other terms and conditions apply.

Cover: LANDCADD renderings courtesy of Estrada Land Planning, San Diego, CA. Background image © 1998 PhotoDisc, Inc.

Staff:
Publisher: Michael McDermott
Acquisitions Editor: Sandy Clark
Editorial Assistant: Christopher Leonard
Art & Design Coordinator: Mary Beth Vought
Production Coordinator: Jennifer Gaines

Printed in the United States of America
For more information, contact:

Delmar Publishers/Autodesk Press
3 Columbia Circle, Box 15-015
Albany, New York USA 12212-5015

International Thomson Publishing Europe
Berkshire House 168-173
High Holborn
London, WC1V 7AA
United Kingdom

Thomas Nelson Australia
102 Dodds Street
South Melbourne, Victoria 3205
Australia

Nelson Canada
1120 Birchmont Road
Scarborough, Ontario
Canada, M1K 5G4

International Thomson Publishing Southern Africa
Building 18, Constantia Park
240 Old Pretoria Road
P.O. Box 2459
Halfway House, 1685 South Africa

International Thomson Editores
Campos Eliseos 385, Piso 7
Colonia Polanco
11560 Mexico D. F. Mexico

International Thomson Publishing GmbH
Konigswinterer Strasse 418
53227 Bonn Germany

International Thomson Publishing France
Tour Maine-Montparnasse
33, Avenue du Maine
75755 Paris Cedex 15, France

International Thomson Publishing -Japan
Hirakawacho Kyowa Building, 3F
2-2-1 Hirakawa-cho Chiyoda-ku
Tokyo 102 Japan

International Thomson Publishing Asia
221 Henderson Road
#05-10 Henderson Building
Singapore 0315

1 2 3 4 5 6 7 8 9 10 XXX 03 02 01 00 99 98

Library of Congress Cataloging-in-Publication Data
Gordon, Kent.
Using LANDCADD / Kent Gordon.
p. cm.
Includes index.
ISBN 0–8273–8626–5
1. Landscape design—Data processing—Programmed instruction.
2. Computer-aided design—Programmed instruction. 3. LandCadd
4. AutoCAD (Computer File) I. Title
SB472.45.G67 1998
712'.3—dc21

97–34723
CIP

contents

acknowledgments

The author would like to gratefully acknowledge the assistance and support of many people who helped make this project a reality.

First, many thanks go to the LANDCADD staff who graciously answered my endless questions and showed enormous patience with my frequent phone calls. These include Chad Bergslein, David Williams, Greg Jameson, and Alan Buckingham. Thanks, guys.

Second, great appreciation goes to the architects, designers and principals at Purkiss-Rose/RSI for their guidance and inspiration: Steve Rose, Dominic Oyzon, Jeff Scott, Matt Farrand, Charlie Prograce, Steve Lang and Jim Pickel.

Third, enormous gratitude goes to my LANDCADD students who found bugs, typos and mistakes throughout the development of the projects in this book. This book would not exist if it were not for your contributions, suggestions and (usually) constructive criticisms.

Fourth, I would like to thank my colleagues at Fullerton College for their patience and support during the process of producing this project. I have been fortunate to work with wonderful coworkers who have all contributed to this effort. Special thanks to Geoff Smith, Peggy Smits, Adrian Erb, Dr. Dave Thomas and Tom Lennon.

Finally, special thanks and appreciation extend to my family, whose love, patience and support made this project possible. To Julia, Erin and Mike—thank you with all my heart. To my brother Lance—thanks for your patience with my late-night "tech support" queries.

dedication

This book is dedicated to the memory of Joyce Boyajian, a wonderful LANDCADD student and an enthusiastic supporter of horticulture, learning and life.

introduction

As a vocational educator with more than twenty years of teaching experience, I have discovered certain truths about learning in general and learning software programs in particular. Specifically, I have discovered that:

People learn best by example.

Almost every software program released has a tutorial exercise which accompanies it. This exercise is designed (hopefully) to show you the capabilities of the program and how best to use them. You **cannot** learn a software program without using it to perform the tasks you want to learn. Theoretical learning and rote memorization will not help you learn to run software.

The more realistic the example, the more focused and motivated the learner.

When you have a relevant, realistic example, you can see the software perform the work that you might have performed by hand or with a less capable software program. Not only do you see the power of the new program, you see how it can make your work easier and faster. This tends to improve your focus and heighten your sense of enthusiasm for learning.

Background helps.

Unfocused, irrelevant tutorials too often reflect a "monkey-see, monkey-do" approach that fails to provide explanation and background for the commands and sequences that are used. It is important to provide a context for the work that you are doing in a tutorial. A running commentary is particularly valuable to give an expanded explanation as to **why** a function is used, as well as how to use it. This can assist you in deciding when to apply the function again as you work with the software after the tutorial is completed.

People don't like jargon (especially computer jargon).

Basic explanations of terminology can help you feel that the process of learning computer programs is less threatening and intimidating. If you can understand computer and software terminology, you can usually gain a clearer picture of how the software and hardware work together. More importantly, you can usually get a better handle on how to control them. A side benefit is the ability to cut through the verbal smokescreen that some people use to make themselves sound more knowledgeable than they really are!

People want to learn how to use software efficiently.

It's really not enough to learn how to use a software program. You also need to know how to use it efficiently. LANDCADD and AutoCAD provide so many ways to perform drawing tasks that some guidance may be required to find the most effective way to use them. As with most other software packages, LANDCADD has very important features that you really **must** know, other features that you **should** know, and some optional features that you can learn at your leisure. Many tutorials fail to emphasize the most important portions of a program and exaggerate the importance of others. We will try to steer toward the important features and show how to take advantage of them to work quickly and intelligently.

People start learning software with all different levels of experience.

Some readers will be fully conversant in computers generally and AutoCAD in particular. Others will not have any great familiarity with either computers **or** AutoCAD. This book assumes that you have some AutoCAD experience, but not great mastery of this enormous program. If you lack fundamental AutoCAD skills, you will find an introduction to the basic commands used in architectural drawing. If you are proficient in AutoCAD, you can skip to the LANDCADD exercises right away.

Learning CAD is not a substitute for learning design skills, graphic skills, or fundamental knowledge of drafting concepts.

Learning LANDCADD will not make you a landscape architect or designer, any more than learning how to use a calculator will make you a mathematician. You need fundamentals in design principles, graphic communications, drafting standards, plant usage, grading and drainage, irrigation design, landscape construction materials, etc. LANDCADD does not provide you with any of this vital background and experience.

Learning any type of CAD is very much like learning to use a pencil, triangle, tee-square and scale. It provides you with the tools to communicate your designs and ideas in an electronic (rather than paper) medium. This medium can be used to improve your visualization skills, enhance your design and drafting efficiency, and increase your productivity so that your work will become more profitable in the long run. But without training in the basics of landscape architecture and design, you will simply be able to draw very clear images of uninspired, unimpressive designs.

CAD designers are often asked to set up a CAD system at architectural and landscape architectural offices.

Learning a CAD program almost never happens in isolation. A CAD designer is often called upon to organize a CAD file system, administer a small network and make recommendations for thousands of dollars of equipment purchases. The line between CAD designer and CAD manager is often very thin. If you want to set up a CAD system for a small company, you will find information and references in *Using LANDCADD* to make this job easier.

Using LANDCADD attempts to address all of these issues, albeit at different times and at different places throughout the book. *Using LANDCADD* attempts to provide a set of realistic drawing exercises to emphasize the important parts of LANDCADD and some suggested ways to get the most out of this leading landscape design software package. Particular attention is also paid to the establishment of a CAD drawing system, including file management and backup, office procedures and network use. Beginning CAD users will be able to learn the fundamentals of using CAD to do landscape design work, while intermediate users will find techniques to use LANDCADD and AutoCAD more effectively. In particular, the book provides experienced users with techniques to customize LANDCADD to make it fit your office graphic style and symbol selection.

Using LANDCADD is intended to guide the beginning LANDCADD user through a series of guided exercises which demonstrate the most important features of LANDCADD in the production of working drawings. The exercises follow the standard sequence of drawings that flow from a traditional residential landscape design project. After developing a blank drawing (to establish basic drawing parameters) we build a sample title block and logo. Then we begin with a base plan for site layout and building location, then follow with a construction and hardscape plan to lay out planters and paved areas. Next, using the plant database, we develop a plant list, and lay out a planting plan. After assigning costs and categories to each plant, we build a full installation cost estimate. Then we create an irrigation plan, and complete it with an irrigation legend and materials schedule. Finally, we build a sample detail sheet to highlight some construction details. Additional exercises are included to provide experience in converting 2-D plant symbols to 3-D symbols for presentation drawings, using AutoCAD's Paper Space, and building customized irrigation symbols.

The intention of these exercises is to illustrate the efficient use of LANDCADD commands and their relationship to fundamental AutoCAD commands. It is **not** the function of this book to discuss landscape design fundamentals, land planning theory or irrigation design practices. In fact the design is deliberately kept at a rudimentary level to highlight the capabilities of LANDCADD, rather than concentrate on the esthetics of landscape architecture. *Using LANDCADD* is meant to highlight the commands and routines found in the Landscape Design, Irrigation Design and Site Planning and Construction Details modules of LANDCADD. These modules are most central to the landscape design process and relate to the production of working drawings for landscape projects. Note that these modules are also tied to the AutoCAD platform, which is the most popular format for sharing drawings with other design and engineering professionals. In addition, these modules offer the most opportunity for customization and adaptation to individual office standards.

For reasons of brevity and focus, information was not included for the Eagle Point standalone modules including Site Designer, Irrigation Designer, Picture Perfect, and Virtual Simulator. The first two programs provide a subset of the commands found in Landscape Design, Site Planning and Irrigation Design. The second two programs provide visualization tools more suited for presentations and animations which are beyond the scope of this volume. In addition, no coverage is given to the civil engineering modules included in the LANDCADD product line. These modules include Surface Modeling, Site Analysis, COGO and Survey Adjustment. While these modules provide great utility to many architects and land planners working on larger developments, they too are beyond the limits of this book.

Using LANDCADD does make certain assumptions about the computer and about the reader. First, the author assumes that you have AutoCAD R13 installed on your computer, and that you run it in Windows 95 or Windows NT 4.0. If you have AutoCAD R12 installed, or you do not use Windows 95 or NT, **LANDCADD runs almost identically**. The menus and command structure of LANDCADD is essentially the same no matter which Intel/Microsoft-based platform you use (Windows 3.X, DOS, Windows NT 3.5). I also assume that you have LANDCADD 13 installed and configured correctly. In addition, I assume that you have a basic understanding of how to save, retrieve and move drawings in the platform that you use. I assume that you can operate the pointing device that your computer uses, whether it is a mouse, digitizer or track ball. It is also important that you have a **basic** knowledge of AutoCAD. You should be able to draw simple objects, and have an understanding of different methods for inputting points, selecting objects and editing them.

A final (and important) assumption is that you have access to an output device. You can use a plotter (inkjet or pen type), a laser printer or an inkjet printer to produce output. Any of these devices will allow you to produce hard copies to compare to the illustrations in the book. You should make test plots frequently to check your work and spot errors before proceeding too far. Since the drawings follow a sequence, a major error at the beginning will make it difficult to complete all of the exercises.

Who Should Read This Book

Whether you are an experienced landscape design professional or a beginning landscape architecture student, you can benefit from this book. *Using LANDCADD* can help you shorten your "learning curve" and help you become productive and efficient more quickly. If you have invested in a computer system, AutoCAD and LANDCADD, you have spent $6000 to $8000 in software and equipment already. Making this investment pay off requires learning how to use LANDCADD and to take advantage of its time-saving tools and routines. This book will provide you with the fastest possible way to learn LANDCADD—through realistic examples of how to use this program for the production of construction drawings.

Beginning and Intermediate LANDCADD users

If you just purchased LANDCADD or have used it for some time, you can benefit from *Using LANDCADD.* LANDCADD is a large and complex program that can make your landscape drawing work more efficient and productive **once you learn how best to use it**. The exercises are intended to show you how to get the most from LANDCADD in the shortest amount of time. The exercises included in this book show you how to produce working landscape drawings without the "trial and error" period required when learning most software. In addition, this book was not written by computer programmers or software engineers. It is written without excessive jargon, and does not assume that you have an extensive technical computer background. *Using LANDCADD* is written for landscape designers who want to use a CAD program, not for computer professionals who want to learn landscape design.

Landscape Architecture Students

You can use the exercises in *Using LANDCADD* to learn LANDCADD and use it for your own design projects and assignments. Knowing LANDCADD can be a persuasive argument for employment once you finish your landscape architecture studies. *Using LANDCADD* can be used as a text book, or as a tutorial book for students. The introductory section on AutoCAD commands can be helpful to those without AutoCAD experience, while the drawing exercises can help you master LANDCADD itself. The final section on setting up AutoCAD and LANDCADD in a design office can be very useful when you begin employment in the landscape design field.

Landscape Architects setting up AutoCAD and LANDCADD for use in a design office

If you have recently purchased LANDCADD and want to get set up in a productive CAD system in your office, *Using LANDCADD* has information to help you get started. You can learn how to set up office standards for title blocks, layering, fonts, dimensions and plotting, as well as how to set up a basic CAD network. In addition, you can find helpful information about organizing and backing up files, creating different kinds of output and working with CAD drawings from consultants and outside contractors. *Using LANDCADD* can help you set up an office system that works properly from the beginning, without experiencing some of the pitfalls that await those who set up systems without a cohesive plan.

How This Book Is Organized

Using LANDCADD is set up in three separate parts. Each part is related to, but not dependent on the others.

Part One: Introduction to AutoCAD for the Landscape Architectural Designer

This portion of the book is intended to provide a brief explanation of the AutoCAD commands most needed to begin using AutoCAD and LANDCADD to produce landscape architectural drawings. The chapters in this section are most appropriate for those who have little experience in running AutoCAD. They can help provide a "crash course" for new users. Chapters 1 and 2 explain the AutoCAD drawing and editing commands and introduce the concept of selection sets. Chapter 3 shows you how to control the cursor and the appearance of the objects in the drawing. Chapter 4 shows you AutoCAD and LANDCADD tools for creating boundaries and hatches. Chapter 5 explains how to handle text in AutoCAD and LANDCADD. Chapter 6 relates to scaling and dimensioning AutoCAD and LANDCADD drawings. Chapter 7 teaches you to handle blocks, attributes and external reference drawings. Chapter 8 deals with file management and organization in AutoCAD, while Chapter 9 explains how to create output on printing and plotting devices.

Part Two: LANDCADD Drawing Exercises

The second part of the book contains the drawing exercises which show you how to use LANDCADD to produce construction drawings. We begin Chapter 10 with the Eagle Point Project Manager, and produce a blank drawing to act as a basic prototype. The first drawing exercise begins in Chapter 11 with the Title Block, which is used in most all of the remaining drawings. We create the lot and building footprint in Chapter 12, and add utilities to complete the Base Sheet. Chapter 13 leads you through the production of a Hardscape and Construction Plan. In Chapter 14 we build a Planting Plan, and use LANDCADD's attribute tools to build a planting Materials List and Estimate. The Chapter 15 drawing exercise leads you through the production of an Irrigation Plan, including how to create a Materials Schedule and Legend. Chapter 16 leads you through the production of a Detail Drawing Sheet to round out the project. Chapter 17 shows you a drawing exercise that illustrates the conversion of 2-dimensional into 3-dimensional symbols, and building shaded and rendered images. The Chapter 18 exercise leads you through the use of AutoCAD's Paper Space and Model Space drawings and title blocks. In Chapter 19, the final exercise shows you how to build customized sprinkler symbols for use in your irrigation plans.

Part Three: Using LANDCADD in a Landscape Architectural Office Environment

The third part of the book deals with setting up and using LANDCADD in a production setting. First, Chapter 20 explains the relationship between AutoCAD and LANDCADD. Following this section, Chapter 21 covers the installation and set-up of LANDCADD in different operating systems. After this, Chapter 22 discusses the available methods for customizing LANDCADD for your office operations. Chapter 23 explains the value of creating standardized office procedures for CAD drawing and file management, handling text files, outside consultant drawings, and hard copy drawings. In addition, Chapter 23 explains how to establish standardized settings and styles within your drawings to make it much easier to integrate your work with other CAD designers. Finally, Chapter 24 teaches you the basics of running CAD on a network, as well as some hints to help you implement a new network.

How To Use This Book

For new users without AutoCAD experience, you can use this book as a guide to learning the basics of the program. The first section of the book explains the fundamentals of drawing objects with AutoCAD. This is not an exhaustive treatment of the program, but will help you get a handle on this large and (often) intimidating program. Use the first part of the book, along with other resources, to begin the process of learning AutoCAD.

Those with strong AutoCAD experience can begin the exercises in the middle section, which traces a small residential landscape design project through the production of a base plan, hardscape and construction plan, planting plan, irrigation plan and irrigation schedule, and detail sheet.

The first three drawings (the Title Block, Base Sheet and Construction and Hardscape Plan) are straightforward drawings which take advantage of both LANDCADD and AutoCAD commands, but basically set the stage for later drawings. It is important to complete these drawings to be able to build the Planting Plan and Irrigation Plan that follow.

The Planting Plan and Irrigation Plan are the largest drawings and take the most time to produce. These drawings thoroughly explain how to use the Landscape Design and Irrigation Design modules in LANDCADD. The Landscape Design and Irrigation Design modules are complex programs with great capabilities for many types of design work, so it is important to examine each of these modules fully. Although you should stop to save your work frequently, you can plan on spending several hours to complete each of these drawings. In addition to producing a landscape plan, the Planting Plan also gives you an introduction to producing custom plant symbols, using LANDCADD's Plant Database, and producing a cost estimate for a planting plan. When you finish the Irrigation Plan, you also use LANDCADD routines to build a Materials Schedule and Legend.

The final three drawings show how to use LANDCADD to produce a Detail Sheet, 3-D concept and presentation drawing and Paper Space plan. These drawings can be completed independently of the Planting and Irrigation Plan exercises, although they do require that you have completed the Title Block exercise in Chapter 11.

The final section is provided to help anyone who expects to use LANDCADD in an office environment to produce work for clients. If you do not yet have LANDCADD installed on your computer, there is a brief guide to installing and configuring the program included, along with an explanation of the relationship between AutoCAD and LANDCADD. There are also important chapters which guide you through the process of establishing and documenting office CAD standards. The final chapter can help you untangle some of the complex problems that arise when you set up a CAD network.

Online Companion

The Online Companion™ is your link to AutoCAD on the Internet. We have compiled supporting resources with links to a variety of sites. Not only can you find out about training and education, industry sites, and the online community, we also point to valuable archives compiled for AutoCAD users from various Web sites. In addition, there are pages specifically for users of **Using LANDCADD.** These include an owner's page with updates, a swap bank where you can share your drawings with other AutoCAD students, a page where you can send us your comments, and chapter Review Questions. You can find the Online Companion at:

http://www.autodeskpress.com/onlinecompanion.html

When you reach the Online Companion page, click on the title **Using LANDCADD**.

Comments or Problems

If you need assistance with the exercises in this book, or find a new bug, you can contact the author by e-mail at k.gordon@fullcoll.edu or through CompuServe at 103261,2476@CompuServe.com.

c h a p t e r

Introduction to AutoCAD for the Architectural Designer

To use LANDCADD effectively, a designer and/or architect needs a firm grounding in AutoCAD. While this fact should seem obvious, too many employers and designers try to ignore this "dirty little secret" in hopes of discovering a short-cut to productive use of LANDCADD without a proper foundation in AutoCAD. For those hoping for a short-cut through the process of learning AutoCAD, I can only apologize in advance. There simply are no short-cuts. The core concepts used in AutoCAD are central to using any add-on CAD program effectively, LANDCADD included. You must learn the basics of drawing, editing and saving files. You must learn how to work with drawing units and dimensions, text and blocks. You must learn how to plot, print and manage drawing files. By learning these basic skills, you gain mastery of AutoCAD and lay a foundation for becoming a more efficient and productive design professional.

Knowing how AutoCAD works and how to navigate through its commands makes using LANDCADD faster and easier. Once you learn to use AutoCAD, you can more easily understand how LANDCADD works and how to take advantage of its many features. Investing the time and effort in learning the basic AutoCAD program pays great dividends later.

A key concept in understanding AutoCAD:

The "User Friendly" Continuum

←	→
Easy to Learn	Hard to Learn
"User Friendly"	"User Surly"
small set of features	large set of features
limited capability	nearly unlimited capabilities

Most computer programs can be placed easily on the continuum shown above. There are many simple, inexpensive programs that take a few days to a few weeks to learn and master. Soon, however, we run into the limitations of the program and begin to wish the program had more features and did more work for us. In fact, we may find ourselves investing time and effort in learning how to overcome the limitations of the program, rather than performing productive work.

At the other extreme, there are other programs that are much more daunting to learn. They have a large number of commands and often, many ways to perform the same task. These programs often provide the user with too many options and can leave the user feeling frustrated and overwhelmed. We try to use these programs and sometimes feel that the programmers have deliberately created a "user surly" program that defies our efforts to learn it. The fact that a program can do so many things may make it difficult to learn to do any one single thing. On the positive side, these programs usually have enormous capabilities. Users can work with the program for many months (or years) and continue to be more and more productive. Frequently these programs have customizing capabilities built in to assist users in automating and streamlining their work.

If you think that AutoCAD belongs on the far right side of the continuum, you are not alone. Many users begin working with AutoCAD and quickly feel overwhelmed. The "learning curve" can be long and bumpy. The key to learning a program like AutoCAD is to concentrate on those sections of the program that you **need** to know first, then gradually branch out into other areas of interest. There will, no doubt, be portions of the program that you may **never** learn (or need). Many users spend years running AutoCAD and really only use 50%—or less—of its features. The program has so many options and capabilities that no one really uses them all. In fact, the programmers at Autodesk have deliberately created a program that has vast capabilities and can be customized and tailored to many different drafting and design disciplines. Third party applications (like LANDCADD) are written to take advantage of AutoCAD's many ways of performing tasks and to present them in commands and routines that complement the user's normal working process. The "open architecture" approach used by AutoCAD has given birth to an entire industry of software developers. AutoCAD functions lie at the center of such programs as ArchT®, LANDCADD®, Auto-Architect® and ArchCAD®. Possessing a strong foundation in AutoCAD makes learning to use these programs much easier.

Another key point to remember is that AutoCAD has become the common denominator for civil engineers, mechanical engineers, architects, landscape architects and even industrial piping designers. Any of these design professionals can exchange drawing files (and ideas) through AutoCAD. Knowledge of AutoCAD has become the "coin of the realm" in the design and construction industry. In today's employment market, knowledge of AutoCAD is not optional for many types of design positions. Being able to "run AutoCAD" is often necessary to even get an interview at many design firms.

With these words behind us, let's examine some of the essential commands used in AutoCAD.

Drawing Commands in AutoCAD

The LINE command

Line (**L**) The **Line** command simply draws a line between the two endpoints you specify. Type **L**, then **<Pick>** the first endpoint. After the first endpoint is chosen, set the second endpoint with the **<Pick>** button to draw the line. You can continue drawing line segments end-to-end if you press the **<Pick>** button again after the second endpoint has been set. This draws a new line segment connected to the first one. You will continue to draw line segments until the **<Enter>** key is pressed.

Lines are not related to each other in any particular way. Each line can be moved or edited without affecting any other adjoining lines.

The POLYLINE command

Polyline (**PL** or **Pline**) A polyline is a group of line segments or arcs treated as a single entity. Because these parts are linked together, line segments (or arcs) within a polyline cannot be moved or edited without affecting the other parts of the polyline. Moving one part of the polyline moves the entire object. More importantly, polylines are very versatile. Unlike line segments, polylines can be edited to create fitted and spline curves. Polylines may be closed to create boundaries for hatches. Polylines can be three dimensional, and can define planes or borders in space.

As you begin the **Polyline** command, proceed as you would when building a line. That is, specify the first endpoint.

`Command:`
PLINE (PL)

`From point:`
Pick a beginning point for the polyline.

`Current line-width is 0'-0"`
`Arc/Close/Halfwidth/Length/Undo/Width/<Endpoint of line>:`
The **Current line-width is 0'-0"** indicates the current setting for the polyline width. Since polylines can have widths greater than zero (unlike lines) it is appropriate to display the setting before we start picking endpoints.

The standard (default) setting for polylines is the **Line** mode. The **Line** mode allows you to create a polyline consisting of different line segments. You can select a series of endpoints for these line segments (as in the **Line** command above) so that the polyline will be a connected series of line segments. Use the **<Pick>** button repeatedly for this effect.

> **Arc**—If **Arc** is selected, the polyline command shifts to **Arc** mode. See below for details on **Arc** mode.

Close—Use the **Close** option to close the polyline back to its first endpoint. This is an important step if the polyline is to be used as a boundary for hatching, or when an area calculation is made. Once a polyline is closed, AutoCAD can keep track of the perimeter length and the area enclosed within the polyline.

Halfwidth—This is the dimension from the center of the polyline to its outer edge. It may be more convenient to specify the **Halfwidth** of the polyline instead of the **Width**. As with the **Width** option, the **Halfwidth** can be entered as a numerical value, or can be selected as the distance between two points chosen on the screen. The initial value or size indicated is the beginning **Halfwidth**. The ending **Halfwidth** must be entered or chosen just as the beginning value. All subsequent polylines or polyline segments will have the same beginning and ending **Halfwidth**s until the settings are changed.

Length—In **Line** mode, you can also specify the **Length** of a polyline segment by using a numerical value. The new segment will be at the same angle and connected to the previous line segment. If the previous piece of the polyline was an arc, **Length** makes the current segment tangent to the end of the arc.

Undo—The **Undo** function allows you to remove each segment of your polyline until you reach the origin point. This allows editing of the polyline as you draw it.

Width—You can adjust the **Width** of a polyline, the dimension from one edge to the other, by selecting **Width**. To set the width, either type in a numerical value at the prompt, or select two points to show the width on the screen. This defines the beginning width of the polyline. The ending width can either be equal to the beginning one (the default), or be chosen in the same manner as the beginning width. All subsequent polylines (or polyline segments) will have the same beginning and ending widths until you change the **Width** setting again.

Endpoint of line—This is the location of the next endpoint of the polyline. The endpoint can be entered in relative, absolute or polar coordinates, or picked on the screen.

Another option for the **Polyline** command is **Arc**. This option changes the **Polyline** command to **Arc** mode. The Arc mode allows you to add arc segments to a polyline. When in Arc mode, you have several options from which to choose:

Endpoint of arc—draws an arc segment starting at a point tangent to the previous polyline segment.

Angle—chooses the angle which will be included in the arc. After entering an angle (a positive number generates a counter-clockwise arc, while a negative number generates a clockwise arc) you must complete the other information needed to create an arc. This can be either the **Endpoint**—which simply draws in the arc segment—or the **Center** or **Radius**. Once **Center** is selected, AutoCAD draws the arc using the included angle and center point you chose. If you select **Radius**, the command waits for you to specify a point on the screen. Then you are asked to pick a second point (the length of the radius) to show the size of the arc segment. Finally, you are prompted to show the **Direction of chord** which rotates the direction of the arc tangent to the first point.

Center—chooses the center of the arc segment. After the center is chosen, you have three options to complete the arc: **Endpoint**—which simply draws in the arc segment; **Angle**—which asks for an included angle for the arc segment; and **Length**—which asks for the chord length (linear distance between the endpoints) of the arc segment.

Close—will close the polyline back to its origin point with an arc segment.

Direction—asks for two points. The first point indicates a starting direction for the arc, while the second point sets an endpoint to the arc segment.

Halfwidth—shows the dimension of a polyline from the center to the outer edge.

Line—returns the command to the **Line** mode.

Radius—allows you to first define the radius of the arc segment by specifying a beginning and ending point for the radius. Once the radius is defined, you then must either indicate the: a) endpoint for the arc; or b) included angle and direction of chord for the arc.

Second pt—permits the selection of the second and third points of a three-point arc segment.

Undo—removes the most recent arc segment from the polyline.

Width—sets the width for the polyline arc. This becomes the default setting for the ending width, and remains the polyline width until it is changed.

Note: Complex polylines with widths can enlarge the disk size of your drawing, and complicate editing changes in the drawing. Avoid using wide polylines if regular lines and arcs will draw what you want.

Polyline Exercise

This exercise is built on a screen with the following settings:

Units:	Architectural
Denominator of smallest fraction to display:	16
Systems of angle measure:	Decimal degrees
Number of fractional places for display of angles:	0
Direction for angle 0:	0
Angles measured clockwise?:	No
Limits:	
Lower left corner:	0'-0",0'-0"
Upper right corner:	250'-0", 185'-0"
Grid:	10'
Snap:	5'

The following exercise illustrates the use of polylines to build a box with arcs at each corner. Note that each arc is built in a different way: Angle-Radius-Chord, Endpoint-Tangent-Endpoint, and Radius-Angle-Chord. The final arc is built with the **Close** option. The Polyline defaults help us to minimize the response information needed.

At the prompt:

`Command:`
Type **Snap** to enter a Snap setting.

`Snap spacing or ON/OFF/Aspect/Rotate/Style <1'-0">:`
Type **10'** to set the Snap distance.

`Command:`
Type **PL** to begin the Polyline command.

From point:
Type (or pick) **30',20'** to set a starting point.

Arc/Close/Halfwidth/Length/Undo/Width/<Endpoint of line>:
Type **W** (or **Width**) to set the polyline width.

Starting width <0'-0">:
Type **6"** for the width.

Ending width <0'-6">:
Press **<ENTER>** to accept the default value.

Arc/Close/Halfwidth/Length/Undo/Width/<Endpoint of line>:
Type (or pick) **230',20'** to set the first endpoint.

Arc/Close/Halfwidth/Length/Undo/Width/<Endpoint of line>:
Type **A** (or **Arc**) to enter the Arc mode.

Angle/CEnter/CLose/Direction/Halfwidth/Line/Radius Second pt/
Undo/Width/<Endpoint of arc>:
Type **A** (or **Angle**).

Included angle:
Type **90.**

Center/Radius/End point:
Type **R** (or **Radius**).

Radius:
Type **10'** to set the radius.

Direction of Chord <0>:
Type **45** to set the direction of rotation.

Angle/CEnter/CLose/Direction/Halfwidth/Line/Radius/Second
pt/Undo/Width/<Endpoint of arc>:
Type **L** (or **Line**) to enter the Line mode.

Arc/Close/Halfwidth/Length/Undo/Width/<Endpoint of line>:
Type (or pick) **240',160'** to set the next endpoint.

Arc/Close/Halfwidth/Length/Undo/Width/<Endpoint of line>:
Type **A** (or **Arc**).

Angle/CEnter/CLose/Direction/Halfwidth/Line/Radius/Second pt/Undo/Width/<Endpoint of arc>:
Type (or pick) **230',170'** to set the arc endpoint.

Angle/CEnter/CLose/Direction/Halfwidth/Line/Radius/Second pt/Undo/Width/<Endpoint of arc>:
Type **L** (or **Line**).

Arc/Close/Halfwidth/Length/Undo/Width/<Endpoint of line>:
Type (or pick) **30',170'** to set the next endpoint.

Arc/Close/Halfwidth/Length/Undo/Width/<Endpoint of line>:
Type **A** (or **Arc**) to enter the Arc mode.

Angle/CEnter/CLose/Direction/Halfwidth/Line/Radius/Second pt/Undo/Width/<Endpoint of arc>:
Type **R** (or **Radius**).

Radius:
Type **10'** to set the radius.

Angle/<End point>:
Type **A** (or **Angle**).

Included angle:
Type **90.**

Direction of Chord <180>:
Type **225** to set the angle.

Angle/CEnter/CLose/Direction/Halfwidth/Line/Radius/Second pt/Undo/Width/<Endpoint of arc>:
Type **L** (or **Line**) to enter the Line mode.

Arc/Close/Halfwidth/Length/Undo/Width/<Endpoint of line>:
Type (or pick) **20',30'** to set the last line segment endpoint.

Arc/Close/Halfwidth/Length/Undo/Width/<Endpoint of line>:
Type **A** (or **Arc**) to enter the Arc mode.

Angle/CEnter/CLose/Direction/Halfwidth/Line/Radius/Second pt/Undo/Width/<Endpoint of arc>:
Type **C** (or **Close**) to close the polyline.

See the results below:

After the border is drawn, try drawing some tapered and curved polylines inside.

Figure 1.1

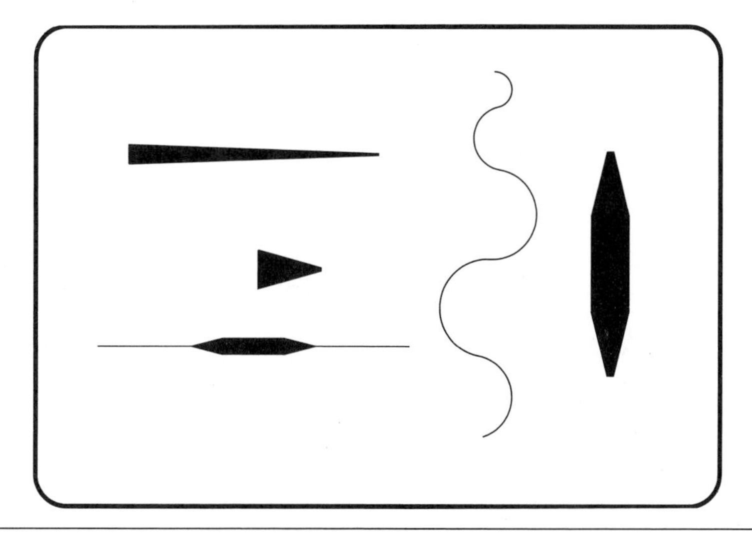

The ARC command

Arc (**A**) The **Arc** command draws an arc (circle segment less than 360°) in a *counter-clockwise direction* using three of several points or indicators. When an arc is constructed, we can specify combinations of the following:

Start point—the beginning point of the arc.

Second point—a point along the path of the arc between the start and end point.

End point—the ending point of the arc.

Angle—a measure of the included angle contained within the arc; using a negative angle will result in a clockwise arc.

Chord—the linear distance between the start point and the end point of the arc; using a negative chord length will result in an arc angle greater than 180°.

Direction—the angular direction tangent from the starting point of the arc; this can be entered from the keyboard or indicated with a pointing device.

Radius—the linear distance of radius of the arc; a negative radius will result in an arc angle greater than 180°.

Using these indicators, there are *eleven* ways to build an arc:

SSE—Start, Second, End

SCA—Start, Center, Angle

SCL—Start, Center, Length of chord

SCE—Start, Center, End

SEA—Start, Angle, End

SED—Start, End, Direction of tangent

SER—Start, End, Radius

SEC—Start, End, Center

CSE—Center, Start, End

CSA—Center, Start, Angle

CSL—Center, Start, Length

To draw a series of connected arcs, press **<Enter>** after completing an arc to repeat the **Arc** command. Press **<Enter>** again to begin the new arc at the endpoint of the previous arc.

The CIRCLE command

Circle (0)—The **Circle** command draws circles five different ways using the following data:

Center point—the center of the circle.

Radius—the distance between the center and the edge of the circle.

Diameter—the distance across the circle through the center.

Tangent—an object on the screen for the circle to be tangent to.

Using these definitions, there are five ways to build a circle:

Center-Radius—specify a center point, then a radius.

Center-Diameter—specify a center point, then a diameter.

2 Points—specify the endpoints of the diameter to define the circle.

3 Points—specify three points for the circle to pass through.

Tangent-Tangent-Radius—pick two objects for the circle to be tangent to; then specify the radius of the circle to be drawn.

Arc and Circle Exercise

One of the best ways to explore the **CIRCLE** and **ARC** commands is to build a series of boxes in a grid pattern on the screen, then use the different ways of building circles and arcs in the different squares. See the example below:

Figure 1.2

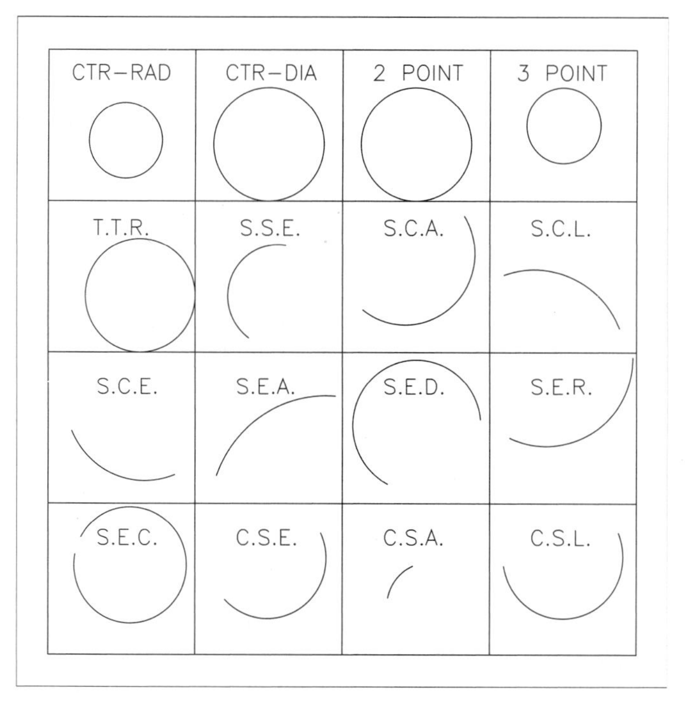

There are more ways to build these objects than you will probably use, but it is an excellent idea to try them all. You may find that you may need to insert circles when you only have tangent objects and the circle radius available. You might also need to locate a circle where the two endpoints of a diameter are known, but not the center point. The large number of ways to build an arc allow you great flexibility in this command, so that nearly any combination of point, center and other instruction will yield an arc.

The POLYGON command

Polygon—The **Polygon** command builds a polygon with up to 1,024 sides. When we choose to build a polygon, we choose the **Number of sides** (three or more—up to 1024), then the **Edge/Center** of the circle which defines the border of the polygon. When we build the polygon, we are also given the choice of building the polygon within the bordering circle (inscribed), or outside of the circle (circumscribed).

The DONUT command

Donut or **Doughnut**—Occasionally, we have the need to define a solid dot, or a widened circle. Thus we use the **Donut** command. (AutoCAD also responds to the proper spelling **Doughnut** as well.) In this command, we simply define the **Inside diameter**, then the **Outside diameter**. After these factors are defined, then we simply pick the **Center of donut**. Donuts will continue to be placed on the screen with each press of the **<Pick>** button until we press the **<Enter>** button. When we want to place solid dots on the screen, we set the inside diameter to a value of **0**. If we want to build an empty (zero thickness) circle with the properties of a polyline (allowing a line width to applied later), we can specify the inside diameter and the outside diameter at the same value.

The POINT command

Point—The **Point** command builds dimensionless points on the screen. These points have no length or width, but they do have location (X, Y and Z values). Points can be placed on the screen to define lines or polylines which can be drawn later. AutoCAD has several options for how points are actually displayed on the screen.

Use the "Point Style" dialogue box (available from the **EP and AutoCAD** menu; pick the **Options** pull-down menu, then **Display ▸** and finally **Point Style...**). See the dialogue box in Figure 1.3.

Once points are placed on the screen, you can use the **Osnap—Node** option to snap to them when building polylines, lines, arcs or other objects. Notice that one of the available options for point display is to make them invisible, so that they can disappear for plotting purposes, but not be lost from the drawing database.

Figure 1.3

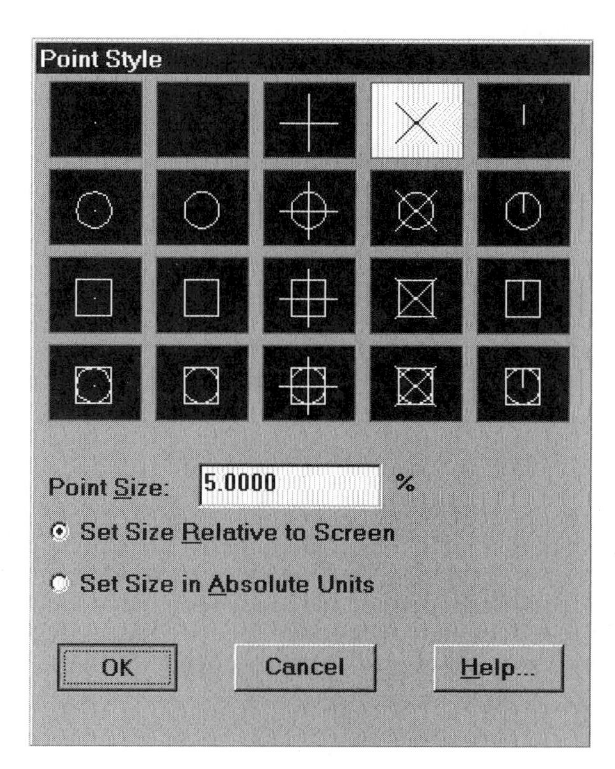
Point Style
Point Size:
5.0000
%
Set Size Relative to Screen
Set Size in Absolute Units
OK
Cancel
Help...

c h a p t e r

Editing Commands in AutoCAD

Editing objects in AutoCAD is really a two-step procedure. First, you need to select the objects to be edited. Second, you actually perform the editing command. Selecting the objects to be edited can be as time-consuming as performing the actual editing. By taking advantage of the various ways of selecting objects to edit, you increase your efficiency and drawing speed. By knowing the various editing commands, you can learn how to perform tasks in slightly different ways. As you become more adept at AutoCAD drawing, you will realize an important truth: There are many ways of doing any single task in AutoCAD, and none of them are wrong. Some methods are more efficient than others, but none of the several choices available are really wrong. We almost always can use our personal preferences when choosing how to work with AutoCAD commands.

Selection Sets

One of the key concepts in editing objects on the screen is that of **Selection Sets**. A selection set is the object or objects that you wish to edit. Objects in a selection set will **highlight** (become dashed) after being chosen.

While using an editing command, we will see the prompt:

```
Select objects:
```

This means that AutoCAD is expecting us to choose a selection set. There are several ways to include objects in a selection set. These include:

1. **Manual Selection** (or **Pick**)—after the cursor turns into a **pick box**, we can move around the screen and select objects by touching them (preferably near an edge) and pressing the **<Pick>** button.
2. **Window**—this option allows us to draw a "rubber band box" around the objects we wish to select. Any objects not completely enclosed within the box will not be selected. We can activate the **Window** option two ways:

`Select objects:`
Type **W** at the prompt. We are then prompted to pick a first and second corner point to define the enclosing box.

Or

while on an empty portion of the screen, press the **<Pick>** button, then drag the cursor to the **RIGHT**. This will automatically activate the **Window** function, and the window will be stretched from the original point where you pressed the **<Pick>** button.

The **Window** option will draw a "rubber band box" using a **solid** line.

3. **Crossing**—this option allows you to draw a dashed "rubber band box" which selects any objects enclosed **within** or crossing **through** the box. The crossing box is one of the most useful tools in selecting objects in AutoCAD, because it selects a large number of objects very quickly. Like the **Window** option, the **Crossing** option can be activated two ways:

   `Select objects:`
   Type **Cr** at the prompt. We are then prompted to pick a first and second corner point to define the dashed **Crossing** box.

   Or

   while on an empty portion of the screen, press the **<Pick>** button, then drag the cursor to the **LEFT**. This will automatically activate the **Crossing** function, and the dashed window will be stretched from the original point where you pressed the **<Pick>** button.

   Remember that a **Crossing** window will always draw a "rubber band box" using a **dashed** line.

4. **Group**—this option selects a group of objects which were linked together using the **Group** command. The **Group** command allows you to assign objects to a selection set that you can name, save and describe in a dialogue box. When this option used, the prompt will ask for the name of the group to be edited.

   `Select objects:`
   Type **G**

   `Enter group name:`
   Type in the name of the group to be edited. You may need to use the **Group** command to display a list of the named groups in your drawing.

5. **Previous**—this option will highlight the selection set that was used in the last editing command. To use this option:

   Select objects:
   Type **P**—the previous selection set will highlight on the screen.

   This can be very convenient when using a large selection set which took some time to create. A bit of careful planning can allow you to use the same selection set for several commands in a row before a new selection set is created.

6. **Last**—this option will highlight the **Last** object placed on the screen. This last object could be any object created with a drawing command, or a block inserted into the drawing.
7. **All**—the **All** option selects all of the objects displayed in the drawing. Objects on **frozen** or **locked** layers are not selected.
8. **Window Polygon**—although not frequently used, this option allows the user to define an irregular polygon with any number of sides to enclose the selection set. As with the **Window** option, only those objects completely enclosed by the polyline will be selected. To use this option:

   Select objects:
   Type **WP** to activate the **Window Polygon** option.

   First polygon point:
   Choose the first corner point of the polygon.

   Undo/<Endpoint of line>:
   Stretch the "rubber band polygon" to surround the desired object(s).

   Undo/<Endpoint of line>:
   Continue choosing corner points until the object(s) are enclosed within the polygon.

9. **Crossing Polygon**—another infrequently used option, **Crossing Polygon** allows the user to define an irregular polygon cross through the selection set. As with the **Crossing** option, any object crossing through the polyline will be selected. To use this option:

   Select objects:
   Type **CP** to activate the **Crossing Polygon** option.

   First polygon point:
   Choose the first corner point of the polygon.

```
Undo/<Endpoint of line>:
```
Stretch the dashed "rubber band polygon" to cross through the desired object(s).

```
Undo/<Endpoint of line>:
```
Continue choosing corner points until the dashed polygon crosses through all of the desired object(s).

10. **Fence**—this option is a variation of the **Crossing** box or polygon. In this case, the **Fence** acts as a **Crossing** line. This means that any object on the screen passing through the **Fence** line will be selected. The **Fence** can consist of any number of connected line segments. To use this option:

    ```
    Select objects:
    ```
    Type **F** to activate the **Fence** option.

    ```
    First fence point:
    ```
    Pick the first endpoint for the **Fence**.

    `Undo/<Endpoint of line>:`

    Pick the second endpoint for the **Fence**. Press **<Enter>** when finished building line segments for the **Fence**. All of the objects crossing through the **Fence** line will be selected.

11. **Remove**—the **Remove** option switches the selection set mode to **Remove** (rather than add) objects from the selection set. Any objects chosen after this option is activated will be removed from the selection set and the dashed highlight will disappear. Any of the previous methods for selecting objects can also be employed to remove objects from the selection set. This is especially handy for removing one or two objects from within a large number of other objects that were added to the selection set.

12. **Add**—the **Add** option returns the selection set mode back to its default **ADD** setting. Objects picked after switching to the **Add** mode will be added to the selection set. It is not uncommon to switch back and forth from **Add** and **Remove** to place the desired objects in a complex selection set. If you wish to define and save a complex selection set, use the **Group** command. See the comments on **Group** later in this chapter.

    After the selection set has been chosen, always press **<Enter>** to indicate that you have finished selecting objects. This causes AutoCAD to return to the editing command.

Editing Commands

The editing commands in AutoCAD can be divided up in several ways. We will look at commands which act to duplicate objects, change the location or size of objects, and change the properties of objects. Finally, we will examine the power of Object Snaps to control the drawing and editing process.

Duplicating Objects

The COPY command

Copy (**C**) The **Copy** command allows you to create a copy (or copies) of a selection set in any new location of your choice. You can make single or multiple copies of your objects.

`Command:`
Type **Copy** (or **C**).

`Select objects:`
Pick the selection set you wish to copy.

`Select objects:`
Press **<Enter>** to complete the selection set and return to the **Copy** command.

`<Base point or displacement>/Multiple:`
Pick a “handle” or “grab” point on the object to indicate where the copy will be made from. This can be done using the freehand cursor, with **Object snaps** or with coordinate input.

Or

press **M** to build multiple copies. If this option is used, you will be asked for a Base point afterward.

`Second point of displacement:`
Pick the point where the copy(ies) will be made to. This second point can also be entered using the freehand cursor, **Object snaps** or with coordinate input.

To add precision to the **Copy** command, you can use some of these options:

1. Use **Snap** to control the cursor and place the objects in precise locations.
2. Use **Object snaps** to pick the Base point and/or Second point. This allows you to “grab” the selection set from a precise spot on an object (midpoint, center, quadrant, etc.) and place it in another precise location.

3. Use **Ortho** mode to place objects in horizontal or vertical alignment.
4. Use relative coordinate input to specify the location of the Second point. This allows you to precisely choose the angle and distance from the First point.

The OFFSET command

Offset (OF) The **Offset** command allows you to make parallel copies of an object at a specified distance away from the original. In this way, you can make a series of parallel lines from one original, or extra copies of an object at known distances apart. The **Offset** command will also make a copy of a circle or arc concentric to the original. If a polyline is selected to **Offset**, the duplicate will be parallel to the original.

`Command:`
Type **Offset** (or **OF**).

`Offset distance or Through <Through>:`
This distance can be specified from the keyboard, or by picking two points on the screen. The distance between the two points will be the **Offset** distance.

`Select object to offset:`
Pick an object to offset. **Offset** can only be used on one object at a time.

`Side to offset?`
Select the side of the object on which to create the copy. Circles and arcs are copied concentric to the original. Closed polylines will be copied either inside or outside the original. Lines and other objects are copied parallel to the original.

`Select object to offset:`
Either choose another object to offset, or press **<Enter>** to complete the command.

The **Offset** command is one of the most efficient in AutoCAD. Parallel copies are used very frequently in architectural and mechanical drawings. Closed polyline copies are created easily and accurately. An irregular curve can also be copied without calculation of new radii or tedious plotting of data points. This command can often substitute for **Copy**, **Scale**, and even **Array**.

The ARRAY command

Array (AR) The **Array** command creates multiple copies of the selection set in either a **Rectangular** or **Polar** coordination. We will discuss each separately.

A **Rectangular array** copies objects in horizontal rows and/or vertical columns. We must specify the number of rows and columns for our array. In addition, we must specify the distance between the rows and the columns. The distances refer to the distance between a particular point on one object to the corresponding point on the copied object.

`Command:`
Type **Array** (or **Ar**).

`Select objects:`
Pick the objects to **Array**.

`Select objects:`
Press **<Enter>** when finished selecting objects.

`Rectangular or Polar array (R/P) <R>:`
Press **<Enter>** to accept the default and create a rectangular array.

`Number of rows (---) <1>:`
Enter the number of horizontal rows to create.

`Number of columns (|||) <1>:`
Enter the number of vertical columns to create.

`Unit cell or distance between rows (---):`
To set the distance from the keyboard, enter a numeric value here.

Or

`Unit cell or distance between rows (---):`
To set the distance between the rows **and** columns on the screen, create a **Unit cell**. A **Unit cell** is a "rubber band box" which indicates the horizontal (between-column) distance and vertical (between-row) distance between the objects in the array. **Pick** a point on the object to be copied. AutoCAD will stretch a "rubber band box" away from this original point.

`Other corner:`
Pick a point which corresponds to the desired spacing in the vertical and horizontal distance and direction for the copies to be made. If this option is used, the array is constructed immediately after selecting this second point.

`Distance between columns (|||):`
Enter a numeric distance from the keyboard. The array will be created at the distances specified.

The **Rectangular array** option can make use of either positive or negative distances when choosing row and column spacing. Positive values build rows and columns up and to the right. Negative values build rows and columns down and to the left.

A **Polar array** will copy objects in a circular fashion around a center point specified by the user. The user can specify how many objects to create around the center, the angle through which the objects are to be rotated (in an arc rather than a circle), and whether the objects are rotated as they are copied.

`Command:`
Type **Array** (or **Ar**).

`Select objects:`
Pick the objects to **Array**.

`Select objects:`
Press **<Enter>** when finished selecting objects.

`Rectangular or Polar array (R/P) <R>:`
Type **P** to switch to polar array.

`Center point of array:`
Pick a point at the desired center of the array. This point can be picked from the screen, or entered as a coordinate value. It may be desirable to use an **Object snap** to specify the center point with precision.

`Number of items:`
Enter the number of objects desired in the array (including the original object).

`Angle to fill (+=ccw, -=cw) <360>:`
Press **<Enter>** for a circular array. For an array which forms an arc, enter a value less than 360°. If the value is negative, the arc will be formed in a clockwise direction. A positive arc value will create a counter-clockwise array.

`Rotate objects as they are copied? <Y>`
If **Y** is selected, the objects will be rotated around the center point as they are arrayed. If **N** is chosen, the objects will retain their original position as they are rotated. These options are shown in Figure 2.1 and Figure 2.2.

The MIRROR command

Mirror (**MI**) The **Mirror** command will create a "mirror image" of a selection set on the opposite side of a "mirror line." The mirror line can be drawn at any angle on the screen, often either horizontal or vertical. This is very easy to do when **Ortho** is activated. You may elect to keep or erase the original selection set.

`Command:`
Type **Mirror** (or **Mi**).

`Select objects:`
Select the objects you wish to mirror.

Figure 2.1

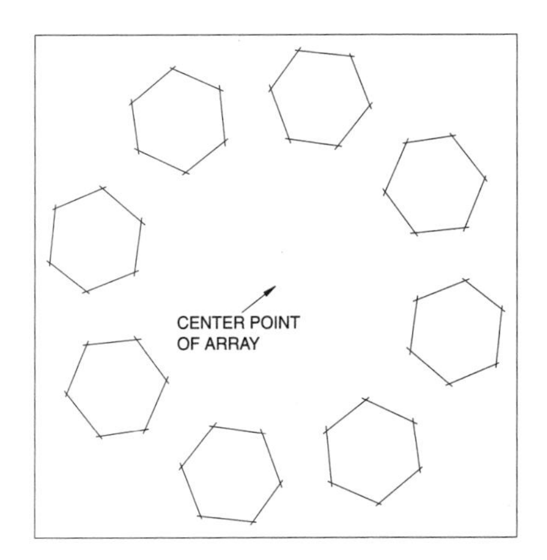

Figure 2.2

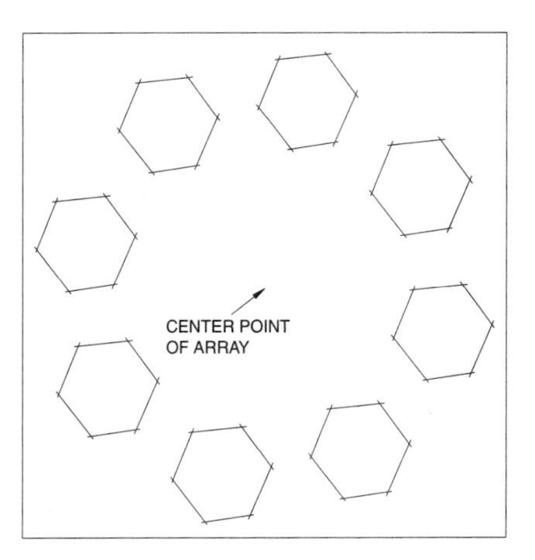

`Select objects:`
Press **<Enter>** to complete the selection set.

`First point of mirror line:`
Pick the first endpoint of the line to mirror across. Select this point with care since it dictates not only the direction but also the distance at which the copy is made.

`Second point:`
Pick the second endpoint of the mirror line. The length of the line is irrelevant.

`Delete old objects? <N>`
Type **Y** if you wish to remove the original selection set, and **N** if you wish to leave the original selection set intact.

Mirror is another very efficient command, since it can greatly speed the drawing of symmetrical objects. If care is taken when one side of an object is created, the other half can be created quickly and easily.

Changing Size or Location of Objects

The MOVE command

Move (M) The **Move** command allows you to select an object (or objects) and move it from one location to another. After choosing a selection set, you specify a **Base point** (another "handle" or "grab" point) on the screen. You then are asked to specify a **second point** which represents where the objects will move to. The orientation of the selection set to the **Base point** remains consistent as the selection set is moved across the screen.

```
Command:
```
Type **Move** (or **M**).

```
Select objects:
```
Pick the selection set you wish to move.

```
Select objects:
```
Press **<Enter>** to complete the selection set and return to the **Move** command.

```
Base point or displacement:
```
Pick a convenient location on the object as a "handle point" by which to drag the objects.

```
Second point of displacement:
```
Pick a new location for the "handle point". The selection set will be moved along with the "handle point."

The TRIM command

Trim (TR) The **Trim** command cuts off a line, polyline, circle or arc back to its intersection with another object ("cutting edge") that you specify. The cutting edge can also be a line, polyline, circle, or arc. Pick the cutting edge(s) first, then select the object(s) to be trimmed. The **Trim** command is a simple but versatile command with complex applications.

To begin with, you can pick two cutting edges intersecting a desired line and trim off the excess from both ends, or eliminate the portion in the middle, leaving the outside pieces.

To use the command:

```
Command:
```
Type **Trim** (or **Tr**).

```
Select objects:
```
Pick the object(s)—using the selection set tools mentioned earlier in this chapter—to be used as the cutting edge(s).

```
Select objects:
```
Press **<Enter>** to complete the selection set and return to the **Trim** command.

```
<Select object to trim>/Project/Edge/Undo:
```
Pick the portion of the object you wish to be trimmed.

```
<Select object to trim>/Project/Edge/Undo:
```
Press **<Enter>** to end the command. **Undo** will undo any accidental trims.

Hint: When prompted for the first selection set (cutting edges) during the **Trim** command in AutoCAD 13, you can press **<Enter>**. This will select all of the objects on the screen as cutting edges. You can then proceed with trimming any object as before. This is a very efficient short-cut for using the **Trim** command, and is not found in any of the AutoCAD documentation.

In addition, the **Trim** command normally requires you to select only one object at a time to **Trim**. If you wish to **Trim** several objects simultaneously, use the **Fence** option (or any other selection set option) to pick the objects to **Trim**. You may select as many objects as you wish in this manner.

Another use of the **Trim** command is to trim objects to an **extended edge**. This means that an object can be trimmed to a cutting edge, even though they might not cross each other. AutoCAD can temporarily (within its coded functions) extend the cutting edge to the object to be trimmed.

Figure 2.3

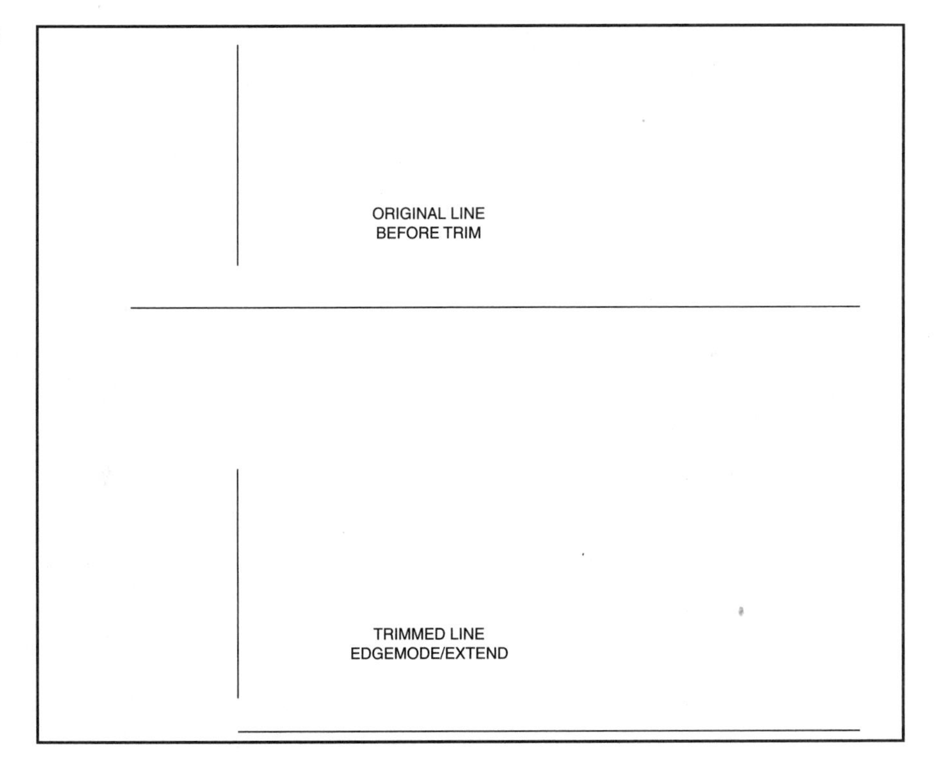

To use this option:

```
Command:
```
Type **Trim** (or **Tr**).

```
Select objects:
```

Pick the cutting edge which does not extend to the object to trim.

```
Select objects:
```

Press **<Enter>** to complete the selection set and return to the **Trim** command.

```
<Select object to trim>/Project/Edge/Undo:
```

Type **Edge** (or **E**).

```
Extend/No extend <No extend>:
```

Type **Extend** (or **E**).

```
<Select object to trim>/Project/Edge/Undo:
```

Pick the object to trim. AutoCAD will invisibly extend the cutting edge to the object, then trim the desired object.

```
<Select object to trim>/Project/Edge/Undo:
```

Press **<Enter>** to leave the command.

One limitation of the **Trim** command relates to its handling of wide polylines. The **Trim** command will always trim to the center of a wide polyline. This may lead to unsightly edges which may require further editing.

Figure 2.4

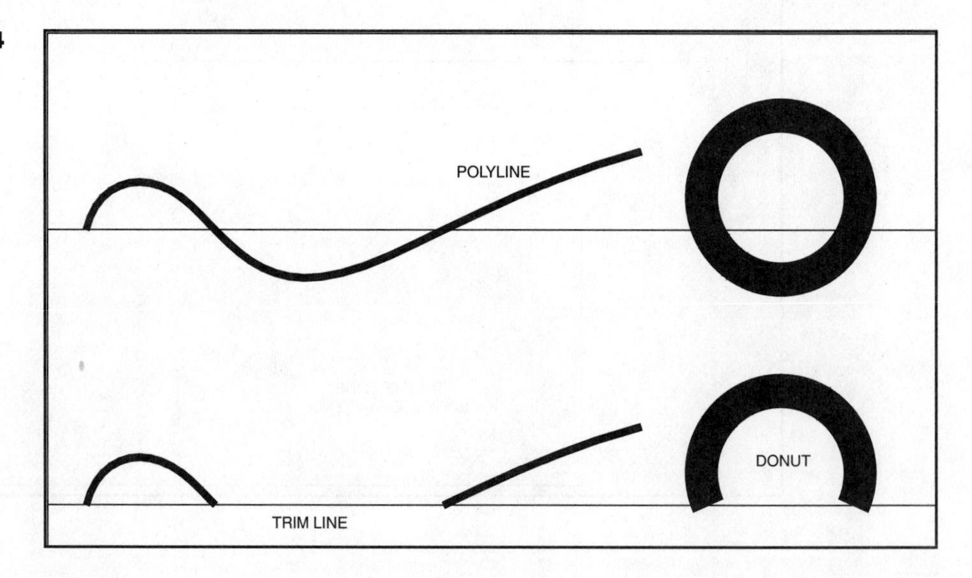

A final option in the **Trim** command relates to trimming objects in three-dimensional space. When **Project** is selected, three options are available:

None—this default value indicates that only objects within the same plane as the cutting edge can be trimmed, or that the object and the cutting edge must actually intersect in space. No projection of cutting edges will occur. This is the normal value when working in two-dimensional space.

UCS—this value indicates that objects will project a cutting edge perpendicular to the current **User Coordinate System**. This way, in the plan view, an object on one plane can act as a cutting edge to trim an object on another plane. A horizontal line at 10' elevation (10 feet on the Z axis) can be used to trim a circle on the base plane (0 feet on the Z axis) elevation. Even though the objects are on different planes and do not intersect in space, the **Project** option allows the cutting edge to **Project** down to the base plane. Remember that this cutting plane is always perpendicular to the current **UCS** X-Y plane.

View—this option allows objects to project cutting edges into space along the **current viewing plane**. This viewing plane may be different than the **User Coordinate System** and so the projected plane may be different as well. The use of this option is dictated by the location of the viewing point in space. Objects which **appear** to intersect from the current view point can be trimmed.

The EXTEND command:

Extend (**X**) The **Extend** command works as an inverse of the **Trim** command. As with the **Trim** command, the first selection set is the "edges." The **boundary edges** are the objects to which the desired object is extended. The second selection set is the object to be extended. As with the **Trim** command above, AutoCAD normally accepts only one object to **Extend** at one time. Use **Fence** or another selection set option to pick several objects simultaneously. To use the command:

`Command:`
Type **Extend** (or **X**).

`Select objects:`
Select the boundary edges that will define the newly extended object.

`Select objects:`
Press **<Enter>** to complete the selection set and return to the **Extend** command.

`<Select object to extend>/Project/Edge/Undo:`
Select the object that you wish to extend. Use the selection set options to allow you to pick more than one object to **Extend**.

`<Select object to extend>/Project/Edge/Undo:`
Press **<Enter>** to complete the command.

The **Project** and **Edge** options function exactly as they do in the **Trim** command. The **Edgemode** option is especially useful when extending objects to boundary edges when the boundary edge is too short to create an intersection.

Figure 2.5

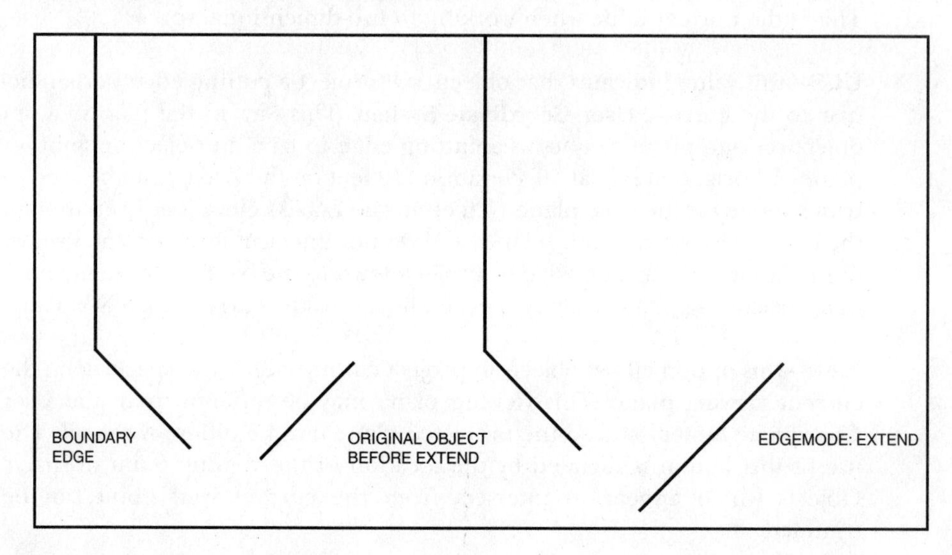

The ROTATE command

Rotate (**Ro**) The **Rotate** command allows you to rotate a group of objects around a specified point. After the selection set and **Base point** are specified, the command asks for a **rotation angle**. The selected object will then be rotated around the **base point** by the indicated angle. A positive angle results in a counter-clockwise rotation, while a negative angle results in a clockwise rotation. Using **Ortho** mode will result in 90° increment rotations.

`Command:`
Type **Rotate** (or **Ro**).

`Select objects:`
Select the objects you wish to rotate.

`Select objects:`
Press **<Enter>** to complete the selection set and return to the **Rotate** command.

`Base point:`
Specify the center point for the rotation. The selection of this point can be made on the screen or from the keyboard.

Care should be taken when selecting the **Base point**. Picking a point too far from the selected objects can result in them being rotated off-screen, even past the current drawing limits.

```
<Rotation angle>/Reference:
```
Enter the desired angle of rotation.

The **Rotation angle** method allows you to enter a rotation angle as a value, from the keyboard. In addition, you can rotate the objects using your pointing device, watching the objects drag through different rotation angles. These methods are useful if an object is at a known angle, and/or the desired angle of rotation is known.

Or

```
<Rotation angle>/Reference:
```
Type **R** to activate the **Reference** method.

```
Reference angle <0>:
```
Pick a point along the selection set, or enter a value as a reference.

```
Second point:
```
Pick a second point along the selection set. (This prompt will not appear if the **Reference angle** was entered as a value in the previous step.)

```
New angle:
```
Enter an angle as a value, or pick a point on the screen to indicate the desired angle.

The ALIGN command

Align—The **Align** command actually combines the functions of the **Move** and **Rotate** commands. After you create a selection set of objects to align, you pick a source point on the selection set, and a corresponding destination point where the objects are to align. This process is repeated for a second source and destination point. If a 3-D object is selected, a third point is required for the alignment.

Using **Align** is most efficient when **Osnaps** are used. Selecting intersections, endpoints or other **Osnaps** for the first and second points make the alignments easy and precise. See the section below on **Osnaps**.

To use the command:

```
Command:
```
Type **Align**.

```
Select objects:
```

Select the objects you wish to align.

```
Select objects:
```

Press **<Enter>** to complete the selection set and return to the **Align** command.

```
1st source point:
```

Pick the first point on the selection set. This point will act as one of the "handle points" by which the selection set will be aligned. This first point actually sets the location of the **Base Point** during the **Move** portion of the command.

```
1st destination point:
```

Pick the new location of the first "handle point." This will be the new location of the **Base point**.

```
2nd source point:
```

Pick the second "handle point" on the selection set. This second point is used to set the **Reference** angle during the **Rotate** portion of the command.

```
2nd destination point:
```

Pick the new location of the second handle point. The selection set objects will then be moved and rotated into position along the destination points. Note that the second destination point is used only to set the rotation angle for the selection set, not to move it away from the first destination point.

```
3rd source point:
```

Press **<Enter>** to stop naming points for 2-D objects. If using **Align** on 3-D objects, a third source and destination point are required.

```
<2d> or 3d transformation:
```

Press **<Enter>** to complete the command.

Figure 2.6

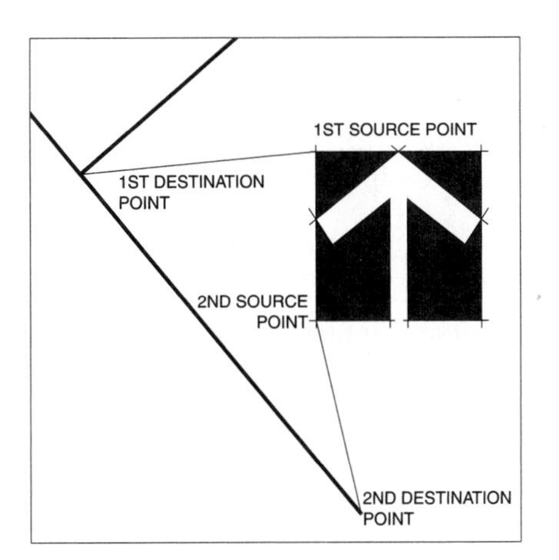

Figure 2.7

The STRETCH command

Stretch—The **Stretch** command is one of the most powerful editing tools in AutoCAD. This command allows you to change the length of objects in a selection set *while leaving them connected to other objects not included in the selection set.* The **Stretch** command changes the locations of object endpoints which are included in a crossing box. The endpoints of those objects which are outside of the crossing box are left in place. Lines, polylines and arcs can all be stretched. In addition, dimensions are stretched as the objects they reference are stretched.

Selecting objects to **Stretch** is somewhat different than creating other selection sets. The only options for selecting objects to **Stretch** include **Crossing box** and **Crossing polygon**. While entire objects highlight as the selection set is created, only those endpoints included in the **crossing box** (or **polygon**) are actually affected. As during the **Move** command, a **Base point** is selected (from the keyboard or on the screen), then a **Second Point of Displacement** is chosen. The second point might be selected from the screen, entered as absolute coordinates from the keyboard, or entered as polar or relative coordinates from the **Base point**.

`Command:`
Type **Stretch**.

`Select object(s) to stretch by crossing-window or -polygon:`
Type **Cr** (to activate a **Crossing box**) or **Cp** (to activate a **Crossing polygon**), then pick the objects to be stretched.

`First corner:`
Pick a corner for the **Crossing box**.

`Second corner:`
Pick the opposite corner of the **Crossing box**. Be sure to include only those object endpoints (including polyline segment endpoints) which you wish to be modified. Although entire objects will be highlighted, only those endpoints included in the box will be affected.

`Select objects:`
Type **<Enter>** to complete the selection set and return to the **Stretch** command.

`Base point or displacement:`
Pick a "handle point" point from which the objects will be stretched. As with other commands, using **Object snaps** will assist in maintaining accuracy and speed in your drafting.

`Second point of displacement:`
Pick a second "handle point" where the first point will be relocated. All of the selected endpoints will be moved the same angle and distance. Since the other endpoints remain stationary, the objects remain anchored, but are stretched the indicated distance.

Figure 2.8

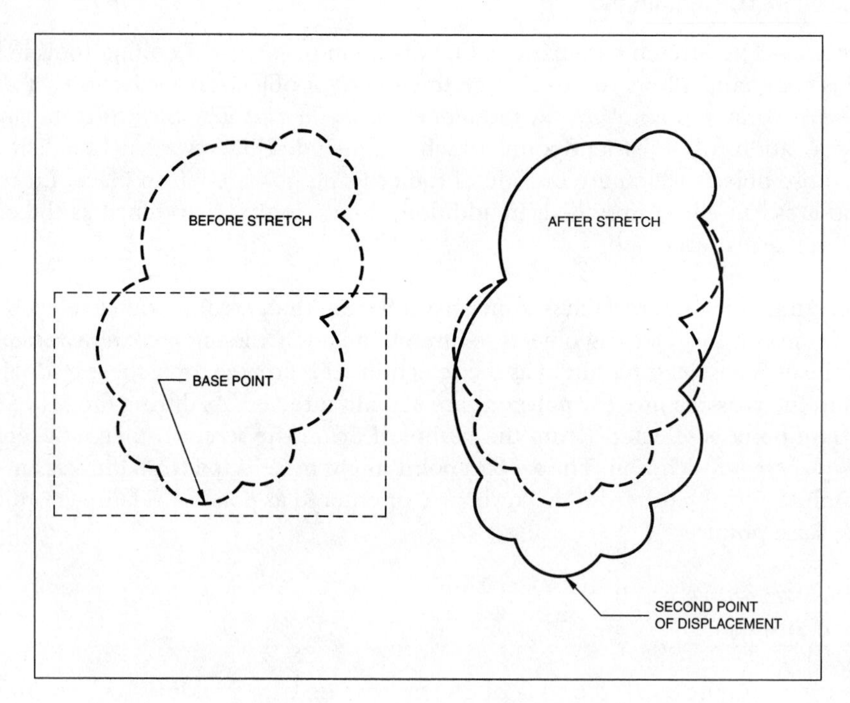

The SCALE command

Scale (**SC**) The **Scale** command allows you to change the size of an object—either larger or smaller. The object will retain its correct proportions at any size. Here you specify the stationary point (the point which will not move as the object is scaled) and the scaling factor. Any scale factor larger than 1 will increase the size of the object. Any scale factor less than 1 (use decimals) will decrease the size of the object.

`Command:`
Type **Scale** (or **Sc**).

`Select objects:`
Pick all of the objects you wish to resize.

`Select objects:`
Press **<Enter>** to complete the selection set and return to the **Scale** command.

`Base point:`
Pick the stationary point on the object(s) to be scaled. The selection set will grow (or shrink) around the stationary point.

`<Scale factor>Reference:`
To use the **Scale factor** option, enter a value from the keyboard. For best results, use decimal values. The selection set will be resized according to the **Scale factor** chosen.

Or

to use the **Reference** option, pick two points on the selection set (or enter a value from the keyboard). This distance (or value) represents the reference length used for scaling. A new value for this reference length can then be entered from the keyboard or stretched across the screen. To use this option:

`<Scale factor>Reference:`
Type **R** (for **Reference**).

`Reference length <1>:`
Enter a value from the keyboard, or pick a point on the selection set. If a first point is selected, you are asked for a second point.

`Second point:`
Pick a second point on the selection set.

`New length:`
Either enter a value from the keyboard or move the cursor on the screen to change the size of the object(s).

The FILLET command

Fillet (**FI**) The **Fillet** command lets you square off or round off corners to any radius you specify. A radius of **0** will yield a squared-off corner, while a radius setting will create a rounded corner. In addition, **Fillet** will trim and extend objects automatically. This means that objects need not intersect to use the **Fillet** command on them. The **Fillet** command also contains a switch to enable or disable automatic trimming. This may be especially useful when performing multiple fillets on the same line.

The most basic form of **Fillet** is to simply round off the intersection of two lines. To begin, set the radius desired:

`Command:`
Type **Fillet** (or **Fi**).

`(TRIM mode) Current fillet radius = 0'-0″`
`Polyline/Radius/Trim/<Select first object>:`
Type **R** (for **Radius**).

`Enter fillet radius <0'-0″>`
Enter a value for the desired radius, or pick two points on the screen to indicate the desired radius. The command ends after the radius is entered.

After the radius is entered, objects (lines, arcs and circles) can then be **Fillet**ed. Note that a polyline can have its own intersections **Fillet**ed, but cannot be **Fillet**ed with another object.

`Command:`
Type **Fillet** (or **Fi**) or press **<Enter>** to repeat the last command.

`(TRIM mode) Current fillet radius = 0'-6″`
`Polyline/Distances/Trim/<Select first object>:`
Pick the first object to fillet. Be sure to pick the object toward the end where the fillet is desired. Picking the object in the middle will yield unpredictable results.

`Select second object:`
Pick the second object to fillet. The command will end after the second object is chosen and the fillet is created.

The **Fillet** command with a radius set to **0** is particularly useful in creating square corners and cleaning up line intersections. Since **Fillet** will perform automatic **Trim** and **Extend** functions, it is a very efficient way to edit lines which might (or might not) intersect. **Fillet** is a good command to lend accuracy and speed to the drafting process.

When **Fillet**ing several objects to the same line, use **Fillet** with the **NOTRIM** option. This causes the desired line to remain unchanged as the other lines are **Fillet**ed.

Figure 2.9

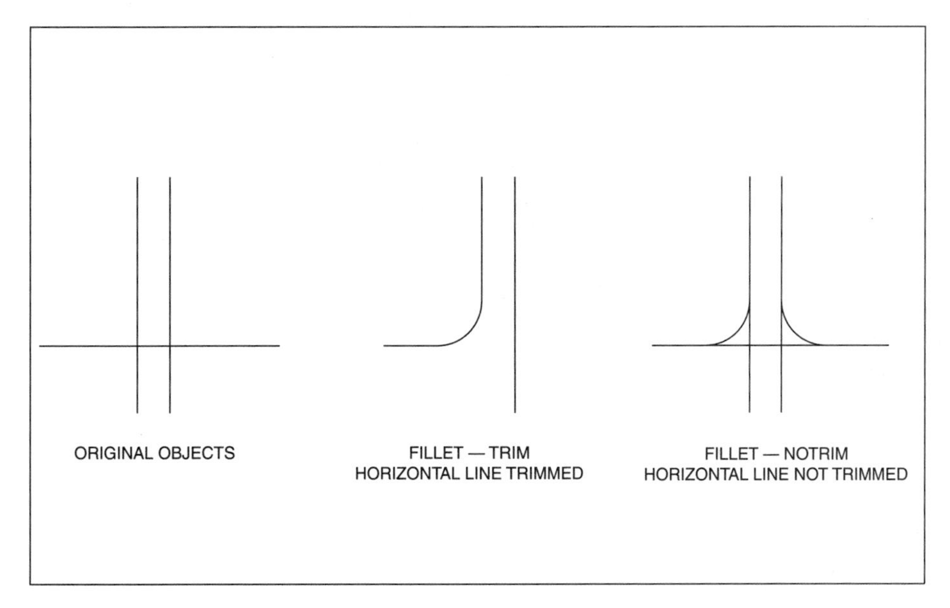

Changing Properties of Objects

The CHANGE command

Change (**Ch**) One of the most flexible commands, **Change** allows you to change the endpoint of a line (or group of lines) or arc, change the properties of an object (color, linetype, layer, elevation, etc.) or the contents of a string of text (style, insertion point, height, and the text itself). Two enhancements of the **Change** command are also available:

> **Ddmodify**—always brings up a context-sensitive dialogue box. After selecting an object to modify, a dialogue box appears which allows the properties of **that specific object** to be modified. There are different dialogue boxes for lines, arcs, polylines, splines (specially edited polylines), circles, hatches and text. Each dialogue box allows you to edit any of the properties of the object you chose.
>
> **Chprop**—works on two or more objects at once. After selecting the objects to modify, a dialogue box appears allowing you to choose which property(ies) of the objects to change. Because many kinds of objects might be chosen, only a few properties can be modified through the **Chprop** command.

Color—changes the color of the selection set to another color, regardless of the layer on which the items reside. You can also designate the color as **Bylayer** which causes all items to be displayed with the color of the layer on which they reside.

Layer—changes the layer on which the items in selection set reside.

Linetype—changes the specific linetype with which the items in the selection set are drawn. Changes made here might not be visible unless the **Linetype scale** is set properly.

Ltscale—changes the linetype scale of the objects in the selection set. Previous versions of AutoCAD have only allowed the linetype scale to be assigned to the entire drawing. No individual linetype scaling was possible. Version 13 has created the possibility of editing individual object linetype scaling. Changes made to the linetype scale in the **Chprop** command will only affect the items selected and not the global setting of **Ltscale**.

Thickness—the dimension of object on the Z axis is altered with this setting. Most 2-D objects are created with a **Thickness** set at **0**.

When the **Change** command is used, the prompt indicates two of the three modes available:

`Command:`
Type **Change** (or **Ch**).

`Select objects:`
Choose the objects you wish to **Change**.

`Select objects:`
Press **<Enter>** to complete the selection set and return to the **Change** command.

For the **Properties** option:

`Properties/<Change point>:`
Type **P** to change the properties of the selection set.

`Change what property (Color/Elev/LAyer/LType/Scale/Thickness)?`
These options are identical to those found under **Chprop**, except for **Elevation**. The **Elevation** of an object is its position (not thickness) on the Z axis.

Or

for the **Change point** option:

`Properties/<Change point>:`
Pick a point which will act as the new endpoint to all of the objects in the selection set.

Or

for the **Text** option:

`Properties/<Change point>:`
Press <**Enter**>. If the object in the selection set was a text entity, these properties can be edited:

`Enter text insertion point:`
Enter the new insertion point for the text.

`Text style: STANDARD`
`New style or RETURN for no change:`

The top line indicates the current style for the text. Enter the name of any existing text style to change the style of the selected text. Press <**Enter**> for no change.

`New height <2'-0">:`
Indicate the new desired height for the text. Press <**Enter**> for no change.

`New rotation angle <0>:`
Indicate the new desired rotation angle for the text. Press <**Enter**> for no change.

`New text <EXISTING TEXT REMARK>:`
Enter any new text to replace the existing text. Press <**Enter**> for no change.

The EXPLODE command

Explode (**Ex**) The **Explode** command is used to break complex objects into simple objects which can be edited with traditional editing commands. Polylines, polygons, hatches, dimensions and blocks are all built of multiple objects, and all cannot be edited by normal editing commands. The command is extremely simple to use.

`Command:`
Type **Explode** (or **Ex**).

`Select objects:`
Pick the objects to explode.

`Select objects:`
Press <**Enter**> to complete the command. The objects will be **Explode**d.

Although the **Explode** command is simple and easy to use, it is powerful and should not be used lightly. Exploded objects can greatly expand the size of the drawing, and change object properties dramatically. Polylines lose their width when exploded, and may consist of a very large number of line segments if they had been converted to splines previously. Exploded dimensions will be placed on the **0** layer and will lose their associativity. Their color designation is **Byblock** and must be changed to **Bylayer** if you want them to behave like other objects in the drawing. Exploded hatches are also placed on layer **0**, and also lose their associativity. They are broken into individual line segments which must be edited individually rather than as a single object. Exploded blocks undergo dramatic—and usually undesirable—changes. Like exploded dimensions (which were actually blocks), exploded blocks are placed on the **0** layer. All attributes become visible text entities, and are no longer associated with the block. Once a block has been exploded, it will no longer be recognized when attributes are searched and listed. If there are other alternatives to exploding blocks, it is best to take advantage of them.

The PEDIT (POLYLINE EDIT) command

Pedit (**Pe**) The **Pedit** command allows you to change the properties of a polyline. With **Pedit**, the user can control many facets of a polylineÍs appearance, location of its vertex points and even its curvature in relation to existing vertex points.

`Command:`
Type **Pedit** (or **Pe**).

`Select polyline:`
Pick the polyline you wish to edit.

```
Close/Join/Width/Edit vertex/Fit/Spline/Decurve/Ltype
gen/Undo/eXit <X>:
```
Each of these options are explained below:

Close (or **Open**)—this option allows you to close the polyline back to its origin point. Closed polylines have no beginning or ending points. A closed polyline is useful since AutoCAD can compute its length (circumference) and the area within the polyline. If the polyline has been closed with the **Close** option, **Pedit** can be used to open the polyline to undo the **Close** command.

Join—this option allows you to add line segments or arcs to an existing polyline. This is only possible when the endpoint of the additional object corresponds to the endpoint of the existing polyline. If there is **any** discrepancy, the object(s) will not be added to the polyline. Closed polylines cannot have objects **Join**ed to them.

Width—this option allows you to assign a new width to the entire polyline. If the width is different in separate segments of the polyline, this option will reset the widths of all segments to the new value.

Edit Vertex—this option introduces an entire set of editing tools to modify the vertex points in the polyline. In response to the **Edit vertex** option, AutoCAD responds with the choices:

```
Next/Previous/Break/Insert/Move/Regen/Straighten/Tangent/Width/
eXit <X>:
```

Next—advances the marker to the next polyline vertex visible on the screen. The marker is displayed as an "X" mark on a polyline vertex point.

Previous—reverses the direction of the movement of the vertex marker, and returns it to the last vertex point.

Break—removes a segment of the polyline between the current vertex point (where this option is invoked) and a selected vertex point. The prompt reads:

```
Next/Previous/Go/eXit <N>:
```

Choosing **Next** moves the vertex marker to the **Next** vertex. You can continue to specify **Next** to advance the marker to the desired vertex if you want to remove several segments at the same time. Choosing **Previous** changes direction of movement of the vertex marker to the **Previous** vertex. If you continue to choose **Previous** you can move the marker back through the vertices until the desired one is highlighted. Choosing **Go** creates the break in the polyline between the current vertex point (where **Break** was invoked) and the selected vertex point (where **Go** was entered). If you specify the current and desired break point as the same vertex, the polyline will be broken in two parts with no visible break between them.

Insert—inserts a point after the marked vertex point. This option should be used carefully to ensure that the point is inserted in the proper sequence to create the desired effect.

Move—allows the relocation of any vertex point. After this option is selected, you are prompted for a new location for the currently selected vertex.

Regen—regenerates the polyline (not the screen) to show the new curvature, width or other edited feature of the polyline.

Straighten—works similarly to **Break**, except that the polyline is straightened between the marked vertex and the desired vertex. The vertex marker can be advanced through the vertices in either direction by repeated use of the **Next** or **Previous** options. After the desired vertex is marked, **Go** actually creates the straightened segment.

Tangent—allows you to specify a tangent direction to any vertex in the polyline. This tangent direction will be stored with the vertex for use in a curve-fit polyline. AutoCAD prompts:

```
Direction of tangent:
```

then allows a rubber band line to be stretched from the vertex to indicate the desired direction of tangency. The tangent value stored with the vertex will only take effect if the polyline is subsequently turned into a curve-fit polyline.

Width—allows the assignment of a width to a specific segment of the polyline without affecting the width of the other segments. The width of the segment immediately after the marked vertex will be modified. The **Width** option also allows for a different width at the beginning and the end of the segment. Use the **Regen** option to display any change in the width of a segment of the polyline.

eXit—leaves the **Edit vertex** mode and returns to the **Pedit** options.

Fit—this converts the polyline from line segments into arcs. This has the effect of creating a pair of arcs from each vertex to the next vertex. The resulting smooth curve will pass through every vertex point in the polyline and will use any tangent directions attached to the vertex points.

Spline—this converts the polyline to a *Bezier spline* (or B-spline). A *Bezier spline* is a curved line whose shape is dependent on **control points** which are located along its path. Although the **spline** does not pass through these control points, they act to "pull" the curves of the **spline** and modify its shape. When a polyline is converted to a **spline**, the vertex points are converted to control points. The **spline** will pass through the first and last control points, while the others "pull" the curved polyline toward them. If the original polyline includes arc segments, they are straightened as the **frame** for the **spline** is calculated. The **frame** is the original polyline (and its vertices) without any curved segments. This **frame** is retained in the drawing database. If the spline is then **decurved**, the frame is restored. It is also possible to display the **frame** and the **spline** at the same time with the **Splframe** AutoCAD system variable.

A **spline** is very different from a **fit** curve in that the **fit** curve will pass through each **control point** (polyline vertex), but the **spline** does not. In addition, the mathematical formula for calculating a **spline** polyline can be performed as a quadratic and a cubic function. Control over this function is built into the **Splinetype** AutoCAD system variable. To change the type of curve function used, set this option before using the **spline** command. A quadratic *B-spline* is pulled more strongly to the **control points** (for a tighter curve-fit) while a cubic *B-spline* is less strongly pulled, yielding "looser" curves. To exercise even more control over the process of building **splines**, use the AutoCAD system variable **Splinesegs**. The larger this setting, the smoother the **splines** will be, but will occupy more drawing space and will take longer to generate on the screen.

Figure 2.10

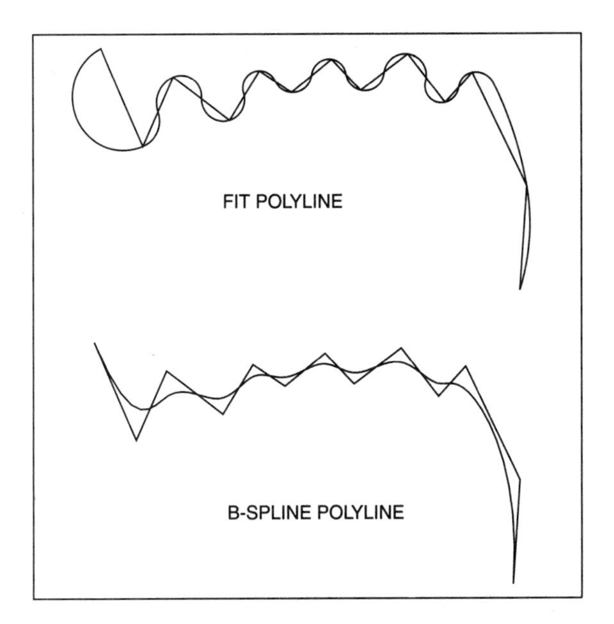

Splines can also be created with the **Spline** drawing command. A **spline** created with this command will pass directly through vertex points rather than being "pulled" by them. The **fit tolerance** can also be adjusted so that the "tightness" or "looseness" of the curve can be set to the needs of the user. The **Splinedit** command brings a large set of advanced editing tools to editing all types of splines.

Decurve—this option acts to undo the effects of the **fit** and **spline** choices. The **frame** of the polyline is restored, while the **fit curve** or **spline** is discarded.

Ltype gen—this option controls how the linetype of the polyline is drawn relative to its vertices. If the **Ltype gen** setting is turned **On**, the polyline is overlaid with a linetype pattern which will ignore the vertices. If the linetype has spaces included, a vertex might have a space over it. The pattern remains consistent through the length of the polyline. If the **Ltype gen** is set to **Off**, the linetype pattern will begin anew at each vertex. In this way, each polyline segment appears separate because its linetype will begin and end at its vertices.

Figure 2.11

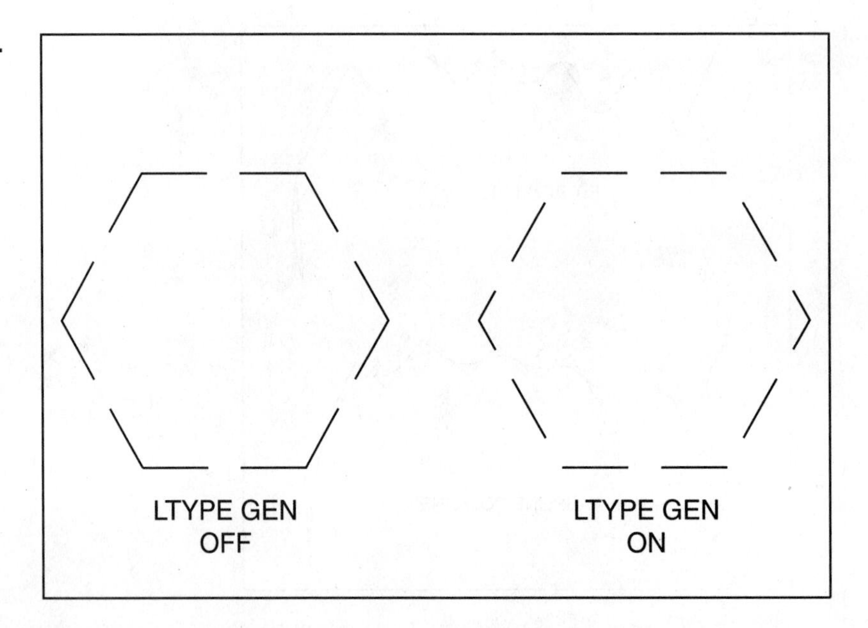

Undo—this will undo the last **Pedit** operation, and can be used to undo changes back to the beginning of the **Pedit** session.

eXit—this option exits the **Polyline** command.

The GROUP command

Group—The **Group** command allows an object (along with a collection of other objects in the selection set) to be assigned to a named group. This group can be then be used as a selection set on which any number of editing commands might be used. Prior to AutoCAD Release 13, selection sets could only be temporary. This often resulted in tedious repetition building the same selection set several times in the process of creating a drawing. By using the **Group** command, objects can be placed in a selection set that can be saved with a name and description. In addition, a **group** can be saved with a status of **selectable** or **non-selectable**. A **selectable group** can be selected as a whole if one of the group members is selected, or if **group** is entered at the "select objects" command prompt. A **non-selectable** group will not be selected as an entire entity if one of its members is chosen during the "select objects" phase of an editing command.

From the command line:

```
Command:
```

Type **Group**. This opens the "Object Grouping" dialogue box. If you type **-group**, the command will be run from the command line.

`?/Order/Add/Remove/Explode/REname/Selectable/<Create>:`

The options shown listed here are described in the functions of the "Object Grouping" dialogue box below.

Figure 2.12

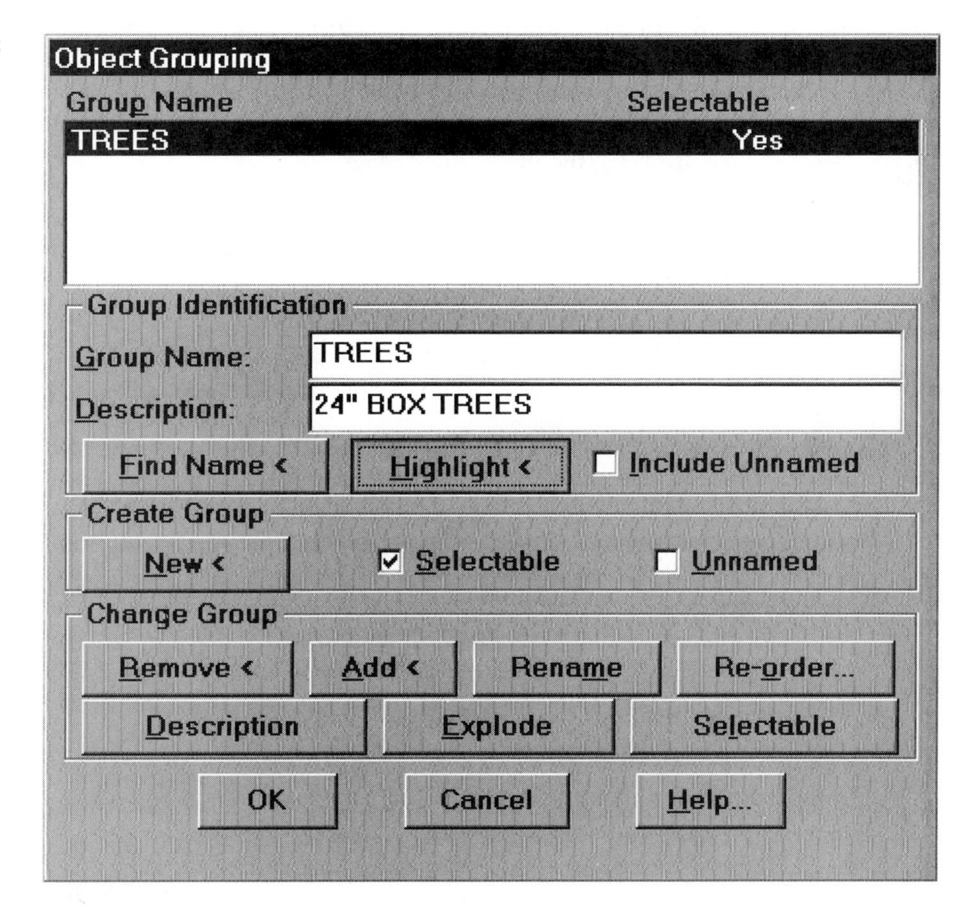

To build a **group**, first enter a **Group name:** in the text box. If you choose not to name a **group**, select the **Unnamed** option. The **group** will be given the default name ***A*n** where ***n*** represents the number of the group and **A** stands for **Anonymous**. Next, enter a **Description** for the **group**. This description can contain up to 64 characters. Finally, choose whether or not the **group** will be selectable, using the **Selectable** toggle box. With these preliminaries out of the way, use the **New <** button to pick the objects you wish to include in the group. Press **<Enter>** when finished selecting members of the group. When you return to the dialogue box, the name of the **group** and its **selectable** status will be added to the list of **groups** at the top of the dialogue box.

The Find Name < button allows you to select an object on the screen and determine which **group** it belongs to.

The Highlight < button will highlight all of the members of the **group** whose name appears in the **Group name:** text box.

The **Include Unnamed** toggle box will cause any unnamed **groups** to be included in the **Group Name** display at the top of the dialogue box.

The **Change Group** options at the bottom of the dialogue box allow you to alter the characteristics of any of the **groups** in the drawing.

The Remove < option is used to remove selected objects from a **group**. After you are returned to the drawing screen, the **group** objects are highlighted, and you are prompted to remove objects from the **group**. Even when all of the objects can be removed from a **group**, it retains its name and description.

The Add < option is used to add selected objects to the selected group.

The Rename option will rename the selected group to the name in the **Group Name** text box. This operation can be performed on both named and unnamed **groups**.

The Re-order ... option allows you to change the sequence in which the objects in the **group** are listed. This sequence might be critical in certain types of operations which move from item to item in a **group**. This option opens another dialogue box which is used to control the ordering of the objects in the **group**.

The Description option allows you to edit the existing description of a **group** or add a description to a **group** which does not yet possess one.

The Explode option removes all of the objects from a **group** and strips the **group** name from the list of **groups** at the top of the dialogue box.

The Selectable button specifies whether a highlighted **group** is selectable. This button is used to change the status of an existing **group** rather than to define its status during creation.

The BREAK command

Break (**Br**) The **Break** command is used to create a gap in an object, break the end off of an object, or make two objects out of one. Once this command is activated, the two points selected will be designated as the endpoints for the gap. If the first endpoint is not at the desired location, it can be respecified during the command. At the prompt:

`Command:`
Type **Break** (or **Br**).

`Select object:`
Select the object to break.

`Enter second point (or F for first point):`
If the first point is not the correct first break point, type **F** to respecify the first point. If **F** is chosen, the command prompt reads:

`Enter first point:`
Respecify the first break point. If **F** is not chosen, the prompt will read:

`Enter second point:`
To create a break in an object, pick the second point to indicate the end of the break region.

To split an object in two pieces without creating a gap between them, type **@** to indicate the second point is identical to the first point. The object will then be split in two pieces.

To break the end off of an object, the first point should be chosen on the object (at the point where the object will be cut off) while the second point is chosen beyond the endpoint of the object.

All of the Windows versions of AutoCAD offer icons which can be used to perform each type of **break**.

Using icons can speed the selection of a specific type **Break** you want to perform. A **1 Point break** will create the break at the point selected, with the second point automatically entered as **@**. This option requires that you select the point of the **break** accurately the first time. A **1 Point Select break** allows you to select the object to break first, then asks you to specify the **break point** next. The second point is automatically entered as **@**. A **2 Point break** is the same as the typed **Break** command. The first point selected on the object will also be the first **break point**, unless you type **F** to respecify the first point. You are then prompted for the second **break point**. A **2 Point Select break** allows you to select the object with the first pick, then select the first **break point** and finally the second **break point**.

c h a p t e r

Controlling the Cursor and the Appearance of Objects

Controlling the Cursor

The OSNAP (OBJECT SNAP) command

Object Snaps or **Osnap** (**Os**) The **Osnap** settings will cause your cursor to find a specified place on an object. If you set the **Osnap** to **Endpoint**, your cursor will snap to the endpoint of any line or arc or polyline that you touch. The **Osnap** feature improves your accuracy and speeds the drawing process. When you can easily move to the precise endpoint, midpoint, or other location on an object, you can be assured that lines will meet correctly, objects will intersect cleanly, and that your drawing will be **accurate** for purposes of engineering, architecture and construction functions. This type of accuracy is an important advantage of CAD drawings over hand drawn plans.

Osnap modes include the following options:

From—allows you to snap to a point at a specified angle and distance from a **Base point** that you pick. This option is only available from the cursor menu.

ENDpoint—allows you to pick the precise **endpoint** of a line, polyline, arc or spline.

MIDpoint—allows you to snap to the precise **midpoint** (between the endpoints) of an object.

INTersection—allows you to snap to the precise **intersecting point** between two objects.

APParent Intersection—in 3-D drawings, this allows you to snap to a point where two objects in different planes **appear** to intersect.

CENter—allows you to snap to the **center** point of a circle or arc.

QUAdrant—allows you to snap to a point at four different positions around a circle. You can snap to a **Quadrant** at 0°, 90°, 180°, or 270° on a circle.

PERpendicular—allows you to snap **perpendicular** to a selected object. This option can also be used to terminate line segments at perpendicular intersections with other objects.

TANgent—allows you to snap to the calculated **tangent** point of an arc or circle.

NODe—allows you to snap to a defined **point** on the screen. This **node** point can be placed individually, or as the result of the **Divide** command.

INSertion—allows you to snap to the **insertion** point of a block.

NEArest—allows you to snap to the object nearest to the cursor. This option does not lend any particular accuracy to a point selection, other than guaranteeing that the point will be **on** a selected object rather than **off** of it.

Osnaps can be activated several different ways during the drawing process. The method used for invoking **Osnaps** will vary depending on the habits of the user, the platform of AutoCAD used, and the particular task at hand. Because **Osnaps** are particularly valuable tools for drawing, each method of using them should be thoroughly understood by every AutoCAD user.

Cursor Menu —This pop-up menu in AutoCAD is activated by the pointing device. When using a two-button mouse, use the **Shift+Rt Button** (or **Shift+#2 Button**) combination to bring the **Cursor Menu** to the screen. When using a three-button mouse, use the **Center Button** (or **#3 Button**) to activate this menu. When using a digitizer, press the **#3 Button** (Figure 3.1).

Osnap Toolbar—In the Windows platforms, a toolbar specifically for **Osnaps** is available either as a "fly-out" toolbar, or as one that can be placed on the screen with all of its options visible at one time (Figure 3.2).

Command Line—With the command line option, any of the **Osnap** modes can be activated while in the midst of performing an AutoCAD command. When prompted for a point during a command, simply typing in **INT** (or any three letter abbreviation) will activate the appropriate **Osnap mode**:

Figure 3.1

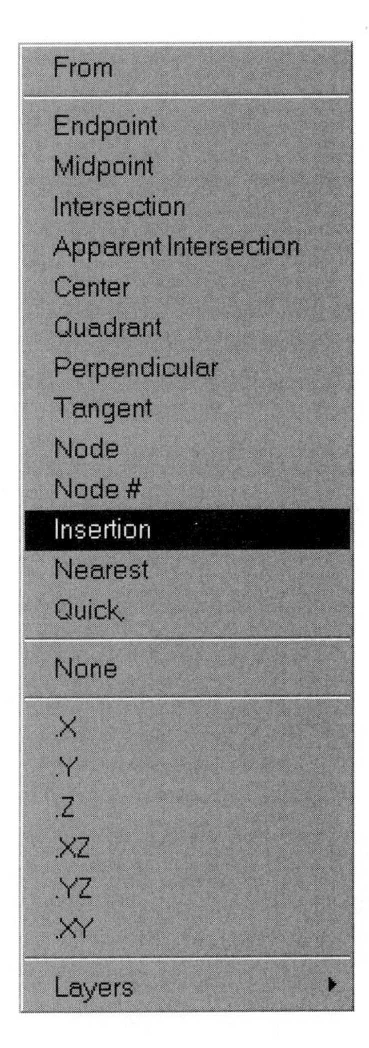

Figure 3.2

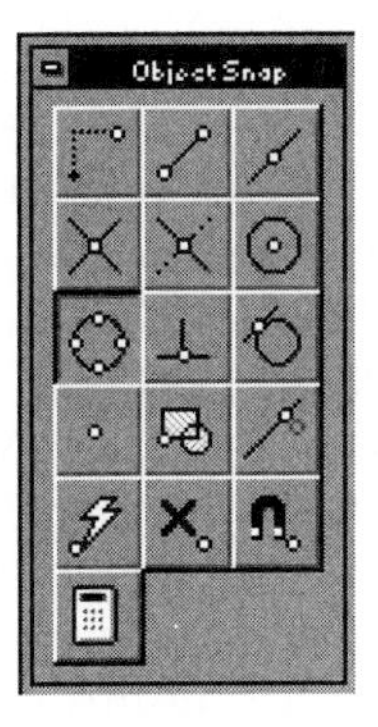

`Command:`
Line

`From point:`
Type **INT.**

`_Int of:`
Pick the desired intersection for the first point of the line.

`End point:`
Type **PER.**

`_Per to:`
Pick the object to which the line should be perpendicular.

All of these methods can be used ***transparently***. This means that they can be chosen during the process of picking a point during a command without interrupting the command itself. While this is desirable in many (even most) situations, these methods work for only a *single point at a time.*

Running Osnaps

There are occasions where you may wish to work with the same **Osnap** active for several commands or steps. You may wish to draw a series of lines from one intersection to another throughout the drawing. In this instance, it is better to activate the **Intersection** snap and leave it active during the entire line building process. This is called a **Running Osnap**. Using this method is generally faster than setting a single **Osnap** each time a point is selected. There are also times where it may be desirable to have more than one type of **Running Osnap** active at a time.

Running Osnaps can be activated in two ways:

1. From the command line :

   `Command:`
   Type **Osnap** (or **Os**).

   `Object snap mode:`
   Type **INT** or any other **Osnap mode** (more than one entry can be typed here—separate them with a comma, but not a space).

   The **Object snaps** will then be active **until they are shut off**. It is important to shut off **running Osnaps** when you have finished performing the desired operations. Allowing **Osnaps** to run inadvertently can cause the cursor to behave unpredictably. It may be difficult to move the cursor into a position **not** specified by a **running osnap** that is active. To shut off a **running osnap**:

   `Command:`
   Type **Osnap.**

   `Object snap mode:`
   Type **Off.**

2. From the "Running Object Snap" dialogue box:

   `Command:`

   Type **Ddosnap** or pick **Running Object Snap...** choice from the **Options** pull down menu (Figure 3.3).

Figure 3.3

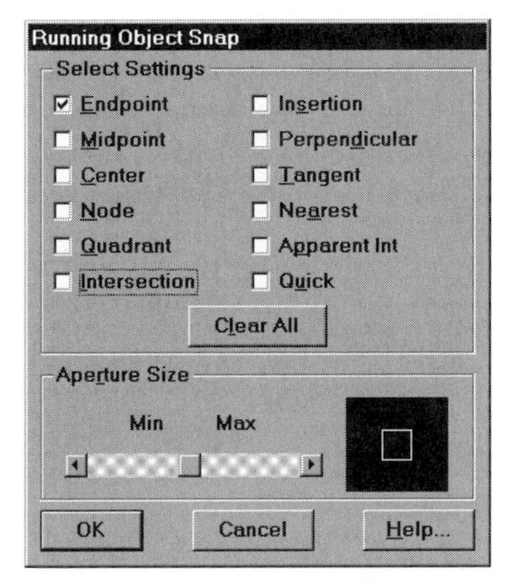

As with the command line method, any combination of **object snap**s can be activated. In addition, the **Aperture Size** can be set in this dialogue box. The **Aperture Size** refers to the size of the box surrounding the cursor crosshairs when an **object snap** is active.

Remember to turn **OFF** any running **OSNAP** modes when you are finished with them. You can shut them off from the command line (see above) or by returning to the "Running Object Snap" dialogue box and selecting **Clear All**.

Be sure that your aperture size is large enough to make it distinct from the normal cursor. That way it will be obvious when an **Osnap** is active, and you will know to shut it off if it is not wanted. If you are unable to control your cursor properly, the odds are good that an **object snap** is running without your being aware of it.

Grips are editing devices which are available on any object created in AutoCAD. There is no command for activating **Grips**. Simply pick a group of objects (picking manually, or using the automatic **Window** or **Crossing box** options) without first specifying a command, and their **grips** will highlight. Any number of objects can have their **grips** activated at one time. **Grips** will appear at the endpoints and the midpoint of lines, the center and quadrant points of circles, the center and endpoints of arcs, and at all of the vertex points of a polyline. Blocks and text each have a **grip** at their insertion point (though text created with **Mtext** will have grips at four corner points). Dimensions typically have **grips** for each dimension and extension line and for the text.

Since **grips** always appear at **object snap** points on an object, they can be used to reduce steps in selecting points when editing. By using **grips** you can actually take advantage of **Osnaps** and editing commands simultaneously. You can invoke **Stretch**, **Move**, **Rotate**, **Scale**, **Mirror** and **Copy** commands without typing or picking them from menus or icons (Figure 3.4).

Figure 3.4

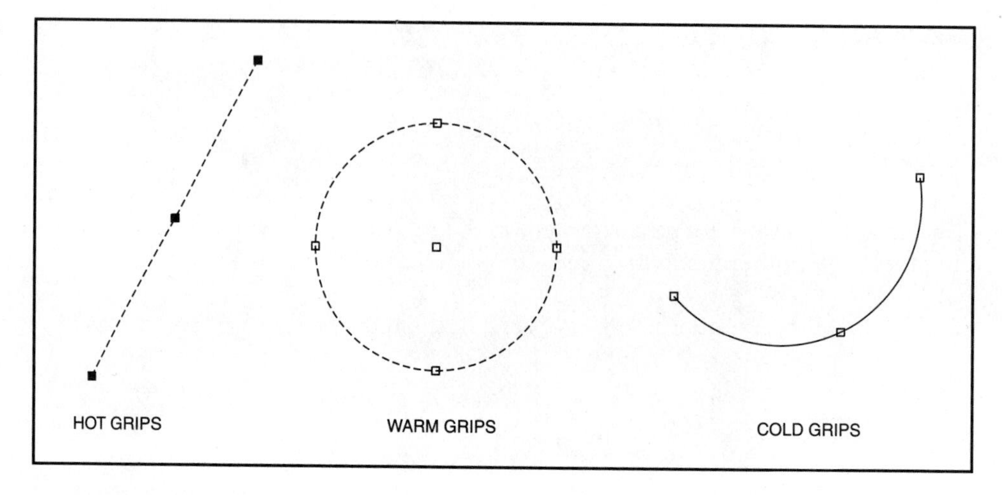

Grips Status

Grips exist in three states. A **grip** initially appears as a blue outline box when an object is picked. As long as the object is highlighted, the blue box represents a **warm grip** that is not being actively edited. When you move your cursor inside this box and press the **<Pick>** button, this box turns solid red. This represents a **hot grip**. Two or more **grips** can be made **hot** by holding down the **<Shift>** key while selecting each **grip**. Once a **grip** becomes **hot**, it can be edited, or used as a base point while editing other objects. If an object is removed from the selection set of highlighted objects, its grips will still remain visible. The **grips** on non-highlighted objects are **cold grips**. These **cold grips** are available as **object snap** points for editing other **grips**, but cannot be edited themselves. Generally, there are no restrictions on the number of **hot**, **warm** and **cold** grips at any given time.

The status of **grips** can be modified by using the **<Esc>** key. The number of times the **<Esc>** key is pressed indicates the number of levels **grips** are reduced. When an object is selected, it becomes highlighted, and its **grips** are always **warm**. Pressing the **<Esc>** key once will reduce the status of **warm grips** to **cold**, and **hot grips** to **warm**. If a set of **grips** is already **cold**, pressing the **<Esc>** key once will remove the **grips** completely. Pressing the **<Esc>** key twice will remove all of the **warm grips**, and turn any **hot grips** to **cold grips**. Pressing the **<Esc>** key three times removes all **grips** from the screen.

Editing with Grips

Once one or more **hot grips** have been selected, the editing functions for **grips** become available. The command prompt disappears, replaced by a version of the **Stretch** command. See below:

`Command:`

Select objects to activate their **grips**. Select one or more **hot grips**.

```
** STRETCH **
<Stretch to point>/Base point/Copy/Undo/eXit:
```

The **grips** version of the **Stretch** command allows you to change the location of the **hot grip** while leaving the other grips anchored. When the endpoint of a line, polyline, or arc is selected, it is relocated to the new point. The **Stretch** option will only affect the objects which have a **hot grip** activated on them. The other options within the **grip** editing tools allow other flexibility.

Base point allows another point (rather than the **hot grip** itself) to be used as the "handle point" as the object is **Stretch**ed.

Copy causes the object(s) to be copied as they are **stretch**ed. This would be a convenient way to build concentric circles.

Undo will undo the previous **Copy** or **Base point** selection.

Reference is used to establish a reference angle or distance when using the **Rotate** or **Scale** functions.

eXit causes the command to terminate.

If you want to perform another editing function using **grips**, press the **Space bar** to advance through the options:

```
** STRETCH **
<Stretch to point>/Base point/Copy/Undo/eXit:
```

Press the **Space bar** to advance to the **Move** option.

```
** MOVE **
<Move to point>/Base point/Copy/Undo/eXit:
```

Press the **Space bar** to advance to the **Rotate** option.

```
** ROTATE **
<Rotation angle>/Base point/Copy/Undo/Reference/eXit:
```

Press the **Space bar** to advance to the **Scale** option.

```
** SCALE **
<Scale factor>/Base point/Copy/Undo/Reference/eXit:
```

Press the **Space bar** to advance to the **Mirror** option.

```
** MIRROR **
<Scale factor>/Base point/Copy/Undo/eXit:
```

Press the **Space bar** to advance to return to the **Stretch** option.

Although it may take some time to acclimate to the use of **grips**, using them to edit objects can represent real time savings. A problem that typically arises in drawing relates to polylines which do not meet at the same endpoint to allow them to join. The two "adjacent" polylines might have their grips activated to allow editing of their endpoints. If one polyline's endpoint is stretched to meet the other polyline's endpoint, they will be precisely connected at their endpoints, allowing the **join** to occur. **Grips** allow this type of editing to be done without stepping through **osnaps**, **pedit**, **edit vertex**, and **move** commands. Using **grips** to edit dimensions also creates significant time savings. Each of the elements of a dimension can be edited accurately and conveniently by using **grips**. This represents real progress over the previous methods of editing dimensions.

Viewing Options

Viewing options are commands and settings that affect the appearance of your drawing, and the speed and convenience with which it is viewed. These commands do not affect the drawing itself but help you to see the drawing more clearly on the display.

The ZOOM command

Zoom (**Z**) The **Zoom** command allows you to increase or decrease the viewing area on the screen. It acts very much like the zoom lens on a video or still camera. Typing in a scale value will cause the drawing to be displayed at a scale multiplier compared to the original limits. If you type **.5X** the objects will be displayed at half their original size compared to the original drawing limits. If you type **2X** the objects will be displayed at twice the size as shown at the original drawing limits.

Zoom Options:

All—(**ZA**) This causes the entire drawing limits to be displayed on the screen. Even though the objects drawn may not fill the screen, the drawing's limits will be fully displayed. If an object extends beyond the limits, the entire object will be displayed.

Center—This option will zoom to fill the screen with the drawn objects, and center them on screen.

Dynamic—(**ZD**) This option causes a large view of the entire drawing to appear, with a few special features: a green dashed box indicating the previous view, and the white "rubber-band" box with an "X" and arrow in it. The rubber band box can be adjusted in size and location with the **<Pick>** button. When it is oriented over the correct part of the drawing, press the **<Enter>** button. This will cause the portion of the drawing that was in the box to be displayed at full-screen size. This is a useful command when you wish to enlarge your drawing frame less than the **ZA** or **ZE** commands would permit.

Extents—(**ZE**) This option causes the objects on screen to fill the entire screen. Even though the objects may not fill the entire limits, ZE will cause them to fill the screen. If the objects extend past the limits, ZE will also display them.

Left—This option allows you to identify the lower left corner of the new window, then specify either the new window height or the magnification factor for the new window. You may never use this.

Previous—(**ZP**) This option causes the last zoom window to be displayed.

Vmax—(**ZV**) This zooms to the largest extents possible without causing the drawing to regenerate. This is a very useful command on slow computers, those with limited memory (RAM) or on very large drawings. On a fast computer with plenty of memory, this command may not be used often.

Window—(**ZW**) allows you to define a "rubber band" box to zoom into a smaller part of the drawing. That part of the drawing will then fill the screen.

In—This option will cause the display to be magnified by a factor of **2X**. The objects will appear larger on the screen (Figure 3.5).

Figure 3.5

Out—This option will cause the display to be reduced by a factor of **.5X**. The objects will appear smaller on the screen (Figure 3.6).

Figure 3.6

Limits—This option will cause the display to zoom to the limits of the drawing, regardless of whether or not objects extend beyond the limits. This option is not available from the command line version of the **Zoom** command.

The PAN command

Pan (**P**) The **Pan** command allows you to move the drawing frame without changing the magnification. This command works differently from the command line than it does from the screen icon.

From the command line:

`Command:`
Type **Pan** (or **P**).

`PAN Displacement:`
Pick a point on the screen or type in the coordinates of a point.

`Second point:`
Pick (or type) a new location for that same point. The drawing window will shift by the difference between the two points you have chosen.

`With the screen icon:`
Pick the icon from the Standard Toolbar. The "fly-out" options allow you to choose which direction the **pan** will occur. The distance that will be covered by the **pan** operation will be about half the size of the current drawing screen. The various arrows dictate the direction of the **pan** operation. This preset **pan** may be useful in certain situations, but lacks flexibility. Scroll bars can be activated at the right and bottom edges of the viewing screen to enhance this ability. The primary drawback of scroll bars is that they consume precious screen real estate. Most AutoCAD users want as much drawing area as possible on the screen.

The VIEWRES (VIEW RESOLUTION) command

Viewres (**Vre**) This is an AutoCAD setting that controls the display of circles, arcs and curved polylines. The term **Viewres** stands for *View Resolution*, indicating a numerical resolution that affects only the **display** of the objects, not the objects themselves. All circles and arcs, regardless of their appearance on the screen, will plot smoothly.

What is actually controlled is called the **zoom percent**, which corresponds to the number of line segments used to approximate the curvature of circles, arcs, etc. The setting used here will affect whether or not the curves appear jagged and rough, as opposed to smooth and even. Not coincidentally, the **Viewres** command also has an affect on the speed of drawing regeneration. A high **Viewres** value will smooth the curves in a drawing, but will also slow down the process of regenerating the display. From the command line:

`Command:`
Type **Viewres** (or **Vre**)

`Do you want fast zooms?<Y>`
Press **<Enter>** to say yes. If you respond with **N** here, AutoCAD will cause the drawing to regenerate after each use of the **Zoom**, **Pan** or **View** commands.

`Enter circle zoom percent (1-20000)<100>:`
Enter the desired value for the zoom percent.

A normal range for the zoom percent might be from 500 to 5000. Values higher than this will slow the regeneration process, while values lower than this can make the display look inaccurate. With extremely large or complex drawings, leave the **Viewres** setting at the AutoCAD default of 100.

The REGEN (REGENERATE) command

Regen—This command causes AutoCAD to recalculate the screen display by examining the actual drawing database. This process is invoked automatically by certain commands (**Zoom Extents** for example) but may be performed manually at any time. You may wish to regenerate a drawing after changing the characteristics of a text style, linetype scale, or other AutoCAD system variable. While older versions of AutoCAD required frequent **regens**, AutoCAD 13 requires very few. Since large drawing files have very complex databases, they cause very slow **regens**. It is best to minimize the use of **regens** when editing these large files.

The REDRAW command

Redraw (**R**)—This command simply refreshes the drawing display by redrawing the objects on the screen. The **Redraw** command does not examine the drawing database and does not consume appreciable amounts of drawing time. It is wise to use **Redraw** frequently when editing objects. When a circle overlies a line, the two objects share some screen pixels in common. When the circle is erased, all of the pixels in the circle will return to the background color. It then appears that gaps exist in the line which are not really there. Use the **Redraw** command to refresh the screen and restore all of the pixels in the line to their correct color. **Blips** (small markers on the screen) left after editing objects are also cleared when using the **Redraw** command. Use this command often.

The VIEW command

View —This command is used to name and save the current display of your drawing. There are many occasions where you wish to zoom into a specific location in the drawing, then move away to perform some other command. When you want to return to the specific zoom window again, you would need to zoom carefully into the original position. Several **Zooms** and **Pans** might be required before you could recreate this position. The **View** command allows you to name a view and save it so that you can restore it in a single step.

There are several options available for the **View** command.

`Command:`
Type **View.**

`?/Delete/Restore/Save/Window:`
Enter the desired operation.

?—displays a list of currently available views.

Delete—removes a named view from the list of available views.

Restore—allows you to retrieve one of the available views and make it the current view.

Save—saves the current view with a name that you specify.

Window—asks you to specify a name, then allows you to pick a window on the screen and save it as a view.

After saving a view, it is wise to use the ? function to be sure that the new view has been added to the list. To make the **View** command more efficient, use numerical or very short names for your views. This makes restoring a view very quick. If you name your views this way, you can restore a view in four or five keystrokes. A dialogue box version of the **View** command is available by typing **Ddview**, or picking Named Views ... from the View pull-down menu in AutoCAD.

Views are also useful during the plotting process. Any saved view can be plotted at any time during the drawing process.

Hint: Save views of detailed areas frequently during the drawing process. You can plot these details on an office laser printer to check the specifics of the drawing, and even FAX them to other offices for comments or approvals.

The FILL setting

Fill—This is an AutoCAD *setting* rather than a command. A setting controls some function in the AutoCAD program while not affecting the specific drawing being edited. The **Fill** setting affects the **display** of solid objects on the screen such as polylines and solids. With **Fill** turned on, wide polylines and solids display normally. With **Fill** turned off, wide polylines and solids will only display as outlined areas. You may want to turn **Fill** off when working with very large drawings. This will speed regeneration and redraw times and make drawings easier to navigate. When **Fill** is turned off, solids and polylines appear as outlines both on the screen and in plotted drawings. Remember to turn **Fill** back on if you want your wide polylines and solids to plot normally.

The TEXTFILL setting

Textfill—This setting affects the appearance of solid-filled fonts. TrueType® and Postscript® fonts are often filled and require significant time to display and regenerate. If you turn **Textfill** off, the solid fill in these letters is removed, which can greatly speed AutoCAD's handling of these complex fonts. When **Textfill** is turned off, the fill does not appear in the display, and will not be plotted. If you use these fonts regularly in your drawings, turn off **Textfill** while drawing, but turn it on before you plot. The **Textfill** setting is actually an AutoCAD *system variable.* This means that you change the **value** of the variable, rather than turning it on or off.

`Command:`
Type **Textfill.**

`New value for TEXTFILL <1>:`

The value of **<1>** indicates that the variable is currently turned on. Type **0** to turn **Textfill** off.

`Command:`

Type **Regen**. This will display the effects of the **Textfill** system variable.

`Regenerating drawing.`

Information Commands

Information commands report data to you about objects or locations on the screen. In AutoCAD, these commands are grouped together as **Inquiry** commands. They can be accessed through the command line or from the **Edit** pull-down menu.

The ID (IDENTIFY) command

ID—This command will return the X, Y and Z coordinates of any point you select on the screen. You can activate any **Osnap** option to determine the precise location of endpoints, centers, intersections, etc. Another use of this command is to allow it to establish AutoCAD's **last point**. This point is stored as the last point located on a drawing so that it becomes available for *relative coordinate* usage. This means that you can set objects relative to this point by using the **@** (or ampersand) symbol.

`Command:`

Type **ID.**

`Point:`

Pick the point on the screen you wish to identify.

`Point:   X = 140'-0"      Y = 80'-0"      Z = 0'-0"`

AutoCAD reports the coordinates of the point you selected.

The DISTANCE command

Distance—This command allows you to determine the distance between any two points you pick on the screen. As with the **ID** command, **Distance** is most effectively used in combination with **Osnaps**. You can determine the distance between specific objects, the length of lines, the radius of circles, etc.

`Command:`

`Type` **Dist.**

`First point:`

Pick the first endpoint of the distance you wish to measure. Use **Osnaps** where appropriate.

```
Second point:
```
Pick the second endpoint of the distance you wish to measure.

```
Distance = 172'-3 7/16",   Angle in XY Plane = 353,   Angle from
XY Plane = 0
Delta X = 171'-0",   Delta Y = -21'-0",   Delta Z = 0'-0"
```

AutoCAD reports the linear distance between the two points, the angle of the line formed by the two points and the angle of the line from the viewing plane. In addition AutoCAD reports the change (**Delta**) in the X, Y and Z axes.

The LIST command

List—The **List** command is used to report the particular data associated with an object on the screen. The type of data (and the amount of it) depends on the object chosen. Typically, you will see a display of the type of object, the layer on which it resides, the color with which it is displayed, its endpoints, angle and length. Polylines typically display data for each vertex. The command is easy to initiate:

```
Command:
```
Type **List.**

```
Select object:
```
Pick the object to **List.**

The listed information will be displayed in text mode. In DOS versions, the display will flip to the text screen, while with Windows versions, a text window will open to display the text information.

Use the **List** command to inquire as to the name of a hatch pattern, the name of a font, the scale of a block, the name of a linetype, or the layer on which some object might reside. The following is the partial result of a **List** command on a LANDCADD tree symbol:

```
BLOCK REFERENCE  Layer: 0293-NEW__-TREES
Space: Model space
Color: 3 (green)    Linetype: BYLAYER
Handle = 1D1D
USERL007
at point, X=101'-1 1/4"  Y=119'-10 9/16"  Z=0'-0"
X scale factor  120.0000
Y scale factor  120.0000
rotation angle       0
Z scale factor   60.0000
```

```
ATTRIBUTE  Layer: 0293-NEW__-TREES
Space: Model space
Handle = 1D1E
Style = STANDARD     Font file = HLTR.SHX
mid point, X=101'-1 1/4"  Y=119'-10 9/16"  Z=     0'-0"
height      1'-0"
value Viburnum japonicus
tag ITEM
rotation angle       0
width scale factor  1.0000
obliquing angle      0
flags invisible
generation normal
```

Changing Your Mind

Some commands in AutoCAD are used when you make a mistake. Thankfully, most mistakes are relatively easy to fix by using one of these three commands.

The UNDO command

Undo—The **Undo** command is used to reverse the effects of a command (or group of commands) used previously. Use **U** to undo the effects of the command immediately preceding its use. **Undo** is used to undo the effects of more than one command at a time. The **Undo** command has a number of relatively sophisticated features which makes it valuable to use when working with complex AutoCAD and LANDCADD command routines.

LANDCADD contains a number of command routines which perform a set of AutoCAD commands in sequence. If you make an error in selecting the command routine (usually a menu selection) or preparing the drawing for the command routine, you will probably see things appear on the screen that you had not planned for. You might choose to cancel the command before it is completed. Alternatively, you might allow the command to finish even though the results are obviously not what you wanted. In either case, you can end up with a number of objects on the screen which must be removed. Erasing a large number of objects on the screen wastes time you might rather spend otherwise. To make matters worse, erasing those objects might not delete them from the drawing database, but simply remove them from the screen. This can lead to large drawing file sizes and frustrated designers. It is far more efficient to use the **Undo** command. Controlling the **Undo** command will save time and keep your drawing file sizes to a minimum.

To understand the way **Undo** works, we need to see how other commands are treated by it. First, a record of each command (and the objects it creates) is stored in the drawing database. Even though a line might be drawn, stretched, rotated and then erased, it is still possible to use the **Undo** command to recover the line in its original form. This explains why drawings can become extremely large even though they have very few items stored in them. Second, each command is treated as a separate step in the **Undo** process. One **Undo** will undo one command. Ten **Undo**s will undo ten commands. Third, It is possible to force AutoCAD to treat separate commands together as a single group. If this is done, a sequence of ten commands can be undone with one operation of the **Undo** command. Finally, you can insert markers at different places in the drawing sequence to stop the process of an **Undo** command that is working back through many steps. Once a marker is placed, an **Undo** command can be used to undo all of the commands issued until the marker is encountered. If you placed a marker five commands ago, then asked the **Undo** command to undo ten steps, it would stop at the marker after five **Undo**s.

`Command:`
Type **Undo.**

`Auto/Control/BEgin/End/Mark/Back/<Number>:`
Each option deserves an explanation:

Auto—this option sets the way that the **Undo** function operates with respect to menu selections. A single menu selection could allow you to pick a tree symbol, select its name from a list, change layers, insert the symbol on the screen and copy it to other locations in the drawing. If **Auto** were turned on, these operations would all be treated as a single step. One **Undo** command could undo all of the steps at one time. If **Auto** were turned off, each step would require its own **Undo** command. The **Control** function may act to override the **Auto** function if the **Undo** command has been limited in any way.

Control—this option will act to limit or turn off the **Undo** command. There are three choices for **Control**:

The **All** option leaves the **Undo** command functioning normally.

None disables the **Undo** command so that none of the previous commands are stored with the drawing database. This means that none of the previous commands can be undone. If you try to use the **Undo** command with **None** selected, the command line automatically reverts to the **Undo/Control** function, allowing you to reactivate the function if you wish.

One limits the **Undo** function to a single command. This means that only one previous command is stored in the drawing database, and that you can only undo one command.

If **None** or **One** is in effect, the **Auto**, **Begin**, and **Mark** functions will not operate. Use **None** or **One** only if controlling the drawing size has extremely high priority.

Begin—this option creates a grouping of commands to be treated as a group by the **Undo** command. The **Begin** option starts the formation of the command group. When a group of commands is grouped together with **Begin** and **End** they can be **Undone** by one **Undo** command. On the positive side, this can save time because the effects of grouped commands can be eliminated immediately. On the minus side, all of the commands are **Undone** at one time, preventing you from stepping back through the commands to stop at the point you want to resume drawing.

End—this option ends the command grouping. Once a **Begin** and an **End** are placed around a set of commands (as parentheses enclose this phrase) they are treated as a single group. One **Undo** step will undo the group of commands.

Mark—this places a marker at a particular phase of the drawing. Once placed, you can **Undo** back to this marker at any time. The **Mark** will stop the **Undo** function from proceeding with any further **Undos**. If there are no **Marks** placed in a drawing, it is possible to **Undo** all of the commands since the drawing was opened, or since the last **Save** command.

Back—this option allows you to step back through the drawing process and **Undo** all of the commands until a **Mark** is encountered. If AutoCAD does not detect any **Marks** in the database, it displays the warning message:

```
This will undo everything. OK? <Y>:
```

Think twice before accepting this default. You will **Undo** any changes in the drawing since you opened it or saved it last.

Once a **Mark** is encountered, you can repeat the **Undo/Back** command to find the next previous **Mark**. Using **Mark** and **Back** are usually better than using the **Begin** and **End** functions since **Mark** and **Back** give you more control of the **Undo** process. LANDCADD programmers usually place a **Mark** at the beginning of complex command routines, so we can make use of the **Undo/Back** command to remove the effects of commands that do not behave as we expect them to.

Number—this sets the number of steps you wish to **Undo**. If a **Mark** is encountered, the **Undo** process will cease, even though the number of the steps indicated were not **Undone**. To **Undo** past the **Mark**, use the **Undo** command again.

The REDO command

Redo—The **Redo** command is used to undo the effects of the last **Undo** command. **Redo** must be used immediately after the **Undo** to negate its effects. The net result of a **Redo** is as if the last **Undo** were never performed.

The OOPS command

Oops— Use the **Oops** command to undo the erasure caused by the last **Erase** command. **Oops** can be used at any time to undo the effects of the last **Erase**. This means that you do not have to use **Oops** immediately after an **Erase** command to undo its effects.

Controlling the Appearance of Objects

One of the most basic AutoCAD concepts relates to controlling the appearance of objects that you create. The appearance of an object will relate to the color in which the object is displayed and the linetype which is used to make up the lines, polylines or arcs in the object. Therefore, in order to control the appearance of the objects in your drawing, you should decide how you want to take control of colors and linetypes. An object on the screen can have a color and linetype assigned to it directly. On the other hand, the colors and linetypes can be assigned to layers rather than objects. Placing an object on a layer can automatically cause that object to acquire the properties of that layer. Layers can also be used to control whether or not objects will appear on the screen, since layers can be turned on or off. When drawings are small and simple, the importance of this type of control is minimal. It does not really matter how colors and linetypes are assigned, because there are few enough objects in the drawing that exerting control over them is easy. When drawings become large and complex, the stakes increase quickly. It becomes progressively more important to control the appearance and visibility of objects in a drawing as its complexity increases.

Most AutoCAD users employ layers to control the appearance of the objects in their drawings. For large drawings, the use of organized, standardized layers is essential. Layers can be named for the functional types of objects that are placed in the drawing. Landscape architects might want to have different layers for trees, shrubs, groundcovers, lighting, utility lines, irrigation lines, hardscape materials, property lines, dimensions, topographic lines, etc. Layers not only help control the appearance of objects, but also help us to organize the objects into categories whose visibility can be turned on or off. In this way, a single drawing could contain all of the information needed for a site plan, planting plan, construction plan and irrigation plan. Turning layers on and off could control which objects are visible on the screen and which objects will be plotted to a hard copy. Most design offices take advantage of a standardized set of layers, colors and linetypes to make electronic drawings easier to understand and hard copy drawings look more legible.

Another aspect of layers to remember is that layers (along with their colors and linetypes) are assigned to your current drawing and will accompany it if your drawing is inserted into another designer's drawing. If you use a drawing as a **prototype** for other drawings, the other drawings will contain the same layers (and colors and linetypes) as the **prototype**. The fact that drawings often interconnect in this way makes layer standardization even more critical.

The LAYER command

Layer—The best way to gain access to the layer control functions is through the "Layer Control" dialogue box. There are five ways to gain access to this dialogue box:

1. Type **Ddlmodes**. This opens the "Layer Control" dialogue box directly.
2. Pick Layers... under the Data pull-down menu in AutoCAD.
3. Pick Layers ▸ under the Tools pull-down menu in LANDCADD, then select AutoCAD Layer Dialog.
4. Pick Layers ▸ from the LANDCADD cursor menu, then select Layer Control ... .
5. In Windows versions of AutoCAD, pick the Layer Control icon. This icon will be visible in the upper left corner if the **Object Properties** toolbar is activated (Figure 3.7).

Figure 3.7

Which of these methods you use to access the dialogue box will depend on which menu is active and which platform of AutoCAD you are using. Regardless of which way you choose to open this dialogue box, it represents the quickest and most direct way to control the layer functions in AutoCAD (Figure 3.8).

Once the "Layer Control" dialogue box is opened, you can immediately see the names of the layers which are present in the drawing. The **Current Layer:** line displays the name of the layer which is now active. Any objects drawn will be on the active or current layer.

Figure 3.8

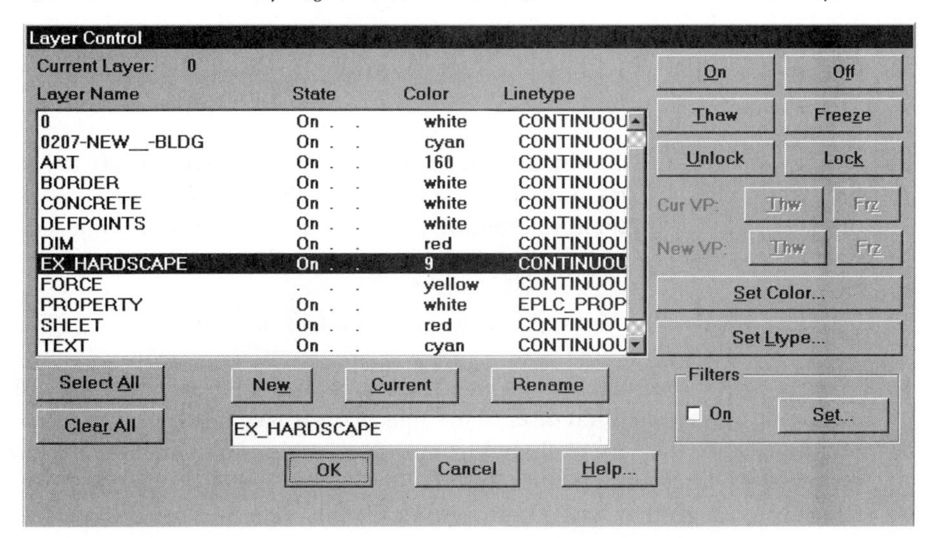

This, however, does not preclude you from moving an object to another layer later during the drawing process. The **Layer Name** box displays the layer names in alphabetical order. If more names are present than are visible at one time, a scroll bar appears at the right margin of this box. If you pick one of the layer names in this box, the entire line will highlight. Several of the options to the right side of the dialogue box will then become active. This allows you to edit the properties of the layer which is highlighted.

Layer Names and Layer Selection

Select All allows you to highlight all of the layers at one time.

Clear All allows you to clear (remove) the highlight from all of the selected layers. These options are useful when you want to perform editing functions on a large number of layers at one time.

New allows you to make the entry in the text box below into a new layer. A layer name can contain up to 31 characters, but cannot contain spaces. To create multiple layers in one step, simply enter each layer name, one after another, separated by commas. After the **New** option is selected, these layers will be listed in the **Layer Name** box. There is no limit to the number of layers which a drawing can contain, although practical considerations might limit the number to under 250 layers.

Current makes the highlighted layer the active drawing layer. This option will function when only one layer name is highlighted.

Rename changes the name of the highlighted layer to that in the text box. Again, this option will function only when one layer name is highlighted.

Layer Properties and Layer Status

On turns on the highlighted layers. Layers which are turned **on** are visible on the display and will plot normally. The objects on these layers require normal regeneration time.

Off turns off the highlighted layers. Layers which are turned **off** are not visible on the screen and will not plot. These layers are still part of the drawing database, and objects contained on them will require normal regeneration time. Objects contained on layers in the **Off** state cannot be edited unless the selection set is **All**.

Thaw will remove highlighted layers from a **frozen** state.

Freeze will freeze highlighted layers. Objects on **frozen** layers are not visible, will not plot, and cannot be edited or erased. Objects on **frozen** layers do not require regeneration time, since they are excluded from the active drawing database. A **frozen** layer has the letter **F** placed in the **State** column of the **Layer Name** box. **Frozen** layers are not visible, even though they have **On** written in the **State** column next to the layer name.

Unlock will remove highlighted layers from a **locked** state.

Lock will lock highlighted layers. Objects on **locked** layers are visible and will plot, but cannot be edited. Objects on locked layers cannot be erased, even when **All** is used as the selection set. Layers which are **locked** can be either **On** or **Off**. Objects on **locked** layers will require regeneration time, whether turned **On** or **Off**.

The **Cur VP** and **New VP** options for freezing and thawing layers are used when **Paper Space** is active. When using **Paper Space** it is possible to look at the same drawing using several different viewports. You can control the status of each layer in each viewport. This means that you can control the visibility of objects independently in different viewports of the same drawing. Using a golf course as an example, it might be necessary to display 3" contour lines in a 10' scale enlargement viewport of a golf green. In a 40' scale viewport of the entire hole, these contour lines would appear too close together and would be illegible. In **Paper Space**, we can look at the same drawing in two (or more) ways. We can create a 40' scale viewport with the 3" contour lines kept invisible on a frozen layer, while at the same time displaying them in a 10' scale enlargement viewport of the same drawing. Each viewport can have a different scale and have different drawing layers visible. This way we can create many different views of the same drawing. We will examine **Paper Space** in an exercise later in this book.

Set Color... this option opens the "Select Color" dialogue box.

Select the desired color for the highlighted layer(s) from one of the displayed colors. The number of colors displayed here will depend on your system video card (or graphics controller), your monitor and the video drivers available for AutoCAD. The minimum standard VGA (or video graphics array) allows 16 colors. Most systems now use SVGA video equipment (Super VGA) and allow 256 colors. Newer video cards can allow a higher number of colors, although AutoCAD will not make use of them in its standard drafting mode.

The selection of layer colors can have more significance than might be immediately apparent. Since most plotting is performed where a specific color will correspond to a specific pen number, the colors on the screen can relate to line widths on paper. This means that the choice of colors on the screen can make it easier to create a line weight hierarchy on plotted drawings.

The colors at the top of the dialogue box are AutoCAD's **Standard Colors** numbered one through nine. The **Full Color Palette** contains colors 10 through 249, while the **Gray Shades** are numbered 250 through 255. If your video equipment supports fewer than 256 colors, the unsupported colors will display as gray shades. They will still retain their number in the AutoCAD Color Index (ACI) system, and can still be assigned to specific pen numbers (and line widths) for plotting.

The **Logical Colors** are not available for assignment to layers. These options are only available if you wish to change the color assigned to a specific object on the screen, or reset the current drawing color.

Set Ltype this option opens the "Select Linetype" dialogue box.

When first started, AutoCAD will only display the **CONTINUOUS** linetype. Any other linetypes must be loaded into the current drawing. As linetypes are loaded, they are stored with the drawing so that they are available to use when the drawing is reopened. To load a new linetype, pick the **Load...** option. This opens the "Load or Reload Linetypes" dialogue box.

The **File...** displayed here will be **acad.lin**. This file contains all of the linetypes which come with AutoCAD. These linetypes are listed below in the **Available linetypes** box. When LANDCADD is loaded onto your system, there are two additional files that contain linetypes you can use. One is found in Eagle Point's **Support** folder, and is called **ep.lin**. The other is found in Eagle Point's **Program** folder, and is called **lc_ltype.lin**. Select as many linetypes as you need to load into your current drawing. These linetypes will become part of the drawing database, so try to avoid loading more linetypes than necessary. Unused linetypes and layers can be removed with the **Purge** command.

Once linetypes have been loaded into your drawing, you can return to the "Select Linetype" dialogue box to select the desired linetype to assign to the highlighted layer(s). The **ISO Penwidth** option is only available if an ISO linetype is selected. The **Linetype Scale** option is only available if you are changing the linetype of an object in the drawing. Any object in the drawing can have a linetype and a linetype scale assigned to it.

The **Filters** section refers to the way that layer names are displayed in the list of **Layer Names**. When there are dozens of names in the list of layers, it can be very tedious to sort through the layer names to find the individual layers you are looking for. Using the **Filters** provided by AutoCAD, you can cause the list to display only those layers meeting your specific **sort criteria**. For example, you can choose to display only those layers which are **locked**, assigned the color **blue**, and only those whose names begin with the characters **C_1**. Each of these is a **sort criterion**. To establish these **sort criteria**, pick the **Set...** option. This opens the "Set Layer Filters" dialogue box.

The illustration shows each of the possible filters (or **sort criteria**). For the top five choices, we have paired options. We can select those layers which meet one or both of the conditions in the pair. As an example, we will see those layers which are either **On** or **Off** (because **Both** is selected), but must be **Locked**. Use the pop-down boxes to select which member of the pair you wish to use. Use the **Reset** to restore all of the pairs to the **Both** setting.

The bottom three options allow you to enter filter options from the keyboard. In this case, only those layers assigned the color **blue** (the filters are not case-sensitive, so either upper or lower case letters can be used), and begin with the characters **C_1** will be displayed (Figure 3.9).

Figure 3.9

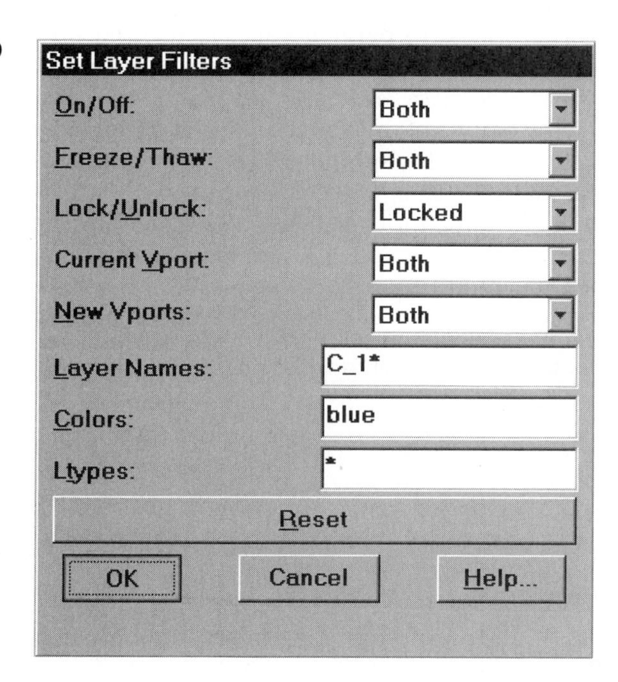

The layer filters support *wild card* characters which can be used to substitute for other characters. * can be used to indicate **any characters**. Thus a layer name **C_1*** will mean any layers beginning with **C_1** followed by any combination of characters. This is a very valuable feature when working with drawings which adhere to **CSI** or **AIA** layering standards. Once these filters have been set, they can be turned **On** or **Off** using the toggle in the "Layer Control" dialogue box.

Special Layers in AutoCAD

There are two layers in AutoCAD that deserve special recognition. The **0** layer will exist in every AutoCAD drawing. It is the default layer when the prototype drawing **ACAD.DWG** is opened. This layer cannot be purged from the drawing database, and it possesses a peculiar property: any object created in the **0** layer and converted into a block becomes a chameleon block. That is, it takes on the properties (color and linetype) of the layer on which it is inserted. Other objects turned into blocks will retain their layer characteristics, and will bring their layer(s) with them when they are inserted into other drawings.

Another layer deserving mention is the layer called **Defpoints**. This layer is created when you build associative dimensions in an AutoCAD or LANDCADD drawing. The purpose of this layer is to hold **definition points** that are associated with the objects that are being dimensioned. For example, when a circle is dimensioned, it has **defpoints** placed at its center and at the end of its radius line. When a line is dimensioned, **defpoints** are placed at its endpoints. If an object is edited, its **defpoints** are normally edited along with it. Objects which are scaled, trimmed, extended, or stretched will normally have their **defpoints** changed as their geometry changes. When this happens, the dimensions are automatically updated to reflect the editing of the objects. For this reason, **defpoints** should be left alone where possible. Keeping the **Defpoints** layer frozen will prevent you from accidentally erasing the **defpoints**. Freezing the **Defpoints** layer does not influence the functionality (or visibility) of the **defpoints** contained on the layer.

Alternative Methods to Control Layers

If you choose to access the **Layer** command from the command line, the prompt will permit you to perform most of the layer control functions mentioned above:

`Command:`
Type **Layer** (or **La**).

```
LAYER
?/Make/Set/New/ON/OFF/Color/Ltype/Freeze/Thaw/LOck/Unlock:
```
? lists the layers on the drawing.

Make allows you to create a new layer and make it the current layer at the same time.

Set allows you to change the current layer to the layer name you enter.

New allows you to create one or more layers at the same time. Type the names of the layers you want to create, separated by commas.

On and **Off** allow you to turn on or off one or more layers.

Color allows you to assign a color to one or more layers.

Ltype allows you to assign a linetype to one or more layers.

Freeze and **Thaw** allow you to freeze or thaw one or more layers.

LOck and **Unlock** allow you to lock or unlock one or more layers.

These commands are quick to operate, but require that you know the names of the layers and the desired colors or linetypes for the target layers. Once the list of layers grows beyond a certain point, you will probably need to use the "Layer Control" dialogue box.

Windows users can take advantage of the pop-down box of layer control functions stored in the **Object Properties** toolbar. The icons indicate the state of the layer, along with its assigned color (Figure 3.10).

Color—The color of an object can be controlled by two methods. First, we can assign the object the unlikely color called **Bylayer**. This "color" indicates that the object will display the color of the layer on which it is placed. This is the only way to control the color of objects purely by layer assignment. Second, we can set the color with the **Color** command or in the **Ddcolor** dialogue box. When the color is set using either of these tools, any subsequently created objects will take on the new, manually set color regardless of their layer assignment. Although there may be some situations where this type of color assignment is desirable, it is generally less flexible and useful than using the **Bylayer** color definition.

At the command prompt:

`Command:`
Type **Color**.

`New object color <BYLAYER>:`
Enter the new color name or ACI (AutoCAD Color Index) number.

To use the dialogue box:

`Command:`
Type **Ddcolor**, or pick **Color** from the AutoCAD **Data** pull-down menu. This opens the "Select Color" dialogue box.

Select the desired color from the displayed set of available colors. In this dialogue box, the **Logical Colors** are available. This means that the "colors" **Bylayer** and **Byblock** can be selected. **Bylayer** forces objects to take on the colors assigned to the layers on which they are drawn or placed. **Byblock** forces objects turned into a block to take on the color

Figure 3.10

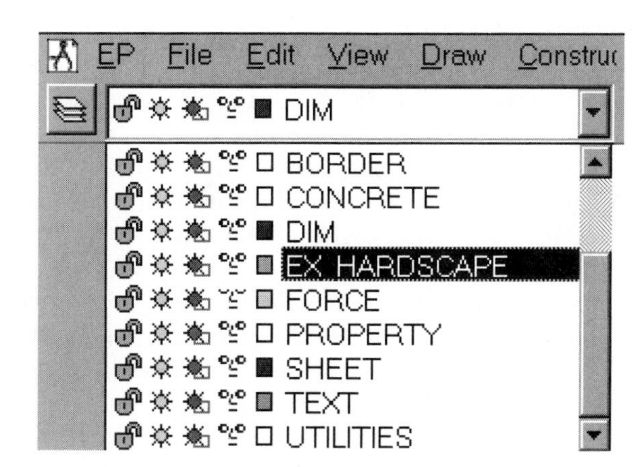

Figure 3.11 

currently active when the block is inserted into a drawing. This means that the color (and linetype) will be dictated by the current color setting, not by the layer on which the block is placed. If **Bylayer** is the active "color" when the block is inserted, the block will assume the **Bycolor** designation.

Windows users can also load the "Select Color" dialogue box from the icon found in the **Object Properties** toolbar (Figure 3.11).

Linetype—The same principles which operate in the selection of colors also operate for selecting linetypes. Linetypes can be assigned to layers just as colors can be assigned to layers. This is the method to use when you want to be sure that objects will always assume the linetype of the layer on which they reside. The other method for linetype assignment is to use the **Linetype** or **Ddltype** command to set the current linetype for objects created from that time forward. This method means that objects will take on the newly assigned linetype regardless of which layer they are built upon. The amount of confusion this can cause is considerable. To make matters even more confusing, it is also to possible to change the **Linetype Scale** of a specific object, of objects created only after this value is reset, or of all of the objects in the drawing at the same time. We will discuss the **Linetype Scale** later in the book. Generally speaking, it is wise to assign linetypes to layers, then reset the **Linetype Scale** globally to suit the scale of the plotted drawing.

The linetype can be set independently of the **Layer** command in three ways:

From the command line:

`Command:`
Type **Linetype.**

`?/Create/Load/Set:`

> **?** lists the linetypes currently loaded in the drawing.
>
> **Create** allows you to create custom linetypes. Refer to the AutoCAD Command Reference or any of the excellent books on AutoCAD customization to learn more about custom linetypes.
>
> **Load** allows you to load an existing linetype into the drawing. After typing in the name of the linetype, you are shown the "Select Linetype File" dialogue box. You must select the file in which the new linetype is stored to load it successfully into your drawing (Figure 3.12).

Figure 3.12

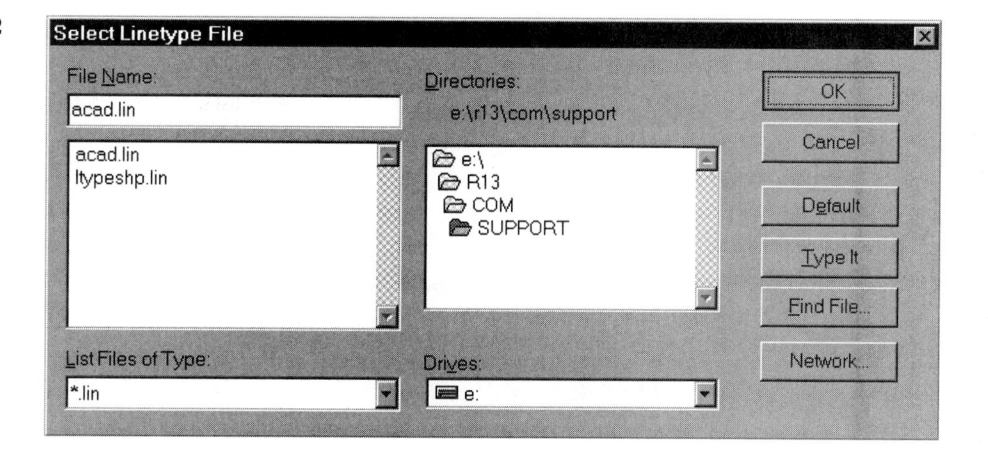

All of the standard AutoCAD linetypes are stored in the file **acad.lin**. Extra "customized" linetypes are stored in the file **ltypeshp.lin**.

Set allows you to set the linetype for all objects created subsequently, regardless of the layer on which they are created.

From the "Select Linetype" dialogue box:

```
Command:
```

Type **Ddltype** or pick **Linetype...** from the **Data** pull-down menu. This loads the "Select Linetype" dialogue box.

Use this dialogue box to select from the available linetypes, or to load new linetypes into the drawing. To load a new linetype, pick the **Load...** option.

This opens the "Load or Reload Linetypes" dialogue box. The **File...** displayed here will be **acad.lin**. This file contains all of the linetypes which come with AutoCAD. LAND-CADD linetypes can be found in two additional files. One is found in Eagle Point's **Support** folder, and is called **ep.lin**. The other is found in Eagle Point's **Program** folder, and is called **lc_ltype.lin**. Select as many linetypes as you need to load into your current drawing. These linetypes will become part of the drawing database, and will be available any time you want to change the current linetype or assign a new linetype to a layer.

The **Linetype Scale:** box will set the linetype scale for those objects created after this value is changed. Changes made in this dialogue to the active linetype and to the linetype scale are not retroactive and will not affect objects created previously. Although it may be convenient to adjust the linetype scale for objects independently of other objects, it can lead to confusion. Avoid using this method until you are thoroughly familiar with the concepts of **LTSCALE** and **CELTSCALE** covered later in this book.

Windows users can also load the "Select Linetype" dialogue box from the icon found in the **Object Properties** toolbar (Figure 3.13).

Figure 3.13

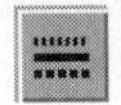

Windows users can also take advantage of the pop-down box of loaded linetypes to control the current linetype. This pop-down box will **not** allow you to change the linetype associated with a particular layer. If you wish to assign linetypes by layer, leave this set to the **Bylayer** linetype (Figure 3.14).

Figure 3.14

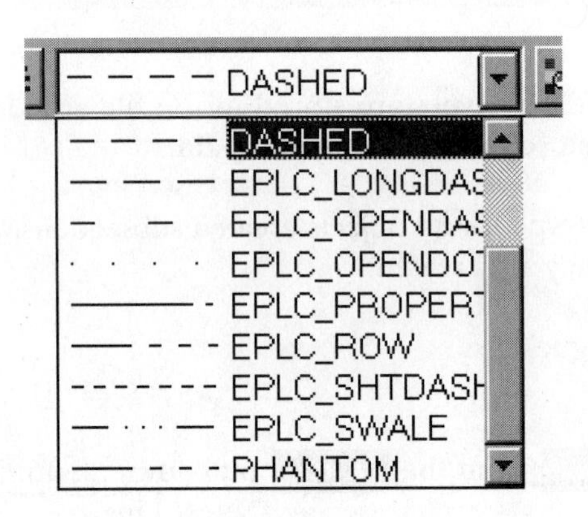

c h a p t e r

Boundaries and Hatches

Boundaries

The BOUNDARY command

Boundary creates an enclosed boundary around an area bordered by lines, polylines, arcs, polygons, circles, etc. The **boundary** that is drawn can be a **polyline** or a **region**. A **region** is a complex and powerful object which can have areas added to it and subtracted from it. **Regions** are especially useful in creating 3-D objects. For the purposes of building a **hatch**, a polyline **boundary** is sufficient to meet your needs.

To create a **boundary**, use the "Boundary creation" dialogue box:

`Command:`
Type **Boundary**, or pick it from the **Construct** menu under **Bounding Polyline**. This will open the "Boundary Creation" dialogue box (Figure 4.1).

The **Object Type:** pull-down box allows you to select either a **polyline** or a **region**.

The **Define Boundary Set** settings allow you to choose the group from which the **boundary** objects will be selected. For all but very large drawings, the radio button **From Everything on Screen** setting is appropriate. When working with very large drawings, you can select

Make New Boundary Set < to pick a selection set from which the **boundary** set objects will be chosen.

If the **Island Detection** toggle is active, the **Ray-Casting** option is not available. If this toggle is removed and **Island Detection** disabled, you can choose which angular direction AutoCAD will use to search for the **boundary**. For relatively simple boundaries, this is not usually an issue. For complex or narrow boundaries, there may be some advantage to being able to specify this angle.

Figure 4.1

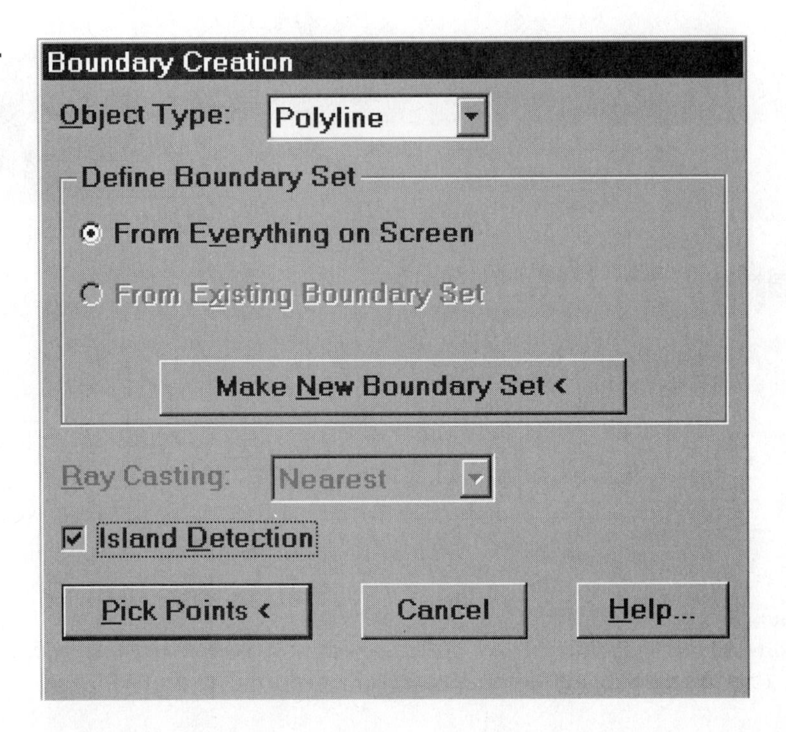

Island Detection activates the AutoCAD mechanism for detecting enclosed areas within the **boundary** area. If such an area is detected, it is outlined to create a "hole" in the boundary.

Pick Points < allows you to pick a point within the **boundary** that you wish to create. The command will cause AutoCAD to search for the **boundary** and create the polyline. The **boundary** will be highlighted as the command prompt changes:

```
Select internal point:
```

Move the cursor inside the desired area, then press the **<Pick>** button.

```
Selecting everything...
Selecting everything visible...
Analyzing the selected data...
Analyzing internal islands...

Select internal point:
```

Press **<Enter>** to complete the process of building the **boundary**.

```
BOUNDARY created 1 polyline
```

Figure 4.2

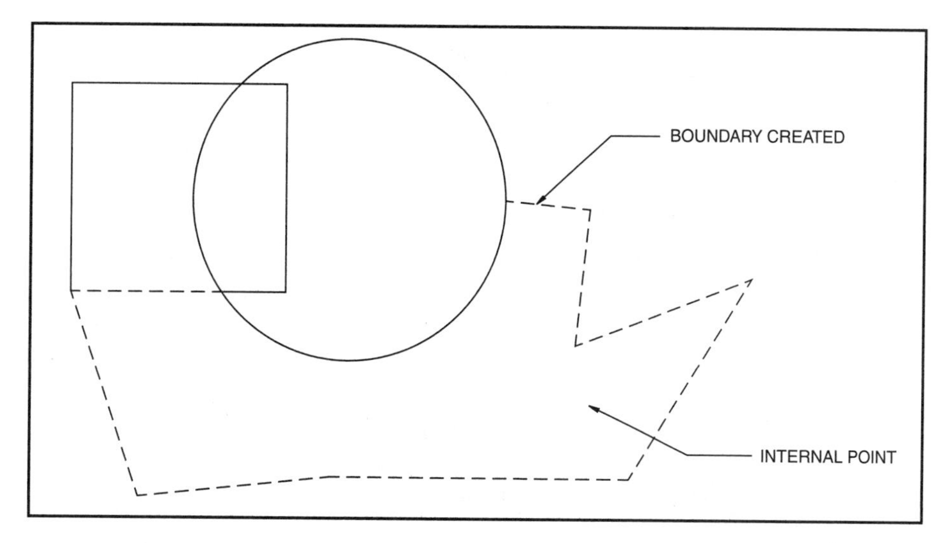

Hatches

Boundary and **Hatch** are related commands in that a **hatch** is created within the limits of a **boundary**. Although you can create a **hatch** without predefining a **boundary**, it makes the job easier.

Hatch spreads a selected pattern over an enclosed area at a defined angle and scale. The actual **Hatch** command is very limited, however, and has largely been replaced by the **Bhatch** command. The **Bhatch** command creates an **associative hatch** which changes as its **boundary** objects are edited, while **Hatch** does not. In addition, there are no **ray casting** functions built into the **Hatch** command. This means that the **boundary** must be manually selected to very exacting standards. Typically, objects which extend past the **boundary** must be broken at their intersection with other objects which make up the **boundary**. This tedious process alone makes **Hatch** a less than desirable option when building most **hatch**es.

The BHATCH (BOUNDARY HATCH) command

Bhatch (**Bh**) opens the "Boundary Hatch" dialogue box which controls the various portions of the **hatch** that is created (Figure 4.3).

At the top of the dialogue box, the user can choose the **Pattern Type**. Both AutoCAD and LANDCADD are equipped with many hatch patterns. However, you can also choose to use a **User-defined** hatch in the **Pattern Type** pop-down box.

In the **Pattern Properties** area, you define the characteristics of the **hatch** itself. The **Pattern:** can be chosen through this pop-down box. After AutoCAD and LANDCADD are loaded, there will be several dozen patterns to choose from. The LANDCADD patterns

Figure 4.3

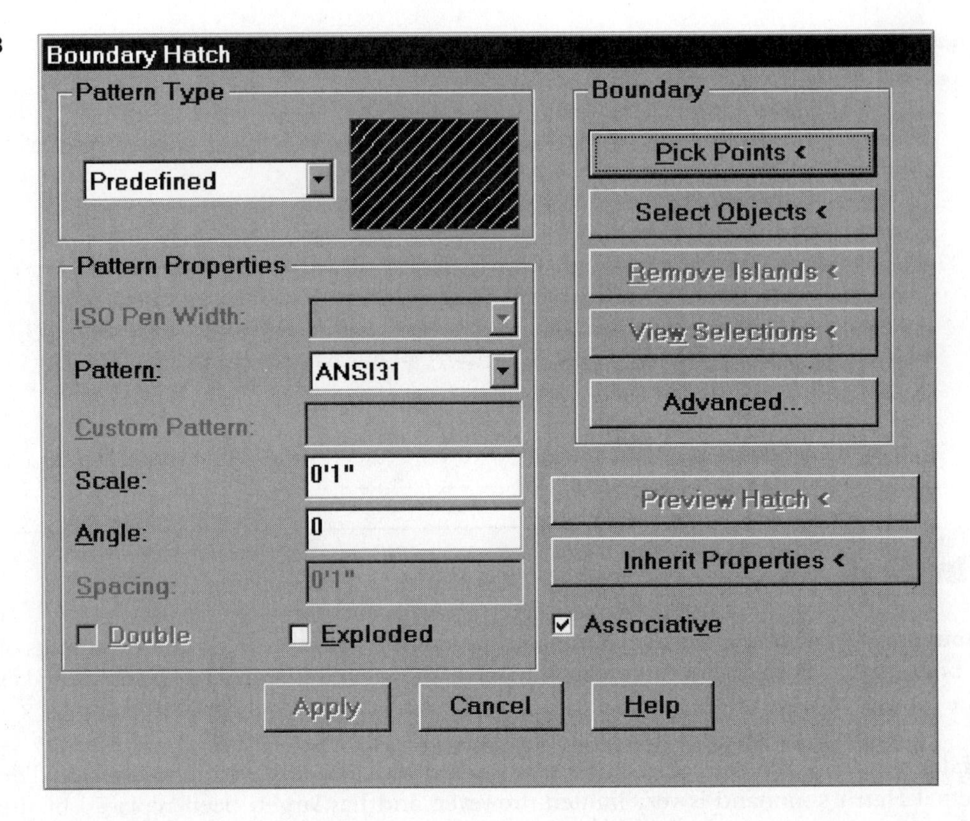

have the prefix **EPLC**. The other way to select a **hatch** pattern is to move the cursor over the image of the pattern and press the **<Pick>** button. This will cause the images to sequence through all of the AutoCAD patterns, and the names of the patterns will be displayed in the **Pattern:** pop-down box below. After moving through the sequence of AutoCAD patterns, the LANDCADD patterns are listed in the pop-down box, but their images are not available. The procedure for viewing and using LANDCADD **hatch** patterns is discussed later in this section.

The **ISO Pen Width** option is only available if an **ISO** pattern is selected. **ISO** patterns are intended for metric drawings, and are not often used in landscape applications. If one is selected, the pen width for **hatch**ing these patterns can be selected here.

The **Pattern:** text box allows you to enter the name of any custom pattern file you have created. This custom pattern is typically created and stored with the extension **.PAT**.

The **Scale** box is where the value associated with the **size** of the **hatch** is placed. Use the scale factor calculated from the relationship of plotted size to actual size. A **1/4" = 1'-0"** drawing scale computes to a scale factor of **48** (1" = 4' or 1" = 48"). A **1/20** scale drawing (1" = 20'-0") computes to a scale factor of **240** (1" = 240").

Hint: If in any doubt as to the correct value of the scale factor, set it to a **large** number. The worst that will happen is that the hatch will be too coarse to display properly on the drawing. If the value is set extremely low, a **very** dense **hatch** pattern will result, slowing your computer to a stand-still, or perhaps even crashing your AutoCAD session. Any time you see this happening, press **<Esc> immediately**. This will terminate the generation of the **hatch** pattern and allow you to re-enter the value before trying again.

The **Angle:** box allows you to enter the desired value for the angle or slope of the hatch. An angle of **0** indicates that the slope of the hatch will match that shown on the image. Any angle entered here will rotate the displayed hatch pattern by that value in degrees. Both the **Scale** and **Angle** factors are entered using the currently selected units.

The **Spacing:** option is only activated if a User-Defined **hatch** pattern is selected. The value entered here will correspond to the spacing between the **hatch** pattern lines.

The two toggle boxes in the **Pattern Properties** area relate to very different properties. The **Double** toggle box is only active if a User-Defined hatch is selected above. This option will create a second layer of hatching oriented at 90° from the original pattern.

The **Exploded** option draws the pattern as a series of lines, rather than a cohesive block. This means that individual lines can be edited without affecting the entire **hatch**. The **hatch** though, will lose its **associativity**. This means that if the **hatch** boundary is modified, the **hatch** will not be modified along with it. There are few occasions where this choice should be used.

Bhatch Boundaries

The **Boundary** options in the "Boundary Hatch" dialogue box allow you to select which areas will be **hatch**ed with the pattern specified previously. There are two methods for defining the **Boundary**: **Pick Points** and **Select Objects**. Either of these methods might be employed, or a combination of both of them might be used.

Pick Points < will take advantage of AutoCAD"s **ray casting** technique in searching for a boundary. Simply pick a point inside the area to be **hatch**ed, and allow the **ray casting** to find the bordering area for the **hatch**. The prompts resemble those given with the **Boundary** command:

```
Select internal point:
```

Move the cursor into the area to be hatched, then press the **<Pick>** button.

```
Selecting everything...
Selecting everything visible...
Analyzing the selected data...
Analyzing internal islands...
```

```
Select internal point:
```

Add another area to **hatch** or press **<Enter>** to complete the process of finding the area to hatch.

Once the point has been picked, the **ray casting** process examines the objects on all sides of the selected point and searches for a **boundary**. A temporary polyline is built and highlighted around this calculated **boundary**. Next, any internal islands are examined and outlined. By default, these will be excluded from the **hatch** area. Although the precise location of the point selected is usually (though not always) unimportant, the characteristics of the **boundary** are always important. The **boundary** must always be completely closed, without any gaps. If gaps are detected in the **boundary**, AutoCAD will return an error message.

If the **boundary** is composed of complex polylines, the **ray casting** process can bog down your system, or may yield incorrect results. The **boundary** might need to be modified, or the area broken into smaller, less complex sections with temporary lines.

Select Objects < allows you to select the **boundary** of your **hatch** manually. The objects which comprise the **hatch** area must form a completely enclosed area, and cannot overlap. When the **boundary** objects are manually selected, AutoCAD does not build a temporary polyline around the area, and it is the temporary polyline which compensates for overlapping objects. When overlapping lines are found, they should be edited (broken, erased, trimmed, filleted, etc.) where appropriate. Where gaps are found, they must be closed with lines, polylines, arcs, etc., or by extending, filleting or stretching existing objects. If these corrections are not made, AutoCAD draws **hatch**es that may be unsatisfactory. See below for examples (Figure 4.4).

It may be appropriate to use the **Select Objects** option to create an outer **boundary**, but allow AutoCAD to find internal areas using the **Pick Points** option. Both options can be used in establishing a **hatch** area.

After a **hatch boundary** has been selected, the "Boundary Hatch" dialogue box returns to the screen with new options available. The **Remove Islands <** option allows you to "pick and choose" amongst the internal islands that AutoCAD finds within the **hatch** area. Use this option to remove any islands from the **boundary** that were found by the **ray casting** technique and permit them to be **hatch**ed. The **View Selections <** option will re-highlight the selected **hatch boundary** in the drawing. This allows you to ensure that all of the desired areas are highlighted before the **hatch** is drawn. The **Advanced ...** option opens the "Advanced Options" dialogue box.

The upper portion of this dialogue box operates in an identical manner to the "Boundary Creation" dialogue box. The **Object Type:** pop-down box is only active if the **Retain Boundaries** toggle box is marked. When this toggle is marked, AutoCAD will build a **boundary** object that will stay with the drawing even if the **hatch** is erased. As with

Figure 4.4

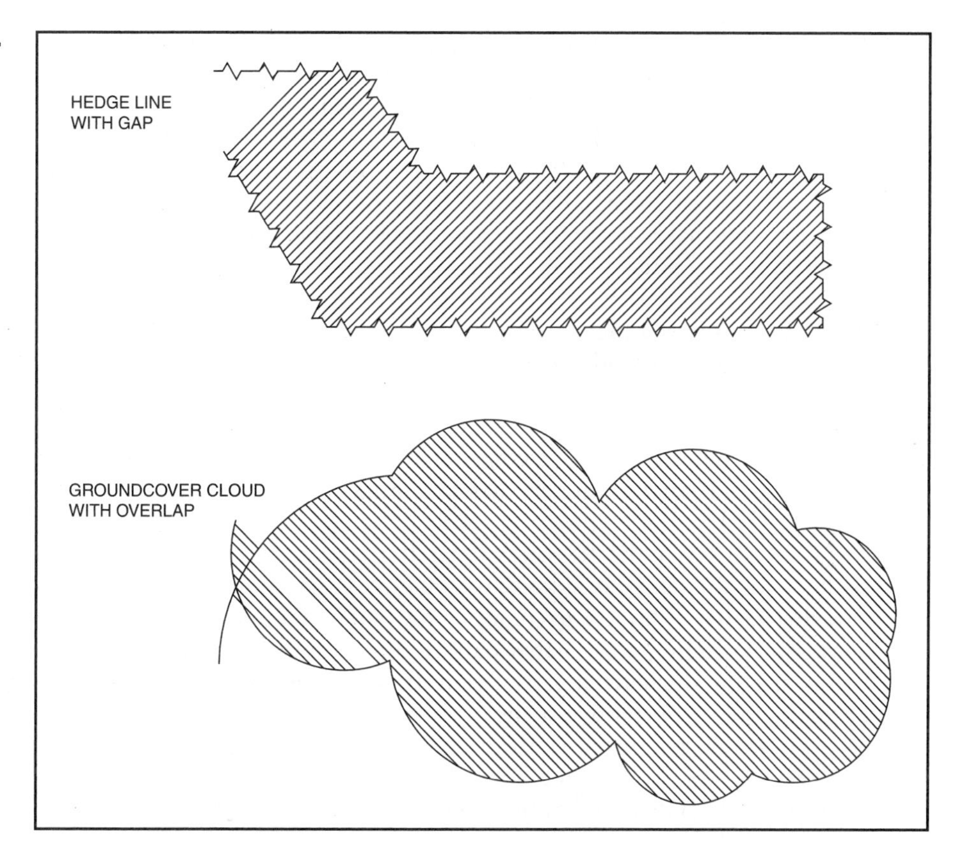

the "Boundary Creation" dialogue box, this **boundary** object can be either a **polyline** or a **region**. The **Define Boundary Set** selections operate identically to those found in the "Boundary Creation" box. In most cases, the radio button selection **From Everything on Screen** is appropriate. This allows AutoCAD to search all of the objects on the screen to find objects to include in a **hatch boundary**. In very large drawings, you can use the **Make New Boundary Set <** to define a smaller set of objects from which to choose for the **hatch boundary**.

The **Style:** pop-down box allows you to select the options for hatching text and internal islands in the **hatch** area. The **Normal** option will avoid hatching over text, and will not hatch in interior islands. If there are objects within the interior islands, they will be **hatch**ed. The hatch pattern will be applied in an alternating pattern between the outer edges, internal islands, and objects within the islands. This default value is correct for most drawings. The **Outer** option will hatch outer areas, avoiding internal islands as before. Those objects **inside** the internal islands will not be **hatch**ed. The final **Ignore** option will ignore internal islands entirely. The entire area within the **hatch boundary** will be hatched regardless of any internal islands.

The **Ray Casting** pop-down box is available if the **Island Detection** toggle is turned off. When **Island Detection** is disabled, you can control how AutoCAD will search for the **hatch boundary**. You can choose which direction AutoCAD will use in **ray casting**. Your choices include **+X**, **-X**, **+Y**, **-Y**, and **Nearest**. The relationship of the pick point and the **ray casting** direction is crucial. If the wrong direction is chosen, the first object encountered might be an internal island rather than a boundary edge. Since **Island Detection** is disabled, the internal island will be sensed as a boundary, and your pick point will be outside it. This will cause AutoCAD to report that your pick point is outside the boundary.

Island Detection is a toggle that controls whether or not AutoCAD will hatch over the top of any internal islands found within the **hatch** area. This has the same effect as using the **Ignore** style above.

The **Retain Boundaries** toggle controls whether or not AutoCAD will build a polyline or region that will survive after the **hatch** command is completed.

The final options in the "Boundary Hatch" dialogue box represent the most convenient aspects of the **Bhatch** command. Once a **hatch boundary** is established, the **Preview Hatch <** option becomes active. **You should always use this option**. This will allow you to examine the **hatch** pattern appearance and the **hatch boundary** integrity **before** making the **hatch** a part of your drawing. Because **hatch**es contain large amounts of data, they can (and usually will) greatly enlarge your drawing. By using the **Preview Hatch <** option, you can make any necessary adjustments to the scale, angle and boundary of your **hatch** before placing it in the drawing. After examining the **hatch** itself, press the **Continue** button to return to the "Boundary Hatch" dialogue box.

Inherit Properties < is used to duplicate an existing **hatch** in the drawing. The **hatch** pattern, scale and angle will be retrieved and used for the next **hatch**. After this option is picked, you are returned to the drawing screen where the prompt reads:

`Select hatch objects:`

Pick the **hatch** area that you wish to duplicate. The hatch properties are retrieved and are available as you return to the "Boundary Hatch" dialogue box.

The **Associative** toggle box is used to control whether or not the **hatch** created will be associated with its **boundary**. An associative **hatch** will be modified as the **hatch boundary** is modified. If this option is disabled, the **hatch** will not change regardless of the status of the **boundary**.

The **Apply** button at the bottom of the dialogue box actually places the **hatch** on the drawing. This ought to be the **last** step in the **hatch**ing process. Do not apply a **hatch** to a drawing before it has been previewed and examined carefully. This will save considerable time in erasing and purging incorrect **hatch**es.

Hatching in LANDCADD

LANDCADD provides over 60 different hatch patterns for use in landscape drawings. Accessing these hatch patterns and placing them in drawings is somewhat different than in AutoCAD. Regardless of which LANDCADD module is active, you can use LANDCADD hatches from the LANDCADD **Tools** pull-down menu. (Remember that AutoCAD also has a Tools pull-down menu with different commands in it.)

Pick the **Hatch** selection to activate the **Hatch** Hatch sub menu (Figure 4.5).

Each option on this sub-menu activates a different dialogue box. The ACAD BHatch option will open the AutoCAD "Boundary Hatch" dialogue box described above. The Bomanite option opens a dialogue box showing the various Bomanite paving choices (Figure 4.6).

The Brick option contains patterns for several popular paving materials (Figure 4.7).

The "Select brick pattern" icon menu has another page with several more choices (Figure 4.8).

The Fill option offers other choices for hardscape surfaces (Figure 4.9).

The Misc. option offers unusual patterns suitable for various fills and solid areas (Figure 4.10).

The Roof option supplies hatches appropriate for elevation or 3-D views of roof areas (Figure 4.11).

Figure 4.5

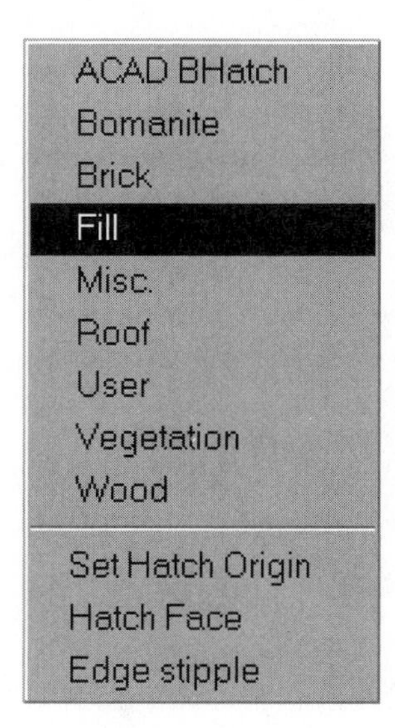

Figure 4.6

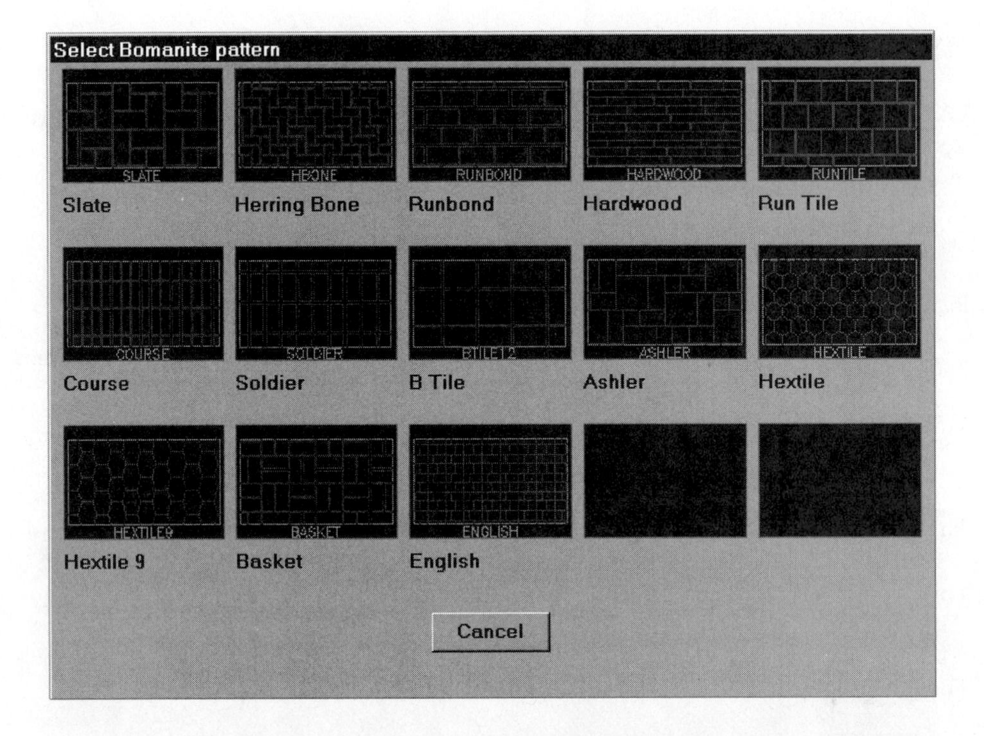

Figure 4.7

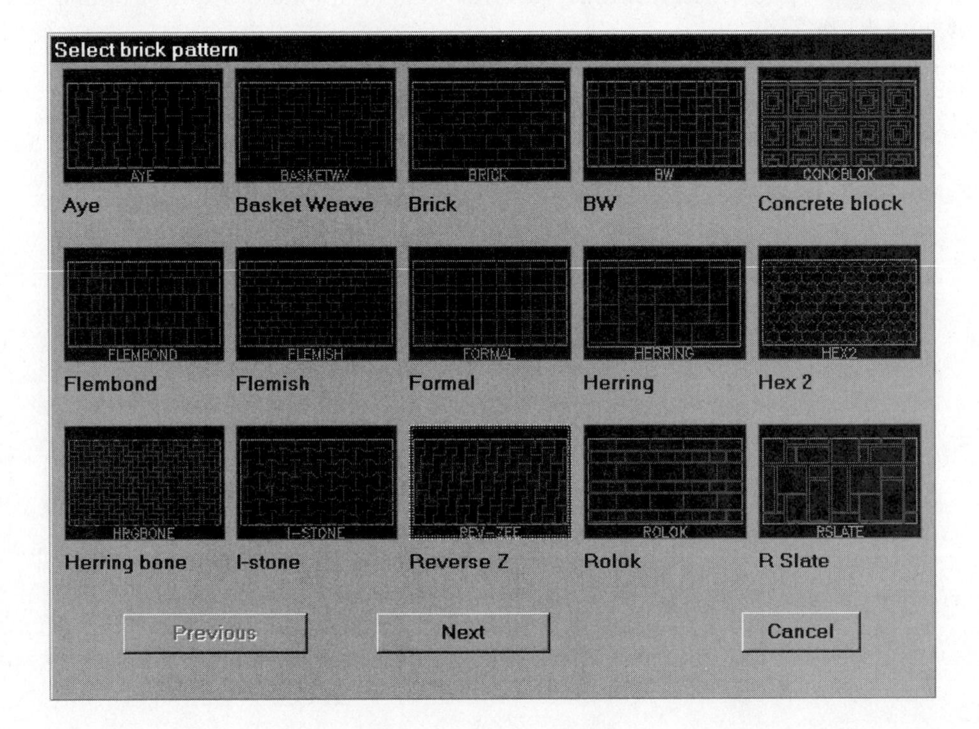

Figure 4.8

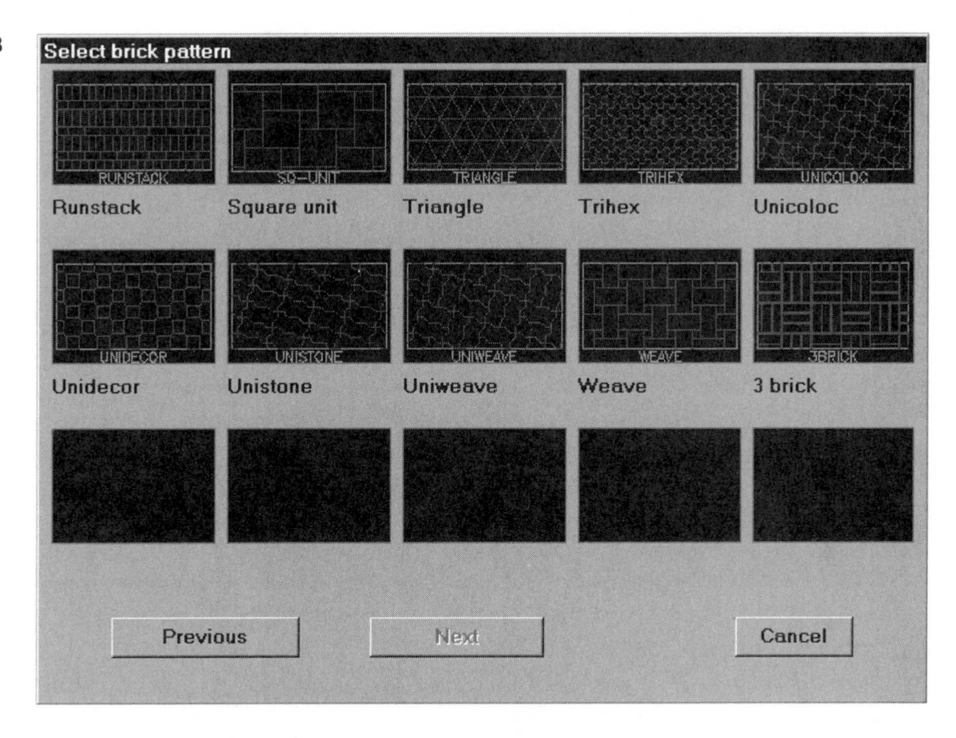

Figure 4.9

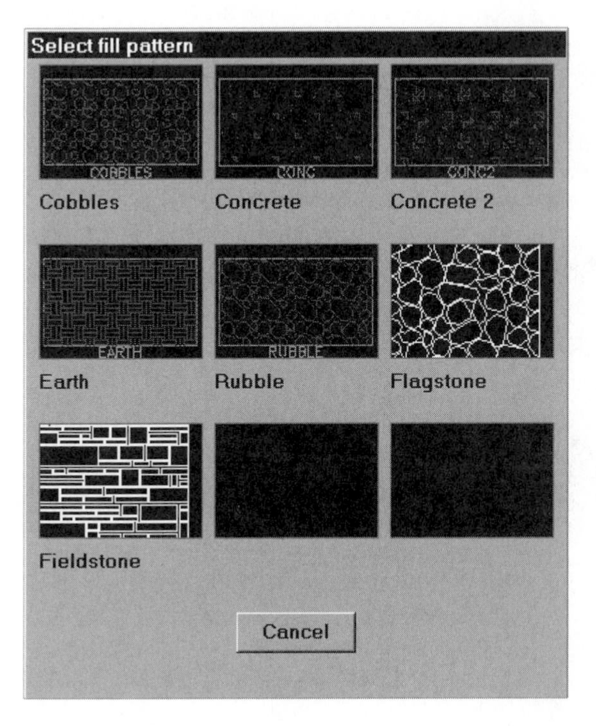

Figure 4.10

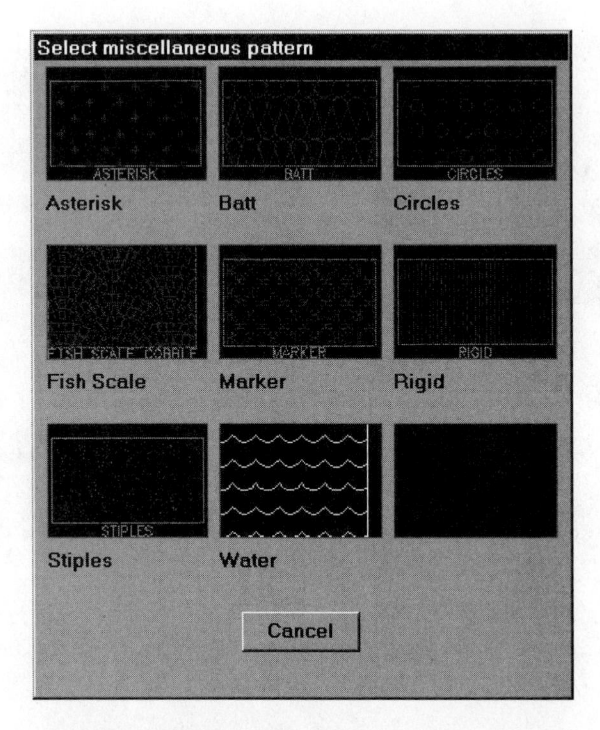

Figure 4.11

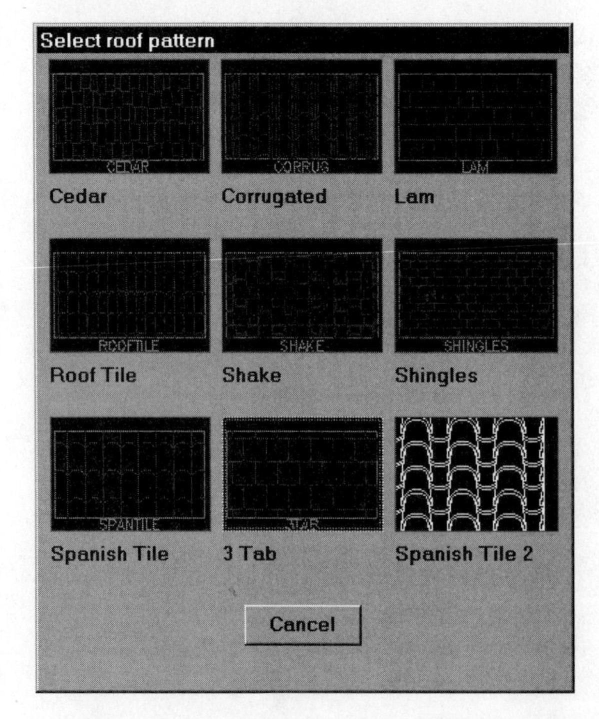

The **User** command uses the AutoCAD command line **Hatch** command to allow you to specify a hatch pattern created to your requirements. After making this selection, you are returned to the drawing screen and the hatch is created from responses at the command line:

```
Command: _hatch
Pattern (? or name/U,style) <U>: U
Angle for crosshatch lines <0>:
```

Enter the desired angle for your hatch lines. This is the first response you can give during this command. The previous responses are built in to this command routine.

```
Spacing between lines <1.0000>:
```

This prompt will be given in decimal units regardless of your drawing units. In architectural units, 1.0 = 1 inch. Enter a value here.

```
Double hatch area? <N>
```

Indicate whether or not to place a second set of hatch lines at 90° from the original hatch angle. Type **Y** or **N**.

```
Select hatch boundaries or RETURN for direct hatch option:
```

Either select the hatch boundary manually (as with the **Select Objects <** button in the "Boundary Hatch" dialogue box), or use the **direct hatch** option.

The **direct hatch** allows you to define a polyline around the hatch area. Instead of selecting objects, you select vertex points for the polyline. Before identifying the points in the polyline, you must decide whether or not to keep the boundary polyline created by this command:

```
Retain polyline? <N>
```

Answer **Y** or **N.**

```
From point:
```

Pick the beginning point of the polyline.

```
Arc/Close/Length/Undo/<Next point>:
```

Use the traditional polyline options to build your polyline. Be sure to **Close** the polyline at the end.

```
From point or RETURN to apply hatch:
```

At this point, you can create another polyline (`From point`) or press **<Enter>** to place the hatch in the defined area. You cannot preview the hatch or edit it before it is placed on the drawing.

If the boundary objects are available on the screen, choose them and press **<Enter>** to place the hatch on the screen.

`Select objects: 1 found`

The hatch is placed in the boundary area.

The Vegetatian option allows the selection of three patterns suitable for filling areas with random or arranged patterns of plants (Figure 4.12).

The Wood option allows you to select hatch patterns suitable for details and elevation plans of wood structures (Figure 4.13).

Figure 4.12

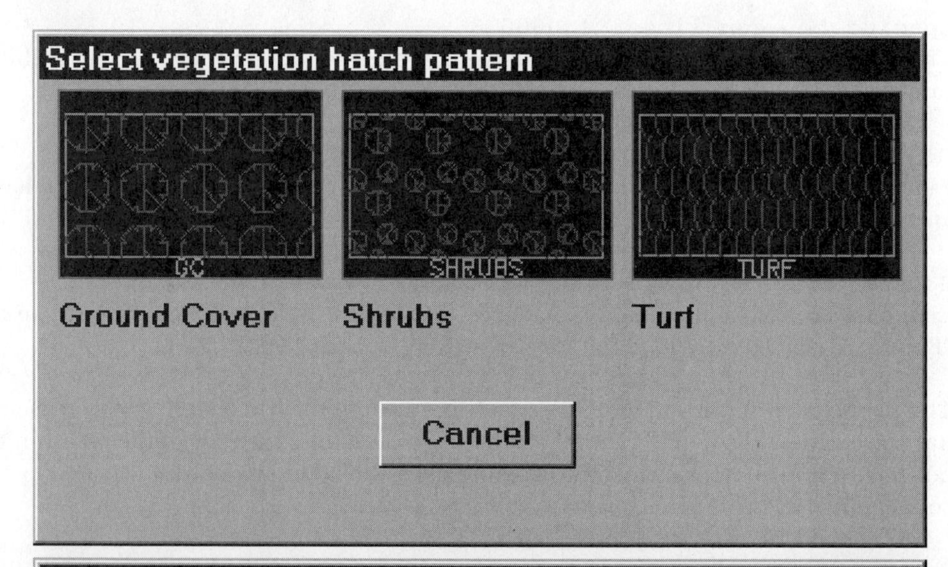

Figure 4.13

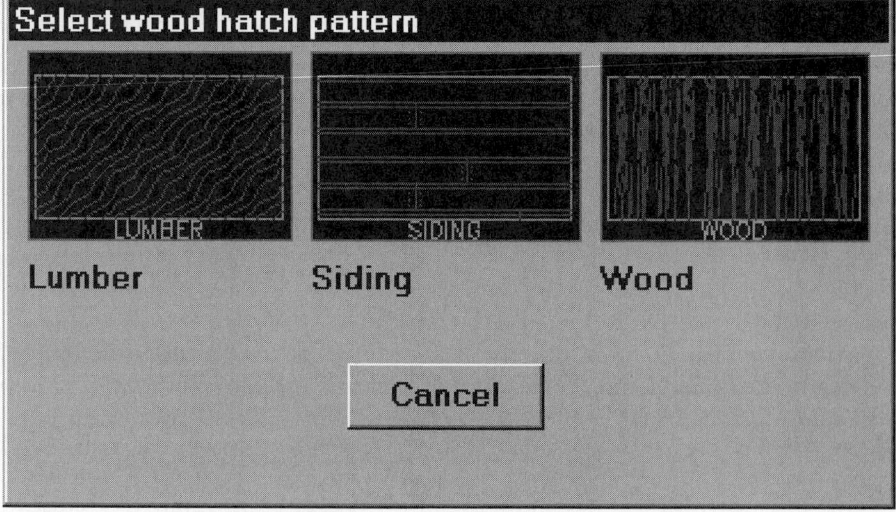

To insert any of the LANDCADD hatch patterns from these icon menus, follow this procedure:

Pick the **Tools** pull-down menu, then select **Hatch**. Select the type of hatch category you need, opening the appropriate icon menu.

From the icon menu selected, pick the desired hatch pattern, then pick **OK**. Follow the instructions at the command prompt:

```
Command: EPLC_texture
Hatch scale <12">:
```

Enter a scale for the hatch pattern. The LANDCADD hatches are created at different scales, so there is no direct correlation between the scale factor and the scale of the hatch. Since there is no way to preview the hatch, there is much trial-and-error associated with these patterns using LANDCADD's hatching command. The decision about scale will depend on whether the hatch is intended to be proportional to the drawing or is simply intended to illustrate a pattern for presentation work. As before, try to enter a **large** value here first, then reduce the scale to suit your needs. The following figures illustrate the LANDCADD hatch patterns, with each shown at the scale listed beneath. Use the relative scales of each square as a guideline when you begin using these hatch patterns (Figures 4.14, 4.15).

```
Rotation angle <0>:
```

Enter a rotation angle for the hatch.

```
Objects to hatch:
```

Select the objects you wish to be hatched. The objects selected must create an enclosed, non-overlapping boundary. Be sure to edit the objects forming the hatch boundary **before** trying to create the hatch.

```
Select objects: 1 found
Select objects:
```

Press **<Enter>** to complete the selection set and return to the LandCADD"s hatching command.

```
Press ENTER to accept or [S]cale/[P]attern/[A]ngle:
```

If the hatch requires editing, type the first letter of the parameter to edit. Both the **Scale** and **Angle** can be edited easily enough, but the **Pattern** is more of a problem. Since this command is carried out at the command line, you cannot pick the hatch pattern from a list or an icon menu. You must know the full name of the new hatch pattern you wish to use. Without access to the list of LANDCADD hatch patterns, it may be difficult to identify the names of all of the patterns. If you type the name of the new pattern incorrectly, the command terminates and you must start over.

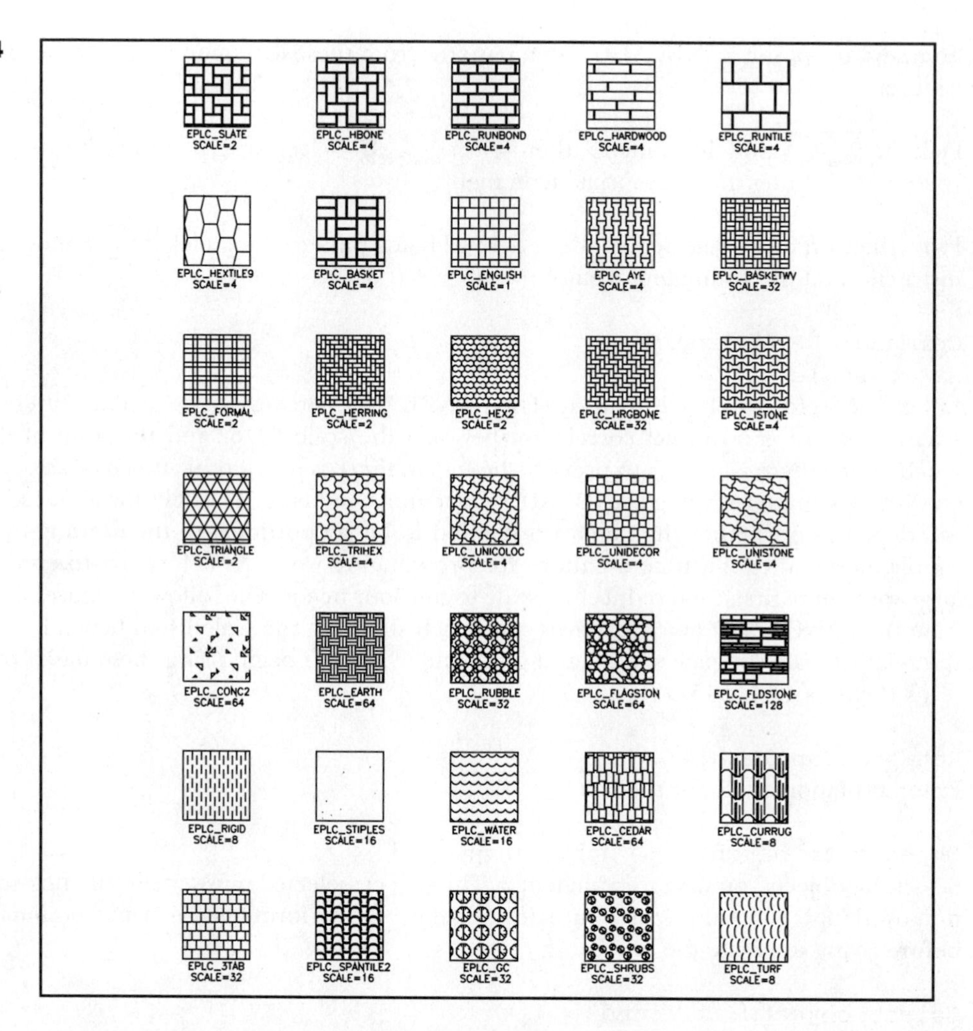

Figure 4.14

After editing these hatch pattern parameters to your satisfaction, press **<Enter>**.

```
Done.
```

From these problems it is clear that LANDCADD"s hatching command is not as capable as AutoCAD's **Bhatch** command. The best solution (until LANDCADD improves its hatching command) is to use LANDCADD's icon menus to identify the hatch patterns, then use AutoCAD's **Bhatch** command to perform the actual hatching. That way, you can see the LANDCADD hatch patterns and identify them (keep a pencil handy) with the icon menus, but still get the benefit of previewing the hatch and controlling it more easily with the **Bhatch** command.

Figure 4.15

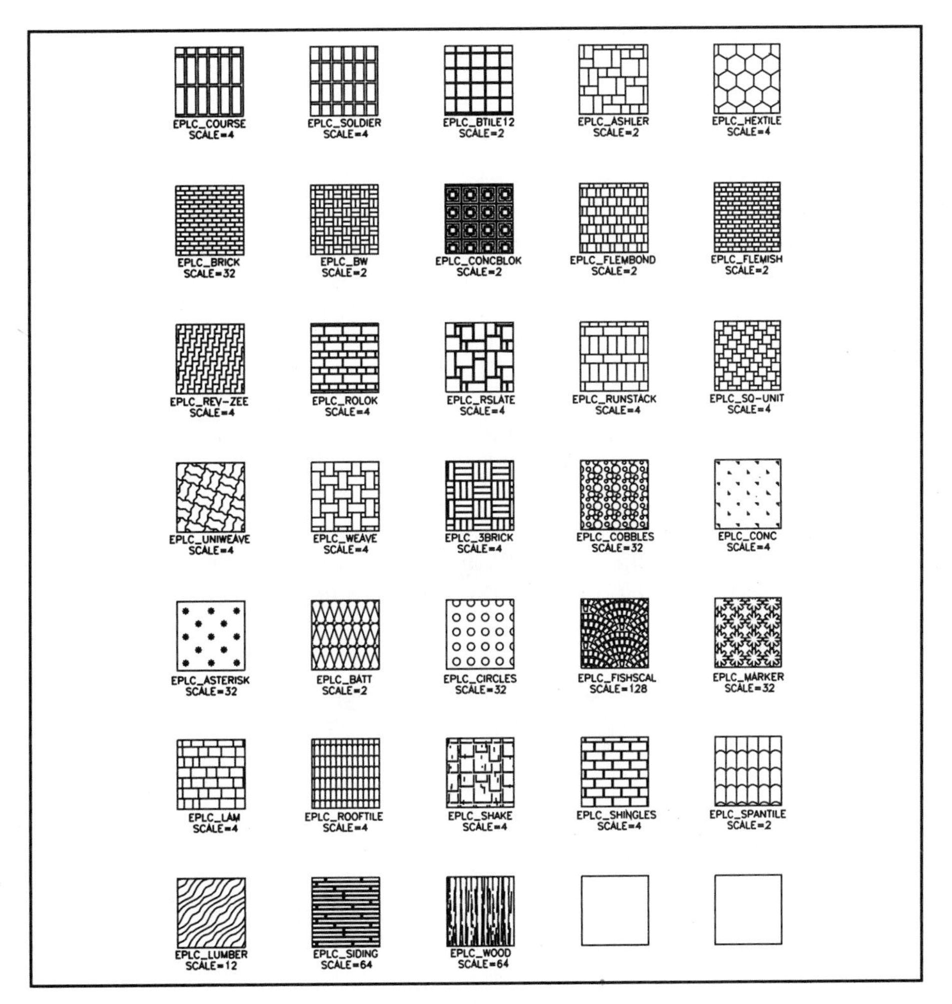

The **Set Hatch Origin** option is used to establish a new origin point when you want a hatch pattern to begin in a specific spot. This is valuable when you need a pattern to begin with a whole unit rather than a partial unit. It is also useful for performing hatches in other UCS (User Coordinate System) planes because it allows the hatch to be performed from a specific spot on the screen. The command prompt reads:

```
Command: _snap
Snap spacing or ON/OFF/Aspect/Rotate/Style <1'-0">: _r

Base point <0'-0",0'-0">:
```

Enter or pick a new base point for the hatch.

Hatch Face permits you to place a hatch on a 3-D face defined in the drawing. 3-D faces can be used to build roof surfaces, walls and other objects in LANDCADD. The command prompt reads:

`Select face to hatch:`

Pick the 3-D face you wish to hatch. The command will obtain the coordinates of the 3-D face and place the hatch on the correct plane.

`Pattern? <ANSI31>:`

Enter the name of the hatch pattern you want to use.

`Scale? <1">:`

Enter the scale of the hatch you wish to use. If the scale or angle are incorrect, use **Undo** to eliminate the hatch. This will keep the size of your drawing as small as possible.

`Hatch Angle? <0>:`

Enter the angle of the hatch. 0° will be oriented along the first side located on the 3-D face. After the angle is entered, the hatch will be placed on the 3-D face.

Edge Stipple is a complex routine that allows you to hatch an area enclosed by a polyline in a gradually fading pattern. The stipple pattern is dense toward the outside of the object, fading to a sparse pattern toward the center of the object. The command creates temporary polylines offset from the outside border, then creates hatches at different scales for each region defined. The polylines are erased but the hatch pattern remains. At the command prompt:

`Select edge:`

Pick the outside edge of the object to be hatched. Make sure that the object is enclosed by a single, non-overlapping polyline.

`Select point at end of stippling:`

Pick a point near the center of the area to be hatched. LANDCADD will then draw several temporary polyline edges to create a graduated hatch pattern. If the polylines spill outside the edge of the area or look incorrect, use **<Esc>** to cancel the command. Use **Undo—Back** to remove the "temporary" polylines, then start over.

`Scaling factor (smaller numbers = more dense <25.000>):`

Set the scaling factor for the hatch. This regulates the density of the stippling.

`Change stipple scale? (no/yes):`

Type **N** if the hatch pattern looks correct. Type **Y** if you want to change the scale factor to adjust the density of the hatch.

chapter

Entering Text in AutoCAD and LANDCADD

There are several ways to place text into AutoCAD drawings. The method you will use will depend on the quantity and size of the text to be entered, and the precision with which the text must be inserted into the drawing. In addition, the platform of AutoCAD that you work on will dictate which of the text handling tools are available to you.

Text in AutoCAD takes on interesting properties. In some ways, text is treated as any other object on the screen. That is, text can be copied, rotated, scaled, moved and erased just as you would a line, an arc or a circle. On the other hand, AutoCAD has unique tools to handle the appearance, size and orientation of the lettering used, and still other tools to edit existing text in the drawing.

We will break up our discussion of text into sections relating to the functions of the commands. There are commands which are used to create text, establish the style (lettering) of the text, edit existing text in a drawing, and import text from outside AutoCAD.

Text Creation Commands

The TEXT command

Text—The **Text** command is the original tool for creating text in the original versions of AutoCAD. As the oldest text creating command, it is the most primitive tool for creating text. To use the **Text** command:

`Command:`

Type **Text** or pick **Text ▸** from the **Draw** pull-down menu, then pick **Single-Line Text**.

`Justify/Style/<Start point>:`

Pick a start point for the text. This will be the lower left corner of the first letter of text. We will examine the functions of the **Justify** and **Style** in the discussion of **Dtext**.

```
Height <1'-0">:
```

Enter the correct height for the text to be entered. To obtain the desired height for your text, multiply the scale factor of the drawing times the desired size of the lettering. For a 1/4"= 1'-0" scale drawing where the desired text size is 1/8" high, the result will be 6". (Scale factor is 1/4" = 1' or 1" =4' or 1"=48" or a factor of 48; text size will be 1/8" times 48 or 6").

```
Rotation angle <0>:
```

Enter the angle of rotation for the text. An angle of **0** indicates text that will be written from left to right horizontally across the screen.

```
Text:
```

Enter the text you want to place on the drawing. The text will appear on the command line. After you have finished creating the line of text, press **<Enter>** to place it on the screen.

The **Text** command only places one line of text on the screen at one time. To place a line of text directly below the previous one, press **<Enter>** again. This restarts the **Text** command.

```
Command:
TEXT  Justify/Style/<Start point>:
```

Press **<Enter>** again. This places the new start point one line below the previous start point.

```
Text:
```

Type in the second line of text. The text will use the same height, rotation, and justification as the text entered previously. Press **<Enter>** to place the text on the screen.

The limitations of the **Text** command are evident. You can only enter one line of text at one time, and there is no way to see the text on the screen until the text line is completed and you press **<Enter>**. If the text is too long for the area, if the lettering size is too large, or if the text style is incorrect, you have no way of seeing it until the text is on the screen. This may have been acceptable during the pioneering days with AutoCAD, but most users today expect more convenience when entering text.

The DTEXT (DYNAMIC TEXT) command

Dtext (**Dt**)—The **Dtext** command inserts one line of text at a time, but does so *dynamically* (**Dtext** stands for Dynamic Text). That is, the text is visible as it is inserted. After the command is started, a square appears on the screen which illustrates the size and location of the first letter in the text line. This makes it possible to check the relative size of the letters before an entire line of text is entered on the screen.

`Command:`
Type **Dtext** (or **Dt**) or pick **Text ▸** from the **Draw** pull-down menu, then pick **Dynamic Text**.

`Justify/Style/<Start point>:`
Pick a starting point (the lower left corner of the first letter) or select one of the other options for this command:

Justify—selecting this option will return the following prompt:

`Align/Fit/Center/Middle/Right/TL/TC/TR/ML/MC/MR/BL/BC/BR:`
Select the type of justification desired. The command will then ask for the height and rotation for the text. The justification options are discussed below:

Align—this option allows you to pick two points on the screen which will act as the starting and ending point (lower left and lower right) for the text line. The angle of the text will align with the line, and the lettering size will shrink or grow proportionally to allow the line of text to fill the space between the two points.

Fit—this option fits the lettering between the two points specified (as above) but the height of the lettering will remain constant. The width of the lettering will shrink or grow to allow the lettering to fill the space between the points. This can lead to very distorted lettering if the space is large and the line of text is short.

Center—this selection centers the line of text so that the selected point is at the *bottom* center point of the lettering. The next line of text will be centered below the first line entered.

Middle—this selection centers the line of text so that the selected point is at the *middle* center point of the lettering. As before, the next line of text will be centered below the first line entered.

Right—this option creates a line of right-justified text. The lettering will be placed so that the selected point is located at the lower right corner.

The next selections adjust the relationship of the selected point to the line of text which will appear on the screen. The lettering will appear on the screen with normal left justification until you press **<Enter>** to complete the line of text. At that point, the text will move into proper justification.

TL—sets the *top left* point of the text.

TC—sets the *top center* point of the text.

TR—sets the *top right* point of the text.

ML—sets the *middle left* point of the text.

MC—sets the *middle center* point of the text.

MR—sets the *middle right* point of the text.

BL—sets the *bottom left* point of the text.

BC—sets the *bottom center* point of the text.

BR—sets the *bottom right* point of the text.

The results will appear slightly different if upper and lower case text is used. When lower case letters are used, the **BL**, **BC**, and **BR** points will correspond to the bottom of letters which extend below the base line of the text. The lower edges of such letters as "p" and "y" will determine the location of the **BL**, **BC**, and **BR** points.

Style—This option allows you to select which of the existing text styles you want to use to display the text to be entered. The style you choose must already exist in the drawing. If you choose, you can examine a list of existing styles in the drawing.

`Command:`
Dtext.

`Justify/Style/<Start point>:`
Type **Style.**

`Style name (or ?) <STANDARD>:`
Type **?.**

`Text style(s) to list<*>:`
Press **<Enter>** to list all of the available text styles.

```
Style name: 1 Font files: romans.shx
Height: 0'-0" Width factor: 1.0000 Obliquing angle: 0
Generation: Normal

Style name: 2 Font files: swissb.ttf
Height: 0'-0" Width factor: 1.0000 Obliquing angle: 0
Generation: Normal
```

```
Style name: 3 Font files: swissk.ttf
Height: 0'-0" Width factor: 1.2500 Obliquing angle: 0
Generation: Normal

Style name: STANDARD Font files: HLTR.SHX
Height: 0'-0" Width factor: 1.0000 Obliquing angle: 0
Generation: Normal

Press RETURN to continue:
```

This list displays the available text styles, the fonts assigned to them and their display characteristics. We will discuss these characteristics later in this section.

Start point—this option places the text justified **Bottom left** to the start point.

The MTEXT (MULTI-LINE TEXT) command

Mtext—This is the newest addition to AutoCAD's text handling tools. **Mtext** was added with Release 13. **Mtext** behaves more like a word processor than a text line editor. Text is added in a paragraph form and can be formatted to allow more sophisticated control over the appearance of the text. The **Mtext** command will allow you to define an insertion point and a "width point" which define the overall width of the paragraph created by the text. The text will wrap onto another line if the width of the text exceeds the width of the line defined on the screen. After the insertion point and "width point" are defined on the screen, **Mtext** switches to a text editing screen which allows you to enter text and apply certain types of formatting to the text. For instance, you can assign colors to individual words in a paragraph of text, underline certain words, change fonts, change text height and create stacked fractions within the text editing function. With **Mtext** (as with many new AutoCAD Release 13 features) the best functionality of the command is found in the Windows platforms. The Windows 'Edit Mtext' dialogue box allows you to see the results of most text formatting as you create it. The DOS version of AutoCAD Release 13 automatically invokes the DOS Edit command, which cannot display any formatted text. Any text formatting must be created by inserting *formatting codes* which define how the text will appear after it is inserted into the drawing. See the AutoCAD documentation book *AutoCAD User's Guide* or one of the AutoCAD reference books mentioned in Appendix R for more information about AutoCAD text formatting codes.

To use the **Mtext** command:

`Command:`

Type **Mtext** (or **Mt**) or pick **Text ▸** from the **Draw** pull-down menu, then pick **Text**.

`Attach/Rotation/Style/Height/Direction/<Insertion point>:`

Pick an insertion point for the text, or choose another option. If you choose an **Insertion point**, the prompt returns:

`Attach/Rotation/Style/Height/Direction/Width/2Points/<Other corner>:`

On the screen, you will see a "rubber band box" stretched across the screen. When you specify the **Other corner**, you are setting the **width** of the paragraph only. The box on the screen only shows the width of the paragraph, but does not indicate two other important features: the length of the paragraph and the orientation of the text relative to the **Insertion point**. There are good reasons for these "shortcomings" but they can lead to confusion when entering text.

Since the "rubber band box" really only sets the **width** of the paragraph, it provides no information about the **length** of the paragraph. The length of the paragraph is entirely dependent on the quantity of text created in the text editing dialogue box and the size of the lettering. There is no guarantee that the text will fit within the defined box. The text may run out past the bottom (or top) of the box, or may even fail to fill a single line.

The other options for the **Mtext** command operate as follows:

Attach—this sets the orientation and justification of the text relative to the original **Insertion point**. The direction of **Attach**ment sets the justification type. If **Attach** is selected, the options are displayed. If **Attach** is not selected, **TL** (Top Left) attachment is always assumed.

`TL/TC/TR/ML/MC/MR/BL/BC/BR:`

These options correspond to those described in the **Dtext** command. The text will always **Attach** to the **first** point selected. This is true regardless of the location of the second point. It is possible to locate the second point above and to the left of the first point (establishing the paragraph width), but still have the paragraph text justified below and to the right of the first point (Figure 5.1).

You can control the justification of the text more easily if you remember that the location of the **first point** is critical. The paragraph will **always** be oriented with respect to this first point, whether it is on the left side, center or right side of the text, at the top, middle or bottom of the paragraph.

Figure 5.1

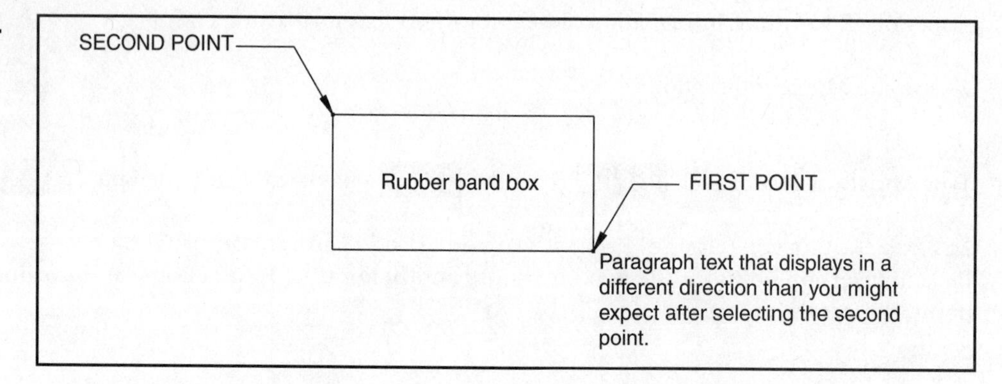

The **Left**, **Center** and **Right** justification of the paragraph will always affect the alignment of multiple lines of text. Unlike the **Dtext** and **Text** commands, **Mtext** treats all of the text as a single object. It is not possible to reset the justification of a single line of text within the paragraph.

Rotation—this option will rotate the text to an angle specified by picking two points, or by entering an angle as a value. Remember that this **rotation** will always occur **with respect to the first point** entered. The location of the second point has no bearing on the **rotation** or justification of the text. The **rotation** will, however, be affected by the method of **attach**ment used. The **rotation** about the center or right **attachment** point would clearly yield different results from a rotation about a left **attach**ment point.

Style—this option allows you to set the style of the text used in the paragraph. This operates in the same manner as the **Dtext** and **Text** commands mentioned earlier. You can, however, change the text style in the "Edit Mtext" dialogue box while the text is being entered, or after the text has already been inserted.

Height—this corresponds to the height setting used in the **Dtext** and **Text** commands. This height can also be modified in the "Edit Mtext" dialogue box.

Direction—this option allows you to orient the text horizontally or vertically. Vertically oriented text will modify the function of the second point selected as the "rubber band box" is created. In the case of vertically oriented text, the second point will define the **height** of the paragraph, rather than the width.

Width—this allows you to enter a value to specify the distance between the first point and the second point. If you know the precise width of the paragraph you want to create, you can enter the **width** with the keyboard.

2Points allows you to specify the width of the "rubber band box" (and the paragraph) by picking two other points on the screen. These two points are chosen independently of the first point chosen. This option is useful when you want to fit the text within a size defined by some other object on the screen.

Once you have defined the characteristics of the paragraph that you want to create, the "Edit Mtext" dialogue box will appear on the screen (Figure 5.2).

The size of the white editing area displayed in the "Edit Mtext" dialogue box will correspond roughly to the proportions of the text and the size of the "rubber band box" which defines the width of the paragraph. In this dialogue box, text can be entered and edited using common Windows conventions. For example, holding down the left button and dragging it across text will highlight it for editing with other functions.

Figure 5.2

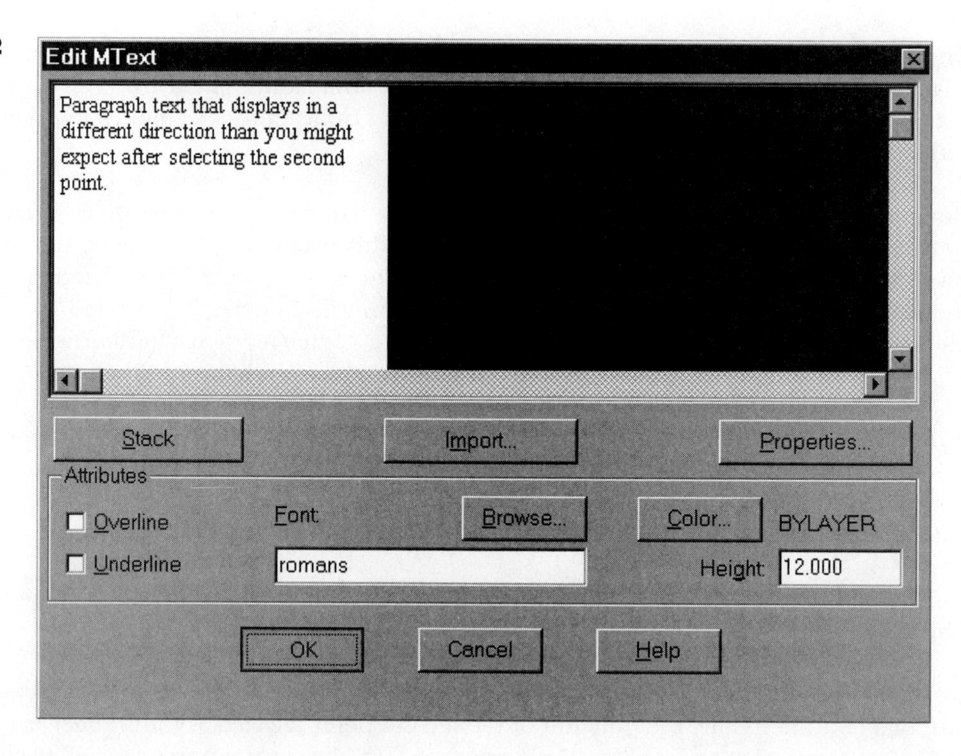

Stack will create a stacked fraction when two highlighted phrases are separated by a slash.

Import... will open a dialogue box to allow you to insert a text file into the "Edit Mtext" dialogue box. This text file can append an existing paragraph, or can be inserted into an empty editing box.

Properties will open the "Mtext Properties" dialogue box (Figure 5.3).

This dialogue box is used to modify the characteristics of the entire paragraph. Changes made in this dialogue box can be used to modify the settings used to create the paragraph originally. This means that you can not only reset the features of the text, but the features of the shape and size of the paragraph itself. This dialogue box can also be opened by typing **Mtprop** at the command prompt.

In the **Attributes** section, you can edit the characteristics of the highlighted text in the editing box. Use the **Overline** or **Underline** toggles to place lines over or under the highlighted text.

While text is highlighted in the editing box, you can change its appearance. To change the font to a different selection, type in the new name in the **Font** text box, or use the

Figure 5.3

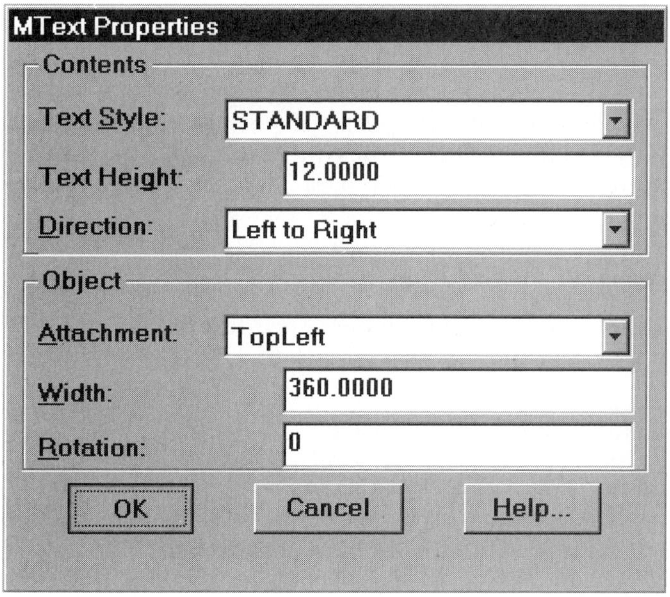

MText Properties
Contents
Text Style: STANDARD
Text Height: 12.0000
Direction: Left to Right
Object
Attachment: TopLeft
Width: 360.0000
Rotation: 0
OK | Cancel | Help...

Browse... button to search for a new font.

To reset the color of highlighted text, pick the **Color** button to make your selection.

To reset the height of the highlighted text, change the value in the **Height** text box.

One of the unique advantages of the "Edit Mtext" dialogue box relates to the fact that changes made to the format of the text will appear in the editing box. While not a perfect indicator of the final appearance of the text in the drawing itself, the visible formatting information can help you see obvious errors or formatting problems before the text is actually placed on the screen.

Once the paragraph text is placed on the screen, using the **Ddedit** text editing command will bring the "Edit Mtext" dialogue box back to the screen. This dialogue box is the only vehicle for editing paragraph text.

Text Style Commands

These commands are used to create and edit text styles. An important distinction exists between a **Style** and a **Font**. A text **style** will contain information which relates to the orientation, size, rotation and slant of the lettering built by the **font**. The **font** is a required part of a **style** but is not the same as a **style**. The **fonts** which are available to AutoCAD have increased with the introduction of AutoCAD Release 12. AutoCAD will now make use of **SHX** (AutoCAD font) files, **TTF** (TrueType® font) and **PFB** and **PFA** (Postscript® font) files. AutoCAD provides many of these files (in the **R13\COM\SUPPORT** folder), and many more may be available if you have any Windows version installed on your computer. Typically, Windows supports many types of TrueType® fonts which are stored in the **WINDOWS\SYSTEM** folder. Other TrueType® fonts may be installed with any newer word processing programs. Be sure to find these files on your computer so that you can use them in your drawings if needed.

The STYLE command

Style (or **St**) This command is used to create or edit text styles.

`Command:`
Type **St** or pick **Text Style** from the **Data** pull-down menu.

`Text style name (or ?) <STANDARD>:`
Enter the name of the style you wish to edit or create. If you want a listing of the existing styles, type **?**. If the style exists, the prompt will respond:

`Existing style.`
You then will see the "Select Font" dialogue box (Figure 5.4).

Use this dialogue box to select a different font if you wish to change an existing style, or pick **OK** to accept the font assigned to this style. You can search through any drives and directories (folders) on your entire computer to find font files. After the font has been selected, the prompt returns with a series of questions:

`Height <0'-0">:`
Enter the desired height of the text for this style. For most applications, **this height should be left at 0**. This allows you to create text of different sizes at different times while working with the same style. If this height is preset, it will affect dimension text sizes and

Figure 5.4

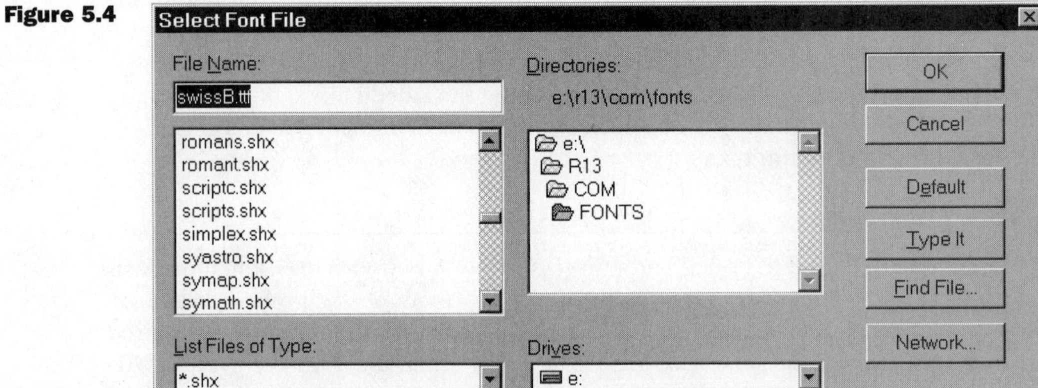

labeling routines in LANDCADD, yielding unpredictable results. Unless you have reason not to do so, leave this value set to **0**.

`Width factor <1.0000>:`
This value relates to the proportion of the width of the lettering to the height of the lettering. Values greater than **1** widen the letters, while values less than **1** tend to make the lettering appear pinched and narrow.

`Obliquing angle <0>:`
This value defines the slant of the lettering. Maintaining a **0** value keeps the lettering vertical, while positive values cause the lettering to slant to the right (forward). Negative values cause the letters to slant to the left (backward).

`Backwards? <N>`
This causes the lettering to be displayed (strangely enough) backward. This might be useful for certain types of reproduction work.

`Upside-down? <N>`
Answer **Y** if you want the lettering to display upside-down when entered. If the style is new, you will be asked:

`Vertical? <N>`
Answer **Y** if you want your text oriented vertically.

It is important to keep track of the names of your styles, and to use style names consistently throughout all of your drawings. This is especially important when drawings are inserted into other drawings. For example, assume a block drawing contains the text style STYLE1, with a simple font like ROMANC. Assume that this block was inserted into a large drawing with the same style name STYLE1, but **this** version of STYLE1 had a complex filled font like SWISSB. The AutoCAD program must then decide which style definition to use. When this occurs, the original text style definition is lost and the text takes on the style characteristics of the new drawing. This could result in illegible text that will mar the appearance of the drawing. If you keep the style names and definitions consistent from drawing to drawing, this problem will not occur.

The DDSTYLE (DIALOGUE BOX STYLE) command

Ddstyle—This command opens a dialogue box which is only available in Windows versions of AutoCAD Release 13 Version C4.

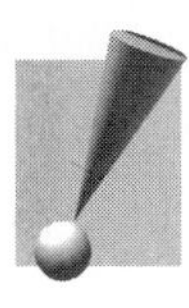

AutoCAD Warning: Although the AutoCAD help screen claims that this command is available through the **Text Style** option of the **Data** pull-down menu, this only activates the **Style** command in most releases of AutoCAD R13. The only way to activate this command is from the command line. There is no mention of the **Ddstyle** command in the AutoCAD documentation or in other AutoCAD reference books (Figure 5.5).

Figure 5.5

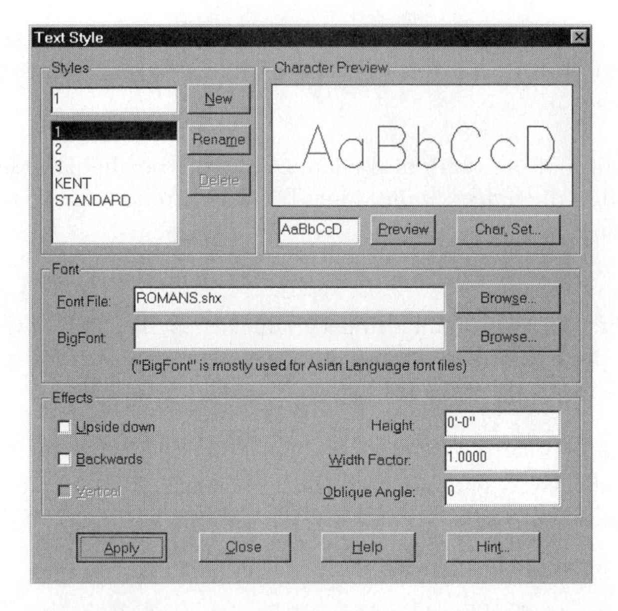

The "Text Style" dialogue box allows you to see the names of the existing styles in your drawing and to preview the font and modifying characteristics applied to it. In addition, you can create new styles, rename existing styles and modify the features of existing styles.

In the **Styles** section, you see a list of styles that exist in your drawing. The top text box shows the active style. When a style name appears in the active style box, it can be renamed. Use the **New** button to create a new style. If no name appears in the active style box, or if an existing style appears there, the new style will be named **STYLE1**. If an existing style name appears in the active style box, you can use the **Rename** button to rename it.

The **Character Preview** area is a unique feature that actually allows you to see the font as it will be displayed with the current style active. In other words, you get to see the effects of the obliquing angle, width factor, etc. on the font you have selected for the text style.

You can also examine two other important features of the font you are using. First, you can modify the characters which are being displayed in the preview area by moving your cursor into the text box to the left of the **Preview** button. Change the text displayed in this text box (the default is always **AaBbCcD**) to the text characters you want to see. You might want to examine the appearance of punctuation or numerical characters. After changing the text in this box, press the **Preview** button to have the new characters displayed in the preview area.

Another important feature of the "Text Style" dialogue box is the **Char. Set...** button. This option opens a display of the entire character set provided by the font.

Because not all fonts contain all of the characters supported by AutoCAD, it is convenient to see which characters are available for the font in question. Non-supported characters are usually displayed as the ? character.

In the **Font** area, you can select the name of the font you wish to assign to the style you have selected. You can assign either a standard font file or a **big-font** file to a text style. Big-font files contain non-ASCII characters found in kanji or other Asian language alphabets. You can even assign both types of font files to the same style. Use the **Browse** buttons to search all of the available drives on your computer for AutoCAD, TrueType or Postscript font files.

The **Effects** section illustrates the choices you have made to modify the font in the existing style. Use the settings shown here to edit an existing style or create different effects in a new style. To see the result of your **Effects** settings in the preview area, pick the **Preview** button. If you create changes to an existing style, then move to examine another style, AutoCAD provides a warning screen to ask you if you want to save the changes to the style.

Select **Yes** or **No** depending on your preferences.

At the bottom of the dialogue box, you can choose to **Close** the dialogue box without saving your changes, or you can **Apply** your changes before you exit.

The **Hint...** button provides information about the size of fonts and the **Textfill** system variable in the display of the font and style in the preview area.

Text Editing Commands

The DDEDIT (DIALOGUE BOX EDIT) command

Ddedit (or **DD**) This command is useful for modifying existing text in your drawing.

`Command:`
Type **Ddedit** (or **Dd**) or pick **Edit Text ...** from the **Modify** pull-down menu.

`Select an annotation object>/Undo:`
Pick the text you want to edit. If the object you selected was not text, this prompt will continue.

If the text was created by either the **Text** or the **Dtext** command, the "Edit Text" dialogue box appears.

Use normal editing tools (text highlighting, **<Backspace>**, **<Delete>**, **<Insert>**, **<End>** keys, etc.) to modify the existing text. After your editing is complete, pick the **OK** button. The prompt

`Select an annotation object>/Undo:`
will continue until you press **<Enter>** to end the command.

If the text was created with the **Mtext** command, the "Edit Mtext" dialogue box will appear. The same tools used to create the text can be used to edit the existing text. DOS users will be returned to the DOS Edit screen.

The DDMODIFY (DIALOGUE BOX MODIFY) command

Ddmodify brings up a context-sensitive editing dialogue box. After selecting an object to modify, a dialogue box appears which allows the properties of the specific object to be modified. There are different dialogue boxes for lines, arcs, polylines, splines, circles, hatches and text. Each dialogue box allows you to edit any of the properties of the object you chose.

To use this command:

`Command:`
Type **Ddmodify** or pick **Properties...** from the **Edit** pull-down menu.

`Select object to modify:`
Pick the object you want to modify.

When you select text which was created by the **Text** or **Dtext** command, the "Modify Text" dialogue box appears (Figure 5.6).

Figure 5.6

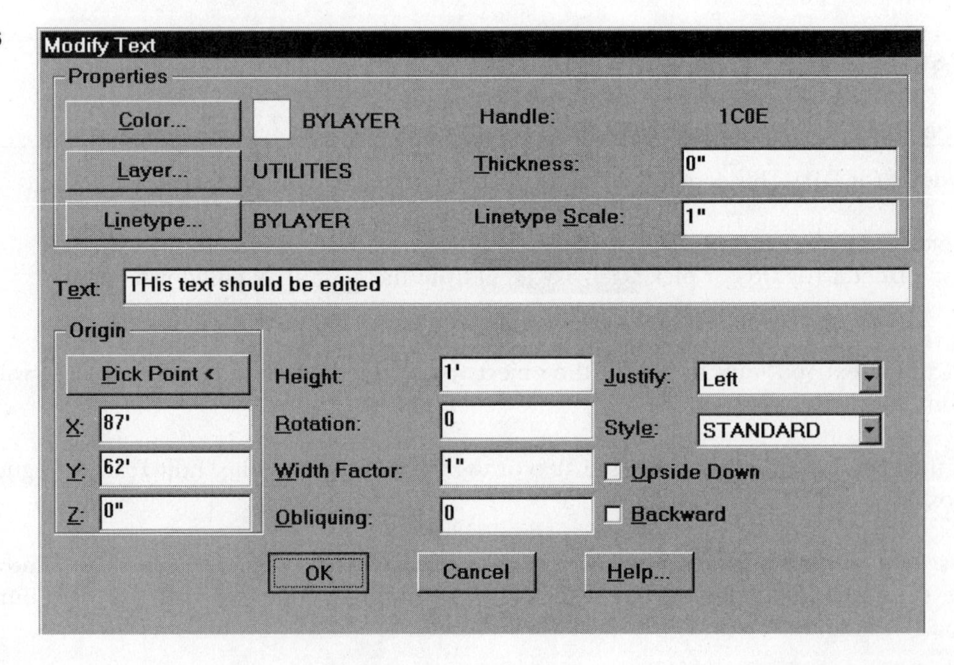

Any properties of the text can be modified in this dialogue box, including the style of the text and the individual appearance of the text. Note that changes in the **Height, Rotation, Width Factor**, **Obliquing**, etc., will only affect the text chosen, and will not affect other text created with the same style.

If the text was created with **Mtext**, the "Modify Mtext" dialogue box will appear. You can change a few features of the paragraph text in this dialogue box. If you choose to edit the text itself, pick **Edit Contents....** This returns you to the "Edit Mtext" dialogue box.

If you choose to edit the characteristics of the paragraph text, pick **Edit Properties....** This will return you to the "Mtext Properties" dialogue box (Figure 5.7).

The MTPROP (MULTI-LINE TEXT PROPERTIES) command

Mtprop—This command will allow you to modify the properties of paragraph text. Any text not created by the **Mtext** command will be ignored by **Mtprop**.

```
Command:
```
Type **Mtprop.**

```
Select an MText object:
```
Pick any paragraph text created with **Mtext.**

In the "Mtext Properties" dialogue box, you can edit the style of the text, its height and direction. The properties of the entire paragraph can also be modified. You can change the attachment type (along with the justification of the paragraph), the width of the paragraph, and the rotation angle of the paragraph.

Figure 5.7

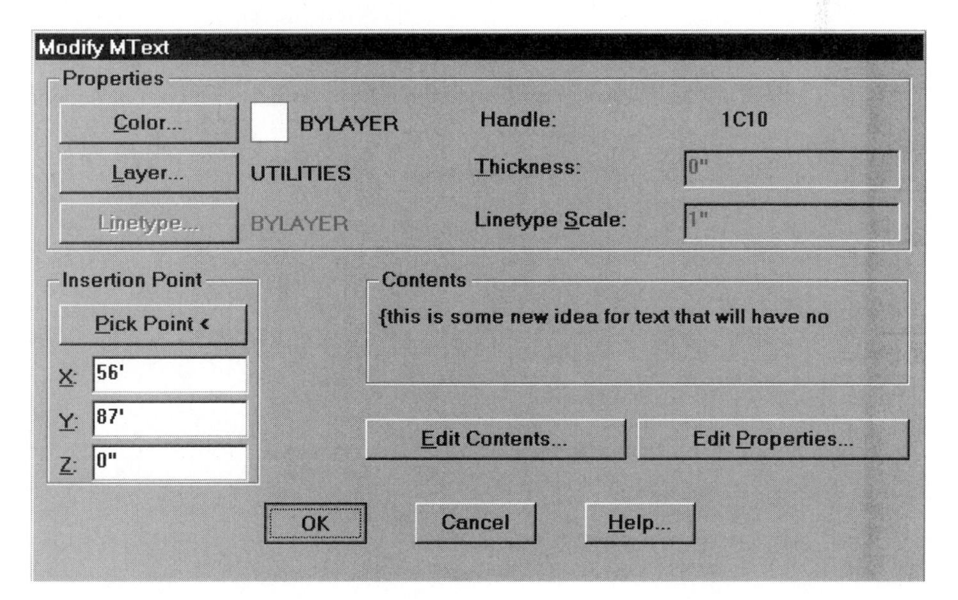

The SPELL command

Spell—This command activates AutoCAD's spell checker. Any text in the drawing, regardless of whether it was created with **Text**, **Dtext**, or **Mtext** will be checked with this command.

`Command:`
Type **Spell** or pick **Spelling...** from the **Tools** menu.

`Select objects:`
Pick the text whose spelling you want to check.

`Select objects:`
Press **<Enter>** to open the "Check Spelling" dialogue box (Figure 5.8).

The spell-checking function can help you correct obvious typographical errors and common misspellings. When a word you use does not appear in AutoCAD's dictionary (names, new words or abbreviations) you can choose to ignore a single occurrence of the new word with the **Ignore** button.

If the word occurs often in the drawing, you can use the **Ignore All** button to ignore every occurrence. If the word in question is truly misspelled, you can change the word to one of the suggested ones by highlighting the correct one, then pick **Change**. To change all of the occurrences of the same word, pick **Change All**.

To check the spelling of a word against comparable words in AutoCAD's built-in main dictionary, type the word in the **Suggestions:** text box, then pick **Lookup**.

Figure 5.8

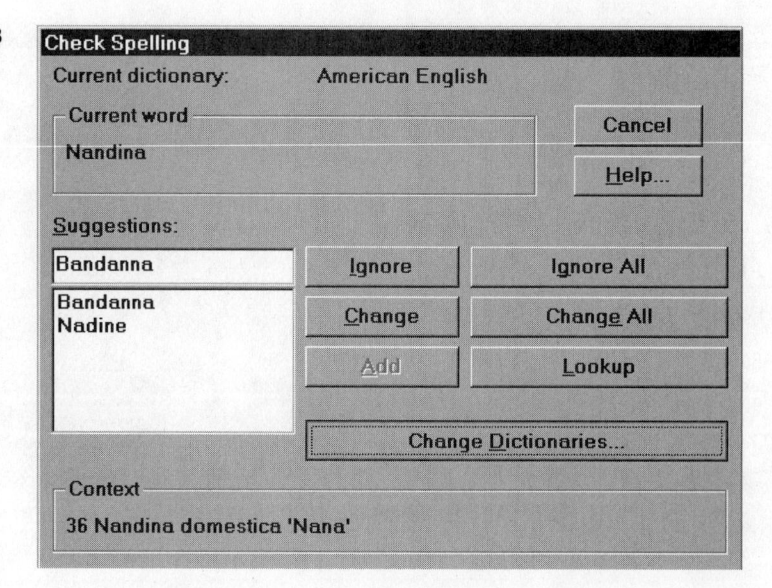

The **Main dictionary** used by AutoCAD depends on the version of AutoCAD you purchased. French, German, Italian, Russian and Spanish versions of AutoCAD will contain dictionaries which reflect those languages. To change **Main Dictionaries** to another dictionary provided by AutoCAD, pick **Change Dictionaries**. This will open the "Change Dictionaries" dialogue box.

You can change the main dictionary by using the **Main dictionary** pop-down box, and selecting any of the other options available. The American version of AutoCAD uses a dictionary called **ENU.DCT** . This dictionary is not intended to be edited, so you cannot add words to it.

AutoCAD"s spell-check becomes much more valuable when you can customize the list of words to include those that you use often. A correctly spelled dictionary of plant common and scientific names can be very helpful if you want to check the spelling of a group of plant labels. To create your own dictionary to which you can add your own collection of words, enter a new dictionary name in the **Custom dictionary** text box. The file name of this dictionary should have no more than eight characters, and must carry the extension **.CUS**. Be sure to specify the drive and full path where you want this file to be located after it is created.

Once a **Custom dictionary** is defined, you can add words to it in two ways. In the "Check Spelling" dialogue box, when a word you want to enter is specified in the **Current word** section, simply pick **Add**. This option is only available if a **Custom dictionary** is active. In the "Change Dictionaries" dialogue box, type in the word in the **Custom dictionary words** text box. The words included in your custom dictionary will be listed below. The **Custom dictionary** file is a simple text file that can be edited in any text editing program including Windows Notepad and DOS Edit.

Settings Affecting Text

Qtext—this setting causes text to be displayed as empty rectangular boxes on the screen. Turn this setting **On** to make the text disappear, replaced by rectangles. These rectangles will plot in the place of text. This is very useful when drawings contain large amounts of text. Text can slow the regeneration of the drawing significantly, and also affect plotting time. Once the text is placed on the screen and edited to your satisfaction, you might want to use **Qtext** to speed the display response as you continue editing your drawing. You might also want to leave this setting active for check plots. Turn **Qtext** off to make the text display normally again.

Textfill—this setting causes AutoCAD to display filled fonts as simple outlines. This setting will also act to speed regeneration times and make the display of objects quicker and more responsive. If the system variable is set to **0**, filled fonts will display as outlined text. If **Textfill** is set to **1**, the fonts will fill in normally. With larger drawings, you should probably leave this setting at **0** until the final plotting process.

Importing Text into AutoCAD

The text insertion commands used in AutoCAD all work under the assumption that you will be entering text into your drawing from the keyboard. This may or may not be a valid assumption. Quite often, you will find that it is easier to create text files in other programs. Word processing programs, DOS Edit and Windows Notepad all can produce simple text files which can imported into AutoCAD at a convenient time. These other text editing programs may provide better tools and more control over the text creation process than AutoCAD can offer. Most all word processing programs allow files to be saved in a simple "text-only" format (usually called ASCII or DOS-ASCII). Thus you can create a text file on another computer, or have another person prepare a text file for you on a computer not equipped with AutoCAD.

There are a number of ways to import text from outside AutoCAD.

Mtext

One of the easiest importing tools to use resides in the AutoCAD **Mtext** command. Once the paragraph size is specified, you can choose the **Import...** button in the "Edit Mtext" dialogue box. This opens the "Import Text" dialogue box.

Simply locate the text file you want to import, then select it and pick **Open**. The text file is then inserted into the "Edit Mtext" dialogue box where the text can be formatted with a specific text style, height, colors, under- and overlines, etc.

Windows File Manager or Explorer

This method is available to Windows users. With both AutoCAD and the File Manager/Explorer programs open at the same time, highlight a text file in the File Manager/Explorer, then (holding the left button of the mouse down) perform a "drag-and-drop" operation into the AutoCAD drawing. The drawing will be inserted as an **Mtext** paragraph object, which can then be edited with the **Mtext** tools.

Windows Clipboard

All Windows programs support some form of "cut and paste" using the Windows Clipboard.

In any of the Windows operating environments, you can have several programs operating at one time. To use the Windows Clipboard, work in a word processing program (or even the Windows Notepad text editor) to prepare a paragraph of text. Highlight the text and then use the **Copy** function under the **Edit** pull-down menu to copy the highlighted text into memory. Switch applications to move into AutoCAD. Use the **Paste Special** function under the **Edit** pull-down menu to place the text into the drawing.

The **Paste Special** option allows you to choose the format in which the text will be imported into your drawing (Figure 5.9).

Figure 5.9

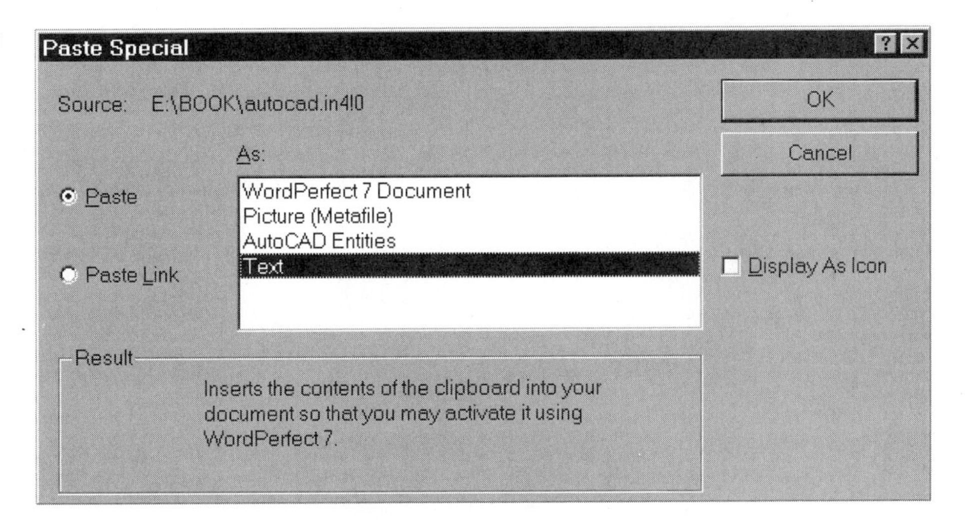

This function can be used to import images (such as GIF, TIF, EPS, and BMP files) as well as pure text files. If you want to maintain a link between the text file and the drawing file (so that the text in the AutoCAD drawing will be updated when the word processor changes it) you can select a **Paste Link**. In this case, you would **Paste** the file as **Text**. This text would be placed in the AutoCAD drawing and handled as text which was entered with **Mtext**.

LANDCADD Insert Text File

LANDCADD contains a text insertion utility which allows the placement of text in a drawing with minimal effort. Pick the **Text ▸** option under the **Tools** pull-down menu.

Select the option **Insert Text File** (Figure 5.10).

This will open the "File to Read" dialogue box, where you can search for the text file that you want to insert into your drawing (Figure 5.11).

After the text file is selected, you are prompted to provide justification, style, text height and rotation angle:

```
Start point or Center/Middle/Right/?:
```

Enter the type of justification for the text. The default value is left justification, which will be used if a **Start point** is selected.

```
Style name <STANDARD>:
```

Pick an existing text style to use for the new text.

Figure 5.10

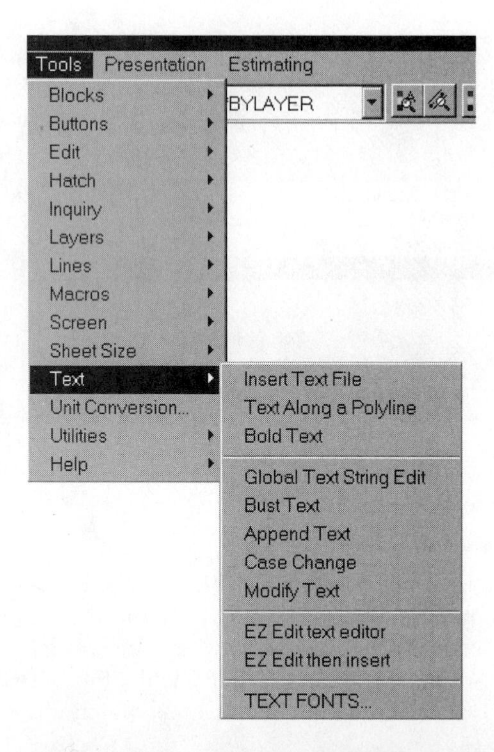

Figure 5.11

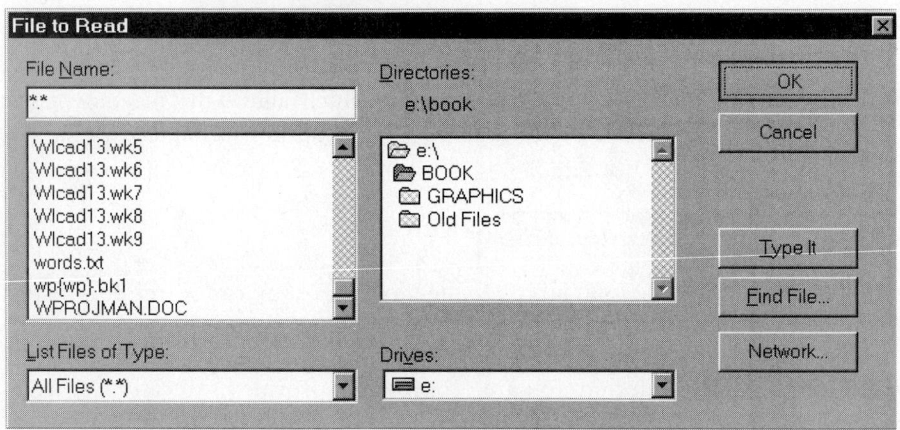

Height <12">:
Enter a height for the new text.

Rotation angle <0>:
Enter a rotation angle for the new text. The new text is then placed in as though it were created with the **Dtext** or **Mtext** commands.

c h a p t e r

Scaling and Dimensioning

When you draw an object on paper, your first decision is how to scale the object so that it fits on your piece of paper. Drawing in AutoCAD is different. There is no fixed paper size on which to draw. Your drawing screen represents **space** not **paper**. The size of the space on your screen can be adjusted to the size of the objects you intend to draw. AutoCAD forces you to work in reverse order from traditional paper drafting. With the different drawing units available, you **always** draw objects at **actual** size. The only time a drawing is actually scaled is when it is plotted. At that time, a plotting scale is imposed so that the drawing can fit to the size of the paper in the plotting device. Because the drawing area (screen) can represent any size, **always** draw your objects at **actual** size. This is crucial for dimensioning, area calculations and other important functions.

Notwithstanding the need for creating objects at actual size, there are important scaling considerations which affect the size of lettering, the scale of non-continuous linetypes, and the appearance of hatch patterns. The display of these items will relate to a concept called "scale factor." The **scale factor** is the relationship between the scale of the objects in the drawing and the paper size they will be plotted on. Consider a 1/8" scale architectural plan:

Drawing scale:	1/8" = 1'-0"
Equivalent scale:	1" = 8'-0"
Equivalent scale:	1" = 96"
Scale Ratio:	1: 96
Scale Factor:	96

This scale factor is used in dimensioning and is the basis for the linetype scale in AutoCAD drawings. **Objects** must be drawn at full size in an AutoCAD drawing, but **text** must be entered at a scale that is appropriate to the plotted drawing. Therefore, the lettering size must be proportionate to the size of the objects **and** the size of the paper. If our architectural standards dictated the use of 1/8" lettering on our plots, we would determine the height of the drawing text this way:

SCALE FACTOR X PLOTTED LETTER HEIGHT = DRAWING LETTER HEIGHT

The Scale Factor of our drawing is **96**, and the plotted letter height is **1/8"** so the drawing letter height is **12"**.

Plotting Scales and Drawing Limits

One of the issues we confront when laying out drawings is how to decide the plotting scale for the drawing. The plotting scale will depend on the size of the objects in the drawing and the size of the paper on which the drawing will be plotted. The size of the objects is determined by the project, and the size of the paper will depend on the conventions established in your business. While these factors are usually set before the drawing begins, the issue of plotting scale can only be decided when the project perimeter and the paper size are laid out on the screen. If you do not know the plotting scale before you start the drawing, one of the easiest approaches to this problem is to follow this sequence:

1. Open a new drawing with a blank prototype drawing. (See **Prototype Drawings.**)
2. Choose the correct units for your drawing (Architectural, Decimal, Metric, etc.).
3. Place the outer geometry of the objects on the screen. Use the lot lines or the limits of the project scope to establish the overall size of the area you will be working on. This drawing should only contain the outer geometry of the project. Do not include **any** details here.
4. Use AutoCAD's **Mvsetup** command or LANDCADD's **Sheet Size** utility to overlay your drawing with a box which represents the paper size at the scale in which your drawing will be plotted. If the outer edges of the sheet do not accommodate the drawing (with room to spare for the title block, legend, labels, notes, etc.) use the **Undo** command to try another paper size or another plotting scale.
5. Examine the object size and the size of the plotted area (determined by the plotting scale and paper size) shown. Record the plotting scale, scale factor and paper size.

6. Save the drawing under a temporary name.
7. Open a new drawing with a normal drawing name, using the prototype with the appropriate title block (paper size and scale).
8. Insert the temporary drawing as a **Block** with the **Explode** toggle active. This inserts the outer geometry into your title block drawing as separate objects that can be edited immediately.
9. Save the drawing and begin work.

While we can manually calculate the size of the drawing limits for any paper size at any drawing scale, the **MVsetup** utility in AutoCAD makes this an easy task.

The MVSETUP (MULTI-VIEW SETUP) command

`Command:`
Type **Mvsetup** (the pull-down option is only available when **Paper Space** is active).

```
Initializing...
Enable paper space? (No/<Yes>):
```

Type **N** since we will perform these operations in **Model Space**.

```
Units type
(Scientific/Decimal/Engineering/Architectural/Metric):
```

Type **A** for architectural units.

Architectural Scales	
(480)	1/40"=1'
(240)	1/20"=1'
(192)	1/16"=1'
(96)	1/8"=1'
(48)	1/4"=1'
(24)	1/2"=1'
(16)	3/4"=1'
(12)	1"=1'
(4)	3"=1'
(2)	6"=1'
(1)	FULL

`Enter the scale factor:`
Enter **96** corresponding to a 1/8" scale drawing.

`Enter the paper width:`
Since we will work with C size (24" W x 18" H) paper, enter **24**.

`Enter the paper height:`
Enter **18**. This places a rectangle for the edge of the paper and resets the drawing limits for the edges of the paper.

The **Sheet Size** routine in LANDCADD performs the same task in a slightly different way:

`Command:`
Pick **Sheet Size ▸** under the **Tools** pull-down menu. Select **Locate**. **Locate**. This presents a dialogue box (Figure 6.1).

Enter the correct **Plotting scale:** in the text box, then select the paper size from the options in the pop-down menu.

After you pick **OK**, a box representing the edge of the paper will appear on the screen. Move this box into the correct position to determine whether or not you have selected the correct plotting scale. This command does not affect the drawing limits. If the drawing limits are incorrect for the sheet box, you can reset them with the **Limits** command or the **Drawing Limits** option under the **Data** pull-down menu.

The **Drawing Limits** refer to the size of the displayed area on your screen **in the current units**. The limits you use will depend on the units employed in the drawing, the paper size and the plotting scale. Limits should be proportional to the paper size, which is usually maintained at a ratio of approximately 4 units wide to 3 units high. The default drawing limits in AutoCAD are 12" W X 9" H in architectural units or 12.0 W X 9.0 H in decimal units. The drawing limits you can use for some commonly used architectural paper sizes and architectural drawing scales are illustrated below:

Figure 6.1

Architectural Paper Sizes and Architectural Drawing Limits

Paper Size	Nominal Size	1/4 Scale	1/8 Scale	1/10 Scale	1/20 Scale	1/30 Scale	1/40 Scale
A	11" X 81/2"	44' X 34'	88' X 68'	110' X 85'	220' X 170'	330' X 255'	440' X 340'
B	17" X 11"	68' X 44'	136' X 88'	170' X 110'	340' X 220'	510' X 330'	680' X 440'
C	24" X 18"	96' X 72'	192' X 144'	240' X 180'	480' X 360'	720' X 540'	960' X 720'
D	36" X 24"	144' X 96'	288' X 192'	360' X 240'	720' X 480'	1080' X 720'	1440' X 960'
E	42" X 30"	168' X 120'	336' X 240'	420' X 300'	840' X 600'	1260' X 900'	1680' X 1200'
F	48" X 36"	192' X 144'	384' X 288'	480' X 360'	960' X 720'	1440' X 1080'	1920' X 1440'

Plotting Scale

1"=1"	1"=4'	1"=8'	1"=10'	1"=20'	1"=30'	1"=40'

Text Size Conversions

1/8"	1/8"	6"	1'	1.25'	2.5'	3.75'	5'
1/4"	1/4"	1'	2'	2.5'	5'	7.5'	10'
1/2"	1/2"	2'	4'	5'	10'	15'	20'

Linetype Scale

12	48	96	120	240	360	480

Title Blocks in Model Space and Paper Space

If your firm or company has a standard title block for each paper size used in hand-drawn work (most do), it is also very useful to develop a standard set of title blocks for each paper size in CAD. This would mean that you should create a separate CAD title block for each paper size used in your work. Once this is done you have two options:

First, you can create copies of these title blocks at each drawing scale that you intend to use. As an example, your firm may have a title block that is used for all D-size paper drawings. If the firm draws in the most common architectural scales (1:1, 1/4", 1/8", 1/10, 1/20, 1/30, and 1/40), you might have seven sheets on file for your use:

Scale	Scale Factor	Default Text Height
1:1 scale	1	1/8"
1/4" scale	48	6"
1/8" scale	96	12"
1/10 scale	120	1.25'
1/20 scale	240	2.5'
1/30 scale	360	3.75'
1/40 scale	480	5'

With the drawing limits set up, the standard drawing layers established, standard text styles and dimension styles preset, and a few essential drawing blocks built into these title blocks, you can begin your drawings with a minimum of redundant work. Following through with this method, you would need to create these seven sheets for each paper size you intended to use in your office. The time invested to create these title blocks is considerable. It may take several hours to build the title blocks and create the necessary copies and install the correct settings for each paper size. But the time saved later is considerable. Your title block can be (see **Issues of Scale** above) early in the drawing process, bringing a large quantity of useful settings and information with it.

Second, you can take advantage of **Paper Space** for your title block drawings. With **Paper Space**, you create only one title block at actual size (1:1 or 1"=1") for each paper size you intend to use. The title block resides in **Paper Space**, in which you create an opening (or openings) to view **Model Space** drawings. Standard drawings are always created in **Model Space**. You can display any drawing you create in **Model Space** within a viewport created in a **Paper Space** title block. The title block acts as a sort of matte frame with a rectangular opening cut in it. The viewport cut in the title block will actually display a model space drawing **at any scale you require**. The matte frame remains the same size while the drawing inside grows or shrinks by the **Zoom Scale Factor** you use to view it. When you plot the drawing, you return to **Paper Space** and always use a plotting scale of 1"=1", since the scaled drawing always fits within the (actual size) title block. Because you can force dimensions in **Model Space** to be scaled proportionally to the **Zoom Scale Factor**, they will maintain a constant lettering and arrowhead size. The title block text is placed on the title block at actual size, since the title block exists in **Paper Space** while the drawing itself exists in **Model Space**.

You should use **Paper Space** for title blocks only, and do all of your drawing work in **Model Space**. In fact, these should be treated as separate drawings until you are: 1) ready to plot them; 2) ready to save them as a file to store in your archives; or 3) ready to send them to a client. To prepare to plot a drawing combining a **Paper Space** title block and a **Model Space** drawing use this procedure:

1. After your **Model Space** drawing is completed, create a new title block drawing (using an existing prototype stored for that purpose) with a **Model Space** viewport already active.
2. Complete the project information required in the title block text. Fill in the appropriate blanks in the title block.
3. Insert the **Model Space** drawing into the viewport in the **Paper Space** title block. Although it is possible to insert the **Model Space** drawing as a **Block**, it is far better to use the **Xref** function to do this (See **Working with Xrefs and Prototypes**). The drawing and the title block are maintained as two separate drawings until the file is actually saved in archive form.
4. Plot the drawing at a **1"=1"** plotting scale.

The text labeling of **Model Space** objects should be performed in **Model Space**. While it is tempting to use **Paper Space** for nearly all text-related functions (the text can be entered at actual size— avoiding the need to calculate its correct height at the scale factor being used), the decision to do so is probably unwise. Most LANDCADD labeling and attribute reading routines and commands only operate in **Model Space**. When LANDCADD creates schedules and legends it will always do so in **Model Space**. While you may want to use **Paper Space** for your title blocks, you still will need to make ample use of all **Model Space** functions.

One of the primary strengths of **Paper Space** is that you can open multiple **Model Space** viewports of the same drawing, **each at a different scale**, all within a single **Paper Space** title block. Multiple scale views of the same drawing are not only possible, they are very easy to implement. This allows you to create details and enlargements of specific areas of a **Model Space** drawing that can be viewed independently of each other. Layers can be controlled independently in each viewport, so detailed information can be included in small scale enlargement views that will be turned off in larger scale views. We will examine **Paper Space** more thoroughly in LANDCADD exercises later in the book.

Ltscale, Celtscale and Dimension Scale

When you use linetypes (other than **continuous**) in AutoCAD drawings, the appearance — or scaling — of the linetypes is regulated by a setting called the **Ltscale** (or **Linetype Scale**). This is a global setting that will affect all of the linetypes in the drawing. As a starting point, most designers set the **Ltscale** to the same value as the **Scale Factor**. This frequently yields linetypes whose dashes and dots are too coarsely spaced. Other users cite a preference for a **ltscale** set at about one-half that of the **Scale Factor**. This is a subjective decision, so you need to try several different values before deciding which proportion you are happier with. Whatever your preference, keep a record of your decision and apply it uniformly to all of your work in AutoCAD and LANDCADD.

Ltscale affects the linetypes of all of the objects in a drawing **except** that individual linetype scales can be assigned to different objects in the drawing. The setting that creates individual linetype scales is called **Celtscale** (for Current Entity Linetype Scale). **Celtscale** allows you to control the linetype scale of one entity (object) separately from those in the rest of the drawing. The flexibility of this setting (like using colors and linetypes not assigned to layers) can lead to great confusion. This is exacerbated by the use of the term **Linetype Scale** in AutoCAD dialogue boxes where it means **Ltscale** in some and **Celtscale** in others. To use the **Celtscsale** setting, remember that the default value is **1**. This allows you to scale the **Celtscale** to values greater than **1** to create a larger (coarser) or smaller (finer) linetype for the entity you are drawing. Like the **Color** command, you will continue to create objects with the current **Celtscale** setting until you change it back to **1**. Be sure you have a thorough familiarity with **Ltscale** before you begin to set the **Celtscale** in your drawings.

Another scale factor deserving attention is that of the **Dimension Scale**. This setting is entered in the drawing from the "Dimension Styles" dialogue box series, or can be entered at the command line as **Dimscale**. The **Dimension Scale** is always the same as the **Scale Factor**. This setting affects the size of the dimension text, the arrowheads, tickmarks and other features used in dimensioning. If this setting is the same as the **Scale Factor**, the size of the text, gaps, arrowheads, etc. will be exactly as entered in the dialogue boxes. If your **Dimension Scale** is incorrectly set, the proportions of the dimensions will be affected. It is easy, but critically important, to set the **Dimension Scale** correctly in dimensioned drawings.

Dimensioning in AutoCAD 13

One of the most important improvements in AutoCAD over the last few versions has been in dimensioning. In older versions of AutoCAD, all dimensioning was performed with command line options. The appearance of dimensions was (and is) controlled by **dimension variables** which were only accessible from the command line. In AutoCAD Release 12 we saw the introduction of the **DDIM** dialogue box, which offered a more visual approach to setting the appearance of dimensions. In Release 13, this dialogue box has been improved and made much easier to use.

Before we open the dialogue boxes used to create dimensions, let's examine the anatomy of a typical architectural dimension (Figure 6.2).

Extension lines extend away from the dimensioned object to make room for the dimension text and dimension line. The gap between the extension line and the dimensioned object can be adjusted, and the display of either the first or second extension line can be suppressed. Finally, the distance that the **extension lines** extend beyond the **dimension line**(s) can be set by the user.

The **dimension line** runs between the extension lines. The **dimension line** will have arrowheads at its ends whose size can be adjusted. The arrowheads can be selected from AutoCAD's available list, or supplied by the user. At your direction, or if the space between the **extension lines** is very small, AutoCAD will split the **dimension lines** and place them outside the **extension lines**. The display of either the first or second **dimension line** can be suppressed.

The **dimension text** may sit above, below or break the dimension line. It can be centered along the **dimension line** or be placed at either side of it. The format of the text (presence or absence of zeros, units of measure, text style, gap between text and dimension line, etc.) can be adjusted in a dialogue box.

Figure 6.2

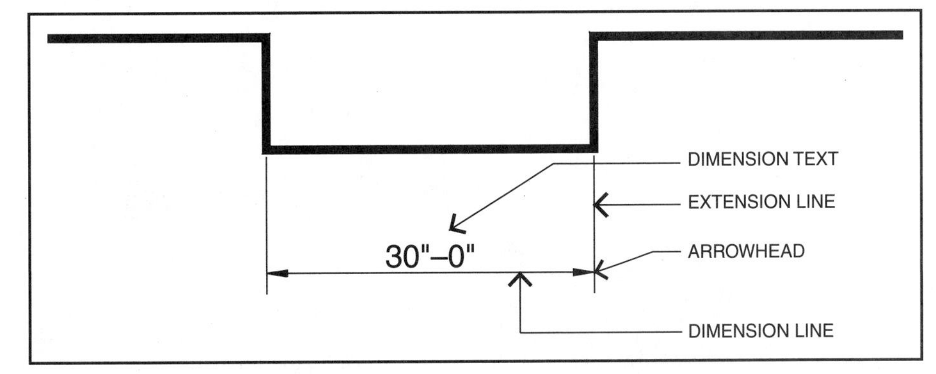

In addition, the colors of each of the elements of a dimension can be independently controlled.

The DDIM (DIALOGUE BOX DIMENSION) command

To operate the **DDIM** command:

```
Command:
```

Type **DDIM** or pick Dimension Style... from the Data pull-down menu.

This opens the "Dimension Styles" dialogue box (Figure 6.3).

The **Dimension Style** area is used to name, rename and retrieve existing dimension styles stored in your drawing. Changes made to a **dimension style** can be saved, or allowed to operate on a temporary basis. The **Family** area is used to control the appearance of dimensions according to the type of dimension being created.

The Geometry..., Format... and Annotation... buttons each call up other dialogue boxes that allow you to adjust the settings contained there with visual cues as to the type and result of your entries.

Geometry... opens the "Geometry" dialogue box (Figure 6.4).

Figure 6.3

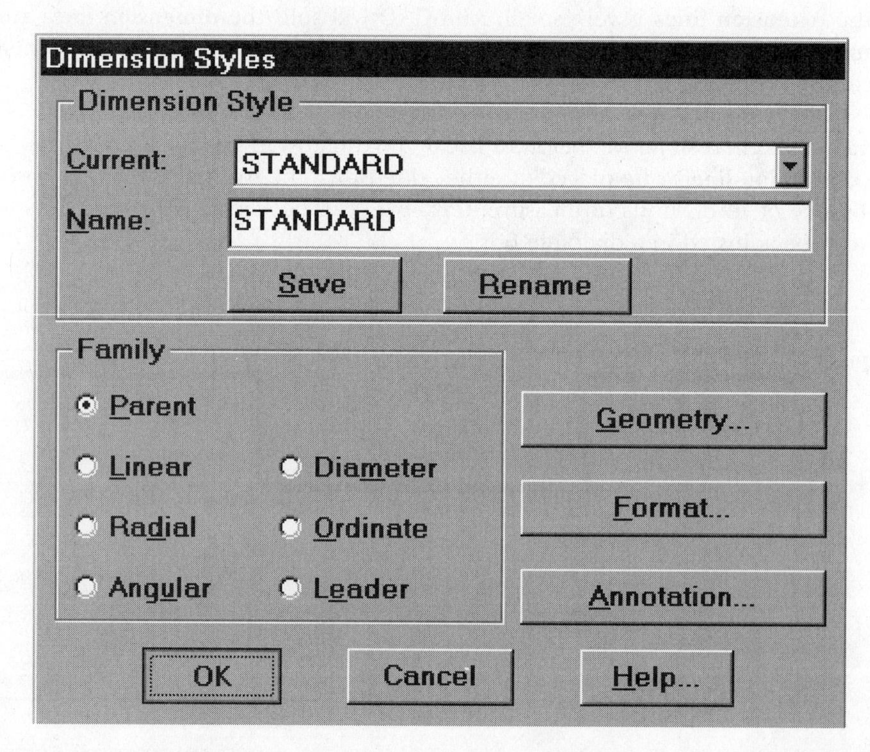

Figure 6.4

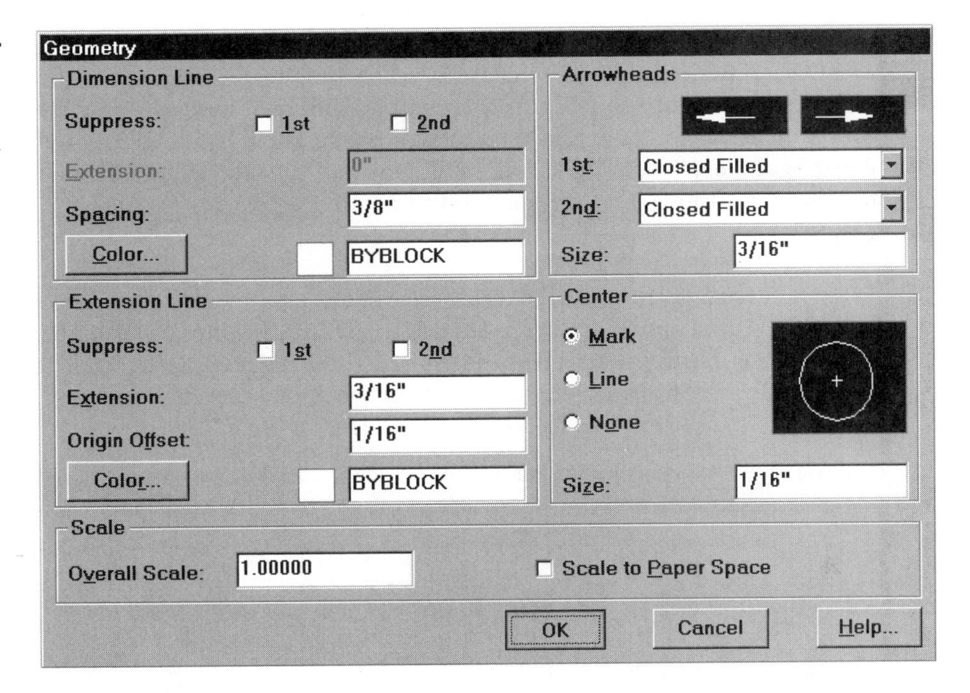

This dialogue box allows you to control the appearance of the dimension lines, extension lines, arrowheads and center marks. **Dimension Line** settings allow you to **Suppress:** the display of either the first or second dimension line. The **Spacing:** setting refers to the distance between the dimension lines when constructing a **Baseline** dimension. See **Dimbaseline** below. Set the color of the dimension lines in the box provided.

Extension Line settings allow you to **Suppress:** the display of either the first or second extension line. You can also regulate the distance that the extension lines run past the ends of the dimension line (**Extension**). In addition, you can adjust the distance between the dimensioned object and the starting point of the extension lines (**Origin Offset**). As before, you can set the display color for the extension lines separately from the other parts of the dimension.

Arrowheads settings allow you to pick the arrowhead style for the first and second arrowhead drawn. The style selected can be from AutoCAD's pop-down list or an arrowhead block that you load into your drawing. The size of the arrowheads can be adjusted according to your standards, and the style of the individual arrowheads can be independently set.

Center settings relate to the control of center marks for circular and radial dimensions. The center marks can be turned off, or set to a mark style or center line style. The size of the center mark or the length of extension past the dimensioned object can be adjusted.

The **Scale** setting is perhaps the most important setting in the "Geometry" dialogue box. The **Overall Scale** text box allows you to set the dimension scale so that the dimensions will be proportional to the **Scale Factor** used to plot the drawing. If you intend to display your dimensions in **Paper Space** (see **Paper Space** title blocks above), mark the **Scale to Paper Space** toggle box. This setting will "gray out" the **Overall Scale** option, and will display the dimensions proportionate to the **Zoom Scale Factor** used for the model space viewport.

Hint: When building your dimension styles, set the **Scale Factor** or **Scale to Paper Space** option first. These settings do not update reliably, meaning that if you change this setting after dimensions using this style are placed in the drawing, you probably will not be able to update the old dimensions to reflect the changes in the style. You will need to erase the dimensions and rebuild them.

Use the **Format...** option to open the "Format" dialogue box (Figure 6.5).

Figure 6.5

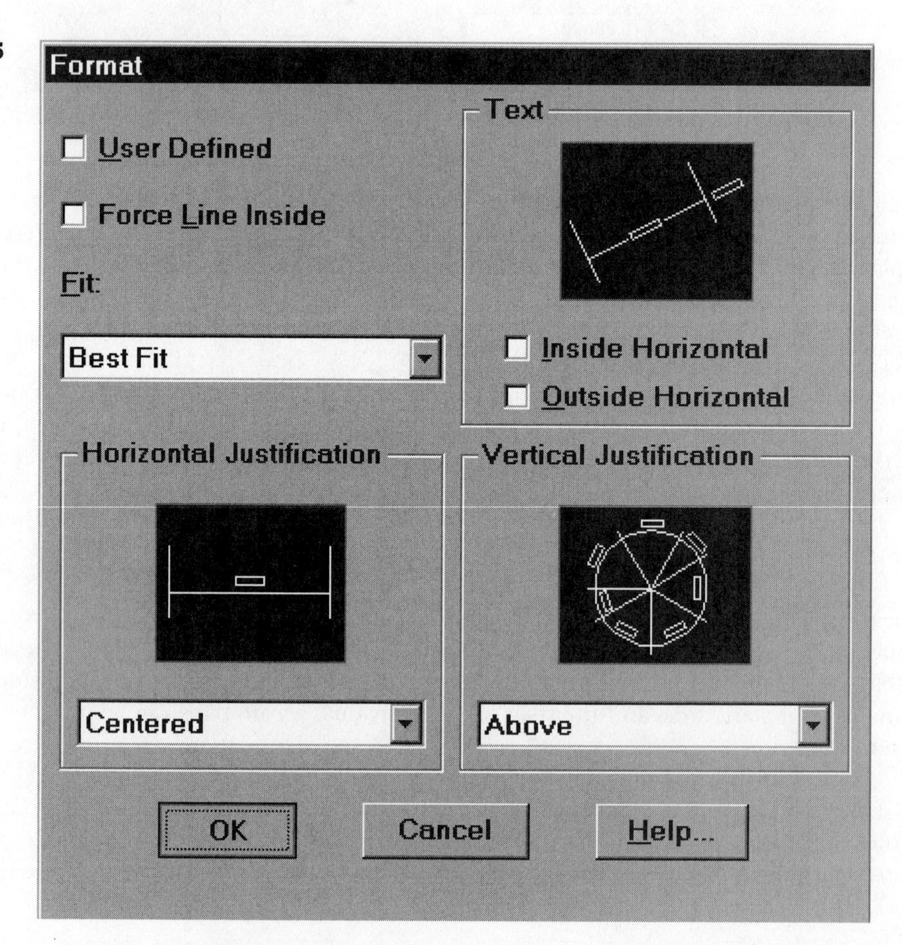

This dialogue box is used to control the placement and display of the dimension text. Since the display of dimension text varies considerably from one industry to the next, these settings are crucial to obtaining the "look and feel" that you want for your drawings. The two toggle boxes at the top of the dialogue box affect different aspects of the text placement. The **User Defined** option allows you to specify an exact horizontal location of the text along the dimension line. Activating this option "grays out" the **Horizontal Justification** pop-down box. The **Force Line Inside** toggle forces the dimension text inside the extension lines if there is enough room to accommodate it.

The **Fit** pop-down box controls the placement of text and arrowheads inside or outside extension lines based on the space between the extension lines. The options provided allow you to place the arrowheads, the text, or a dimension line between the extension lines if space is limited. The **Best Fit** option will automatically make these decisions. The **Horizontal Justification** area allows you to control the position of the text relative to the extension lines. Text can be centered, left- or right-justified, or placed at some **User defined** location (see above). In the **Text** section, you can select where the lettering will be aligned to the dimensioned object or oriented horizontally. Text placed inside and outside of the extension lines can be adjusted separately. The **Vertical Justification** section allows you to select the placement of the text between the extension lines. You can place the text so that it is centered, breaking the dimension line. Other options include placing the text above the dimension line, outside the dimension entirely, or to the **Japanese Industrial Standard (JIS)** location.

Use the **Annotation...** button to open the "Annotation" dialogue box (Figure 6.6).

Figure 6.6

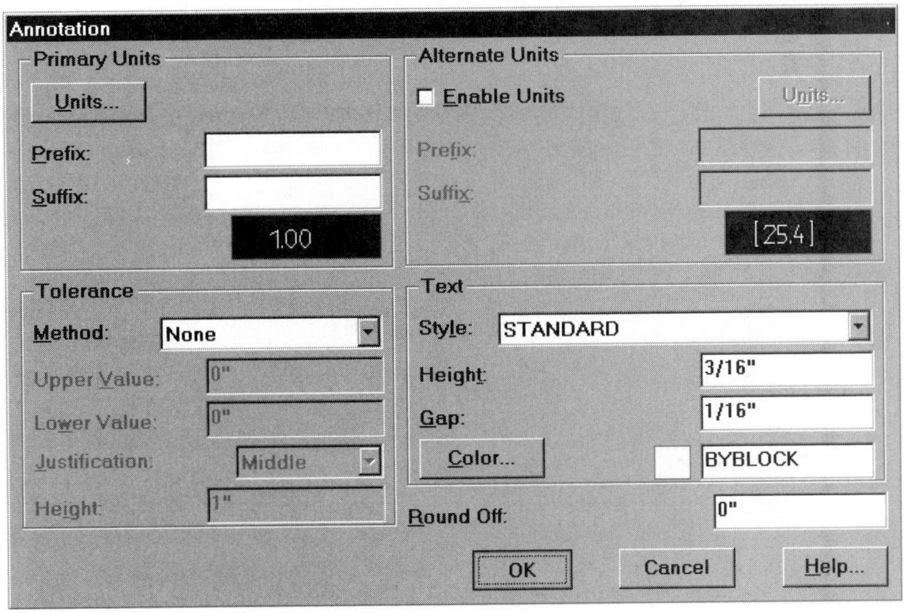

The **Annotation** settings are used to control the appearance (rather than the location and justification) of the text used in dimensions. Amongst its other features, this dialogue box allows you to use a different units system when dimensioning than the units system used when building the objects in the drawing. In this way, an engineering firm using decimal units in its work could provide an architect with drawings using architectural dimensioning units. To activate this option, pick the **Units...** button. This activates the "Primary Units" dialogue box (Figure 6.7).

In the "Primary Units" dialogue box, we can establish the type of units we wish to use in the dimensions. The pop-down box allows us select from a number of options, some of which are new for AutoCAD Release 13.

These options permit the use of stacked fractions, as well as other typical unit systems. The **Angles** section permits the selection of a system for angular measurements. This set of choices parallels the choice of units in the **Units** or **Ddunits** commands. In the **Dimension** area, you can set the number of decimal places of precision, and can control the presence or absence of leading and trailing zeros. This allows you to decide if a dimension should read:

6'-0" or **6'** (with or without a trailing zero)

0'-6" or **6"** (with or without a leading zero)

In the **Tolerance** area, you can establish what level of precision and zero suppression you want to use on a tolerance.

Figure 6.7

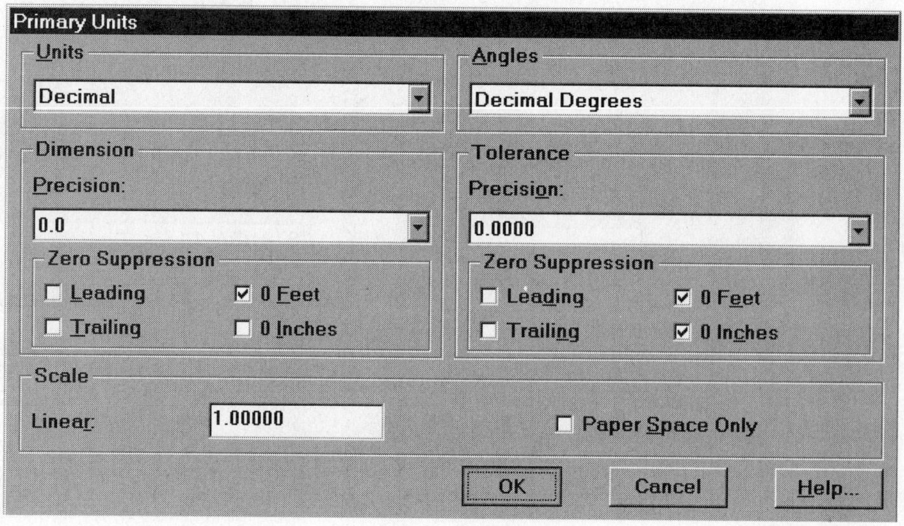

Back in the "Annotation" dialogue box, the **Tolerance** area is used to select the **Method** of tolerancing from the pop-down box. After the **Method** is established, you can set the upper and lower limits of the tolerance, select the justification of the tolerance text, and set the size of the tolerance text. While not often used in architectural drawings, these settings are crucial in mechanical applications of AutoCAD.

In the **Alternate units** section, you can allow an alternate display of units to go along with your primary units display. Some clients require dimensions to be displayed in both metric and English units, so this feature may be very helpful.

The **Text** area allows you to select the style of text used in the dimensions, as well as the text height. Be sure to only use those styles whose text height has been set to **0** if you want to control the height of the dimension text in this dialogue box. The **Gap** refers to the distance between the text and the dimension line(s). If the text is entered above the dimension line, it sets the vertical distance between the two. If the text is centered in a broken dimension line, it sets the gap between the text and the edge of the dimension line. Finally, the color of the dimension text can be set up in this dialogue box.

Back in the "Dimension Styles" dialogue box, we have the opportunity to create separate, named dimension styles that have different characteristics. Better yet, not all of the individual dimension types (child dimensions) must have the same features as the entire style (parent dimension). This means that you can have filled arrowheads for your leader lines, tick marks for your dimensions and dots for your angular dimensions. As long as each "child" dimension is created and saved within a particular dimension style, each can have its own arrowhead, text location, units, etc. The best sequence to establish a dimension style, complete with individual child dimensions, is the following:

1. Set the name for the dimension style. Make it descriptive if you plan to have multiple dimension styles available.
2. With the **Parent** radio button selected, build the common features of the style. These will be the features that will remain common to **all** of the dimension types. Typically these will include such matters as text size, text color, dimension line color, extension line color, units, zero suppression, extension and dimension line suppression, etc. These features define the **dimension family**.
3. Save the dimension style (or family). This style is now made a part of the drawing itself. If you want to retrieve this style, it must be built into a prototype drawing or a block that will be inserted into your drawing before you dimension it.
4. Create the different "child" dimensions, each with its own variations. Now is the time to define each distinctive type of dimension. If you set the **Parent** dimension to contain filled arrowheads, you can now change the **Linear** dimension style by selecting the **Linear** radio button, then selecting the

Geometry... button. Change the arrowhead style to **Tick**, then return to the "Dimension Styles" dialogue box and save the style again. This will store the **Linear** "child" dimension as having a different definition than the rest of this dimension "family."

After you have completed a dimension style, be sure to **Save** it. This will save the **Parent** dimension and all of the **Child** dimension variations. Once you have saved a dimension style, you can still make temporary modifications to it. On occasion you will need to create an individual dimension that does not adhere exactly to the style you defined. You can temporarily **override** the dimension style by creating a change in the style, then picking **OK** **without saving**. This causes a new, temporary style to be created (the original style name with a "+" in front) with the changes you choose. The dimensions you create subsequently will show the properties of the overridden style. One warning: you can only override a style when you make changes to the **Parent** dimension. You will see an error message if you try to create an override to a style by modifying a **Child** dimension.

Once dimension styles are built, the process of dimensioning becomes fairly easy **if the objects in the drawing are drawn at actual size**. This is because the dimensioning in AutoCAD will automatically record the size of the object to be dimensioned and report it as the dimension text is placed. The process of placing dimensions revolves around correctly defining the dimension style you wish to build and accurately creating the objects in the drawing. If both of these tasks are completed correctly, the most difficult part of dimensioning is already over.

As noted earlier, one of the best features of AutoCAD dimensioning is that of **associativity**. This means that the dimensions are **associated** with the objects they refer to. When objects are modified with the **Stretch**, **Scale**, **Extend**, **Trim**, or other such commands, their dimensions will be modified with them. Associativity can be turned on or off, but under all but the most unusual circumstances, you should leave the associativity turned **on**.

Dimensioning Commands

The commands for dimensioning reflect the type of objects being dimensioned. That obvious-sounding statement explains both the name and the function of these commands:

The DIMLINEAR (DIMENSION - LINEAR) command

Dimlinear (**Dl**) This command creates vertical, horizontal or rotated dimensions, depending on the orientation of the object chosen. To create a linear dimension:

`Command:`

Type **Dimlinear** (or **Dl**) or pick **Dimensioning ▸** from the **Draw** pull-down menu, then select **Linear**.

`First extension line origin or RETURN to select:`
Here you have two choices. If you want to control the sequence of endpoint selection (to set differing arrowhead types, suppress dimension or extension lines, etc.) select one of the endpoints of the object you want to dimension. Use **Osnaps** to make this selection accurate. After you select the first endpoint you are prompted for the second:

`Second extension line origin:`
Pick the second endpoint. The next prompt asks for the location of the dimension line.

Or

`First extension line origin or RETURN to select:`
If the sequence of endpoint selection is not important, press **<Enter>** to move to the next prompt.

`Select object to dimension:`
Select the object that you wish to dimension. This is simpler and faster because you only need to use one **<Pick>** to select the object. The only disadvantage is that you cannot control the sequence of the endpoint selection. After either choice, you are next prompted for a location of the dimension line:

`Dimension line location`
`(Text/Angle/Horizontal/Vertical/Rotated):`
This location represents how far away from the object you would like to place your dimension. If **Snap** is active, it becomes much easier to keep all of your dimensions in alignment, even if they are built at different times. Other options include:

Text — use this option if you want to make modifications to AutoCAD's measured length for your object. This command opens the Mtext dialogue box. The <> brackets indicate AutoCAD's measured length for the object. You can place text outside the brackets to add text (like **TYP.**). You can alter the measured length by entering a new value in place of the brackets. **This is not a good idea!** Changing the text to create a desired value is usually a sign of sloppy drafting technique, and should be discouraged. In addition, this cancels the associativity of the dimension since the measured value will not change as an object is scaled, stretched, trimmed, etc.

Angle — use this option to alter the angle of the text as it is placed on the screen. Although the text can be forced to be displayed horizontally, or aligned to the dimensioned object, you can set a defined text rotation angle here.

Horizontal — use this option to force horizontal text orientation if a vertical line or angled line might cause the text to be displayed vertically or at another angle.

Vertical — use this option to force the text to be placed vertically.

Rotated — use this option to cause the **dimension** line to be rotated to some specific angle. Use this option only if you have a specific angle that must be used for the dimension line. If you only need the dimension line to be aligned with the dimensioned object, use **dimaligned**.

After you select an option and/or a dimension line location, AutoCAD's measured value will appear as **Dimension text**.

`Dimension text <30'-0">:`
Although the value here can be altered, it is not advisable to do so. Press **<Enter>** to place the dimension on the drawing.

The DIMALIGNED (DIMENSION - ALIGNED) command

Dimaligned (**Da**) This command creates dimensions that are **aligned** with the dimensioned object. The dimension will automatically be placed at the same angle as the dimensioned object, regardless of the sequence of the endpoints picked, or if the object was selected with a single **<Pick>**.

`Command:`
Type **Dimaligned** (or **Da**) or pick Dimensioning ▸ from the Draw pull-down menu, then select Aligned.

`First extension line origin or RETURN to select:`
Pick the first endpoint of the object or press **<Enter>** to pick the object with one **<Pick>**.

`Second extension line origin:`
Pick the second endpoint of the object.

`Dimension line location (Text/Angle):`
These options are the same as those described in **Dimlinear**.

The DIMCONTINUE (DIMENSION - CONTINUE) command

Dimcontinue (**Dc**) This command is used to build a dimension in line with the dimension created last. The new dimension line will be placed at the same distance from the dimensioned object as the previous dimension. **Dimcontinue** dimensions are always built from the second dimension line created in the previous dimension (Figure 6.8).

`Command:`
Type **Dimcontinue** (or **Dc**) or pick Dimensioning ▸ from the Draw pull-down menu, then select Continue.

`Second extension line origin or RETURN to select:`
To continue a dimension from the last one drawn, simply select the next extension line

Figure 6.8

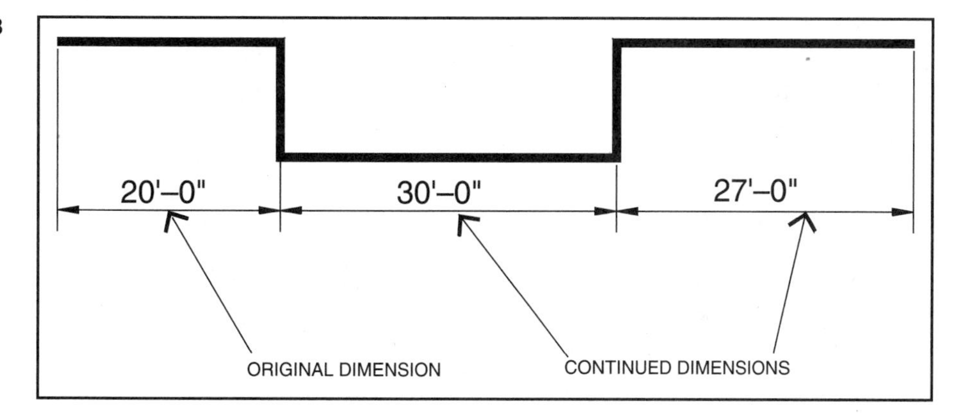

origin you want to use for the next dimension. If the next dimension line you want to create continues from a different dimension than the one just created, press **<Enter>** to select any existing dimension.

The DIMBASELINE (DIMENSION - BASELINE) command

Dimbaseline This command uses a previously placed dimension as a basis for more dimensions placed in the same direction as the original. **Dimbaseline** uses the **first** extension line origin from the previous dimension as the first extension line origin for the new dimension. You are not asked to specify the dimension line location, since the distance between dimensions is defined in the "Geometry" dialogue box (Figure 6.9).

```
Command:
```

Type **Dimbaseline** or **Dimensioning ▸** from the **Draw** pull-down menu, then select **Baseline**.

Figure 6.9

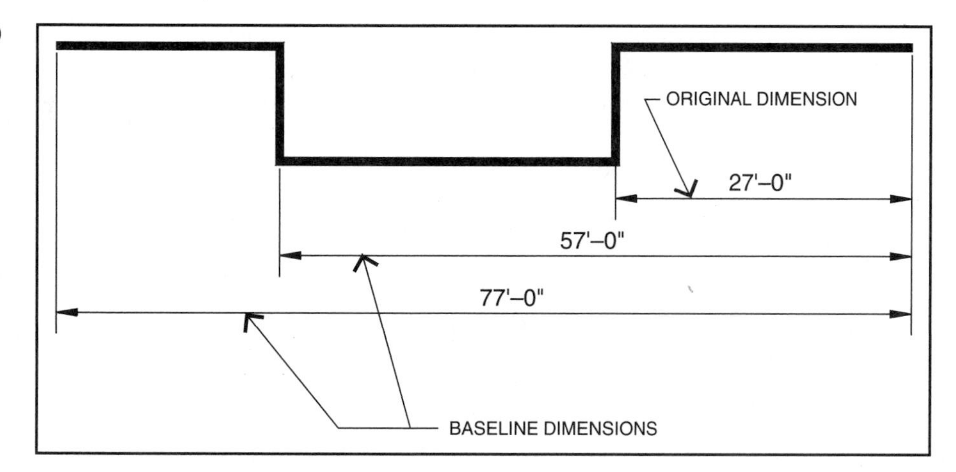

`Second extension line origin or RETURN to select:`
Pick the second extension line origin if the previous dimension is the one you want to use for the original baseline dimension. If you want to pick another dimension to use, press **<Enter>**.

`Second extension line origin or RETURN to select:`
Pick another second extension line origin to continue to build baseline dimensions.

As with the **Dimcontinue** command, you can choose which dimension to use as a basis for the **Dimbaseline** command by choosing to press **<Enter>** rather than picking a second extension line origin immediately.

The LEADER command

Leader (**Le**) **Leaders** are notations placed in a drawing which provide instructions or information about some dimensional feature. The **leader** is actually a type of dimension and has associative properties. When you build a **leader**, you select an endpoint for the arrowhead, then one or more endpoints for the line segments leading to the annotation. The last line segment, placed automatically by AutoCAD, is oriented horizontally.

`Command:`
Type **Leader** (or **Le**) or pick **Dimensioning ▸** from the **Draw** pull-down menu, then select **Leader**.

`From point:`
Pick the first point of the leader line. This is the point where the arrowhead will begin, so it is a good idea to use **Osnaps** to anchor the arrowhead correctly. The next point(s) will define the endpoint(s) of the leader line segments.

`To point:`
Pick the next point in the leader line.

`To point (Format/Annotation/Undo)<Annotation>:`
Pick a third point for the leader line, select **Format** to modify the leader, press **<Enter>** to place text on the command line, or **Undo** the last vertex point.

Format opens another group of options:

`Spline/STraight/Arrow/None/<Exit>:`
Spline allows you to build a spline leader.

STraight produces a normal straightened leader. This is the default setting and does not need to be reset unless you change to the **Spline** option, then change your mind.

Arrow places an arrowhead at the end of the leader. This is the default setting, and does not need to be reset unless you had specified **None**.

None does not place any arrowhead at the end of the leader.

Exit returns you to the previous options.

Annotation places you at the command line to enter a line of text.

```
Annotation (or RETURN for options):
```

Type in the leader text or press **<Enter>** for the available options:

```
Tolerance/Copy/Block/None/<Mtext>:
```

Tolerance opens a *feature control frame* used for geometric tolerancing, a feature used in mechanical drafting.

Copy allows you to select another text block in the drawing and use it as the text for the new leader. This existing text can be created with **Mtext**, **Dtext**, or **Text**.

Block allows you to use a predefined block at the end of the leader. You are prompted for the insertion information as you would be for any inserted block. The block could be a "balloon" containing attribute information, or any other block.

None completes the leader with no text.

Mtext opens the **Mtext** dialogue box (or DOS Edit command) to allow you to create paragraph text to accompany the leader.

One of the hidden features of the **Leader** command is AutoCAD's automatic "hook line" feature which automatically places a short horizontal line at the end of any leader line more than 15° from horizontal. If you are in the habit of placing a horizontal line at the end of your leader lines, this "convenient" feature can be frustrating. You will often find an angled line and a hook line placed at the end of your leaders. To avoid this, simply finish your leaders on an angle and allow AutoCAD to build the hook line horizontally. The size of the hook line will be the same as the arrowhead specified in the **Geometry...** dialogue box for the leader child dimension (if specified) or for the parent dimension for that dimension style (Figure 6.10).

The DIMANGULAR (DIMENSION - ANGULAR) command

Dimangular This command will place a dimension which measures the angle between two lines, the included angle of an arc, or the angular distance between two points oriented to the center point of a circle. This is a **dynamic** command, and will create different types of measurements depending on where the dimension arc line is located.

Figure 6.10

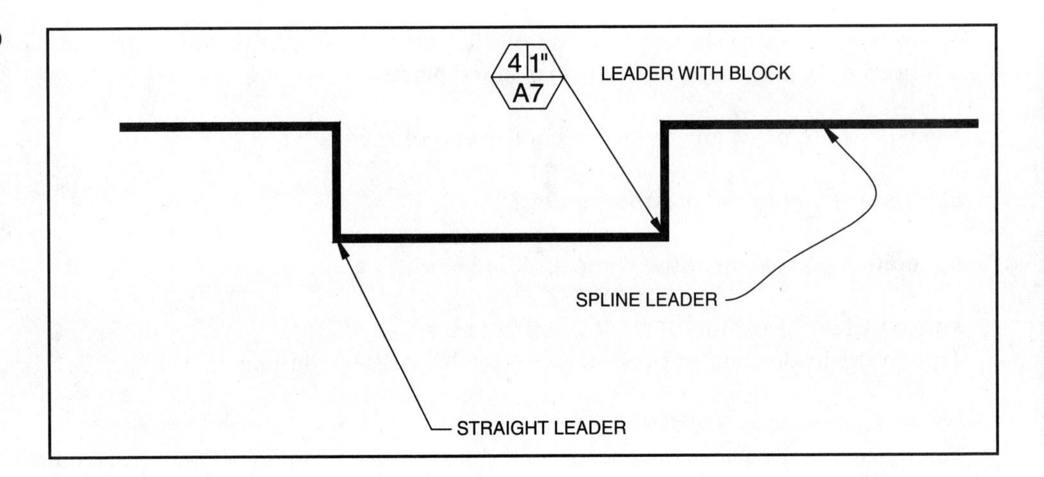

`Command:`
Type **Dimangular** or pick Dimensioning ▸ from the Draw pull-down menu, then select Angular.

`Select arc, circle, line, or RETURN:`
The response of the command will depend on the type of object selected. If you pick a line, you will be prompted for the second line to measure the angle between them.

`Second line:`
Pick the second line of the included angle.

`Dimension arc line location (Text/Angle):`
Choose the location of the dimension line arc. The angle measured and the distance from the vertex will be adjusted as you move the cursor. The options **Text** and **Angle** are explained below:

Text opens an **Mtext** paragraph editing window. As before, AutoCAD's measured value is represented by the symbols <>. You can place text outside this value or in place of this value to remove the measured value.

Angle allows you to enter a rotation angle for the text entered. This value will override the horizontal setting in the "Format" dialogue box, or can be used to enter an angle different from the one displayed as the dimension line is stretched across the angle.

`Dimension text <78.7>:`
The angle measurement will be displayed. The ° symbol is automatically inserted if it is available in the font you have used for your dimensioning text.

If you select an arc as the first object, you are not prompted for a second object. Instead, AutoCAD uses the endpoints of the arc to create the angle about the center point of the arc. Depending on the location of the dimension line, you can measure the inside or outside angle of the arc selected. If you select a circle as the first object, the first extension line origin is the first point selected on the circle. The extension line origin can be any other location on the drawing. The center point of the circle will act as the vertex for the angle.

The DIMDIAMETER (DIMENSION - DIAMETER) command

Dimdiameter—The **Dimdimeter** command will measure the diameters of circles and arcs of more than 180°. Diameters are automatically preceded by the ϕ (phi) symbol in Release 13. This setting can be overridden in the "Annotation" dialogue box if another **Prefix:** is specified. In mechanical drafting, the specifications for diameter dimensions are more exacting than those for architectural work. The angle for drawing diameter measurements is prescribed, along with the location of arrowheads and dimension lines when the dimensions are too small to fit inside the circle.

`Command:`
Type **Dimdiameter** or pick Dimensioning ▸ from the Draw pull-down menu, then select Diameter.

`Select arc or circle:`
Pick the arc or circle to dimension.

`Dimension line location (Text/Angle):`
Move the cursor to choose the location and/or rotation of the dimension line. As the dimension is placed, AutoCAD will place a center mark in the circle or arc if the dimension line is forced outside. This setting can be overridden in the "Geometry" dialogue box if desired.

Text opens the **Mtext** dialogue box to allow you to attach paragraph text to the measured angle. This adds any text you desire, but does not prevent the ϕ symbol from appearing as a diameter prefix. AutoCAD's measured value appears as <> symbols in the **Mtext** screen.

Angle allows you to enter a rotation angle for the diameter text.

The DIMRADIUS (DIMENSION - RADIUS) command

Dimradius This command is used to measure the radius of arcs less than 180°. Where the radius is large enough, the dimension line, the text and the arrowhead are all placed inside the arc. The dimension line will run from the center point of the arc to the edge of the radius. The tip of the arrowhead will touch the arc. If the radius is too small, the dimension line may stay inside the arc, but the dimension text will be forced outside the arc. The text will be connected to the dimension line with a leader line. If the radius is very small, the entire dimension will appear outside the arc, and will resemble a leader. In the case of small radii, the distance of the dimension text from the radius is only changeable if you check the **User defined** box in the "Format" dialogue box. This permits you to drag the leader and text out to a location of your choice.

As with the **Dimdiameter** command, all **Dimradius** dimensions begin with a prefix. In this case **R**. In addition, a center point will be placed as a default. If you want to eliminate the center point, you can shut it off in the "Geometry" dialogue box.

`Command:`

Type **Dimradius** or pick Dimensioning ▸ from the Draw pull-down menu, then select Radius.

`Select arc or circle:`

Pick the arc or circle to dimension.

`Dimension line location (Text/Angle):`

Pick the radius you wish to dimension.

Text opens the **Mtext** dialogue box to allow you to attach paragraph text to the radius dimension. This does not prevent the **R** from appearing as the radius prefix. AutoCAD's measured value appears as <> symbols in the **Mtext** screen. You can enter text inside to replace the measured value (not advised) or add characters to the beginning or the end of the dimension.

Angle allows you to enter a rotation angle for the radius text.

The DIMCENTER (DIMENSION - CENTER) command

Dimcenter — This command places center marks on a circle or arc. The type of center mark placed will depend on the setting in the "Geometry" dialogue box. Unlike those placed with the **Dimdiameter** and **Dimradius** commands, the center marks placed with **Dimcenter** are **not** associative. They can be stretched, trimmed, or erased without affecting associative functions.

`Command:`

Type **Dimcenter** or pick Dimensioning ▸ from the Draw pull-down menu, then select Center Mark.

`Select arc or circle:`

Pick the circle or arc where you wish to place the center mark. The center mark defined in the "Geometry" dialogue box in the **Ddim** (Dimension Styles) command is placed on the object.

Editing Dimensions

As discussed earlier, AutoCAD creates **associative** dimensions by default. This means that the dimensions are tied to the objects they measure. If the object changes size, the dimensions will change with them. This is accomplished because AutoCAD places **definition points** at the crucial locations of objects when they are dimensioned. These are points that appear on the **Defpoints** layer but will not plot on your drawing. It is wise to freeze this layer to prevent these points from being inadvertently erased or modified. Even though these definition points reside on a frozen layer, they can still be accessed by dimensions when objects are modified.

The associativity built into AutoCAD presents both opportunities and challenges in editing dimensions. Dimensions are treated like complex blocks, so modifying them can only be accomplished by using tools designed for the task. If none of the tools described here will allow you to create the effect you need, you may have to **Explode** the dimension in question. This has several effects you should understand. First, the dimension is converted into objects which are no longer associative. They will not change as the objects they were once attached to are edited. Second, the exploded dimension objects will lose their connection to the **dimension style**. Changing the dimension style will have no effect on exploded dimensions. Third, all of the dimension objects from the exploded block can be edited as normal objects. The dimension text is now simply text and can be modified, moved, scaled or erased. Fourth, all of the dimension objects will be placed on the **0** layer, and will assume the **Continuous** linetype and **White** color. Last, if you wish to modify the properties of the exploded dimension objects, you will need to change their color and linetype, which will be defined as **Byblock**. Probably the best color to assign is **Bylayer** so that the objects will assume the color and linetype of the layer on which they are placed.

Grips editing

Grips are one of the most effective ways to edit dimensions. Different types of dimensions have grips at different places, but they all will have a grip on the dimension text, one on each of the extension line origins, and one on each of the dimension line endpoints. Turning one of these grips into a "hot grip" allows you to edit the location of the object the grip is connected to. Most dimension editing revolves around simply shifting the location of one of these basic dimension objects. Stretching a dimension along with the object to which it is linked is easy with the **Stretch** command. Remember to use a **Crossing** box to define the objects to be stretched. The **Stretch** command is the easiest way to edit leaders to move both the text and the leader line at the same time. Another way to edit with grips is to highlight the objects to be stretched and the dimensions attached to them. When you pick a corner of an object to stretch with grips, the dimension stretches too, because the **defpoint** assigned by the dimension will move too.

Dimension editing commands

The DIMTEDIT (DIMENSION TEXT EDIT) command

Dimtedit — This command is used to change the **location** of dimension text.

`Command:`

Type **Dimtedit** or pick Dimensioning ▸ from the Draw pull-down menu, then pick Align Text.

`Select dimension:`
Pick the dimension whose text you wish to relocate.

`Enter text location (Left/Right/Home/Angle):`
Left moves the text to the far left side of the dimension line, allowing for the arrowhead and a very short piece of dimension line at the end.

Right moves the text to the far right side of the dimension line, as above.

Home is used to rotate dimension text back into its original rotation angle. Text moved to the left or right location will maintain that position.

Angle is used to rotate the dimension text to a different angle. Using **Home** will undo the effects of this option.

The DIMEDIT (DIMENSION EDIT) command

Dimedit — This command is used to modify the dimension text. Several of its functions are duplicated in **Dimtedit**. The most significant portion of this command is the **New** and **Oblique** options.

`Command:`
Type **Dimedit**. The only portion of this command accessible from the pull-down menu is **Oblique**. Pick it from the Dimensioning ▸ option under the Draw pull-down menu, then select Oblique.

`Dimension Edit (Home/New/Rotate/Oblique) <Home>:`
Select the option to use to edit the text characteristics.

Home duplicates the function in **Dimtedit** by restoring the original rotation angle of the text in a dimension.

New allows you to create new text in the **Mtext** dialogue box (or on the DOS command line), then choose a dimension in which to place the new text. Remember that the <>

characters represent AutoCAD's measured value. A value in place of these brackets (or even inside them), replaces the measured value, creating non-associative text.

You can restore AutoCAD's original measured value by issuing the **Dimedit** command again, then using **new**. Do not enter any value, but leave the <> symbols. Pick the dimension whose text you wish to restore, and the original value will return.

Rotate duplicates the **Angle** function in the **Dimtedit** command by allowing you to rotate the text to an angle you specify.

Oblique allows you to establish a new angle for the extension lines in one or more dimensions. This command can help you separate a group of dimensions from the others on a drawing by creating a strong visual distinction between them or when dimensions are very crowded. This may also be useful in dimensioning isometric drawings.

The DDMODIFY command

Ddmodify This versatile command opens a different dialogue box for each type of item you select. When you select a dimension, the "Modify Dimension" dialogue box is opened.

This dialogue box allows access to the "Geometry", "Format" and "Annotation" dialogue boxes found in the "Dimension Style" dialogue box. The changes made here will create a **dimension override** which permits you to create an exception to the dimension style found in the **Style:** pop-down box. Be sure that the style in this box is the correct style for creating your temporary **override**. You can also edit such properties as the color, layer and linetype of the dimension. To modify the text in the dimension, pick the **Edit** button which opens the **Mtext** editing screen.

chapter

Working With Blocks, Attributes, Xrefs and Title Blocks

Blocks

A block is a group of objects combined together and treated as a single object by AutoCAD. Once a block is created, it has been given a **block definition** within the drawing. Each block in a drawing is stored in the **block definition table**. There are no limits as to the type of objects that can be turned into blocks. In LANDCADD we see plant symbols, hardscape and architectural elements, irrigation equipment, various markers, tags, labels and call-outs which are all defined, stored and inserted as blocks. An entire directory within LANDCADD containing over 1000 files is devoted to blocks.

Why should we use blocks in drawings? There are a number of advantages in using blocks:

1. Blocks make drawing faster. You can insert a block as often as you need to without drawing the same objects repeatedly.
2. Blocks save disk space. Once a block is defined in the block definition table, it can be repeated over and over with almost no increase to the drawing file size.
3. Blocks improve uniformity and make drawing standards easier to implement and maintain. Using blocks allows the use of consistent title blocks and drawing symbols for each drawing produced by a firm.
4. Blocks improve efficiency. Once a complex object is created by one designer, it can be used by other designers in the firm without tedious redrafting of the same object in another drawing.
5. Blocks can be stored with **attributes**. Attributes are additional information attached to a block that can be read later. Lists of attributes can be compiled (or *extracted* in AutoCAD language) to create a materials list and even a construction estimate.

6. A block can be used to store text attributes in your title block. This allows you to place the same style text in the same location each time when building a title block. This can save considerable time.
7. Blocks are easy to manipulate in a drawing. A block is always selected as a single unit, which makes it very easy to move, copy, scale, etc.

Although LANDCADD has a huge number of drawing blocks, you will probably end up building your own custom blocks for your drawings. LANDCADD's creators have anticipated this and have made allowances for inserting custom blocks into several of their modules. There are exercises later in this book which lead you through the process of creating customized planting symbols and customized irrigation symbols for use in the LANDCADD program.

The BLOCK command

Block (**B**) The **Block** command is used to create blocks in AutoCAD. Remember that the new block will only exist within the drawing you are working on.

```
Command:
```

Type **Block** (or **B**) or pick **Block** from the **Construct** pull-down menu.

```
Block name (or ?):
```

Type in the name of the block or enter **?** to see a list of the blocks that are present in the drawing. The block name can be up to 31 characters in length, and can contain letters, numbers and the characters "$" (dollar sign), "_" (underscore), and "-" (hyphen). AutoCAD will automatically convert the block name to upper case letters.

If you enter the name of a block that is already present in the drawing, AutoCAD will return the message:

```
Block BLOCKNAME already exists.
Redefine it? <N>:
```

If the existing block is outdated or contains information that is no longer correct, you can redefine the block at this point. All occurrences of the block will be updated to the new definition that you create. If you do not want this to happen, type **N** so that AutoCAD will cancel the command. Use a new (unique) name for the block and start over.

If the block name is unique, you can proceed with the command:

```
Insertion base point:
```

Specify a point on the drawing. The **base point** will become the **insertion point** for the block, so select it with an eye toward how you will insert it later. Picking a central point or a lower left corner of the new block is typical. The block can be rotated during insertion, so the **base point** will also become the center point during rotation.

`Select objects:`

Pick the objects you want to include in the block. You can select any visible object to place in this new block. A block can contain another block, which is said to be **nested** inside the new block. As an example, you could place block symbols for a bathtub, sink and toilet in a bathroom of an apartment building. If you created a new bathroom block which included these other blocks, they would be nested within it. This would allow you to insert the bathroom block as many times as you wished, and the other blocks would be carried with it.

The properties of the objects in the block are reflected when the block is inserted later. If you create a block on layer **0**, it will assume the linetype and color of any layer onto which you insert it. If the block objects are assigned the color and linetype **byblock**, they will assume the color and linetype which are in effect when the block is inserted. Be sure to keep these in mind when you create blocks.

If you want to attach attributes to the block (see below), build them before the block is created, then include them as the block objects are selected.

After you finish selecting the objects in the block, press **<Enter>**. The objects in the block then disappear from the screen. The objects have been assigned to the block. You can type **Oops** to return them to the screen.

Warning: Never name a block with the same name as the drawing that will contain it. This creates a *"circular reference"* which will cause problems later. If you name a drawing ABC and then save a block within it called ABC, you create such a reference. If you attempt to insert this drawing into another drawing, AutoCAD will abort the insertion. There is no way for AutoCAD to make a distinction between the overall drawing ABC and the block ABC within the drawing if they both have the same name.

Adding Attributes

Blocks are much more valuable if they contain information. The information that can be attached to a block is almost unlimited. You can attach a name, a description, a part number, a price, even an entire database of information to a block. The mechanism for attaching this data is the **attribute**. We can create attributes on the screen, then attach them to blocks as we define them. To create an attribute, you can use either the **attdef** (attribute definition) command or the **ddattdef** (dialogue attribute definition) command. **Attdef** is a command line command which can be difficult to use. **Ddattdef** works through a dialogue box and is much easier to use.

The DDATTDEF (DIALOGUE BOX ATTRIBUTE DEFINITION) command

Ddattdef is a command to create attribute definitions and insert them into the drawing. Once an attribute is inserted into the drawing, it can be included in a block. An attribute can have numerous properties. The first property to consider is the attribute **mode**. The mode controls the way that the attributes are set and displayed. There are four possible modes.

Invisible—this indicates whether or not the attribute is visible on the screen. Many attributes are left invisible so that extra clutter is not added to the screen each time a block is inserted. Plant names attached to plant symbols are typically set to be **invisble**.

Constant—this attaches an attribute that cannot be changed. When an attribute is set this way, you are not prompted for the information to attach, and you cannot edit the information after the block is inserted.

Verify—this mode sets whether or not you are prompted to verify (as well as enter) the information in the attribute before it is inserted into the drawing.

Preset—this mode sets a default setting for the attribute when it is inserted. This attribute is different from a **constant** mode attribute because it can be edited after the block is inserted.

Once you have set the modes for the attribute, they cannot be edited after the block is inserted. If you make a mistake setting the modes for the attribute, erase it and start over (Figure 7.1).

Figure 7.1

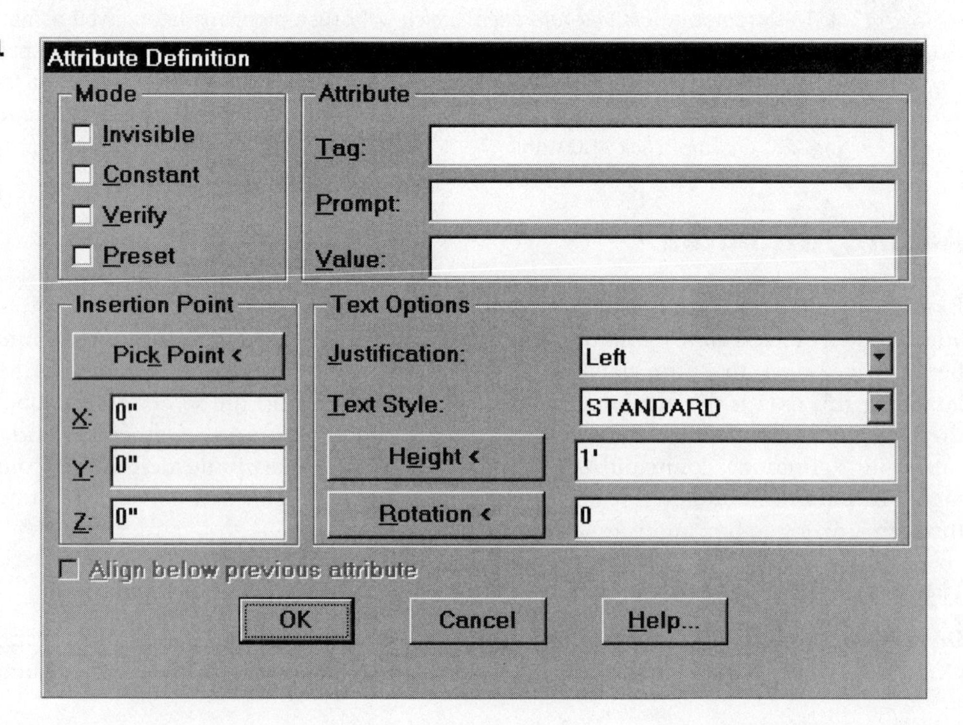

The **Attribute** section allows you to establish the portions of the attribute that will display, or that the user will edit as the block is inserted.

Tag—this is the **name** of the attribute. The tag is displayed as a line of text after you have completed the attribute definition. All attributes **must** have a tag. The tag name can be up to 256 characters long, but cannot contain any spaces.

Prompt—this is the text you want to appear when the attribute is inserted. The **prompt** may take the form of a question (What is the PLANT NAME?) or a category (Planting Size:) or any other form that you want the user to answer when the block is inserted.

Value—this is the default text that will appear opposite the prompt when the block is inserted. Use an entry here when there is an expected answer to the prompt. This will save time if there are no changes as the block is inserted.

Once these entries have been completed, the location of the tag and the characteristics of its display can be entered. Remember that the attribute will simply appear as a string of text on the screen. The following entries simply establish the appearance and location of the text.

Justification —this sets the manner in which the text is justified as it is placed. For invisible attributes, this is not important, but when attributes are part of the visible drawing you will want to control the appearance of the text. Choose from the various options from the pop-down box. These options correspond to the options found in the **Text**, **Dtext**, and **Mtext** commands.

Text Style—this sets the style of text. Choose from any of the existing styles in the drawing.

Height—allows you to specify the lettering height.

Rotation—allows you to specify a rotation angle for the text. **This option may not work properly in AutoCAD 13**.

Insertion Point allows you to select the location where the text will be located. You can specify the **X**, **Y**, and **Z** coordinates in the text boxes, or select **Pick Point <** to select the point on the drawing.

Regardless of whether or not the attribute was **invisible** when defined, the **tag** will be placed on the drawing as a piece of text. If it is not located correctly, you can move it (or scale, copy, rotate or mirror it) to satisfy your requirements. When you use the **Block** command, select the attribute along with the block objects and they will disappear together. When the block is inserted into the drawing, the attribute will display (or not) according to your definition.

When creating a block with multiple attributes, pay attention to the sequence you use to select the attributes. **Always pick the attributes in the order you want them to display**. If the prompts lead you through a sequence of questions, be sure to select the attributes in the order you want to ask them.

Attribute Dialogue Boxes

After you have created blocks with attributes, you can control the appearance of the prompts which the user sees when the block is inserted. Specifically, you can control whether the prompts are seen at the command line or are visible in a dialogue box. This is set with the AutoCAD system variable **ATTDIA.** This variable is set at **O** (Off) or **1** (On).

```
Command:
```

Type **Attdia.**

```
New value for ATTDIA <0>:
```

Type **1** to turn on the attribute dialogue box display (Figure 7.2).

Dialogue boxes are easier to read and easier to edit than command line prompts. The user has an opportunity to review the entries before placing them on the screen. In addition, dialogue boxes are more in keeping with the modern graphical interface employed by most current programs.

Figure 7.2

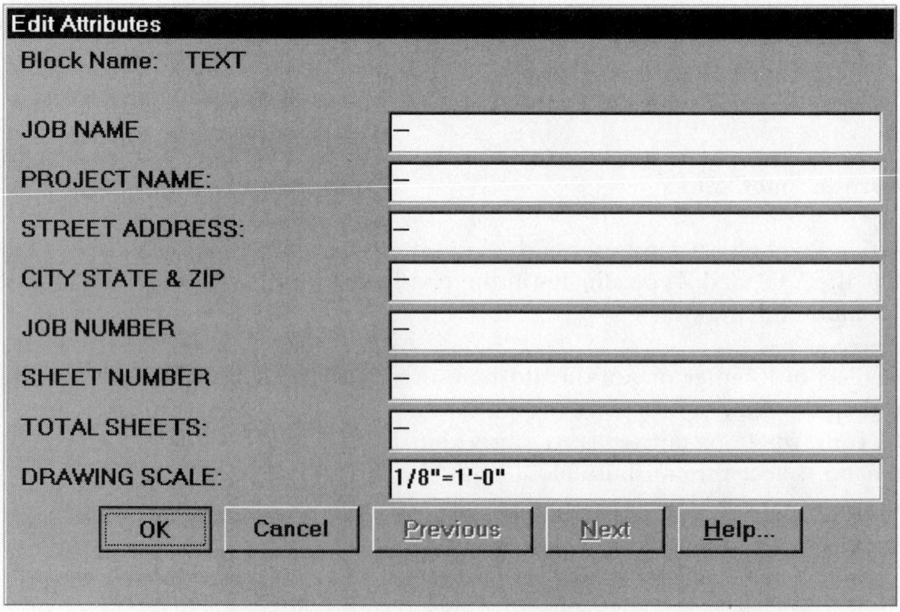

Edit Attributes
Block Name: TEXT
JOB NAME
PROJECT NAME:
STREET ADDRESS:
CITY STATE & ZIP
JOB NUMBER
SHEET NUMBER
TOTAL SHEETS:
DRAWING SCALE: 1/8"=1'-0"
OK | Cancel | Previous | Next | Help...

Inserting Blocks

Insert and **Ddinsert** are commands which allow you to insert blocks into your drawing. The **Insert** command works from the command line while **Ddinsert** opens a dialogue box.

`Command:`
Type **Ddinsert** or pick **Insert ▸** from the **Draw** pull-down menu, then select **Block**. This opens the "Insert" dialogue box (Figure 7.3).

When this dialogue box is used, you can choose from any of the existing blocks that are included in the block definition table, or you can search for a drawing file to insert as a block. If you select the **Block...** button, you are presented with the "Defined Blocks" dialogue box (Figure 7.4).

This listing of the block definition table allows you to select one of the existing blocks and place it in the drawing. After you have selected one of the blocks here, you are returned to the "Insert" dialogue box. The block name you selected will appear in the text box next to the **Block...** button. Once the block name appears on this line, it will appear as the default the next time you open this dialogue box.

Two important toggle boxes appear in the "Insert" dialogue box. If you select **Specify Parameters on Screen**, the **Insertion Point**, **Scale**, and **Rotation** options are "grayed out"

Figure 7.3

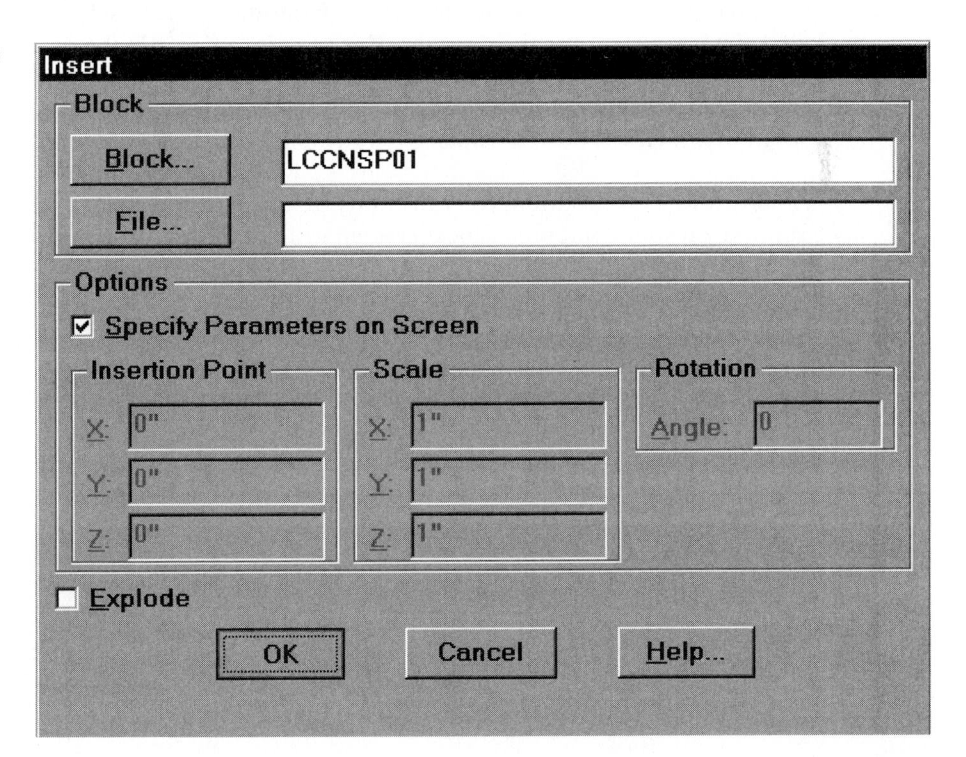

Figure 7.4

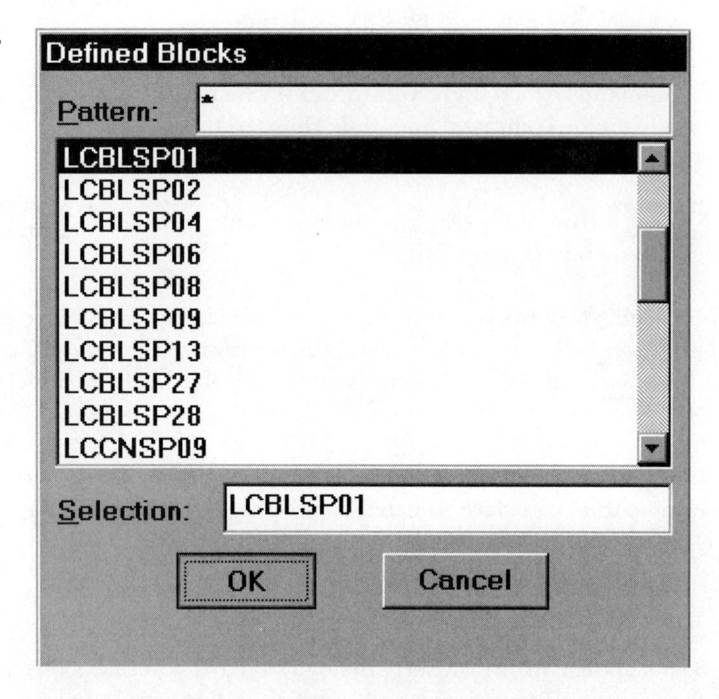

and are entered as the block is inserted. If the **Explode** option is selected, the block is exploded as it is inserted into the drawing. This permits you to edit the objects that make up the block immediately. This causes any attributes previously attached to the block to be converted into text and drawn independently of the block. Use this option only if you are certain of the effects of your choice.

If you choose the **File...** button, you can choose to insert a block that is stored as a drawing elsewhere on your computer. Blocks created with the **Block** command are only a part of your current drawing. If you want to gain access to them from another drawing, you must store them as separate drawings on your hard disk. See the **Wblock** command below. When you select the **File...** button, you will open a "Select File" dialogue box.

You can navigate through all of the folders and drives on your computer to search for AutoCAD drawings which can be inserted as blocks.

Once you have selected a block or file to insert, pick **OK** to complete the insertion.

`Insertion point:`
Pick the insertion point for the block.

`X scale factor <1> / Corner / XYZ:`
Enter the correct **X** scale factor for the block, or move the cursor to specify both the **X** and **Y** scale factors. This supercedes the **Corner** and **XYZ** options.

```
Y scale factor (default=X):
```

Specify the correct **Y** scale factor for the block or press **<Enter>** to use the same scale factor as was used for **X**.

```
Rotation angle <0>:
```

Enter the correct rotation angle for the block. If **Ortho** (F8) is active, you can control the rotation to 90° increments.

Exploding Blocks

Once a block is present in a drawing, it cannot be edited. If any of the variable attributes stored with the block are incorrect, they can be edited with the **Ddattedit** command. The appearance of the block is tied to the block definition table and cannot be modified. The only way to change the appearance of a block is to **Explode** it. When a block is exploded, it reverts to its previous form—a collection of circles, lines, arcs, polylines, etc.—along with the attributes which were originally tied to the block. Each of these objects can be edited independently of the others, because they no longer are tied together in a block. If you want to modify **all** of the occurrences of a block in the drawing, you can edit the objects and/or attributes, then create the block with the same name. This will redefine the block and will modify the appearance of **all** of the comparably named blocks in the drawing.

Creating Blocks and Drawings

As we saw earlier, a block only exists within the confines of a current drawing. It is not accessible from another drawing. Using the **Block** command does not help us to create a library of symbols that can be used over again, or by coworkers on other drawings. What we really need to be able to do is to create blocks that are stored in a way that anyone can access at a later time. In fact, we need to be able to turn blocks into drawing files. Drawing files can be inserted into other drawings at any time. This is where the **Wblock** command is used.

The WBLOCK (WRITE BLOCK) command

Wblock—This command is used as its full name implies: Write Block. The **Wblock** writes a block to the hard disk as a full-fledged drawing. This means that anyone can use this block (now a drawing) at a later time because it exists on the hard disk under its own name.

```
Command:
```

Type **Wblock** or pick **Export...** under the **File** pull-down menu.

If **Wblock** is entered, the "Create Drawing File" dialogue box appears (Figure 7.5).

Enter the desired name of the new drawing in the **File Name:** text box, being sure to place it in the appropriate drive and folder. Pick **OK** when you have entered the correct name and folder.

Figure 7.5

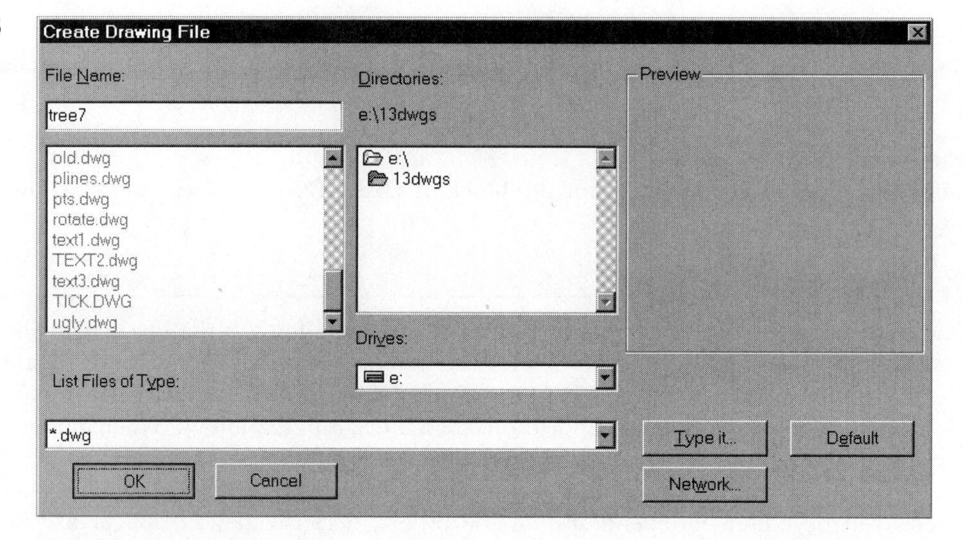

If you pick **Export...** you will open the "Export data" dialogue box. As the "Export Data" dialogue box appears, be sure to select the ***.dwg** file format. This is the same format that is used by AutoCAD to save its drawing files (Figure 7.6).

Enter the desired drawing name in the **File Name:** text box. Be sure to select the correct drive and folder where you want the drawing to be stored.

The naming process for naming blocks deserves special attention. It does not take long to accumulate dozens, even hundreds of blocks to be used in your drawings. Try to plan

Figure 7.6

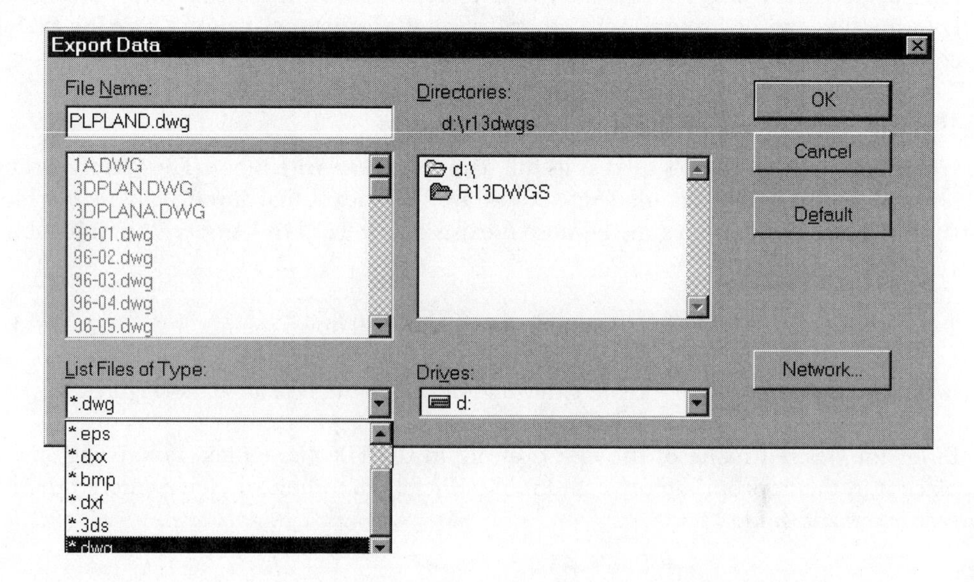

out a naming scheme for your blocks ahead of time. If possible, keep a record of your drawing blocks, including a printed copy of each file. This can be invaluable in organizing your work and setting up standardized office procedures.

After entering the name of the file to be saved, pick **OK**.

`Block name:`

Your response here will depend on the status of the objects that you want to include in the new drawing. If the object is an existing block, enter the name of the block here. This will have the effect of taking an existing block in the current drawing and converting it into a drawing on its own. If the desired drawing name entered in the last dialogue box is the same as the existing block, you can enter a "=" at this prompt.

If the desired drawing is not an existing block in the drawing, press **<Enter>**. This will create the new drawing outside of the current drawing. It will not be listed as a block in the current drawing after the **Wblock** command is completed.

`Insertion base point:`

Pick the base point for the new block. Pick an insertion point that will allow you to locate the block easily when it is inserted into another drawing. The centers of circular objects are always a good choice, while the lower left corner of square or rectangular objects is sometimes used. If the block is used in the same location every time you insert it, you could specify the insertion point as **0,0**. This will help you place the new block in the same location each time it is used.

`Select objects:`

Pick the objects you want to include in the new drawing. These new objects can include anything on the drawing screen, including other blocks. Be careful in the naming of your blocks to avoid duplication. If an inserted drawing (**Test1.dwg**) shares the same block name (**XYZ**) with a block in the current drawing (**Test2.dwg**), the inserted (**Test1**) drawing version of **XYZ** will be lost, overwritten by the current (**Test2**) drawing definition.

`Select objects:`

Press **<Enter>** to complete the command. The objects will disappear from the screen (as they did with the **Block** command). Type **Oops** to make them return to the screen so that you can continue working with the objects.

Inserting drawings with nested blocks has a number of interesting applications. Some LANDCADD drawings may appear blank, but have numerous block definitions stored in their block definition tables. When these drawings are inserted, their blocks are inserted into the current block definition table even though there is no evidence of them on the screen. Another option to insert a block definition without inserting any screen object is to use this procedure:

While working in a drawing, use **Ddinsert** to insert a drawing file. When asked for an insertion point, press **<Esc>** or **<Ctrl>-C** to cancel the command. The block definition will be inserted even though nothing appears on the screen. The block can be inserted as though you had built the block in the current drawing.

When you build a drawing that you intend to become a block to insert into other drawings, you may decide not to convert it into a block, but simply save it as a drawing. This does not require any special steps, but does prevent you from specifying an insertion point for the drawing. AutoCAD automatically assumes a default value of **0,0** for the insertion point of drawings inserted into other drawings. If you want to specify another point as the insertion point of your new drawing/block use the **Base** command.

The **BASE** command:

`Command:`
Type **Base**.

`Base point <0'-0",0'-0",0'-0">:`
Pick a new base point for the drawing, or specify a point from the command line.

Removing Unused Blocks

Sometimes your drawing can accumulate "baggage" that is not needed. Blocks, linetypes, layers, text styles, etc., that are defined, but not used in a drawing can add to the size of drawing file without adding any benefit. You can eliminate these unused blocks, linetypes, text styles and layers from the drawing by using the **Purge** command. With the release of AutoCAD 13, it is now possible to use the **Purge** command at any time during the drawing process. (Previous versions of AutoCAD confined use of the **Purge** command to immediately after opening or saving a file.)

The PURGE command

`Command:`

Type **Purge** or pick **Purge ▸** under the **Data** pull-down menu. This opens another menu which has the same options as the command line.

```
Purge unused Blocks/Dimstyles/Layers/Ltypes/Shapes/STyles/
Mlinestyles/All:
```
You can select which of these categories you wish to purge, or pick **A** for all.

The command prompt will respond by listing all of the unused entities that it finds in the drawing. If you responded with **All**, you might see a response like this:

```
No unreferenced blocks found.
No unreferenced layers found.
Purge linetype EPLC_PROPERTY? <N>
Purge linetype EPLC_ROW? <N>
Purge linetype EPLC_SWALE? <N>
Purge linetype EPLC_LONGDASH? <N>
Purge linetype EPLC_SHTDASH? <N>
Purge linetype EPLC_OPENDASH? <N>
Purge linetype EPLC_OPENDOT? <N>
Purge linetype DASHED? <N>
Purge linetype PHANTOM? <N>
Purge linetype CENTER? <N>
```

After the listing of an item, answer **Y** if you wish to **Purge** it from the drawing file.

Another way to eliminate unused information from the drawing file is to use the **Wblock** command in an unconventional way. You can delete all unused blocks, layers, linetypes, etc., in a single step with the **Wblock** command. In this case, the file name of the drawing you create will be the same as that of the existing drawing.

```
Command:
```
Type **Wblock**.

When the "Create Drawing File" dialogue box opens, accept the default file name—that of the current drawing. Pick **OK**.

AutoCAD will warn you that you are preparing to overwrite your drawing file.

Pick **Yes**.

```
Block name:
```
At this prompt type *. This will select all of the objects on the screen and in frozen layers. Any unreferenced blocks, layers, linetypes, etc. will be removed in one step. **There are no prompts before removing these definitions**. Be sure to use this procedure with caution.

A more radical use of this command is to press **<Enter>** at this prompt (no block name), then select only the items on the screen to include in the drawing. This will eliminate any objects in frozen layers, and the frozen layers themselves. This procedure will greatly shrink the size of over-bloated drawing files.

Working With Xrefs and Prototypes

A very important advantage in the use of CAD (and AutoCAD in particular) is the shared use of drawing information between several designers simultaneously. The best way to create this kind of relationship is through the use of the **Xref** command. **Xref** stands for *External Reference.* **Xref** creates a dynamic linkage between drawings because one drawing makes reference to another.

As an example, a Base Plan can be created from surveying and civil engineering data for a particular landscape development, then placed in the drawing **BASEPLAN.DWG**. The information included on this plan (street locations, setbacks, spot elevations, lot lines, building footprints, utility locations, etc.) would be needed in all subsequent drawings for the project. Several designers in a firm might want to include this information in their plans, and would certainly need to have this information updated if the Base Plan were modified. The **Xref** command can create the links between the Base Plan and the other plans as they are developed (Figure 7.7).

Figure 7.7

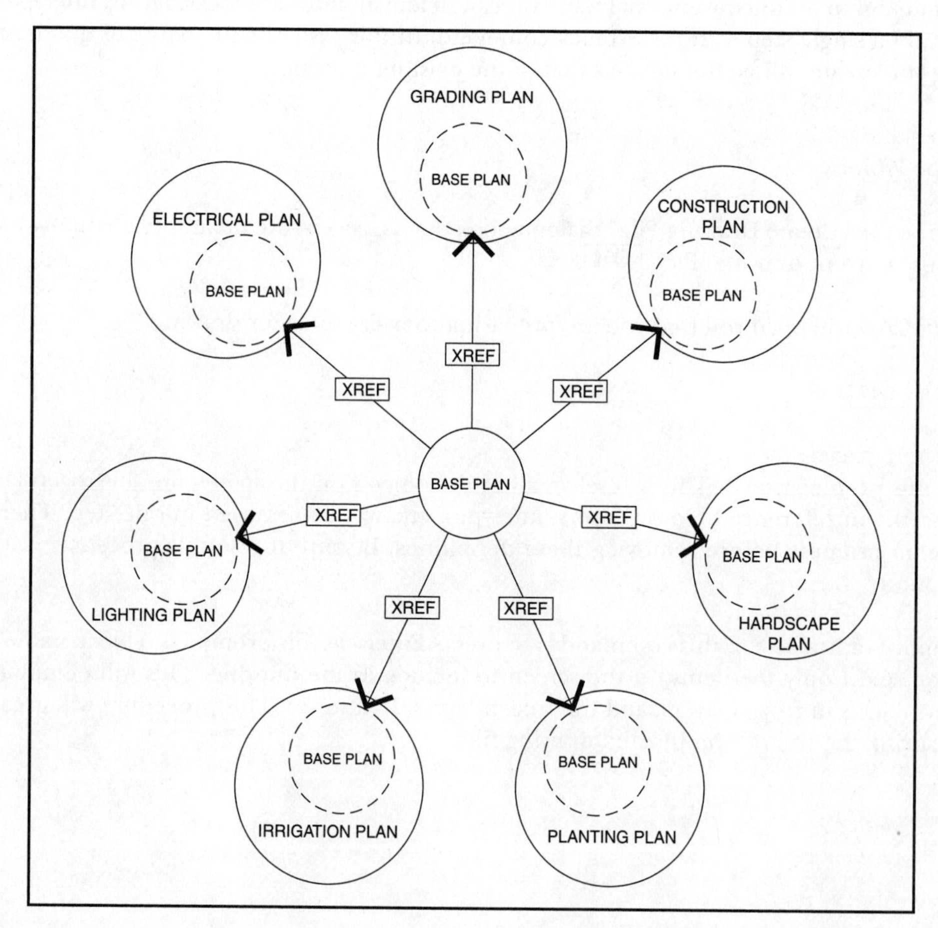

Xrefs and **Blocks** are similar in that they can be inserted into an existing drawing. There are no limits to the number of **Xrefs** and **Blocks** that can be inserted into a drawing. Like **Blocks, Xrefs** will bring their layers into the drawing with them. **Xref** layers will appear in the "Layers" dialogue box accompanied by their drawing name. If your current drawing had an **Xref** to the drawing **BASEPLAN**, the **Text** layer from the Base Plan would appear as **BASEPLAN|TEXT**. These layers can be controlled in the same way as other layers, so the appearance of the **Xref** can be changed to suit your needs.

As with a **Block**, an **Xref** can have blocks nested within it. The blocks contained in an **Xref** will show up in the Block Definition Table in the same manner as the layers appear in the "Layers" dialogue box. A **block** called **GRATE** from the **BASEPLAN Xref** would be listed as **BASEPLAN|GRATE** in the Block Definition Table. While the layers and blocks retain their own identity, they are still tied to the **Xref**. These are known as *dependent layers* and *dependent blocks,* respectively. Other items (linetypes, layers, text styles, etc.) will also accompany an **Xref** as *dependent objects.*

Xrefs can also have other **Xrefs** nested inside them—even the same **Xref**. In the previous example, the designer working on the Planting Plan could **Xref** the Hardscape Plan in order to place plants within the boundaries of the planting areas defined there. The definition of the **BASEPLAN**, the shared **Xref**, would be duplicated but AutoCAD would ignore it (Figure 7.8).

Figure 7.8

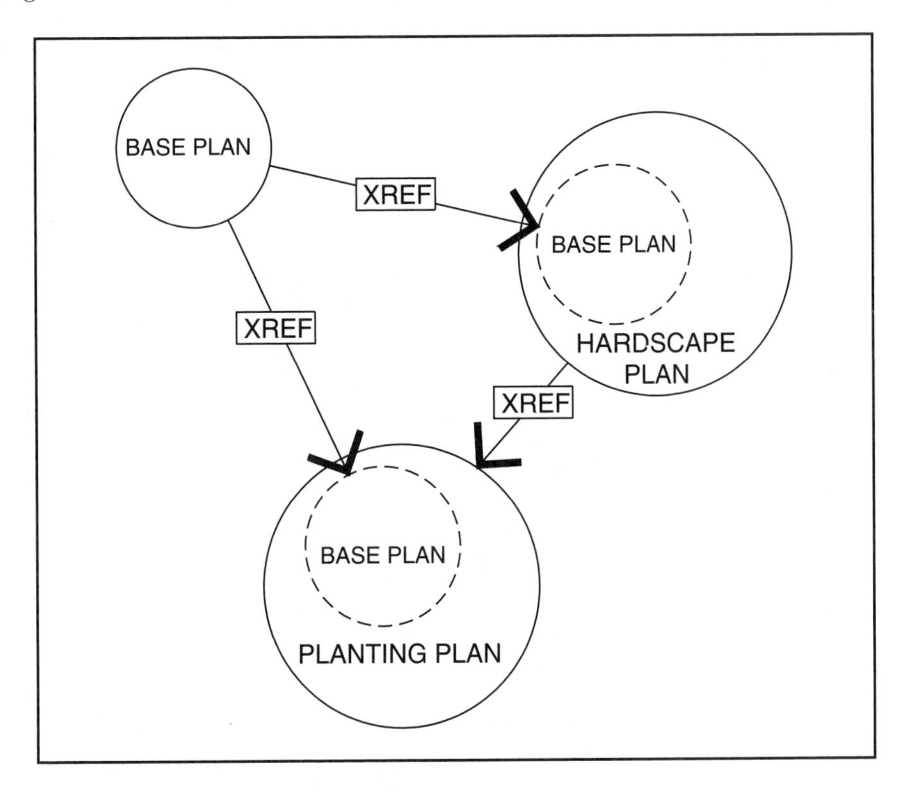

There are three primary differences between **Xrefs** and **Blocks:**

1. An **Xref** is updated each time the drawing containing it is loaded. Any changes made to the **BASEPLAN** drawing would be reflected in any of the other plans which include it as an **Xref** each time they are opened. This means that the linkage between the drawings is renewed each time the drawing containing the **Xref** is opened and edited.
2. The objects in the **Xref** cannot be modified or edited. What appears in the drawing is simply the **image** of the **Xref** to allow the designer to use its information, but not edit it in any way.
3. An **Xref** file, no matter how large, does not enlarge the drawing that references it. A **block** inserted into a drawing will enlarge the file by approximately the file size of the block. **Xrefs** do not add substantially to a drawing file size.

Though **Xrefs** can be nested and shared between drawings, they cannot be circular. This means that the Hardscape Plan cannot reference the Planting Plan if the Planting Plan references the Hardscape Plan. These linkages can only occur in one direction. If this were true, the Planting Plan, through the Hardscape Plan, would reference back to itself, causing an error in AutoCAD. The program would refuse to display this **Xref** as either the Planting Plan or the Hardscape Plan (Figure 7.9).

Figure 7.9

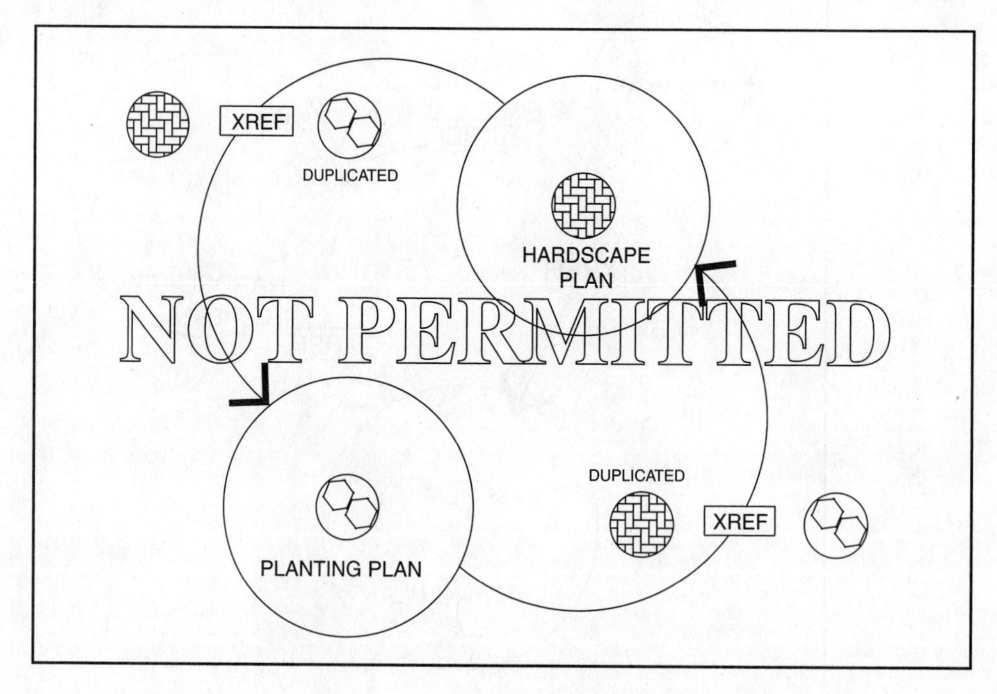

To avoid circular **Xrefs** in cooperative work—where **Xrefs** may link several drawings together at the same time—AutoCAD has added the **Overlay** feature. An **Overlay** is a special type of **Xref** which will **not** allow nesting **Xrefs**. This is important when drawings are **Xref**ed in a circular manner. In the previous example, the linkage between the Hardscape Plan and the Planting Plan drawings could be only one-way. The planting designer could see the hardscape designer's work, but not vice versa. If an **Overlay** were used, the Planting Plan could have an **Overlay Xref** to the Hardscape Plan, and vice versa. The Planting Plan would not carry the Hardscape Plan as a nested **Xref**, and the Hardscape Plan would not carry a nested **Xref** of the Planting Plan. This would allow each designer to refer to each other's work without creating a circular **Xref**. This can be a tremendous asset in large firms where several designers work on the same project simultaneously (Figure 7.10).

The XREF (EXTERNAL REFERENCE) command

Xref—Use the **Xref** command to create **Xref** attachments and control their activities.

`Command:`

Type **Xref** or pick **External Reference ▸** from the **File** pull-down menu.

`?/Bind/Detach/Path/Reload/Overlay/<Attach>:`

The options are displayed below:

Figure 7.10

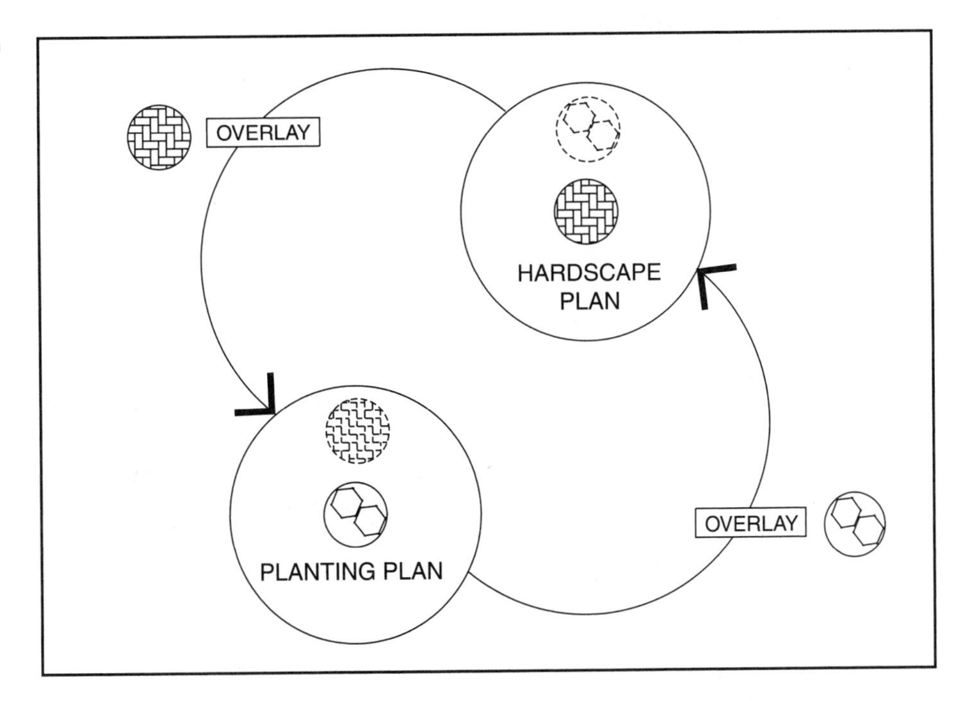

? displays a list of all of the **Xrefs** currently attached or overlayed in the drawing.

`Xref(s) to list <*>:`
Press <Enter> to list all of the **Xrefs**.

```
Xref name           Path                    Xref type
---------------     ---------------         ---------------

TEST1               e:\r13dwgs\test1.dwg    Attach

TEST2               e:\r13dwgs\test2/dwg    Overlay

Total Xref(s): 2
```

Bind makes an **Xref** part of the drawing. This converts the **Xref** into a traditional **Block** and places it in the drawing file.

`Xref(s) to bind:`
Enter the name of the **Xref** to bind.

Type in the name of the **Xref** to bind to the drawing. Enter the * character to bind all of the **Xrefs** at once.

```
Scanning...
Regenerating drawing.
```

Detach removes an **Xref** from the drawing. This option has the same effect as purging a block.

`Xref(s) to detach:`
Enter the name of the **Xref** to detach from the drawing. Use the * character to detach all of the **Xrefs** at once.

`Scanning...`

The **Xef** is removed from the drawing and disappears from the screen.

Path allows you to change the location where AutoCAD will search for the **Xref** drawing. This is especially important when drawings are moved from one location in the computer to another. If the drawing has been moved, AutoCAD will be unable to find the file and will not be able to display ("resolve") the **Xref**. Use the **Path** option to respecify the location of the file.

`Edit path for which xref(s):`
Name the **Xref** which has been relocated.

`Scanning...`
`Xref name: TEST1`—the name of the **Xref** named above
`Old path: TEST1.DWG`—the old location of the TEST1.DWG
`New path:`
Enter the new path or location for the drawing.

E:\R13DWGS\TEST1.DWG—

`Reload Xref TEST1: E:\R13DWGS\TEST1.DWG`
`TEST1 loaded. Regenerating drawing.`
Reload forces the **Xref** drawing to load so that a fresh version of the drawing will be displayed. This may be important if someone has edited the **Xref** file since you opened your drawing.

`Xref(s) to reload:`
Enter the name of the **Xref** to reload.

`Scanning...`
`Reload Xref TEST1: e:\r13dwgs\test1.dwg`
The command will reload the **Xref** from wherever the current **Path** specifies.

`TEST1 loaded. Regenerating drawing.`
Overlay creates an **Overlay Xref** which will not carry nested **Xrefs**. Use this feature when several drawings will refer to each other and the potential exists for circular **Xrefs**. Selecting this option opens the "Select file to Overlay" dialogue box (Figure 7.11).

Figure 7.11

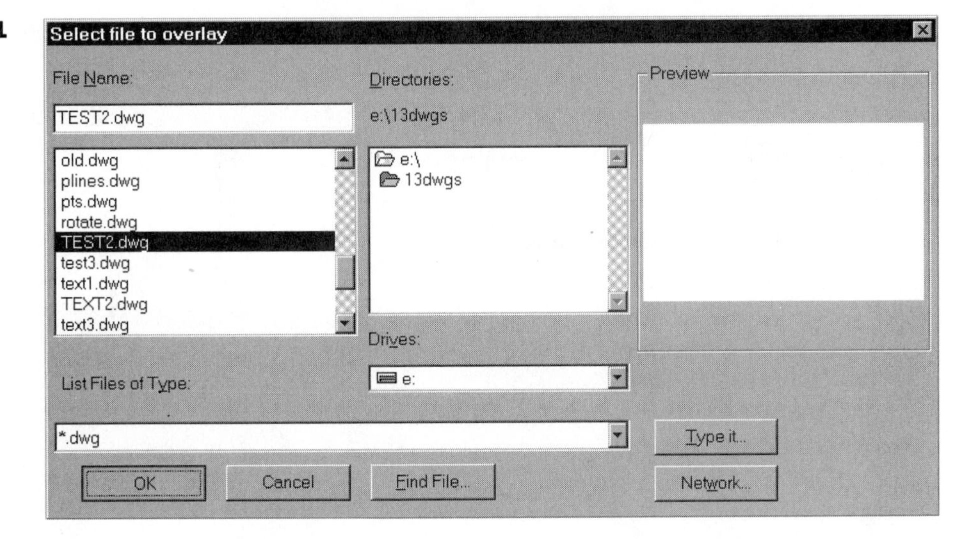

Select the file you wish to overlay.

`Overlay Xref TEST2: TEST2.dwg`
This is the file name selected to overlay.

`TEST2 loaded.`
`Insertion point:`
Pick an insertion point for the overlay file.

`X scale factor <1> / Corner / XYZ:`
Enter a scale factor or move the cursor to indicate the scale for the overlay.

`Y scale factor (default=X):`
Press **<Enter>** or type a Y scale factor for the overlay file.

`Rotation angle <0>:`
Enter a rotation angle for the overlay file.

Note that these prompts are identical to those used in the **Insert** command. This is due to the similarity to a **Block** and an **Xref**. The same prompts are used in the **Attach** option, shown below:

Attach creates the link between the current drawing and the target drawing. Choose this drawing from the "Select file to attach" dialogue box (Figure 7.12).

Select the file you wish to attach as an **Xref**.

`Attach Xref TEST4: TEST4.dwg`

Figure 7.12

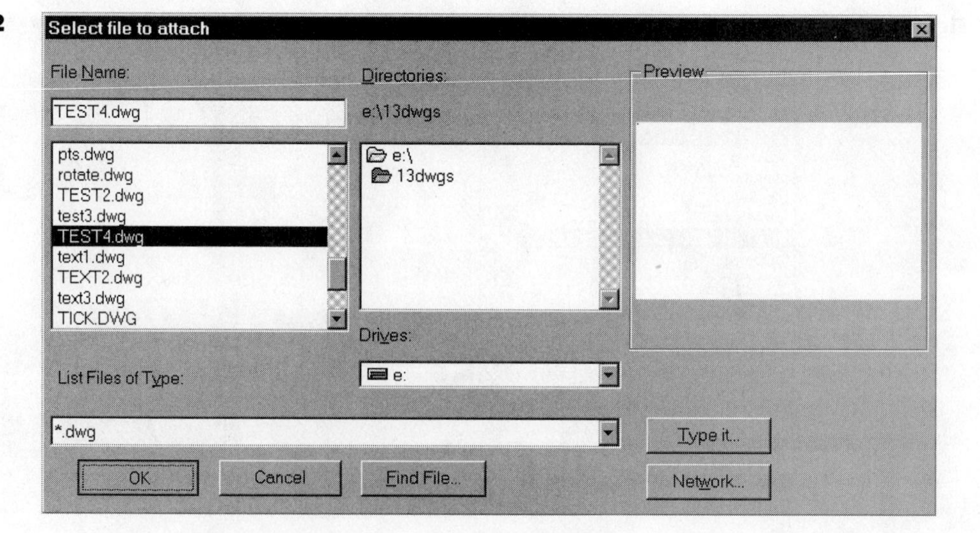

This is the file name selected to overlay.

```
TEST4 loaded.
Insertion point:
```
Pick an insertion point for the attached **Xref** file.

```
X scale factor <1> / Corner / XYZ:
```
Enter a scale factor or move the cursor to indicate the scale for the attached **Xref**.

```
Y scale factor (default=X):
```
Press **<Enter>** or type a Y scale factor for the attached **Xref**.

```
Rotation angle <0>:
```
Enter a rotation angle for the attached **Xref**.

Prototype Drawings

Another AutoCAD tool which can improve your efficiency and help you standardize your drawing procedures is that of **Prototype Drawings**. Prototype drawings are drawings which house your most frequently used settings, text and dimension styles, linetypes, layers, and blocks. When you use a prototype drawing as the basis for a new drawing, all of the ingredients of the prototype drawing are copied into your new drawing—without taking the risk of altering the original file. This is accomplished when you start a **New** drawing in AutoCAD, or when you open a **New** project in LANDCADD's Project Manager.

New—When you want to begin a new drawing in AutoCAD, you use the **New** command.

```
Command:
```
Type **New** or pick **New** from the **Files** pull-down menu. This opens the "Create New Drawing" dialogue box (Figure 7.13).

Figure 7.13

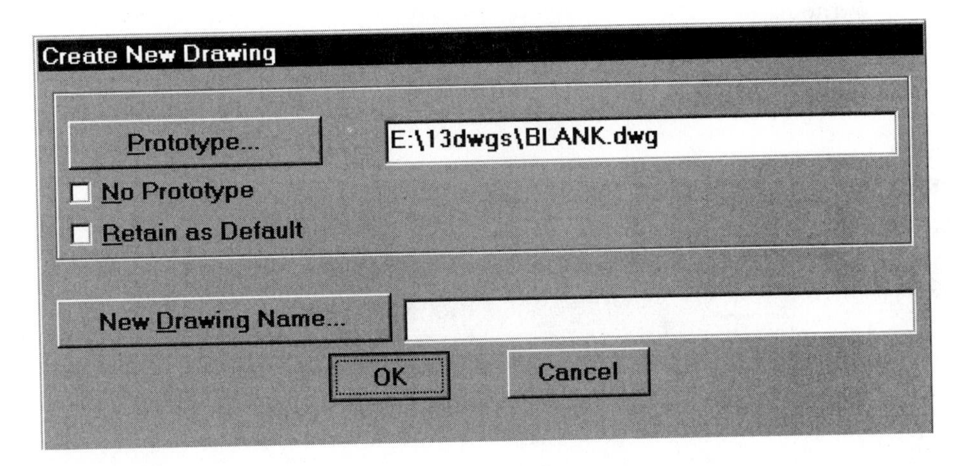

This dialogue box allows you to specify the name of the drawing you will create, the name of the prototype drawing (if any), and whether you wish to retain the current prototype to use the next time you use the **Open** command.

If you pick the **Prototype...** button, you can pick your prototype by navigating through the "Prototype Drawing File" dialogue box.

You can select a prototype drawing from any drive or folder on your computer. If you maintain many prototype drawings, consider creating a folder devoted strictly to their storage. If you are familiar with DOS commands, Windows Explorer or Windows File Manager, you can make your prototype drawings "Read-Only" to prevent any accidental overwriting.

If you already know the exact location and name of the prototype file you wish to use, you can enter it in the text box to the **Prototype...** button. If you choose not to use a prototype drawing, you can use the toggle box labeled for that purpose. If you wish to keep the current prototype as the default prototype drawing, you can also indicate this in the toggle box labeled **Retain as Default**.

You can specify the name of your new drawing in the text box next to the **New Drawing Name** button.

If you want to create a drawing in a specific folder other than the default location on your computer, you can pick the **New Drawing Name** button. This will open the "Create Drawing File" dialogue box.

If you select an existing drawing to overwrite with your new drawing, you receive a warning message (Figure 7.14).

Since overwriting a file is a serious decision, be sure to consider the consequences of doing this. You probably will not be able to recover the drawing if you choose to overwrite it.

If you choose not to name your drawing in this dialogue box, your drawing will be given the name **UNNAMED**.

Figure 7.14

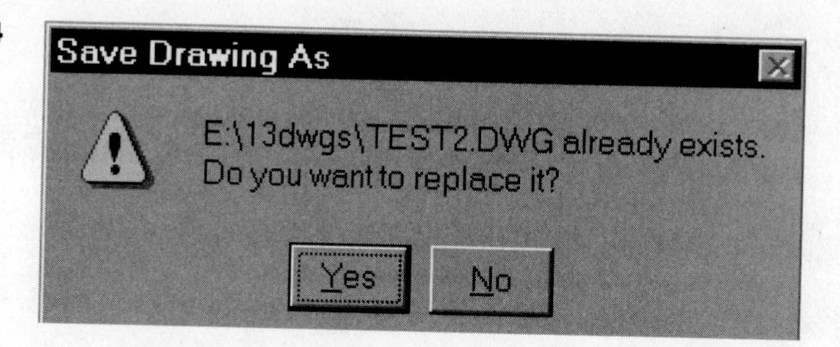

After you finish entering data in the "Create New Drawing" dialogue box, pick **OK**. This will open your drawing and begin your drawing session.

Eagle Point Project Manager

The Project Manager portion of LANDCADD offers utilities for opening new drawings and assigning them to *Projects.* When the Project Manager is on the screen, you can begin a new drawing by using the **New...** button (Figure 7.15).

The **New...** menu selection opens the "New Project" dialogue box (Figure 7.16).

Figure 7.15

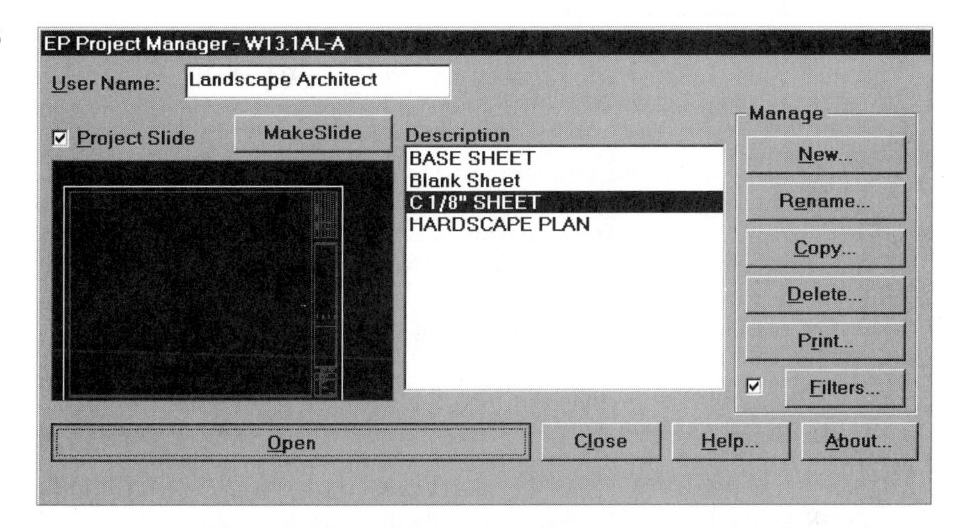

Figure 7.16

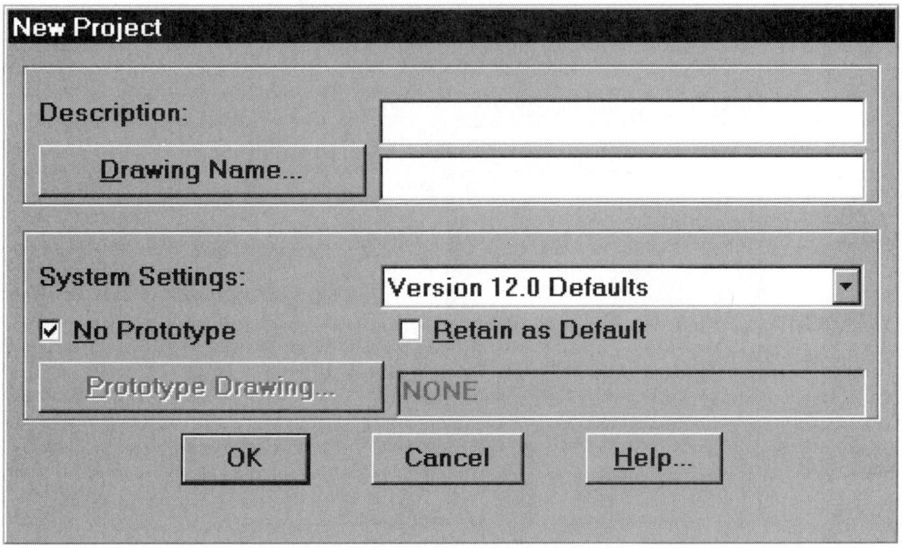

This dialogue box allows you to build a new project by entering a **Description**, and entering a **Drawing Name**. There are two ways to use the **New...** option.

First, you can create a new project using an existing drawing. To do this, pick the **Drawing Name . . .** box. Use the "Select Drawing" dialogue box to navigate through the different directories (or folders) to find the correct location and name of the existing drawing (Figure 7.17).

After you select the desired drawing, click on **OK**. You are then returned to the "New Project" dialogue box with the correct drawing name, along with the full path, in the **Drawing Name . . .** text box.

Enter a unique description for the project in the **Description:** text box. After you click on **OK**, the project name will appear on the Description list in the File manager screen.

Second, you can create a new project and open a new drawing at the same time. To open a new drawing, type in the full name (including drive and path) of the drawing that you wish to create in the **Drawing Name . . .** text box. Be sure to type in the **full** name of the drawing, including the extension **.DWG**.

When you create a new drawing, you can also specify a **Prototype Drawing** of your choice. To activate the prototype feature remove the mark from the **No Prototype** toggle box. To make this feature function properly, place a mark in the **Retain as Default** toggle box. This will cause the Project Manager to call the specified drawing as the prototype in the same way that we saw in AutoCAD's "Create New Drawing" dialogue box. When you wish to change this default drawing, simply specify your new prototype drawing (or choose **NONE**), then toggle the **Retain as Default** on again. The new prototype then becomes the default selection.

Figure 7.17

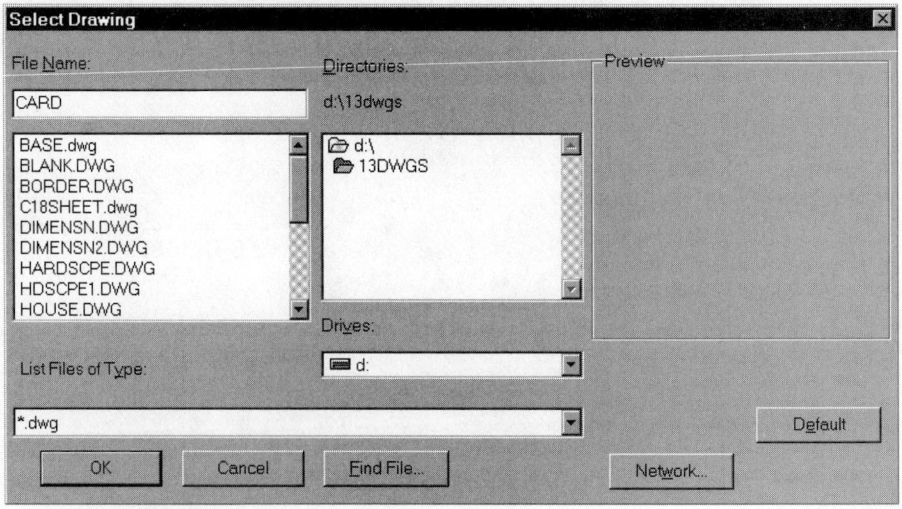

When you open a new drawing in either AutoCAD or the LANDCADD's Project Manager, and call for a prototype drawing—here is what happens:

1. A copy is made of the prototype drawing—including all of its styles, layers, linetypes, blocks, etc.
2. The copy of the prototype is renamed to the name you specified when you opened the new drawing.
3. The copied drawing (now the current drawing) is opened for you to begin your drawing session.

There are important reasons why this is the best method to begin your new drawings:

- It saves time because three functions (**Copy**, **Rename** and **Open**) are performed in one step.
- It assists in standardization since all of the office default styles, layers, linetypes, blocks, etc. are built into the prototype.
- It saves time because the important settings, style, layers, linetypes, blocks, etc. are already available and need not be created again.
- It preserves the original (prototype) drawing and protects it from being accidentally overwritten during **New** and **Save** functions.

One of the best ways to build prototype drawings is to combine them with title blocks so that all of the information needed to begin a drawing is present. If you build your prototype drawings this way, the following should be present when you open the drawing:

- Units (including angular measurement)
- Drawing Limits—set according to Paper Size, Plotting Scale and Scale Factor
- Snap—set appropriate to Drawing Limits
- Text Styles—use fonts and style settings appropriate to your drawing discipline
- Dimension Styles—use settings appropriate to the Drawing Limits and Scale Factor
- Linetypes—load those which you will use in most every drawing
- Linetype Scale—set according to Scale Factor and personal preference
- Layers—set the names, colors and linetypes according to your drawing discipline, your project or your office standards
- Title Block—set according to the Paper Size, Plotting Scale and Scale Factor

- Title Block Text Attributes—set to the correct text style, size and location for your title block
- Custom Symbols—incorporate blocks that you use in dimensions, notes or other standard features directly into the Block Definition Table of your prototype
- Preferred settings for AutoCAD system variables—preset **Textfill**, **Fill**, **Viewres**, **Attdia**, **Blipmode** and other variables to meet your (or your company's) preferences.

Your use of prototype drawings will depend on your system of drawing, and your use of **Paper Space** and **Model Space**. There are three alternatives:

1. Use **Model Space** only. This is the "old-fashioned" way, and is appropriate when you only work with one scale at a time, work with fairly simple drawings, and do not work much in 3-D. If you do not use **Paper Space**, consider the following guidelines:

   a. Build a title block for each paper size you intend to use.

   b. Create scaled title blocks for each plotting scale used. If you use five paper sizes (B, C, D, E and F) at seven drawing scales (1:1, 1/4, 1/8, 1/10, 1/20, 1/30, 1/40), this would require the creation of 35 title block drawings.

   c. Build your layers, text styles, and load your often-used linetypes and blocks into each title block.

   d. Set your text height, linetype scale, and dimension style according to the final plotting scale of each title block.

   e. Name the title block drawings in an easily recognizable format, and place them in a directory/folder specifically for that purpose.

   f. Use these title blocks as prototype drawings when you begin work on a new drawing. You may insert other blocks into this drawing as you begin to lay out the project. (see "Title Blocks in Model Space and Paper Space").

   g. Create a blank prototype drawing with "normal" settings for limits, snap and grid. The blank prototype is used when you want to determine the drawing scale and paper size for the perimeter objects in the drawing. See "f." above.

2. Use **Paper Space** occasionally. This may be a good option if you want to work with two or more scales in the same drawing, if you work in 3-D occasionally, or if you want to take advantage of the 1:1 plotting scale used by **Paper Space**. Consider these guidelines:

a. Build a title block for each paper size you intend to use.

b. Create a **Model Space** viewport in each title block.

c. If you use multiple **Model Space** viewports frequently, build a title block for each configuration of viewports used.

d. Build your layers, text styles, and load your often-used linetypes and blocks into each title block. Your linetype scale and text height will be set after the drawing scale factor is determined.

e. Set your dimension style(s) to **Paper Space** scaling.

f. Name the title block drawings in an easily recognizable format, and place them in a directory/folder specifically for that purpose.

g. Create a blank **Model Space** prototype drawing with "normal" settings for limits, snap and grid. You may have to reset these later. The blank **Model Space** prototype is used when you want to determine the drawing scale factor and paper size for the perimeter objects in the drawing.

h. Build your **Model Space** drawing as you normally would. Set the linetype scale and the text height after you have determined the drawing scale factor.

i. Use **Xref** to attach the **Model Space** drawing to the **Paper Space** title block when you are ready to plot.

j. Bind the **Xref** to the **Paper Space** title block when sending the file to the client and/or saving the file to an archive disk, cartridge or tape back-up.

3. Use **Paper Space** exclusively. This is a good strategy when you know the intricacies of **Paper Space** and need it for your multi-scale drawings, 3-D drawings and frequently used complex viewport configurations. Follow the recommendations above with the following exceptions:

   a. Place all of your text in **Paper Space**. This will save the effort of determining the proper text height for each scale factor used in **Model Space**.

   b. Be cautious about multiple **Xrefs**. Each layer and block from each **Xref** is accounted for separately. You can end up with a tangle of layer and block names to contend with.

   c. Reset your **MAXSORT** AutoCAD system variable to display the large number of layers that this option frequently creates. The **MAXSORT** system variable controls the number of names that can be alphabetized in the "Layers", "Load File", "Create File" and similar dialogue boxes. If the number of names exceeds the **MAXSORT** value (default = 200) the names will not be alphabetized. Increasing this value may slow AutoCAD performance slightly.

d. Thoroughly investigate the control of such factors as viewport layer control, **Paper Space** linetype scale, and regenerations of individual viewports.

See Chapter 6 for more information regarding the set-up of title blocks and plotting scales for drawings.

chapter

File Management in AutoCAD

When working with any computer program, the **most** important skill to learn first is how to save and retrieve your work. The basis for getting any constructive work done with any computer is knowing how to save your working file with an appropriate name, then locate and open it the next time you need it. There are a number of underlying, DOS-based principles in file management that affect your work regardless of the operating system that you use. Although you may use Windows, Windows 95 or Windows NT as your operating systems, you are still affected by the old-fashioned world of DOS.

File Names

Files are named with a **file name** and an **extension**. The file name used should adhere to what is called the **8.3** format:

FILENAME.EXT

"FILENAME" can be any combination of three to eight characters **without** spaces or punctuation.

"." acts as a separator between the filename and the extension.

"EXT" is the extension name. All AutoCAD drawings carry the extension name **DWG**.

While Windows 95 and Windows NT now support extended file names up to 255 characters (including spaces and punctuation), other platforms (Windows 3.1 and DOS) do not. When you try to use these long file names in a DOS application, the name will be truncated to eight letters. This may cause problems if different files are named with the same eight letter abbreviation. To avoid difficulty with file names it is probably best to adhere to the **8.3** standard for the near future.

File Directories/Folders and CAD Organization

The files in a computer are usually stored in **directories** (now called **folders** in Windows 95 and Windows NT). The file directories are typically arranged in a hierarchy that reflects the categories of files and the required separation of files. When we examine the hard disk that contains CAD files and programs, we can see that the AutoCAD program will occupy a specific directory—in this case **R13**. LANDCADD occupies a separate directory—**EPWIN13**. Even though the AutoCAD and Eagle Point programs run simultaneously, their major files are separated into directories. One of the easiest ways to understand the relationships between directories is to use the analogy of an outline:

1. **C:** C: Drive Root Directory
   - A. **R13** AutoCAD Directory
     - 1) **COM** Common files to all AutoCAD versions
       - a) **FONTS** SHX, TTF, PFB and PFM font used in AutoCAD
       - b) **SUPPORT** Commands, dialogue boxes, slide libraries, etc.
       - c) **SAMPLE** Sample drawings and commands
     - 2) **WIN** Files specific to Windows versions of AutoCAD
   - B. **EPWIN13** Eagle Point Windows Release 13 files
     - 1) **BLOCKS** Drawing blocks used by LANDCADD
     - 2) **DETAILS** Detail drawings used by the Details module
     - 3) **DWG** Drawings and slides used by Eagle Point modules
     - 4) **LANDCADD** Files used by LANDCADD modules
     - 5) **PROGRAM** Menus and dialogue boxes used by LANDCADD
     - 6) **SUPPORT** AutoCAD and Eagle Point support files
     - 7) **TEXTURES** Texture files used by LANDCADD Virtual Simulator
     - 8) **TUTORIAL** LANDCADD and Eagle Point tutorial exercises
   - C. **LIBRARY** Library of files used in many different drawings
     - 1) **SHEETS** Directory of drawing sheet files
       - a) **TITLEBLK** Stored title blocks of different sizes
       - b) **COVER** Stored cover sheets of different sizes
       - c) **DETAIL** Stored detail sheets of different sizes
     - 2) **DETAILS** Directory of details categories

a) **PLANTING** Stored planting details

b) **IRRIGATION** Stored irrigation details

c) **PAVING** Stored paving and hardscape details

d) **ADA/OSHA** Stored details related to ADA and OSHA standards

e) **PLANTERS** Stored walls and planters details

f) **SEATING** Stored seating and facilities details

ETC.

3) **DWGBLOCKS** Directory of common drawing blocks

a) **CALLOUT** Stored call-out and note balloons

b) **LEGENDS** Stored legends commonly used in drawings

c) **DIG ALERT** Stored Dig Alert® symbols

d) **PLANTS** Stored custom plant symbols (not stored with LAND-CADD)

e) **SPRINKLERS** Stored custom head symbols (not stored with LAND-CADD)

ETC.

4) **TXTBLOCKS** Directory of common text blocks

a) **GRLNOTES** Stored general notes used in drawings

b) **IRGNOTES** Stored irrigation notes used in drawings

c) **SLSNOTES** Stored soils notes used in drawings

d) **PLTNOTES** Stored planting notes

e) **SPECS** Stored specifications commonly used in drawings

ETC.

D. **CLIENTS** Library of clients listed by name

1) **LSANGLES** City of Los Angeles

a) 96-110 Project #96-110

i) **SITEDEV** Site development plan

ii) **BASE.DWG** Base plan

iii) **PLANTPL.DWG** Planting plan

iv) **HARDSCP.DWG** Hardscape plan

v) **IRRIG.DWG** Irrigation plan

ETC.

b) 94-893 Project #94-893

i) **SITEDEV** Site development plan

ii) **BASE.DWG** Base plan

iii) **PLANTPL.DWG** Planting plan

iv) **HARDSCP.DWG** Hardscape plan

v) **IRRIG.DWG** Irrigation plan

ETC.

2) **PETERSON** Peterson Development Co.

a) 97-085 Project #97-085

i) **SITEDEV** Site development plan

ii) **BASE.DWG** Base plan

ETC.

b) 95-450 Project #94-450

i) **SITEDEV** Site development plan

ii) **BASE.DWG** Base plan

ETC.

While the AutoCAD and LANDCADD directories are placed on the hard disk as the programs are installed, the remaining organizational structure is up to the user. What is usually called for is a structure which permits the organization and storage of two types of drawing files: those files which are part of the company "library," and those which are drawn specifically for clients.

The Library portion of files is divided into four or more categories. First, drawing sheets for each paper size (and potentially each plotting scale), featuring the standard company title block, are stored in a separate directory. The drawing sheets are broken into categories for drawing title blocks, cover sheets and detail sheets. There may be as few as two or three files or as many as 50 files in each category. Second, detail drawings of assemblies, techniques, materials, or other enlargements are stored according to the type of area or plan in which they might be used. In instances where many drawing files are stored in the same directory, descriptive file names are crucial. A third section of the library directory is for commonly used drawing blocks. The number of potential categories here is nearly unlimited, and will depend on the designer's drawing discipline and on the number of repeated elements in the drawings. In those instances where there are many repeated elements, there is good reason to store many blocks to save redundant drawing. A final section is used for commonly used text blocks used in drawings. General and specific notes can be stored here, along with commonly used specifications.

Client drawings should be separated from the standard library drawings to allow easy access and storage. Keeping client files in directories separated from the Library files makes sense for many reasons. It is easier to locate projects when you can use the client name to search for them. It is also easier to back up files using a back-up program when the files you want to back up are stored together. Finally, keeping clients separated allows you easily to move (or remove) drawing projects when they are no longer active. Once this occurs, you can move files to a tape cartridge, a portable drive disk or other archival storage medium.

CAD Drawing File Naming

Organizing individual CAD drawing file names is less simple. The method of naming and organizing client files will depend on the quantity of drawings created for each project, and the number of plotted drawings produced for each drawing file. As long as there is only one plotted drawing produced per file, and there are only a few standard drawings produced for each project, the system above is satisfactory. Each project is organized by project number, while drawing files are descriptively named and placed in the project number directory. Using this type of system also allows the mixture of drawing files, database files, word processing documents, and workflow management data files together if this becomes desirable in the firm's approach to computer network management.

If the number of files per project grows too large, or if the number of plotted drawings per file grows to three or more, it will become necessary to rethink this approach. Firms often develop their own system for file naming and organization that establishes prefixes, drawing numbers and suffixes to indicate the particular subject matter of the drawing by category, the sequential number in the category and the number of plots included in the drawing.

Project:	—	Thomas	
Project Number:	—	96-101	
Drawing Subject:	—	Irrigation plan	(IR)
Drawing Sequence:	—	Second irrigation drawing	(02)
Plots in Drawing:	—	3	(03)

Directory: **THOMAS**

Project: **96-101**

File Name: **IR_02_03**

Probably the most essential part of file naming and organization boils down to **preplanning**, **standardization** and **good record keeping**. Try to anticipate your requirements before accumulating so many drawing files that renaming and reorganizing them becomes an ordeal not unlike cleaning out an attic. Be sure that the naming system is used by all designers in the firm to avoid the use of individualized, inconsistent file naming systems. Keep a record of your system, and develop a CAD manual to be sure that everyone understands how the file system is organized and how to use it. This improves communication and can help shorten the training time necessary for new personnel.

Saving Drawings

Release 13 in AutoCAD has four different commands to save your file, and a few variations on those commands. This causes a great deal of confusion (and frustration) amongst users. We will examine the different options available here:

The QSAVE (QUICK SAVE) command

Qsave (**Qs**) The **Qsave** command performs the simple action of saving your file under its current name into its current directory (or folder). There are no prompts to answer for this action, the command simply saves the file.

`Command:`
Type **Qsave** (or **Qs**).

QSAVE

The only exception to this operation is when the drawing has not yet been saved. If the drawing does not have a name, AutoCAD will display the "Save Drawing As" dialogue box. This will give you the opportunity to name the file and show where it should be stored in the hard disk.

The SAVEAS command

Saveas This command will automatically invoke the "Save Drawing As" dialogue box, shown below (Figure 8.1).

This dialogue box allows you to enter the new file name of the file you are saving, and the location where you want to place it on the hard disk. This command can be used to

Figure 8.1

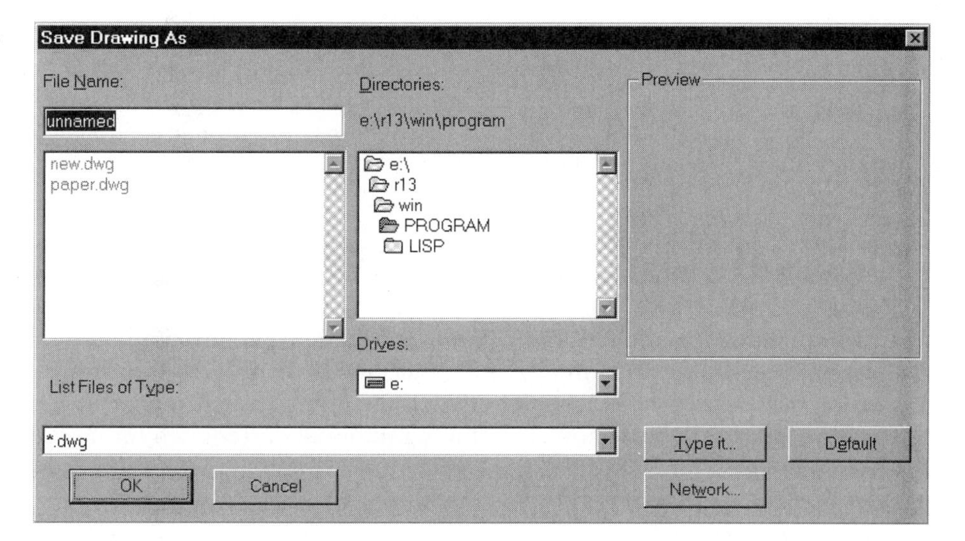

name a new drawing or to create a copy of the current file under a new name and/or location. An important feature of this command is that the newly named version becomes the current drawing immediately.

Do not use this command to save your drawing to a floppy disk. This will have the unintended effect of making a floppy disk drawing the current, active drawing. When this occurs, AutoCAD will begin to write its temporary swap files onto the floppy disk, rather than to the hard disk. If these temporary files fill the floppy disk (they usually will in a few minutes) AutoCAD will crash and you may lose all of the data on the floppy disk. Copying files to a floppy disk should be performed with the **Files** command or with the File Manager or Windows Explorer.

The SAVE command

Save—This command functions differently depending on how the command is invoked. If you type **Save** from the command line, it has the same effect as the **Saveas** command, opening the same dialogue box.

If you pick Save from the File pull-down menu, or pick the Save icon from the standard toolbar, it has the same effect as the **Qsave** command. The Save As... option is available from the File pull-down menu, and works as the **Saveas** command.

The SAVEASR12 (SAVE AS RELEASE 12 DRAWING) command

Saveasr12—This command saves your file in AutoCAD Release 12 format. While several types of 3-D objects and some important 2-D objects do not translate properly, this is a way to create compatible files with consultants or clients who want Release 12 drawings. Associative **Leaders**, true **Splines**, **Ellipses** and **Mlines** (multi-lines) will not translate (they did not exist in Release 12) during this process, as well as all 3-D solid objects. Be sure to name or locate R12 drawings in such a way as to be able to identify them and separate them from your R13 work.

Saving to a Floppy Disk

The traditional method of transporting a drawing from one computer system to another has been through the floppy disk. While 5 1/4" disks were popular in the 1980s, most users have switched to high-density 3 1/2" disks for such purposes. These disks can accommodate 1.44 megabytes of drawing information (enough for most drawings) and are well-suited for CAD drawings. Writing information to these disks can be done in safe and reliable ways if the proper methods are used.

DO NOT use standard file saving commands in AutoCAD. These methods work well for saving drawings to the hard disk, but are not suitable for saving drawings to a floppy disk. Using these tools may result in crashing AutoCAD, losing your work and ruining the existing data on the floppy disk.

The first method to use exists in AutoCAD, in the command **Files**. This command contains several utilities which can help you copy and manage files on either floppy or hard disk drives.

The FILES command

Files—This command is used to manage files.

```
Command:
```
Type **Files**. This opens the "File Management" dialogue box (Figure 8.2).

Each button opens a different dialogue box, or combination of two dialogue boxes.

Figure 8.2

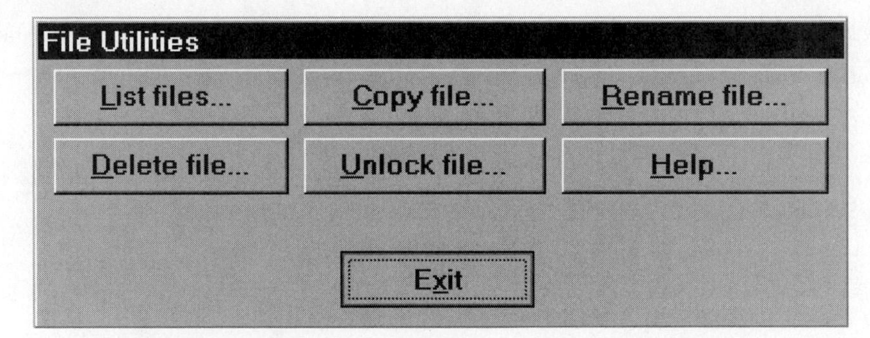

List files ... opens the "File List" dialogue box which displays the files existing on a particular drive in a selected directory. Files shown here cannot be copied, deleted or renamed, nor have any other action performed on them. Use this option to examine the contents of a floppy disk or hard disk directory. You can select the type of files to display by changing the entry in the pop-down box **List Files of Type:**. The default value is ***.*** or all files (Figure 8.3).

Copy file... allows you to copy files from a **Source File** to a **Destination File**. A separate dialogue box appears to select each file name. The "Source File" dialogue box asks you to select the file to be copied (Figure 8-4).

Choose from the available files, or navigate to the drive and directory which contains the file you want to copy. Highlight the file, then pick **OK**. This opens the "Destination File" dialogue box (Figure 8.5).

Figure 8.3

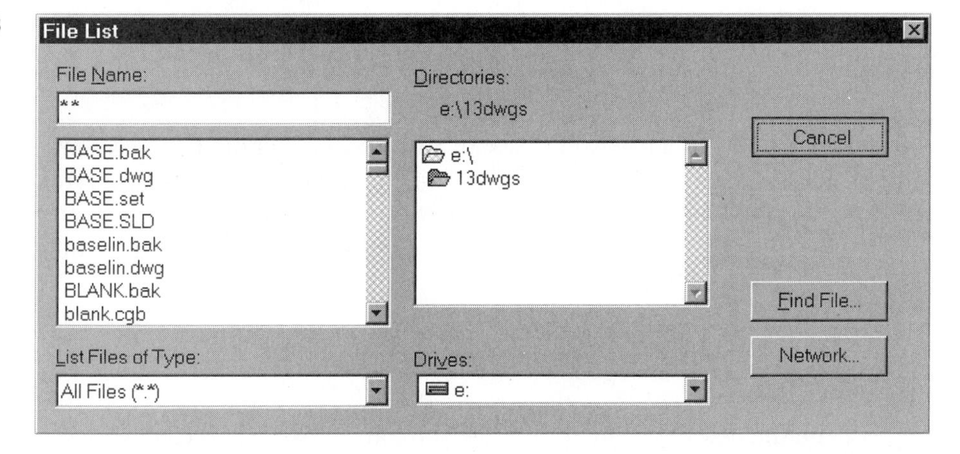

Figure 8.4

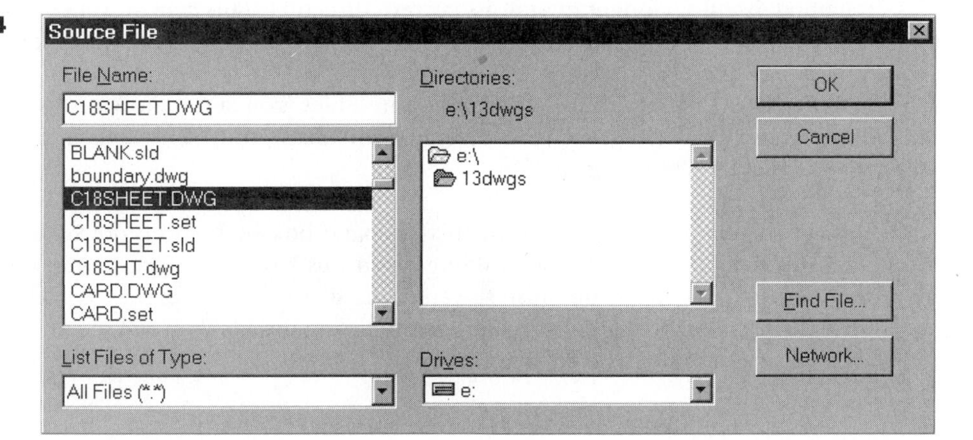

Figure 8.5

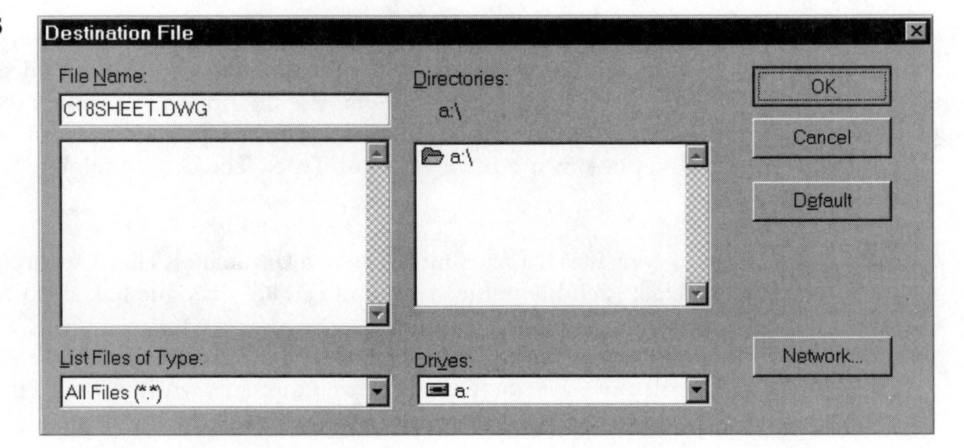

Select the location where you want to place the copy of the file. You can select amongst any of the drives or directories available on the computer. AutoCAD will not allow you to overwrite the file back in its original directory. **This is the best way to copy a file to a floppy disk from within AutoCAD.**

Rename file... allows you to assign a new name to a file that exists on the computer. When **Xrefs** are used, files can have complex relationships, so be careful when renaming files. You may need to provide a new path to the file if it is used as an **Xref** in another drawing. This selection will open the "Old File Name" dialogue box, resembling the other dialogue boxes above.

Select the file you wish to rename. Highlight the file and pick **OK**. This will open the "New File Name" dialogue box.

Create the new file name for the file you have chosen. This can be in the same directory or in another directory, or even on another drive. Remember that when a file is renamed, it will no longer exist in its current directory. This may prevent you from loading the **.SET** and **.SLD** files created and used in the Eagle Point Project Manager.

Delete file... allows you to remove unwanted files from a directory or disk. More than one file can be highlighted and deleted with this option. This opens the "File(s) to Delete" dialogue box (Figure 8.6).

You can highlight multiple files in this dialogue box by holding down the **<Ctrl>** key when picking from the file list. Holding down this key is not necessary in the DOS version of AutoCAD. After you finish selecting the files to delete, pick **OK**. This will bring up a warning box asking if you really want to delete the file. You will see a separate warning box for each file selected (Figure 8.7).

Figure 8.6

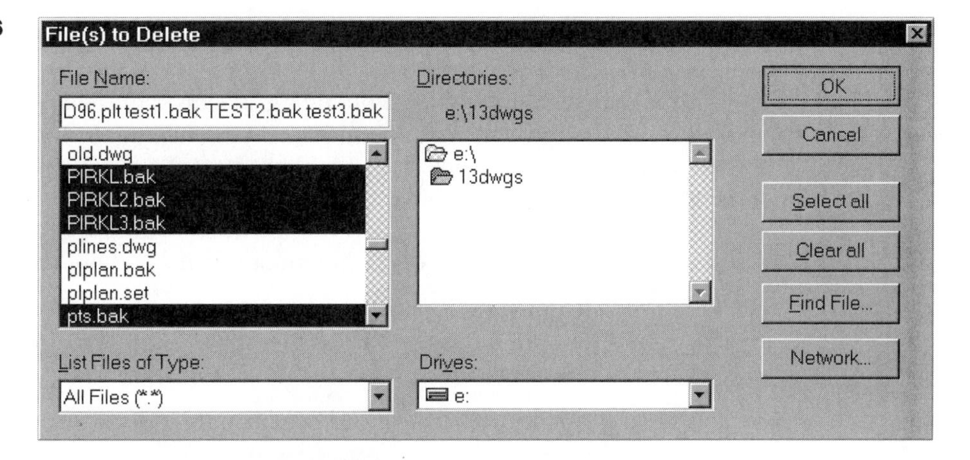

Figure 8.7

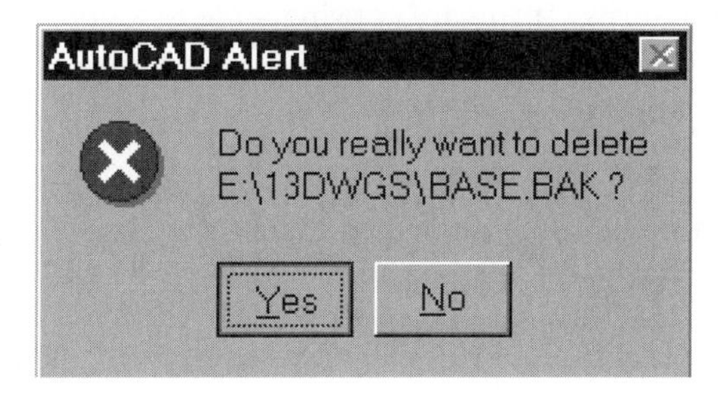

This will give you an opportunity to double check before the deletion will proceed.

Unlock file... allows you to choose a file to unlock. This feature allows you to remove a drawing lock that may have been placed on a file in a network system or when File Locking has been enabled in the AutoCAD configuration file. Lock files are normally attached to drawing files to prevent more than one designer working with the file at one time. A lock file is normally deleted when a drawing is closed normally. When AutoCAD terminates unexpectedly (crashes) the drawing lock file may not be removed properly. This option will remove the drawing lock—named with the extension **DWK**. You can also use the **Delete file...** option to remove a drawing lock file. This option opens the "File(s) to Unlock" dialogue box.

Select the file(s) that you wish to unlock, then pick **OK**.

If you prefer to perform file management chores outside of AutoCAD, there are several tools at your disposal. The tools you can use for copying files will depend on the operating system you are working in.

Windows 95 and Windows NT

The best tool for file management for these operating systems is the Windows Explorer. The Explorer allows you to navigate through a map of your computer drives and directories. Once you locate the file to copy, then highlight it. Once the file is highlighted, you have three ways to copy it (Figure 8.8).

1. Press the right mouse button to open the options available there (Figure 8.9).

   Select the **Copy** function, then move the highlight to the drive and directory where you want the file copied. Press the right button again to open the options available (Figure 8.10).

   Use the **Paste** option to place the file into the target directory.

2. Pull down the **File** menu to use **Send To ▸**, allowing you to copy the selected file to the target drive. If you are copying a file to a floppy drive, this is probably the fastest option (Figure 8.11).

3. Pull down the **Edit** menu to use **Copy** to copy the file into the Windows Clipboard.

   Once you have located the destination drive and directory, use the **Paste** option to place the file in the destination.

Figure 8.8

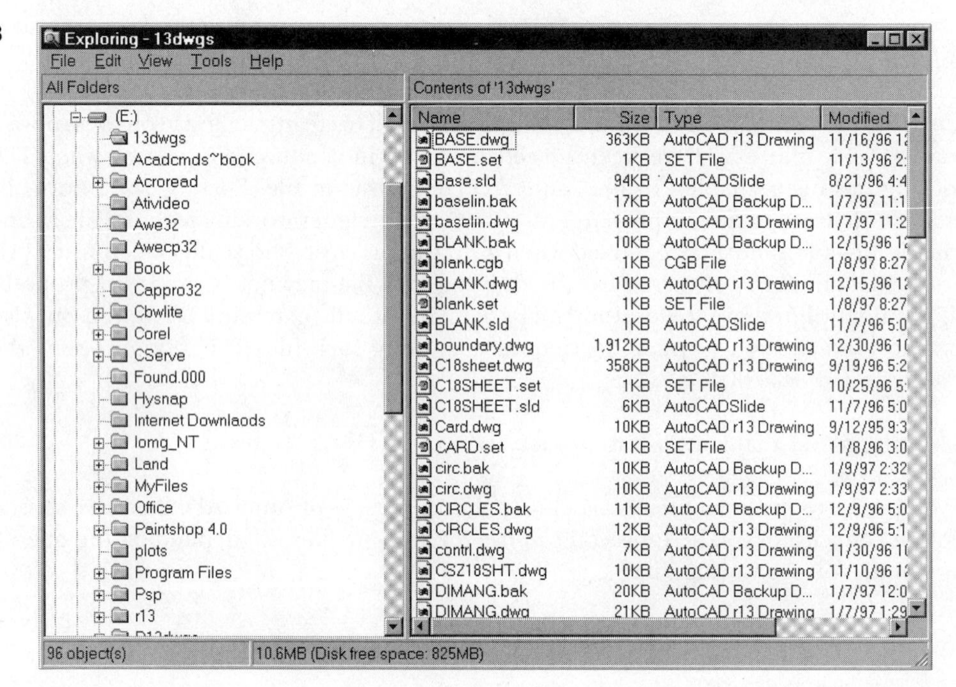

Figure 8.9

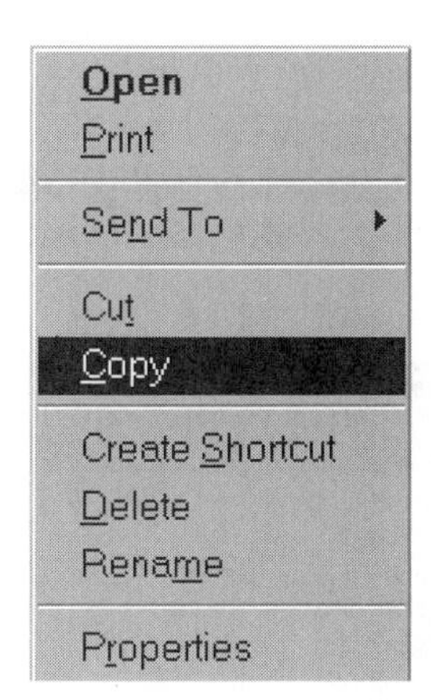

Figure 8.10

Figure 8.11

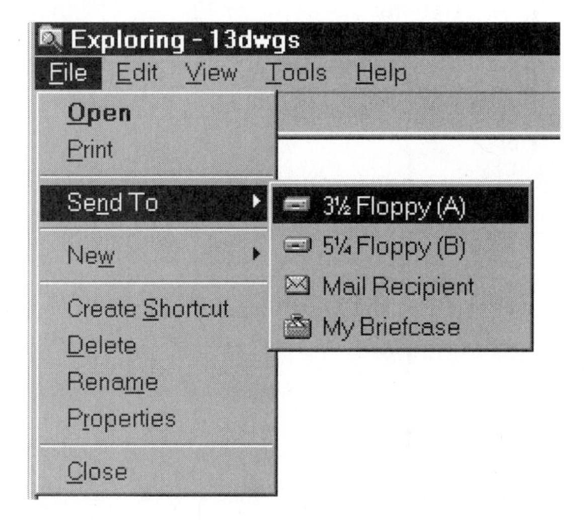

Windows 3.1

When working in Windows 3.1, use the File Manager to perform the **Copy** and **Paste** functions. Open the File Manager to search for the file you want to copy, then highlight it (Figure 8.12).

Figure 8.12

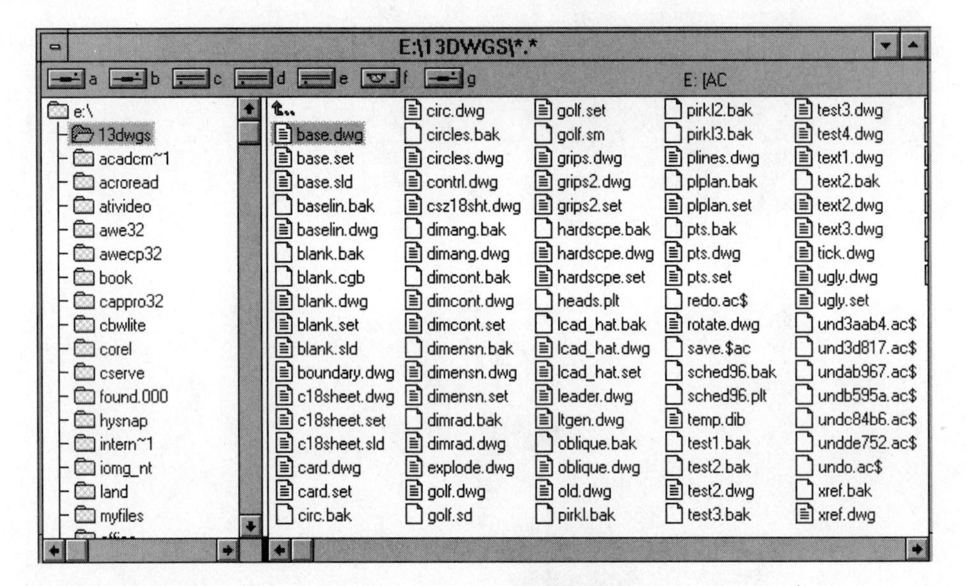

Move to the **File** pull-down menu, then pick **Copy** to copy the file to the desired drive and directory (Figure 8.13).

Figure 8.13

File Disk Tree View Options

Open	Enter
Move...	F7
Copy...	F8
Delete...	Del
Rename...	
Properties...	Alt+Enter
Run...	
Print...	
Associate...	
Create Directory...	
Search...	
Select Files...	
Exit	

This opens the "Copy" dialogue box. Enter the destination for the file (including the file name, drive and full path) in the **To:** text box (Figure 8.14).

The file will be created in the drive and directory you specify. Note that you can assign a new name during this copy process in Windows 3.1.

DOS

When working in DOS, you can use the **COPY** command to create a copy of the file in any location you choose. Use the syntax:

COPY <source file name> <destination file name>

When copying a file in this manner, remember:

1. You can leave off the full path name to the source file if you are currently working in that directory (i.e., within the AutoCAD drawing directory).
2. You must include the full path (including drive, path and full file name) when specifying the destination file name.
3. You can assign a new file name for the destination file name.
4. Remember to use a single space between the command **COPY**, the source file name and the destination file name.

Auto Save

Auto Save is a feature that can be activated with the **Configure** command. This option sets a timer to automatically back up your working drawing at a time interval you set. A standard value might be ten to 20 minutes. This is a way of preventing a program crash or system failure from destroying many hours (or days) of work. It is a convenient feature that is easy to set and easy to use. The back-up file created will usually have a name

Figure 8.14

like **AUTO.SV$**. This file must be renamed as a **.DWG** file before it can be opened and used. To activate Auto Save, use the following procedure:

```
Command:
```

Type Config or pick **Configure** under the **Options** pull-down menu. After displaying the current configuration settings, AutoCAD will display these options:

```
Configuration menu
0. Exit to drawing editor
1. Show current configuration
2. Allow detailed configuration
3. Configure video display
4. Configure digitizer
5. Configure plotter
6. Configure system console
7. Configure operating parameters
Enter selection <0>:
```

Type **7** to configure the operating parameters.

```
Configure operating parameters

0. Exit to configuration menu
1. Alarm on error
2. Initial drawing setup
3. Default plot file name
4. Plot spooler directory
5. Placement of temporary files
6. Network node name
7. Automatic-save feature
8. Speller dialect
9. Full-time CRC validation
10. Automatic Audit after DXFIN, or DXBIN
11. Login name
12. File locking
13. Authorization
14. Use long file names

Enter selection <0>:
```

Type **7** to set the Automatic-save feature.

```
Interval between automatic saves, in minutes
(1 - 600, or 0 for no automatic saves) <0>:
```

Type in the desired interval. The recommended value is between 10 and 20 minutes.

```
Enter filename for automatic saves
<e:\r13\auto.sv$>:
```

Enter the desired drive, path and name for the automatic save file name. Press **<Enter>** to return to the Configuration Menu, then **0** to return to the drawing editor.

Remember that this feature only protects your current drawing against lost work during a short-term crash of the AutoCAD program or of the computer system. This is **not** a substitute for a complete back-up system to protect all of your files and all of your programs.

Backing Up Files

When your business relies on computers, you place your firm's economic life in the hands of small mechanical devices that are prone to many forms of damage and failure. The AutoCAD drawings that are stored on your hard disks represent assets which may be valued at hundreds or thousands of dollars. The software you buy (actually lease) can also represent a considerable investment. Leaving these assets unprotected is at best optimistic and at worst suicidal. Businesses that rely on computers simply **must** back up their work files and programs to protect their investment. Fortunately, backing up data has never been easier, cheaper or allowed so many options from which to choose.

Types of Back-Ups

There are three types of back-ups that you can create. You should consider using some variation of all three types.

- A complete system back-up which contains all of your office programs, settings, drivers and program configurations. This type of back-up need only be performed when additional programs are added, new hardware is installed or when a major change takes place in your operating system. This type of back-up should be stored off-site in a secure, fire-proof location. Many businesses use a safe-deposit box in a nearby bank.
- A complete back-up of all of the data files used for projects—including drawing files, correspondence, word processing files, project databases, spreadsheets, etc. These files are typically not stored with the program applications, but stored by client and project. This type of back-up could be performed on a daily or weekly basis, depending on what other types of back-ups were being performed.
- An incremental back-up of only those files changed since the previous back-up was performed. Typically, this type of back-up is performed daily.

If your computer system consists of individual non-networked computers, each station should have a complete system back-up stored off-site, and a separate back-up strategy for program files. This might consist of a weekly back-up of all project files, and a nightly incremental back-up of all files changed during that working day. This allows you to restore a complete work station if it is stolen, damaged, or if a hard disk fails. A complete system can be restored with current data up until the last incremental back-up. If done properly, a maximum of one day's work could be lost from a work station.

If you work on a network (with or without a server), the individual work stations (and the server) should be backed up to make a safe copy of all of the office programs, settings, drivers and operating systems. As before, this system-wide back-up should be stored off-site. This type of back-up is only needed when hardware is added, new programs installed, or a change is made to the operating system.

A separate back-up should be done with the drawing and data files for all of the projects stored on the system. An incremental back-up of the project data should be performed daily, while a full back-up of all of the project data should be performed each week. A network system makes it very convenient to perform back-ups since the back-up itself can be performed by a single computer that can save files from all of the other computers on the network. If the system has a server which stores all of the project data files, the back-up can be preformed on the server hard drive(s) while the other work stations can be shut down for the night.

When devising a strategy for daily and weekly back-ups of project data, you can choose from several options:

- Full back-ups weekly (Friday) on a single tape or cartridge; incremental back-ups nightly using the same tape or cartridge. The nightly back-up overwrites the previous night's back-up. The weekly back-up could be stored off-site.
- Full back-ups nightly; a different tape or cartridge is used on each day of the week. The nightly back-up overwrites the previous week's backup. Friday's tape or cartridge is taken off-site.
- Use the same tape or cartridge to perform daily and weekly back-ups. Remove the tape or cartridge only when it becomes worn and unreliable. (When you discover this, it may be too late!)

Whatever your strategy, follow the same guidelines as with file organization: **pre-planning**, **standardization** and **good record-keeping**. These management practices may seem unrelated to AutoCAD and LANDCADD, but they are essential to keep a design business operating in the long run.

Back-Up Media

Back-up copies can be created on individual floppy disks, on a tape drive with high capacity tape cartridges, or on a special disk drive with high-capacity disks or cartridges.

Floppy disks are the traditional method for backing up data, but are no longer satisfactory for that purpose due to the increased size of drawing and data files. Even with file compression (a potentially risky choice with CAD files) a high-density floppy disk might only hold one or two drawing files. The number of floppy disks needed to back up an entire CAD system could number in the thousands.

Tape cartridges are used for internal or external tape drives which can hold up 350 megabytes for some styles, while others hold 1.5 gigabytes or more. When very large hard drives are backed up, it will be necessary to use multiple tapes for a complete back-up. Tape drives often come with software which can be scheduled to back up data at night or even in the background when other programs are in use.

Portable, removable disk drives made by such companies as Iomega and SyQuest also provide an excellent back-up option. These drives use removable disks that can hold up to a gigabyte of information. They are extremely fast (nearly as fast as a standard hard disk) and can be used to restore data quickly. Because these disks provide non-linear storage of data, you can find lost files very quickly and restore them to your system without waiting for the tape drive to wind through to the location of your file, then play it back to restore it to your computer.

chapter

Creating Drawing Output

Creating hard copies of drawings from AutoCAD requires that some basic configuration chores be completed:

1. AutoCAD must be properly configured for all of the output devices which are connected to the computer system. AutoCAD supplies drivers for many different output devices which can be selected during the **Configure** command. Each device must be defined and configured separately. You can also configure AutoCAD for plotters that are not connected to your system. If you use a plotting service with a particular brand and model number, you can configure AutoCAD for that plotter. This is useful when you want to plot to a file, then send the plot file to the plotting service at a later time.
2. The network and operating system must be set up so that AutoCAD can plot to the output devices. AutoCAD running in Windows 95 and Windows NT requires special settings and a plotting batch file to work properly. See Appendix 2 **Plotting in Windows 95 and Windows NT**. Other network operating systems may require separate configuration.
3. The output devices must be functioning properly, connected to the computer, and loaded with paper and ink or toner.

Creating hard copy output in AutoCAD is usually done through use of the **Plot** command.

The PLOT command

`Command:`
Type **Plot** or pick **Print** from the **File** pull-down menu. This opens the "Plot Configuration" dialogue box. If this dialogue box fails to open, reset the AutoCAD system variable CMDDIA. When this system variable is set to **0**, the plotting process takes place from the command line.

Once this dialogue is on the screen, set the various controls to produce the plot you want (Figure 9.1).

Device and Default Selection... This option sets the output device which you want to use for your plot. You can connect laser printers, ink jet plotters, pen plotters, and desktop color printers to your system and produce AutoCAD output. Since each device can be configured in different ways to handle different colors and line-weight hierarchies, you can create different options for multiple devices and multiple device set-ups.

If different clients or consultants use different colors to represent different line weights, you can maintain separate *plotter configuration parameter* files (with the extension **.PCP**) for each client. When you are ready to plot a specific project, you can retrieve the settings from the file rather than setting them manually in the "Plot Configuration" dialogue box. A plotter configuration file can be created and stored using the **Save Defaults to File...** button, and retrieved using the **Get Defaults From File...** button.

The **Show Device Requirements...** button allows you to see the "time-out" value for the plotting device. A plotting device has an amount of memory storage called a **buffer**. Once the buffer fills, it stops accepting information from AutoCAD. This value represents the time interval that AutoCAD will hold plotting information in computer memory while waiting for the plotter to empty its buffer by plotting the information already received. When this value is exceeded, AutoCAD asks you if you want to abort the plot. If your plots are very large or your plotter is slow, you may want to increase the value shown here.

Figure 9.1

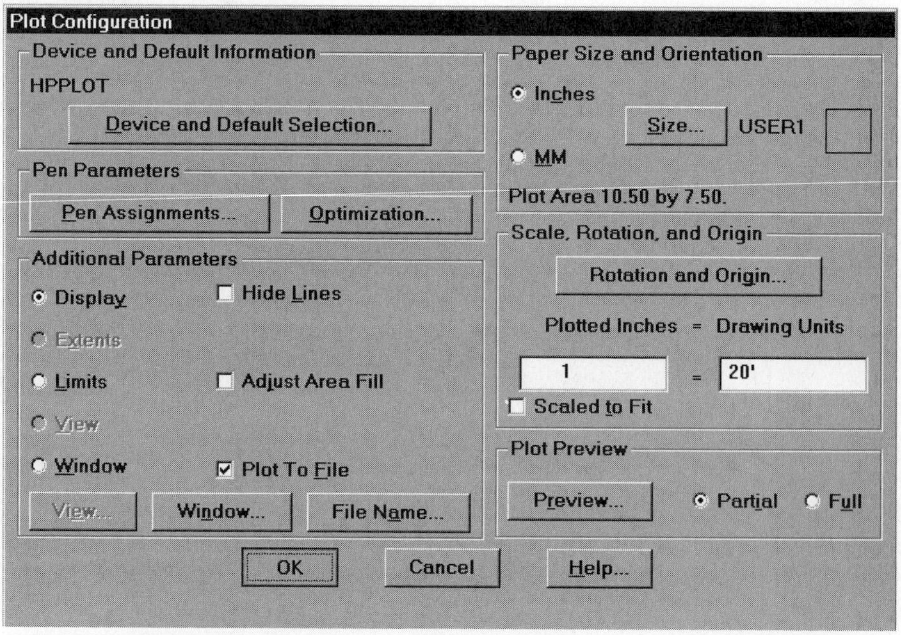

To increase the "time-out" value, select the **Change Device Requirements...** button to set the value desired.

The **Pen Parameters** section is used to set how the plotter pens are assigned to the colors on the screen, and special instructions for the plotter pen movement.

The **Pen Assignments...** button is used to open the "Pen Assignments" dialogue box.

Use this dialogue box to set the pen number, linetype, pen speed and pen width assigned to each color used in your drawings. The manner in which you use the flexibility in pen assignments will depend on whether you use colors in your plots, the number of pens available on your plotter, and your desired line-weight hierarchy.

Whether you use colors in your plots will depend on whether you make plots for blueprint reproduction or for presentation work. Although colors do not translate well in reproducibles, gray tones (AutoCAD Color Index numbers 250 through 255) can be used in reproducible plots to provide a "half-tone" effect for base plan information, building footprints, etc. Color inkjet plotters allow the use of 256 "pens" (the plotters contain three color ink cartridges, not pens)—one for each color available in AutoCAD. Pen plotters allow only the number of physical pens present on the plotter. Remember that one pen number can have more than one color assigned to it.

Linetype settings are used to assign a color to a specific linetype in the plotter itself. This setting is normally left at **0** to allow AutoCAD to control the linetype in the drawing rather than have the plotter or printer control the linetype. The **Feature Legend...** button allows you to see which printer or plotter linetypes are available.

Pen speed is set when pen plotters are used. If the pen speed is set too high, the pens may skip or produce fuzzy lines. When modern ink-jet plotters are used, these settings are normally irrelevant.

Pen widths are set to provide you with a hierarchy of very narrow to very wide lines. The effect of these widths is to provide more weight and depth to some lines, and less to others. The result is to control the emphasis on specific areas or items in the drawing.

Remember that colors are assigned to pen numbers and pen numbers are assigned line widths. The final result is that when working with reproducible plots, screen colors translate into line widths in your plotted drawings. The "Pen Assignments" dialogue box allows you to control that translation.

The **Optimization...** button opens the "Optimizing Pen Motion" dialogue box.

The features set here affect pen plotters and their pen movements in creating the plotted drawing. The optimization settings here are normally set by the plotter driver supplied by AutoCAD or the manufacturer of the plotter.

The **Paper Size and Orientation** settings allow you to choose between the various paper sizes allowed by the plotter you have selected. The radio buttons allow you to specify the size of the drawing in millimeters or inches, and the **Size...** button opens the "Paper Size" dialogue box (Figure 9.2).

You can choose between the various paper sizes allowed by the plotter, and define your own sheet sizes, so long as they do not exceed the maximum allowed by the plotter.

The **Additional Parameters** section actually allows you to specify what will be plotted, and the manner that the plot will proceed.

Display will cause the screen display to be plotted. The centering and enlargement of the image on screen will be reflected in the plot produced.

Extents will plot the extents (outer edges) of the objects included in the drawing, no matter how far they are away from the main body of the drawing. Use **Zoom/Extents** to see the results of this option before you plot.

Limits will plot the area you specified with the **Limits** command.

View will plot a named view created with the **View** command. If there are no views saved with the file, this option will not be available. Pick a view to plot with the **View...** button at the bottom of the "Plot Configuration" dialogue box to activate this option. Whichever view is selected will be plotted.

Window will allow you to plot a specific portion of the drawing that you can define by a "rubber band box" or with coordinates. Selecting this radio button will plot the window

Figure 9.2

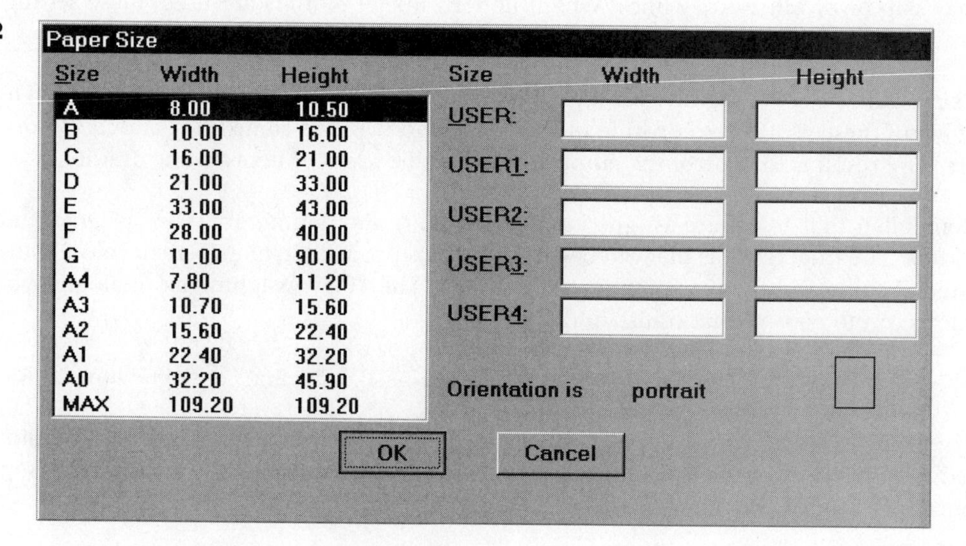

which is specified with the **Window...** button at the bottom of the dialogue box. If you do not specify the corners of the window to plot, the window last defined will be plotted.

Hide Lines allows you to specify whether or not hidden lines in 3-D drawings will be hidden or plotted.

Adjust Area Fill causes AutoCAD to adjust the width of filled areas so that the pen width does not increase the outer edge size of the object. This adjustment is seldom needed in architectural plots.

Plot to File is used when you want to create a file version of a plot rather than sending the information directly to the plotter. The plot file will have the extension **PLT**, and will contain the plotted information specifically for the type of plotter which was specified when the plot was done. This can be useful if you want to take the plot file to a plotter service or another device for plotting later. Plotting to a file is generally required when plotting in Windows 95 or Windows NT. See Appendix **Plotting in Windows 95 and Windows NT**.

The **Scale, Rotation and Origin** section sets the plotting scale, and the relationship of the plotted area to the paper available. To set the plotting scale, specify the relationship between the **Plotted Inches** to the **Drawing Units**. The drawing units can be expressed as inches or feet. A 1/20 drawing scale could be expressed **1" = 20'** or **1" = 240"**. If the **Scaled to Fit** toggle is active, the drawing scale will be calculated by AutoCAD to fill the available paper size, regardless of the awkwardness of the scale.

The **Rotation and Origin...**button opens the "Plot Rotation and Origin" dialogue box (Figure 9.3).

Figure 9.3

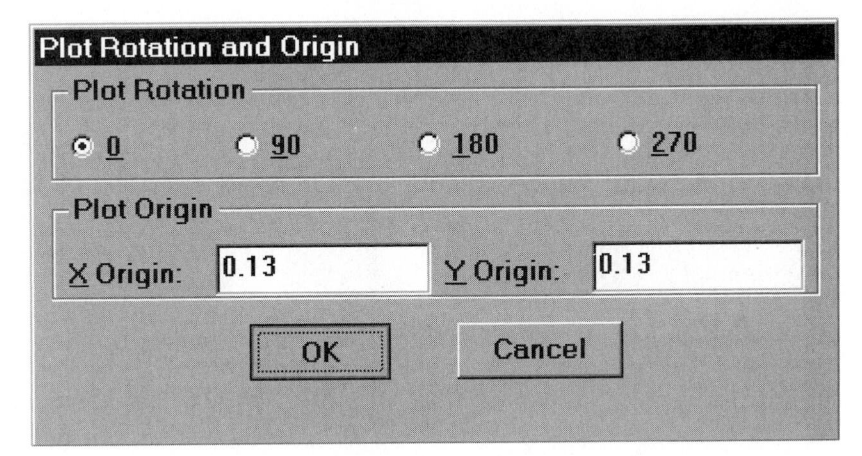

Set the rotation angle of the plot compared to the paper at 90° increments. The **Plot Origin** refers to the distance from the device's starting position to the corner of the plot. The starting position of plotters is their lower left corner in landscape orientation, while the starting position of printers is generally their upper left corner in portrait orientation.

The **Plot Preview** area actually gives you a preview of the appearance of your plotted area related to the paper you will be plotting on. If the plotting scale is incorrect for the paper size, you can see it before you actually waste the time and materials necessary to create the plot. The preview can be set as a **Partial** or **Full** preview (Figure 9.4).

A **Partial Preview** gives a generalized view of the plot related to the paper size. This is an excellent option if the plotted image is very large or complex, and would take an excessively long time to regenerate in the **Full Preview** mode.

If the plotted drawing is too small or too large for the paper size, a warning message is displayed below. A red box displays the paper edge while a blue box represents the plotted "Effective area." The paper size is defined in the **Paper Size and Orientation** area, while the "Effective area" is defined in the **Additional Parameters** and plotting scale areas.

Figure 9.4

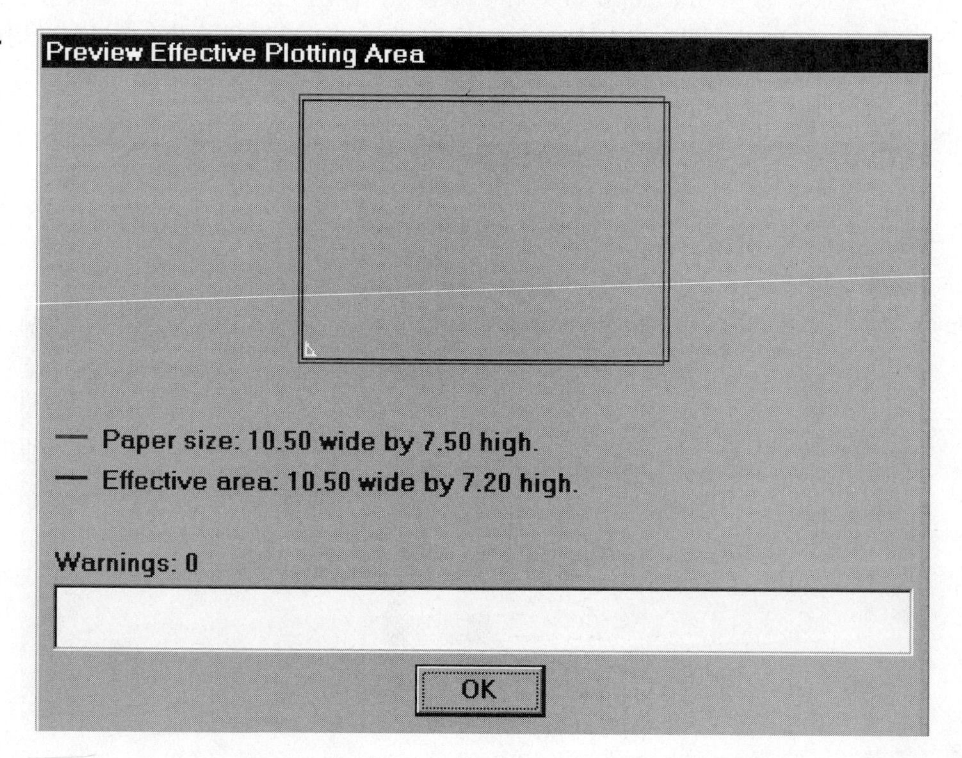

A **Full Preview** is used when you want to examine the plotted "Effective area" more closely to ascertain whether or not specific areas will be included in the plot. Use the Pan and Zoom features to move around in the window to check on specific areas. The Pan and Zoom controls do not affect the outcome of the plotted drawing (Figure 9.5).

In the design and architectural arena, there are three primary types of plotter devices in popular use:

Inkjet plotters plot with a moving head that disperses ink through a series of tiny holes. The ink is supplied in cartridges, typically one for black ink and the another for cyan, yellow, and magenta inks. The plots are actually *raster* images composed of tiny dots (from 300 to over 700 per inch), but the resolution is so fine that the images look smooth and even. Inkjet plotters are very fast, quiet and reliable. Their prices are falling rapidly; E-sized roll feed color plotters are now available for about $7000, while E-sized single sheet black and white plotters are available for less than $3000. The ease of maintenance, speed of plots and relatively low cost have made these devices the preferred choice in many design firms.

Pen plotters use individual pens to produce plots. The pens are relatively expensive (up to $7 each for quality pens), and can clog easily if not maintained. Pen plotters produce continuous smooth lines and circles, but will show double strike lines (like those found in some extension lines in dimensioning work) differently than single stroke lines. Pen

Figure 9.5

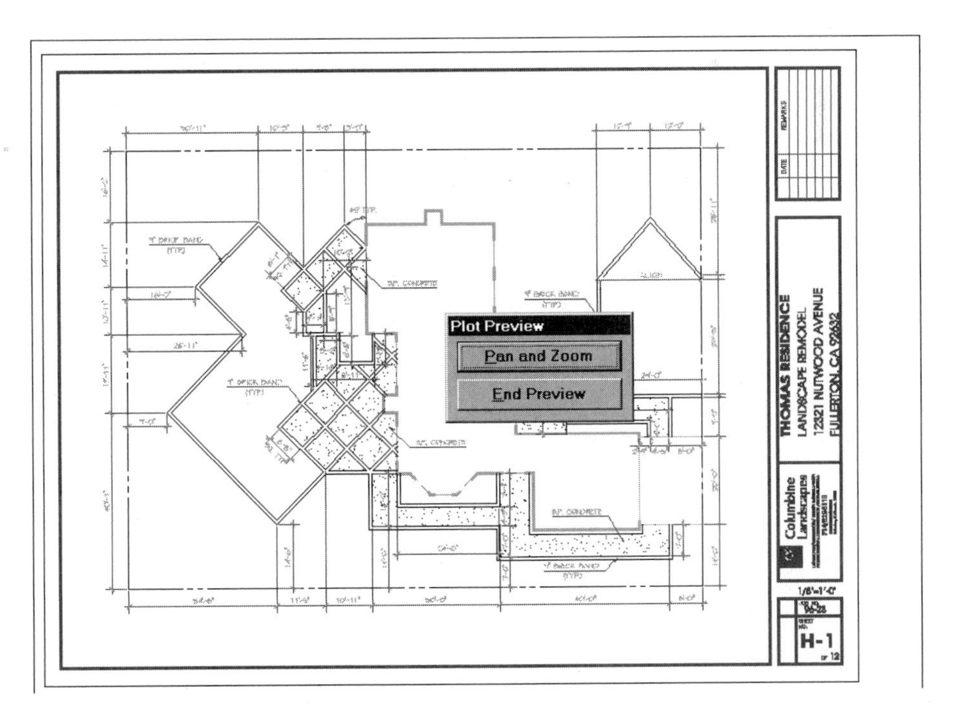

plotters are generally quite slow, requiring 45 minutes or more for a complex E-sized plot (a "final" quality comparable inkjet plot might require 10 minutes). In addition, they can be noisy, and their pens difficult to maintain. Pen plotters are beginning to become the equivalent of dot-matrix printers: noisy curiosities found unsuitable for most business purposes.

Laser printers are excellent for producing fast, low cost check plots. The standard office laser printer can be used to produce enlargements or cut-sheet drawings that are very useful in examining drawings during their creation and revision phases. In addition, small size plots can be used to FAX details to clients or consultants quickly and efficiently. An office quality laser printer can be purchased for $1500 or less, and can serve as a printer for normal office use, as well as an AutoCAD output device.

Plotting to a File

Some CAD designers prefer to send their plots to file, then actually plot the file at a later time. These plot files typically are built with specific instructions for the brand and model of plotter, and contain data that cannot be read in AutoCAD. Unlike standard drawing files, plot files are not opened in AutoCAD, and do not need to be accompanied by special font files, **Xref** files, or any other third party software. Plot files contain only instructions about how to create the specific plot. Plot files are ideal to send to plotting services who have compatible plotting machines to those configured on your computer.

chapter

Preparing LANDCADD for the Thomas Residence Landscape Remodel Project

Before we actually begin building drawings for the landscape project, we will first complete some preparatory work to make our project easier. The settings and adjustments we make here can be performed once, then applied to our other drawings later. This principle underlies most CAD operations: DO IT ONCE. After these preparatory steps are completed, they never have to be performed again. We simply use them over and over in copies of our drawings.

The first "drawing" is really a blank sheet which contains invisible settings. In the blank sheet (**BLANK.DWG**) we will complete the following steps:

- Begin a new drawing using the Project Manager
- Modify major settings for **UNITS, LIMITS, GRID** and **SNAP**
- Change minor settings for **VIEWRES, BLIPMODE** and **UCSICON**
- Set Eagle Point project settings for **SCALE** and **TEXT**
- Save the Eagle Point project settings to the Project Settings Library

Using the Project Manager to Begin a New Drawing

The Eagle Point Project Manager is intended to provide a convenient way to access existing drawings, load Eagle Point and LANDCADD menus and create new drawings using existing prototype drawings. Using this interface properly makes loading and running LANDCADD much simpler.

Hint: The Project Manager is always present when LANDCADD is started. If you choose to close the Project Manager rather than to load a drawing shown there, the process to load LANDCADD menus becomes more complicated. In addition, LANDCADD modules often fail to work properly unless Project Manager is used. Unless there is a compelling reason not to do so, always use **Open** in Project Manager to begin running LANDCADD.

Building a Blank Sheet

One of the ways to make Project Manager easier to use and more convenient to the drawing process is to create a blank drawing which contains settings that you expect to use in most all of your drawings. Your most frequent settings for UNITS, LIMITS, SNAP, and GRIDS can be stored here, along with other AutoCAD features such as BLIPMODE, UCSICON and VIEWRES. See your AutoCAD documentation or reference books for complete explanations of these settings. In addition, you can store your most frequently used layers, linetypes and blocks in this blank sheet.

To build a blank sheet follow these steps:

Open LANDCADD normally. After the program opens and AutoCAD begins, the Project Manager appears in the drawing area (Figure 10.1).

Build a new project by picking **New . . .**. This opens the "New Project" dialogue box (Figure 10.2).

Enter a description (which will appear in the Project Manager description list) in the **Description:** text box. Type **Blank Sheet**.

Enter a drawing name, with the complete path and file extension (**.DWG**), in the **Drawing Name . . .** text box.

Specify the location (drive and directory) where you store your LANDCADD drawings. Do not click on the **Drawing Name . . .** button in the dialogue box. This option is only used if you wish to attach an existing drawing to a project or to overwrite an existing AutoCAD or LANDCADD drawing with a new one.

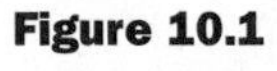
Figure 10.1

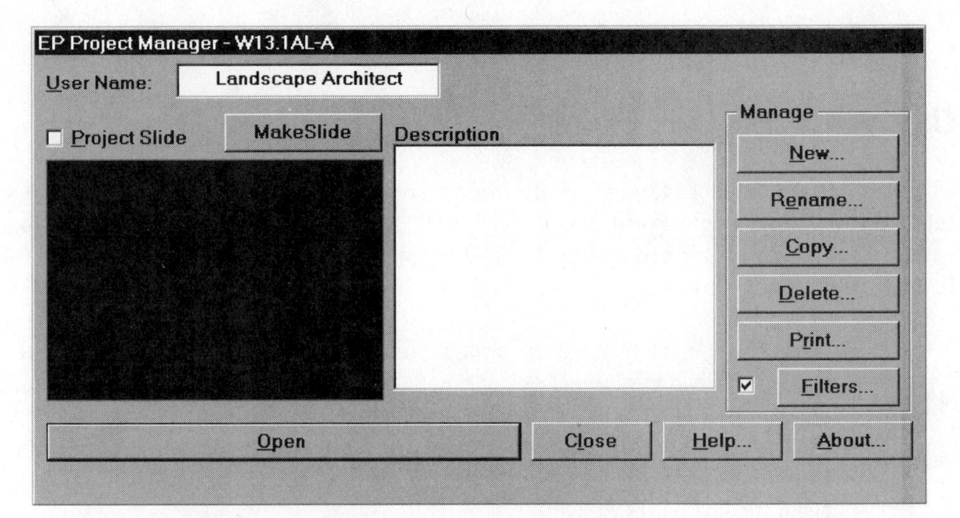

Figure 10.2

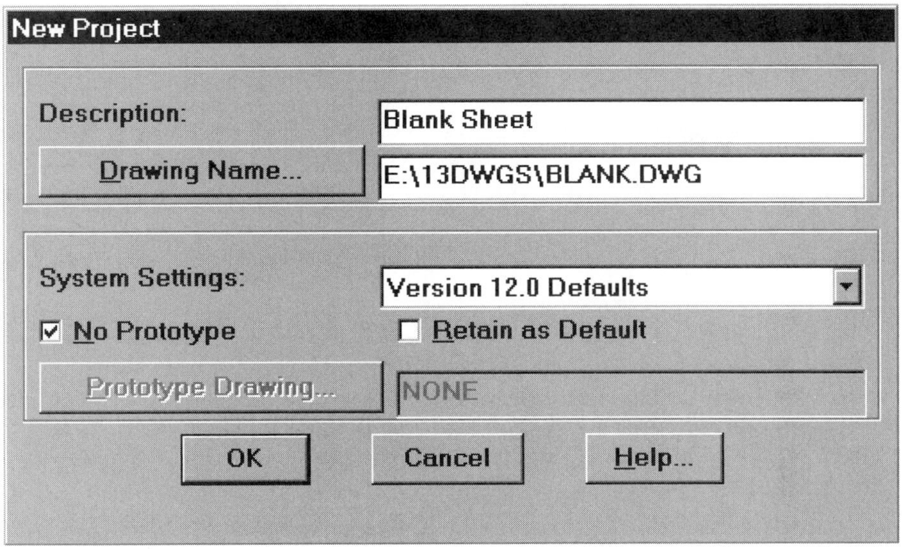

New Project
Description: Blank Sheet
Drawing Name... E:\13DWGS\BLANK.DWG
System Settings: Version 12.0 Defaults
No Prototype
Retain as Default
Prototype Drawing... NONE
OK
Cancel
Help...

Click on **OK**. You are now returned to the Project Manager screen, with your new project listed in the **Description** list (Figure 10.3).

Pick **Open**. The Project Manager dialogue box disappears, and your AutoCAD drawing screen is now visible.

We are now ready to create the settings that will be used throughout the exercises in this book.

Figure 10.3

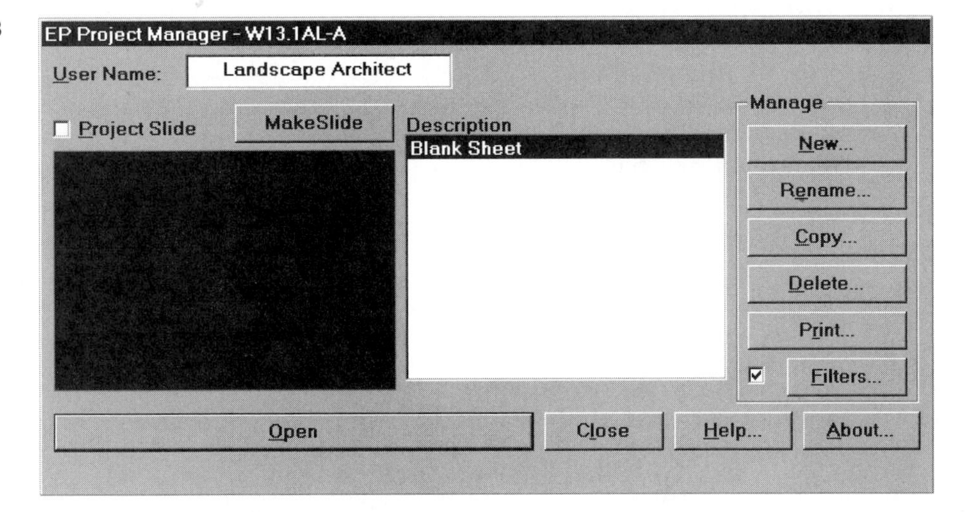

We will begin by establishing architectural, rather than decimal units. At the command prompt:

```
Command:
```

Type **UNITS**.

```
Report formats:         (Examples)
1.  Scientific          1.55E+01
2.  Decimal             15.50
3.  Engineering         1'-3.50"
4.  Architectural       1'-3 1/2"
5.  Fractional          15 1/2
```

With the exception of Engineering and Architectural formats, these formats can be used with any basic unit of measurement. For example, Decimal mode is perfect for metric units as well as decimal English units.

```
Enter choice, 1 to 5 <2>:
```

Type **4** to use architectural units.

```
Denominator of smallest fraction to display
(1, 2, 4, 8, 16, 32, 64, 128, or 256) <64>:
```

Type **16** to establish 1/16" as the smallest fraction of an inch to display.

```
Systems of angle measure:     (Examples)
1.  Decimal degrees           45.0000
2.  Degrees/minutes/seconds   45d0'0"
3.  Grads                     50.0000g
4.  Radians                   0.7854r
5.  Surveyor's units          N 45d0'0" E
Enter choice, 1 to 5 <5>:
```

Type **1** to use decimal degrees.

```
Number of fractional places for display of angles (0 to 8)
<4>:
```

Type **0** to use no decimal places for degree measurement.

```
Direction for angle 0:
East      3 o'clock  =    0
North    12 o'clock  =   90
West      9 o'clock  =  180
South     6 o'clock  =  270
Enter direction for angle 0 <0>:
```

Press **<Enter>** to accept this value. This way, 0 degrees is to the right of the screen.

`Do you want angles measured clockwise? <N>`
Press **<Enter>** to accept this.

The **UNITS** command establishes our system of measurement, along with the amount of accuracy we need while drawing. While these settings are appropriate for the exercises in this book, they may not serve your needs for much larger (or smaller) projects.

Next, we will establish the size of the blank drawing area.

`Command:`
Type **LIMITS**.

`Reset Model space limits:`
`ON/OFF/<Lower left corner> <0'-0",0'-0">:`
Press **<Enter>** to accept the lower left corner value.

`Upper right corner <0'-12",0'-9">:`
Type **250',185'**. Be sure to include the foot marks (') and the comma between the numbers. This represents the size of the area of our new drawing. Our screen will show a space 250' by 185'. This is an appropriate size for a small residential landscape.

This sets the outer limits of the drawing area, but this area is not displayed on the screen yet. We now need to zoom out to display the entire area.

`Command:`
Type **Z** (or ZOOM), then press **<Enter>**.

`ZOOM`
`All/Center/Dynamic/Extents/Left/Previous/Vmax/Window/<Scale(X/XP)>:`
Type **A** (or ALL) to zoom out to the limits of the drawing. You probably will not see any visible effects of this step until later.

`Regenerating drawing.`
The screen now displays the entire drawing area. If you move your cursor to the upper right corner of the screen, your coordinate dial will indicate the distance from the origin point in the lower left corner.

Next, we will add grids to the screen to more clearly indicate the size of the screen distances.

`Command:`
Type **GRID**.

Grid spacing(X) or ON/OFF/Snap/Aspect <0'-0">:
Type **10'**. This places dots on the screen at 10 foot intervals.

We will now set the snap value to restrict cursor movement to the nearest foot when SNAP is active.

Command:
Type **SNAP**.

Snap spacing or ON/OFF/Aspect/Rotate/Style <0'-1">:
Type **1'**. This sets the cursor movement to the nearest 1 foot when SNAP is activated.

This completes the major settings for the blank drawing. We can now set some minor adjustments to the drawing screen which are appropriate to working on landscape plans. Since all of our work will be performed on 2-D plans initially, we do not need the UCS icon at the lower left corner. This icon reminds us that we are working in a plan view mode, which will be obvious here.

To remove this icon, at the command prompt:

Command:
Type **UCSICON**.

ON/OFF/All/Noorigin/ORigin <ON>:
Type **OFF**. The UCS icon is removed from the screen.

To prevent the small "+" objects from forming on the screen after drawing or erasing an object, we will turn off an AutoCAD setting called BLIPMODE.

Command:
Type **BLIPMODE**.

ON/OFF <On>: OFF
Finally, we can ensure that circles will appear round and lines will appear less jagged if we reset the AutoCAD setting **Viewres**. This setting controls how precise circles appear on screen. The default setting is 100, but we can increase the resolution of circles greatly if we change this value.

At the command prompt:

Command:
Type **VIEWRES**.

```
Do you want fast zooms? <Y>
```

Press **<Enter>** to accept this.

```
Enter circle zoom percent (1-20000) <100>:
```

Type **3000** to set this value higher. If the **VIEWRES** setting is too high, the speed of regenerating a drawing can be greatly affected.

```
Regenerating drawing.
```

Whatever is on the screen will be regenerated at the new view resolution.

Once these settings are complete, we can save the drawing to the hard disk.

```
Command:
```

Type **QS** (or QSAVE) to save the drawing.

```
QSAVE
```

Since the drawing has already been named, it is automatically saved under its current name and location (drive and directory).

We have now created a blank page with settings that will be useful in all of the drawings developed for the Thomas residence landscape remodel project.

Editing Project Settings

While most of the Eagle Point module settings operate without affecting the LANDCADD modules, some of the settings affect the defaults for text size. Unless settings are changed here, all text will default to **5"** regardless of changes made in LANDCADD or even AutoCAD. To modify these text defaults, we use the **Edit Project Settings . . .** selection in the **Project Manager** section of the **EP** pull-down menu.

Pull down the **EP** menu, then pick **Edit Project Settings . . .**. This will open the "Project Settings" dialogue box (Figure 10-4).

In the "Project Settings" dialogue box, the options are listed down the right side. The crucial settings are **Scales . . .** and **Text Size . . .**. These two factors combine to influence the setting of the default text height for plant labeling, pipe sizing, and other text functions.

Hint: It is important to remember that the settings created here relate to **DECIMAL UNITS,** and must be converted to **ARCHITECTURAL UNITS** to correctly build the desired settings. The **Scales ...** settings will be 12 times larger than normal. This is because the base unit in decimal units is 1 foot, while the base unit in architectural units is 1 inch. We must multiply the **Scales ...** value by 12 to obtain the correct result.

Figure 10.4

Project Settings

Current Values

```
Drawing Units_____________________________Feet
Angular Units_______Degrees, Minutes, Seconds
Area Units_________________________Square feet
Degree of Curvature Units________100.0000 Arc

Distance Precision___________________________4
North, East Coordinate Precision____________2
Elevation Precision__________________________3
Station Precision____________________________2
Area Precision_______________________________4
Angular Precision______________Nearest Second
```

Units... | Precision... | Format... | Scales... | Text Size... | Options... | Nodes DB... | Tablet...

OK | Cancel | Help...

Pick **Scales . . .** to open the "Scales" dialogue box (Figure 10.5).

The **Horizontal** scale refers to the plan view scale factor, while the **Vertical** scale refers to Z- axis scaling for 3-D functions used in terrain mapping and surveying.

For a 1/8"=1'-0" architectural scale drawing, we know that our scale converts to 1"=8'-0" or 1"=96". Therefore, for a 1/8"=1'-0" drawing, the scale factor will be **96**.

Set the **Horizontal** box to read **96.00000**.

IGNORE THE "ft" UNIT IN THIS DIALOGUE BOX. The "ft" unit relates to decimal scale and is not applicable to architectural scale. Leave **Vertical** scale set at **5.0**.

Figure 10.5

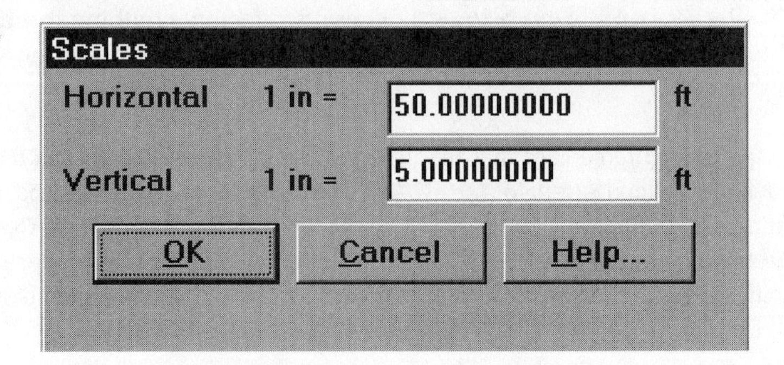

Select **OK** when finished (Figure 10.6).

Figure 10.6

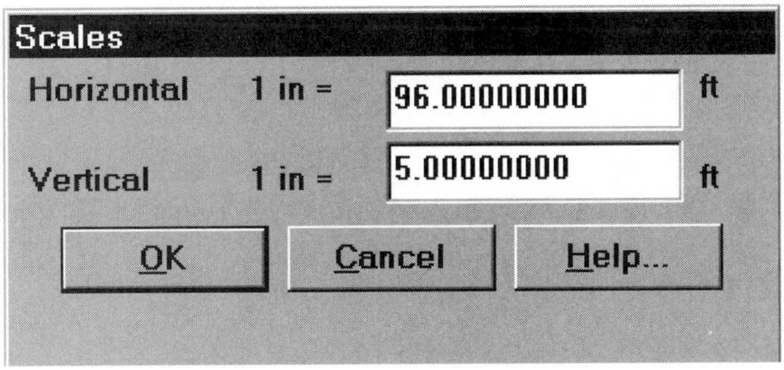

Scales
Horizontal
1 in =
96.00000000
ft
Vertical
1 in =
5.00000000
ft
OK
Cancel
Help...

With the "Scales" box set properly (correcting for the decimal scale), we can then set our text height to its correct value.

Pick **Text Size . . .** to bring up the "Text Size" dialogue box (Figure 10.7).

Figure 10.7

Text Size
0.12500000
Inches
100
Leroy
OK
Cancel
Help...

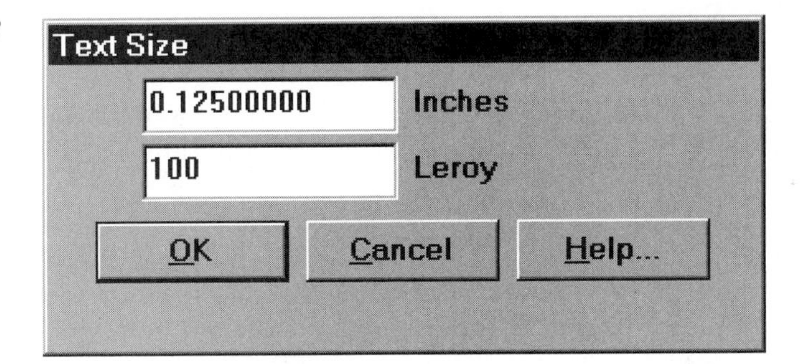

Set the text height to **.1250**. This creates text which will plot out at 1/8" high. Because we corrected for the decimal scale problem previously, we can enter the actual desired text height here.

REMEMBER: SET THE **SCALE** FOR DECIMAL UNITS CORRECTION, BUT MAKE THE TEXT HEIGHT VALUE READ ACTUAL (PLOTTED) SIZE. This will pertain to all architectural drawing scales from 1"=1" to 1"=100' or more.

Pick **OK** when finished.

For a 1/8 scale drawing, the "Project Settings" dialogue box should have entries that resemble those shown below (Figure 10.8).

Figure 10.8

Project Settings
Current Values

Horizontal Direction Format__________Bearing
Vertical Direction Format________Zenith Angle
Station Format___________________________1+00

Horizontal Plot Scale_______1 in = 96.0000 ft
Vertical Plot Scale__________1 in = 5.0000 ft

Text Size___________________________0.1250 in

Auto Plan Load_____________________________ON
Auto Menu Load_____________________________ON
Menu Style__________________________Swapping

Units... | Precision... | Format... | Scales... | Text Size... | Options... | Nodes DB... | Tablet...

OK | Cancel | Help...

Managing Project Settings

Once the project settings have been modified to suit the text and scaling for 1/8 scale drawings, we can save these settings to a library. The settings in the library can, in turn, be retrieved for any 1/8 scale drawings we create. This avoids repeating the steps listed above for each 1/8 scale drawing.

Hint: Each time you build a project at a different drawing scale, edit the Project Settings to meet the needs of the project, then save the settings to the Project Settings library. Be sure to use a descriptive name that includes the drawing scale. Storing these settings to the library means that you can simply retrieve them when you want to, rather than resetting them each time a project is begun.

To save a group of settings in a library, begin by using the **EP** pull-down (Figure 10.9).

Pick **Manage Prototype Settings...**.

This opens the "Prototype Settings Library" dialogue box. This box lists the existing settings prototypes which are included with Eagle Point and LANDCADD. These include **Version 12.0 Defaults** and **METRIC DEFAULTS** (Figure 10.10).

Pick **New . . .**.

This opens the "New Prototype Settings From Project Settings" dialogue box. This box displays any projects which have been created with Project Manager (**Blank Sheet** is the

Figure 10.9

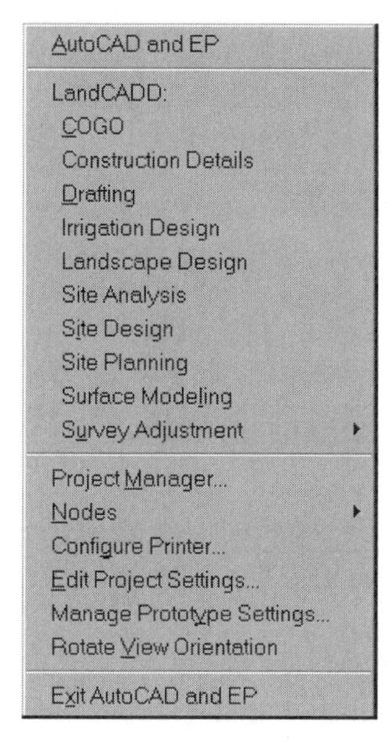

only project shown here). Move your cursor to highlight the project which contains the settings that you wish to save as a prototype for other projects (Figure 10.11).

In this case, we will use the project settings just created for the Blank Sheet. Since these settings would be suitable for all 1/8 scale drawings, we should save them with an appropriate name.

In the **Prototype Name:** text box, type **1/8 SCALE DRAWINGS**, then pick **OK** (Figure 10.12).

Figure 10.10

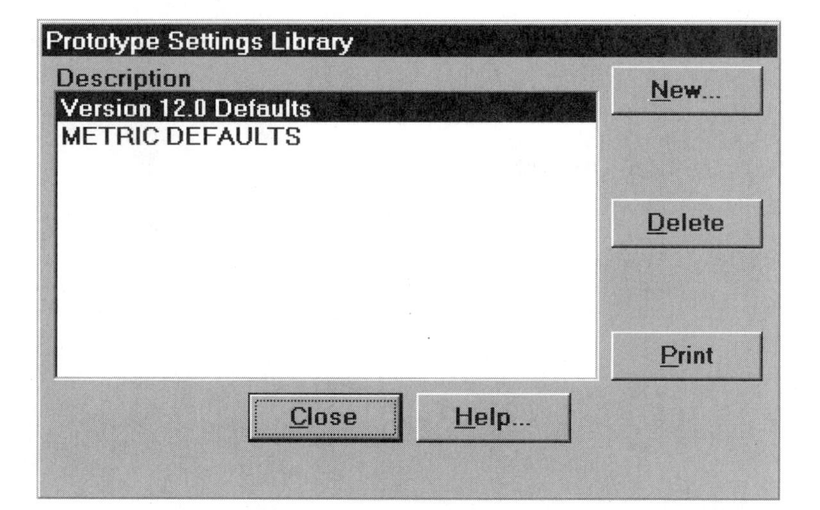

Figure 10.11

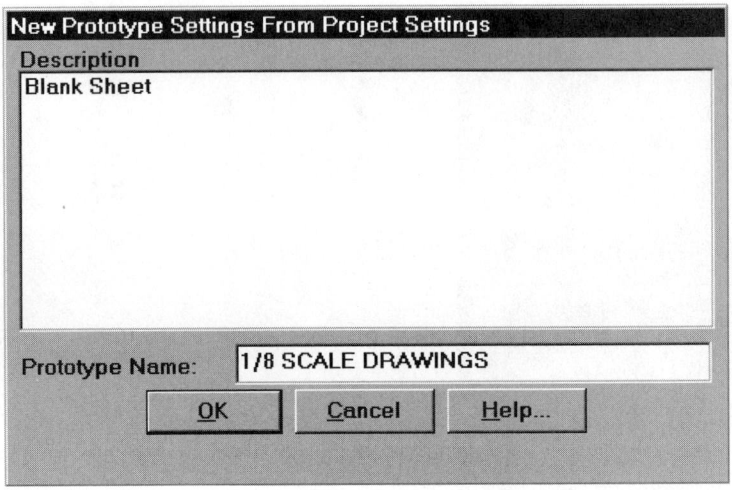

New Prototype Settings From Project Settings

Description

Blank Sheet

Prototype Name: 1/8 SCALE DRAWINGS

OK Cancel Help...

You are then returned to the "Project Settings Library" dialogue box with the new entry at the bottom. The **1/8 SCALE DRAWINGS** selection can now be used every time a new 1/8 scale project is started in the Project Manager.

Pick **Close** to exit the **Manage Prototype Settings...** option to return to the drawing screen.

Remember to create and store settings for other drawing scales that you normally use. This makes beginning new drawings very efficient, since the settings can be picked from a library rather than being created from scratch.

Figure 10.12

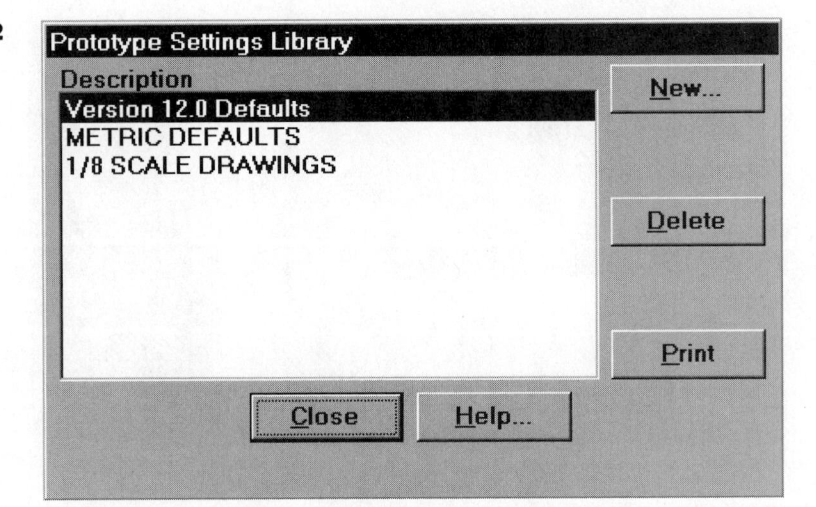

chapter

Drawing One—The Title Block

The first drawing we will create for the Thomas Residence landscape remodel project is the Title Block. For this drawing, we will complete the following steps:

- Create a new project in the Project Manager
- Use project settings from the Project Settings library
- Build the C18SHEET drawing to use as a prototype to create the remainder of the drawings
- Define and use three text styles
- Create a text block with displayed attributes to control the title block information
- Build a customized logo to use in the title block
- Insert the logo into the title block

Title Block Scaling and Prototype Status

The first task to begin the landscape drawings of the Thomas residence remodel project is to create a title block for a C-size sheet of paper scaled at 1/8" per foot. Architectural C-size paper is 18" x 24", so we need to create a trim line at 18" x 24" which corresponds to those dimensions at 1/8" scale. Since we know that 1/8" per foot scale corresponds to 1" per 8', this means that a C-size page would really be 144' (18 x 8') by 192' (24 x 8') at actual size. If our title block has a 1/2" margin around the outside, the working area inside will be reduced to 17" by 23" (or 136' x 164'). We will build a title block that takes these factors into consideration.

It is also possible to build a title block in **Paper Space**. Paper Space is a powerful tool that actually allows us to display a title block at real size (1:1), while displaying a drawing within the title block at a different scale (1/8" per foot). We will examine Paper Space in Chapter 18.

The C-size title block also can be constructed in a way that allows us to use it as a **prototype**. A prototype drawing contains important settings, blocks and layers that can be used in many drawings. If a prototype drawing is specified when a new drawing is opened, the information contained in a prototype drawing is automatically forwarded to the new drawing. This is convenient for two reasons. First, the original prototype is not actually opened and modified in any way. This prevents the prototype drawing from accidentally being altered with unwanted drawing objects. Second, it saves time. The steps of building a title block, inserting text, creating text styles, building layers and establishing settings need only be performed once for an entire set of drawings. The prototype settings are available any time a new drawing is started.

Prototype drawings can be created for any sheet size (and drawing scale) that you are likely to use. In all of the Thomas Residence plans, we will use this C-size prototype drawing.

Opening the Project Manager

After LANDCADD opens the Project Manager, pick the **Blank Sheet** project to begin. The following settings should be active after you open the **Blank Sheet** project:

Units	Architectural	
	Smallest fraction	1/16"
	Angle measure	Decimal degrees
	Direction for 0°	0
	Clockwise?	No
Limits	Lower Left	0'0",0',0"
	Upper Right	250',185'
Grids	10'	
Snap	1'	

Beginning a New Project

Begin a new drawing by selecting the EP pull-down menu, then pick Project Manager... .

In the Project Manager dialogue box, select New . . . . (Figure 11.1).

In the New Project dialogue box, you can specify the new project **Description:** (use **C 1/8" SHEET** or something like it).

In the Drawing Name . . . text box, enter the name of your new drawing (use **C18SHEET**). You will also need to specify the correct path where all of your drawing files are stored. See the section on Project Manager for more details.

To modify the information in the **System Settings:** area, remove the check from the **No Prototype** toggle box. This activates the Prototype Drawing . . . option. Click on this box, which opens the "Select Prototype" dialogue box (Figure 11.2).

Find the **BLANK.DWG** drawing created in the last section, then select it. Place a checkmark in the **Retain as Default** toggle box. Click on OK when finished.

This returns us to the "New Project" dialogue box. We now must retrieve the project settings created in the Project Manager for 1/8 scale drawings.

Pick the **System Settings:** pop-down box to access the settings stored in the Project Settings library (Figure 11.3).

Figure 11.1

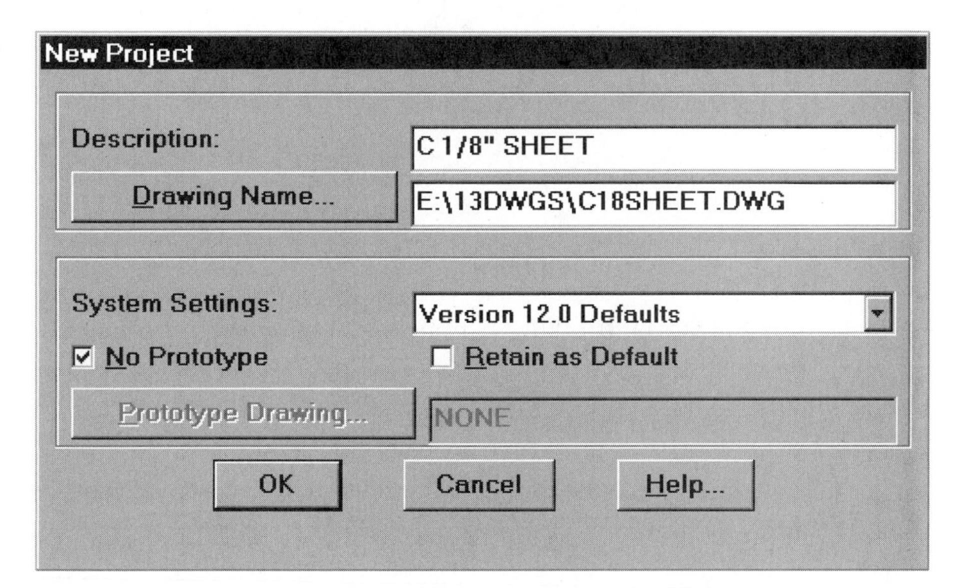

Figure 11.2

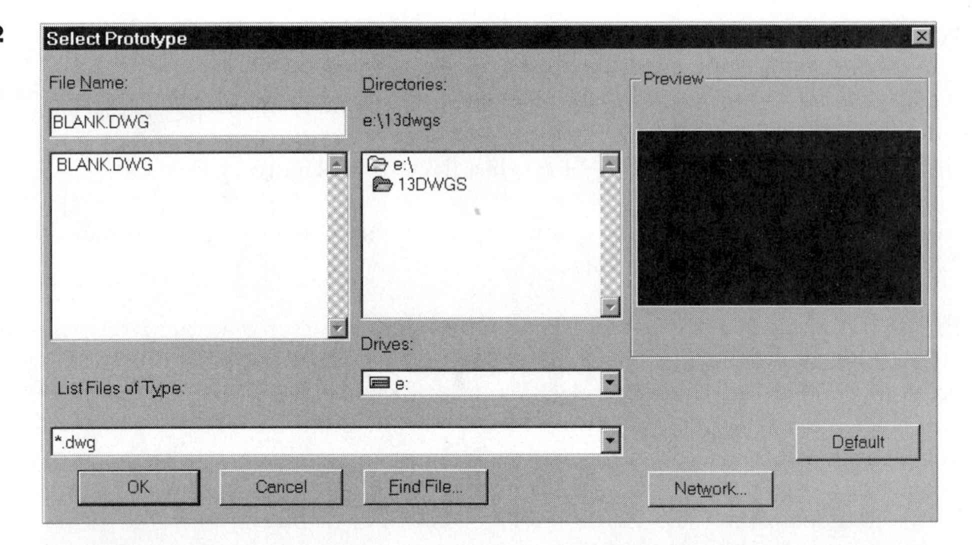

Figure 11.3

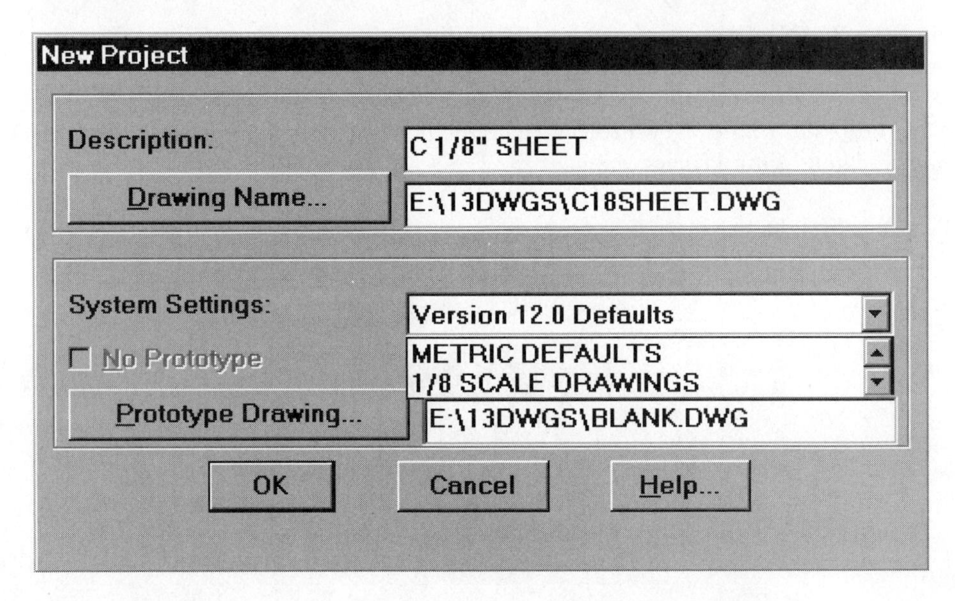

Pick the **1/8 SCALE DRAWINGS** from the list of settings. The final result should look like that shown below (Figure 11.4).

Click on **OK** when finished.

After you have completed the New Project information, you are returned to the Project Manager screen, but you will notice that the **C 1/8" SHEET** has been added to your available projects.

Figure 11.4

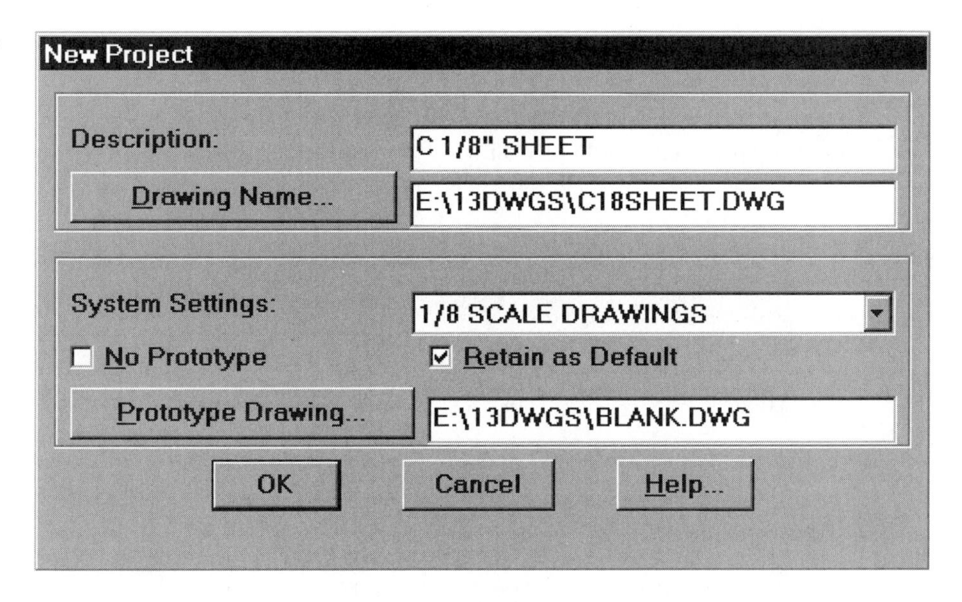

Highlight **C 1/8" SHEET** then select **Open**.

You then see a blank drawing titled **C18SHEET.dwg** on the screen.

Building the Layers

After moving through the Project Manager settings and opening your drawing, you are now ready to add the layers on which the various parts of the title block will be placed.

Move to the Pull-down menu above the graphics area, and pull down the **EP** menu, then select **LANDCADD**.

Select the **Landscape Design** option. This changes the selections above the graphics area to LANDCADD functions and choices.

From the **Tools** menu, select **Layers ▸**. (Windows and Windows 95 users can also pick the layers icon near the top left corner of the screen.)

Note: The **AutoCAD and EP** menu has a **Tools** pull-down menu which does **not** have a **Layers ▸** option available. If you do not see this selection, return to the **EP** pull-down to load the LANDCADD menu.

From the Layers ▸ menu, select AutoCAD Layer Dialog.

In the text box, type **SHEET, TEXT, BORDER** (be sure to include the comma to make three separate names) (Figure 11.5).

Figure 11.5

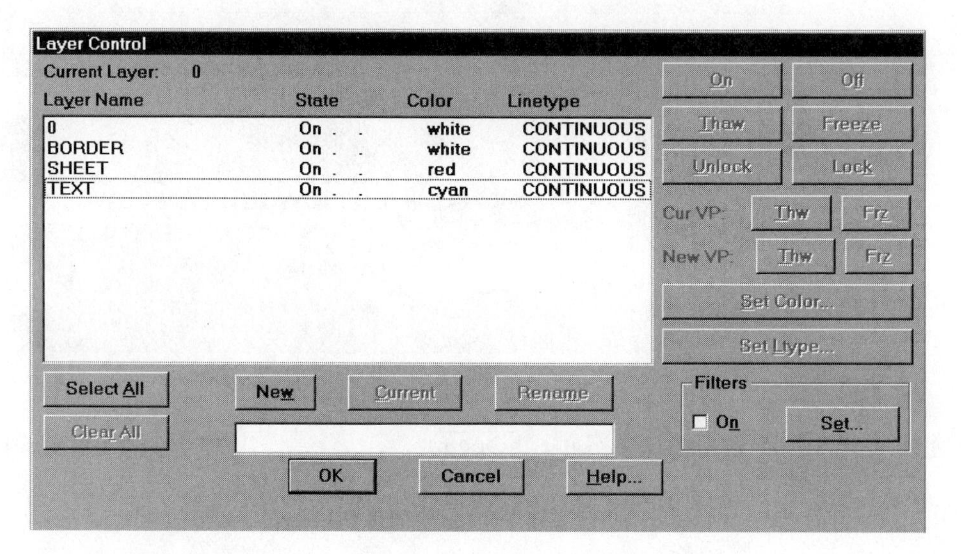

Select New—this adds **SHEET**, **TEXT** and **BORDER** to the list of layers.

Highlight the layer **TEXT**, then select Set Color . . . (Figure 11.6).

Select the top **cyan** (turquoise) square to use cyan as the color for this layer. Pick OK to return to the “Layer Control” dialogue box.

Click on the **TEXT** layer to remove the highlight.

Move the cursor to **SHEET** and highlight it, then select Set Color . . ..

Select the top **red** square to use red as the color for this layer. Return to the “Layer Control” box.

Click on the **SHEET** layer to remove the highlight.

Click on the **BORDER** layer to highlight it.

Select Current — this makes **BORDER** the current layer.

Figure 11.6

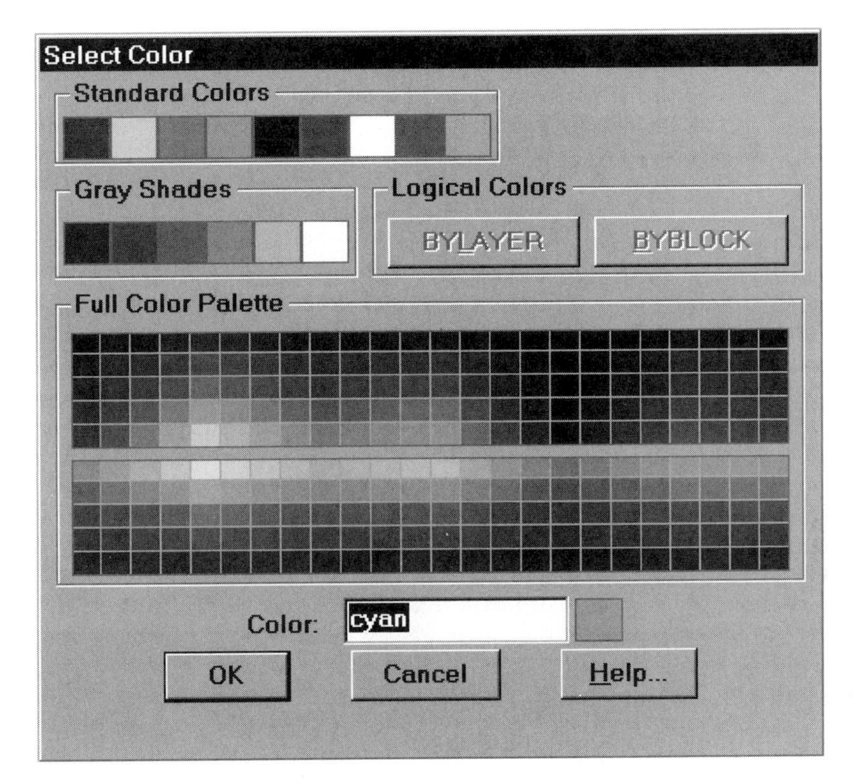

Drawing the Sheet Border

Draw the border for a C-size drawing sheet:

At the prompt:

`Command:`
Type **PL** to start a polyline.

`From point:`
Type (or pick) **0',0'**

`Arc/Close/Halfwidth/Length/Undo/Width/<Endpoint of line>:`
Type (or pick) **192',0'** for the first endpoint.

`Arc/Close/Halfwidth/Length/Undo/Width/<Endpoint of line>:`
Type (or pick) **192',144'** for the second endpoint.

Arc/Close/Halfwidth/Length/Undo/Width/<Endpoint of line>:
Type (or pick) **0',144'** for the third endpoint.

Arc/Close/Halfwidth/Length/Undo/Width/<Endpoint of line>:
Type **C** to close the polyline.

This establishes an imaginary paper edge that helps us center and locate other objects on the drawing. This border will act as the trim line on the final drawing (use it to trim the paper on the paper cutter). It is useful in establishing proper locations for an actual **border**, **title block**, **revision list**, and **sheet box**.

Creating the Title Block

Build the title block for the drawing:

Return to AutoCAD Layer Dialog (through the Tools and Layers ▸ menus), and remove the highlight from the **BORDER** layer. Highlight the **SHEET** layer, then select Current to make it the current layer.

Turn on **Snap** with the F9 key. The word **SNAP** should be black (not gray) at the bottom line of the screen. Turn on **Ortho** with the F8 key. The word **ORTHO** should also be active (black) in the lower portion of the screen.

At the prompt:

Command:
Type **PL** to begin the **Polyline** command.

From point:
Type (or pick) **173',23'**. This is the first point of the rectangle.

Arc/Close/Halfwidth/Length/Undo/Width/<Endpoint of line>:
Type **W** to set the polyline width.

Starting width <0'-0">:
Type **6"** to set the starting width at 6".

Ending width <0'-6">:
Type **6"** again. This sets the entire polyline width to 6".

Arc/Close/Halfwidth/Length/Undo/Width/<Endpoint of line>:
Type (or pick) **173',106'** to set the next endpoint of the polyline.

`Arc/Close/Halfwidth/Length/Undo/Width/<Endpoint of line>:`
Type (or pick) **188',106'** to set the next endpoint.

`Arc/Close/Halfwidth/Length/Undo/Width/<Endpoint of line>:`
Type (or pick) **188',23'** to set the next endpoint.

`Arc/Close/Halfwidth/Length/Undo/Width/<Endpoint of line>:`
Type **C** to close the polyline.

Build the last segment in the title block.

At the prompt:

`Command:`
Type **PL**.

`From point:`
Type (or pick) **173',50'** to set the first endpoint.

`Arc/Close/Halfwidth/Length/Undo/Width/<Endpoint of line>:`
Type (or pick) **188',50'** to set the second endpoint.

Press **<Enter>** to leave the command.

Build the revisions list box using the same **PL** command (and with the same polyline width):

At the prompt:

`Command:`
Type **PL.**

`From point:`
Type (or pick) **173',110'.**

`Arc/Close/Halfwidth/Length/Undo/Width/<Endpoint of line>:`
Type (or pick) **173',140'.**

`Arc/Close/Halfwidth/Length/Undo/Width/<Endpoint of line>:`
Type (or pick) **188',140'.**

`Arc/Close/Halfwidth/Length/Undo/Width/<Endpoint of line>:`
Type (or pick) **188',110'.**

Arc/Close/Halfwidth/Length/Undo/Width/<Endpoint of line>:
Type **C** to close the polyline.

Build the sheet number box by using the **PL** endpoints:

At the prompt:

Command:
Type **PL.**

From point:
Type (or pick) **173',4'.**

Arc/Close/Halfwidth/Length/Undo/Width/<Endpoint of line>:
Type (or pick) **173',19'.**

Arc/Close/Halfwidth/Length/Undo/Width/<Endpoint of line>:
Type (or pick) **188',19'.**

Arc/Close/Halfwidth/Length/Undo/Width/<Endpoint of line>:
Type (or pick) **188',4'.**

Arc/Close/Halfwidth/Length/Undo/Width/<Endpoint of line>:
Type **C** to close the polyline.

Put two polylines in the sheet number box:

At the prompt:

Command:
Type **PL.**

From point:
Type (or pick) **173',15'.**

Arc/Close/Halfwidth/Length/Undo/Width/<Endpoint of line>:
Type (or pick) **188',15'.**

Press **<Enter>** to repeat the **PL** command

From point:
Type (or pick) **177',19'.**

`Arc/Close/Halfwidth/Length/Undo/Width/<Endpoint of line>:`
Type (or pick) **177',4'.**

Finish the revisions list by filling in **lines** (<u>not</u> polylines).

At the prompt:

`Command:`
Type **L** to begin the Line command.

`LINE from point:`
Type (or pick) **176',110'.**

`To point:`
Type (or pick) **176',140'.**

Press **<Enter>** to end the command.

Create additional lines with the **Offset** command, which duplicates lines parallel with the original at on offset distance away.

At the prompt:

`Command:`
Type **OF** to begin the Offset command.

`Offset distance or Through<Through>`
Type **2'** to set the distance between the lines.

`Select object to offset:`
Move your cursor (now a box) over the line you just made. Press the **Pick** button.

`Side to offset?`
Move your cursor to the right, then press the **Pick** button.

This duplicates the line 2' away from your original line. Offset is a chain command, so you can continue to offset lines 2' apart for as long as you want.

Build a total of 5 new lines, then press **<Enter>** to end the command.

Finish the revisions list lines by adding a final line:

At the prompt:

Command:
Type **L** to begin the Line command.

LINE from point:
Type (or pick) **173',115'.**

To point:
Type (or pick) **188',115'**.

Press **<Enter>** to end the command.

Offset this line by the distance of **5'6"** and place the line **above** the original.

At the prompt:

Command:
Type **OF** to begin the Offset command.

Offset distance or Through<Through>
Type **5'6"** to set the distance between the lines.

Select object to offset:
Move your cursor (now a box) over the previous line. Press the **Pick** button.

Side to offset?
Move your cursor **above** the original line, then press the **Pick** button. This duplicates the line above the original. Press **<Enter>** to end the command.

Build a final polyline rectangle to define the drawing area.

At the prompt:

Command:
Type **PL<Enter>** to begin the Polyline command.

From point:
Type (or pick) **171',140'** .

Arc/Close/Halfwidth/Length/Undo/Width/<Endpoint of line>:
Type (or pick) **4',140'.**

Arc/Close/Halfwidth/Length/Undo/Width/<Endpoint of line>:
Type (or pick) **4',4'.**

Arc/Close/Halfwidth/Length/Undo/Width/<Endpoint of line>:
Type (or pick) **171',4'.**

Arc/Close/Halfwidth/Length/Undo/Width/<Endpoint of line>:
Type **C** to close the polyline.

Save your drawing

Type **Saveas** from the command line, keeping the drawing name **C18SHEET**. (This indicates a C-size sheet at 1/8" scale.)

This will save the drawing to the hard disk. This is an important first step, but you should create a back-up copy of the drawing onto a floppy disk.

Use one of the file copying techniques discussed in Chapter 8 to copy your file to a floppy disk.

Entering Title Block Text

The next step in creating the title block is to add the necessary text for the sheet box and revisions list. To enter text on a drawing, we must define a **text style**. The **style** command lets us define a number of factors about how text appears on the screen. These factors include:

Font — the actual appearance of the text itself. This is like a typestyle.

Height — this predefines the height of the style. It is a good idea to leave this set at **0** because the text height can be reset whenever the need arises.

Width factor — the proportion of width to height of the letters. We will adjust this according to how we want to make the letters look.

Obliquing angle — a fancy way of describing the slant of the letters. Leave it at **0**.

Backwards, Upside-down, Vertical — self-explanatory. Answer all of these with the **<Enter>** key which means **No**.

At the prompt:

Command:
Type **Style** (or **ST**) to define the new style.

```
STYLE Text style name (or?) <STANDARD>
```
Type **1.**

```
New Style
```
You then see the Select Font File dialogue box below (Figure 11.7).

To find the correct font, be sure to look in your AutoCAD fonts directory (something like **C:\R13\COM\FONTS**). Move your cursor to the scroll bar to the right of the list of fonts. Click on the arrows or use the page-up or -down keys until the font you want is visible.

Highlight the **romans.shx** font to place its name in the **File Name:** box, then click on **OK.**

```
New style. Height <0'-0">
```
Press **<Enter>.**

```
Width factor <1.0000>
```
Press **<Enter>.**

```
Obliquing angle <0>
```
Press **<Enter>.**

```
Backwards? <N>
```
Press **<Enter>.**

```
Upside-down? <N>
```
Press **<Enter>.**

Figure 11.7

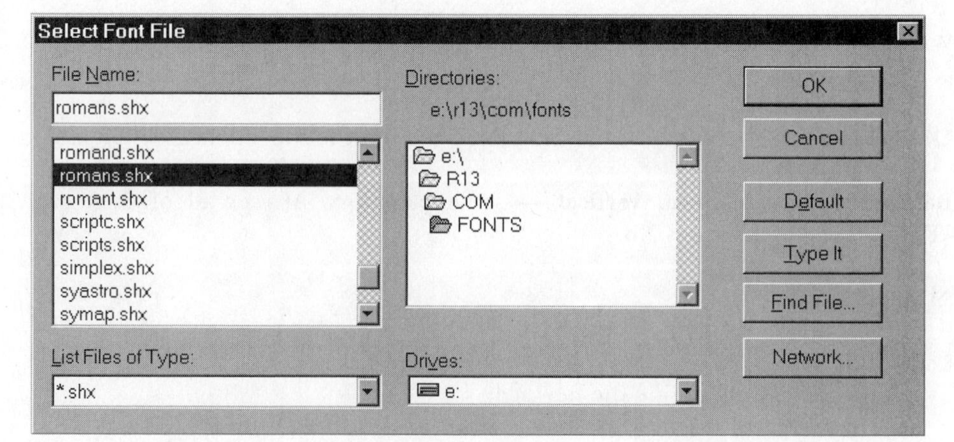

`Vertical? <N>`
Press **<Enter>.**

`1 is now the current text style`
We have just completed building the first text style used in the Title Block drawing. Now you should build the two additional styles for use in the title block. Follow the directions below to build the two new styles. The table below shows the style names and their important characteristics:

Style Name	Font File Name	Width Factor
1	ROMANS.SHX	1
2	SWISSB.TTF	1
3	SWISSK.TTF	1.25

Hint: The solid fill fonts, **SWISSB.TTF** and **SWISSK.TTF** used in this exercise will not display as solid unless the AutoCAD system variable **TEXTFILL** is set correctly. If these fonts do not show on screen as filled, change the system variable as shown:

`Command:`
Type **TEXTFILL.**

`New value for TEXTFILL <0>`
Type **1.**

This allows the fonts to be displayed correctly.

To define the remaining text styles, you will need to use two fonts which are Windows TrueType® fonts. These fonts are provided with AutoCAD Release 13, and can be used in any AutoCAD platform (Windows 3.1, Windows 95, Windows NT, DOS or other operating systems). The "Select Font File" dialogue box allows you to view these TrueType fonts (with the file extension **.ttf**) with an option shown at the bottom of the dialogue box (Figure 11.8).

Change the listing shown in the **List Files of Type:** pop-down box to show the **.ttf** files. Pick the appropriate font names to define styles **2** and **3**.

Figure 11.8

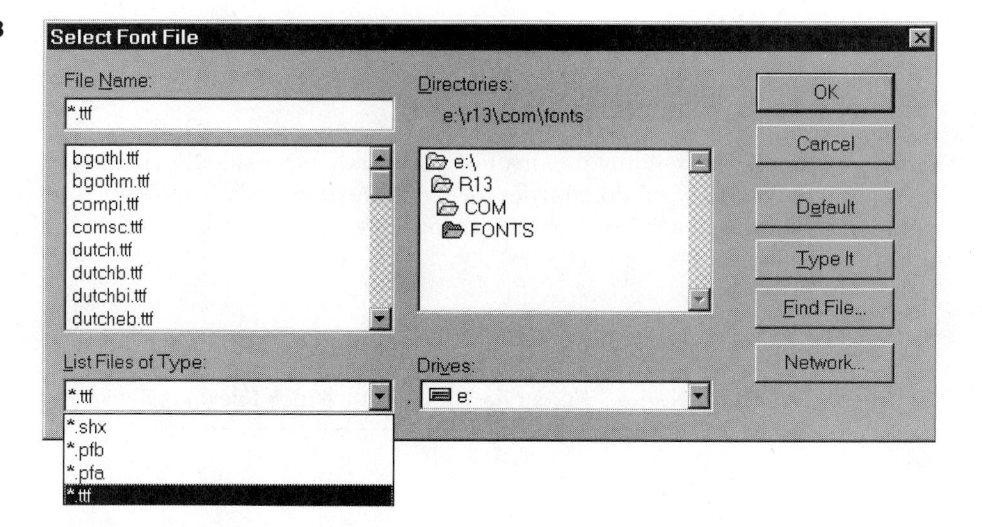

We will place the first text by using the command **DTEXT** (or DT). This is for Dynamic Text, which displays the text as it is typed.

`Command:` **`DT`**
`DTEXT Justify/Style/<Start point>:`
Type **S** (for Style).

`Style name <3>:`
Type **1**. The previous style (3) was kept as the default value, so we must update it to the style we named 1.

`DTEXT Justify/Style/<Start point>:`
Type (or pick) **178',13'** for the starting point of the first text.

`Height <1'-0">`
Type **9"** to set the text height.

`Rotation angle <0>`
Press **<Enter>** to accept a 0° text rotation.

`Text:`
Type **SHEET <Enter> NO. <Enter><Enter>**

This places the words on the screen, then tells the command to end.

Press **<Enter>** to begin the **DTEXT** command again.

`DTEXT Justify/Style/<Start point>:`
Type (or pick) **183',5'** for the text starting point.

`Height <0'-9">`
Press **<Enter>** to accept the default value.

`Rotation angle <0>`
Press **<Enter>** to accept the default value.

`Text:`
Type **OF** then press **<Enter><Enter>** to end the command.

Press **<Enter>** to begin the **DTEXT** command again.

`DTEXT Justify/Style/<Start point>:`
Type (or pick) **178',17'** for the text starting point.

`Height <0'-9">`
Press **<Enter>** to accept the default value.

`Rotation angle <0>`
Press **<Enter>** to accept the default value.

`Text:`
Type **JOB NO.** then press **<Enter><Enter>** to end the command.

Use the **MOVE** command to move the text closer into the corners of the sheet box. Turn off **Snap** and **Ortho** to make it easier to adjust the location of the text (Figures 11.9 and 11.10).

Figure 11.9

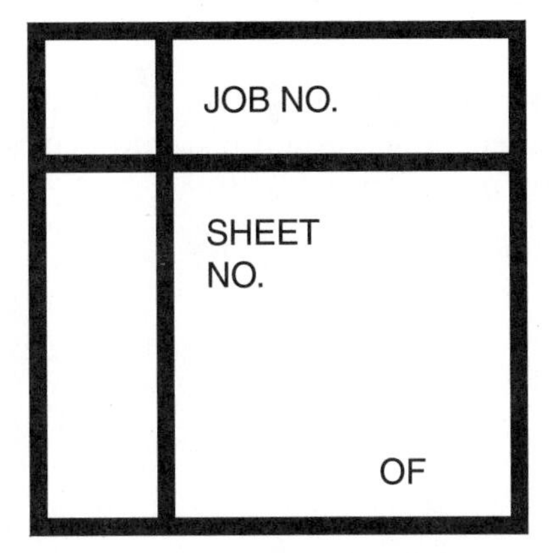

Figure 11.10

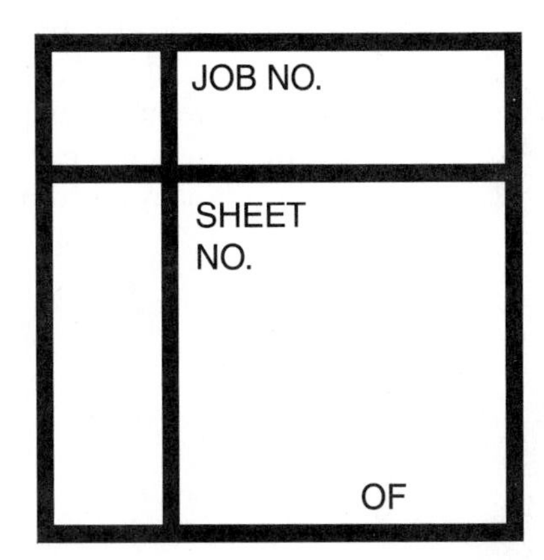

Now insert the text **DATE** and **REMARKS** labels with a height of **12"**, and with the rotation angle set to **90**.

At the prompt:

`Command:`
Type **DT** to begin the Dtext command.

`DTEXT Justify/Style/<Start point>:`
Type (or pick) **175',116'** for the text starting point.

`Height <0'-9">`
Type **12".**

`Rotation angle <0>`
Type **90.**

`Text:`
Type **DATE** then press **<Enter><Enter>** to end the command.

Press **<Enter>** to begin the Dtext command again.

`DTEXT Justify/Style/<Start point>:`
Type (or pick) **175',126'** for the text starting point.

`Height <1'-0">`
Press **<Enter>** to accept the new default.

`Rotation angle <90>`
Press **<Enter>** to accept the new default.

`Text:`
Type **REMARKS** then press **<Enter><Enter>** to end the command (Figure 11.11).

We have now built a C-size title block which can be used for any architectural drawing. We now can create an easy way to enter the important text information which should accompany any drawing we create and plot (Figure 11.12).

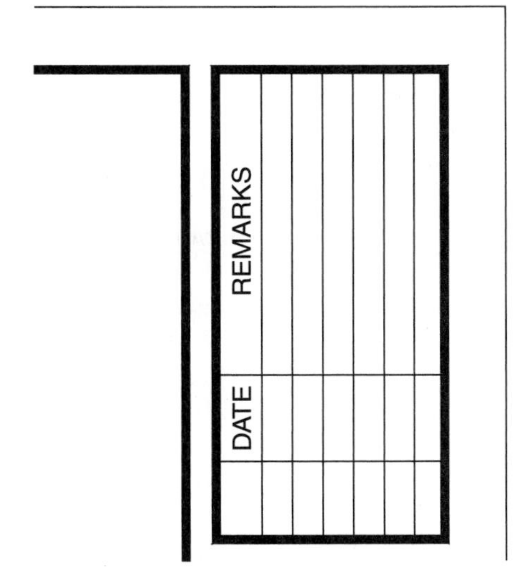

Figure 11.11

Figure 11.12

Building the Text Attributes

The lines of text information we enter can be stored in a special form where each are called **Attributes**. We will create a grouping (called a **Block**) consisting of 8 attributes. Each attribute will require the following information:

Tag — this is the name that AutoCAD uses to keep track of the attribute.

Prompt — this is the actual prompt that the user sees next to the text box where the attribute information is typed.

Style — the font style, size, justification and direction in which the text will be displayed.

Insertion Point — the actual location where the text will start on the drawing.

In addition, other information may be entered with the attribute, such as whether or not to make the attribute invisible, and whether or not there should be a constant (or default) **value** associated with the attribute.

By using the properties of **Blocks**, we can create a sort of **template** that allows us to enter text data easily, accurately and quickly. We will create a block with 8 attributes contained in it. These are the attributes and their properties:

Tag	Prompt	Style	Height	Rotation	Insertion Pt	Justification
JOB	**JOB NAME:**	3	2'	0	177',56'	Left
PROJECT	**PROJECT NAME:**	2	2'	0	177',53'	Left
ADDRESS	**STREET ADDRESS:**	2	2'	0	177',50'	Left
CITY	**CITY, STATE & ZIP:**	2	2'	0	177',47'	Left
NUMBER	**JOB NUMBER:**	2	1.5'	0	179',16'	Left
SHEET	**SHEET NUMBER:**	3	3'	0	178',8'	Left
TOTAL	**TOTAL SHEETS:**	2	1.5'	0	185',5'	Left
SCALE*	**DRAWING SCALE:**	2	1.5'	0	175',20'	Left
*use a **Value:** of **1/8"=1'-0"** for the scale of this title block						

Begin by making TEXT the current layer:

Pull down the **EP** menu, then pick **Layers ▸**

Pick **AutoCAD Layer Dialog** then highlight **TEXT** and make it the current layer.

Click on **OK**.

Bring up the AutoCAD "Attribute Definition" dialogue box.

Pull down the **EP** menu, then pick the **AutoCAD and EP** menu.

Pull down the **Construct** menu, then select **Attribute...**.

We could also type **DDATTDEF** at the command prompt to open the dialogue box for our required information. This dialogue box is illustrated below (Figure 11.13).

We will use the dialogue box to create the attributes, one at a time, in the order presented in the list.

Fill in the text boxes as indicated for the **JOB** attribute. Change the text style in the **Text Style** pop-down box. This box will display any text styles that are currently present in the drawing (Figure 11.14).

Figure 11.13

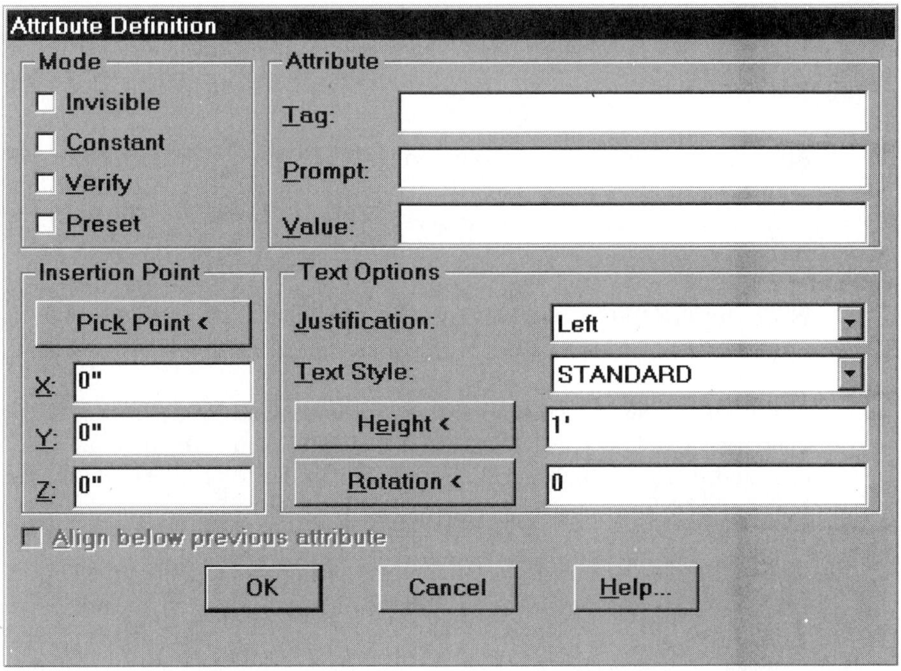

Attribute Definition
Mode
Invisible
Constant
Verify
Preset
Attribute
Tag:
Prompt:
Value:
Insertion Point
Pick Point <
X: 0"
Y: 0"
Z: 0"
Text Options
Justification: Left
Text Style: STANDARD
Height < 1'
Rotation < 0
Align below previous attribute
OK
Cancel
Help...

Figure 11.14

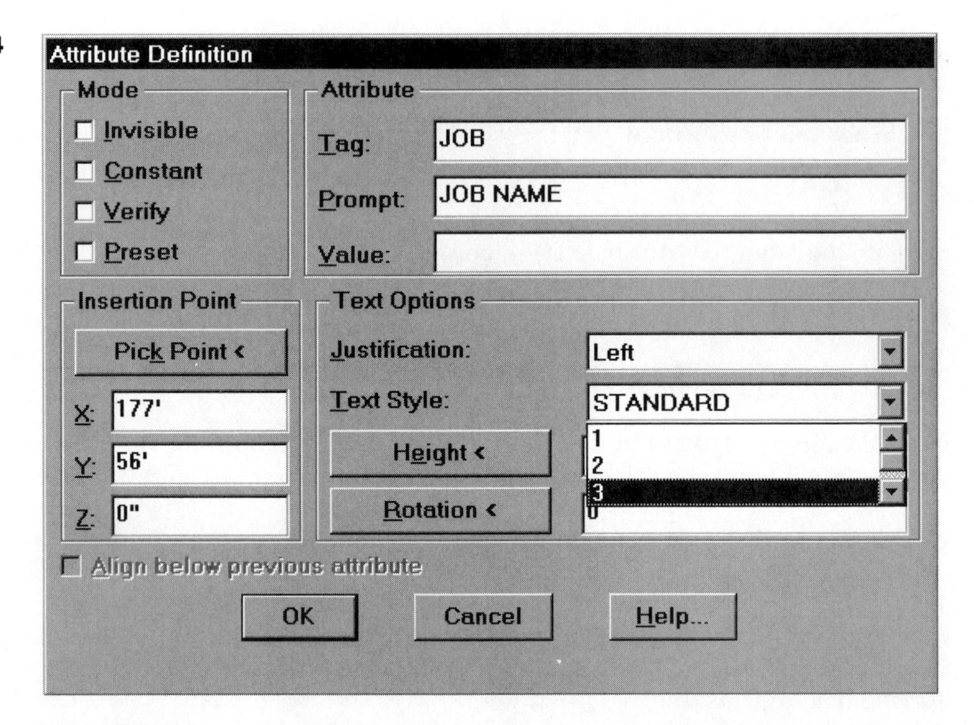

AutoCAD Warning: The Rotation < function in the "Attribute Definition" dialogue box does not work properly. Rather than enter rotation angles here, leave the rotation angle at 0, then rotate the attribute tags after they are inserted (Figure 11.15)

After the information is complete in the "Attribute Definition" box, pick **OK**.

This places the tag **JOB** on the screen. Because of the bug in AutoCAD, we will enter these attributes first, then rotate them later (Figure 11.16).

Place the **PROJECT**, **ADDRESS** and **CITY** attributes on the screen, then rotate them into position as shown below (Figures 11.17 and 11.18).

The final tag, **SCALE**, has an extra attribute called **Value** in the "Attribute Definition" dialogue box. The **Value** attribute is a default value which remains in effect until it is changed. It is reasonable to keep a default value of **1/8"=1'-0"** since this title block is built for 1/8 scale drawings (Figure 11.19).

Notice that as you create the attributes the tag names remain visible. These names are important, but **will not** show on the drawing. After you have completed all of the attributes, and have located them on the drawing, you can check them to be sure they are cor-

Figure 11.15

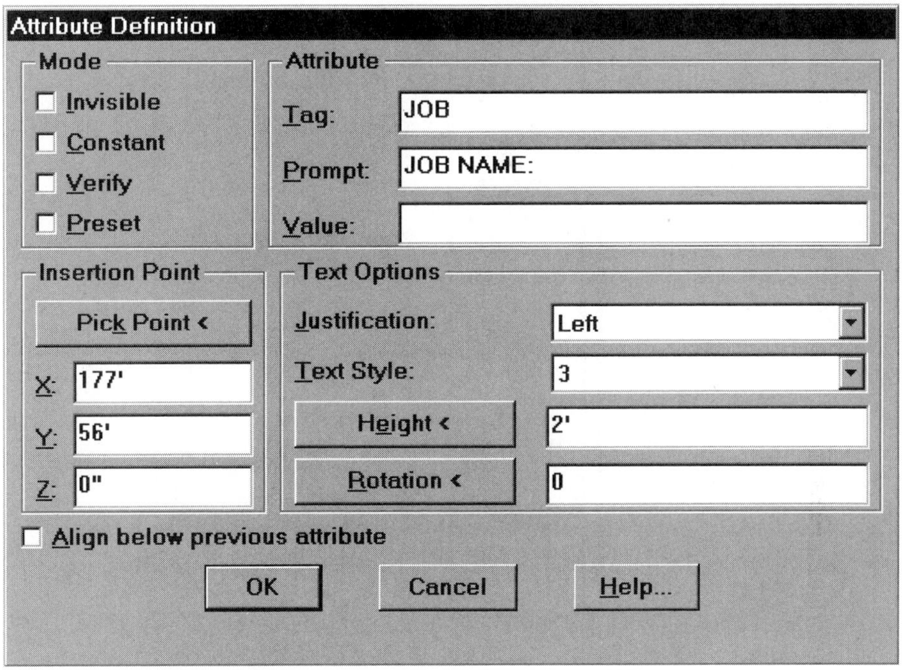

Attribute Definition
Mode
Invisible
Constant
Verify
Preset
Attribute
Tag: JOB
Prompt: JOB NAME:
Value:
Insertion Point
Pick Point <
X: 177'
Y: 56'
Z: 0"
Text Options
Justification: Left
Text Style: 3
Height < 2'
Rotation < 0
Align below previous attribute
OK
Cancel
Help...

Figure 11.16

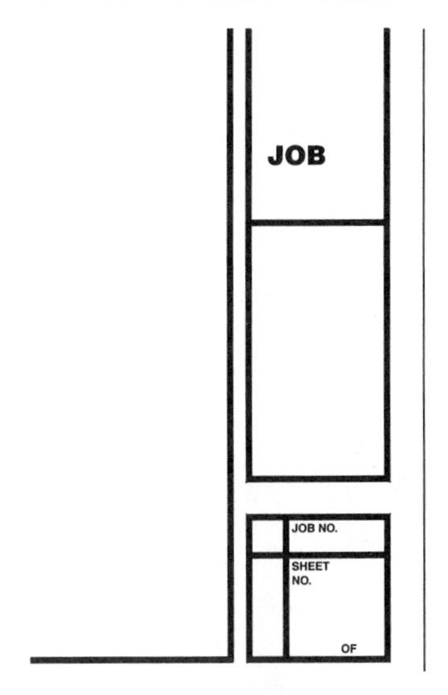

rect. Use the command **DDEDIT** (abbreviated **DD**) to pick an attribute to examine the tag name, prompt and default value. If these are incorrect, you can change them in the dialogue box. If the locations don't look correct, you can still move them easily with the **Move** command. Make sure that you are completely satisfied that all of these are correct before moving to the next step (Figure 11.20).

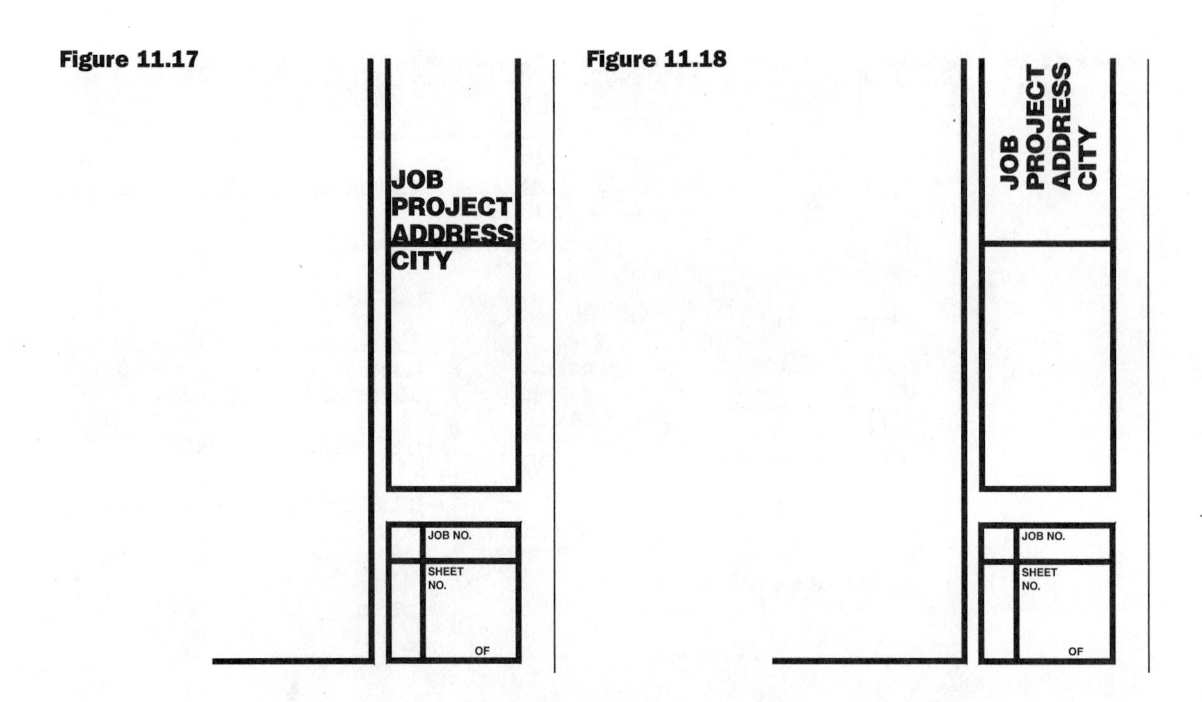

Figure 11.17

Figure 11.18

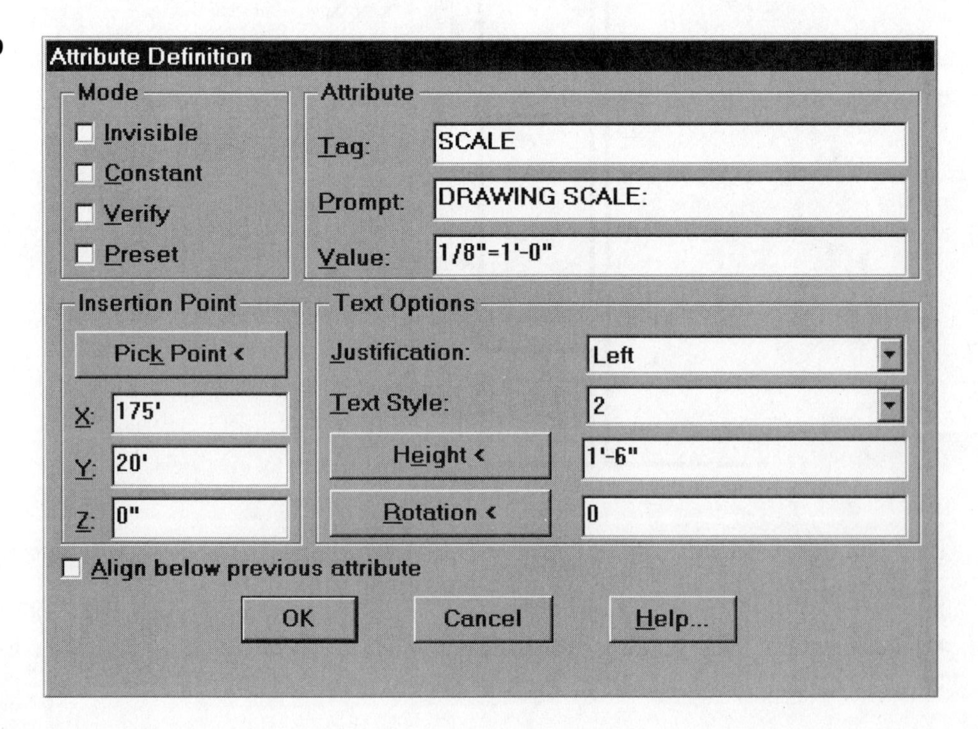

Figure 11.19

Figure 11.20

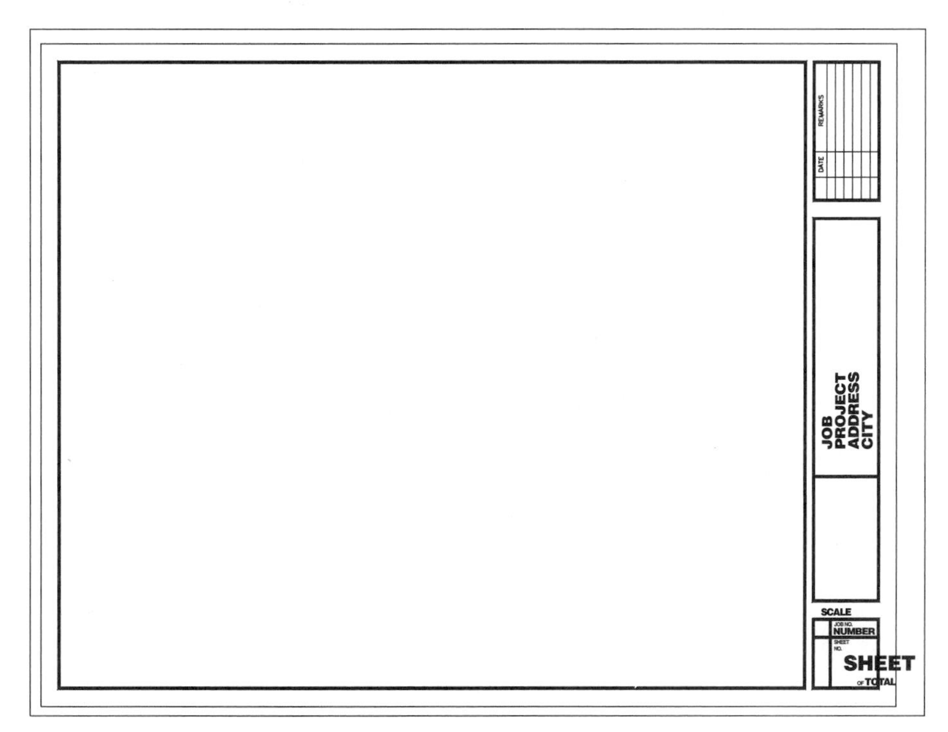

Turning the Text Attributes Into a Block

At the prompt:

`Command:`
Type **B** to activate the Block command.

`BLOCK`
`Block name (or ?):`
Type **TEXTB** to give the block a name.

`Insertion base point:`
Type (or pick) **0,0**.

`Select objects:`
PICK THE OBJECTS IN THE SAME ORDER AS YOU CREATED THEM. In this way, you will be able to answer the prompts in the correct order. If you perform this step incorrectly, you may have to insert the block, explode it and pick the objects all over again.

After you have completed picking the attributes, press **<Enter>** to complete the command. All of the attributes disappear! This is ok. We only want to see the words entered in response to the prompts.

The drawing **C18SHEET** now has the block **TEXTB** nested inside it. It is stored with the drawing and is not visible unless you actually insert it. It is good practice to actually have this text block visible on the drawing, with its presence indicated by place holders. In this case, we will use dashes to indicate the block exists, but must be filled out before the drawing is completed.

Inserting the Text Block

We will now insert the text block into the drawing.

At the prompt:

`Command:`
Type **Insert** to insert a block.

`Block name (or ?):`
Type **TEXTB** as the block name.

`Insertion point:`
Type (or pick) **0,0.**

`X Scale factor <1>/Corner/XYZ`
Press **<Enter>.**

`Y Scale factor (default=x):`
Press **<Enter>.**

`Rotation angle <0>`
Press **<Enter>.**

You now will see a box which shows the prompts to the attributes.

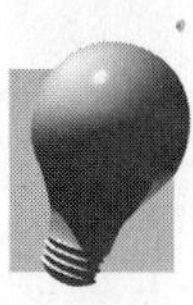

Hint: If a dialogue box does not appear on the screen, change the AutoCAD setting **ATTDIA**.
Command:
Type **ATTDIA.**
New value for ATTDIA <0>:
Type **1.**
This changes the AutoCAD setting and allows the display of the "Edit Attributes" dialogue box.

Fill in all of the prompts with two dashes. The dashes are placed on the drawing in the correct location and on the correct layer (Figure 11.21).

Hint: Whenever you want to attach a text block to a drawing as a permanent feature, leave place-holder characters on the screen. This assures that the block will be scaled, moved and inserted along with the rest of the drawing. If the block is not visible on the screen, it can be removed during the **PURGE** or **WBLOCK** commands, and would not be scaled or moved when these operations were performed on the drawing itself.

Figure 11.21

Edit Attributes

Block Name: TEXT

JOB NAME	--
PROJECT NAME:	--
STREET ADDRESS:	--
CITY STATE & ZIP	--
JOB NUMBER	--
SHEET NUMBER	--
TOTAL SHEETS:	--
DRAWING SCALE:	1/8"=1'-0"

OK Cancel Previous Next Help...

Save your drawing to the hard disk and to your floppy disk as **C18SHEET** (Figure 11.22).

Adding a Logo to the Title Block

In order to personalize and complete your title block, you need to build a logo/business card to place in the blank section set aside for this purpose. You will first create a rectangle the size of an actual business card, then create the appropriate text and graphics to suit your business image (or fantasy business image). Once this card has been created, you will save it as a drawing. You will then insert it into your title block, scale it to the correct size, and rotate it to fit into the appropriate space. Once this is complete, you can resave the title block as the finished version of **C18SHEET**.

Figure 11.22

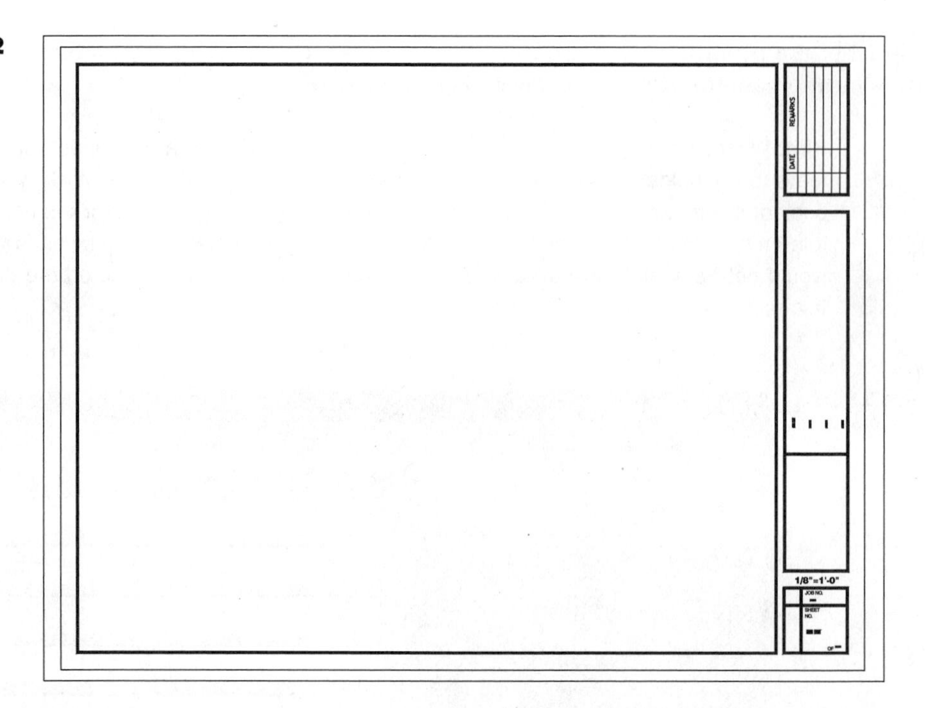

Setting Up the Page

In the Project Manager, begin a new project, called **Business Card**. Use **BLANK** as your prototype drawing, and name your new drawing **CARD.DWG**.

Under the EP pull-down menu, use the Edit Project Settings . . . selection to change the **Horizontal** scale to **1.000000**. Leave the text size at **0.125000**. These settings are appropriate for a **1:1** drawing. After the **CARD** drawing is completed, the project settings can be saved to the Project Settings library. Any later projects which use a **1:1** scale can use these same settings.

Pull down the EP menu, select LANDCADD, then select the Landscape Design option. Use the Tools pull-down menu to select the Layers ▸ option. Pick the option AutoCAD Layer Dialog then immediately build a layer called **CARD**. Make **CARD** the current layer.

We will resize the drawing area by using the **LIMITS** command. The **LIMITS** command defines the outside edges of the “paper” on which you are drawing. Since the original limits were designed for a residential landscape design, they will be too large for the design of a business card. Change the limits as follows:

At the prompt:

```
Command:
```
Type **LIMITS.**

```
Reset Model space limits:
ON/OFF/<Lower left corner> <0'-0",0'-0">
```
Press **<Enter>** to accept this default.

```
Upper right corner <250'-0",185'-0">
```
Type **12,9 <Enter>** to set the drawing area to a 12" X 9" rectangle.

```
Command:
```
Type **Z<Enter>E<Enter>** (or use the keyboard macro **ZE**) to reduce the size of the viewing area to the new one specified in the limits you just typed.

Hint: In architectural units, the base unit is the inch. This means that if you type a distance without units, AutoCAD interprets this distance as inches. When you wish to indicate **inches**, units are not needed. When you wish to indicate **feet**, the ' sign is required after each distance.

You will also need to reset the values for **Snap** and **Grids**. To adjust these features, we will use the AutoCAD "Drawing Aids" dialogue box. We can access this box in two ways:

Pick the **EP** pull-down menu, then select **AutoCAD and EP**. Find and pick the **Options** pull-down menu, then select **Drawing Aids . . .**

Or

type **DDRMODES.**

Either choice will bring up this dialogue box (Figure 11.23).

Figure 11.23

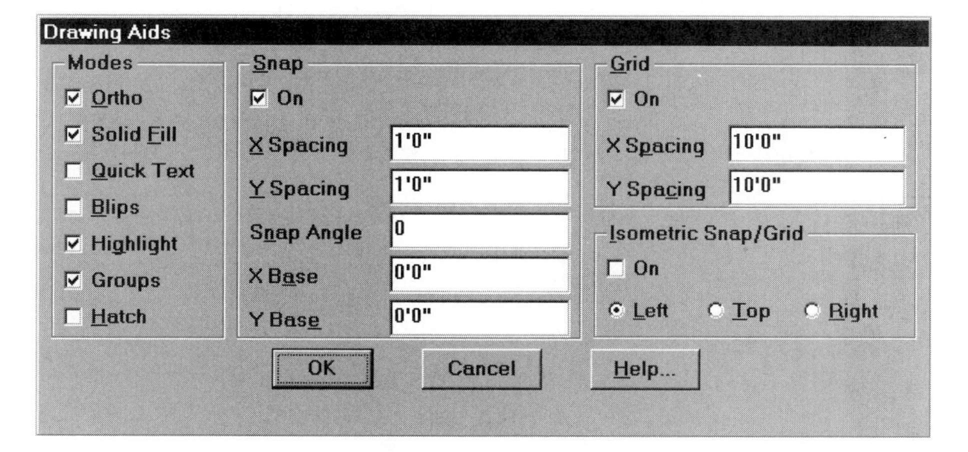

Adjust the settings in the box to the following values:

Snap: 1/8" (note that if you change the X spacing value, the Y spacing changes with it)

Grid: 1/2" (note that if you change the X spacing value, the Y spacing changes with it)

Click on **OK** to accept the rest of the values.

Building the Rectangle

We will use lines to build a business card to this rectangular size and shape (Figure 11.24).

Figure 11.24

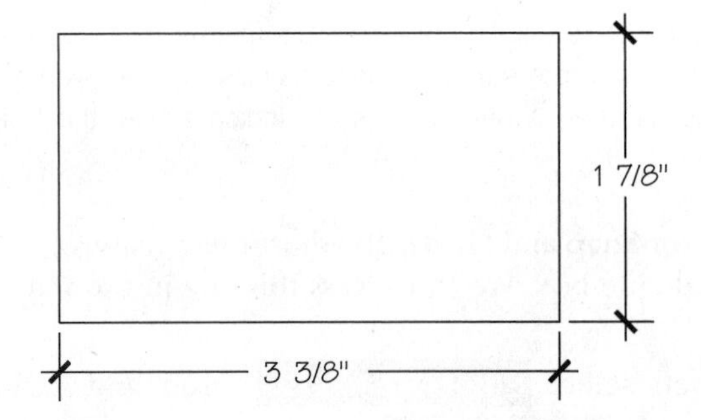

One of the easiest ways to create simple geometry is to use the **Offset** and **Fillet** commands.

To begin, draw a vertical line near the center of the screen.

At the prompt:

`Command:`
Type **OF** to begin the Offset command.

`Offset distance or Through <Through>:`
Type **3-3/8** to give the offset distance.

`Select object to offset:`
Pick the vertical line you just created.

`Side to offset?`
Move your cursor to the right side, then press the **Pick** button.

`Select object to offset:`
Press **<Enter>** to end the command (Figure 11.25).

Figure 11.25

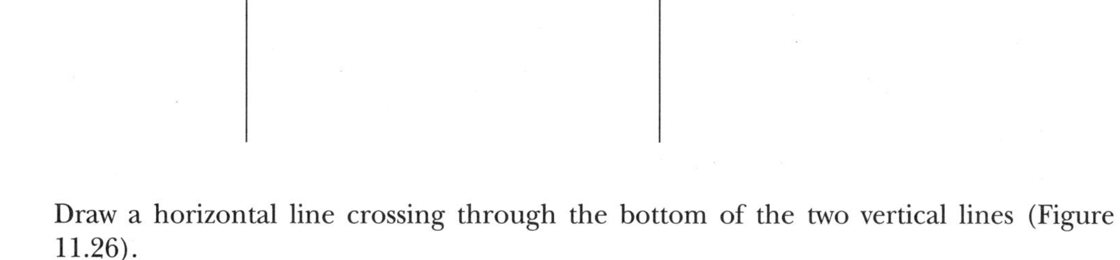

Draw a horizontal line crossing through the bottom of the two vertical lines (Figure 11.26).

Figure 11.26

At the prompt:

`Command:`
Type **OF** to begin the Offset command.

`Offset distance or Through <Through>:`
Type **1-7/8** to give the offset distance.

`Select object to offset:`
Pick the horizontal line you just created.

`Side to offset?`
Move your cursor above the original line, then press the **Pick** button.

`Select object to offset:`
Press **<Enter>** to end the command (Figure 11.27).

Figure 11.27

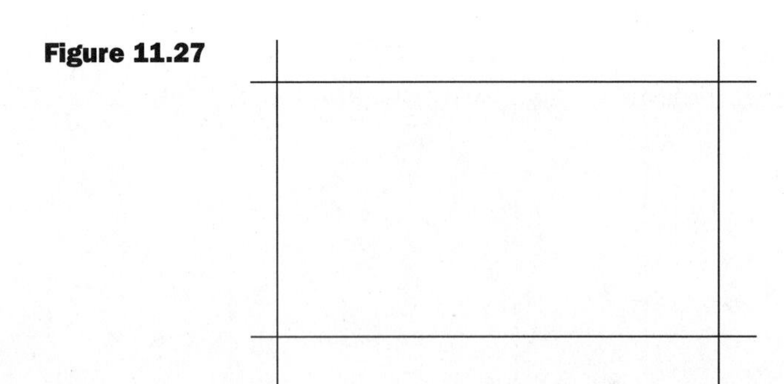

You can use the **Fillet** to eliminate the overhanging line segments.

At the prompt:

```
Command:
```

Type **FI** to begin the Fillet command.

```
FILLET
(TRIM mode) Current fillet radius = 0'-0"
Polyline/Radius/Trim/<Select first object>:
```

Pick the **interior** portion of one of the line segments.

```
Select second object:
```

Pick the **interior** portion of an adjacent line segment. Notice that the exterior portion of the line segments disappear.

Hint: The **FILLET** command has several uses. To use it to trim and terminate lines, be sure that the **Radius** setting is at **0**. When using the command, remember to pick the side of the line segment that you wish to **keep**. The other side of the segment (beyond the intersecting line) will be removed (Figure 11.28).

Figure 11.28

```
Polyline/Radius/Trim/<Select first object>:
```

Repeat this process until all of the unwanted line segments are removed. Remember to **Undo** if you accidentally remove any desired lines.

After you have finished cleaning up the intersections, use the **Move** command to move the entire rectangle so that the bottom left corner is located at **0,0**.

Adding the Text and Graphics

Once the rectangle has been relocated, you can insert text and graphics. Remember that the size of the text will need to be that of the actual text on a business card—1/8" or so for the average text size. Be sure to include your business name, address, telephone and FAX number and your name. Don't worry about using colors, since we will plot to create reproducible drawings—and will use black pens only.

Experiment with different fonts and typestyles. Here are some examples of fonts to try:

Handwritten fonts: **HLTR.SHX***, **HANDLET.SHX****, **CHISEL.SHX****, and **STYLU.TTF**.

Helvetica Solid fonts: **HVBS.SHX***, **HBOLD.SHX***, **HVMS.SHX***, **HVLS.SHX***, **SOLID.SHX****, **SCREEN.SHX****, **SWISS.TTF**, **SWISSB.TTF**, and **SWISSK.TTF**.

Helvetica Outline fonts: **HOUTL.SHX***, **HVLO.SHX***, **HVBO.SHX***, **SWISSBO.TTF**, and **SWISSKO.TTF**

* Included in the **Using LANDCADD** disk

** Included in LANDCADD program

(Remember that the fonts with the **.TTF** extension are TrueType fonts. These will only be filled in when the system variable TEXTFILL is set to 1.)

Hint: There are many sources for extra fonts for your computer system. Many extra AutoCAD fonts are also available. One excellent source of AutoCAD fonts is Technical Software, which sells literally thousands of fonts (called T.C. Fonts™) on a single CD.

To insert different symbols and other drawings, pull down the **EP** menu, select **LANDCADD**, then choose **Landscape Design** Pull down the **Symbol** menu. Notice the choices of **Broadleaf**, **Palms & Cactus**, **Shrub**, and **Flower**. Each of these choices may reveal several options in the box toward the bottom of the screen.

With the Shrub Symbols dialogue box on the screen, you can see the options of **Simple Plan**, **Detailed Plan**, **Simple 3D**, **Detailed 3D,** and **Section View**. The symbols in the dialogue box will change according to which of these you select (Figures 11.29 and 11.30).

Use these drawings as you would any type of "Clip art". If the drawings are too large, you can use the **Scale** command to reduce the size of the drawing to fit onto your business card. See the Editing Commands section to review the **Scale** command. Remember that a scale factor with a value over 1.0 **increases** the size of the symbol, while any value under 1.0 **decreases** the size of the symbol.

Once you have finished creating the logo/business card, save it to the hard drive (and a back-up floppy disk) under the current drawing name **CARD**.

Inserting the Card into the Title Block

Open **C18SHEET** drawing.

Once this drawing is on the screen, type **DDINSERT**—which brings up this dialogue box (Figure 11.31).

Choose File . . . , then move to the drive and directory where your **CARD** drawing is stored to examine the files there.

Figure 11.29

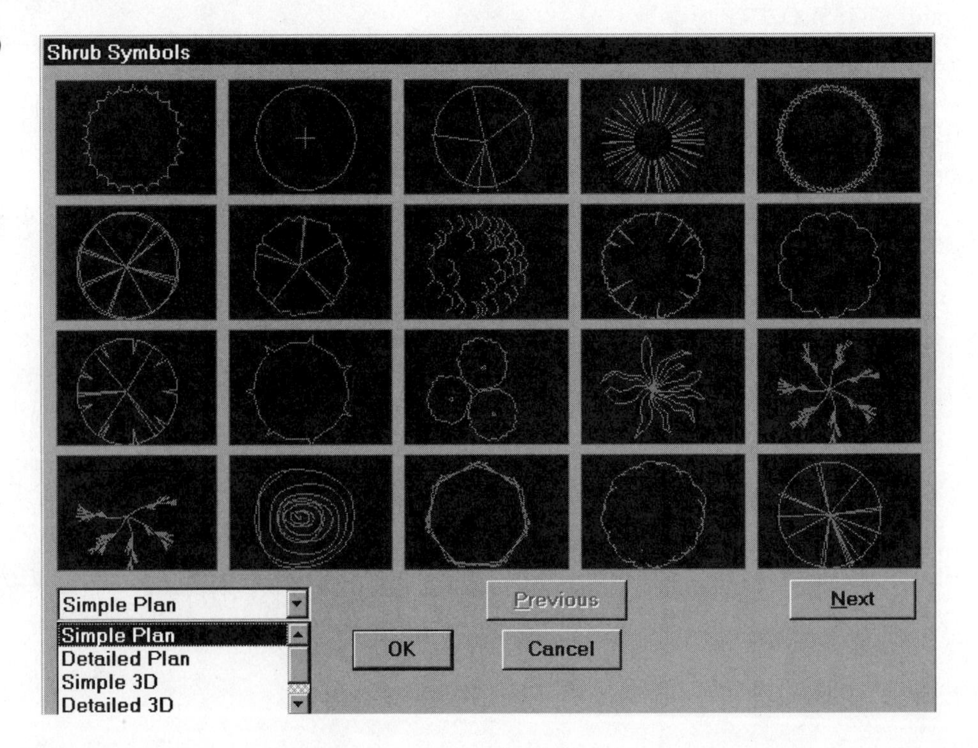

Figure 11.30

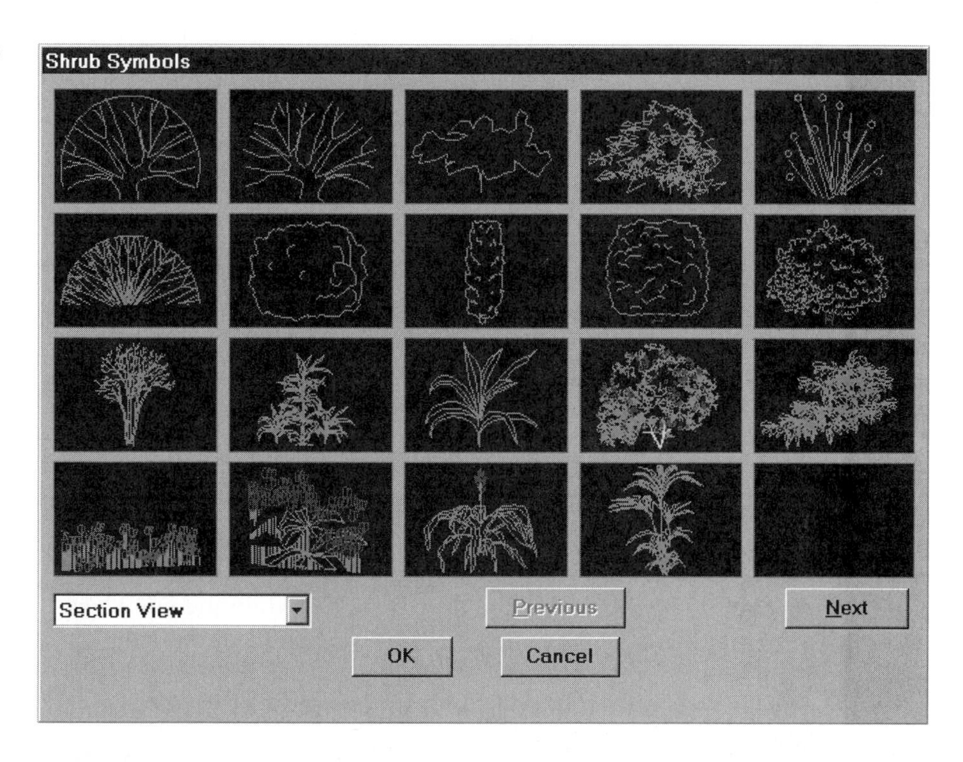

Figure 11.31

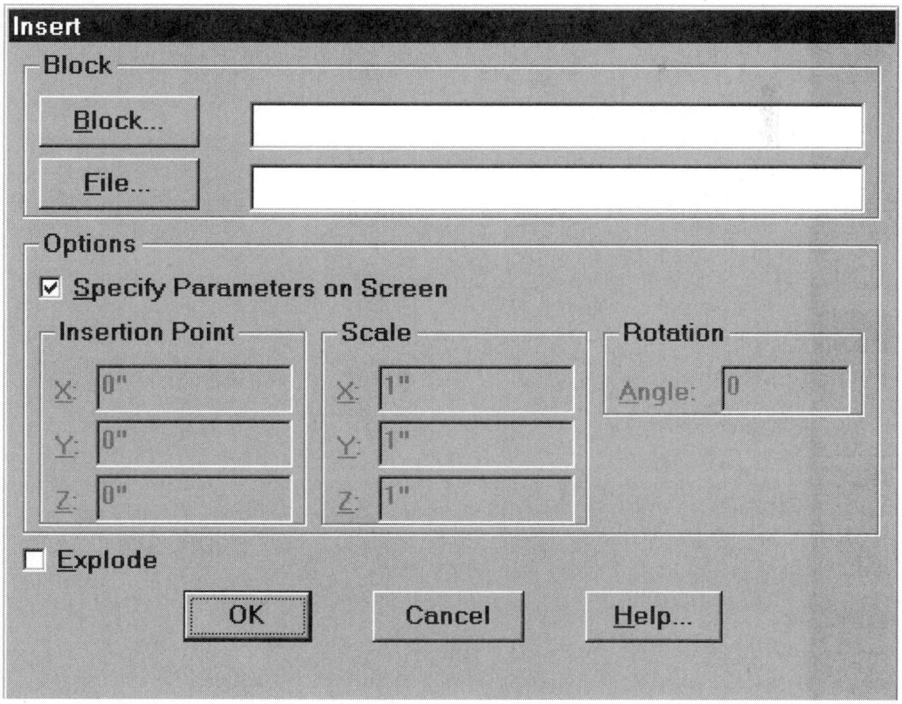

Insert

Block

Block...

File...

Options

☑ Specify Parameters on Screen

Insertion Point

X: 0"

Y: 0"

Z: 0"

Scale

X: 1"

Y: 1"

Z: 1"

Rotation

Angle: 0

☐ Explode

OK Cancel Help...

Once you locate the **CARD** drawing, highlight it and choose **OK**.

You will be asked to select an **Insertion point** in the drawing, then for a scale for the block itself. We will use **96** as the scale factor, since our drawing is at 1/8 scale or **1"=96"** scale. The rotation angle will be **90°**, since we wish to locate the card in the rectangle provided for it.

`Insertion point:`
Choose a point in the center of the screen.

`Insertion point: X scale factor <1> / Corner / XYZ:`
Type **96** to scale the card to the same scale as the title block.

`Y scale factor (default=x):`
Press **<Enter>** to accept the default value.

`Rotation angle <0>:`
Type **90** to rotate the card to the proper angle to fit in the title block.

Finally, use the **Move** command to move the lower left corner of the card (use the **cursor menu**, to highlight **Intersection**), to the appropriate intersection of the title block (Figure 11.32).

Figure 11.32

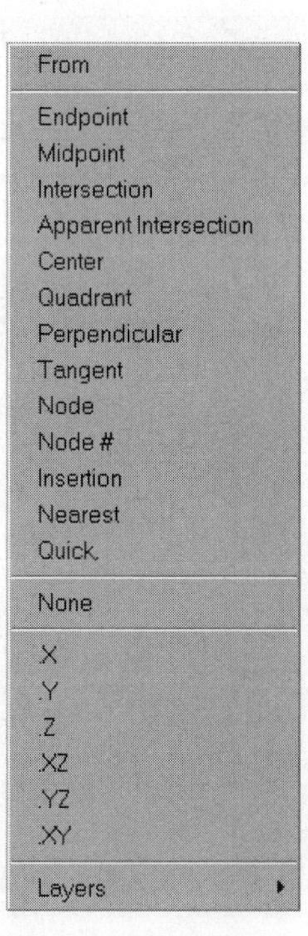

Be sure to use the cursor menu and the **Intersection** mode to make sure that the corner of the card matches with the corner of the title block box. After completing this step, save your drawing as **C18SHEET** one final time. This will be the last modification to this sheet. We will use it as a prototype for most of the other drawings we create in this book (Figures 11.33 and 11.34).

Figure 11.33

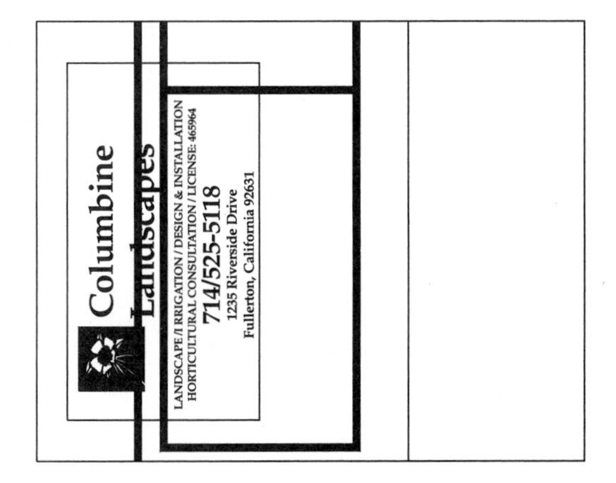

Figure 11.34

chapter

Drawing Two—The Base Sheet

The next part of the Thomas Residence project is the Base Sheet. In the first drawing specifically for the Thomas Residence, we will complete these steps:

- Build a new project in the Project Manager, using the C18SHEET drawing as a prototype
- Use existing project settings from the Project Settings library
- Load a linetype from the LANDCADD linetype file
- Set the linetype scale to display the linetype properly
- Build a property line using the LANDCADD Site Planning module
- Create a building footprint using the LANDCADD Site Planning module
- Stretch the building footprint to correct for measurement errors
- Build a roof using gable and hip styles
- Build a perspective view of the Thomas house
- Show utility lines for the residence
- Create a stipple hatch for the existing driveway
- Show an existing tree (with invisible attributes) on the property

Beginning a New Project

We will begin the new project in the Eagle Point Project Manager. The Project Manager allows us to display projects, assign Project Settings and select prototype drawings for new projects. In addition, the Project Manager provides the most convenient way to load the LANDCADD menu structure (Figure 12.1).

Figure 12.1

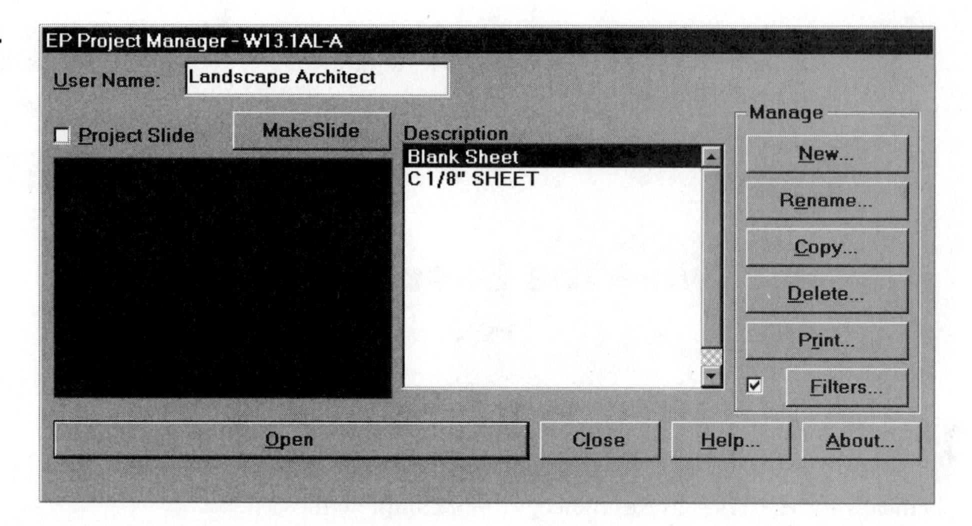

Begin the process by choosing the **New . . .** option. This brings up the "New Project" dialogue box.

In the text box for **Description:** type in **BASE SHEET**. In the text box for **Drawing Name . . .** enter the drawing name **BASE.DWG**.

Be sure to include the path where your LANDCADD drawings are stored before the file name. The full entry might look like **E:\13DWGS\BASE.DWG**. The path name will vary depending on your computer and how AutoCAD and LANDCADD are set up. See your LANDCADD documentation for details on configuration and set up (Figure 12.2).

Figure 12.2

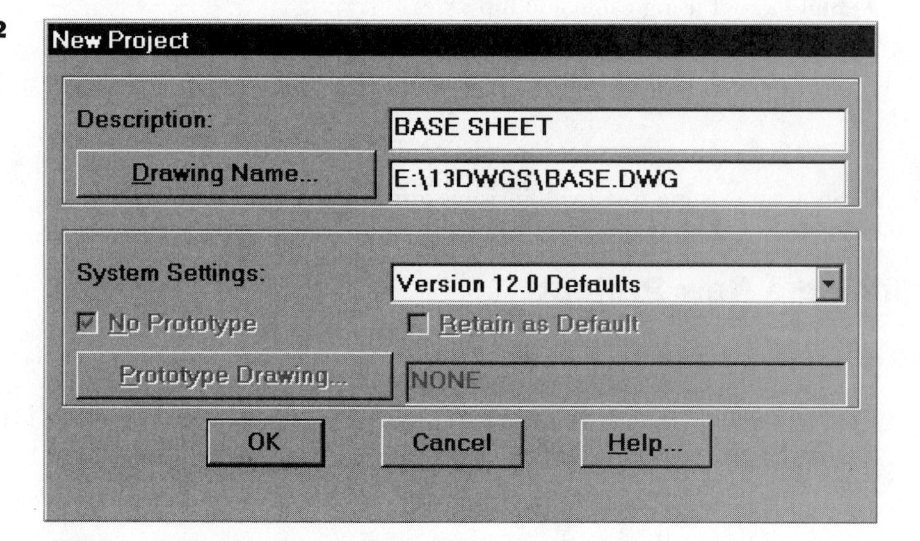

Next, remove the check from the **No Prototype** box. This makes the **Prototype Drawing . . .** box active.

This allows you to specify which drawing you want to use as a **prototype drawing**. When a prototype drawing is used, its features are "borrowed for use by the new drawing, but the original prototype drawing is never changed or altered in any way. We will use the **C18SHEET** prototype for each of our drawings in the Thomas Residence landscape remodel project.

Pick the **Prototype Drawing . . .** box to open the "Select Prototype" dialogue box.

Pick the **C18SHEET** drawing that you finished before. Click on **OK**.

This returns you to the "New Project" dialogue box with the **Prototype Drawing . . .** text box filled in (Figure 12.3).

Figure 12.3

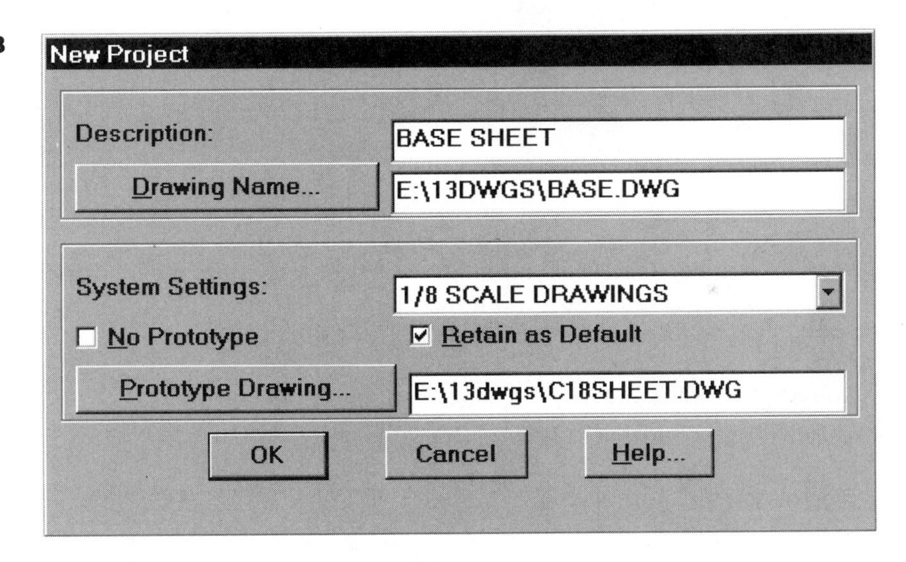

Place a check mark in the **Retain as Default** box. This will set the **C18SHEET** drawing as the default prototype for other projects in the Thomas Residence drawing set.

Be sure that the **1/8 SCALE DRAWINGS** selection is visible in the **System Settings:** box.

These settings are the ones we created for the BLANK drawing, then saved in the Project Settings library. This means that the text size and scale settings are correctly set for the Base Plan.

If this information looks correct, click on **OK**.

This will return you to the Project Manager dialogue box with the **BASE SHEET** project added to your project list (Figure 12.4).

The **BASE SHEET** project will be available any time you use the Project Manager.

Figure 12.4

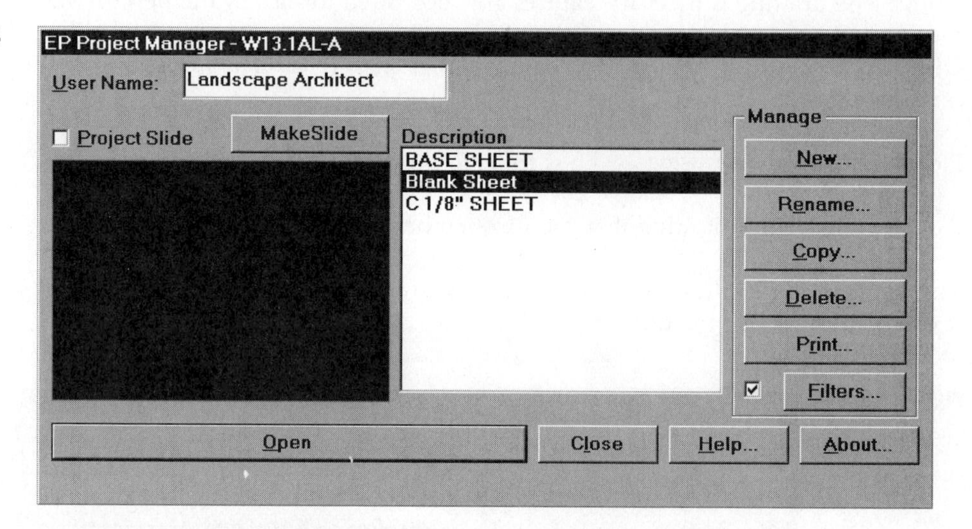

Use the **Open** option to begin work on the base sheet.

We will begin our base sheet drawing by filling out the title block with the appropriate information. To edit the text within this block, use the DDATTEDIT command (the keyboard macro is **AE**). This allows us to edit attributes within a block.

Type **AE**, then pick any of the text block dashes. Replace the dashes with the following information (Figure 12.5).

Next, move through the EP, LANDCADD, Landscape Design and Tools menus to select the AutoCAD Layer Dialog (or simply select the Layers icon in the Standard Toolbar, or type **DDLMODES**). Build a new layer called **PROPERTY** .

Enter the layer name, then select **New**. Highlight the **PROPERTY** layer, then select **Set Ltype . . .**. This activates the "Select Linetype" dialogue box.

At this point the linetype **EPLC_PROPERTY** is not available, so must be loaded from LANDCADD's linetype file.

Pick **Load . . .** to find the linetype file. This brings up the "Load or Reload Linetypes" dialogue box (Figure 12.6).

Figure 12.5

Edit Attributes

Block Name: TEXT

JOB NAME	THOMAS RESIDENCE
PROJECT NAME:	LANDSCAPE REMODEL
STREET ADDRESS:	12321 NUTWOOD AVENUE
CITY STATE & ZIP	FULLERTON, CA. 92832
JOB NUMBER	96-101
SHEET NUMBER	L-1
TOTAL SHEETS:	15
DRAWING SCALE:	1/8"=1'-0"

OK Cancel Previous Next Help...

Figure 12.6

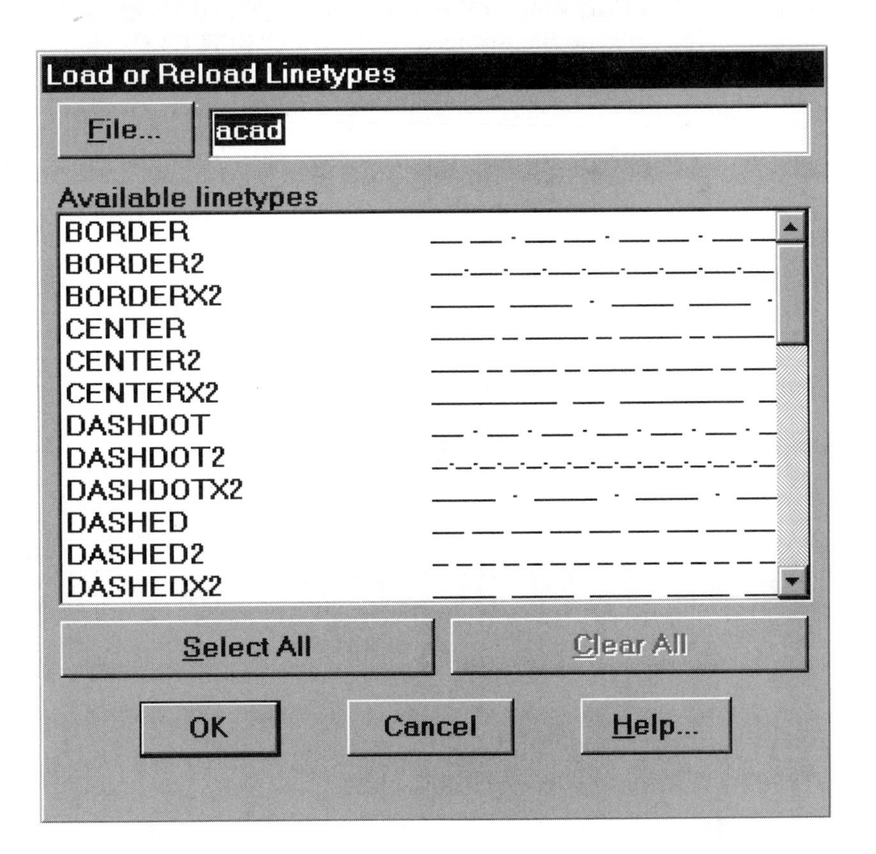

The default linetype file is called (naturally) **acad.**

Since we need to load a different linetype file, pick the **File...** box to find and load the correct file (Figure 12.7).

Figure 12.7

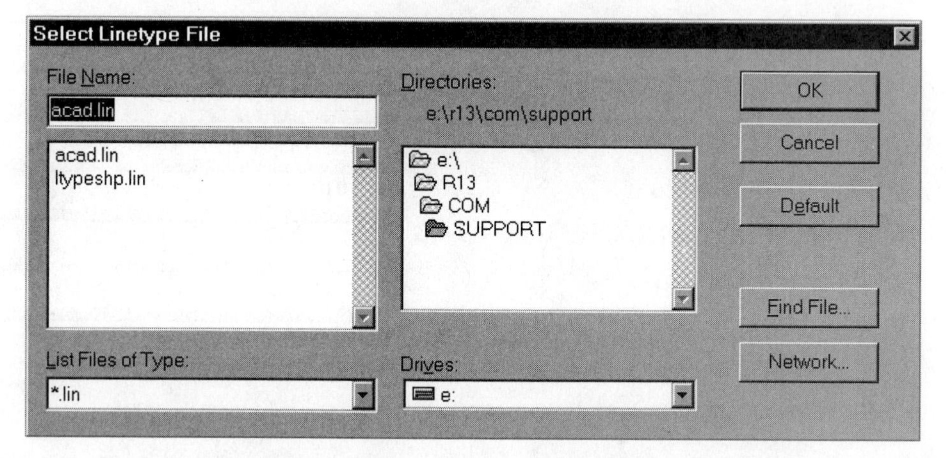

Move from the AutoCAD support directory into the LANDCADD directory to find the linetype file. The linetype file we need is called **LC_LTYPE.LIN**, and is stored in the **EPWIN13\PROGRAM** directory of AutoCAD. If your system has the LANDCADD files stored in another directory, it should still be in a subdirectory labeled **PROGRAM.**

Highlight the **LC_LTYPE** file, so that appears in the **File Name** box, then pick **OK** (Figure 12.8).

Figure 12.8

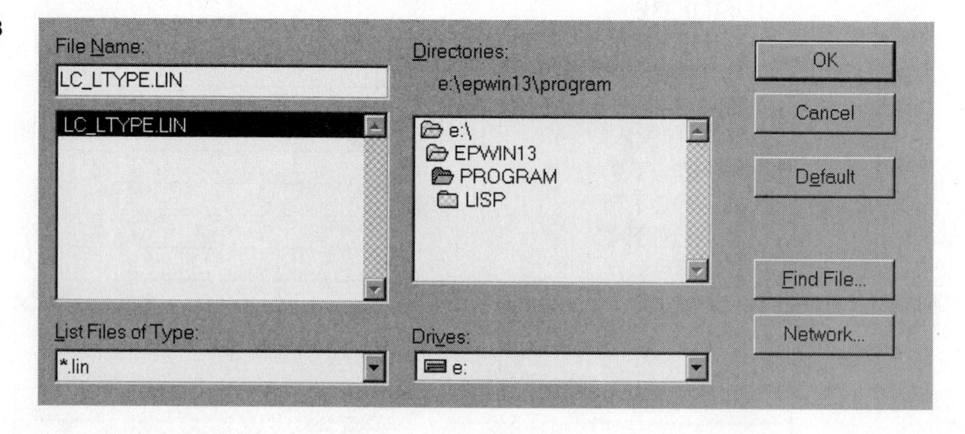

After the "Load or Reload Linetypes" dialogue box reappears, highlight **EPLC_PROPERTY**, then pick **OK** (Figure 12.9).

Figure 12.9

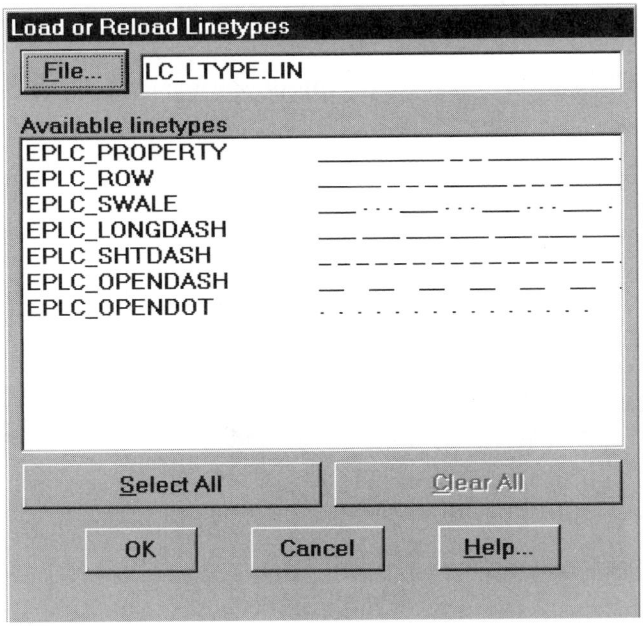

Load or Reload Linetypes
File...
LC_LTYPE.LIN
Available linetypes
EPLC_PROPERTY
EPLC_ROW
EPLC_SWALE
EPLC_LONGDASH
EPLC_SHTDASH
EPLC_OPENDASH
EPLC_OPENDOT
Select All
Clear All
OK
Cancel
Help...

You are returned to the "Select Linetype" dialogue box, but it now allows you to choose between two linetypes (Figure 12.10).

Figure 12.10

Select Linetype
Loaded Linetypes
CONTINUOUS
EPLC_PROPERTY
ISO Pen Width:
Linetype Scale:
0'1"
Linetype:
EPLC_PROPERTY
OK
Cancel
Load...
Help...

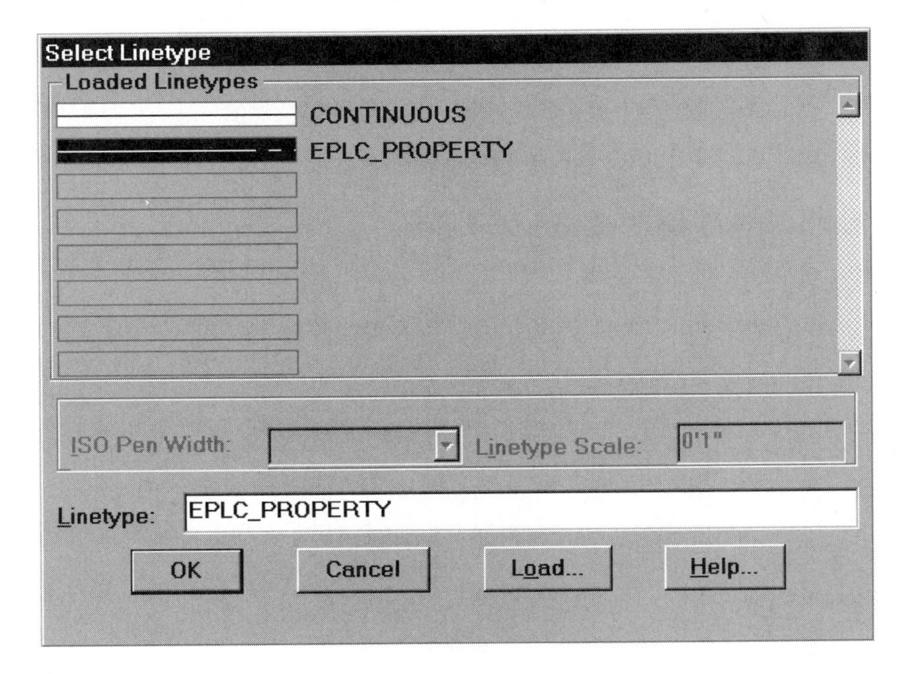

Highlight **EPLC_PROPERTY**, then pick **OK**.

The "Layer Contro"l dialogue box now contains the correct linetype for the layer **PROPERTY**.

After you choose this linetype, select **PROPERTY** as the current layer, then select **OK**.

At the command prompt, type **LTSCALE**. Then type in the value **96**.

This will show the property line at the correct proportions with the rest of the drawing. If the property line appears solid rather than dashed, type **REGEN** to regenerate the drawing with the correct linetype scale on screen (Figure 12.11).

Figure 12.11

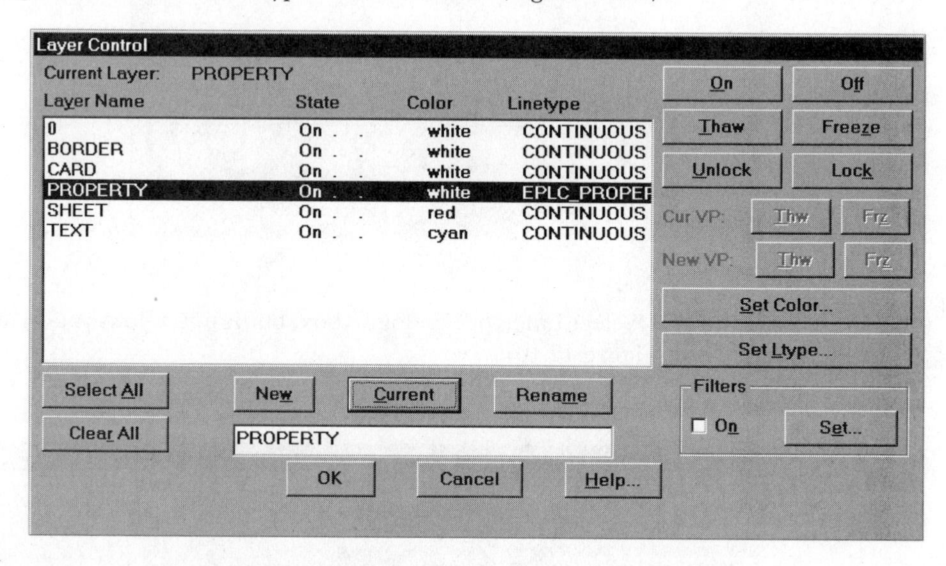

Building the Property Line

Next, we will use LANDCADD routines for building property lines. The property line commands allow us to enter the precise length and orientation of each property line segment. This means that we could duplicate property lines from civil engineering data, survey reports or local plat maps. Although we have the ability to create complex curves, elevation changes and detailed angular geometry, we will build a simple rectangular lot for the Thomas Residence.

Begin the property line by selecting the **EP** pull-down menu, picking **LANDCADD**, and finally **Site Planning**. Select the **Layout** pull-down menu, then choose the **PropLine** option. This opens the "Define property lines" dialogue box.

In the "Define property lines" dialogue box, we choose the annotation included with the property lines, along with the layers on which the labels, lines and point nodes are located. We can also choose the method of inputting the point data. In the **Traverse** option, we enter points in a sequence that builds lines from one endpoint to the next. The **Side Shot** option allows us to enter a series of lines at a known angle and distance from the original point (like spokes on a wheel).

Because our property lines form a simple rectangle for an existing home, we will omit point nodes and property line labels from our plan.

In the "Define property lines" dialogue box, enter the following settings:

For the **Labels** selection, pick the radio button marked **None**.

Remove the check mark from the **Point Nodes** toggle box.

In the **Mode** section, pick the radio button marked **Traverse** (Figure 12.12).

Figure 12.12

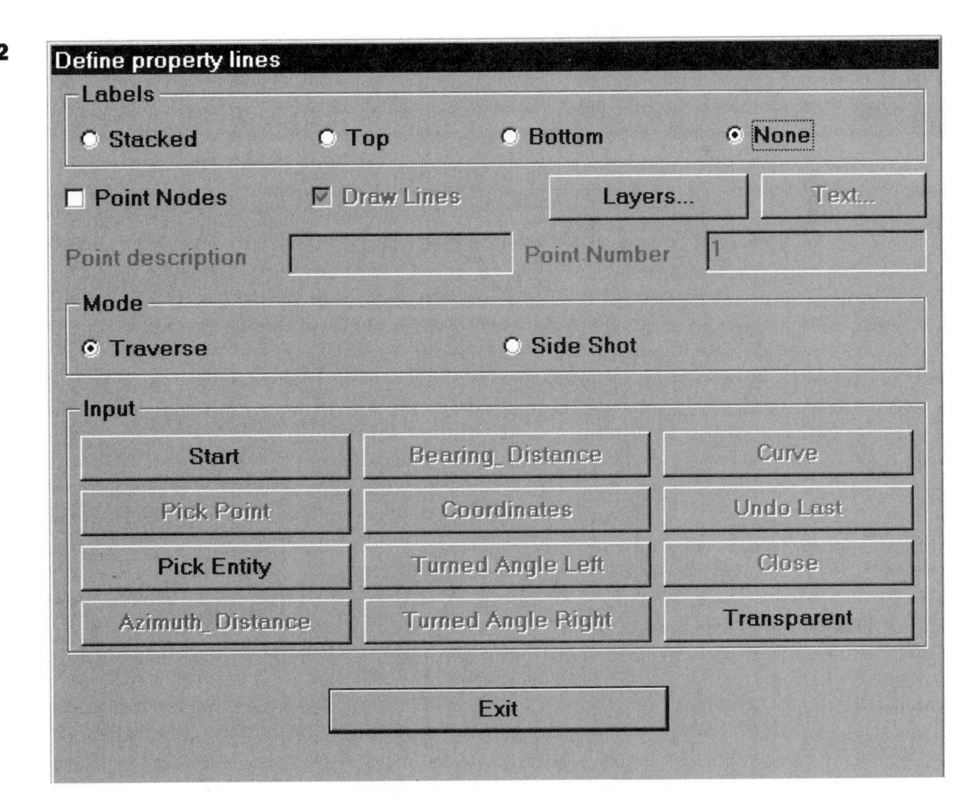

Pick the Layers. . . option to indicate the layer on which the property line will be built (Figure 12.13).

Figure 12.13

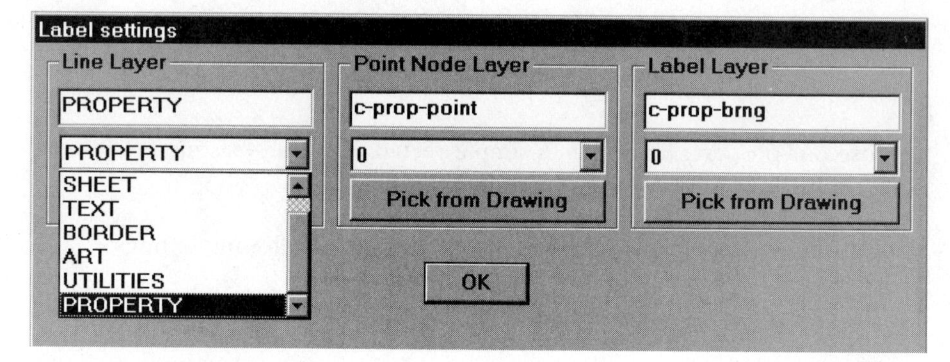

Since we chose not to display point nodes or labels, we can ignore those settings here. We must, however, choose the layer to display the property line. We just created a layer for that purpose, so we will use the new **Property** layer.

Pick the **PROPERTY** layer under the **Line Layer** pop-down box. Pick **OK** when finished.

After you are returned to the "Define property lines" dialogue box, notice that the methods for inputting data are "grayed out." None of these input options are available until a first point is chosen.

Begin building the property line by picking **Start**.

```
Command: EPSP_property
Select start point or number: Number/<point>:
```
Type (or pick) **20',22'**.

After you are returned to the "Define property lines" dialogue box, pick **Bearing Distance**. This input method allows us to define the lines by their angle and distance from our starting point (Figure 12.14).

In the "Bearing Distance" dialogue box, we can enter the quadrant for the orientation of the angle for the property line. We can choose to orient from **North-East**, **North-West**, **South-East** or **South-West**. After choosing a quadrant, we then specify the **Bearing** (angle) and **Distance** from the last point. As a quadrant selection and bearing are entered, the red line in the box above will orient in the direction that the line will be drawn. This is very helpful for double checking the line orientation before it is drawn.

For the ending point of the first line, pick **North-East**, and a **Bearing:** of **0**. Enter a **Distance:** of **100'**.

Figure 12.14

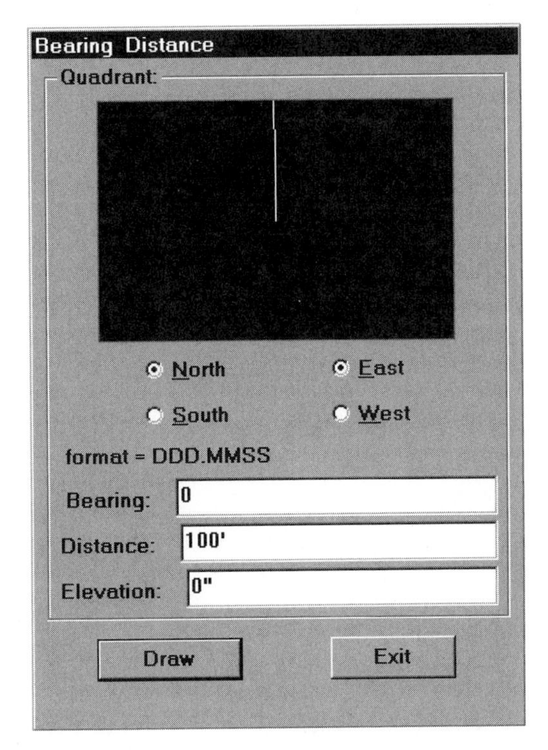

For the second line endpoint, pick **North-East**, and a **Bearing:** of **90**. Enter a **Distance:** of **135'**.

For the third line endpoint, pick **South-East**, and a **Bearing:** of **0**. Enter a **Distance:** of **100'**.

To return to the original point, pick **South-East**, and a **Bearing:** of **90**. Enter a **Distance:** of **135'**.

This lot will then be 13,500 square feet (about one-quarter acre) (Figure 12.15).

The Building Footprint

Begin the building footprint by entering the **Site Planning** module under the **LANDCADD** section of the **EP** pull-down menu. Select the **Bldgs** pull-down menu, then choose the **Footprint** option. This opens the "Building Footprint" dialogue box (Figure 12.16).

Figure 12.15

Since we are working mostly in plan (overhead) view, the **Entity Type:** will be a **Pline** (or polyline).

For **Entity Type:** pick the radio button for **Pline**.

We could also build this house with three-dimensional faces, but these aren't really needed here. For a **Base Elevation** accept the default of **0"**. If our building were to appear in a topographic drawing, we could specify an elevation, or pick a contour line with an elevation suitable for the building pad.

The polylines representing the walls will have a width and a height. (LANDCADD **height** is synonymous with AutoCAD **thickness.**) This makes them suitable for 3-D viewing when we examine perspective views later in this exercise.

If we assume 2"x4" wood wall construction with 1" thick stucco outside, and about 1" thickness for drywall or plaster, we end up with a final wall width of 6".

Type **6"** in the **Width of Walls** text box.

If we assume a raised foundation of 2' and a wall height of 8', this yields a wall height of 10'.

Figure 12.16

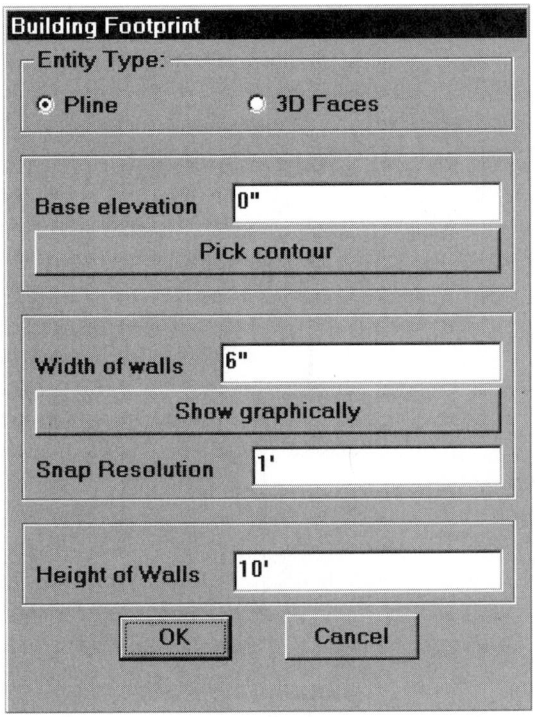

Building Footprint
Entity Type:
Pline
3D Faces
Base elevation 0"
Pick contour
Width of walls 6"
Show graphically
Snap Resolution 1'
Height of Walls 10'
OK
Cancel

Type **10'** in the **Height of Walls** text box.

Once you have entered the values for the construction of the building, you are ready to begin entering in points which define the walls and window and door openings. The footprint you build will be a continuous polyline; the walls have a width of 6", while the openings have a width of 0". The LANDCADD program continues to build walls until you tell it to create a window (or door). The footprint creation routine does not distinguish between windows and doors.

To create a window or door at the end of a wall segment, simply type **W**. Then enter the ending point of the window. The program assumes that you want to begin the window at the point where you typed **W**, and want to end it at the next named point. After you enter the ending point of the window, the program starts building walls again.

The first point of the footprint should be at the bottom opening of the garage door (Figure 12.17).

At the command prompt, begin:

```
Command: EPSP_footprint
```

Figure 12.17

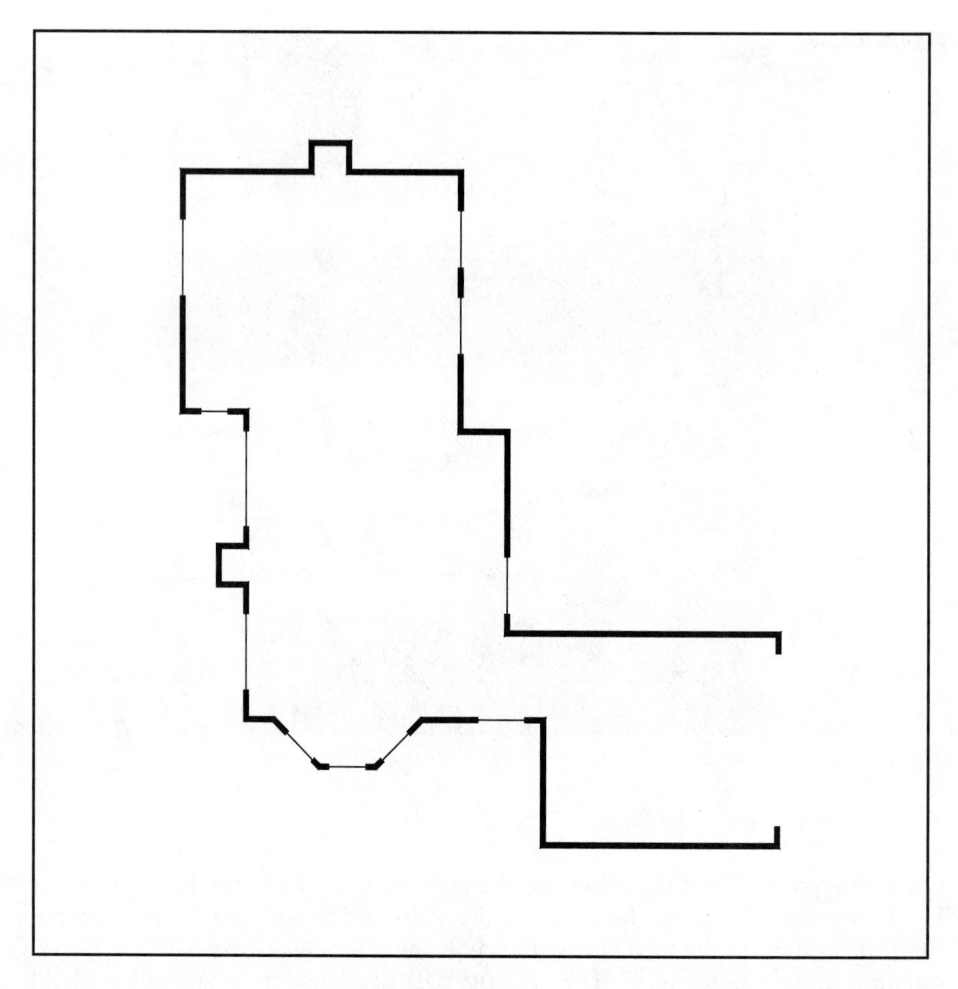

`Start point: Type (or pick)` **`140',37'`**`.`
`Select next point (Close,Window,Undo)` (Figure 12.18):
Type **@2'<270**.

`Select next point (Close,Window,Undo):`
Type **@25'<180**.

`Select next point (Close,Window,Undo):`
Type **@13'<90**.

`Select next point (Close,Window,Undo):`
Type **@2'<180**.

Figure 12.18

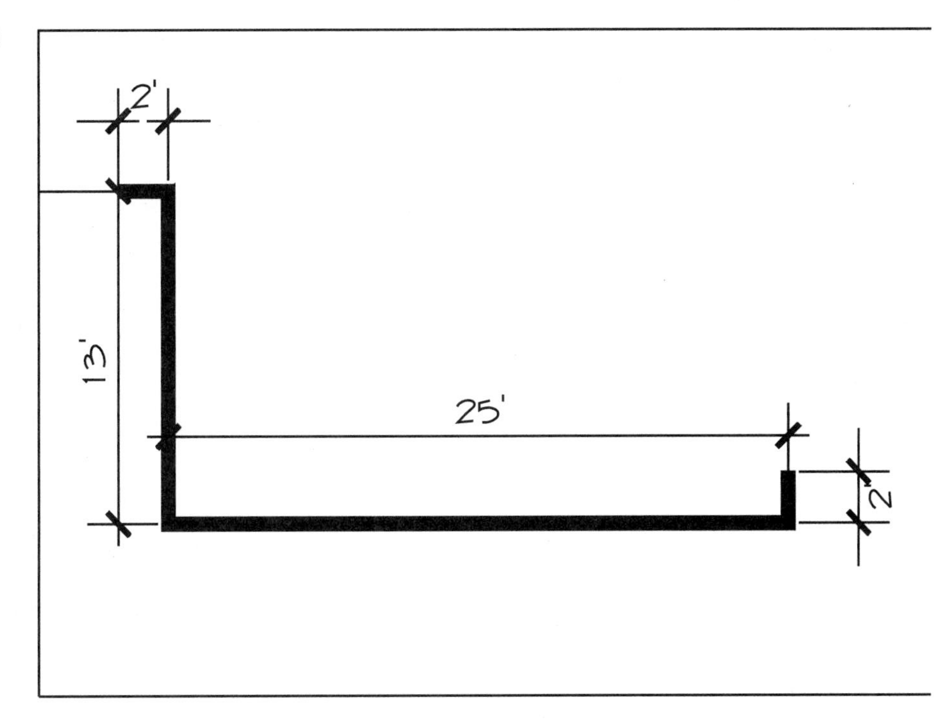

At this point, we build the first opening:

`Select next point (Close,Window,Undo):`
Type **W** to build a window (actually a door) opening(Figure 12.19).

`Pick end point of window/door:`
Type **@5'<180** to mark the far edge of the door.

`Select next point (Close,Window,Undo):`
Type **@6'<180** to build more wall.

`Select next point (Close,Window,Undo):`
Type **@2'<225**.

`Select next point (Close,Window,Undo):`
Type **W** to build a window opening.

`Pick end point of window/door:`
Type **@4'<225** to mark the far edge of the window.

Figure 12.19

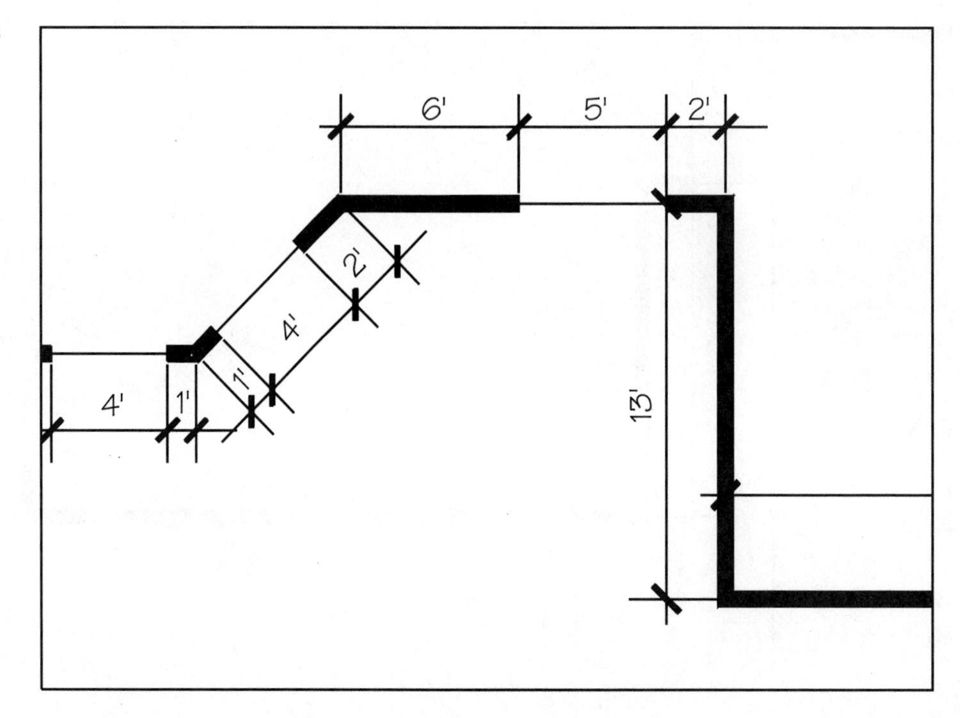


```
Select next point (Close,Window,Undo):
```
Type **@1'<225** to build more walls.

```
Select next point (Close,Window,Undo):
```
Type **@1'<180**.

```
Select next point (Close,Window,Undo):
```
Type **W** to build a window opening.

```
Pick end point of window/door:
```
Type **@4'<180** to mark the far edge of the window.

```
Select next point: (Close,Window,Undo)
```
Type **@1'<180** to build more walls.

```
Select next point (Close,Window,Undo):
```
Type **@1'<135** (Figure 12.20).

```
Select next point (Close,Window,Undo):
```
Type **W** to build a window opening.

Figure 12.20

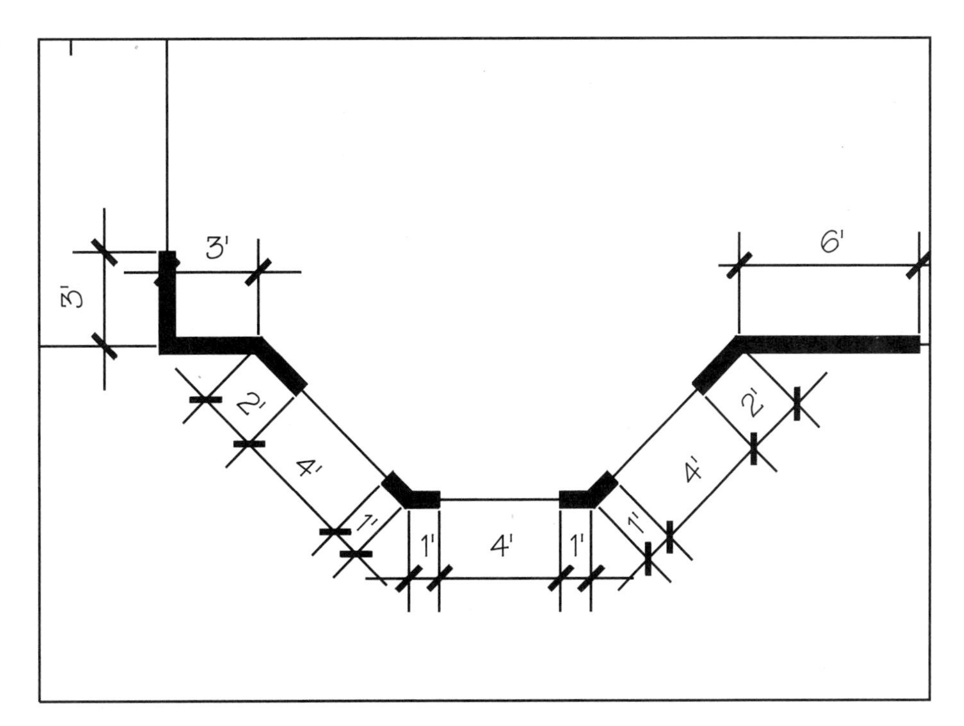


```
Pick end point of window/door:
```
Type **@4'<135** to mark the far edge of the window.

```
Select next point (Close,Window,Undo):
```
Type **@2'<135** to build more walls.

```
Select next point (Close,Window,Undo):
```
Type **@3'<180**.

```
Select next point (Close,Window,Undo):
```
Type **@3'<90**.

```
Select next point (Close,Window,Undo):
```
Type **W** to build a window (actually a door) opening (Figure 12.21).

```
Pick end point of window/door:
```
Type **@8'<90** to mark the far edge of the door.

```
Select next point (Close,Window,Undo):
```
Type **@3'<90** to build more walls.

Figure 12.21

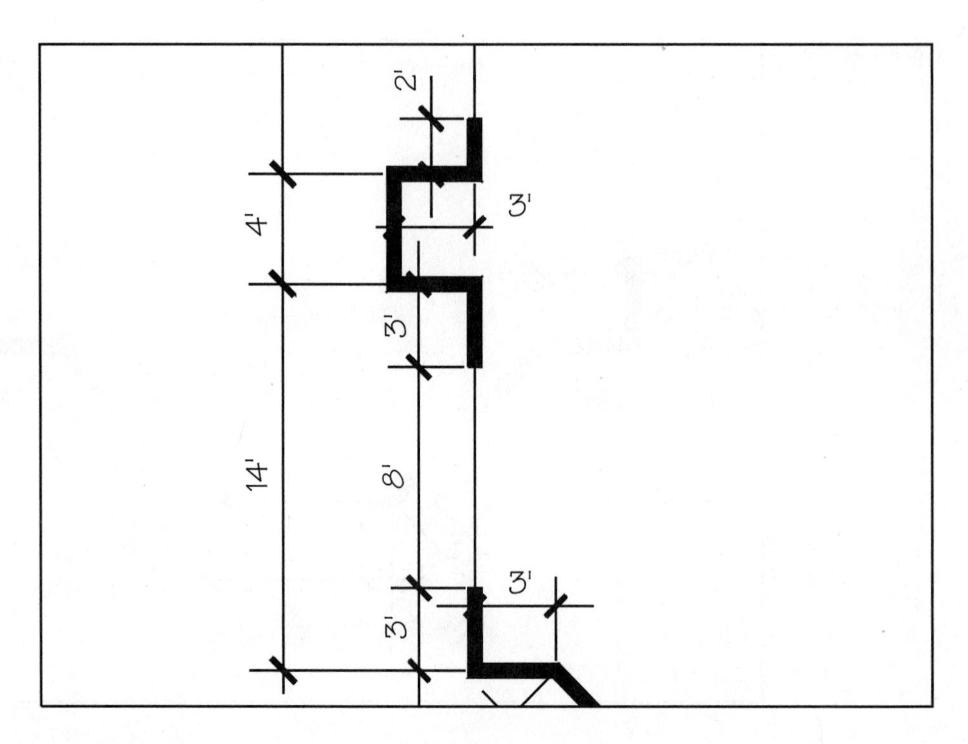


```
Select next point (Close,Window,Undo):
```
Type **@3'<180**.

```
Select next point (Close,Window,Undo):
```
Type **@4'<90**.

```
Select next point (Close,Window,Undo):
```
Type **@3'<0**.

```
Select next point (Close,Window,Undo):
```
Type **@2'<90**.

```
Select next point (Close,Window,Undo):
```
Type **W** to build a window opening (Figure 12.22).

```
Pick end point of window/door:
```
Type **@10'<90** to mark the far edge of the window.

```
Select next point (Close,Window,Undo):
```
Type **@2'<90** to build more walls.

Figure 12.22

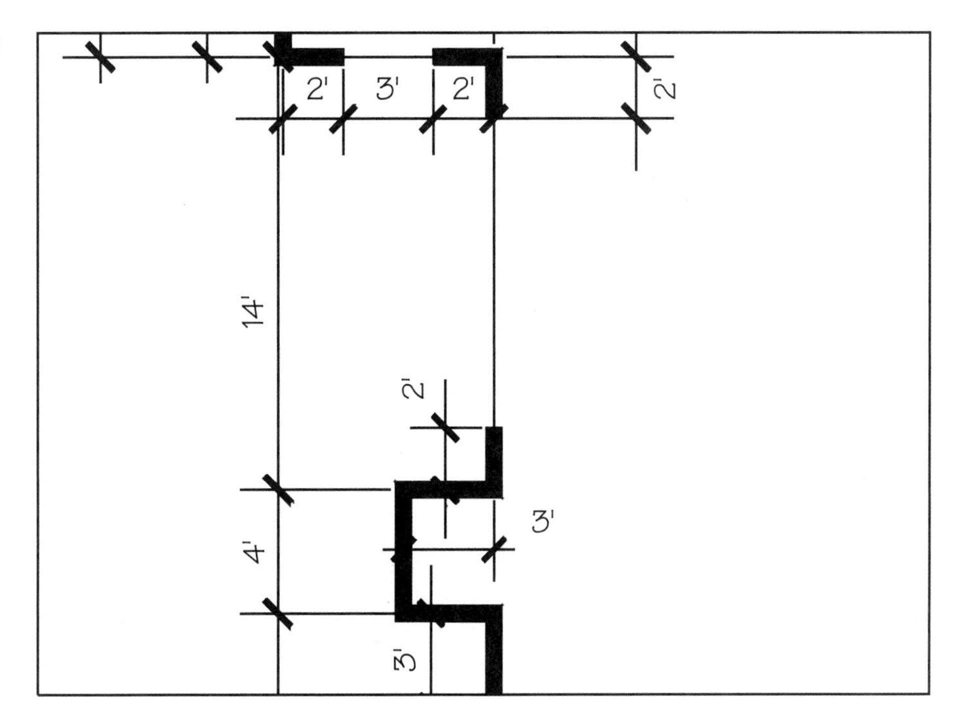

`Select next point (Close,Window,Undo):`
Type **@2'<180**.

`Select next point (Close,Window,Undo):`
Type **W** to build a window opening.

`Pick end point of window/door`
Type **@3'<180** to mark the far edge of the window.

`Select next point (Close,Window,Undo):`
Type **@2'<180** to build more walls (Figure 12.23).

`Select next point (Close,Window,Undo):`
Type **@12'<90**.

`Select next point (Close,Window,Undo):`
Type **W** to build a window (door) opening.

`Pick end point of window/door:`
Type **@8'<90** to mark the far edge of the door.

Figure 12.23

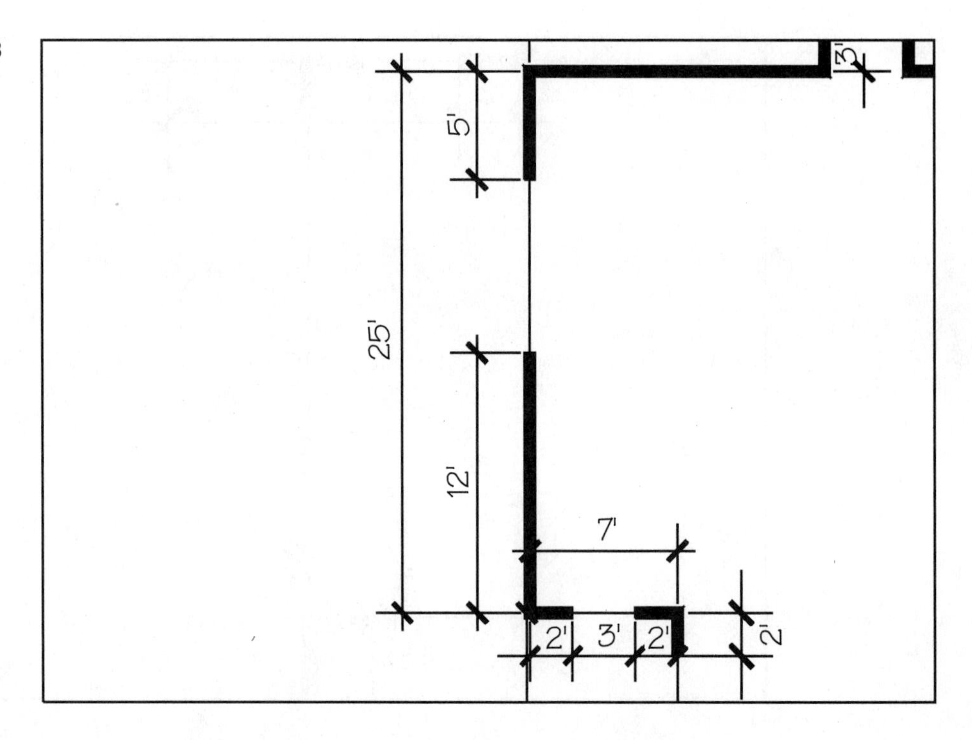


```
Select next point (Close,Window,Undo):
```
Type **@5'<90** to build more walls.

```
Select next point (Close,Window,Undo):
```
Type **@14'<0** (Figure 12.24).

```
Select next point (Close,Window,Undo):
```
Type **@3'<90**.

```
Select next point (Close,Window,Undo):
```
Type **@4'<0**.

```
Select next point (Close,Window,Undo):
```
Type **@3'<270**.

```
Select next point (Close,Window,Undo):
```
Type **@12'<0**.

```
Select next point (Close,Window,Undo):
```
@4'<270 (Figure 12.25).

Figure 12.24

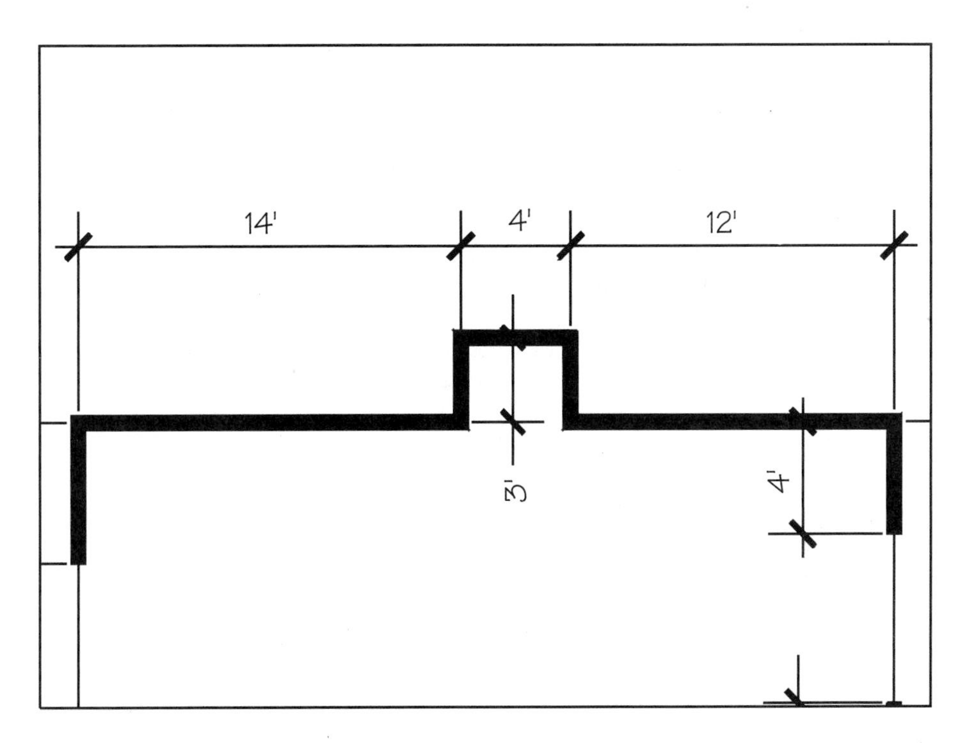

Figure 12.25

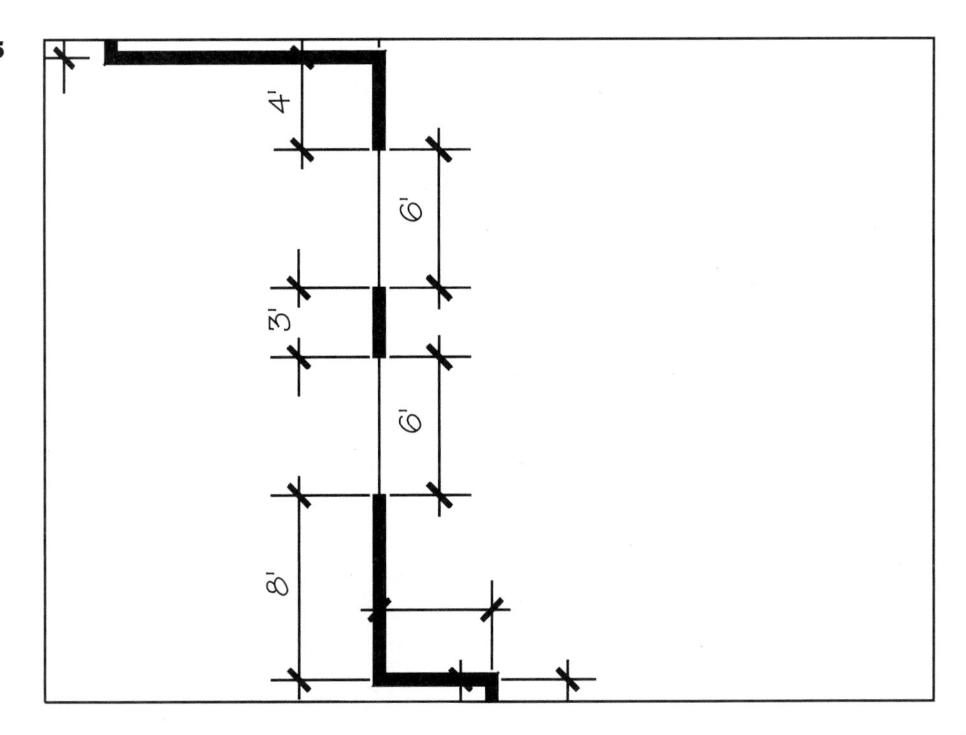

`Select next point (Close,Window,Undo):`
Type **W** to build a window (door) opening.

`Pick end point of window/door:`
Type **@6'<270** to mark the far edge of the door.

`Select next point (Close,Window,Undo):`
Type **@3'<270** to build a wall.

`Select next point (Close,Window,Undo):`
Type **W** to build a window opening.

`Pick end point of window/door:`
Type **@6'<270** to mark the far edge of the window (Figure 12.26).

Figure 12.26

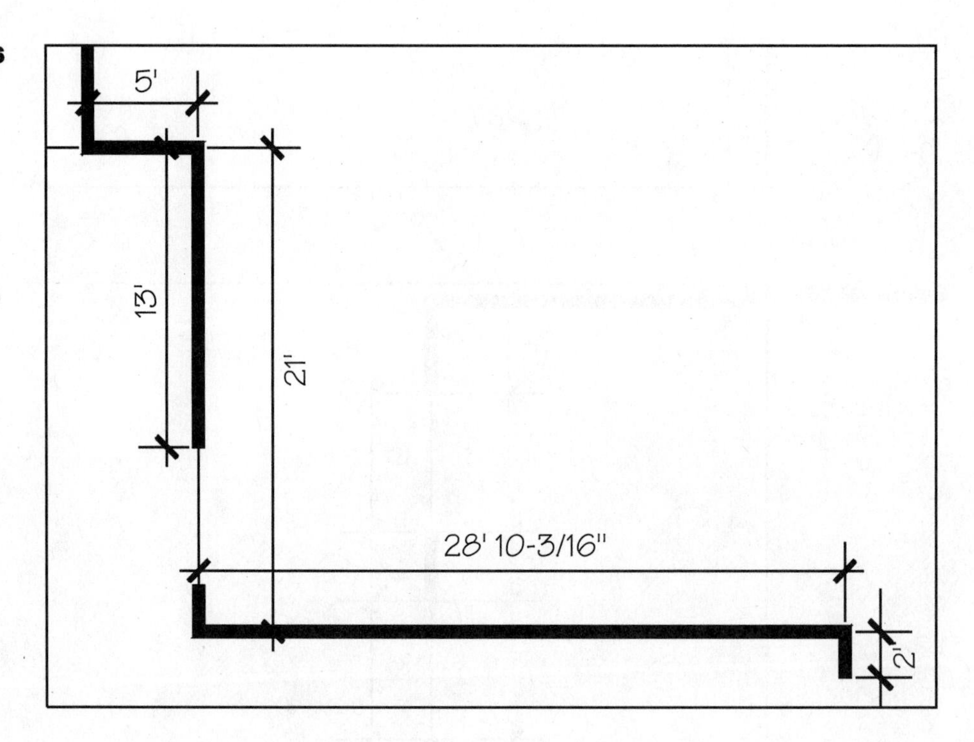

`Select next point (Close,Window,Undo):`
Type **@8'<270** to build more walls.

`Select next point (Close,Window,Undo):`
Type **@5'<0**.

`Select next point (Close,Window,Undo):`
Type **@13'<270**.

`Select next point (Close,Window,Undo):`
Type **W** to build a window (door) opening.

`Pick end point of window/door:`
Type **@6'<270** to mark the far edge of the door.

`Select next point (Close,Window,Undo):`
Type **@2'<270** to build the final walls.

`Select next point (Close,Window,Undo):`
Type **@28'10-13/16"<0**.

`Select next point (Close,Window,Undo):`
Type **@2'<270**.

`Select next point (Close,Window,Undo):`
Type **<Enter>** to leave the command.

Check to be sure that your drawing looks like the House Footprint shown at the beginning of this section.

If you make a mistake, you can use the **UNDO** option to undo the last point you entered. If you made a mistake several points ago, you can either use **UNDO** several times, or you can edit the polyline later, using the **PEDIT** command. You may also find, as a result of errors, that your footprint is actually made up of different polyline segments. To convert them back into a single object, use the **PEDIT** option **JOIN**. This should add the segments back onto the original polyline. If segments do not join properly, they may not share common endpoints. This can be remedied by moving the endpoints of the polyline segments so that they coincide precisely. Use the **ENDPOINT** object snap to assist in this process.

Stretching the Footprint

Once the footprint is completed, you may notice that the corner points of most of the house do not precisely fit to snap points on the drawing. The construction of the house footprint actually leaves most of the corner points of the house 1-3/16" off of the nearest snap points. To edit the misalignment, the best command to use is **STRETCH**. The **STRETCH** command is especially helpful in editing polylines, since it moves only those vertices (corner and end points) which are selected, and leaves the others in place. The polyline is then stretched to make up for the change in locations. Use the process shown below.

At the command prompt, begin:

```
Command:
```
Type **STRETCH**.

```
Select objects to stretch by crossing-window or -polygon...
Select objects:
```
Type: **CR** (for **CROSSING**).

```
First corner:
```
Pick a point to begin the crossing box. Be sure to include the indicated corner points of the footprint, but exclude the others.

```
Other corner:
```
Stretch the crossing box across the screen as shown. The entire footprint will highlight, but only those points included in the crossing box will be relocated. If the crossing box leaves off some of the corner points, or includes too much of the footprint, cancel the command and begin again (Figure 12.27).

```
Select objects:
```
Press **<Enter>** to finish selecting objects.

```
Base point of displacement:
```
Pick any point on the screen (inside the house is probably best).

```
Second point of displacement:
```
Type **@1-3/16"<180**. This instructs the program to move those corner points selected at a distance of 1-3/16" to the left. This places them in alignment with the snap points on the screen.

To verify that the corner points are correctly located, use the **ID** function to examine the location of the corner (intersection) points shown below (Figure 12.28).

If your corner points do not correspond to those shown above, the instructions for the Hardscape, Planting and Irrigation Plans will be inaccurate. Try to correct any errors before proceeding.

Remember that these polyline walls are not flat, but also have a 3-D height. We will put a roof on the walls, and actually show a house in a 3-D perspective view.

This should complete the building footprint. The final step in the completion of the base sheet will be the addition of roof lines, utility lines, and existing plantings and hardscape to remain during the landscape remodel. While the utility lines, hardscape and existing plant information will be useful only in plan view, the roof information will be inserted into the drawing file so that we can see the entire building in a full perspective view.

Figure 12.27

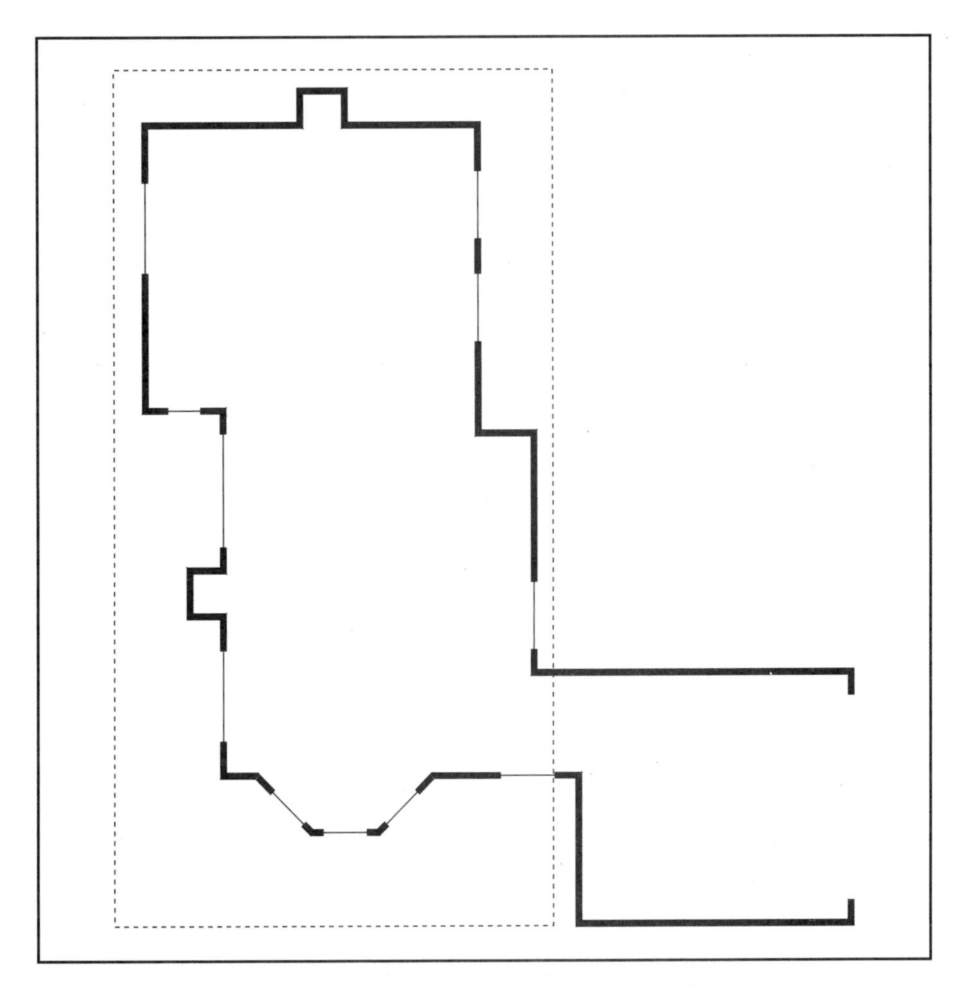

Figure 12.28

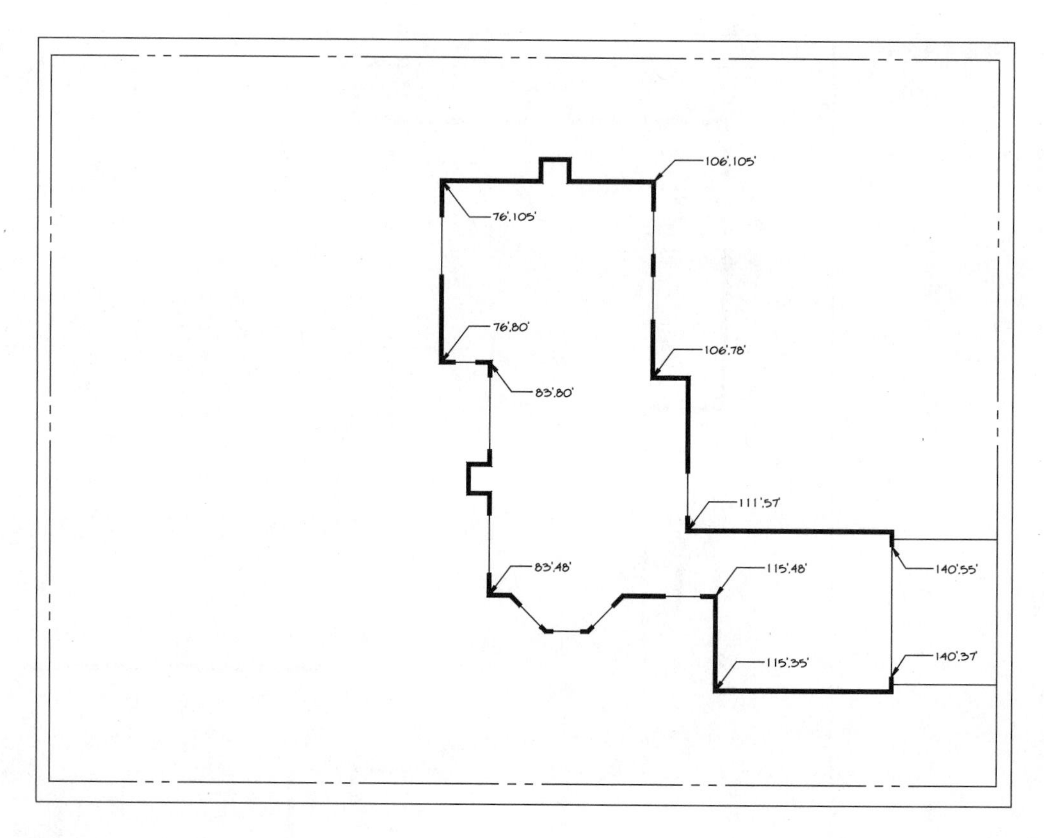

Building the Roof

First, we will put a complete roof on the walls of the building. Building a standard gable roof on a building requires that we establish several fixed points and a pitch (slope ratio) for the roof. The fixed points include the gutter (or eave) height, the ridge height, the corners of the roof and the end points of the ridge. The roof pitch is typically expressed as RISE:RUN. A roof pitch of 6:12 or 5:12 is typical for residential construction.

We can proceed in one of two ways. Using a manual method, we specify the height of the gutter, then lay out the corner points. Once these are established, we calculate the location of the ridge, and calculate the height of the ridge, so that the roof pitch meets a particular ratio. Each of these are entered into the computer as it builds the roof. Using a more automated method, we can specify the gutter height and roof pitch, then allow the computer to generate the ridge height after we choose the corner points. This second method is faster and easier.

Begin by loading **Site Planning**, then use the **Bldgs** pull down menu. Choose **Roofs . . .** (Figure 12.29).

Figure 12.29

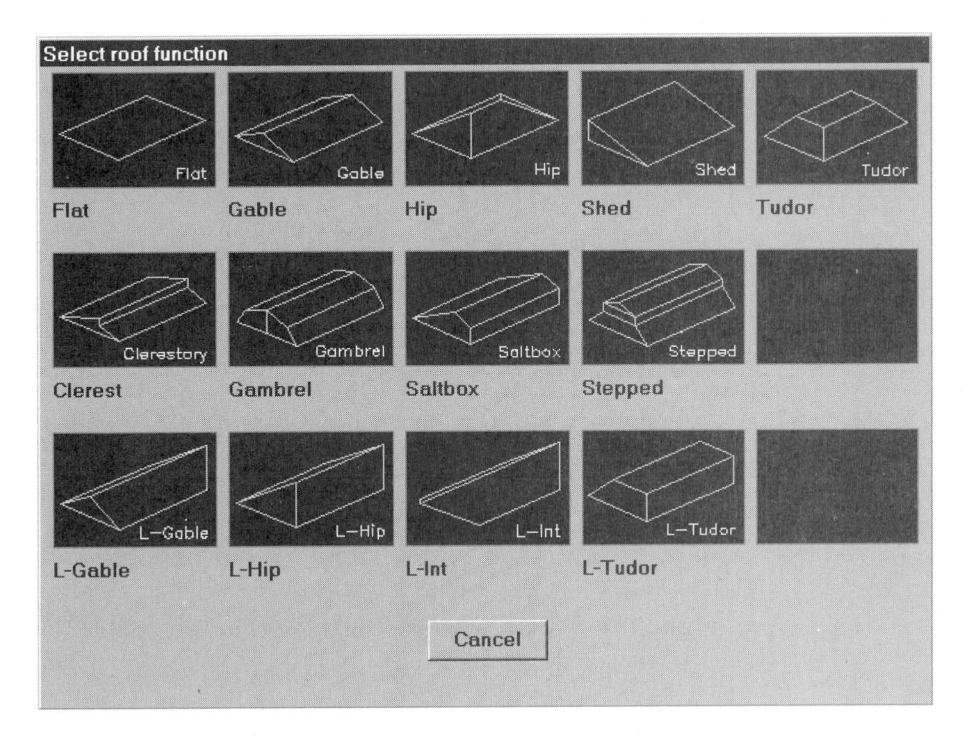

This brings up an icon menu of different roof styles. Of all the options shown, we will use only **Gable**, **L-Gable** and **Hip**.

Move your cursor into the box marked **Gable**. Use the **Pick** button on that option.

Once we choose the **Gable** roof style, the **Gable roof settings** box appears (Figure 12.30).

In this box, we define the **Gutter height** as **10'-0"**. This is the default setting. The peak height can either be set **By elevation** or **By pitch relative to 12**.

We will use a pitch of 5:12, so change the value in the box to **5.00**.

This setting will allow the computer to calculate the ridge height according to our corner points and roof pitch. If we use the setting **By elevation**, we would need to make this calculation ahead of time.

This will bring up the prompt:

```
Show first corner of roof:
```

Type or pick **74',107'**.

Figure 12.30

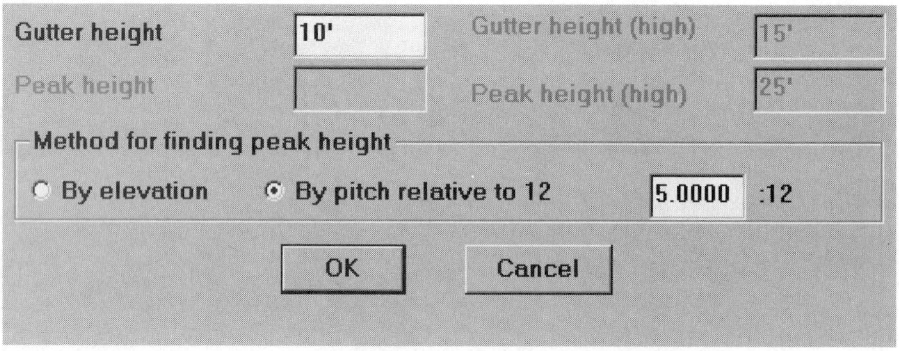

Gutter height 10'
Peak height
Gutter height (high) 15'
Peak height (high) 25'
Method for finding peak height
By elevation
By pitch relative to 12 5.0000 :12
OK
Cancel

`Show gutter ANGLE and longest side of roof:`
Type or pick **74',79'**.

`Opposite corner:`
Type or pick **108',107'**. LANDCADD then builds a complete gable roof over the points you specified (Figure 12.31).

Figure 12.31

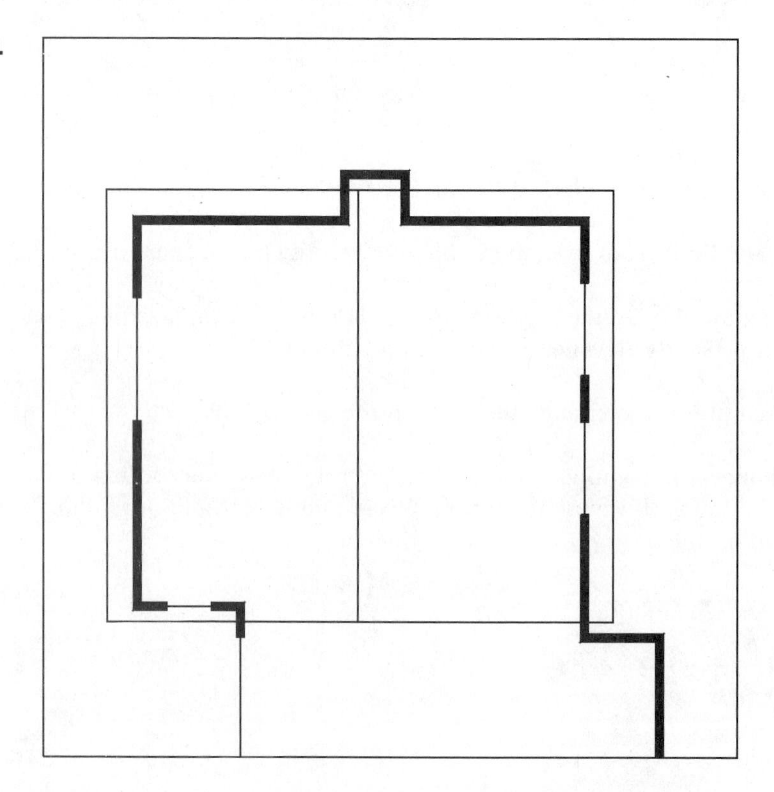

`Command:`
Press **<Enter>** to repeat the command.

Select **Gable** from the **Select roof function** icon menu. Accept the original values for **Gutter height** and **By pitch relative to 12** as the ridge height calculation method.

`Show first corner of roof:`
Type or pick **81',79'**.

`Show gutter ANGLE and longest side of roof:`
Type or pick **81',46'**.

`Opposite corner:`
Type or pick **114',79'** (Figure 12.32).

Figure 12.32

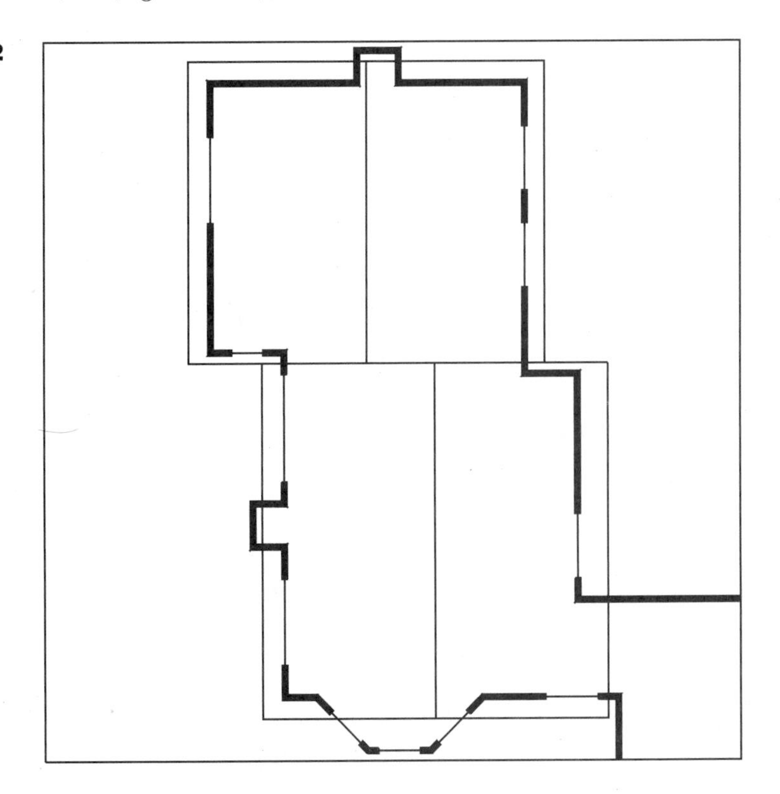

The second type of roof to draw is the **L-Gable**.

As before, pick this from the **Select roof function** icon menu. Again, accept the original values for **Gutter height** and **By pitch relative to 12** as the ridge height calculation method.

```
Show first corner of roof (intersecting point with other
roof):
```
Type or pick **108',101'**. This is the point where the new roof will intersect the old roof.

```
Show gutter ANGLE (perpendicular to other roof):
```
Type or pick **111',101'**.

```
Opposite corner:
```
Type or pick **111',86'**. This builds an L-shaped gable roof that intersects the original roof at a 90° angle (Figure 12.33).

Figure 12.33

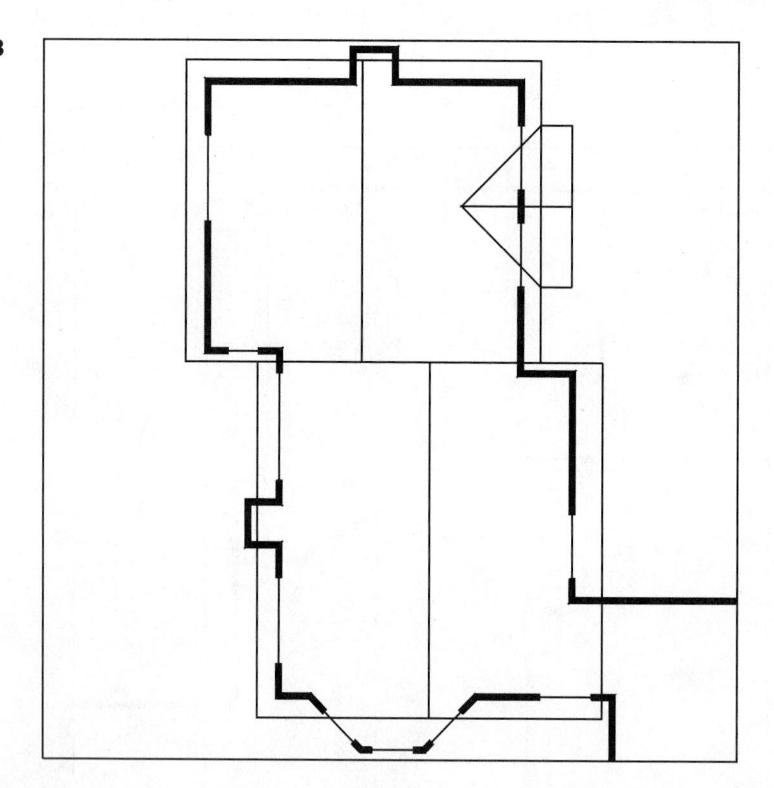

The final roof we will build will be a Hip roof over the garage. Select this style of roof in the same manner as the gable roofs.

Accept the original values for **Gutter height** and **By pitch relative to 12** as the ridge height calculation method.

```
Show first corner of roof:
```
Type or pick **114',59'**.

```
Show gutter ANGLE and longest side of roof:
```
Type or pick **142',59'**.

```
Opposite corner:
```
Type or pick **114',33'**. This will complete the roof for the entire house (Figure 12.34).

Figure 12.34

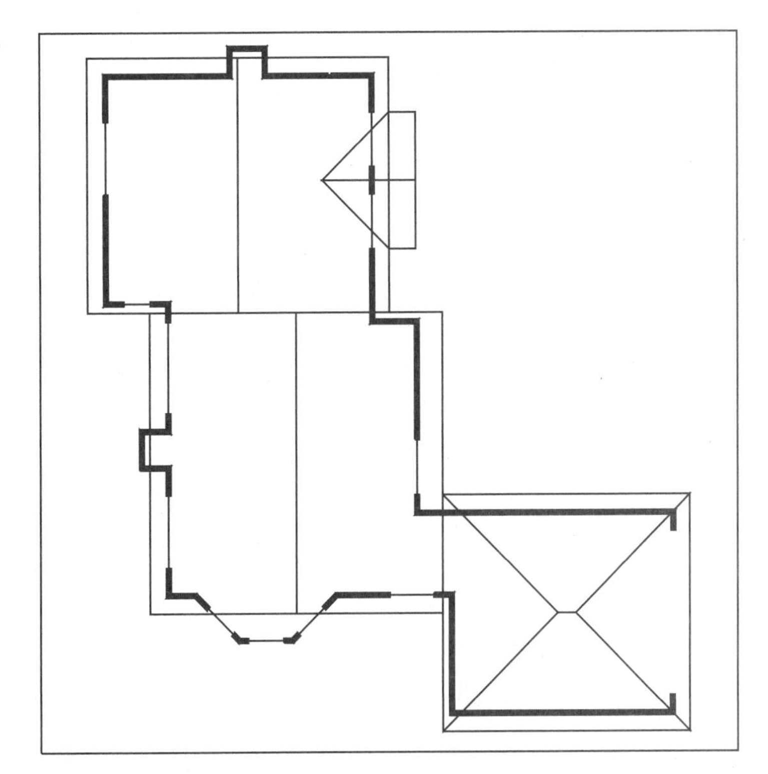

Use the Save command to save your drawing on the hard disk and your floppy disk.

Building a Perspective View

To get a 3-D perspective view of your house, select the Site Planning option, then pick the Presentation pull-down menu. Under the Perspective: line, choose Create to build the perspective view.

In the "Create Perspective View" dialogue box, you can specify where an observer will be situated (**Observers Loc**), where the observer will be looking (**View point**) and the focal length of the "lens" which will view the scene. A smaller focal length indicates a "wide angle" lens, while a larger (longer) focal length indicates a "telephoto" lens.

You can pick an **Observers Loc**ation by either typing in coordinates or using the **Pick Point <** option. Choose a point somewhere on the revisions box near the top of the title block. A good observer viewing height (in the Z axis) is **20'**.

You then must indicate a **View point** which indicates the location where you wish to look. Make sure **Ortho** is turned off. Move your cursor toward the front of the house and pick. You can accept a **View point** Z value of **0'**. For the focal length of the lens, enter the value of **25** (Figure 12.35).

Figure 12.35

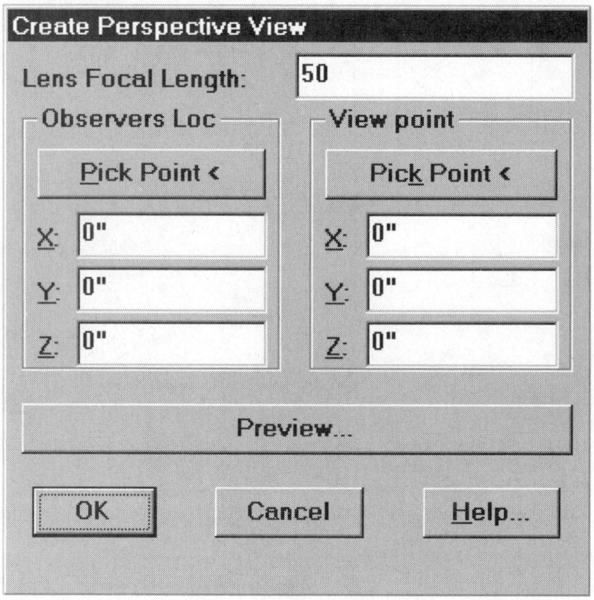

Create Perspective View
Lens Focal Length: 50
Observers Loc
Pick Point <
X: 0"
Y: 0"
Z: 0"
View point
Pick Point <
X: 0"
Y: 0"
Z: 0"
Preview...
OK
Cancel
Help...

The preview option shows you where you will be looking. If the view looks good, select **OK** (Figures 12.36 and 12.37).

If this view looks confusing, type the command **HIDE**. This will cause all of the lines hidden by the walls and roof to disappear (Figure 12.38).

Do not try to modify your drawing in the perspective mode. If you see obvious mistakes, return to the **PLAN** view to do your corrections.

To return to the original Plan view, follow these steps:

`Command:`
Type **PLAN <Enter>**.

<Current UCS>/Ucs/World:

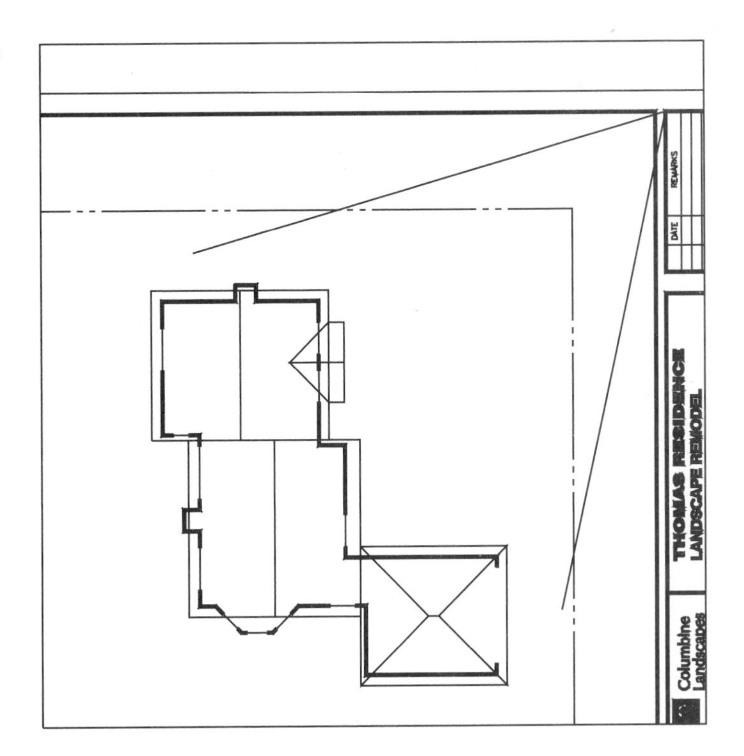

Figure 12.36

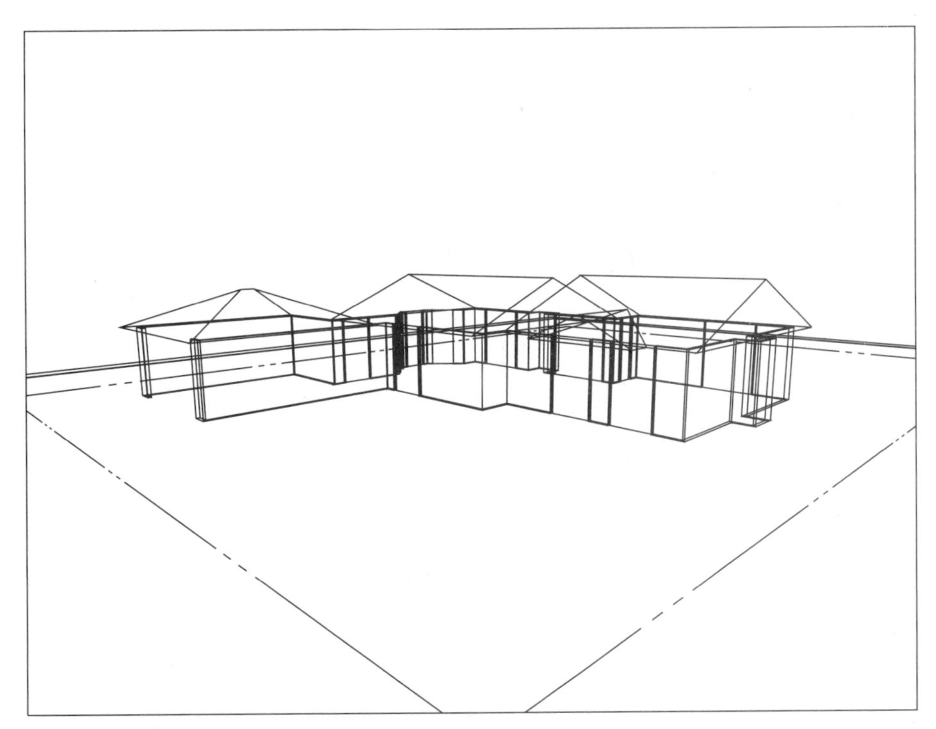

Figure 12.37

Figure 12.38

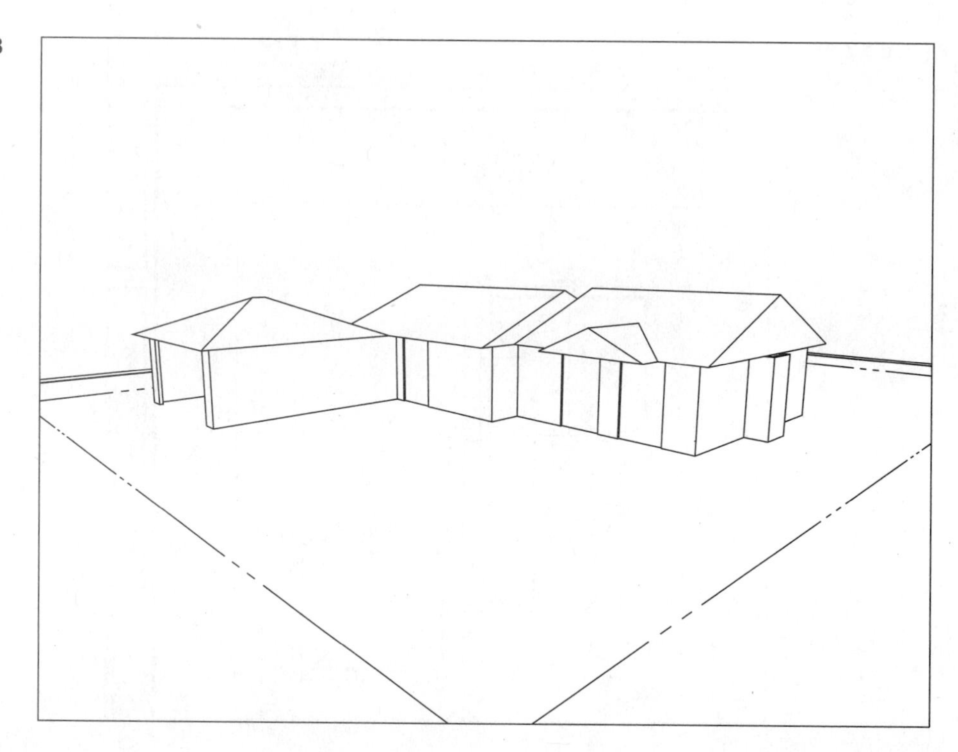

Press <**Enter**> to accept the Current UCS option. These other options refer to the User Coordinate System choices available for various two-dimensional and three-dimensional functions. We will discuss these options later.

This should return you to your original view. If you want to try looking at other views of your house, repeat the Perspective commands, but use other viewing heights, observer points, and focal lengths. You could even try to look outside from some place inside the house. Experiment with this command until you feel comfortable with it. Views can be saved and retrieved at any time. See your AutoCAD documentation on how to use the **View** command (Figure 12.39).

Figure 12.39

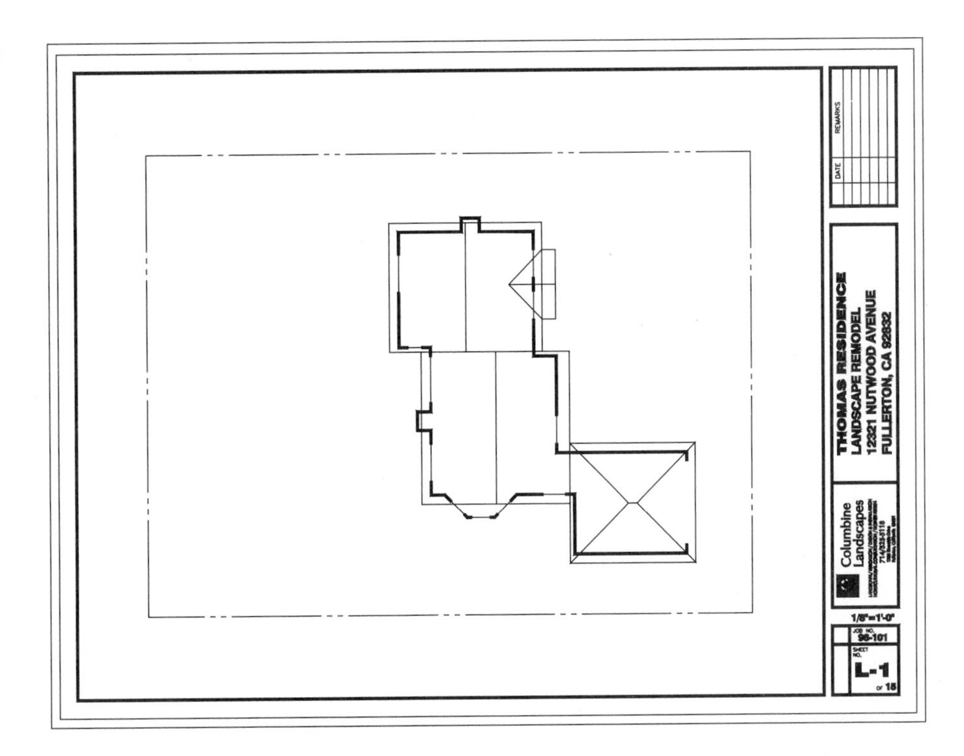

Utility Lines

Before we build the utility lines, we need to return the current text style back to STANDARD. This assures us that the font will match the other text used in the drawings.

`Command:`
Type **ST** to activate the **STYLE** command.

`STYLE Text style name (or ?) <3>:`
Type **STANDARD**.

`Existing style.`
You will then see the "Select Font File" dialogue box (Figure 12.40).

Pick the font **HLTR.SHX** to place it in the **File Name:** text box. Click on **OK**.

`Existing style. Height <0'-0">`
Press **<Enter>**.

Figure 12.40

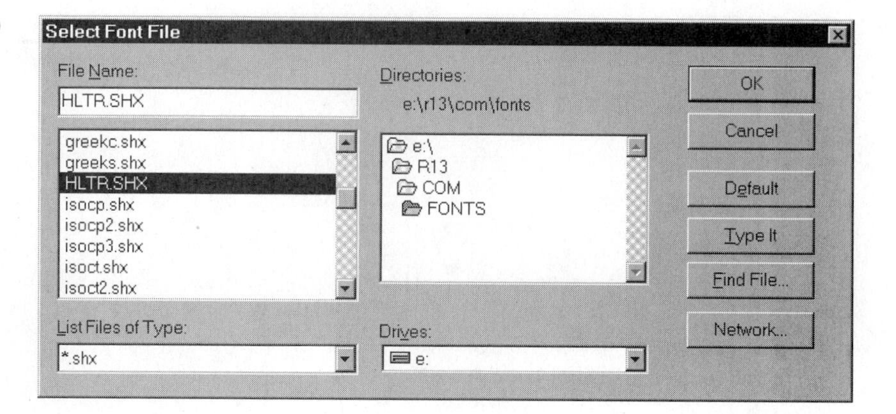

`Width factor <1.0000>`
Press **<Enter>**.

`Obliquing angle <0>:`
Press **<Enter>**.

`Backwards? <N>:`
Press **<Enter>**.

`Upside-down? <N>:`
Press **<Enter>**.

`STANDARD is now the current text style.`
We have redefined the **STANDARD** text style to show the font **HLTR**. This gives us a more "hand-lettered" look in the notes, dimensions and labels on the drawing.

Utility lines are defined by following four steps:

1. Locate the endpoints where the utility line runs
2. Indicate the spacing for the labeling letters (segment length)
3. Select a text height
4. Choose an abbreviation for the utility line being shown

In order for this command to work properly, you must define your line from *left to right* otherwise the text will be placed upside-down.

To enter the utility lines, we use the **Site Planning** menu.

Pull down the **Layout** menu, then select **Utility Lines ▸**. Choose the option **Pick Points**. We begin with the electrical line.

`Show 1st point of utility line:`
Type or pick **20',103'**.

`Show the end point of the line:`
Type or pick **73',103'** where the line touches the edge of the house.

`Enter the length of each segment:`
Type **10'**.

`Enter text height <1'>`
If the default text height is **1'** or **12"**, press **<Enter>**. If this is not the default value, type **1'**.

`Enter the text abbreviation to use:`
Type **E**. You should then see the electrical line being broken into 10' segments with **EL**'s being inserted in between.

`Show endpoint of the line:`
Press **<Enter>** to exit the command (Figure 12.41).

For the water line, choose **Utility Lines ▶** as above.

Figure 12.41

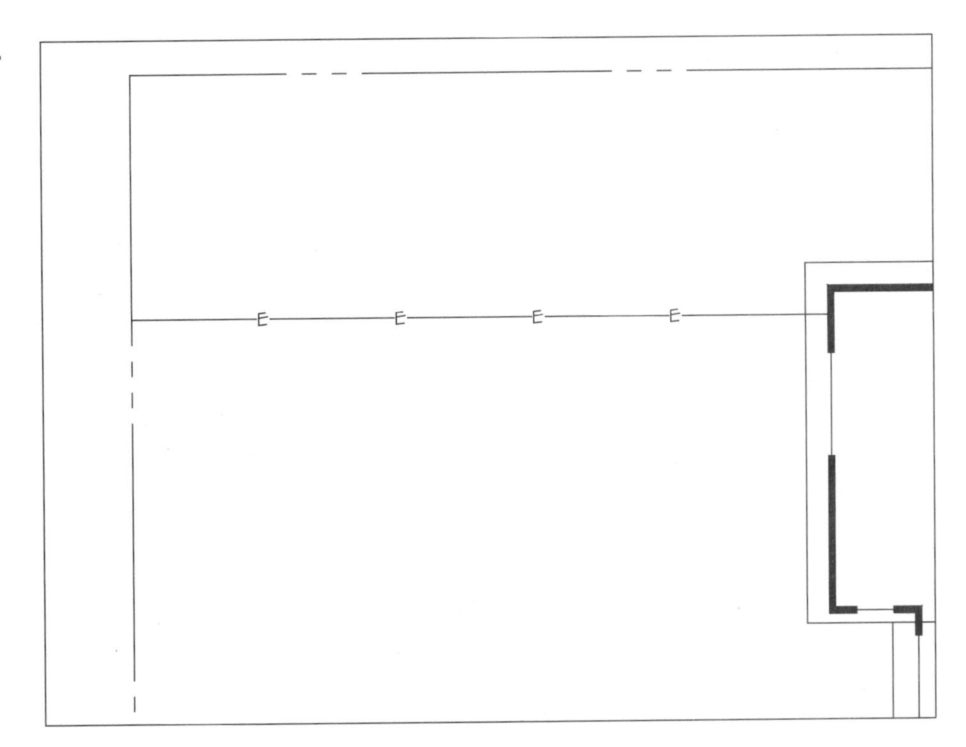

Show 1st point of utility line:
Use the cursor menu to find the endpoints of the water line. Choose the **Midpoint** of the 13' long wall next to the front door as the starting point.

Show the end point of the line:
Pick **Perpendicular** from the cursor menu then touch the property line in the front yard with your cursor. This will draw a line from the midpoint of the wall perpendicular to the front yard property line.

Enter the length of each segment:
Type **10'**.

Enter text height <1'>:
Press **<Enter>** to accept the default.

Enter the text abbreviation to use:
Type **W** for water line.

Show the end point of the line:
Press **<Enter>** to finish the command (Figure 12.42).

The Gas line will have a bend in it, so it will be trickier.

Begin the **Utility Lines ▶** command as usual.

Show the 1st point of the utility line:
Type or pick **105',48'**.

Show the end point of the line:
Type or pick **105',24'**.

Enter the length of each segment:
Type **10'**.

Enter the text height <1'>
Press **<Enter>** to accept the default.

Enter the text abbreviation to use:
Type **G** for your abbreviation.

Show the end point of the line:
Type or pick **155',24'**.

Show end point of the line:
Press **<Enter>** to complete the command (Figure 12.43).

Figure 12.42

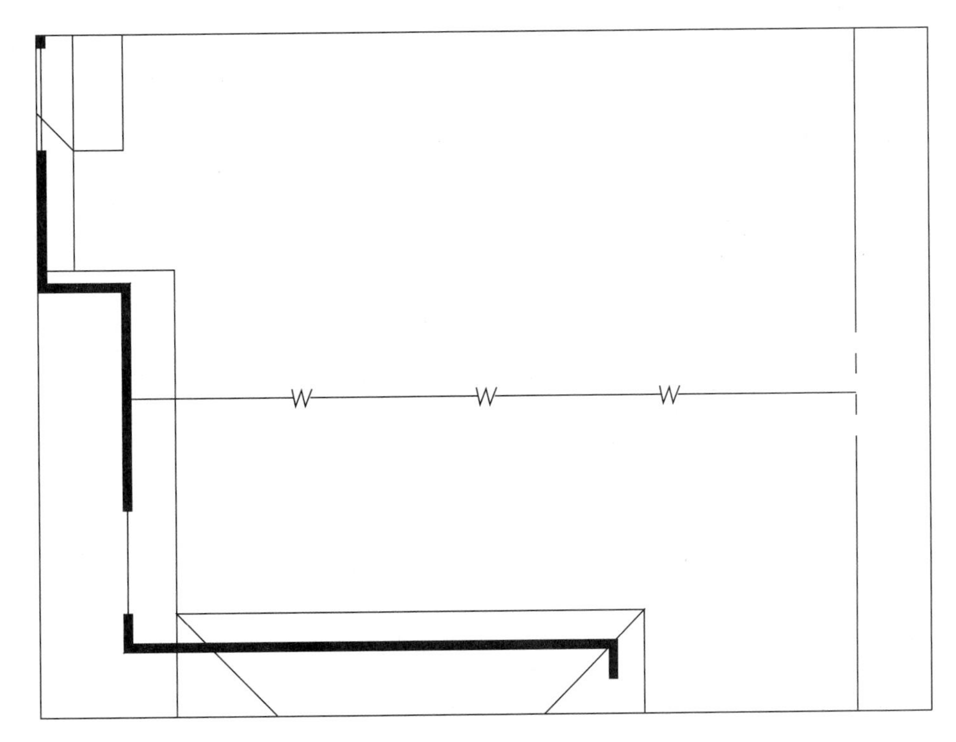

Figure 12.43

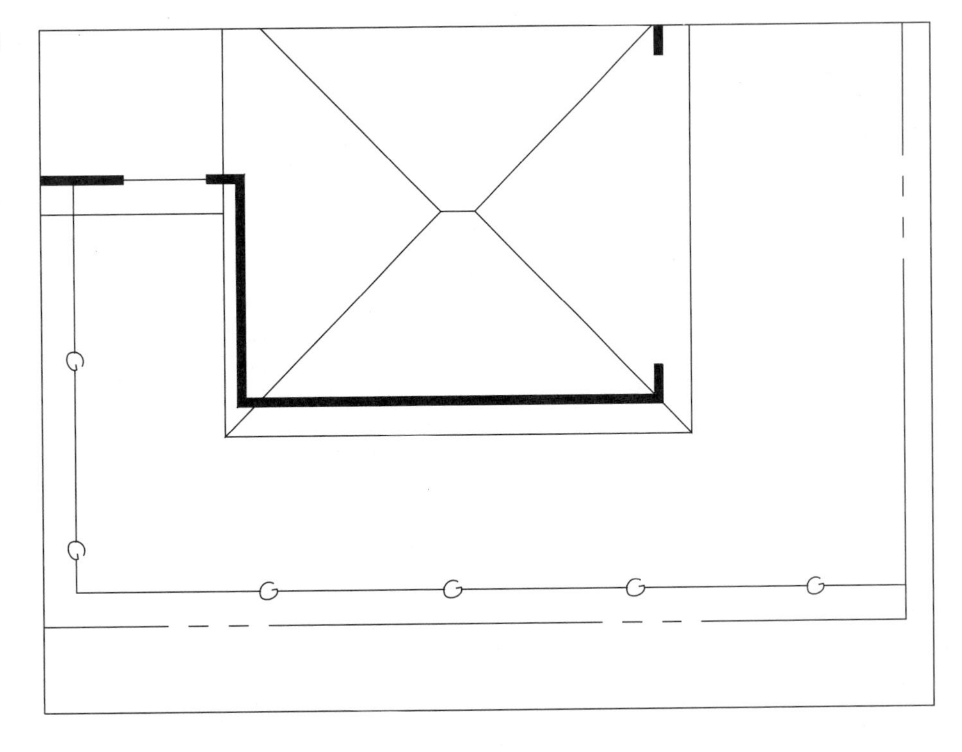

Building the Driveway

We will now build the existing hardscaping to remain (the driveway). Use LANDCADD's cursor menu to bring up the AutoCAD "Layer Control" dialogue box (Figure 12.44).

Figure 12.44

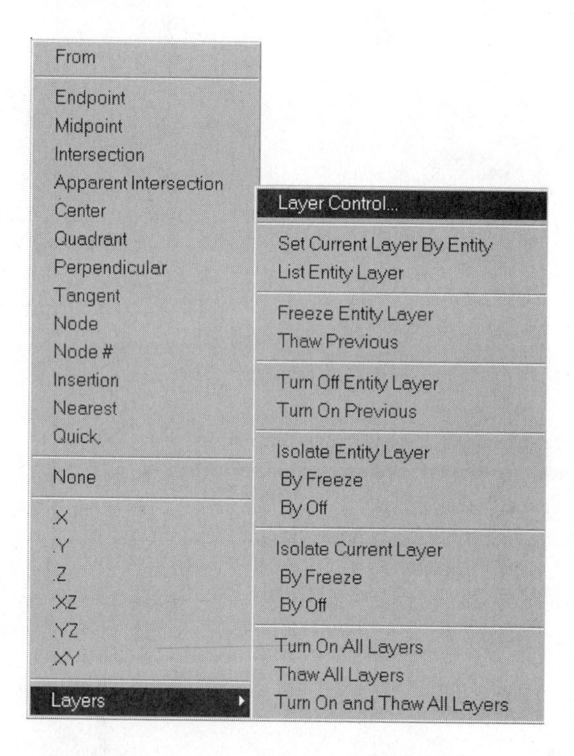

Use the "Layer Control" dialogue box to create a new layer called **EX_HARDSCAPE**. Use the **Set Color . . .** box to choose **gray** as the color (color 9) for this layer. After moving your cursor into the **Color:** text box, you can simply type **9** to assign this color to this new layer. You can also select the gray color box from the top color bar. Use the **Current** box to make **EX_HARDSCAPE** the current layer (Figure 12.45).

Draw the boundary of the driveway as a polyline.

At the prompt:

`Command:`
Type **PL** to begin the **Polyline** command.

`From point:`
Type or pick **140',56'**.

Figure 12.45

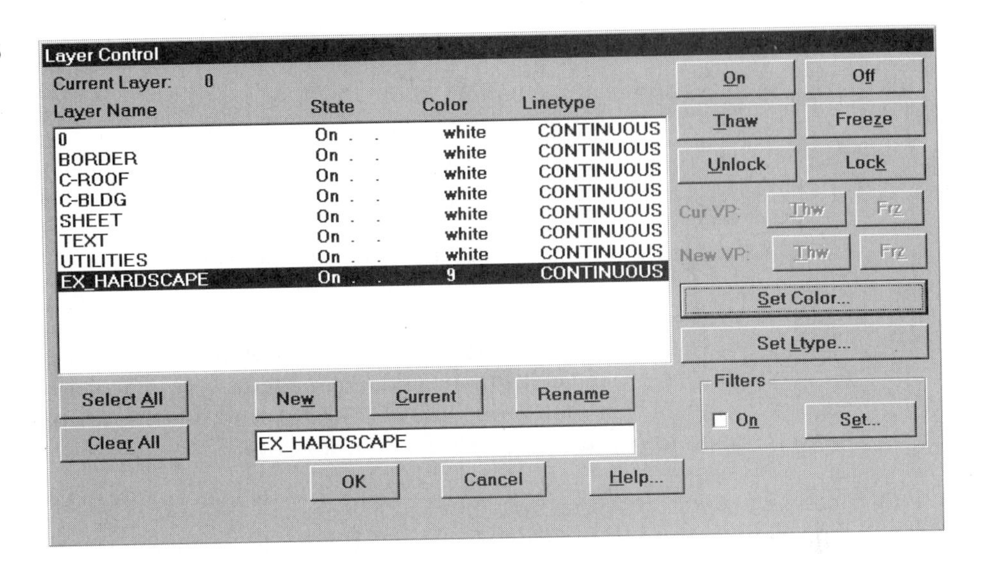

`Arc/Close/Halfwidth/Length/Undo/Width/<Endpoint of line>:`
Type **W** to change the polyline width.

`Starting width <0'-6">`
Type **0** to set the width back to 0.

`Ending width <0'-0">`
Press **<Enter>** to accept the default.

`Arc/Close/Halfwidth/Length/Undo/Width/<Endpoint of line>:`
Type or pick **140',36'**.

`Arc/Close/Halfwidth/Length/Undo/Width/<Endpoint of line>:`
Type or pick **155',36'**.

`Arc/Close/Halfwidth/Length/Undo/Width/<Endpoint of line>:`
Type or pick **155',56'**.

`Arc/Close/Halfwidth/Length/Undo/Width/<Endpoint of line>:`
Type **C** to close the polyline.

Hatching the Driveway

To fill in the driveway with a stipple effect for concrete, we will use a very nice LAND-CADD routine called Edge Stipple. This command fills an enclosed polyline with a graduated stipple pattern which is dense at the outer edge becoming more sparse toward the center.

To use this command, pull down the **Tools** menu, then select **Hatch ▸**. Choose the **Edge stipple** routine at the bottom of the list.

At the prompt:

```
Select edge:
```

Pick the right edge of the driveway closest to the title block.

```
Select point at end of stippling:
```

Pick a point near the center of the driveway. LANDCADD will then draw several temporary polyline edges to create a graduated hatch pattern. If the polylines spill outside the edge of the driveway or look wrong, use **<Ctrl-C>** to cancel the command. Use the **Undo — Back** option to remove the polylines, then start over.

```
Scaling factor (smaller numbers = more dense) <43.275>:
```

Type **50** to set the scaling factor. This regulates the density of the hatch.

```
Change stipple scale? (no/yes):
```

Type **N** if the hatch pattern looks correct. Type **Y** if you want to change the scale factor to adjust the density of the hatch (Figures 12.46, 12.47, 12.48, 12.49, and 12.50).

Figure 12.46

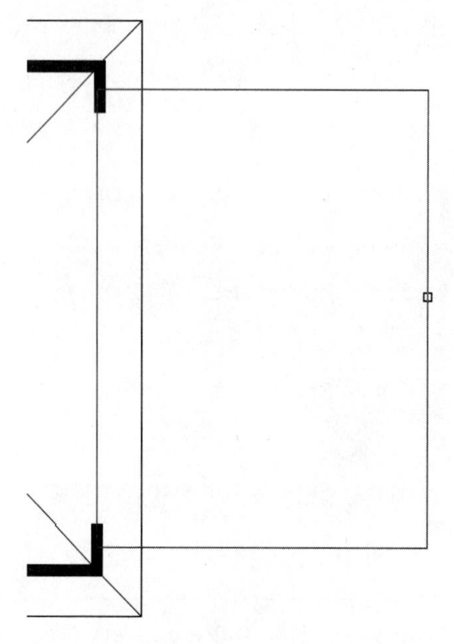

Figure 12.47

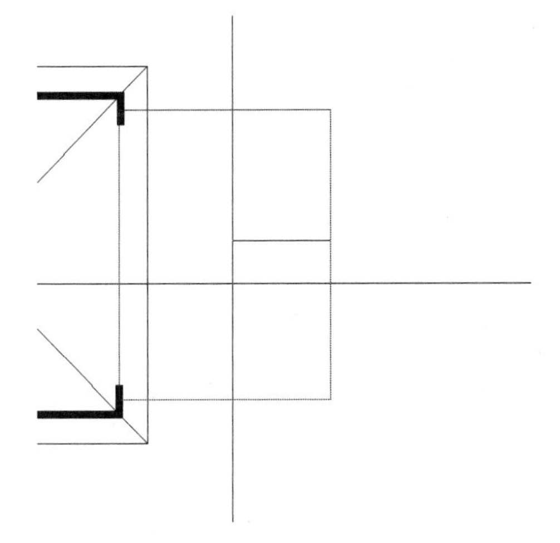

Figure 12.48

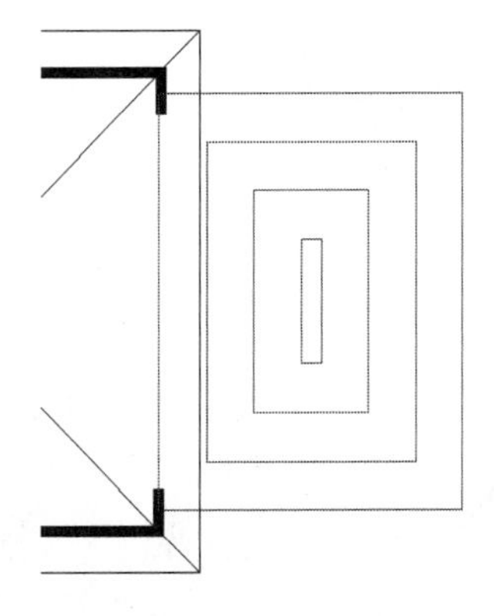

Figure 12.49

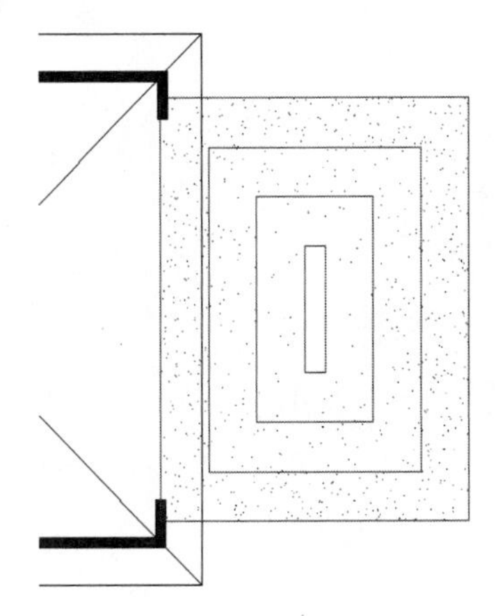

Figure 12.50

Placing the Existing Tree

Place the existing tree by selecting the Landscape Design option under LANDCADD. Use the Planting pull-down menu, and select the Broadleaf option. You will then see the Broadleaf Tree Symbols icon menu. The first symbols displayed are the Simple Plan symbols (Figure 12.51).

Figure 12.51

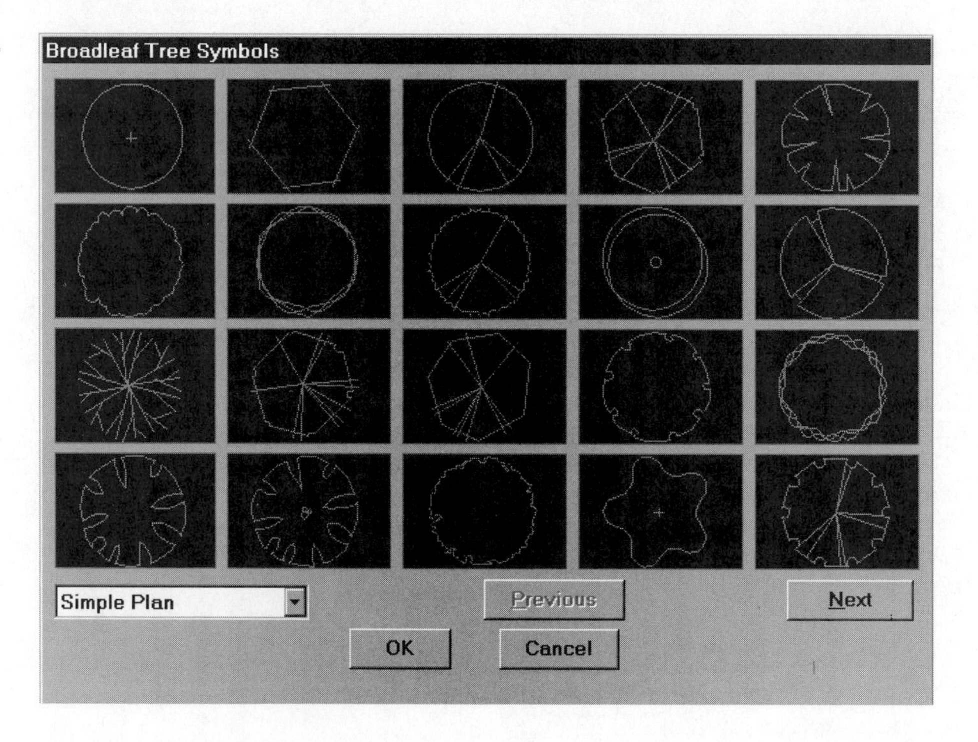

Figure 12.52

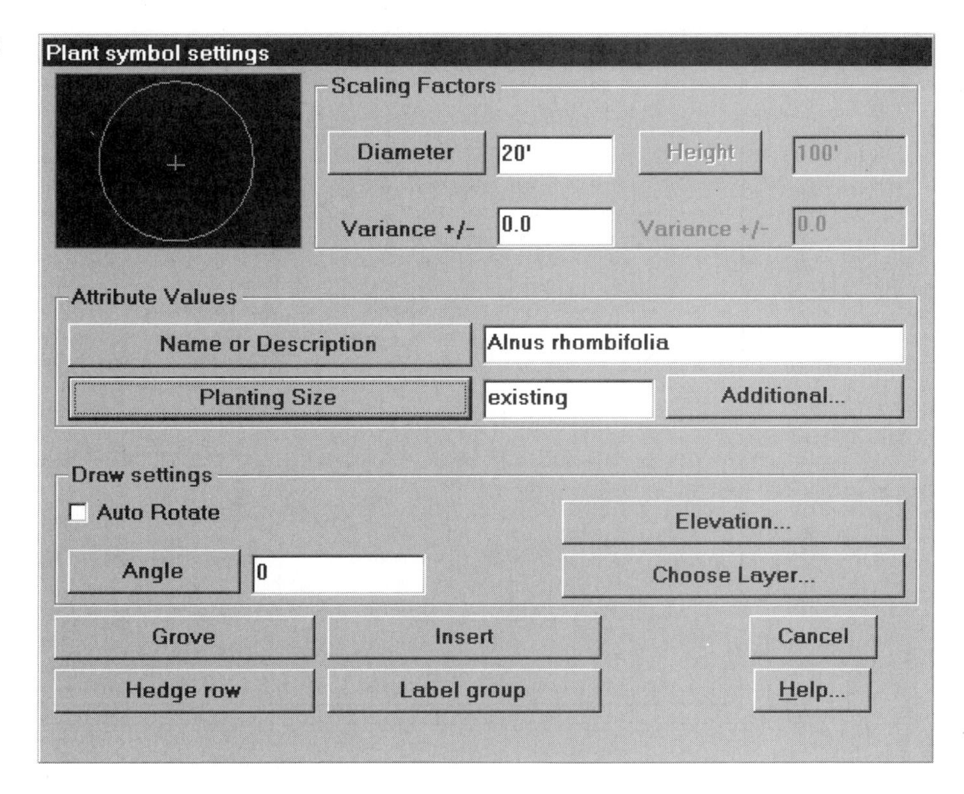

Choose the upper left tree symbol, then pick **OK**. This will bring up the "Plant Symbol Settings" dialogue box (Figure 12.52).

In the **Name or Description** text box, type **Alnus rhombifolia**.

In the **Planting Size** text box, accept the default value **existing**.

In the **Diameter** text box, type **20'**.

Pick **Insert** to place the tree in the drawing.

At the prompt:

```
Locate plants:
```

Move the tree symbol to the location on the drawing as shown. Press **<Enter>** to end the command (Figure 12.53).

The tree is now located in the proper spot, and the information about the name of the tree and its status as an existing tree is stored with the symbol. We will take advantage of

Figure 12.53

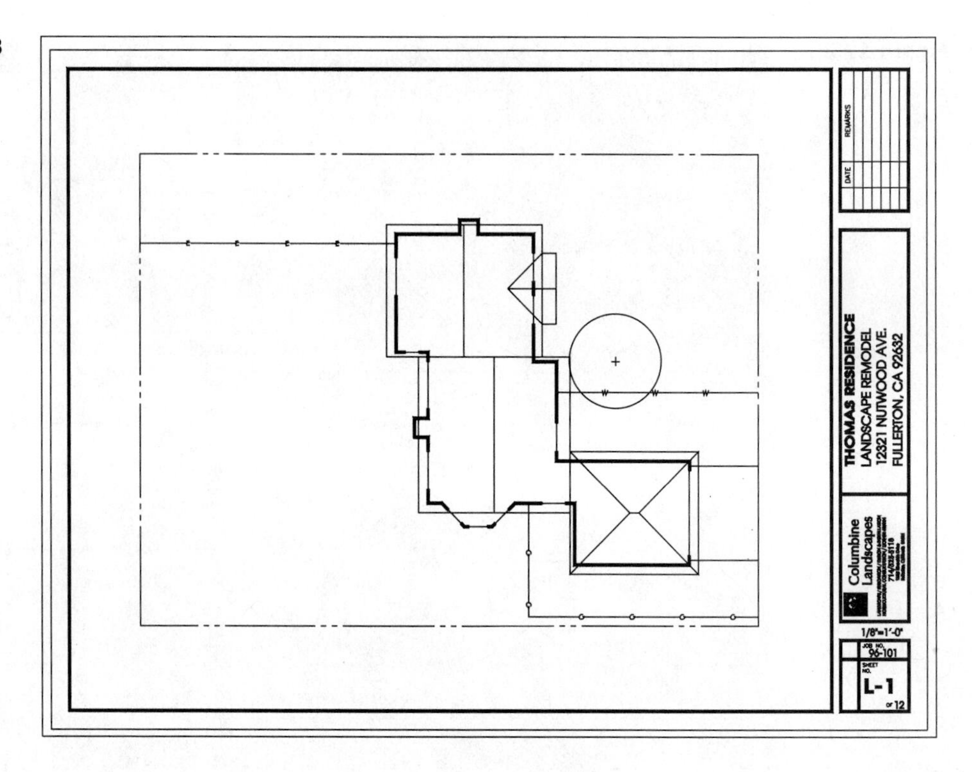

this stored information when we build a planting schedule and project estimate later in this exercise.

This completes the base sheet. Save the drawing as **BASE.DWG** to your hard disk and floppy disk.

chapter

Drawing Three—The Hardscape and Construction Plan

To complete the next part of the Thomas Residence project, we will build a plan which shows the design and location of the hardscape. We will use standard AutoCAD drawing tools to build the patio pads, mow strips and walkways, then make extensive use of AutoCAD dimensioning capabilities to provide dimensions for construction of the project.

In the Hardscape and Construction Plan, we will complete the following steps:

- Build a block of the house footprint from the Base Sheet
- Use existing project settings from the Project Settings library
- Build another new project in Project Manager, using C18SHEET as a prototype
- Import the house footprint block into the Hardscape and Construction Plan
- Build Lines of Force to guide our hardscape design
- Create and control new layers on the drawing
- Create hardscape patios, mow strips and walkways
- Hatch the concrete areas on the plan
- Import a tick mark drawing to use as a dimension arrowhead
- Create and save a dimension style, using parent and child dimensions
- Use colors to control the screen appearance and plotted appearance of dimensions
- Use linear, aligned and angular dimensions
- Use snap to align dimensions

- Use grips and/or dimension editing commands to change the location of dimension text
- Add leaders to show construction notes

We will begin the hardscape plan by creating a BLOCK from the base plan drawing. A block is a portion of a drawing named and stored as a separate entity. We created a text block to use in the Title Block drawing **C18SHEET**. In that case, we used the block to enable us to add text to the tile block in a predefined location and style. This saves time in starting up new drawings.

In this case, we need to "borrow" the building footprint and property line from the Base Sheet in order to create a Hardscape and Construction Plan. However, we do not need the utility lines, the existing tree and the roof lines which appear on that drawing. How can we use the building footprint without the unwanted drawing information? We **could** open the Base Sheet drawing, erase the unwanted portions, then save the drawing under a new name. This is a crude solution and does not permit us to save the building footprint portion of the Base Sheet for use in other plans which might require it.

A more elegant solution is to save the building footprint and the property lines (the portion of the Base Sheet we want to use again), and write them to a block. We can then insert this block into the new Hardscape and Construction plan. This is a more common practice in CAD operations, since it saves time, avoids the potential problem of overwriting an old file with a new one, and allows the use of the block in other drawings.

There are two options for creating a block. The first uses the **BLOCK** command. When **BLOCK** is used, the drawing block becomes a separately defined part of the original drawing. This type of block can be "rubber-stamped" repeatedly on a drawing. This might be useful for repeatedly used symbols or text labels. This **internally** created drawing block is only available in the drawing in which it was built. It cannot be accessed from another drawing.

The second option to create a drawing block is to use the **WBLOCK** command. This option creates the same type of drawing as the **BLOCK** command, but the drawing is saved as a separate file on the hard disk. An **externally** created block built with the **WBLOCK** command is actually saved as an independent drawing. Because the block is written to the hard disk as a new drawing file, it is then available for use in other drawings.

Once a block is inserted into a drawing, it becomes part of that drawing's database. It can be reinserted if erased, inserted repeatedly if necessary, and will be displayed in a list of blocks for that drawing.

Creating the HOUSE Block

To begin, use the Project Manager to open the **BASE SHEET** drawing again. Once the drawing appears on the screen, you are ready to build the new block.

At the prompt:

`Command:`
Type **WBLOCK.** After you type this command, you will have the "Create Drawing File" dialogue box on screen for you to choose a name of your drawing. In the **File Name:** text box, type the name **HOUSE** (Figure 13.1).

Select **OK**.

`Block name:`
Press **<Enter>. Do not enter a drawing name here.**

`Insertion base point:`
Type **0,0**.

`Select objects:`
Pick the building footprint, property line and driveway polyline as the objects to block—pick these objects **only**. Do not select any other portion of the drawing (such as the roof parts or stipples in the driveway). It may be necessary to use the transparent **'ZOOM** command to avoid picking these unwanted objects (Figure 13.2).

Figure 13.1

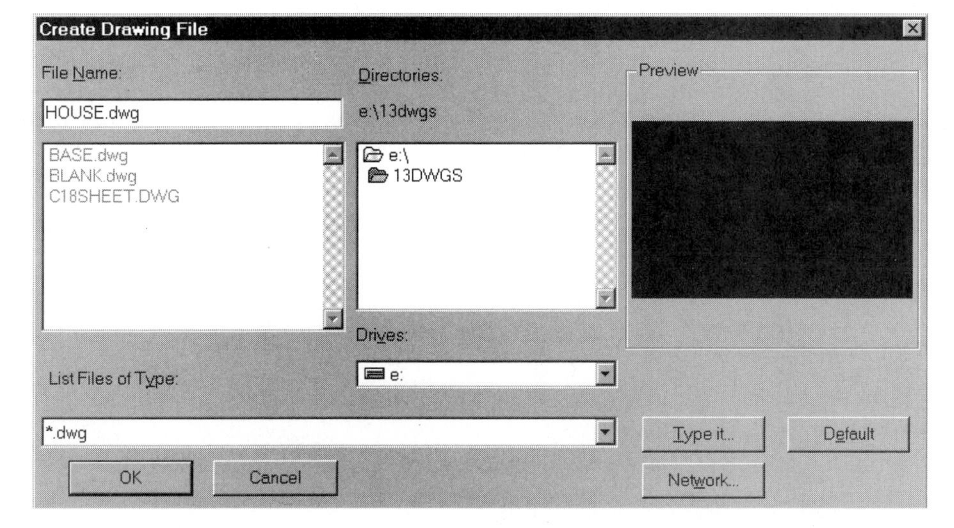

Figure 13.2

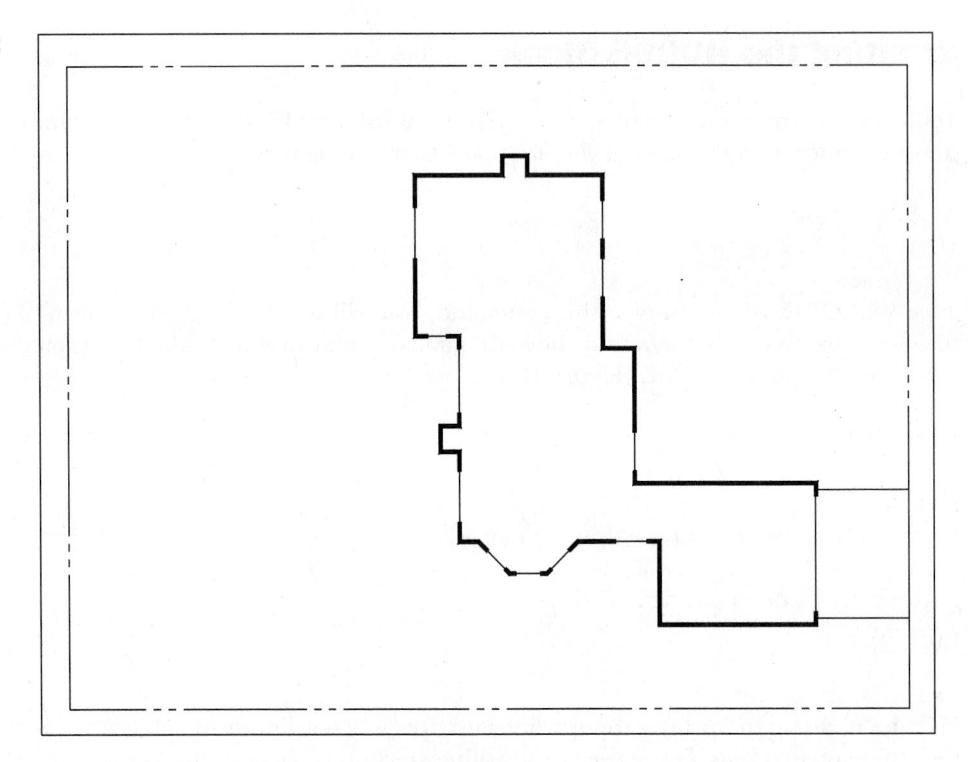

After you have finished highlighting these objects, press **<Enter>**. The building footprint and driveway border will then disappear from the screen. This is supposed to happen. Although the objects are no longer on the screen, they may be retrieved with the **OOPS** command.

`Command:`
Type **OOPS**. This will restore the image back to the screen.

This block is now saved as a drawing to the hard disk. Copy it to your floppy disk. You could use this drawing in any other Thomas Residence drawings that were created later.

When leaving this drawing, do not save any changes. There are no changes to the Base Sheet that we need to save. We have simply copied a piece of this drawing into a new file.

Beginning the Hardscape Drawing

Begin a new drawing by selecting the **EP** pull-down menu, then pick **Project Manager**.

In the Project Manager dialogue box, select **New . . .** (Figure 13.3).

Figure 13.3

In the "New Project" dialogue box, you can specify the new project **Description:** (use **Hardscape Plan** or something like it). In the **Drawing Name . . .** box, enter the name of your new drawing (use **HARDSCPE.DWG** or some other acceptable file name). You will also need to specify the correct path where all of your drawing files are stored. See the section on Project Manager for more details.

In the **Prototype Drawing . . .** box, use the **C18SHEET** drawing you created earlier. Be sure to indicate the correct path to where this drawing is stored. The prompt at the bottom of the dialogue box will tell you if you entered the path incorrectly.

This will bring up a new drawing with your chosen drawing name, and with the information from **C18SHEET** already on the screen.

Use the **AE** keyboard macro (or type **DDATTEDIT**), to edit the text in the title block. Pick any part of the title block text to edit the attributes (Figure 13.4).

Return to the **EP** pull-down menu, and select **EP and AutoCAD**. Select the **Draw** pull-down menu, and choose **Insert ▸**. Then select **Block . . .**. (Or you can simply type **DDINSERT** at the command line) (Figure 13.5).

Pick **File . . .**, then find **HOUSE.DWG** that was created earlier. Select the drawing in the "Insert File" dialogue box. After this file appears in the text box pick **OK**. You are then returned to the "Insert" dialogue box with the **Block . . .** text box and the **File . . .** text box filled in. Pick **OK**.

Figure 13.4

Edit Attributes

Block Name: TEXT

JOB NAME	THOMAS RESIDENCE
PROJECT NAME:	LANDSCAPE REMODEL
STREET ADDRESS:	12321 NUTWOOD AVE.
CITY STATE & ZIP	FULLERTON, CA 92632
JOB NUMBER	96-101
SHEET NUMBER	H-1
TOTAL SHEETS:	12
DRAWING SCALE:	1/8"=1'-0"

OK Cancel Previous Next Help...

Figure 13.5

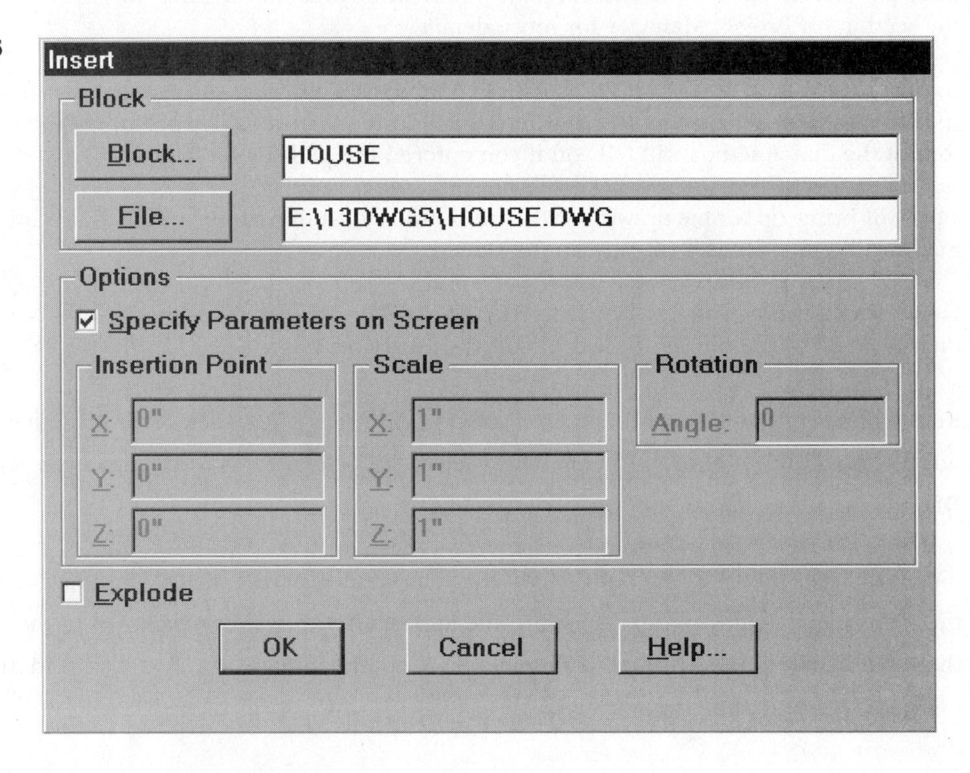

At the prompt:

Insertion point:
Type **0,0**. Standardizing on this base point makes it very simple to insert drawings in exactly the location they were in the previous drawings.

X scale factor <1>/Corner/XYZ:
Press **<Enter>** to accept the default value of 1.

Y sale factor (default=1):
Press **<Enter>** to accept the default value.

Rotation angle <0>:
Press **<Enter>** to accept the default value.

The **HOUSE** drawing is now inserted at the correct location on our drawing sheet. Note that because the **HOUSE** drawing is defined as a block, it cannot be edited or used to trim lines in its current form. In order to use the **HOUSE** block to trim and extend lines, we need to separate the **HOUSE** block into its component polylines and lines again. Once this is done, the components of the building footprint and polyline are no longer part of a block and can no longer carry attributes. The **EXPLODE** command separates a block back into its component parts. Avoid using **EXPLODE** on blocks which have attributes attached to them.

Command:
Type **EX** or (**EXPLODE**). This command can be used to separate blocks, hatches, dimensions and other complex objects. It also converts attributes into text and writes them onto the screen.

Select objects:
Pick any place on the block to highlight it. Press **<Enter>**. The block will then be broken into its previous parts, which can then be used to trim and extend lines.

Although the layers in the drawing accompany the block, the linetype scale does not. We will need to reset the linetype scale to **96**.

At the prompt:

Command:
Type **LTSCALE**.

LTSCALE New scale factor <1.0000>:
Type **96**.

This drawing will contain the building footprint information from the Base Sheet along with the key benchmark information about the hardscape design and construction. The hardscape in the back yard contains an angled grid of concrete pads with brick ribbons running through them. There is also a planter in the center, and a large turf area surrounded by a brick mowing strip. The front yard hardscape consists of simple walkways on either side of the driveway, and a brick mow strip surrounding a front lawn. Although the geometry of this plan may appear intimidating, it is really fairly simple to draw, using the **OFFSET**, **COPY**, **TRIM** and **FILLET** commands. Remember to use the keyboard shortcut macros **OF**, **C**, **TR**, and **FI** to save time and keystrokes. Windows (3.x, 95 and NT) users may wish to use screen icons as an alternative to keyboard macros.

Building the Lines of Force

We will begin this drawing by drawing a series of **lines of force** on a new layer called **FORCE**.

Create this new layer using the **AutoCAD Layer Dialog** selection under **Layers ▸** in the cursor menu. Type **FORCE** in the text box, then use the **Set Color . . .** box to define **YELLOW** as the layer color. Make **FORCE** the current layer. Examine the lines of force on the drawing below (Figure 13.6).

Horizontal Lines

Build the first horizontal line:

Figure 13.6

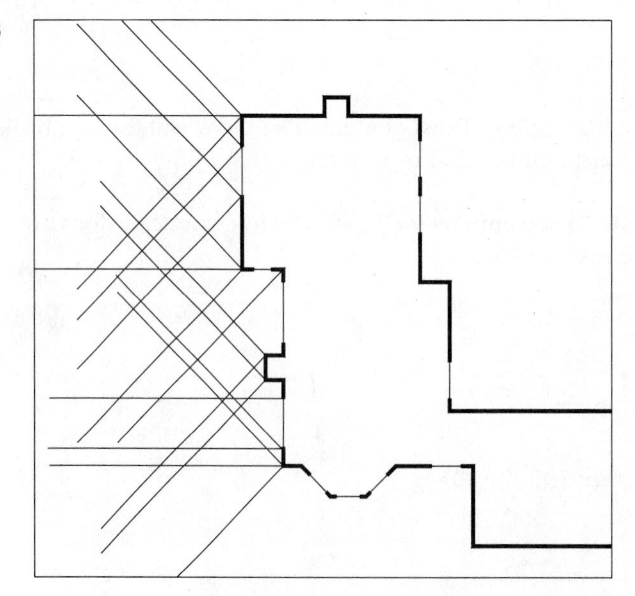

`Command:`
Press **<F8>** to turn **Ortho** on. Type **L** for **Line**.

`LINE From point:`
Type or pick **76',105'**.

`To point:`
Move your cursor about **40'** toward the left (into the back yard) to create a horizontal line. Press **<Pick>** to set the endpoint. (Type **@40'<180** for the same result.) Press **<Enter>** to leave the command.

`Command:`
Type **CM** to activate the **Copy/Multiple** command. This command performs multiple copies of any object you select.

`Select objects:`
Pick the horizontal line you just created as the object to copy.

`Select objects: 1 found`
Press **<Enter>** to indicate that there are no more objects to select.

`Base point:`
Use the cursor menu to highlight **Endpoint**, then choose the endpoint closest to the building as the base point.

`Second point of displacement:`
Place copies of the line at the following locations: (either type or pick)

76',80'

83',59'

83',51'

83',48'

Press **<Enter>** to leave the command.

Angled Lines

`Command:`
Type **L** or **Line**.

`LINE From point:`
Type or pick **80',62'**.

`To point:`
Type **@50'<225** to draw a line angled at 225° (toward the lower left corner of the screen) 50' long.

Use the **CM** command to make copies of the line:

`Command:`
Type **CM**.

`Select objects:`
Pick the diagonal line you just created as the object to copy.

`Select objects: 1 found`
Press **<Enter>** to indicate that there are no more objects to select.

`Base point:`
Use the cursor menu to highlight **Endpoint**, then choose the endpoint closest to the building as the base point.

`Second point of displacement:`
Place copies of the line at the following locations: (either type or pick)

76',105'

76',100'

76',92'

83',80'

76',80'

80',66'

83',48'

Press **<Enter>** to leave the command.

Repeat this process with lines pointing to the upper left corner of your screen (at a 135° angle).

`Command:`
Type **L** or **Line**.

`LINE from point:`
Type or pick **83',51'**.

`To point:`
Type **@50'<135** to set the distance and angle of the line.

Use the **CM** command to make copies of the line:

`Select objects:`
Pick the diagonal line you just created as the object to copy.

`Select objects: 1 found`
Press **<Enter>** to indicate that there are no more objects to select.

`Base point:`
Use the cursor menu to highlight **Endpoint**, then choose the endpoint closest to the building as the base point.

`Second point of displacement:`
Place copies of the line at the following locations: (either type or pick)

76',105'

76',100'

76',92'

76',80'

80',66'

80',62'

83',48'

Press **<Enter>** to leave the command.

Your lines of force are now complete, so we can begin drawing in the edges of the patio.

Building the Patio Edge

We will use the lines of force to help us create the outer edge of the patio. These lines help suggest a shape for the patio, and greatly assist in drawing it. The intersections created by the lines of force will act as endpoints for the lines at the patio edge.

Using the same procedure as before, create a new layer called **CONCRETE**, leaving the color **WHITE**. Make this the current layer.

Begin the outline of the patio with lines connecting points and intersections:

`Command:`
Type **L** or **Line**.

`LINE from point:`
Type or pick **76',100'**.

`To point:`
Pick **71',105'**.

`Continue to the points:`
56',90'

71',75'

55',59'

66',48'

83',48'

Press **<Enter>** to leave the command.

Return to the "Layer Control" dialogue box (using any convenient method). Use the dialogue box to turn the **FORCE** layer **OFF** (Figure 13.7).

Since this represents the perimeter of the patio, use the **OFFSET** command (set at **9"**—the width of a brick ribbon) to offset lines *back toward the house.* This creates the inside edge of the brick ribbon.

At the prompt:

`Command:`
Type **OF** or **OFFSET**.

Figure 13.7

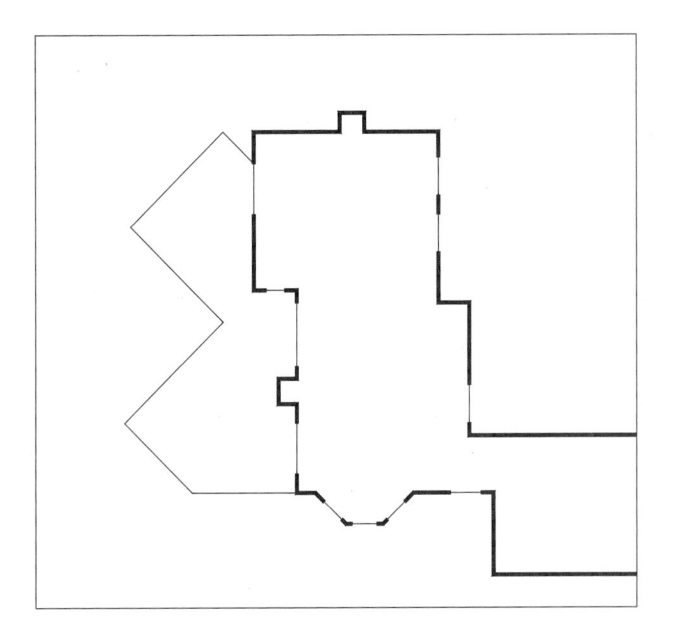

```
Offset distance or Through<Through>:
```
Type **9"**.

```
Select object to offset:
```
Pick any of the patio lines just drawn.

```
Side to offset:
```
Choose the side back toward the house footprint.

```
Command:
```
Press **<Enter>** to repeat the command. Offset each of the patio edge lines just drawn (Figure 13.8).

Use **FILLET** or **TRIM** to clean up each of the corners. See the section Editing Commands to illustrate these commands. You results should resemble the illustration (Figure 13.9).

Building the Bedroom Patio

To build the top bedroom patio, locate the patio edge that runs down from **59',90'** to **71',75'**. The line inside of this (LINE 1 in the illustration) will be used to create the rest of the pads.

Figure 13.8

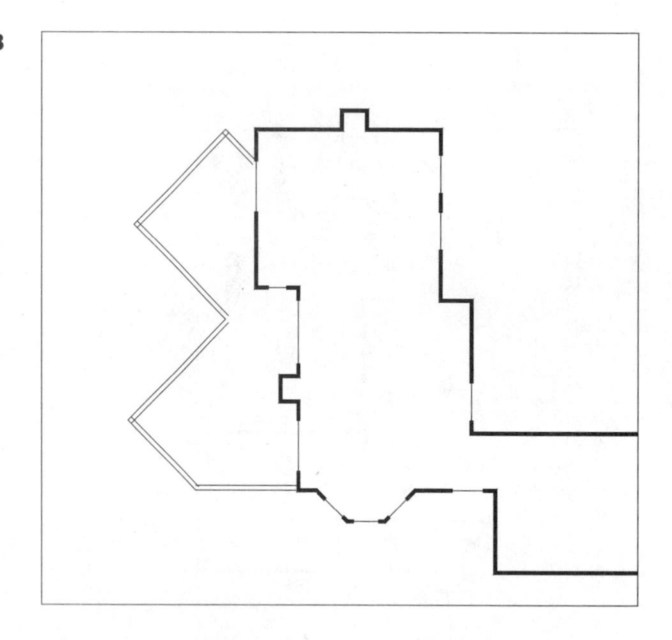

Figure 13.9

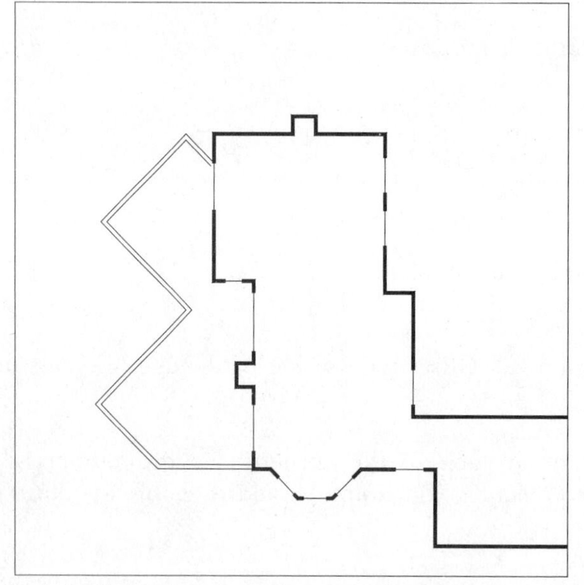

OFFSET this line **73"** then **9"**, then **73"**, then **9"** again. This will begin to create equal sized pads. See below (Figures 13.10 and 13.11).

Find the inside brick edge which runs from about **71',104'** to **57',90'** (LINE 2 in the illustration).

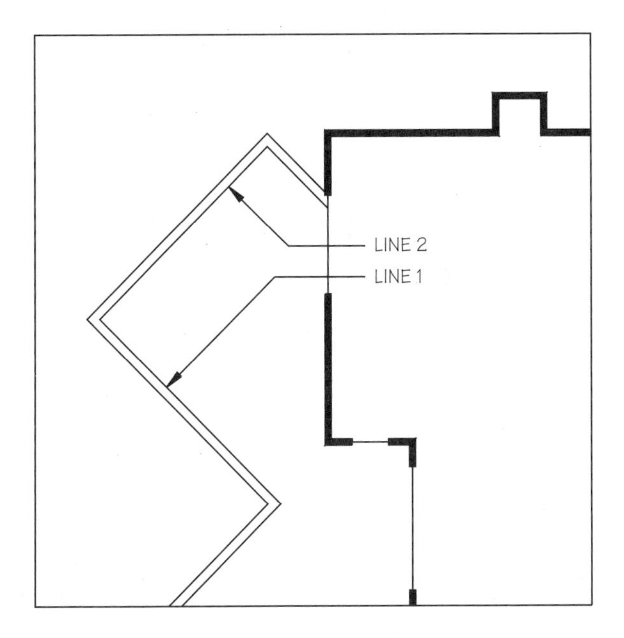

Figure 13.10

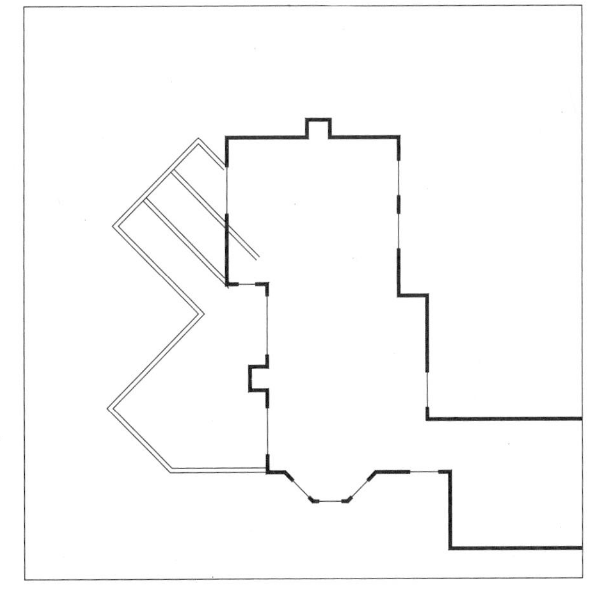

Figure 13.11

OFFSET this line by **73**" towards the house, then **9"** more (Figure 13.12).

Use the **FILLET, TRIM** and **EXTEND** commands to clean up the corners (Figure 13.13).

Figure 13.12

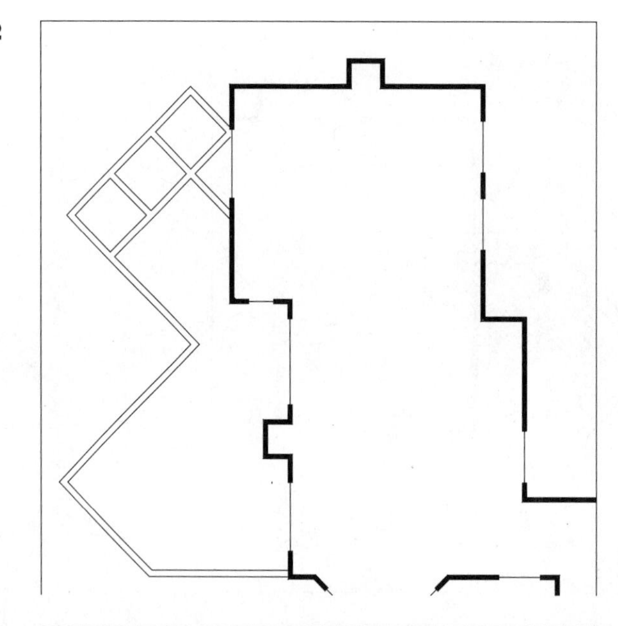

Figure 13.13

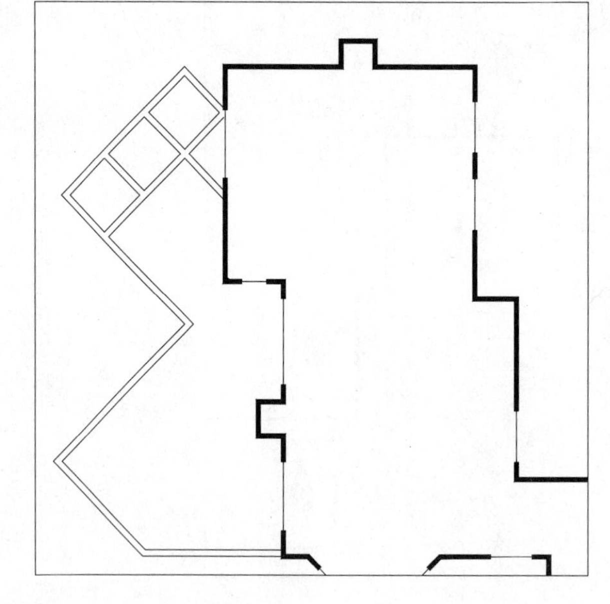

Building the Connecting Pads

To build the connecting pads, we will build temporary lines which will act as the center lines for the 9" brick ribbons. After drawing the original three lines, we will offset these center lines by a distance of 4 1/2". This creates the 9" brick ribbon we need. We will erase these temporary lines when we are finished.

First, draw a horizontal line toward the back yard from the point **76',80'** (since we will trim this line, make it longer than necessary).

Next, you will draw a vertical line using the intersection of two other lines as the start point.

`Command:`
Type **L** or **Line**.

`LINE from point:`
Bring up your cursor menu and highlight **INTERSECTION**, then pick the corner **near 66',91'**. The cursor should snap to this intersection.

`To point:`
Choose a point which brings the line down about **15'** toward the bottom of the screen.

Repeat this procedure for a vertical line from the corner of the pad **near 61',86'**.

Offset each of these lines **4.5"** in each direction. See the results below (Figures 13.14 and 13.15).

Find the first vertical line you drew (from **66',91'**), and offset it **24.5"** to the left and **33.5"** to the right. These lines represent the inside edge of the pads, so offset them each **9"** away from the center of the pad.

Figure 13.14

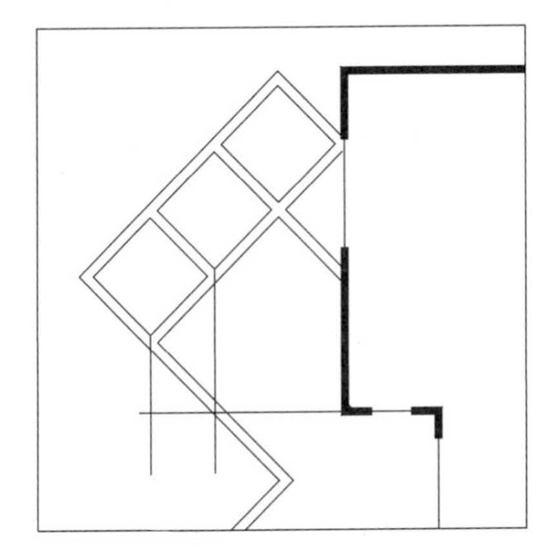

Figure 13.15

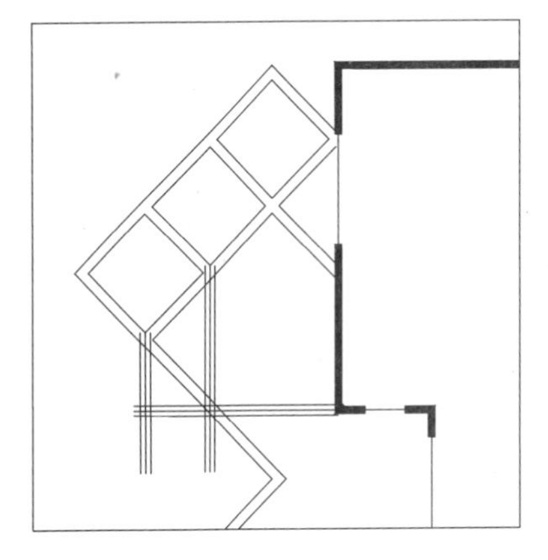

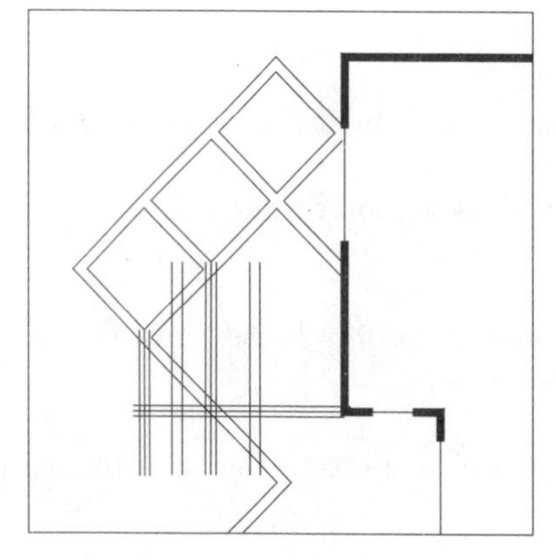

Figure 13.16

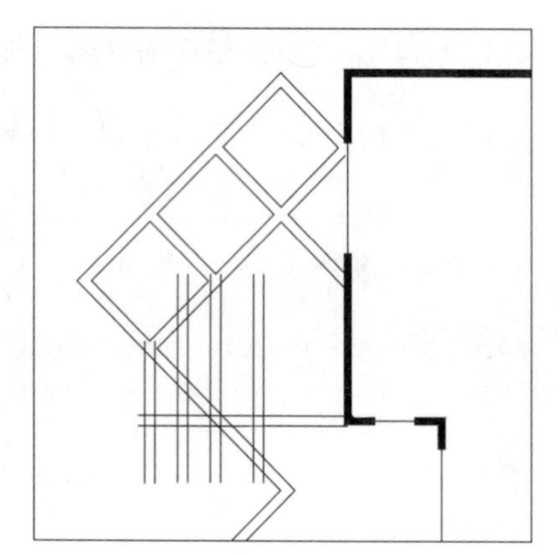

Figure 13.17

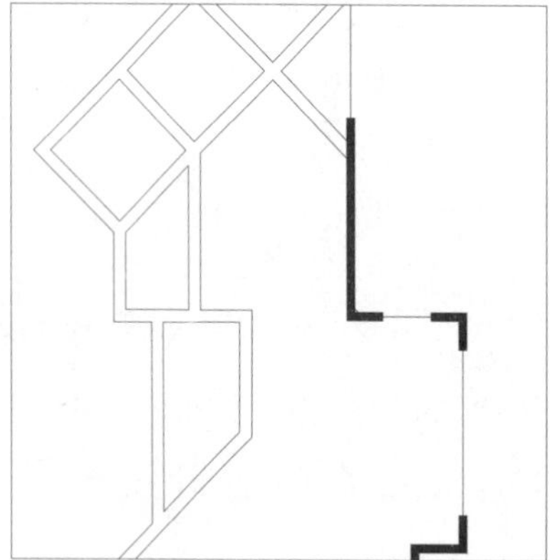

Figure 13.18

Erase the temporary center lines from your offsets. See the figures below for your results:

Use **FILLET**, **TRIM**, and **EXTEND** to clean up and extend these lines to create the following look (Figures 13.16, 13.17, and 13.18).

Complete the border to the patio planter by extending a line down from **78',80'** to **78',73'**, then over to **68',73'**.

Offset the vertical line **9"** to the left, then offset the horizontal line **9"** above.

Clean up the intersections to complete the bedroom patio and the connecting pads. See the illustrations below (Figures 13.19 and 13.20).

Once the bedroom patio and the connecting pads are complete, begin building the main patio.

Figure 13.19

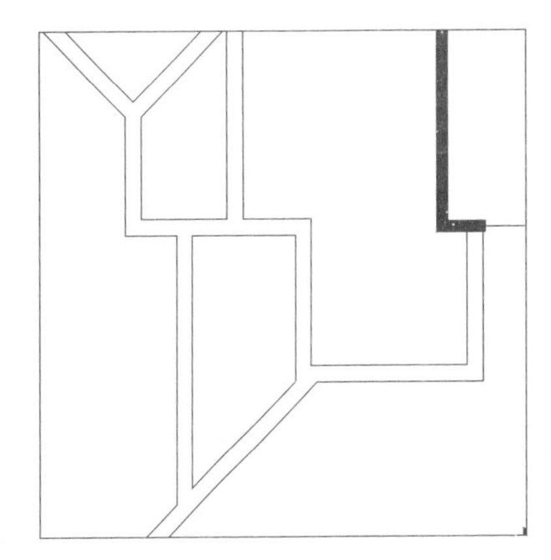

Figure 13.20

Building the Main Patio

Use the **OFFSET** command to offset the line running diagonally down from the point **56',59'** (LINE 3 in the diagram) by a distance of **80.5"** toward the house. Then **OFFSET** this line by **9"** toward the house (Figure 13.21).

Repeat this process three more times to put all of the brick bands running diagonally downward. Use the same process to create the lines running diagonally upward.

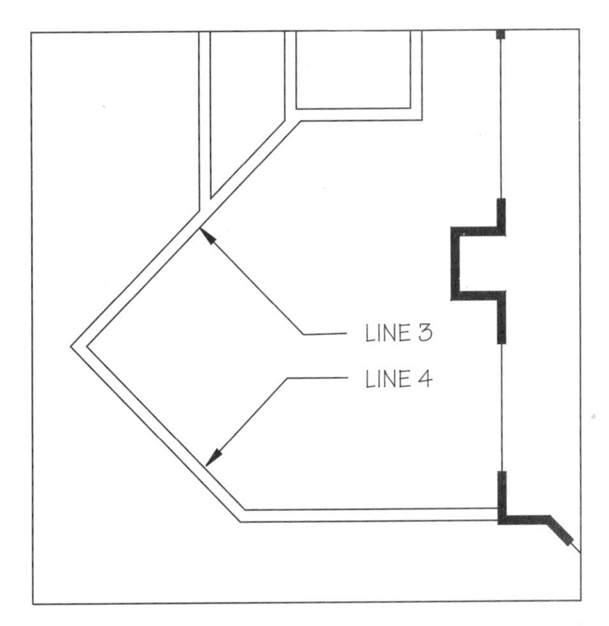

Figure 13.21

OFFSET the line running upward from the point **56',59,'** (LINE 4 in the diagram) by a distance of **80.5"**. **OFFSET** this line by **9"**.

Repeat this sequence twice more. **EXTEND** these lines out to the edges of the patio. Your results should look like the following diagrams (Figures 13.22 and 13.23).

Clean up these lines using **FILLET**, **TRIM**, and other editing commands you know. Remember that **FILLET** terminates two lines at an intersection. **TRIM** is used to remove unwanted pieces by using other lines (or polylines) as cutting edges. Use **ERASE** to eliminate any unwanted line and remember to **UNDO** any accidental erasure (Figure 13.24).

Figure 13.22

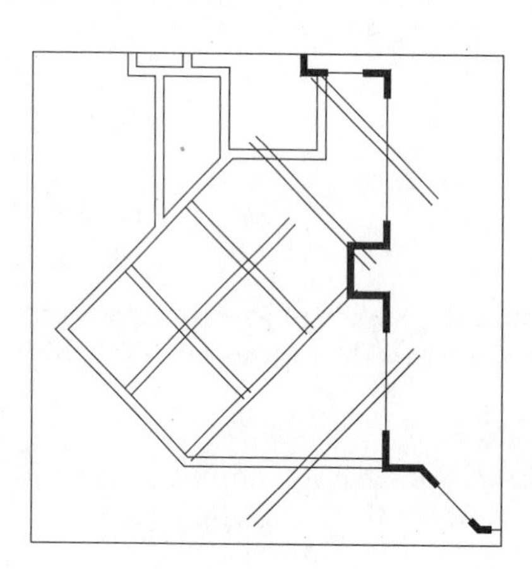

Figure 13.23

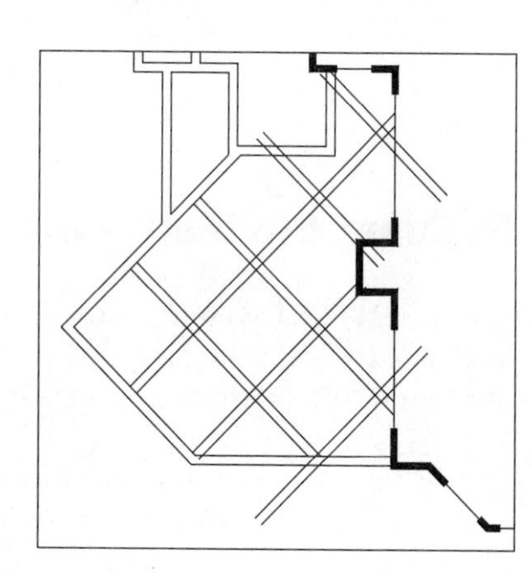

Figure 13.24

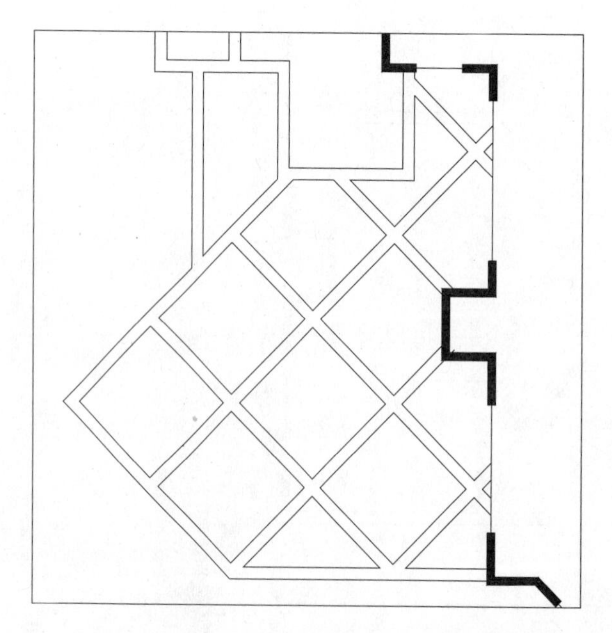

Building the Mow Strip

The final part of the back yard hardscape plan relates to the location of the brick mowing strip which surrounds the back lawn.

Construction Lines

Begin by drawing some construction lines around the perimeter of the lawn.

Draw LINE 1 from **49',37'** to **61',37'**.

Draw LINE 2 from **29',57'** to **29',71'**.

Draw LINE 3 from **36',85'** to **36',95'**.

Draw LINE 4 from **46',105'** to **56',105'**.

Draw the final construction line—LINE 5—from **42',80'** to **52',80'**.

See the illustration for the correct location of the construction lines (Figures 13.25 and 13.26).

Figure 13.25

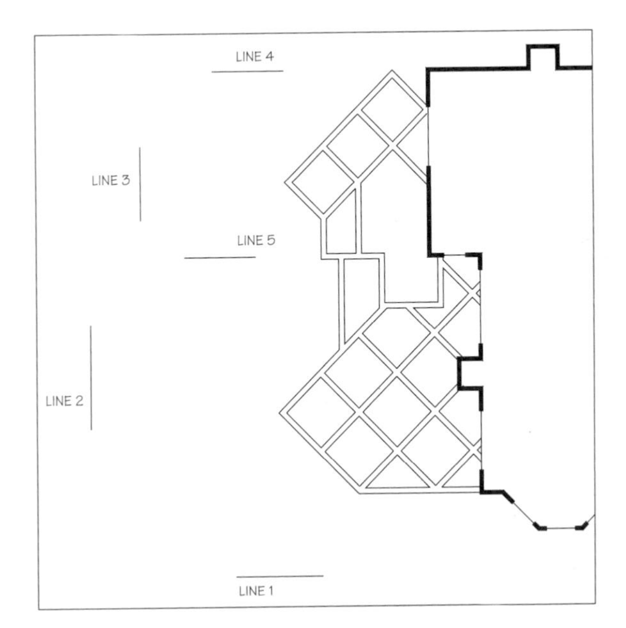

Figure 13.26

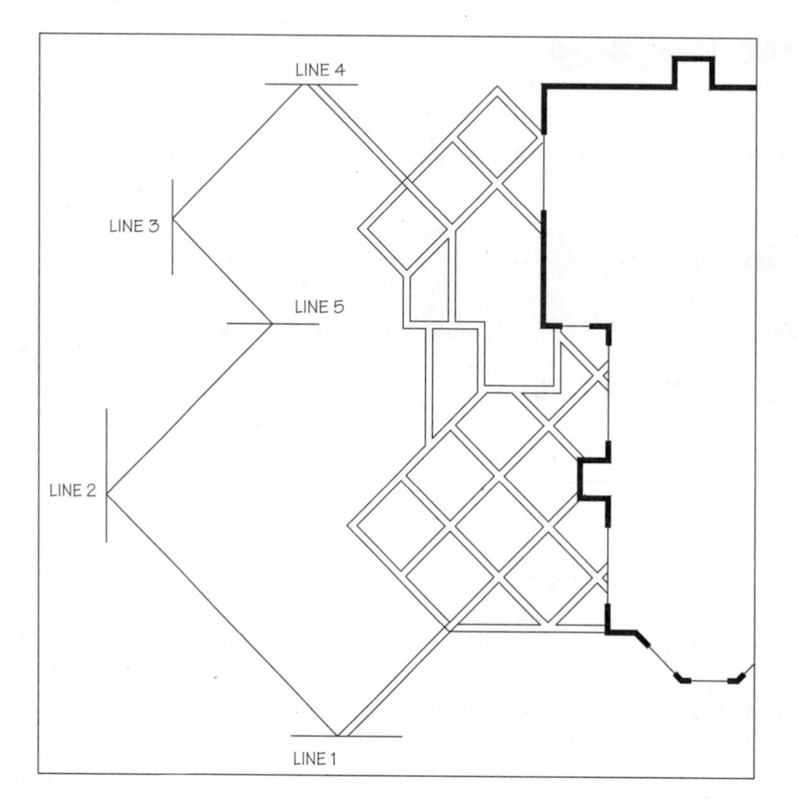

Mow Strip Perimeter

Find the lower diagonal brick edge line which runs down toward the lower left corner of the main patio, and **EXTEND** it down to LINE 1.

Repeat this for the upper diagonal brick edge which runs parallel, 9" away.

From the point where the second (upper) line intersects LINE 1, draw a 10' line at 135°. **EXTEND** it up to LINE 2. Draw a short line from the intersection point at LINE 2 to about 10' at 45°. Extend this line to LINE 5.

From the point where this line intersects LINE 5, draw a 10' line at 135°. Extend this line up to LINE 3. Draw a line from the intersection point on LINE 3 about 10' long at a 45° angle. Extend this line up to LINE 4.

Find the lower brick edge line which runs diagonally up from the bedroom patio toward LINE 4. **EXTEND** it up to LINE 4. Repeat this procedure for the upper brick edge, 9" away. See the illustration to check on the location of your lines.

Fillet the corners at LINE 1 and LINE 4. See the illustrations for details (Figures 13.27 and 13.28).

Figure 13.27

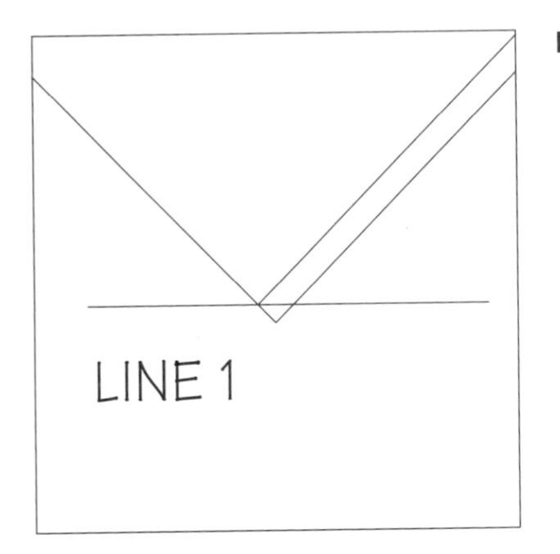

Figure 13.28

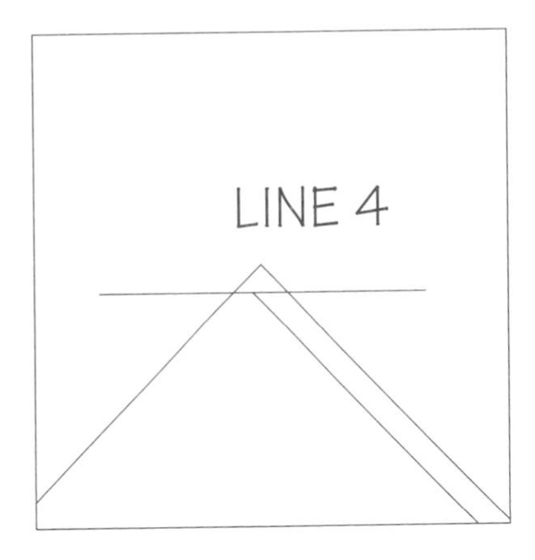

ERASE the construction lines (LINE 1 through LINE 5), then **OFFSET** the remaining perimeter lines by 9" back toward the house (Figures 13.29 and 13.30).

Use **FILLET** and **TRIM** to clean up the inner edge of the mow strip.

Figure 13.29

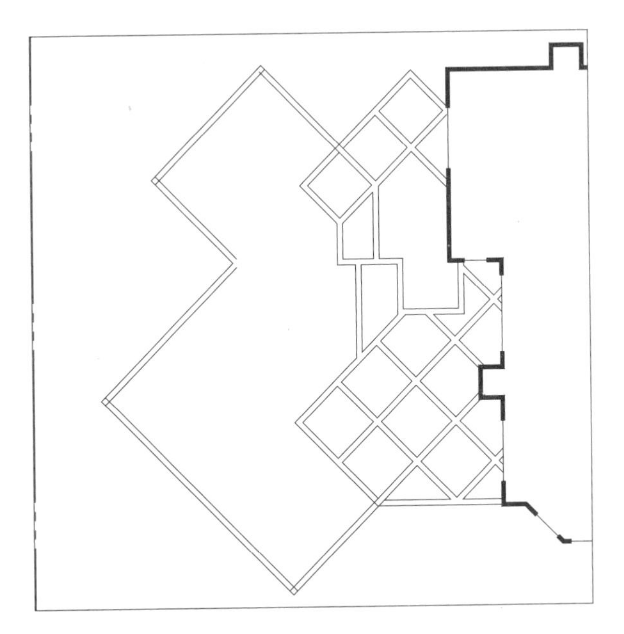

Figure 13.30

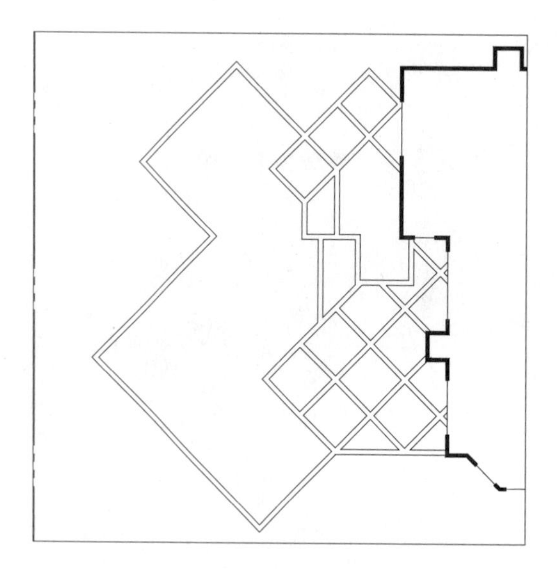

Side and Front Yard Hardscaping

We will complete the hardscape portion of the plan by drawing the side and front yard hardscape. In addition, we will hatch the concrete pads using an attractive stipple pattern to simulate a salt-finish surface.

Side Yard Hardscape

Begin the side yard walkways at the bottom of the yard by drawing a 20' vertical line down from the concrete pad corner **near** the point **77',49'**. **OFFSET** this line by **4.5"** on either side. Erase the center line.

Draw another 20' vertical line down from the point **83',48'**. **OFFSET** it to the right side by **9"**(Figure 13.31).

Add other 20' verticals from the points **107',48'** (LINE A)**,** **114',48'** (LINE B), **140',36'** (LINE C), and **147',36'** (LINE D). **OFFSET** LINE A to the left by 9", LINES B, C and D to the right by 9". See illustration (Figures 13.32 and 13.33).

EXTEND all of these lines to the property line at the bottom edge of the property

Draw horizontal lines from **107',29'** to **147',29'** (LINE E), from **114',34'** to **141',34'** (LINE F), from **76',36'** to **107',36'** (LINE G), from **83',41'** to **107',41'** (LINE H). **OFFSET** LINE E, LINE G and LINE H by **9"** to the bottom of your screen (Figures 13.34 and 13.35).

TRIM, **ERASE** and **FILLET** the lines to build the side walkway (Figure 13.36).

Figure 13.31

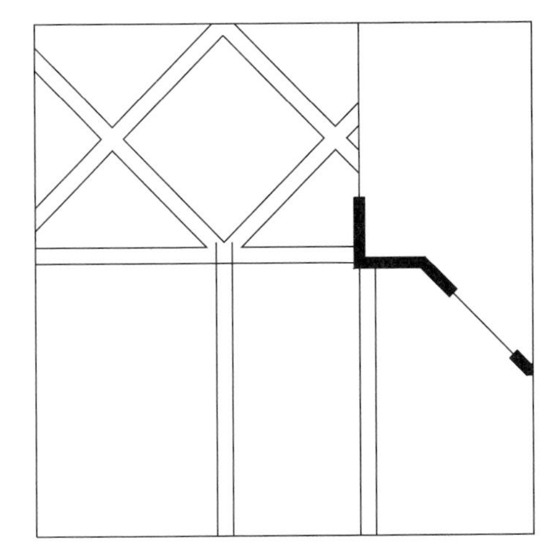

Figure 13.32

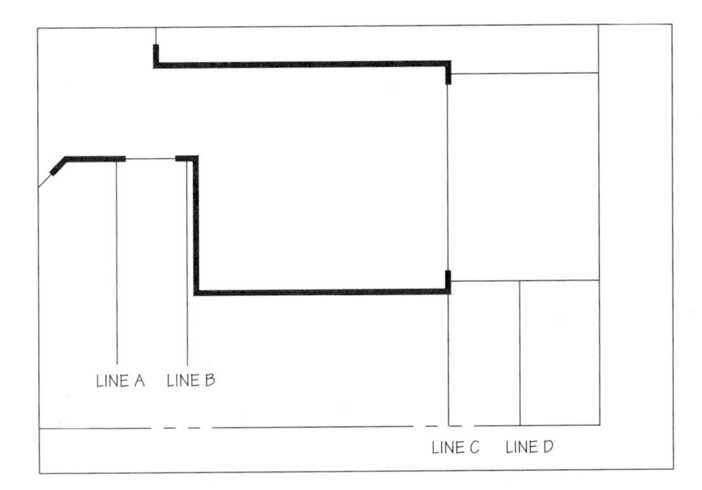

Figure 13.33

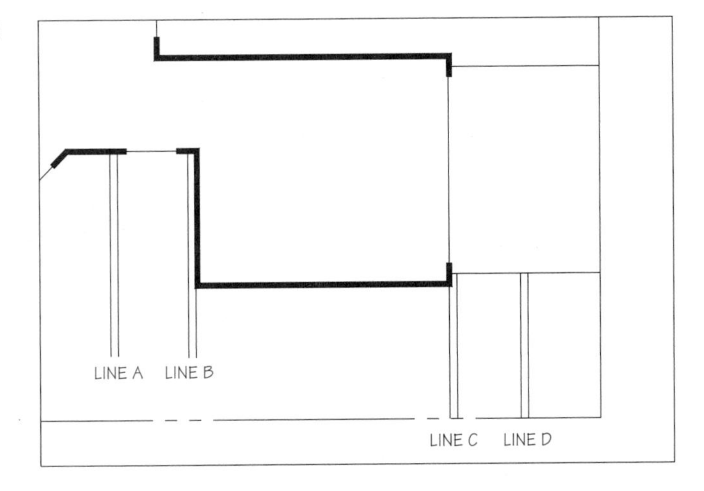

Figure 13.34

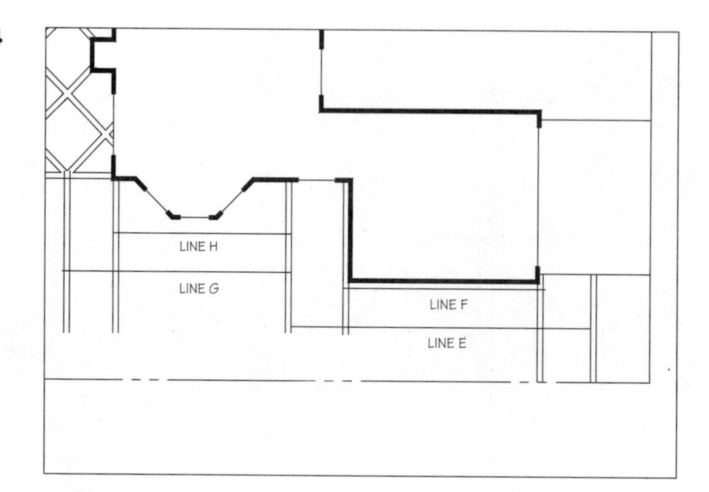

Figure 13.35

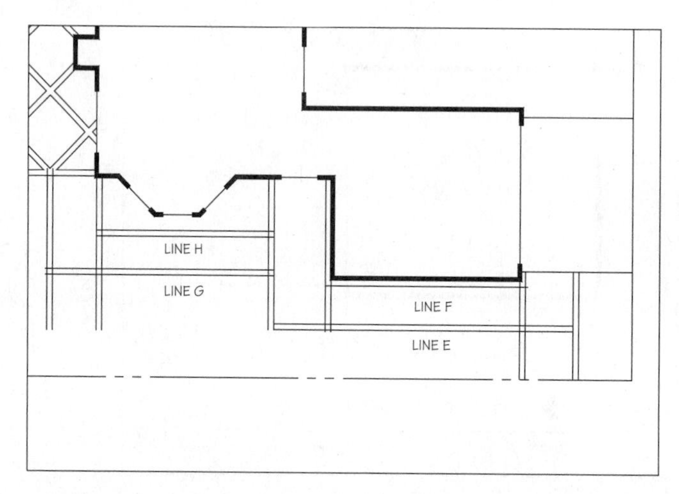

Figure 13.36

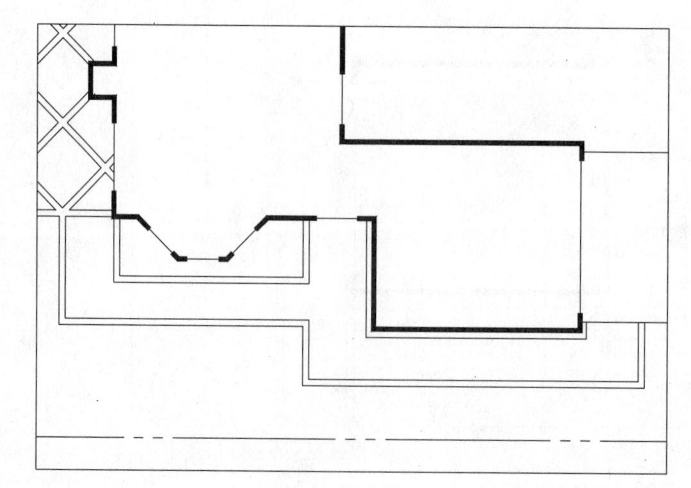

Front Yard Hardscape

For the front yard hardscape, we will build a simple porch, walk and brick mowing strip around the front lawn, porch and walkway.

Begin by drawing a line from **111',67'** to **123',67'** then down to **123',57'**. This completes the front porch (so far). Draw a line from **123',65'** out to the front property line. Repeat this procedure for a line from **123',60'**. These are the inner edges of the walkway, so **OFFSET** them each outward (away from the center of the walk) by **9"**.

Draw two vertical lines down to the driveway from **142',60'** and from **147',66'**. **OFFSET** these lines to the right (towards the front property line) by **9"**.

Using the illustration as a guide, clean up the edges of the walkway and brick mow strip (Figure 13.37).

Mow Strip

Begin the brick mow strip around the front lawn by drawing three lines on the **FORCE** layer (make **FORCE** the current layer when drawing these lines).

Start a line from **106',86'**, and bring the endpoint horizontally out to the front property line. Repeat this procedure from the point **106',92'** and **106',105'** (Figure 13.38).

Returning to the **CONCRETE** layer, draw a vertical line from **131',65'** to **131',92'**. Continue up to **143',105'**, then terminate the line at **155',92'**. These lines will be the inside edges of the brick, so offset them out away from the center of the grass by **9"**. Turn off the **FORCE** layer (Figure 13.39).

Use **TRIM** and **FILLET** to clean up the lines for the front hardscape.

Figure 13.37

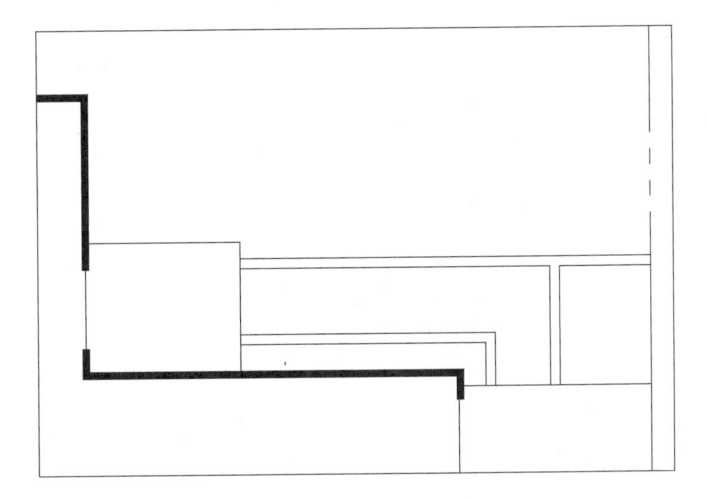

Figure 13.38

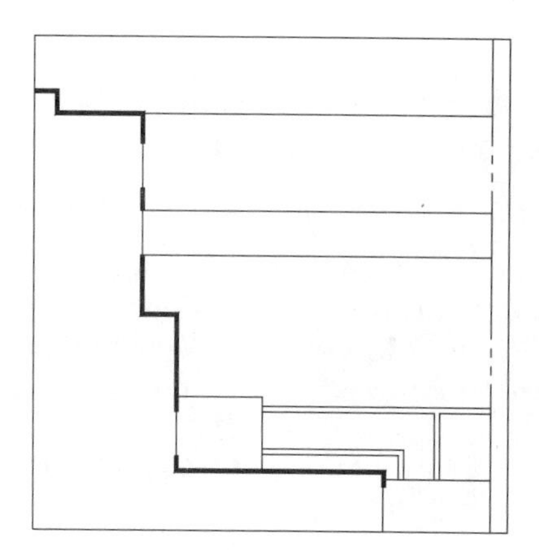

Figure 13.39

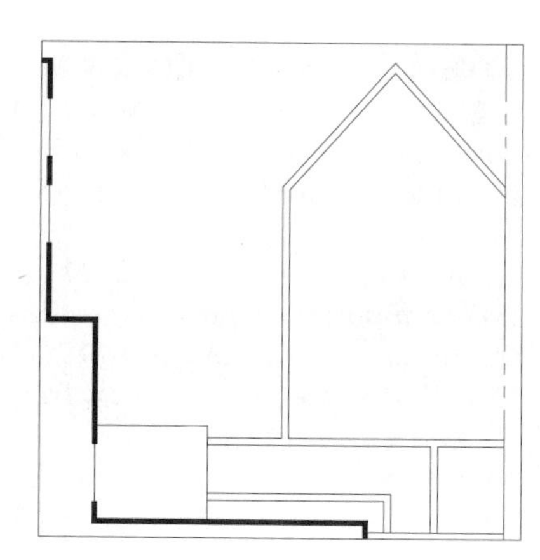

Hatching the Concrete

Once the hardscape is drawn in, hatch in the concrete pads using the **BHATCH** (keyboard macro: **BH**) command. **BHATCH** is different than traditional hatching in that it automatically seeks the boundary of an area to hatch, so you will not need to select the boundary manually (Figure 13.40).

Figure 13.40

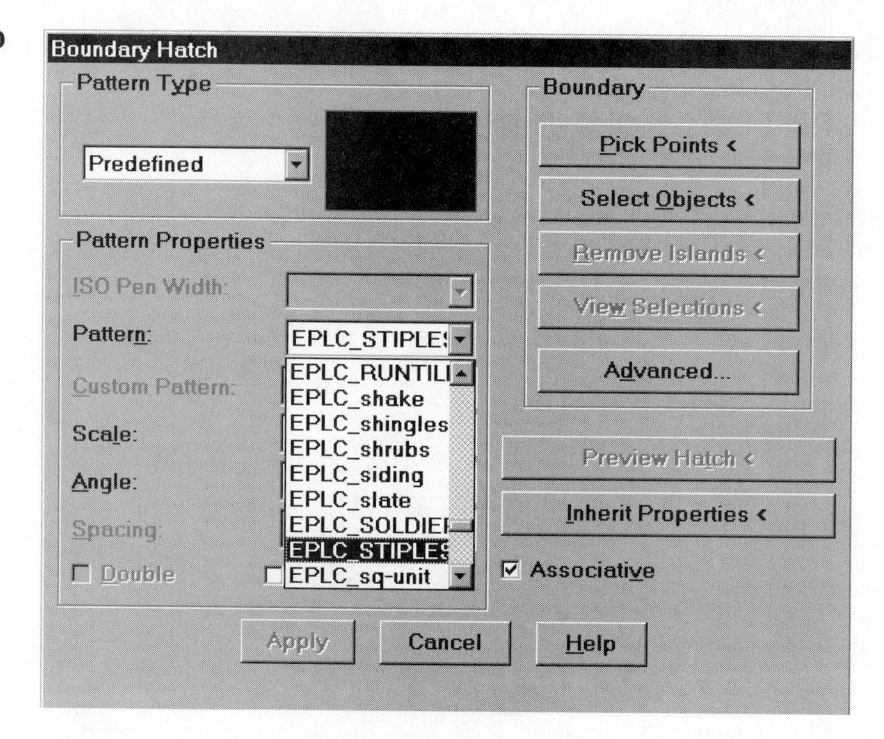

The **BHATCH** command brings up a dialogue box which first asks you to select a hatch pattern (along with its scale and angle), then allows you to choose from two methods of selecting the boundaries of areas to hatch.

In the **Pattern:** pop-down box, move down through the AutoCAD and Eagle Point hatch patterns until you reach **EPLC_STIPLES**. Highlight this pattern to select it.

In the **Scale:** text box, use a scale factor of **60"** or **5'-0"**. This creates a pattern that will give enough density to be visible, but not overpower the drawing. A hatch pattern that is too dense may greatly increase the disk space required for the drawing, and frequently yields muddy plotting results.

The **Angle:** box setting can be left at the default of **0**.

Once these factors have been established, you can choose an area to hatch either by picking a point inside the area, using the **Pick Points <** box, or by selecting the objects which make up the boundary using the **Select Objects <** box. Once you have selected an area to hatch, it will highlight on the screen. This allows you to see if the area is enclosed properly.

If the command behaves improperly (i.e., all the lines are highlighted at once), you probably need to extend a line to another edge to create an enclosed boundary. In the back yard, the **BHATCH** command will not work correctly unless your brick edges actually connect to the house footprint. Use the **EXTEND** command to correct any of these errors (Figure 13.41).

Once you have elected an area to hatch using the **Pick Points <** or the **Select Objects <** option, you can then actually preview the hatch before it is applied to the drawing. Use the **Preview** button for this purpose. If you like what you see, you can use the **Apply** button to place the hatch in the drawing. If there are errors in the hatch-

Figure 13.41

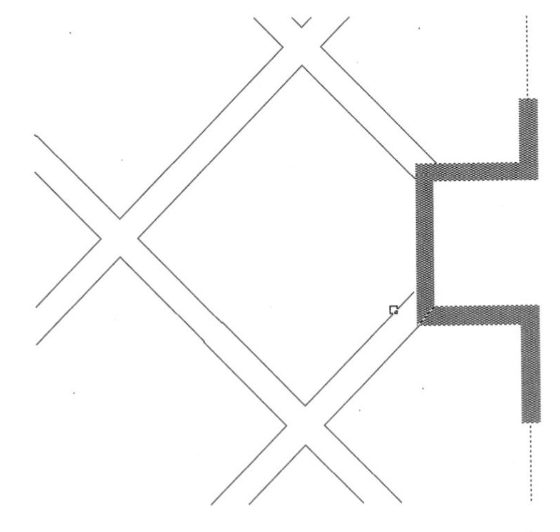

ing, return to the **BHATCH** dialogue box to cancel the command. Make any necessary changes with editing commands before using the **BHATCH** command again.

If you inadvertently use the **Apply** button to place an incorrect hatch on the screen, **DO NOT ERASE IT**. Use the **UNDO** command instead. This will prevent the hatch becoming a permanent part of the drawing file and greatly expanding its size.

Hatch the squares of the patio and all of the walkways.

Examine Figure 13.42 to verify the location of corner points on the house and hardscape.

To determine the actual coordinates of a point, use the **ID** command:

`Command:`
Type **ID**.

`Point:`
Move your cursor to the point you wish to identify. Use the cursor menu or keyboard to activate the appropriate object snap (**Intersection**, **Endpoint**, etc). At the lower left corner of the house, the program reports:

X = 83'-0" Y = 48'-0" Z = 0'-0"

Hint: When using **ID** with polylines, remember that AutoCAD will identify the vertex of a wide polyline (i.e., the intersection of two walls) at the **center** of the polyline.

Use the **DISTANCE** command to determine the distance between the house and the upper property line. Remember that AutoCAD will report the distance from the **center** of the wall to the property line.

`Command:`
Type **DIST** (for **DISTANCE**).

`First point:`
Type **MID** (or choose **Midpoint** from the cursor menu).

`_of`
Pick the chimney wall at the top of the residence.

```
Second point:
```
Type **PER** (or choose **Perpendicular** from the cursor menu).

```
_to
```
Pick the property line at the top of the screen.

```
Distance = 14'-0", Angle in XY Plane = 90, Angle from XY Plane
= 0
Delta X = 0'-0", Delta Y = 14'-0", Delta Z = 0'-0"
```
Continue to use **ID** and **DISTANCE** to verify the locations of the various points on your drawing.

Save your drawing to the hard disk and to a floppy disk back-up.

The following illustrations are intended to help you check the accuracy of your drawings and verify the locations of different elements of your Hardscape Plan (Figures 13.42 and 13.43).

Figure 13.42

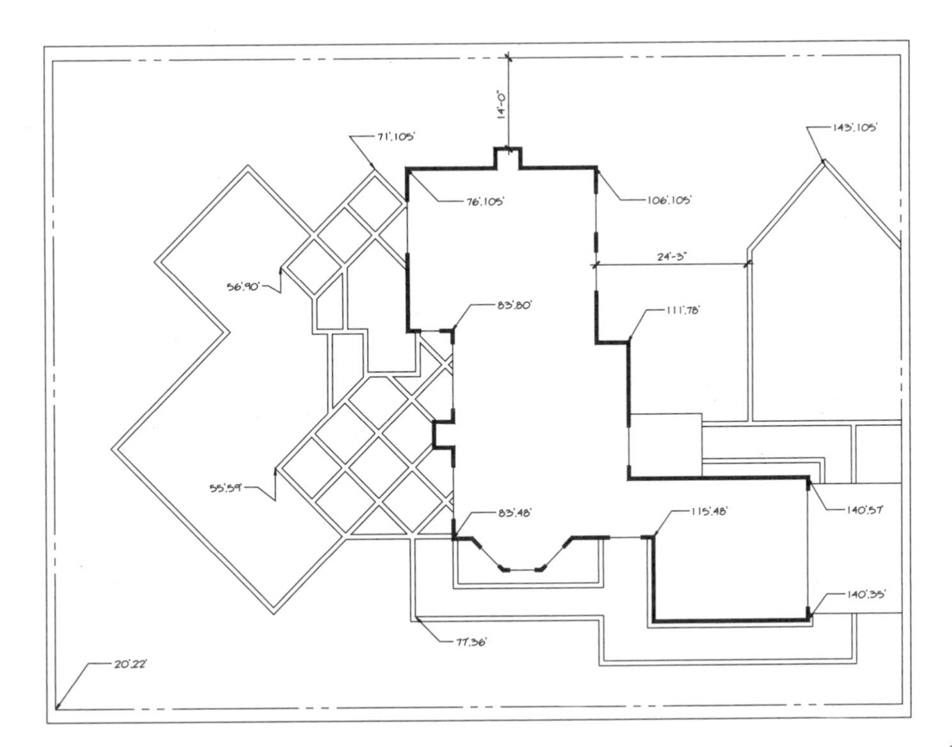

Figure 13.43

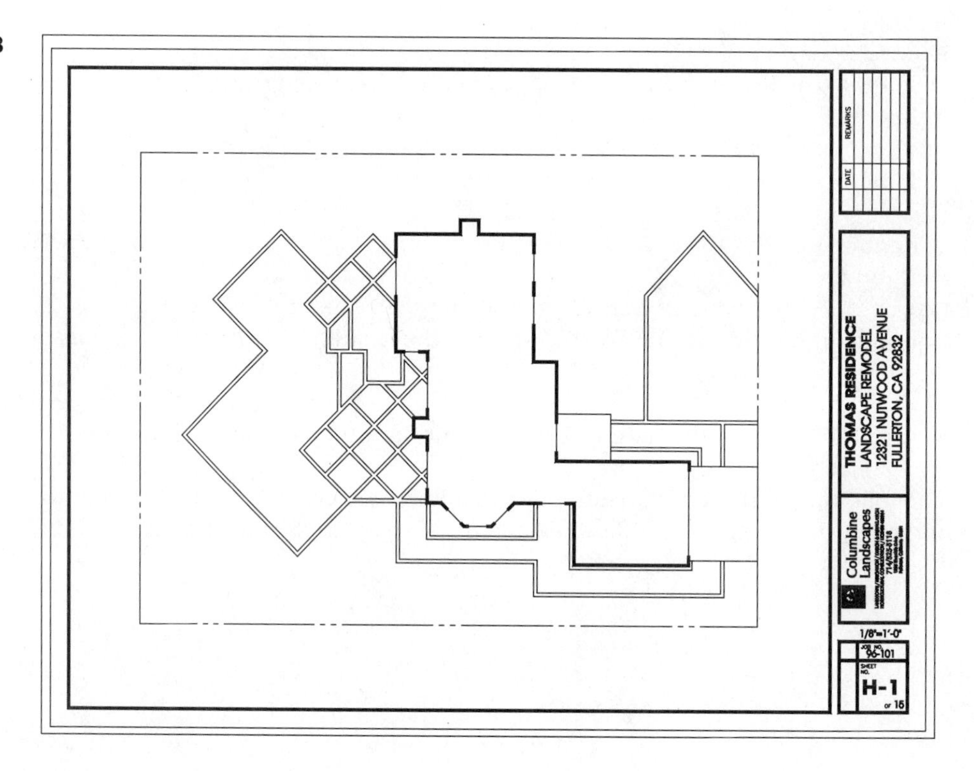

Dimensioning the Hardscape Plan

To complete the Hardscape Plan, a landscape designer will typically include some dimensions in the plan to assist the contractor in locating and installing the hardscape features. While the exact appearance of dimensions is flexible in landscape architecture, some features are fairly standard (Figure 13.44).

Figure 13.44

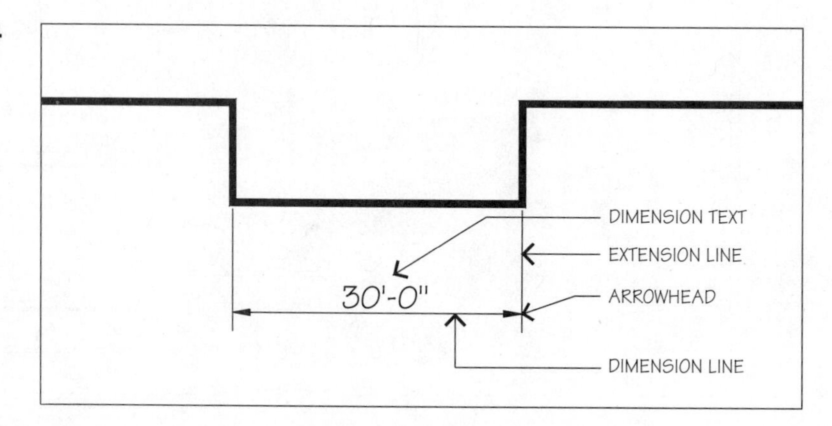

Typically the **dimension text** is located above the **dimension line**, aligned with the dimension line. Dimensions are usually rounded off to the nearest inch, and text is normally forced between the **extension lines** wherever they will fit. The arrowhead style used is usually a **tick mark**, which is usually fairly bold and prominent.

There are several type of dimensions that are used in landscape architecture including Linear, Angular, Radial and Leader dimensions. See the illustration below (Figure 13.45).

Figure 13.45

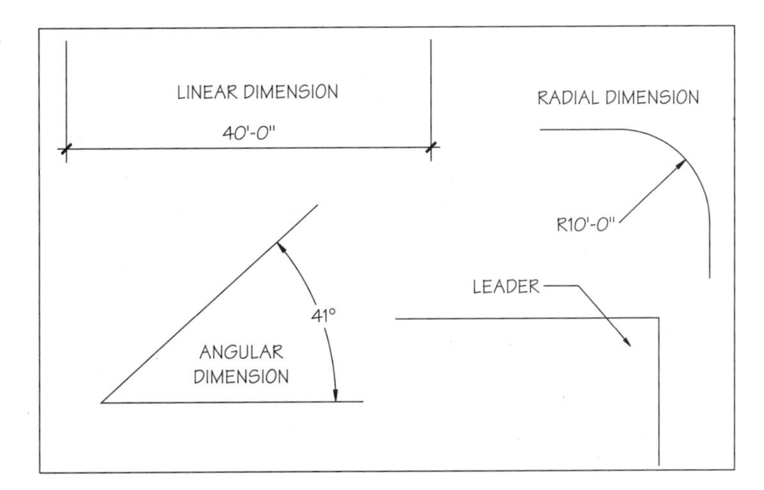

Notice that each dimension type is different here. The Angular, Radial and Leader dimensions do not use a tick mark, but use an arrow instead.

Inserting the Tick Mark

The tick mark provided by AutoCAD is not heavy enough to be used in an architectural drawing. Therefore, we will bring a new tick mark into the hardscape plan to be used in dimensioning. In order to do this, we will use the **INSERT** command to insert the tick mark drawing, but **CANCEL** the command when we are asked to locate the tick mark. This way, we insert the *definition* of the tick mark without actually inserting the image. The *block definition* is stored in the drawing even though the block is not visible. When the block definition is present, we can use it as an arrowhead in our dimension style.

From the command line:

```
Command:
```
Type **DDINSERT**.

This brings up the "Insert" dialogue box (Figure 13.46).

Pick **File . . .**, then find the directory where your drawings are stored (Figure 13.47).

Figure 13.46

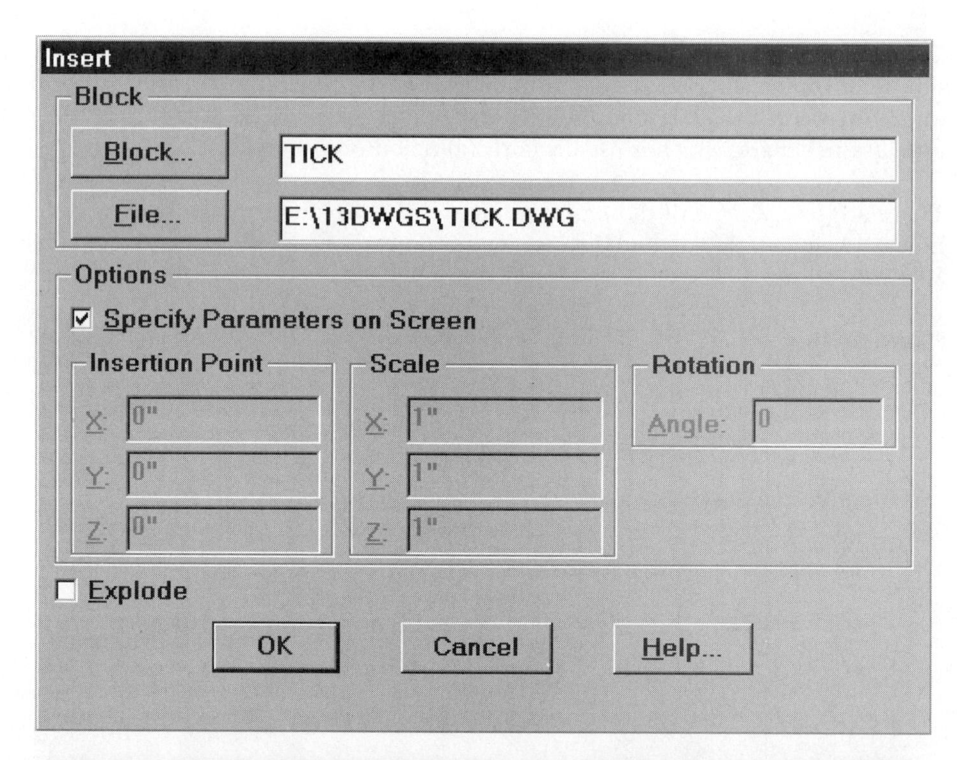

Figure 13.47

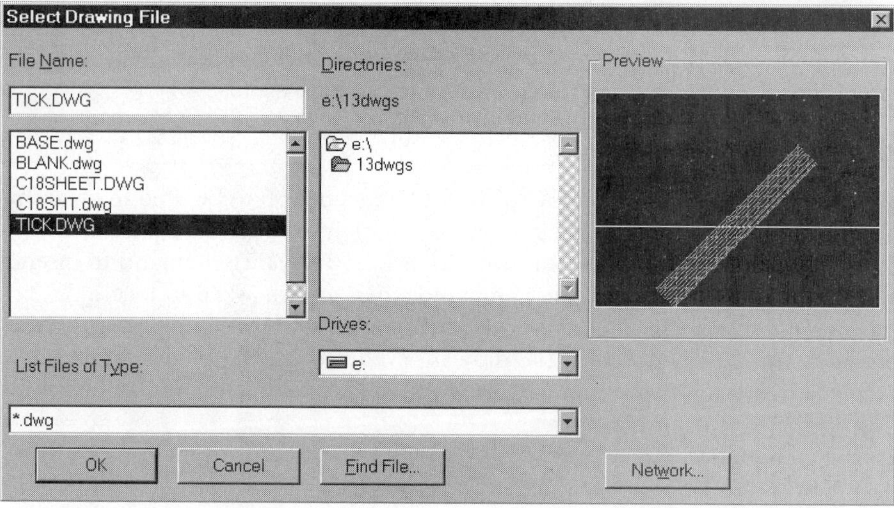

Select the drawing **TICK**. Pick **OK**.

```
Insertion point:
```

Press **<CTRL>-C** (or **<Esc>** in Windows platforms) to cancel the command. Remember that the definition has been inserted, even if the image has not.

Dimension Styles

Because AutoCAD can perform so many types of drafting (mechanical, architectural, engineering, etc.) its dimensioning capabilities are quite flexible. The units, accuracy, text size, arrow size, dimension and extension line location, text orientation and a large number of other dimension settings can be adjusted or modified. To establish the appearance of our dimensions, we need be able to store dimension settings (text size, arrowheads, etc.) without having to reconstruct them for every dimension we create. We can incorporate all of these important settings into a **Dimension Style**, which can be created and saved with the drawing (Figure 13.48).

A Dimension Style is a group of dimensioning settings which are saved under a name that you assign. The dimension settings can be addressed in two different ways:

1. Via the Command Line—here you set the dimension settings by changing AutoCAD's *dimension variables.* There are dozens of variables such as DIMALTZ, DIMTIC, DIMCLRT, etc. Each of these control a single aspect of the dimension's appearance and can be set—assuming that you can keep track of what they are and what they control. See your AutoCAD documentation for details.

Figure 13.48

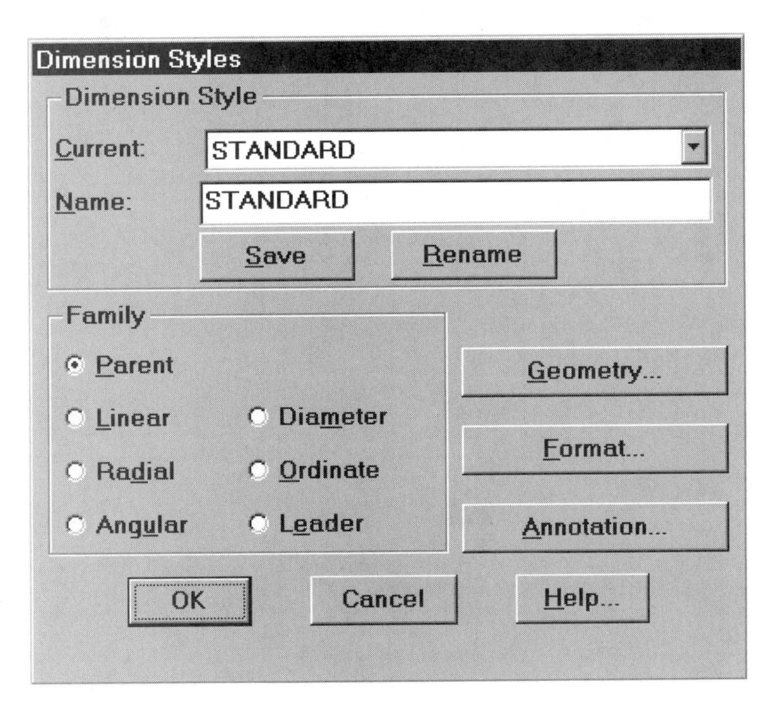

2. Via the "Dimension Styles" dialogue boxes—here you navigate through a series of dialogue boxes to set your dimension variables with pop-up menus, pull-down menus and text boxes to set up the dimension style to your needs. This method is generally simpler and faster than using Dimension variables. It is the method that we will use in our drawing.

Creating a Dimension Style

To begin to create a dimension style, move to the AutoCAD menu, then select the pull-down option **Data**. Pick **Dimension Style . . .**. This opens the "Dimension Styles" dialogue box.

From this dialogue box, we can establish several important settings for the new dimension style. First, we will name our new style **TEMPLATE**. Naming a new style allows us to save the original settings (stored with the style name **STANDARD**) if we should need to reset them. Type **TEMPLATE** in the **Name:** text box.

Pick **Save**. The **TEMPLATE** style is created from the original style **STANDARD** (Figure 13.49).

Geometry

In the "Dimension Styles" dialogue box, we see several categories and options. We will begin with the boxes to the right side.

Figure 13.49

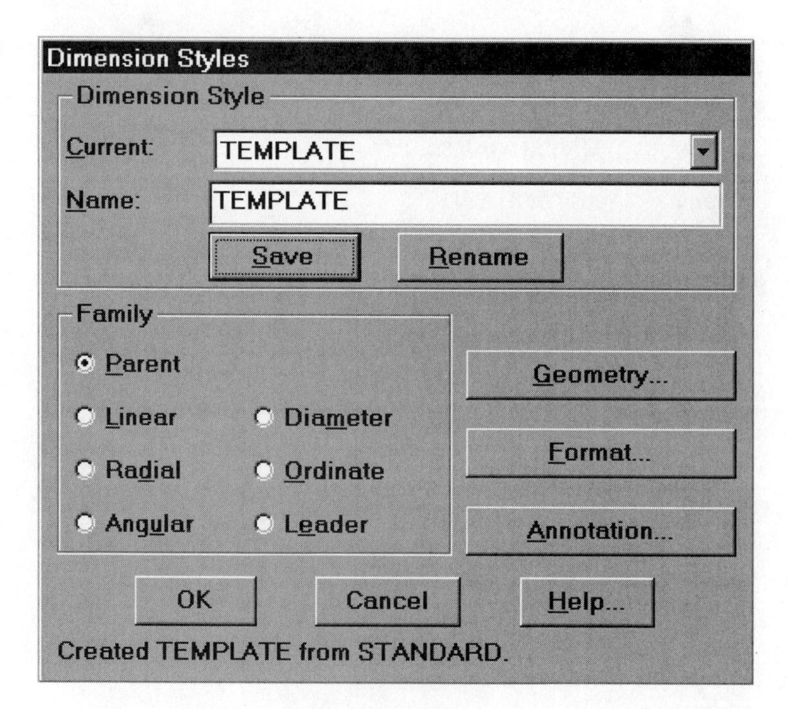

Pick **Geometry . . .** (Figure 13.50).

In the "Geometry" dialogue box, we can set up various features of the Arrowheads, Dimension Lines and Extension Lines. In the case of the dimension style **TEMPLATE**, we want to redefine the type of arrowhead used and its size. Pick in the pop-down box labeled **1st:** **Closed Filled** in the scroll arrow area. This brings down a box of available arrowheads.

Select **Save** **User Arrow . . . .** (Figure 13.51).

Since the program needs to know which user arrow to use, we need to specify the name. In the "User Arrow" dialogue box, type **TICK**. Choose **OK**.

In the Arrowheads section **Size:** box, change the size to **1/8"**.

In the **Overall Scale:** box, change the value to **96**. Pick **OK** at the bottom of the "Geometry" dialogue box (Figure 13.52).

This returns us to the "Dimension Styles" dialogue box.

Use the **Save** option to save your work thus far. Next, pick **Format . . .**. If you have failed to save your work, you will see this dialogue box (Figure 13.53).

Figure 13.50

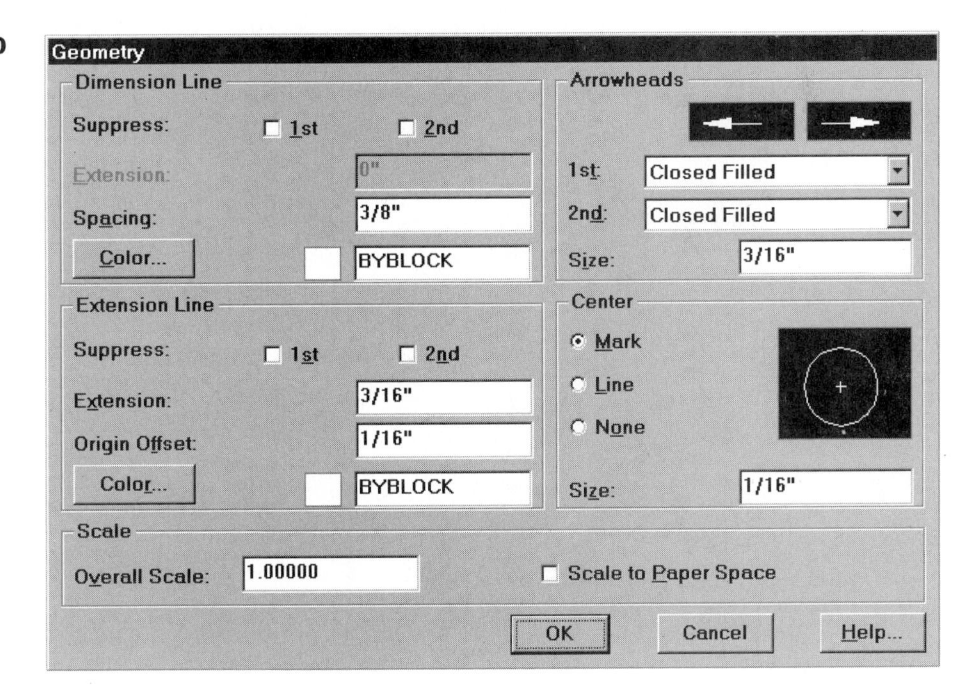

Figure 13.51

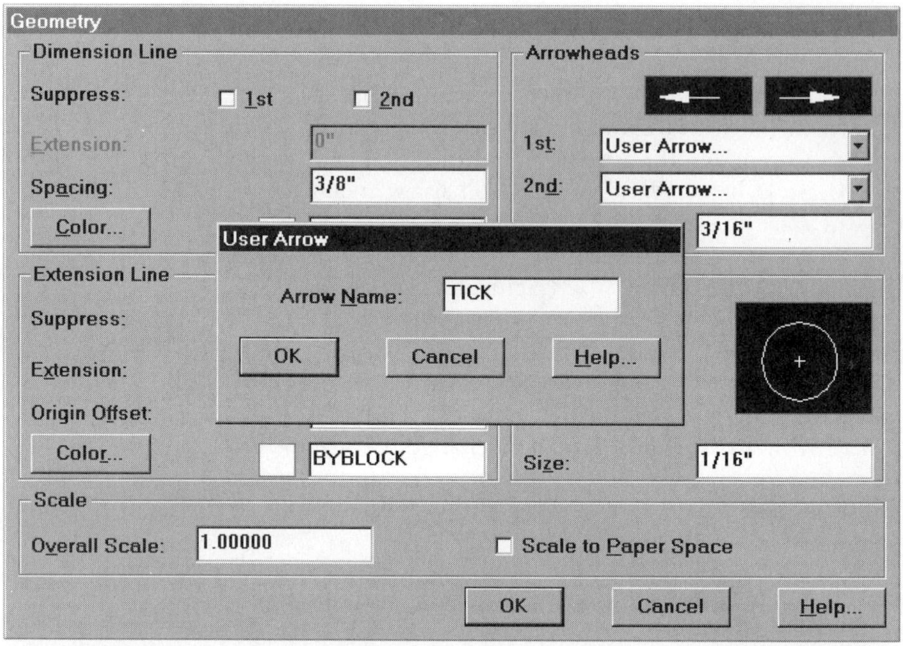

Geometry

Dimension Line

Suppress: 1st 2nd

Extension: 0"

Spacing: 3/8"

Color...

Arrowheads

1st: User Arrow...

2nd: User Arrow...

3/16"

User Arrow

Arrow Name: TICK

OK Cancel Help...

Extension Line

Suppress:

Extension:

Origin Offset:

Color... BYBLOCK

Size: 1/16"

Scale

Overall Scale: 1.00000

Scale to Paper Space

OK Cancel Help...

Figure 13.52

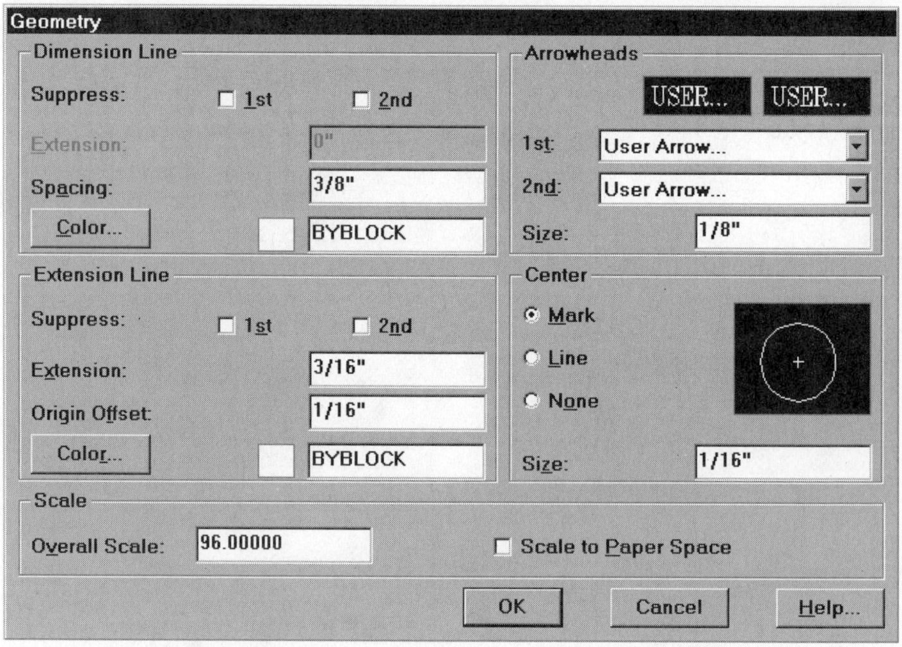

Geometry

Dimension Line

Suppress: 1st 2nd

Extension: 0"

Spacing: 3/8"

Color... BYBLOCK

Arrowheads

USER... USER...

1st: User Arrow...

2nd: User Arrow...

Size: 1/8"

Extension Line

Suppress: 1st 2nd

Extension: 3/16"

Origin Offset: 1/16"

Color... BYBLOCK

Center

Mark

Line

None

Size: 1/16"

Scale

Overall Scale: 96.00000

Scale to Paper Space

OK Cancel Help...

If you have not saved your changes before, do so with this dialogue box.

Figure 13.53

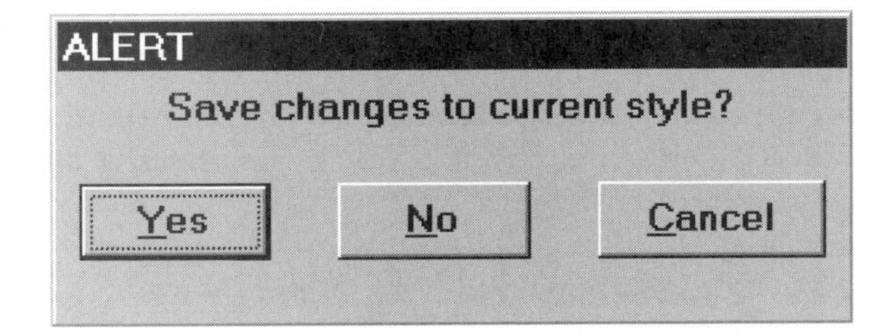

Format

In the "Format" dialogue box, we can set up characteristics of the dimension relating to the location of the text and arrows, and the type of dimension line used (Figure 13.54).

First, make sure the toggle box marked **Force Line Inside** is marked. This places a dimension line between the extension lines regardless of where the text and arrows are placed.

In the **Fit:** pop-down box, select the choice **No Leader**. This allows AutoCAD to place text away from the dimension if there isn't enough room between the extension lines. This text will have no connecting leader associated with it, which makes it easier to move into position later.

Figure 13.54

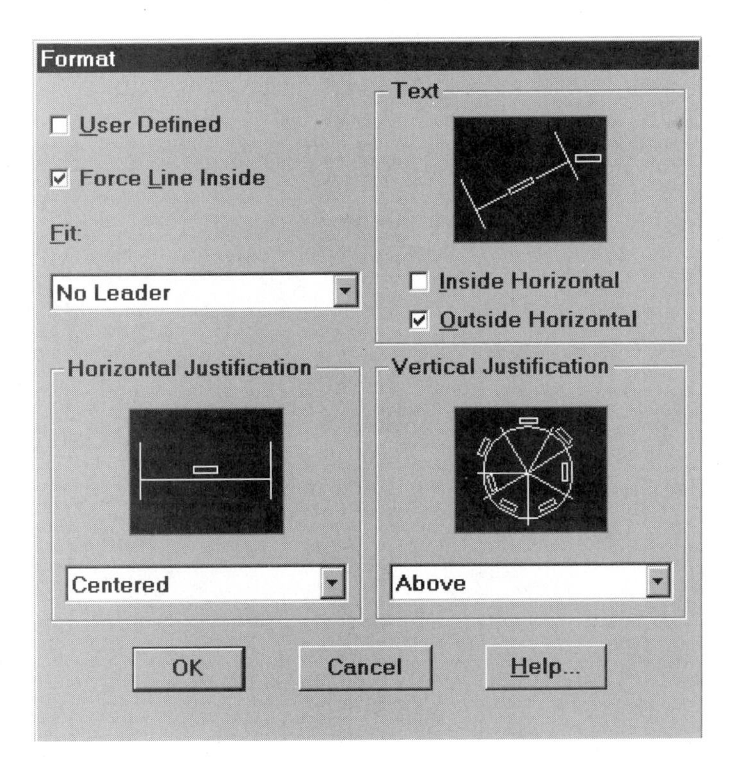

Next, remove the mark from the toggle box marked **Inside Horizontal**, while leaving in the **Outside Horizontal** toggle box. This permits the text inside the extension lines to be aligned with the dimension line, while the text outside of the extension lines remains horizontal. Since we will force the text inside the extension lines in almost every case, this setting will only affect a few dimensions.

Finally, in the **Vertical Justification** pop-down box, choose **Above**. This places the text above the dimension line, permitting an uninterrupted dimension line.

After these changes are made, select **OK** at the bottom of the dialogue box.

Annotation

After returning to the "Dimension Styles" dialogue box, select **Annotation . . .** (Figure 13.55).

The "Annotation" dialogue box refers us to the display of the text in the dimension. We can set the units of measure, allow the display of an alternate unit system (i.e., metric), show tolerances and other features. Most of these features relate to other types of drafting, so are not set here.

Begin by picking **Units . . .** under **Primary Units**. This brings up the "Primary Units" dialogue box (Figure 13.56).

Figure 13.55

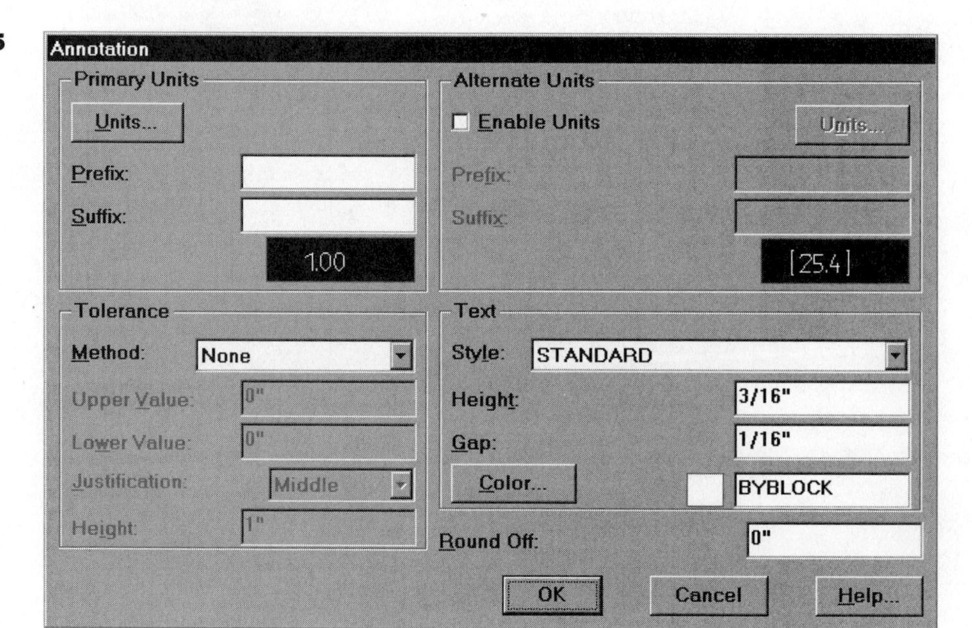

Figure 13.56

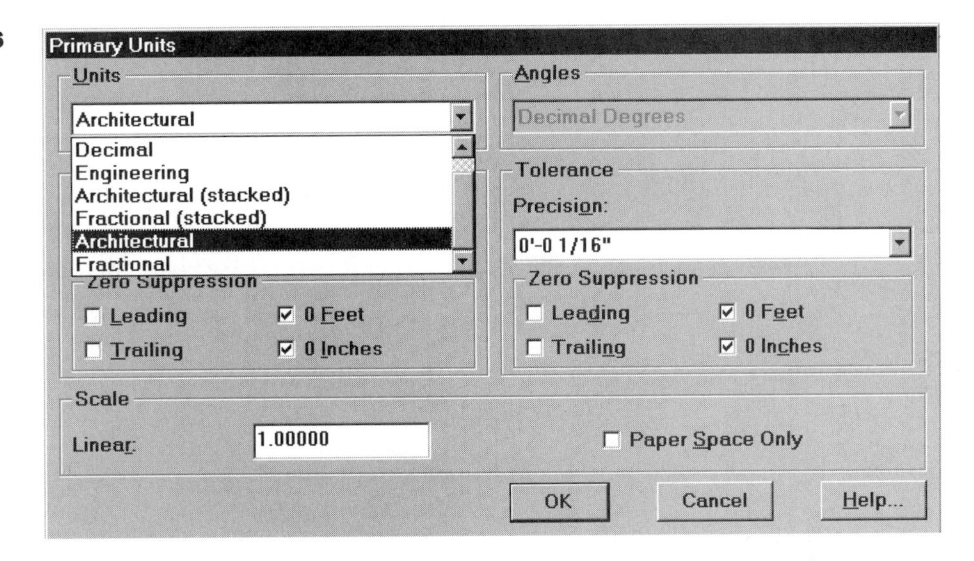

Find the **Units** pop-down box and choose **Architectural**. This will cause the measurements to be displayed in feet and inches (Figure 13.57).

Find the **Dimension** area, and the **Precision** pop-down box. Choose **0' - 0"**. This causes all dimensions to be rounded to the nearest inch. This is appropriate for landscape construction since accuracy to units smaller than one inch is unnecessary.

Next, examine the area labeled **Zero Suppression** (Figure 13.58).

Figure 13.57

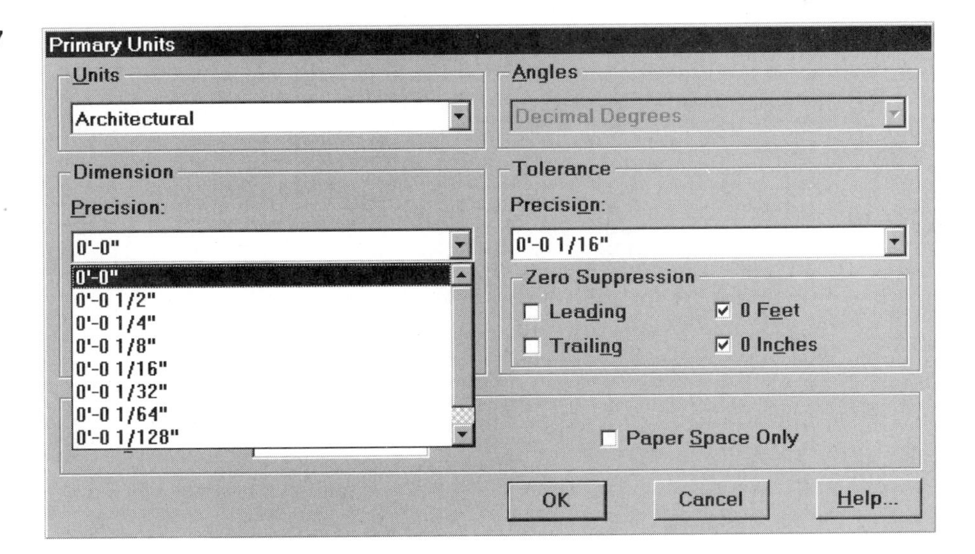

Figure 13.58

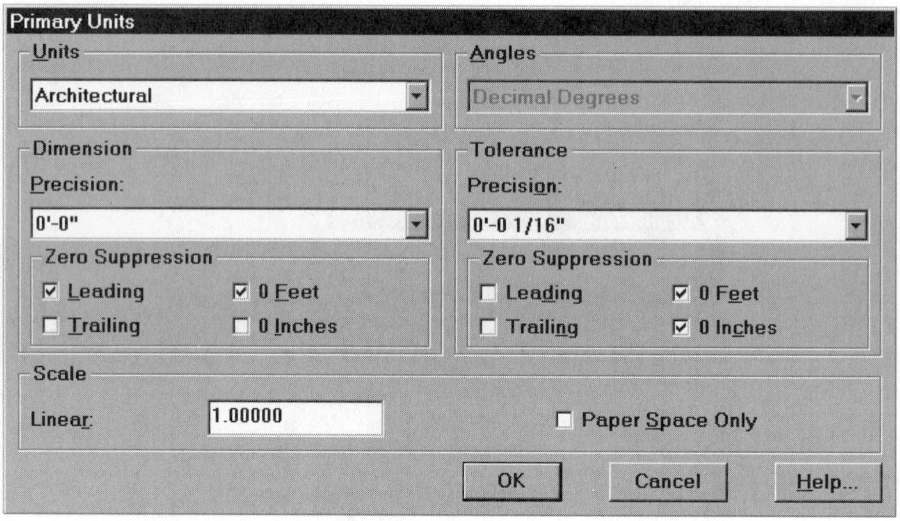

In this area, we can specify whether or not zeros are displayed with a dimension. We will suppress the presence of a leading zero in our feet and inches measurements. This makes a small dimension read **6"** rather than **0' - 6"**. We will allow the presence of a trailing zero in our feet and inches measurements, so that a large dimension would read **30' - 0"** rather than **30'**.

Mark the toggle boxes for **Leading** and for **0 Feet**. This will prevent (suppress) them from being displayed.

Pick **OK** to exit the "Primary Units" dialogue box (Figure 13.59).

We will not use **Alternate Units**, so make sure that the **Enable Units** box does not have an X in it. Additionally, we will not need tolerance measurements, so make sure that the **Method:** pop-down box reads **None**.

We will make two changes in the **Text** area. First, change the value in the **Height:** text box to **1/8"**.

Second, pick the **Color . . .** box. This brings up the "Select Color" dialogue box. Select the color **CYAN**. We will take advantage of the color differences (the text and tick marks are both cyan—while the dimension and extension lines will both be red) when we plot the drawings.

Pick **OK** to close the "Annotation" dialogue box.

After we return to the "Dimension Styles" dialogue box, we need to make a few more changes (Figure 13.60).

Figure 13.59

Figure 13.60

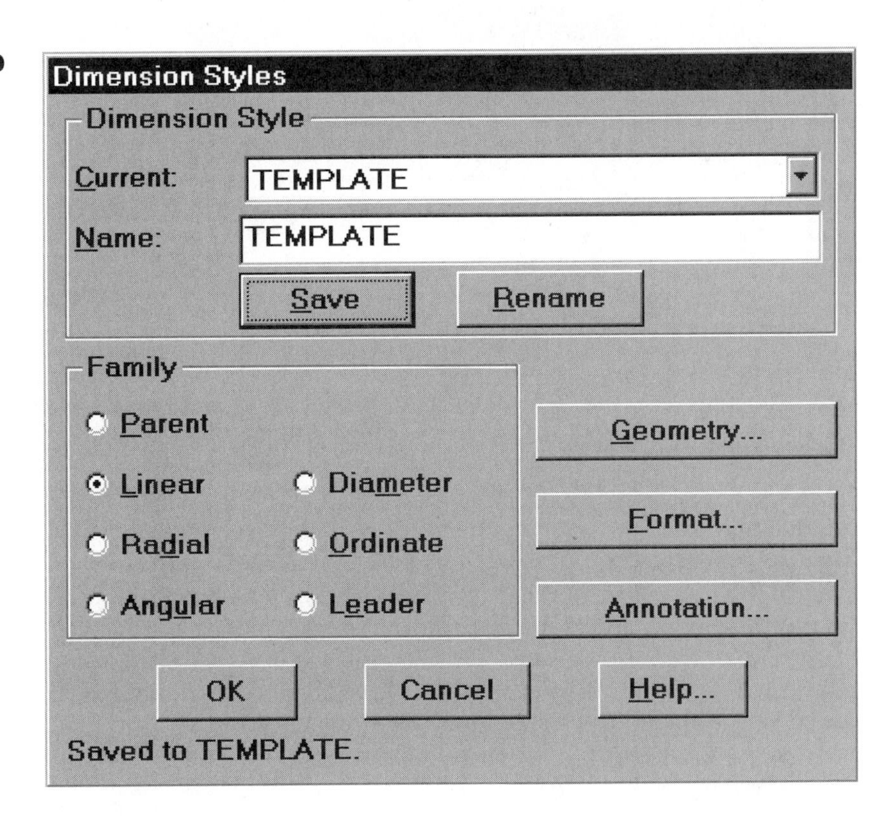

Dimension Families

In AutoCAD Version 13, dimensions are organized into **Families**. There is a **Parent** style, which will share its characteristics with all of the **child** dimensions. Each child dimension (**Radial**, **Angular**, **Diameter**, **Ordinate** and **Leader**) will originally share all of its traits (text style, location, arrowhead style, etc.) with the parent dimension. Since we want to allow some of these child dimensions to differ from the parent style, we can set them here.

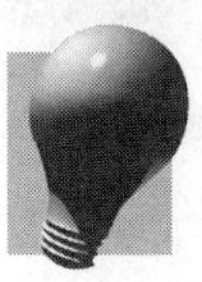

HINT: *IMPORTANT:* The sequence of defining dimension styles is crucial. Name the parent style first, then define its characteristics. Finally, change the child dimensions to meet your needs.

Select the radio button for **Radial**. Move to the **Geometry . . .** dialogue box. In this box, set the **Arrowheads** back to **Closed Filled**, then click on **OK**.

Select the radio button for **Angular**. This brings up the AutoCAD Alert box (Figure 13.61).

Figure 13.61

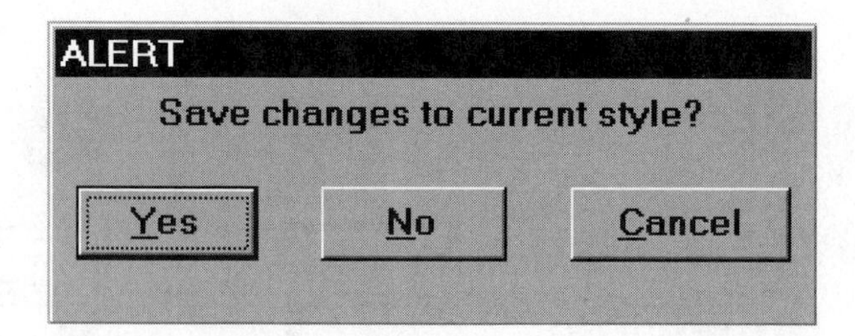

When you change from one child dimension to the next, AutoCAD asks you if you want to save your previous changes.

Pick **Yes**. Move to the **Geometry . . .** dialogue box, and set the **Arrowheads** back to **Closed Filled**. Click on **OK**. Next, move to the **Format** dialogue box, then change the **Text** settings so that both the inside and outside text is displayed horizontally for angular dimensions.

Repeat this procedure for the child dimension **Leader**. After you finish, be sure to **Save** in the "Dimension Styles" dialogue box before you press **OK** to exit.

If you see this AutoCAD Alert box, you have changed a child dimension and have not saved before exiting the "Dimension Styles" dialogue box (Figure 13.62).

Figure 13.62

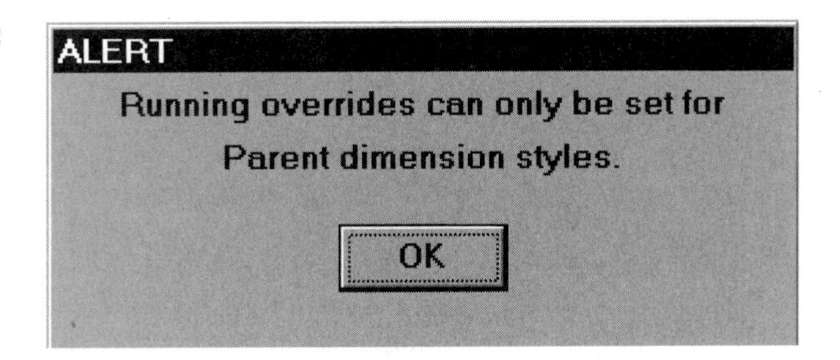

A **Running Override** is a temporary modification to a dimension style which only affects the dimensions created **after** the modification. (When a dimension style is modified and saved, it affects all of the dimensions created **before** the modification.) In addition, a running override is not a permanent change to the dimension style. In any case, if you save the last change before exiting the "Dimension Styles" dialogue box, this Alert box will not appear.

Dimensioning Commands

Once we have finished creating the dimension style we need, we can begin the actual dimensioning process. AutoCAD comes with a sizeable group of commands to create and edit dimensions. Although there are many dimensioning tools available, we will concentrate on five dimensioning commands and three dimension editing commands.

Creating Dimensions

DIMLINEAR (DL) finds the distance (either horizontal or vertical) between two points you specify.

DIMALIGNED (DA) will find the distance between two points at the angle shown between the points. The dimension line and text will also be displayed at this angle.

DIMCONTINUE (DC) will continue building dimensions from the previous point used. This allows us to chain dimensions together while keeping them in alignment.

DIMANGULAR allows us to measure the angle between two objects.

LEADER (LE) allows us to build a leader note to an object on the screen.

Editing Dimensions

DIMTEDIT (DE) will "edit" text by allowing you to move the text where you want it to go. Although it appears that the entire dimension entity (dimension lines, extension lines and text) moves, only the text will move when you click the pick button.

DIMEDIT-NEW will allow you to create a new line of text to replace the one that appears in a dimension. You can then substitute the replacement text for the existing text after you pick the dimension. Although this is available, don't use it to "correct" dimensions that are not accurate. If you select dimension points properly, the dimensions should be correct. Modified dimensions can be a sign of sloppy work and bad drafting.

GRIPS-STRETCH will allow you to move the components of a dimension to a new location. This is one of the best ways to move dimension text where you want it. While no other command is active, simply pick a dimension to modify. The blue grips for the dimension are activated, allowing you to select one grip to make it a red "hot grip." If you make the dimension text grip "hot", you can then move it to a new location without moving the other parts of the dimension.

The Dimension Menu

To use the dimensioning commands, use the EP/AutoCAD menu, and pull down the Draw option. Pick Dimensioning ▸. The major dimensioning commands are listed here.

Dimensioning the Plan

We will begin dimensioning around the outside of the landscape, marking the corner points of the concrete slabs and brick mow strips. Where the brick surrounds concrete pads, we will dimension to the concrete slabs, since the brick trim would be added after the concrete is poured. In the areas where brick trim surrounds planters or turf, we will dimension to the outside corners.

To begin dimensioning, create a new layer **DIM**. Make its color **Red**. Make it the current layer.

Be sure **SNAP** is turned off, but is set to **1'**.

Begin in the lower right corner with the **DIMLINEAR** command.

```
Command:
```

Use the pull-down command Dimensioning ..., then Linear or type **DL** (**DIMLINEAR**).

```
DIMLINEAR
First extension line origin or RETURN to select:
```

Use the cursor menu to highlight **Intersection**, then pick the lower right corner of the property.

```
Second extension line origin:
```

Use the cursor menu to highlight **Intersection**, then pick the corner of the driveway shown below:

Figure 13.63

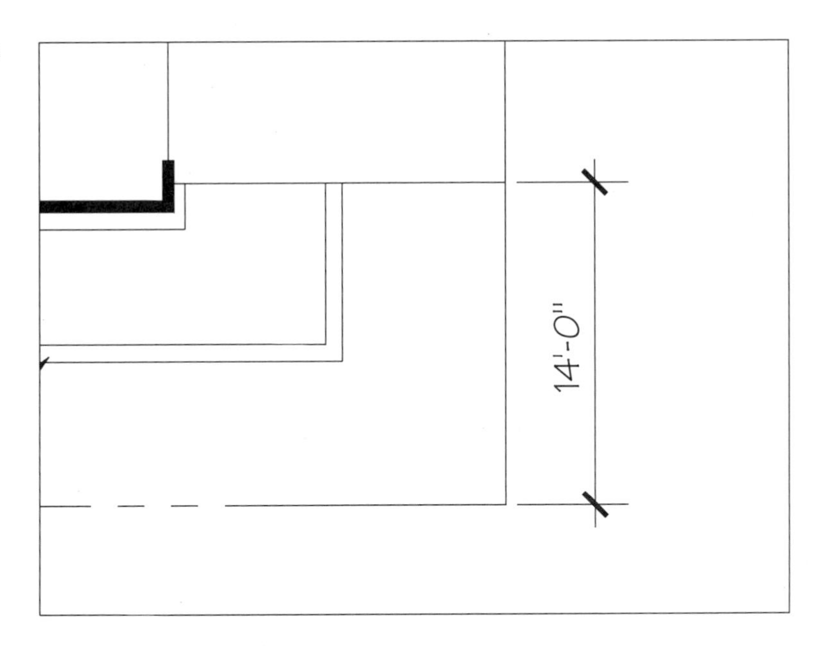


```
Dimension line location
(Text/Angle/Horizontal/Vertical/Rotated):
```

Press **F9** to turn on **SNAP**, then move the cursor so that coordinate dial reads an X value of **159'** (the Y value will not change). This moves the dimension line and text to the area shown. It will also help us keep the dimensions in proper alignment with each other (Figure 13.63).

We will continue to dimension across the front yard.

```
Command:
```

Use the pull-down **Dimensioning ...**, then **Continue** or type **DC** (**DIMCONTINUE**).

```
DIMCONTINUE
Second extension line origin or RETURN to select:
```

Use the cursor menu to highlight **Intersection**, then pick the upper corner of the driveway.

```
Dimension text <20'-0">
```

The dimension is placed automatically, and will be aligned with the first dimension you placed. See below (Figure 13.64).

Press **<Enter>** to repeat the **DIMCONTINUE** command.

```
Second extension line origin or RETURN to select:
```

Figure 13.64

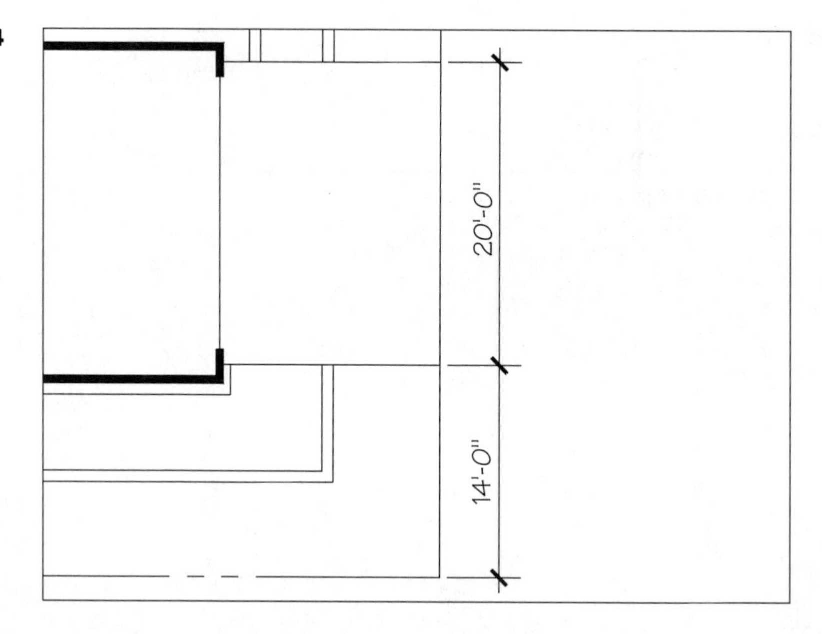

Use the cursor menu (as above) to set the object snap, then pick the corner of the planter. See below (Figure 13.65).

`Dimension text: <9′-0″>`
Press **<Enter><Enter>** to complete the **DIMCONTINUE** command.

Figure 13.65

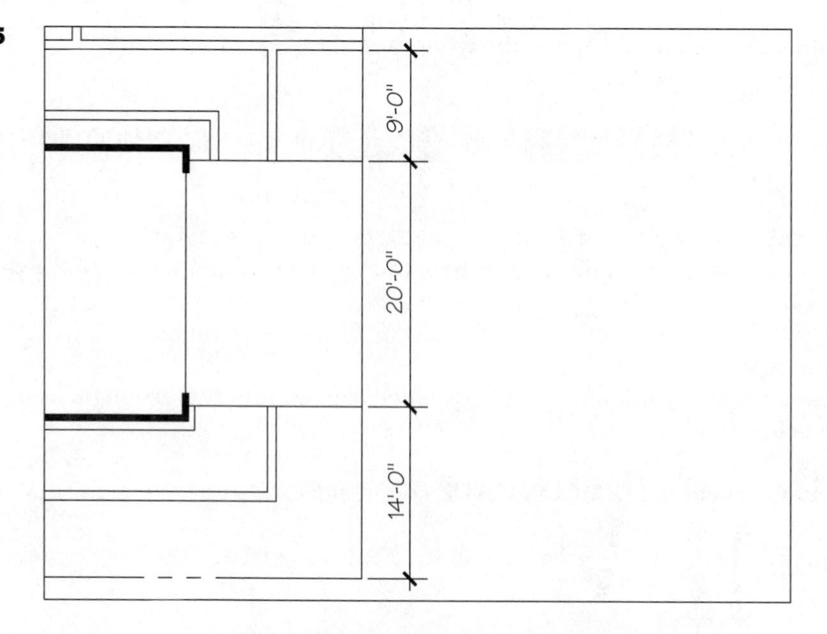

```
Command:
```

Use the pull-down command **Dimensioning ...**, then **Linear** or type **DL** (**DIMLINEAR**).

```
DIMLINEAR
First extension line origin or RETURN to select:
```

Use the cursor menu to highlight **Intersection**, then pick the inside corner of the lawn area.

```
Second extension line origin:
```

Use the cursor menu to highlight **Intersection**, then pick the upper corner of the lawn.

```
Dimension line location
(Text/Angle/Horizontal/Vertical/Rotated):
```

Press **F9** to turn on **SNAP**, then move the cursor so that coordinate dial again reads an X value of **159'**. This places the dimension in exact alignment with the previous ones.

Use the **DIMLINEAR** command to finish the front dimensions as shown(Figures 13.66, 13.67, and 13.68).

Figure 13.66

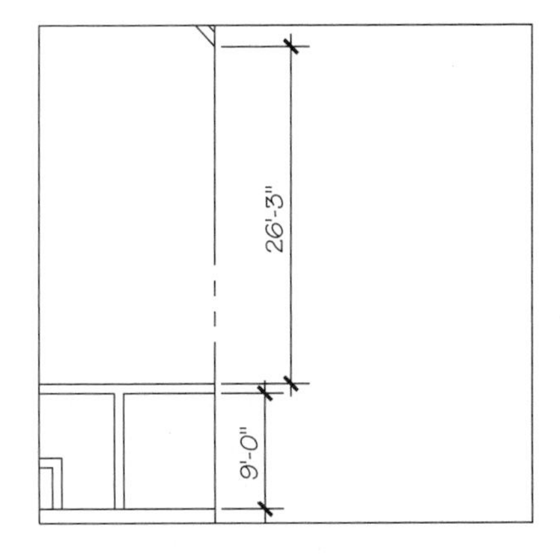

Figure 13.67

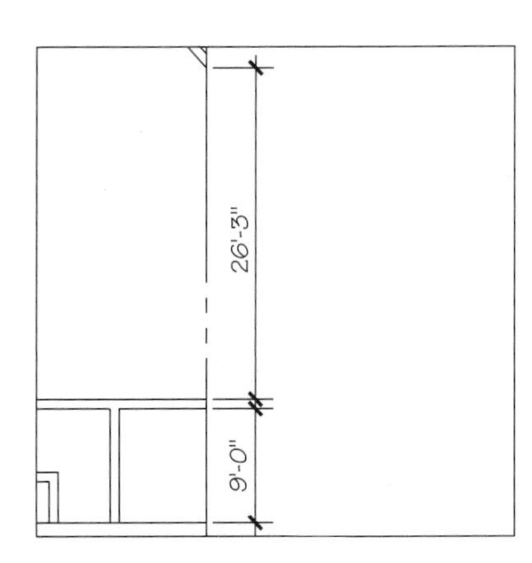

Use **DIMLINEAR** and **DIMCONTINUE** to place the other perimeter dimensions in the side yard. Be sure that **SNAP** is used so that the **Y** coordinate of the dimensions is set to **18'**. This will place the dimensions as shown (Figure 13.69).

For the back yard perimeter dimensions, be sure that the **X** coordinate is **16'** (Figure 13.70).

For the perimeter dimensions at the top of the drawing, be sure that the **Y** coordinate is **126'** (Figure 13.71).

Figure 13.68

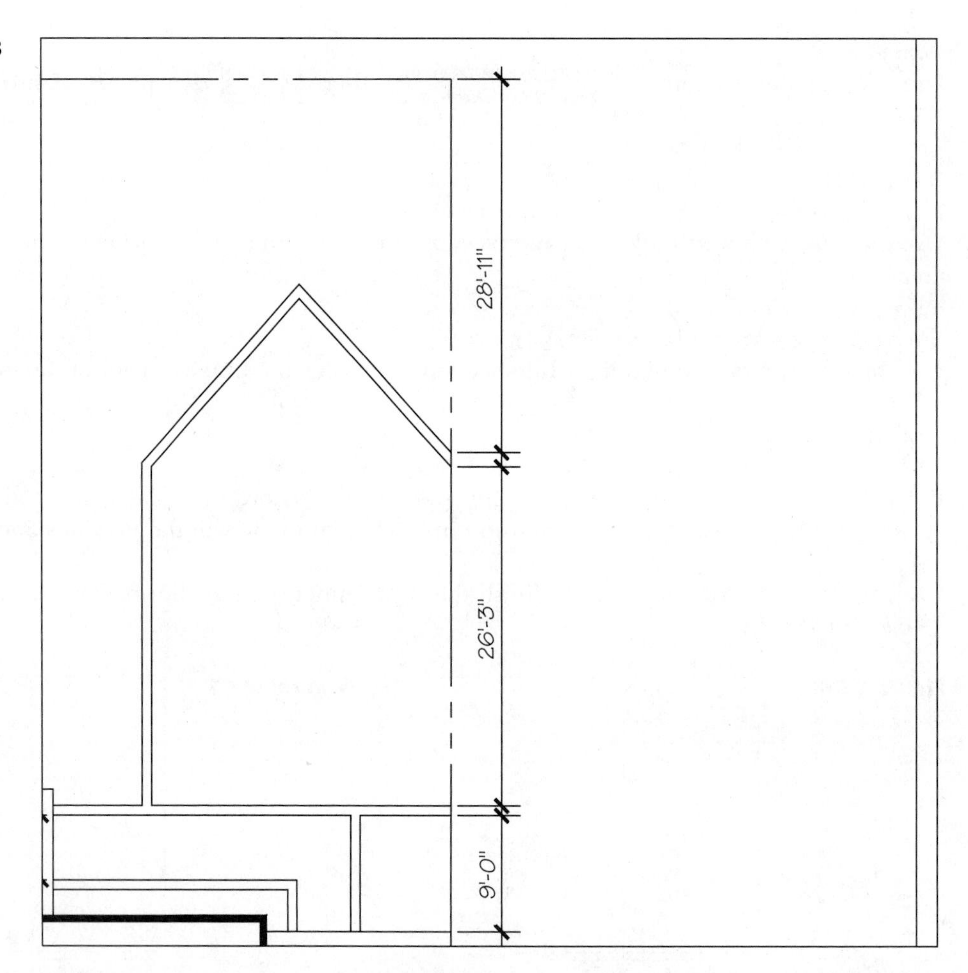

Figure 13.69

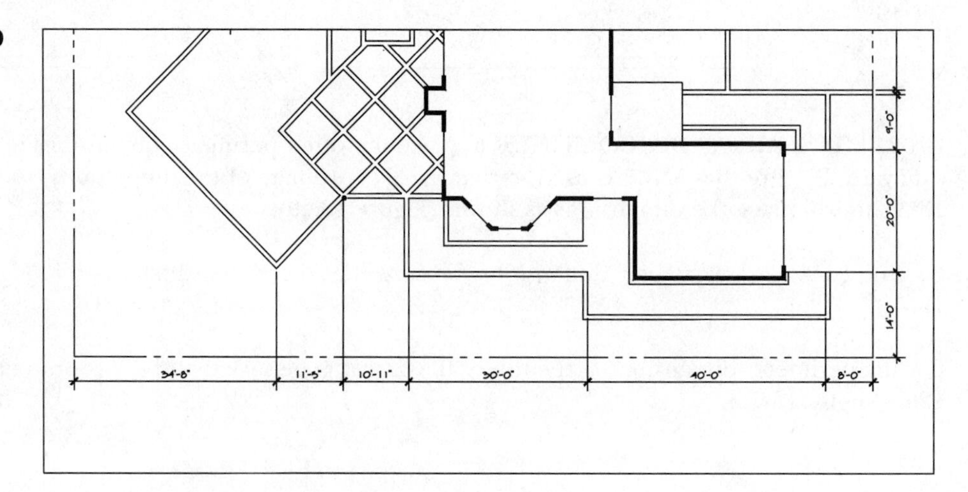

Figure 13.70

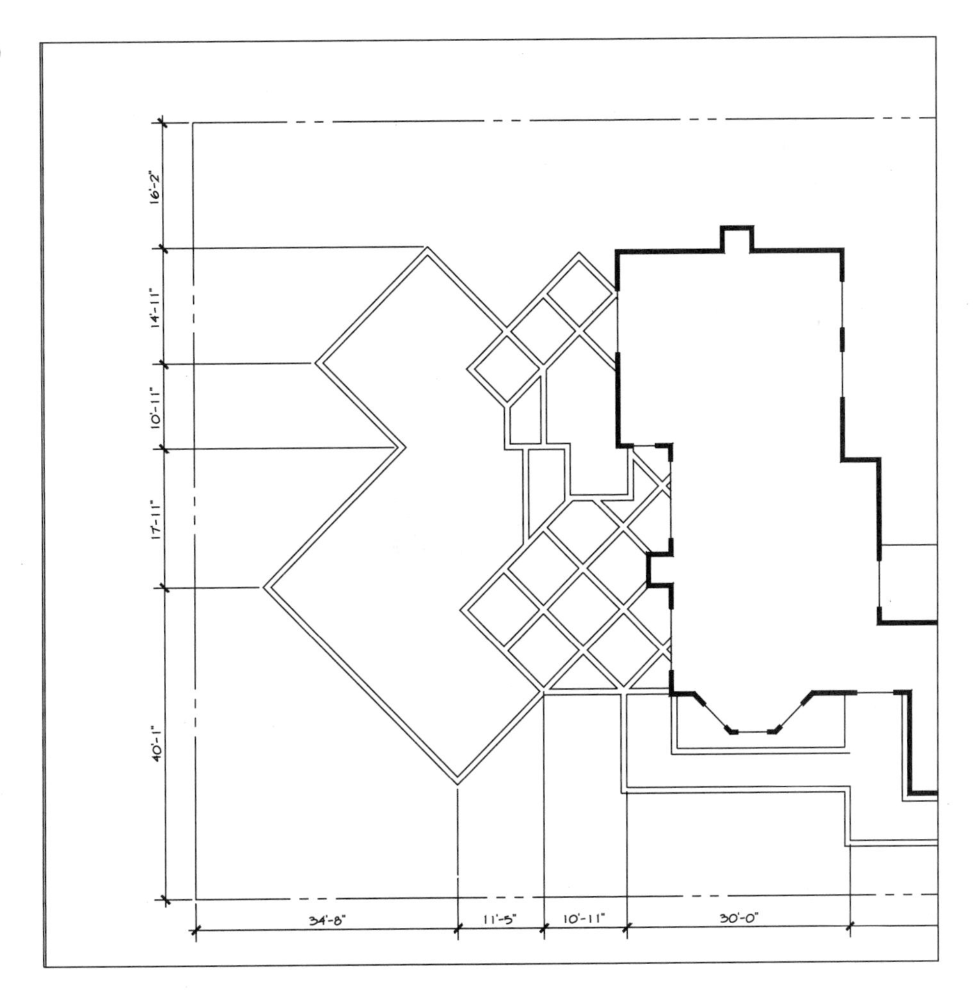

Figure 13.71

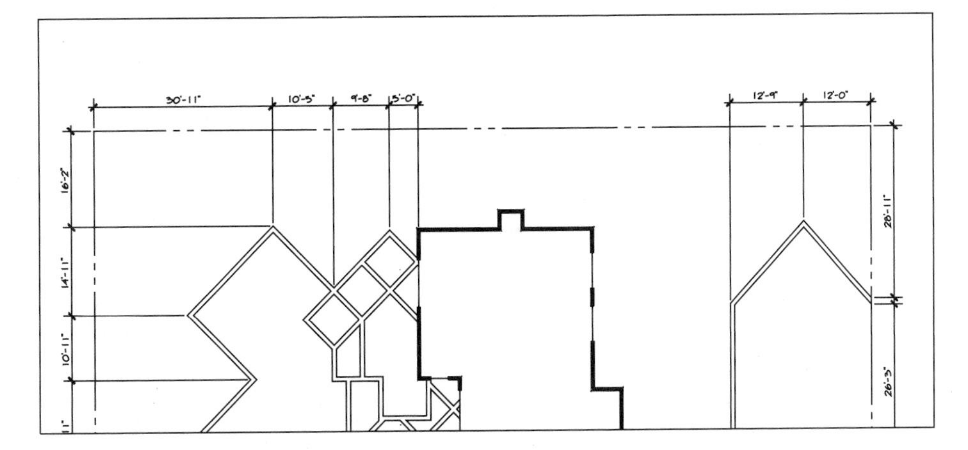

After the perimeter dimensions are completed, we are ready to begin to locate the dimensions perpendicular to our original dimensions. Since a contractor would need two perpendicular dimensions to locate the corner points shown on the drawing, we will provide them here.

We will begin near the side yard door:

Use the **DIMLINEAR** command (either from the pull-down menu or the keyboard macro) to begin a group of linear dimensions (Figures 13.72, 13.73, 13.74, and 13.75).

Use **DIMCONTINUE** to build the dimensions down toward the property line. Notice that the **4'-3"** dimension appears out of line with the other dimensions. This can be modified in two ways:

Figure 13.72

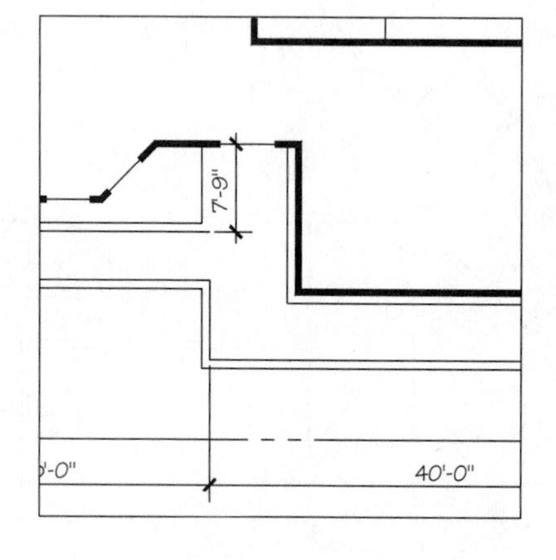

Figure 13.73

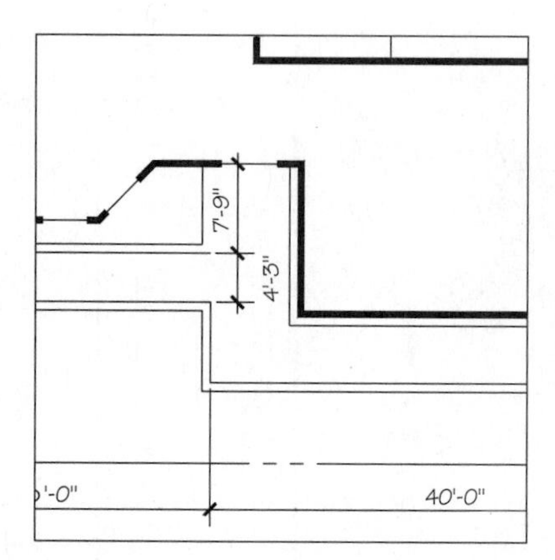

Figure 13.74

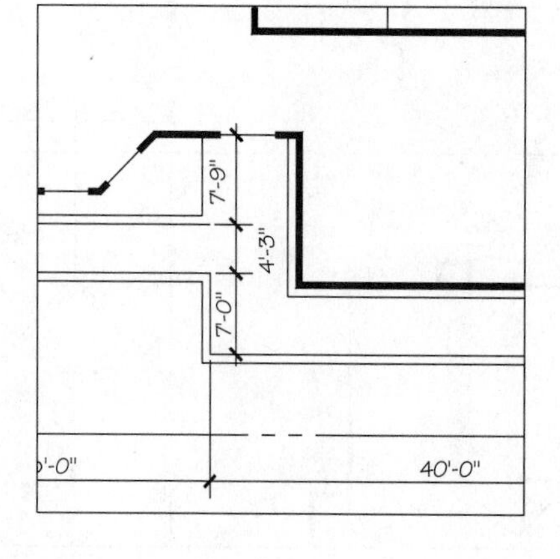

Figure 13.75

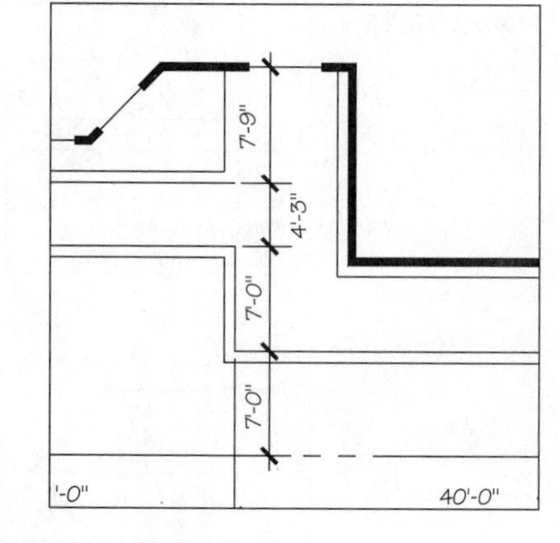

1. Use **grips** to highlight the dimension. Pick the dimension (without any command). This highlights the grips (blue boxes) for the dimension lines, extension lines, and text. Once one of these grips is picked (turning the grip into a red *HOT GRIP)* it can be moved or stretched. Find the grip for the **4'-3"** dimension text, turn it into a hot grip, then move it to relocate the text. The main advantage of this method is speed. You do not need to use any pull-down menus, since the grips are activated whenever you pick an object without issuing a command. Use **<Esc>** (or **<Ctrl>-C** for DOS users) to deactivate the grips and return to normal drawing.
2. Use **DIMTEDIT** (pull-down **Dimensioning ...**, then select **Align Text**).

`Command:`
DIMTEDIT

`Select dimension:`
Pick the **4'-3"** dimension.

`Enter text location (Left/Right/Home/Angle):`
Pick the new location for the text. Repeat the command until the text is located where you want it.

The resulting edited dimensions should look like this (Figure 13.76).

There may be several cases where the text location or dimension line location must be modified to obtain the desired appearance for the dimensions.

Figure 13.76

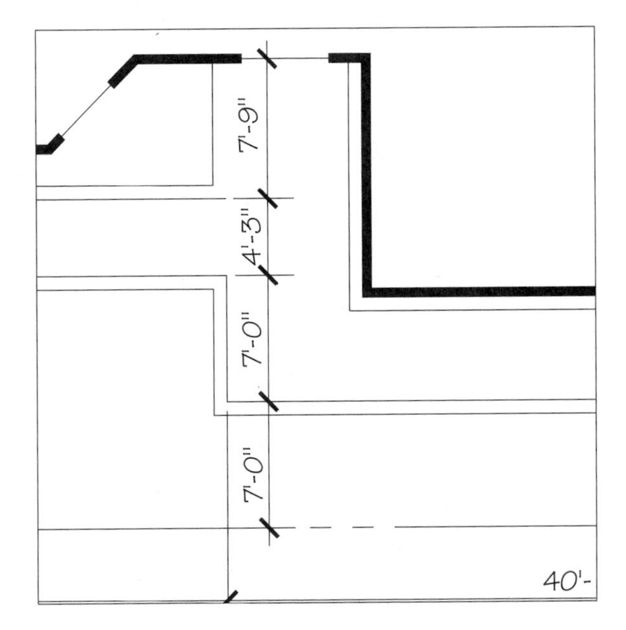

Continue to locate the dimensions around the drawing. We will pay particular attention to those areas where the dimensions are close together and need to be located very carefully to keep the drawing easy to read.

In the front porch area, we need to locate several corner points.

Begin by building a line of dimensions along the porch from the house to the lawn. Add a dimension to the length of the walk (Figures 13.77 and 13.78).

Build a linear dimension from the upper left corner of the driveway to the corner of the walkway slab (see the **2'-9"** dimension below). Continue the dimension across the walk (the **4'-3"** dimension), and to the end of the driveway. Notice that the dimension text is misaligned.

Use the **GRIPS** feature to move the **8'-0"** text below the dimension line (Figures 13.79 and 13.80).

Move the other text into alignment with the **8'-0"** text, then move the **2'-9"** text to the left to make it easier to read (Figures 13.81 and 13.82).

Add the other required dimensions, and see the final results below (Figure 13.83).

In the back yard, we will examine the dimensions around the bedroom and main patio areas.

The first dimension is aligned with the bedroom patio squares. We will label one side of one square, then make a note that these are typically square pads, **6'-1"** per side (Figures 13.84 and 13.85).

To place this note, use the **DTEXT** command, using **1'** for the text height, and **45** as the text angle. Move the text **SQ. TYP.** into position below the dimension line.

This same procedure can be used to dimension the patio squares of the main patio, which are **6'– 8"** per side.

Begin dimensioning the slabs around the patio and planter areas (Figure 13.86).

Notice that the dimension in the center of the planter **15'-9"** continues from the **6'-8"** dimension below. Notice that as more dimensions are added, some dimension text and dimension lines must be moved to make the drawing more legible (Figure 13.87).

As we add dimensions, sometimes the text must be moved from side to side between the extension lines in order to avoid crossing lines on the drawing. See below for an example (Figures 13.88, 13.89, 13.90, and 13.91).

Figure 13.77

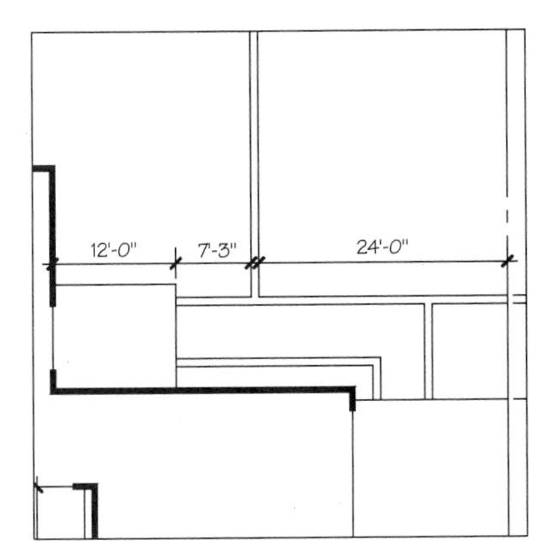

Figure 13.78

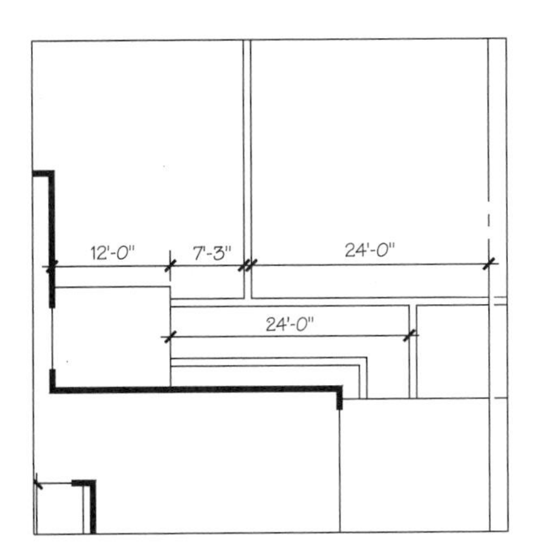

Figure 13.79

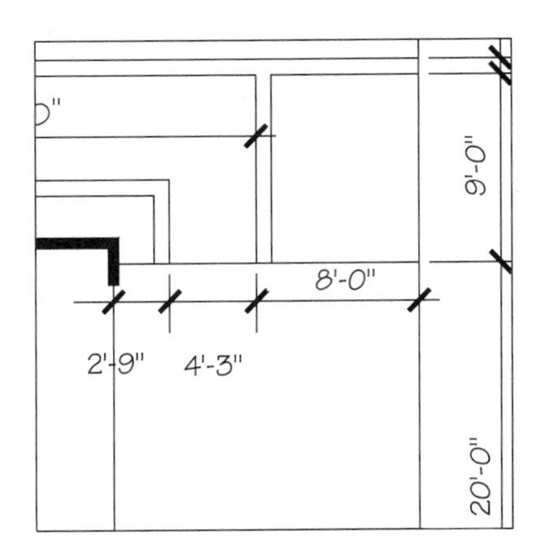

Figure 13.80

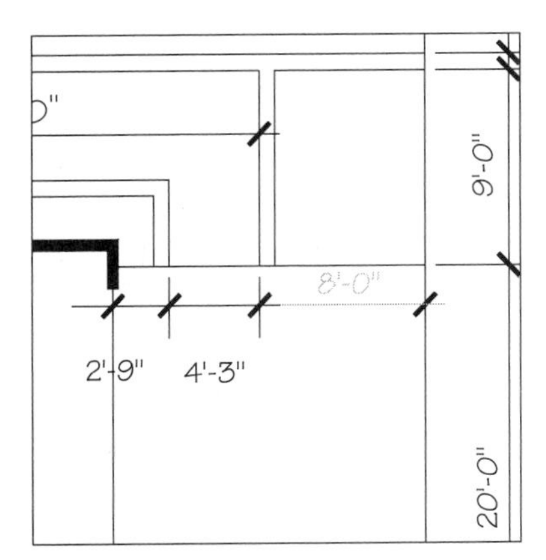

Figure 13.81

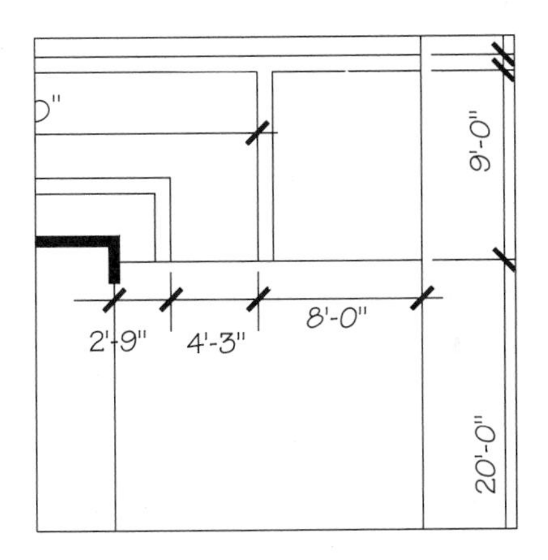

Figure 13.82

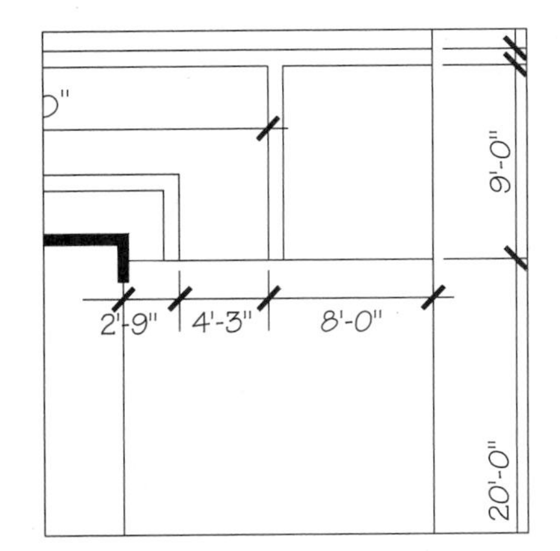

Figure 13.83

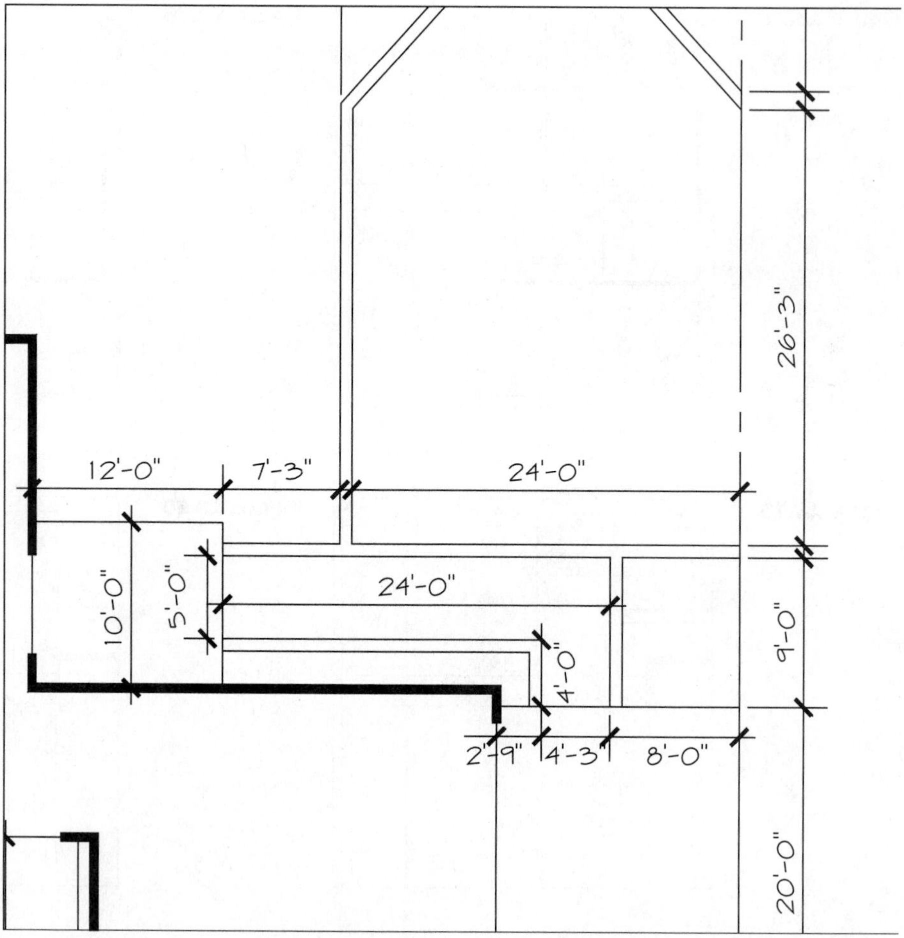

Figure 13.84

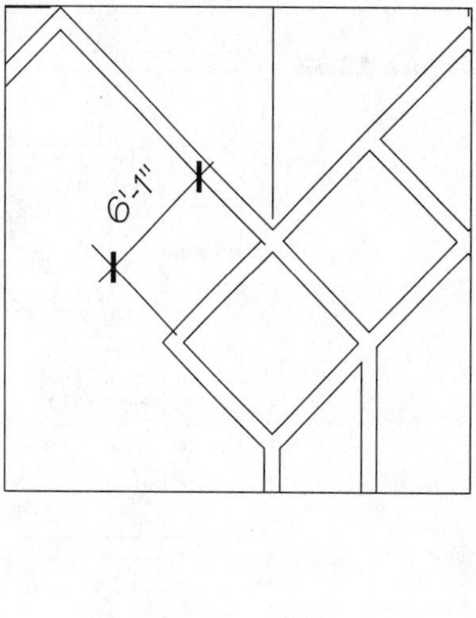

Figure 13.85

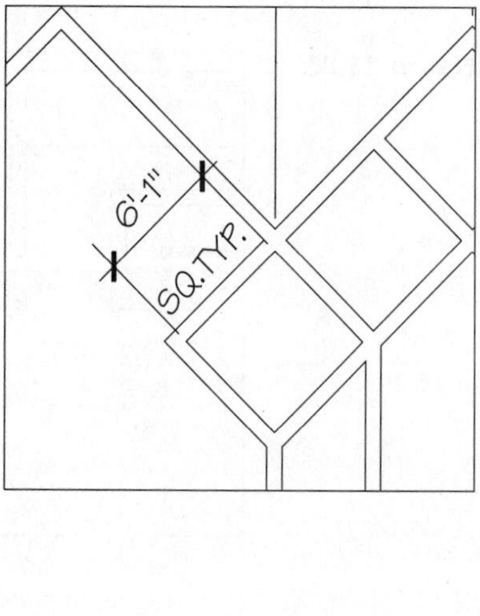

Figure 13.86

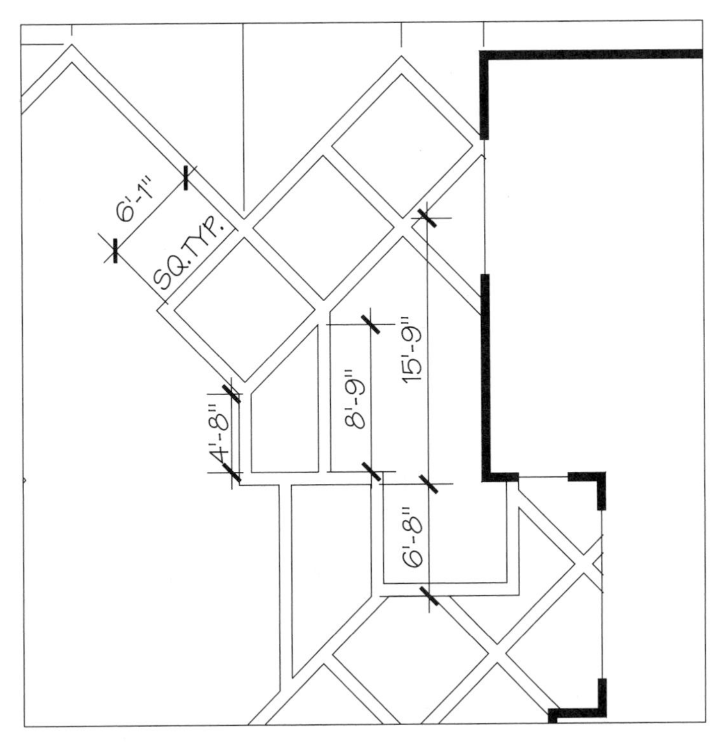

Figure 13.87

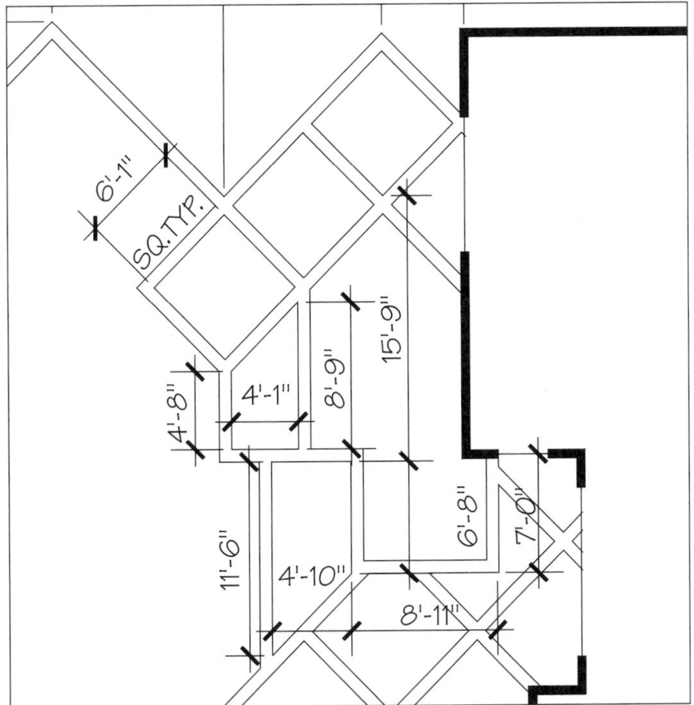

Figure 13.88

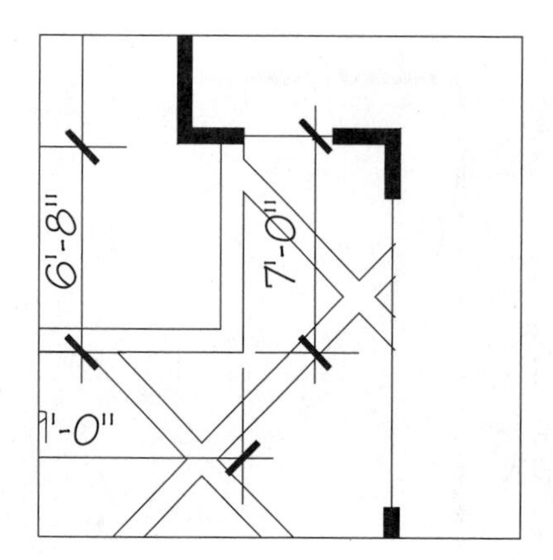

Figure 13.89

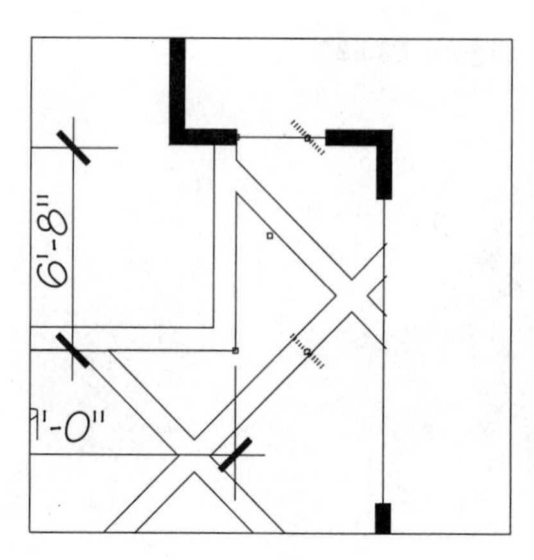

Figure 13.90

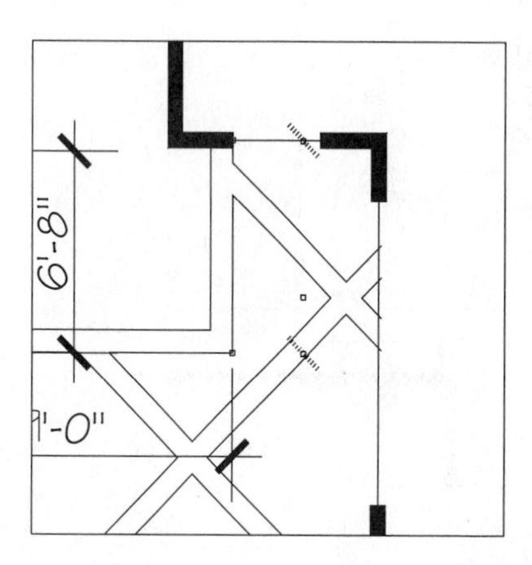

In addition to the dimensions of the slabs, we must also note the angle which is used to form the sides. This is where we use the **Angular** dimension.

`Command:`
Use the Dimensioning ... option Angular, or type **DIMANGULAR**.

`Select arc, circle, line or RETURN:`
Pick the angled brick strip line.

`Second line:`
Pick the house wall.

`Dimension arc line location (Text/Angle):`

Figure 13.91

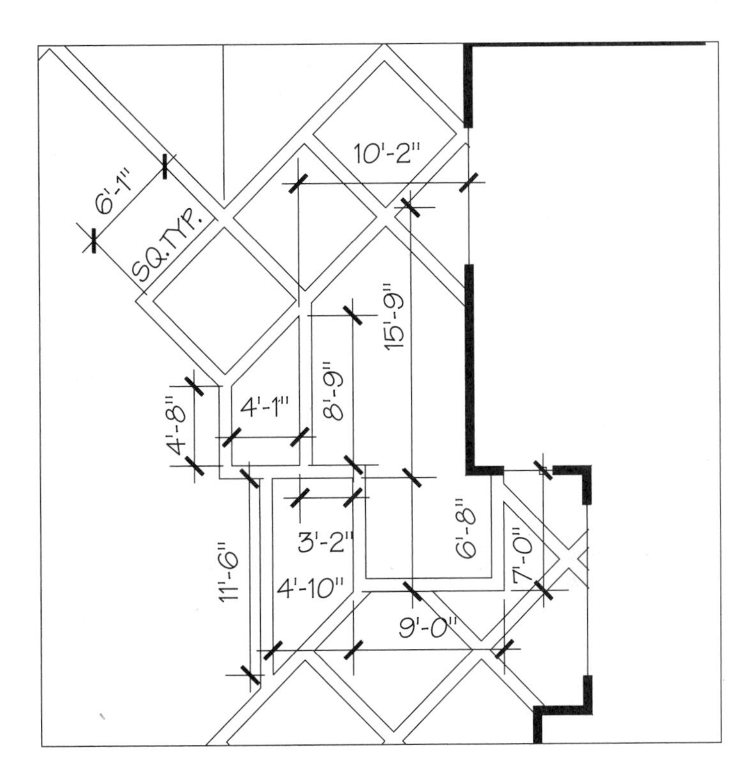

Move the cursor away from the patio. See below (Figures 13.92 and 13.93).

Figure 13.92

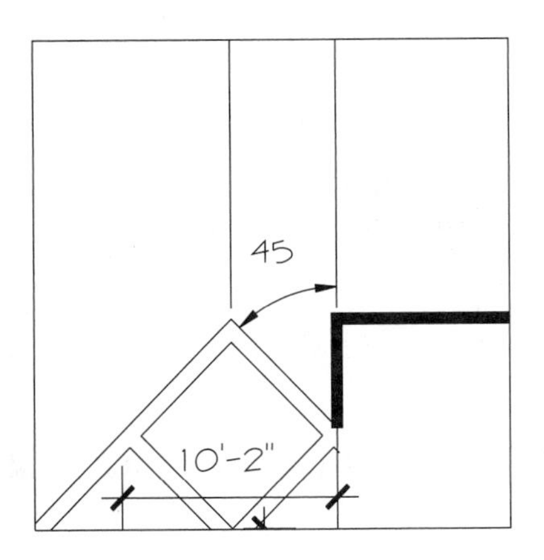

Figure 13.93

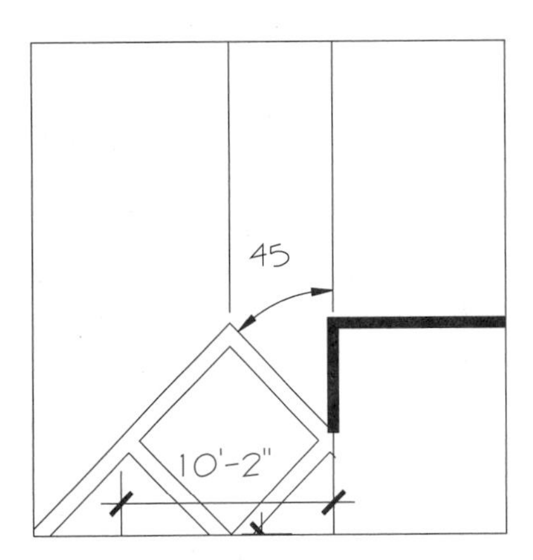

Note that the dimension text is placed far away from the arc because it doesn't fit easily. Use the **grips** function to reposition the text closer to the arc. AutoCAD would normally place a ° symbol next to the **45**, but the HLTR font does not contain this symbol. We can place a simple circle next to the number to simulate this symbol.

Use the **DTEXT** command with **1'** text to place the text **TYP.** next to the **45°**. This tells the contractor that all of the patio and mow strip angles are set on a 45° angle system (Figure 13.94).

We will next place leader notes on the plan. The leaders shown here have text notes to refer to a specific feature of the drawing—namely the concrete finish and the brick strips. Typically these leaders would be inserted with a call-out symbol which would refer to a detail drawing on another sheet. Although we omit those here, they can easily be incorporated into any professional plan.

`Command:`
Use the **Dimensioning ...** option **Leader**, or simply type **LE** (**LEADER**).

`From point:`
Pick a point for the end of the arrowhead.

`To point:`
Pick a point where the arrow will end and the text will begin. Press the **<Enter>** button.

`To point (Format/Annotation/Undo)<Annotation>:`
Press **<Enter>**.

`Annotation (or RETURN for options):`
Type **S. F. CONCRETE** (S. F. indicates salt finish) then press **<Enter>.**

`Mtext:`
Press **<Enter>**.

The leader text then appears over a line connected to the arrow you placed. This style places the text over the leader line, but other choices in the Dimension Style menus can change the location of the text (Figure 13.95).

Placing other leaders with matching text is even easier. Place another leader as follows:

`Command:`
Use the **Dimensioning ...** option **Leader**, or simply type **LE** (**LEADER**).

`From point:`
Pick a point for the end of the arrowhead.

Figure 13.94

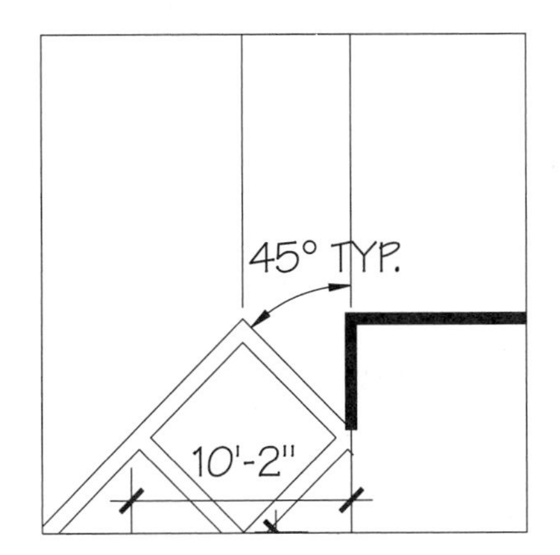

Figure 13.95

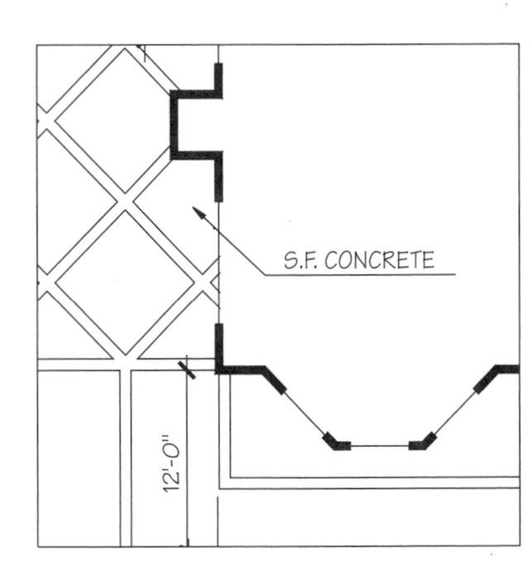

`To point:`
Pick a point where the arrow will end and the text will begin. Press the **<Enter>** button.

`To point (Format/Annotation/Undo)<Annotation>:`
Press **<Enter>**.

`Annotation (or RETURN for options):`
Press **<Enter>**.

`Tolerance/Copy/Block/None/<Mtext>:`
Type **C** for Copy.

`Select an object:`
Pick the text from the last leader. This same text will then appear in the new leader.

Make copies of the **S. F. CONCRETE** leader in the places shown on the drawing.

Place the **9" BRICK BAND** leaders in the places shown on the drawing, and place the **(TYP.)** text under the leaders.

Place the linear dimension across the front lawn, but do not accept the default text entry:

`Command:`
Type or pick **DIMLINEAR**.

`First extension line origin or RETURN to select:`
Pick the intersection on one side of the lawn.

```
Second extension line origin:
```

Pick the intersection on the opposite side of the lawn.

```
Dimension line location
(Text/Angle/Horizontal/Vertical/Rotated):
```

Type **T** (for **TEXT**).

```
Dimension text <24'-0">:
```

Type **ALIGN**—this will replace the default text (Figure 13.96).

See Figure 13.97 for a complete version of the Hardscape and Construction Plan.

Figure 13.96

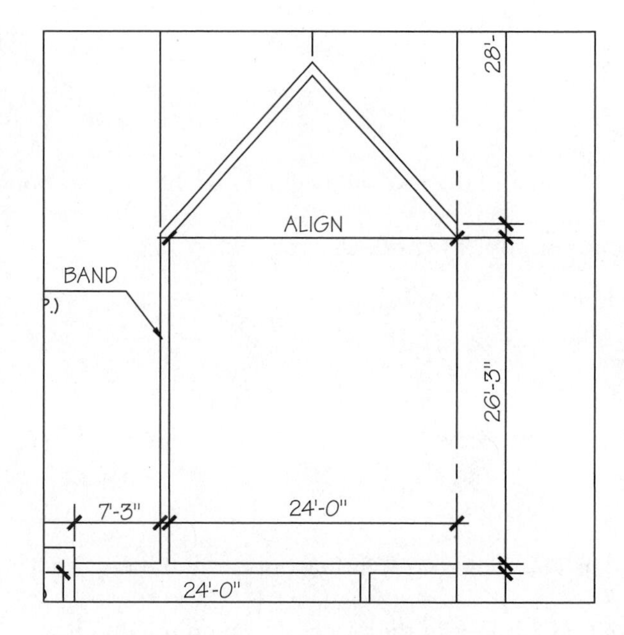

Figure 13.97

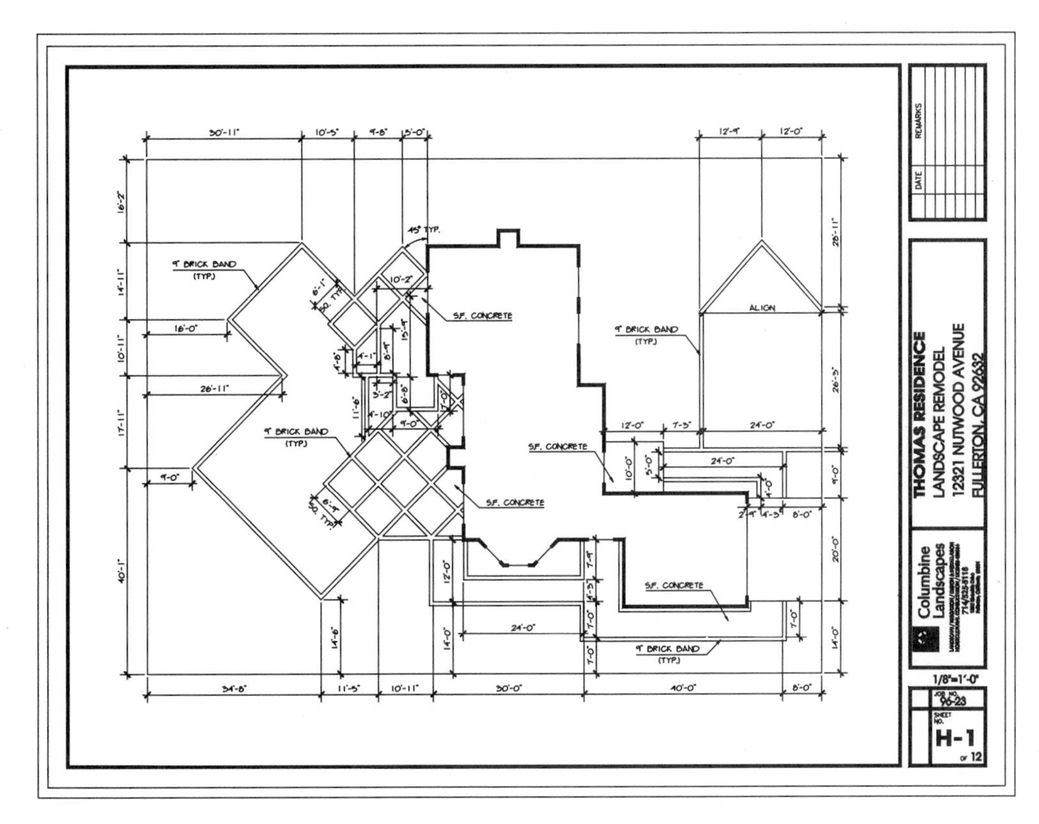
THOMAS RESIDENCE
LANDSCAPE REMODEL
12321 NUTWOOD AVENUE
FULLERTON, CA 92632
Columbine Landscapes
1/8"=1'-0"
H-1
9" BRICK BAND (TYP.)
S.F. CONCRETE
DATE
REMARKS

chapter

Drawing Four—The Planting Plan

While completing the first drawings, we made extensive use of AutoCAD commands, using almost none of the LANDCADD tools. As much as anything, this illustrates the necessity of the LANDCADD user being competent in running AutoCAD. Bear in mind that AutoCAD is integral to all LANDCADD functions, and that the experienced LANDCADD user should be able to move seamlessly from AutoCAD commands to LANDCADD commands and back. The drafting and design process requires that both systems of commands be used regularly.

In the Planting Plan, we will make extensive use of LANDCADD drawing tools and utilities. After creating a new drawing block and inserting it into the Title Block, we will build some simple plan view symbols to use along with the extensive library of symbols provided with LANDCADD. This type of customization is important since it allows the designer to personalize the appearance of the drawing to meet the needs of his or her own graphical style. Once the plan view symbols are completed, we will then make use of LANDCADD's plant database.

While using LANDCADD's plant database, we will modify and create plant records. Database programs typically store information about large numbers of objects. Each object has its information stored in a **record**. Learning to manipulate and use these records is the key to being able to make a database program accomplish real work. Once the database work has been completed, we will examine LANDCADD's options for inserting plant symbols into the project. After the plant symbols are placed on the plan, we will label the symbols and groundcover areas. Finally, we will use LANDCADD's Estimating routines to develop a detailed materials take-off for the Planting Plan.

While developing the Planting Plan, we will complete the following steps:

- Build a block of the house and hardscape outline from the Hardscape and Construction Plan
- Use the Project Manager to open another project, using the C18SHEET drawing as a prototype

- Insert the new house block off to one side of the title block to allow for easier plant symbol labeling
- Build ten new plant symbols for use in LANDCADD's User Library
- Use the Plant Database feature to examine ways to browse and search for plants
- Use the Plant Database feature to edit an existing plant record, copy a plant record, and compose a new plant record from scratch
- Insert plan view plant symbols into the Planting Plan, with plant name and plant size attributes (invisibly) attached
- Build groundcover clouds to show the size and location of the areas to be planted.
- Hatch the groundcover areas
- Label the planting symbols using LANDCADD's Plant Labeling commands
- Label the groundcover areas using LANDCADD's Plant Mix commands
- Use LANDCADD's Estimating routines to build a Report Format for a Planting Plan estimate
- Assign prices and categories to the plants used in the Planting Plan
- Using the Fast Take Off routine, develop a quick listing of the plants used in the front and back yards
- Produce a final Planting Plan estimate, reporting an adjusted price of about $12,000

Creating the HOUSE1 Block

We will begin the planting plan by creating a new block from the Hardscape and Construction Plan. As before, this block will allow us to use a piece of the Hardscape and Construction Plan in the new Planting Plan drawing. The new block will consist of the house footprint, the property line, the driveway and the outer edges of all of the hardscape areas in the front and back yards. The objects to be chosen are illustrated below (Figure 14.1).

After you open the Hardscape and Construction Plan drawing (previously named **HARDSCPE.DWG**) use the following commands to create the block.

`Command:`
Type **WBLOCK**. This brings up the "Create Drawing File" dialogue box.

Figure 14.1

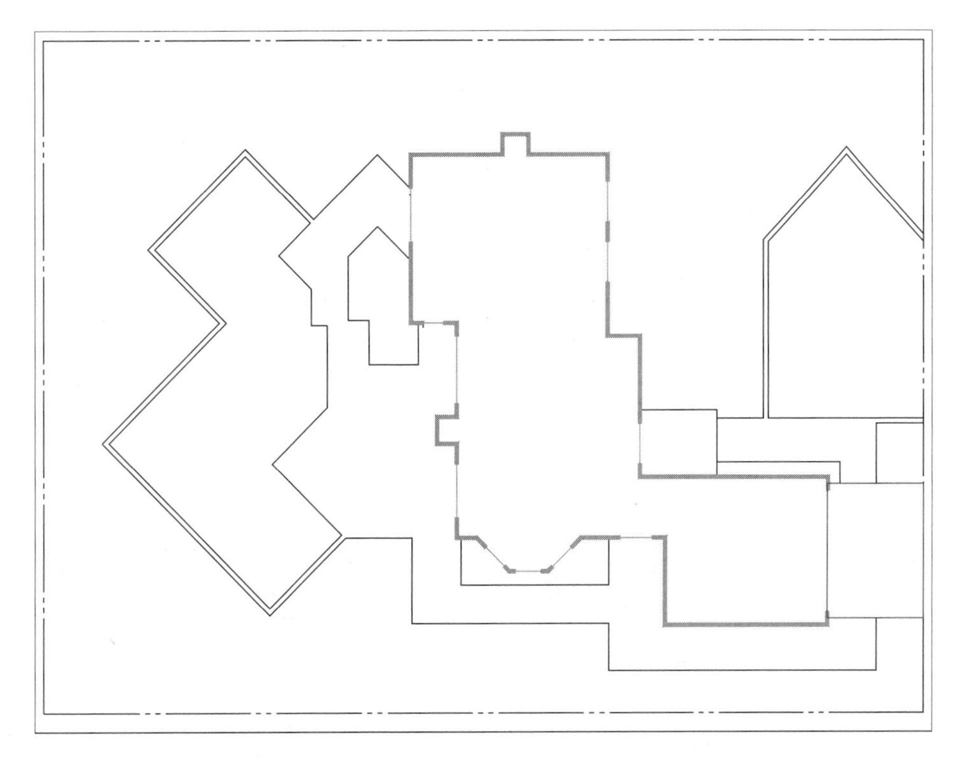

Fill in the name of the **File: HOUSE1** then pick **OK**.

`Block name:`
Press **<Enter>.** Do **not** enter a name at this prompt!

`Insertion base point:`
Type **0,0**.

`Select objects:`
Select all of the objects indicated above. Press **<Enter>** when you are finished picking the objects. All of the objects disappear (again!). This is supposed to happen (again).

HOUSE1 is now saved to your hard disk. If you desire, you can also save it to your floppy disk. Do not save the changes to your **HARDSCPE.DWG**.

Begin a new drawing by selecting the **EP** pull-down menu, then pick **Project Manager** (Figure 14.2).

In the Project Manager dialogue box, select **New . . .** (Figure 14.3).

Figure 14.2

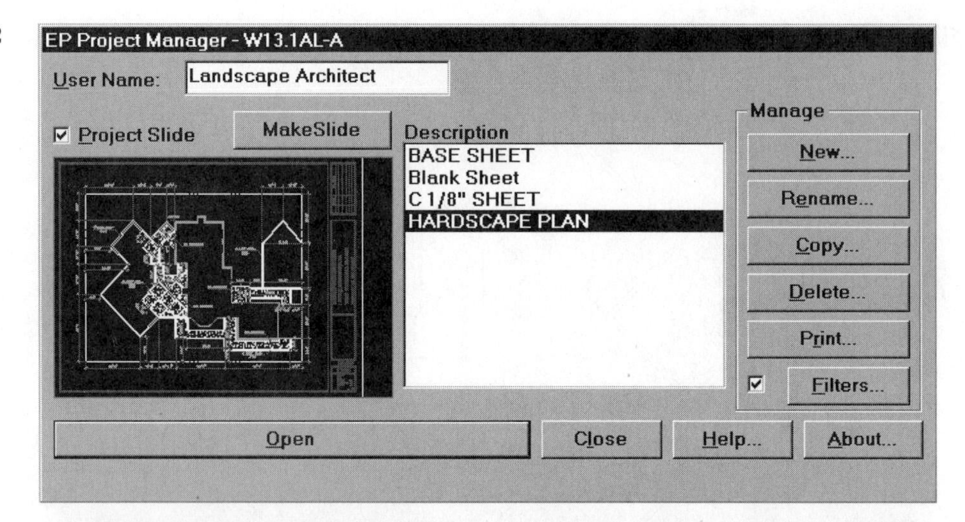

Figure 14.3

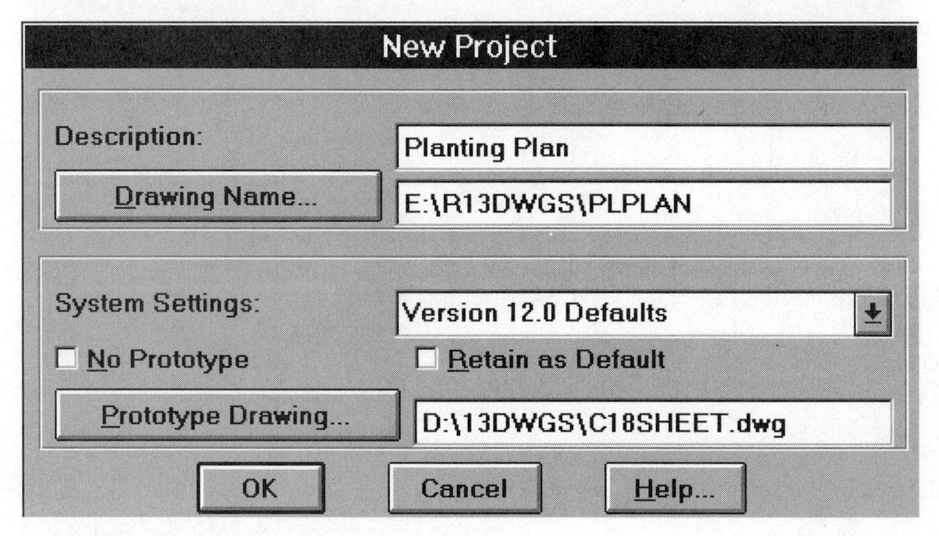

In the "New Project" dialogue box, specify the new project **Description:** and **Drawing Name . . ..** . Be sure to specify the correct path where you want the drawing to be stored. In addition, be sure to set the **Prototype Drawing . . .** to use your **C18SHEET** drawing.

After the drawing is opened, use **AE** (or type **DDATTEDIT**) to edit the text in the title block (Figure 14.4).

After this is complete, you are ready to insert the **HOUSE1** block into your Planting Plan.

Figure 14.4

Type **DDINSERT** at the command prompt. This brings up the Insert dialogue box.

Select **File . . .** then find the **HOUSE1** drawing. After this file appears in the dialogue box, pick **OK**.

At the prompt:

```
Insertion point:
```

Type **-12',0** . This will move our house to the left side of the paper, which will give us more space on the right side for plant labels.

```
X scale factor <1>/Corner/XYZ:
```

Press **<Enter>** to accept a default value of **1**.

```
Y scale factor (default=1):
```

Press **<Enter>** to accept the default value.

```
Rotation angle <0>:
```

Press **<Enter>** to accept the default value.

The **HOUSE1** drawing is now inserted toward the left side of the page (Figure 14.5).

Use **EXPLODE** to separate the **HOUSE1** drawing into separate lines and polylines.

Figure 14.5

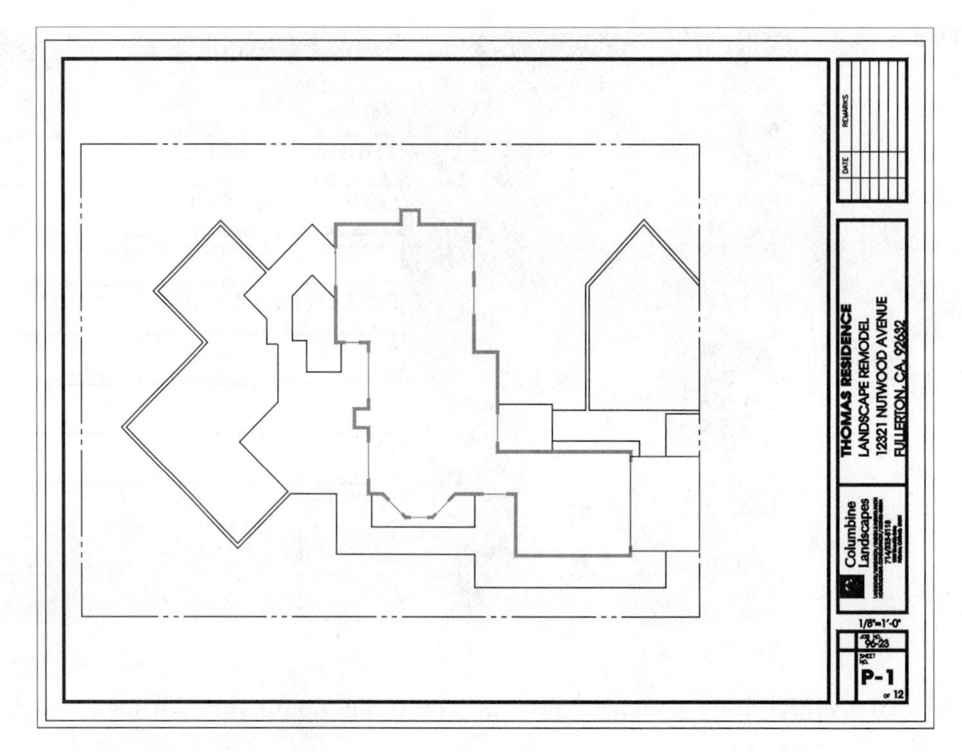

We are now ready to prepare new plant symbols to use in our Planting Plan

Building New Symbols

LANDCADD provides us with the ability to build new symbols and store them in a separate area of symbols called a **User Library**. The symbols stored in this library can be used in the same manner as those stored in LANDCADD's own symbol libraries. This means that we can assign plant names, sizes and other data to the symbols as we insert them. We will build 10 plan view symbols for use in the Planting Plan. We could also construct 3-D symbols for other types of drawings.

Hint: When LANDCADD stores the symbols created with the **Create your own** routine, it places them in a directory called USERL found under the BLOCKS directory in LANDCADD. The drawings and slides created here follow a numbering sequence beginning with USERL001. If you should change workstations, you may want to copy these drawings and slides and install them on the new computer. This will allow you to have access to the customized symbols without having to recreate them.

Begin by using the Project Manager to open a new blank drawing, then choosing the Eagle Point program. Select the **Landscape Design** module. Choose the **Symbols** pull-down menu, then select **Create your own**. If there is anything contained in the current drawing, you will see a warning at the bottom of the screen:

```
WARNING!! Not an empty drawing. Erase everything? [Y/N]:
```

If you answer with **Y** everything on the screen is erased and you are shown the "Custom Tree Settings" dialogue box (Figure 14.6).

Pick the **2D Tree** radio button, which grays out the **Trunk Styles** options. Pick **OK**.

Next, a template appears on the screen, which enables you to draw your symbols at the correct size and scale for LANDCADD to use (Figure 14.7).

Figure 14.6

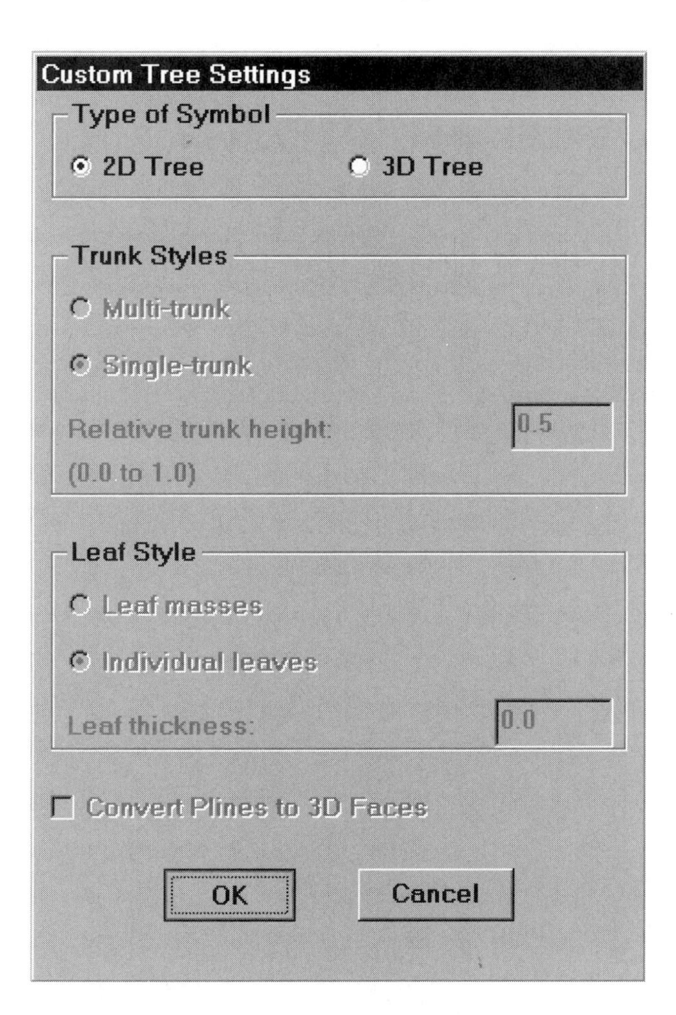

Remember that this template is only a **slide** and does not appear as a part of your new symbol.

The command prompt now allows you to build **Branches** and **Leaves**. The branches and leaves are always drawn as **POLYLINES**. This enables the user to incorporate curves, widths and other complex features in both 2-D and 3-D symbols. Because these objects are polylines, they must be drawn using the **POLYLINE** commands. This is more awkward when drawing a circular boundary for the symbols, so special attention must be paid to this process.

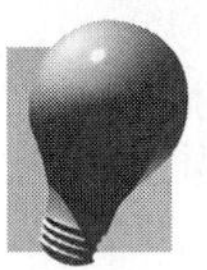

Hint: When building plant symbols, the size of the template slide (and your drawing area) is only about 1" high. If your cursor does not move properly, reset your **SNAP** value (or turn **SNAP** off) to reflect the small drawing size. If the screen fills with color when you draw the polylines, reset your **POLYLINE WIDTH** to **0**.

Building the First Symbol

We will build the first symbol as a green circle (Leaves) with two brown (Branch) lines running through the center. Follow the instructions through to the end without interruption. If you make a mistake and interrupt the command, it is easiest to erase your work and start over (Figure 14.8).

HINT: Use the cursor menu (or type the abbreviation on the keyboard) to activate the indicated **Object Snap** during the selection of points when building the symbols. This speeds the process of building the symbols and greatly enhances your accuracy.

Figure 14.7

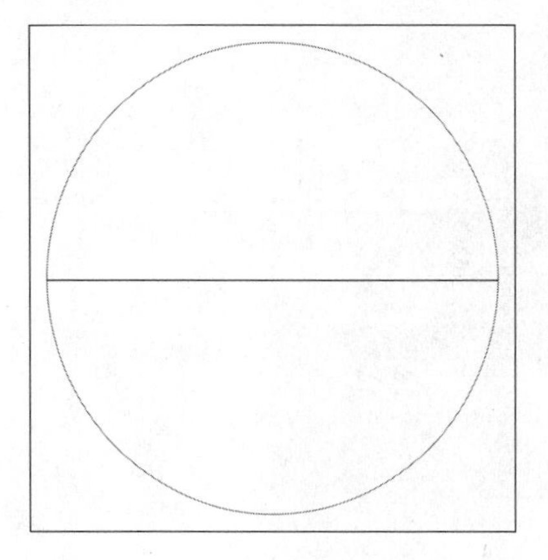

Figure 14.8

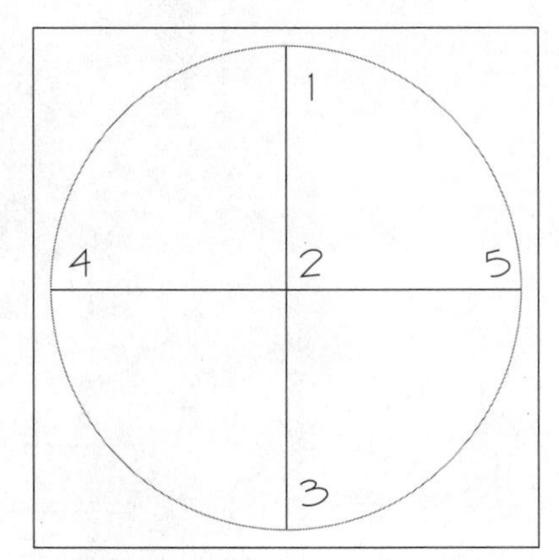

Command:
Draw Branch, Leaves, eXit [B L X]? <B>:
Press **<Enter>** to accept **Branch.**

_.pline
From point:
Choose the **Quadrant** of the circle at **5.**

Arc/Close/Halfwidth/Length/Undo/Width/<Endpoint of line>:
Choose the **Quadrant** of the circle at **4.**

Arc/Close/Halfwidth/Length/Undo/Width/<Endpoint of line>:
Press **<Enter>** to finish the polyline.

Draw Branch, Leaves, eXit [B L X]? <B>:
Press **<Enter>** to accept **Branch.**

_.pline
From point:
Choose the **Quadrant** of the circle at **1.**

Arc/Close/Halfwidth/Length/Undo/Width/<Endpoint of line>:
Choose the **Quadrant** of the circle at **3.**

Arc/Close/Halfwidth/Length/Undo/Width/<Endpoint of line>:
Press **<Enter>** to finish the polyline.

Draw Branch, Leaves, eXit [B L X]? <B>:
Type **L** for **Leaves.**

_.pline
From point:
Pick the **Intersection** at **1.**

Arc/Close/Halfwidth/Length/Undo/Width/<Endpoint of line>:
Type **Arc.**

Angle/CEnter/CLose/Direction/Halfwidth/Line/Radius/Secondpt/
Undo/Width/< Endpoint of arc>:
Type **CE** for **Center.**

Center point:
Pick the **Intersection** at **2.**

```
Angle/CEnter/CLose/Direction/Halfwidth/Line/Radius/Second
pt/Undo/Width/<Endpoint of arc>:
```

Pick the **Intersection** at **3**.

```
Angle/CEnter/CLose/Direction/Halfwidth/Line/Radius/Second
pt/Undo/Width/<Endpoint of arc>:
```

Pick the **Intersection** at **1**.

```
Angle/CEnter/CLose/Direction/Halfwidth/Line/Radius/Second
pt/Undo/Width/<Endpoint of arc>:
```

Press **<Enter>** to complete the polyline.

```
Draw Branch, Leaves, eXit [B L X]? <B>:
```

Type **X** to exit the command and remove the template from the screen.

After the symbol has been drawn, you can save it to the **User Library**.

Under the **Symbols** pull-down menu, select **Save to Library**. The screen prompt then reads:

```
Is this a 2D or 3D symbol?
```

Type **2D**.

```
Enter a number for this symbol <1>:
```

Press **<Enter>** to accept this as your first symbol name.

The screen flickers briefly as the symbol is saved and a slide is created for the symbol. Then the symbol reappears with two attribute tags located on the symbol. These attribute tags are created each time you activate the **Save to Library** function (Figure 14.9).

ERASE these tags before proceeding to the next steps in creating the rest of the symbols.

Figure 14.9

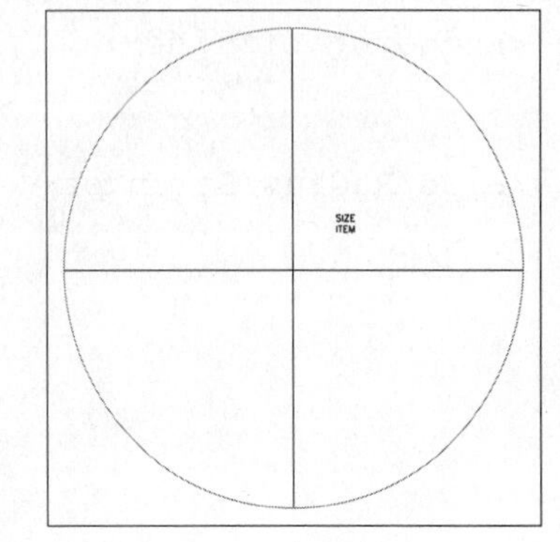

ROTATE the circle and lines by an angle of **45°**, then save the symbol to the library as **2**.

Use the various editing and drawing commands to create the following symbols for your library, while being sure to **erase the attribute tags** before saving each symbol (Figure 14.10).

After you complete these symbols, you can see your completed work in the User Library selection under the Symbols pull-down menu (Figure 14.11).

Figure 14.10

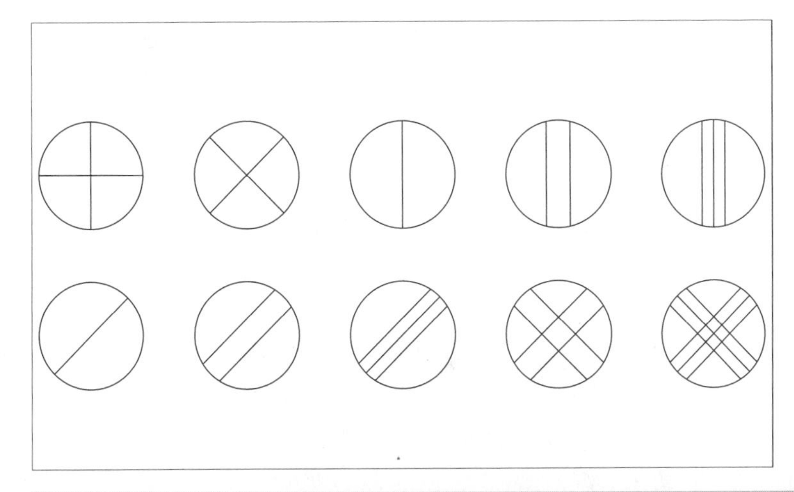

Figure 14.11

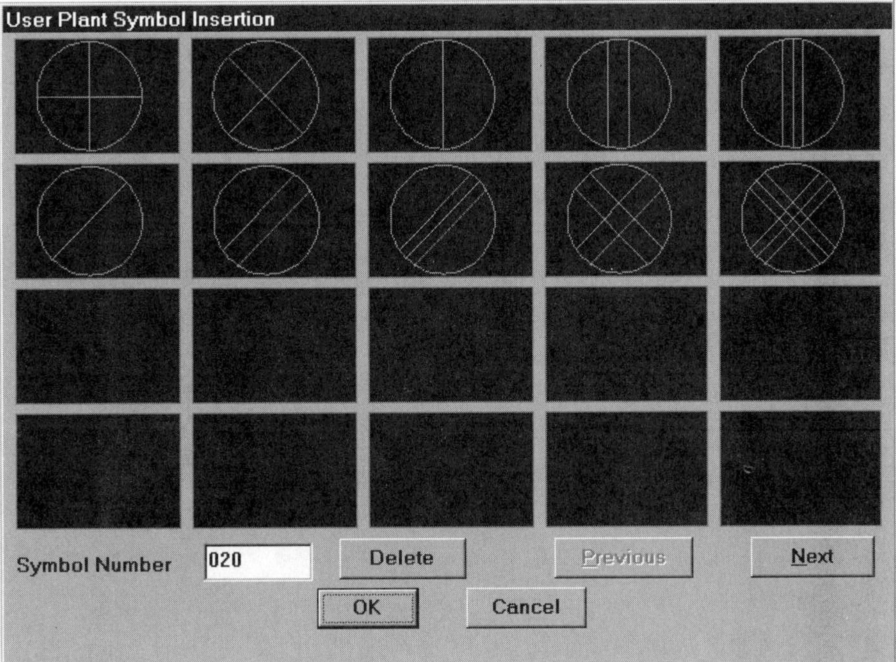

HINT: If you expect to create large numbers of customized symbols, set up a numbering system to categorize them. Symbols 1 through 99 might be plan view trees, 100–199 plan view shrubs, etc. This will make it much easier to locate your custom symbols when you need them.

Selecting a Database File

LANDCADD has a very extensive plant database which is included as part of the Landscape Design module. This database can be used to examine the traits of individual plants, search for plants with certain traits in common, or develop lists of plants for us to draw from as we place plants in the landscape. To access the Database portion of LANDCADD, load the **Landscape Design** module, then select the **Database** pull-down menu (Figure 14.12).

To begin using the Database function, select **Configuration**. This allows us to choose the database file which will be loaded into LANDCADD (Figure 14.13).

Figure 14.12

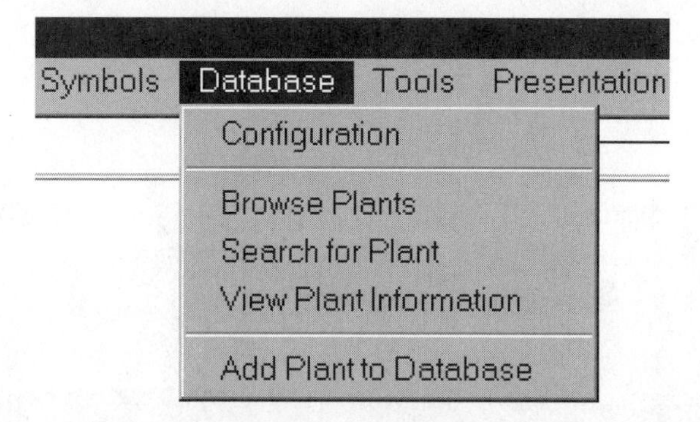

Figure 14.13

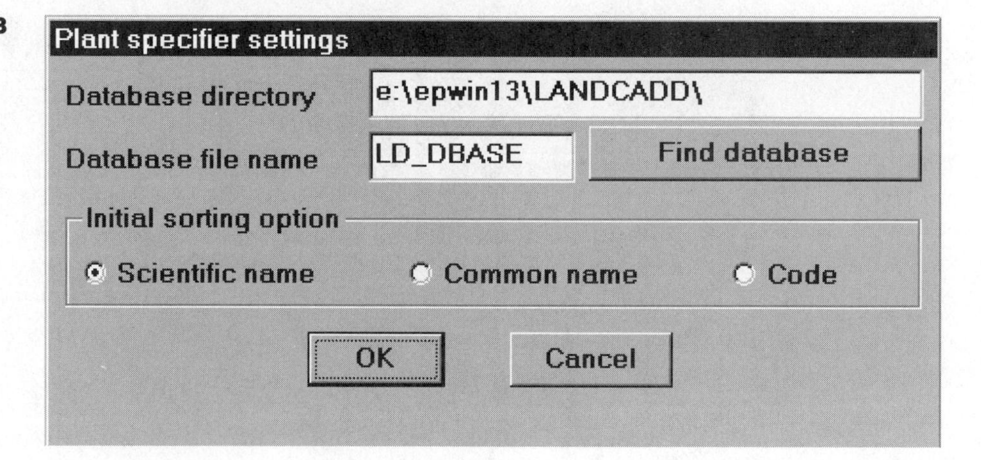

We can select any **.DBF** file with properly structured plant records. Any program capable of creating .DBF files may produce plant lists which can be used by LANDCADD.

To choose which database file to load, pick the **Find database** box (Figure 14.14).

The "Find database file" dialogue box shows us the .DBF files provided with LANDCADD. As you begin to work with the Plant Database program, you may end up building your own database files. If you have .DBF files created in other programs, you may change drives or directories to find them.

Select the file **LD_DBASE.DBF**. This is LANDCADD's own large plant database.

In the **Initial sorting option** box, choose the method of sorting the plant names as they first appear on the screen. The radio button selection should be **Scientific name**.

Browsing Plant Names

After the correct database file is selected, you can then choose the **Browse Plants** option from the **Database** pull-down menu. This option allows us to move through the list of plants and to examine information about any of them. The prompt will show that the program is loading the database:

```
Reading database...............Done
Read in 1161 records
```

This indicates that **LD_DBASE** has 1161 plant entries listed in it.

The first dialogue box we see is "Browse plants" (Figure 14.15).

In the "Browse plants" dialogue box, we see an alphabetical listing of each of the plant species contained in the active database. Any of the plants in the database can be highlighted (selected) by clicking on the plant name.

Figure 14.14

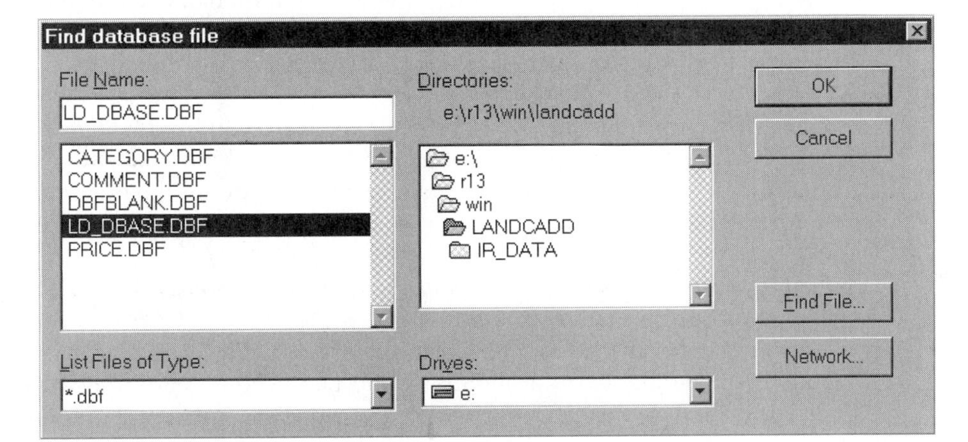

Figure 14.15

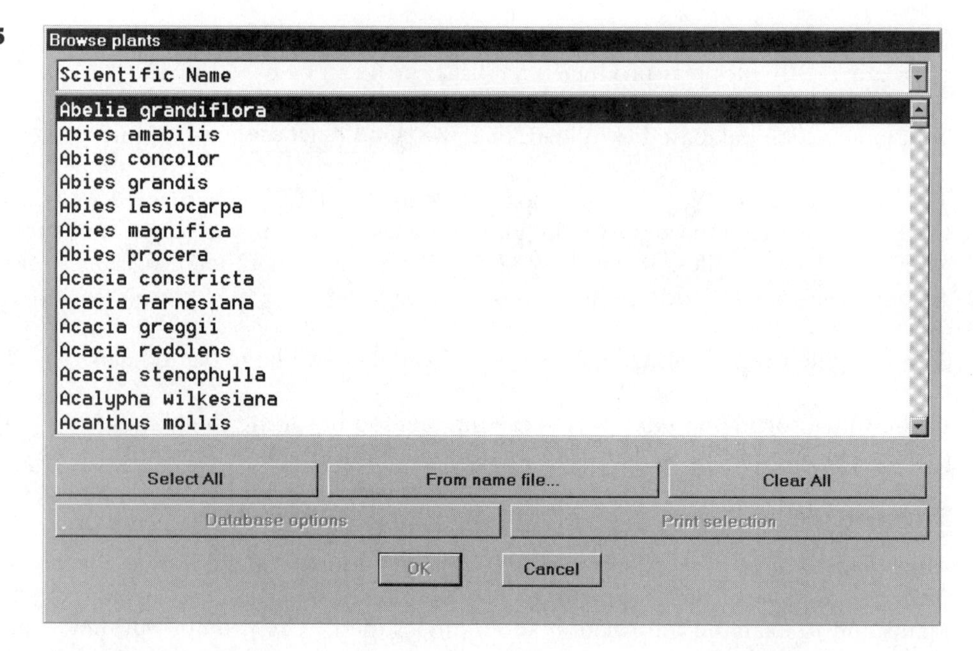

The active buttons below give us options to **Select All**, or to **Clear All** from the group of selected plants. In addition, we can choose to load a list of plants **From name file . . .**.

A **name file** is a list of plant names that we choose to use in a project. While the plant database contains large amounts of information regarding plant usage, environmental requirements and ornamental features, a name file list is much simpler. The name file list does not include a full listing of information about each plant, but instead simply lists the scientific name, common name and plant code for each plant chosen. It is convenient to use a name file list when labeling plants in a drawing since the list may include only 20 or 30 plant entries with three fields (scientific name, common name and plant code) rather than the thousand or more entries (with more than 20 fields each) which might be held in a complete database.

Once a group of plant names (or even a single plant name) is highlighted, two other buttons become active at the bottom of the screen. The **Print selection** option allows us to print the list of plants — either to a text file or directly onto the drawing itself. The latter choice may be especially helpful when developing planting schedules or planting legends.

The **Database options** button opens another set of choices (Figure 14.16).

First, we can save the list as a new plant name file. This list of plant names should have some file name that helps us identify it and tie it to the appropriate project. The plant name files all have the extension **.NAM** so ours will be called **THOMAS.NAM**.

Figure 14.16

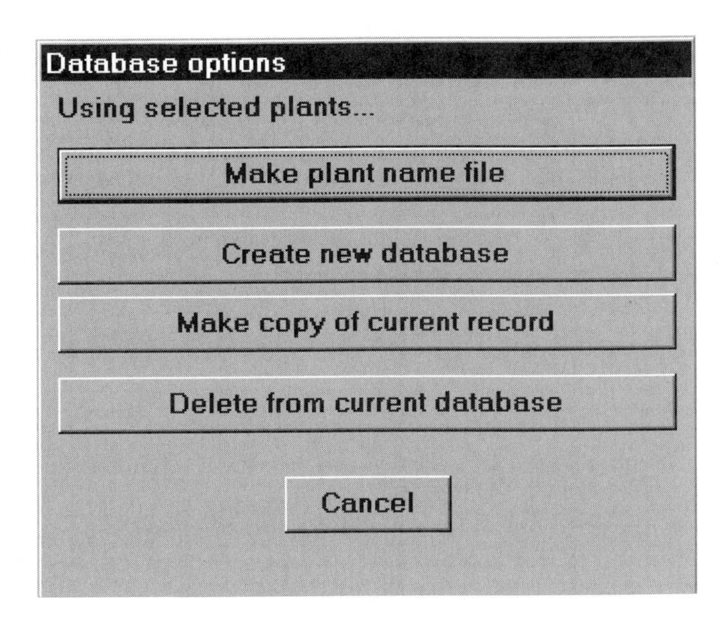

Second, we can create a new database from a selection of plants in the original database. This new, smaller group may reflect a group of plants that are some logical subdivision of the large database, such as those plants suitable in tropical theme, Oriental theme or desert theme landscapes. These smaller databases might help isolate plants which are most suitable for particular projects.

Third, we can make a copy of a particular plant entry (called a record). This might enable us to quickly create a record of some new variety of a listed plant. If a dwarf variety of an existing plant were introduced, we could copy the existing record, then edit the name (to add the variety name) and the size (to reflect the dwarf variety) to complete the new record. This process is much quicker than adding a new plant from a blank record.

Last, we can delete a record from the current database. This is a permanent change to the database file and should only be done with great care. It may take considerable time and research to replace a plant record once it is deleted. The best method for keeping track of plants is to maintain a master database for **all** plants, then create smaller ones for specific uses. Delete records from small databases as required, but leave the master database as complete as possible.

Searching for Plants

One of the primary uses for a database is that of searching. We can search the database for plants which meet certain criteria, then list them on a screen. The list can then be saved, printed or even turned into another database.

The search for plants involves the selection of categories and qualifications which allow us to narrow the range of plants that we want to examine. The categories and qualifications are assembled into a **search expression**. This search expression is a programming representation of all of the parameters we specify, linked together by "logical operators" AND, OR, AND NOT, and OR NOT. This has the effect of excluding some plants and including others. As an example, we may wish to find all of the listed patio trees with pink flowers and which do not have shallow roots and are not poisonous. In LANDCADD, such a search expression would look like this:

(((USE_ARCHIT|4) .AND. (COLR_BLOOM = 0)) .AND. .NOT. (MAIN_TOLER|8)) .AND. .NOT. (MAIN_DETER|131072) (Figure 14.17).

This search expression is built as we pick through the different search buttons, selecting the criteria we want, and the "logical operators" to use in separating the criteria.

The first criterion, patio tree, is found under **Architectural uses** with the **Uses** button.

The second, pink flowers, is found under **Bloom effect** with the **Color** button. These two criteria are connected with the "logical operator" AND since we want the plants to have both of these features. The next two criteria will use the "logical operator" AND NOT since we want the plants **not** to have the next two features.

The shallow root criterion is found under **Tolerances** with the **Maintenance** button. The final criterion, non-poisonous, is located under **Detriments**, also with the **Maintenance** button.

Figure 14.17

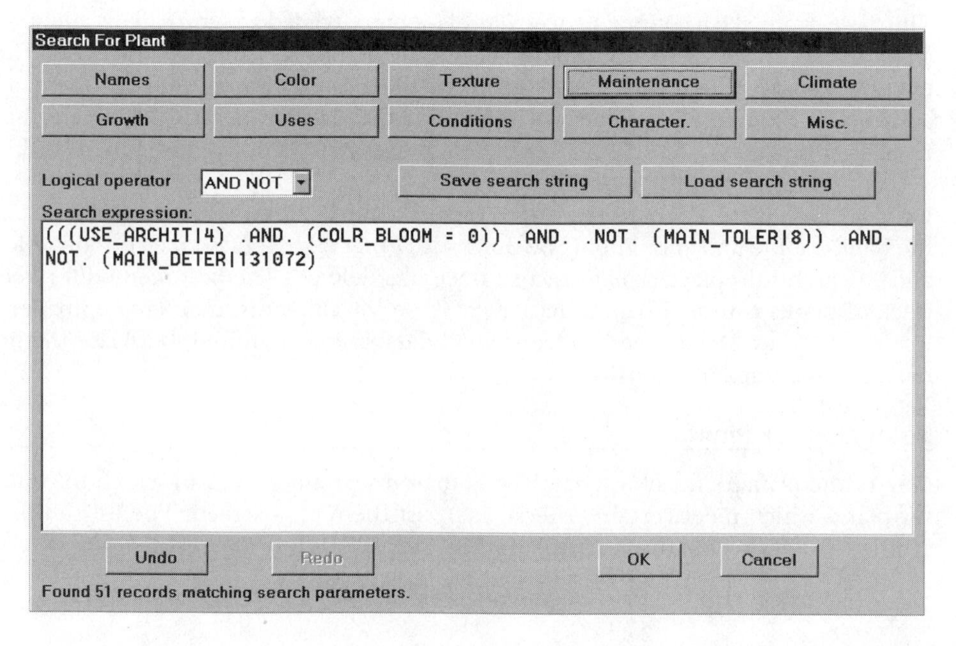

The **Undo** and **Redo** buttons at the bottom of the screen allow us to remove the most recent addition to the search expression, or to place it back in. As the size of the search expression area shows, a search expression can be quite lengthy and complicated.

Once a search expression has been assembled (using the ten buttons available at the top of the screen) it can be saved and/or loaded into the program for use with another database.

Editing a Plant Record

One of the great challenges in landscape design and horticulturally-related careers deals with plant names. Mastering several hundred scientific and common plant names takes a considerable commitment of time and effort. If learning these names weren't challenge enough, the landscape professional must also keep up with names that are changed over the years. A current example is with the old genus **Raphiolepis**, which has been recently given the new spelling of **Rhaphiolepis**. Since we will use this plant in our current project, we will examine how to modify this plant name in the **LD_DBASE** file.

Begin by using the **Database** pull-down menu, then selecting **Search for Plant**.

This opens the "Search for Plant" dialogue box (Figure 14.18).

Pick the **Names** box. This brings up the "Name search" dialogue box (Figure 14.19).

Type in the plant name **Raphiolepis indica**. Click on **OK**.

Figure 14.18

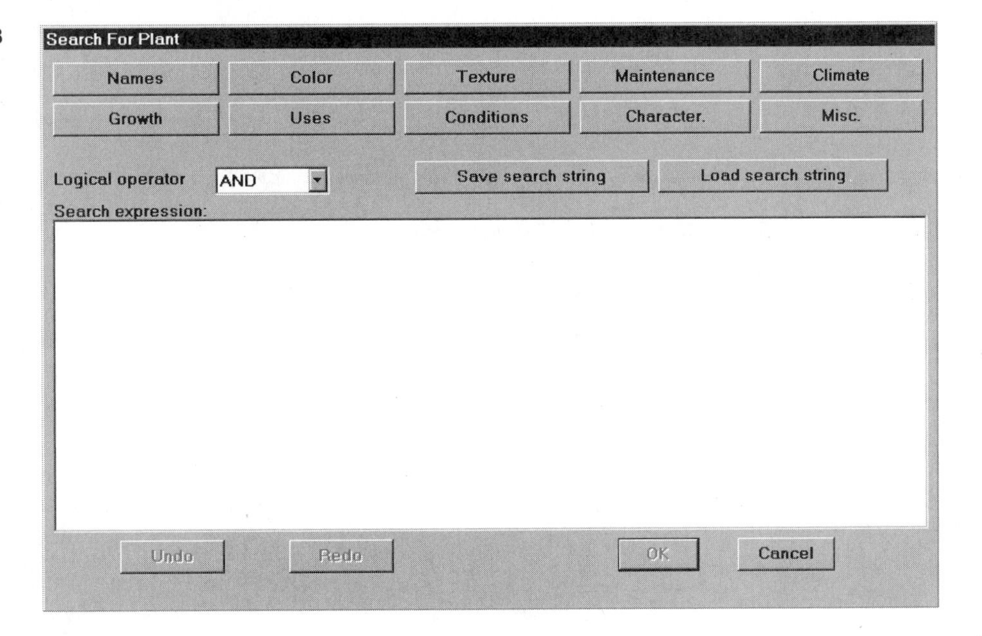

Figure 14.19

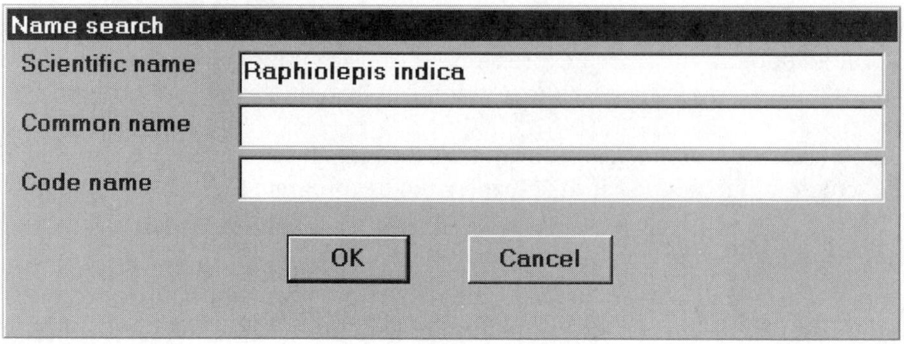

Name search
Scientific name Raphiolepis indica
Common name
Code name
OK Cancel

This creates a "search expression" which tells the program what to look for. When this process is complete, we are returned to the "Search for Plant" dialogue box. At this point the database program searches through all the plant records to find any names which match this one. At the bottom of the dialogue box, we see that LANDCADD finds 1 record which contains the name. If there were different varieties of **Raphiolepis indica** in the database, they would also be listed since they too contain the search expression (Figure 14.20).

Click on **OK** to continue.

This opens the "Browse plants" box, but only one plant name is shown here (Figure 14.21).

Figure 14.20

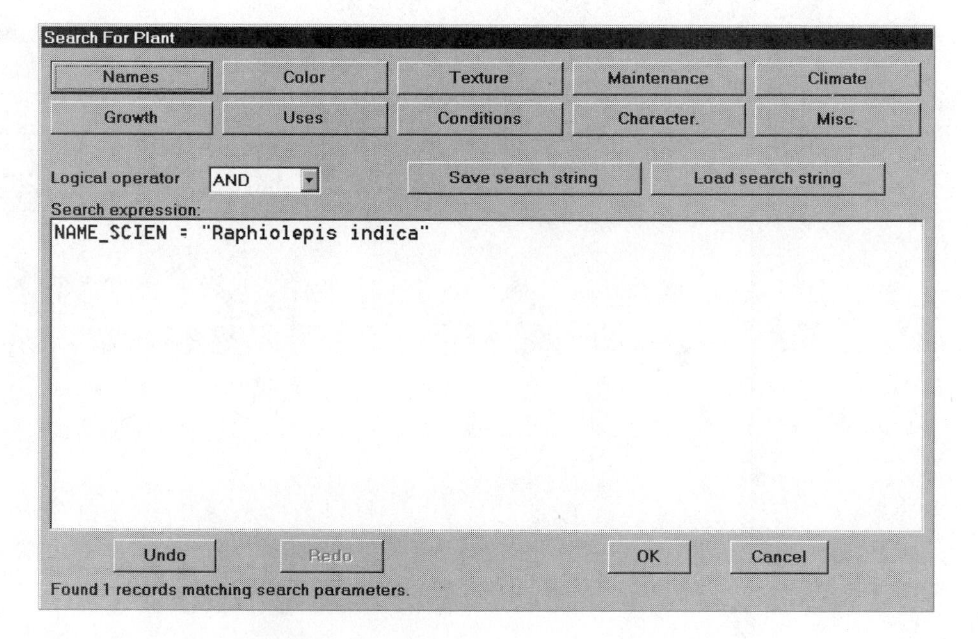

Figure 14.21

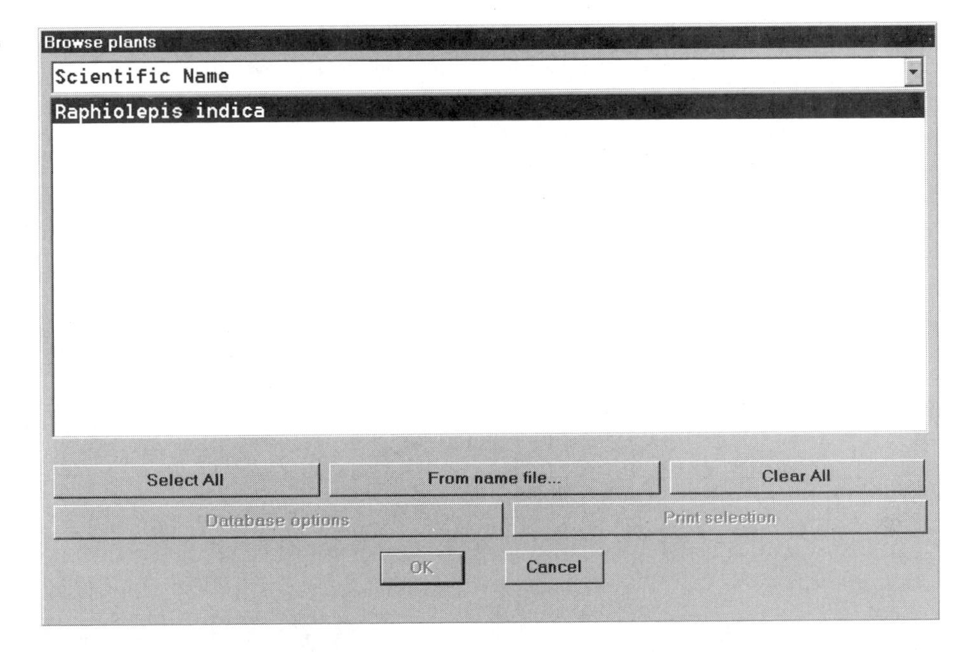

Pick this entry to highlight it. Click on **OK**.

This will open the "View plant information" dialogue box, which will allow us to examine and modify any information about this plant (Figure 14.22).

Figure 14.22

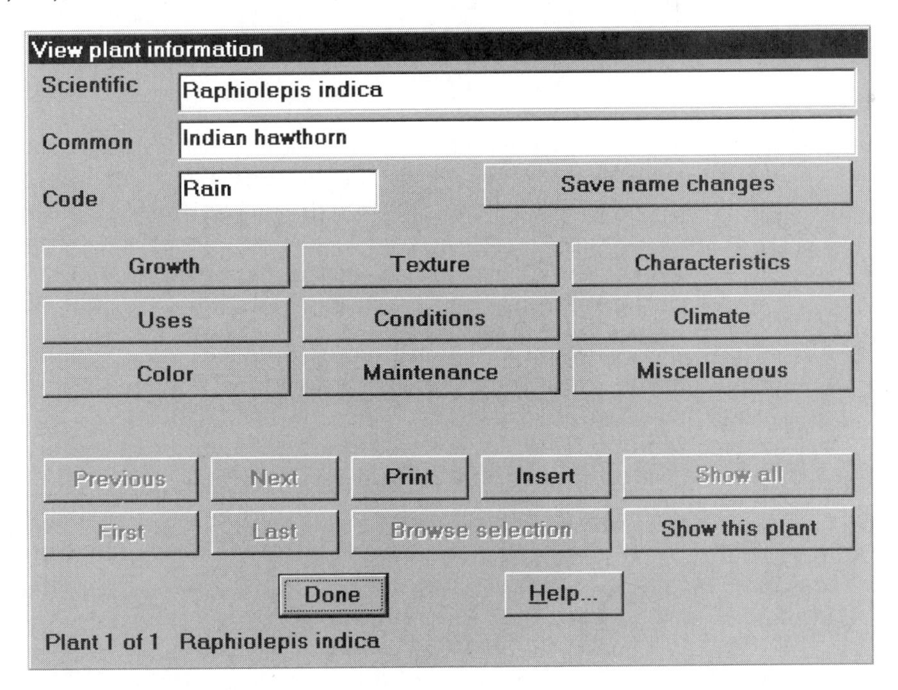

Change the genus name from **Raphiolepis** to **Rhaphiolepis** on the **Scientific** line, then move the cursor to the end of the line and press **<Enter>**. Click on the **Save name changes** box. The line at the bottom of the dialogue box changes to reflect the changes made (Figure 14.23).

Figure 14.23

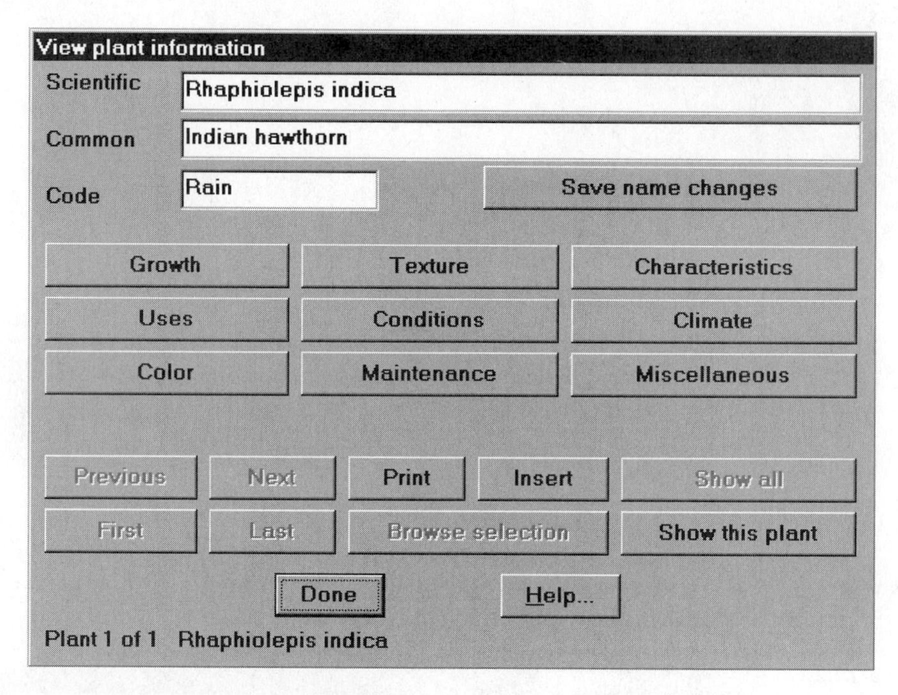

To leave this dialogue box, pick **Done**.

Hint: After editing a plant name you may find that the plant name is not correctly saved, even though you tell the program to save the name change. To be sure that the name is properly saved, pick the **Growth** option, then examine the name of the plant record at the bottom of the dialogue box. If the name change is forwarded to this box, it will be correctly saved when you pick **Done** to leave the "View plant information" dialogue box.

After you have finished successfully updating the plant name, return to the database (either through the **Browse Plants** or **Search for Plant** option), to verify that the change to **Rhaphiolepis indica** has been made to the database.

Copying and Editing an Existing Plant Record

In those instances where we wish to add a new variety of an existing plant to the database, we can simply copy the existing record, then edit it to insert the correct information about the new variety. In this case, we will create a new variety for **Pittosporum crassifolium**.

First, use the **Database** pull-down menu, then select **Browse**. Scroll down the plant list to find **Pittosporum crassifolium**. Highlight the plant name, then pick **Database options**. From the list of available choices, pick **Make copy of current record**. This will create a second entry for **Pittosporum crassifolium**.

Once the copy has been created, you will be returned to the "Browse plants" dialogue box.

Highlight the second of the **Pittosporum crassifolium** entries, then pick **OK**. This will bring up the "View plant information" dialogue box. Edit the **Scientific**, **Common** and **Code** boxes to appear as below (Figure 14.24).

Figure 14.24

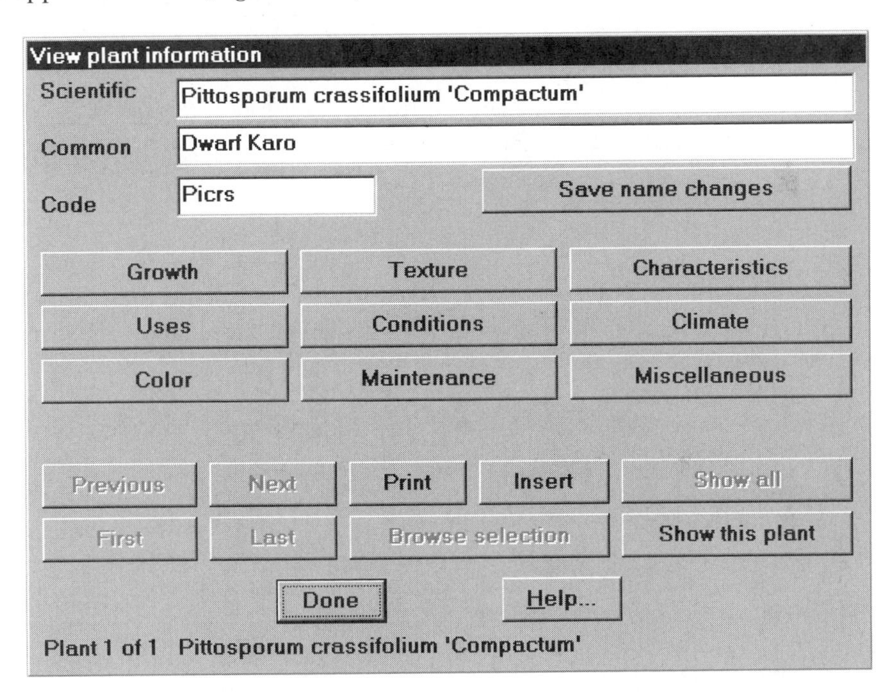

Be sure to press **<Enter>** at the end of each line, then click on the **Save Name changes** box.

Once the changes are made in this dialogue box, you can move on to the **Growth** button. This brings up the "Growth parameters" dialogue box.

Hint: The numbers in **Height** and **Width** in the "Growth parameters" dialogue box use **decimal** units. This means that the base unit is one foot rather than one inch. An entry of **3** indicates a size of **3 feet**. Because all of these records are built with the decimal — rather than architectural — unit system, it is easier to adapt to it than to change it. Be aware that all the plant heights and widths will appear at 1/12 of their intended value when viewed using architectural units (Figure 14.25).

Figure 14.25

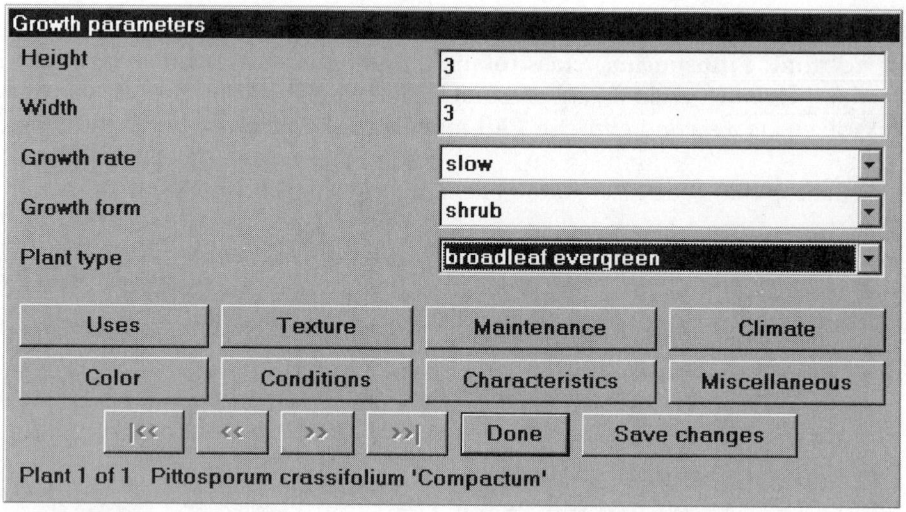

Growth parameters
Height 3
Width 3
Growth rate slow
Growth form shrub
Plant type broadleaf evergreen
Uses | Texture | Maintenance | Climate
Color | Conditions | Characteristics | Miscellaneous
|<< | << | >> | >>| | Done | Save changes
Plant 1 of 1 Pittosporum crassifolium 'Compactum'

Change the entries in the boxes to resemble those above. Be sure to pick the **Save changes** box before you pick **Done**.

Once you have completed this box and selected **Done**, you are returned to the "View plant information" dialogue box.

Once there, select **Done** to finish work on the new record.

If you wish, search or browse the database again to verify that the new plant has been added.

Adding a Plant Record

In addition to editing and copying existing records (plants) in the database, we can also add plants to it. We will create a new plant record for **Sageleaf rockrose**, **Cistus salvifolius**.

Begin by selecting the **Database** pull-down menu, then pick **Add Plant to Database**.

This brings up the "Add plant to database" dialogue box (Figure 14.26).

Fill in the box to match the one below (Figure 14.27).

Figure 14.26

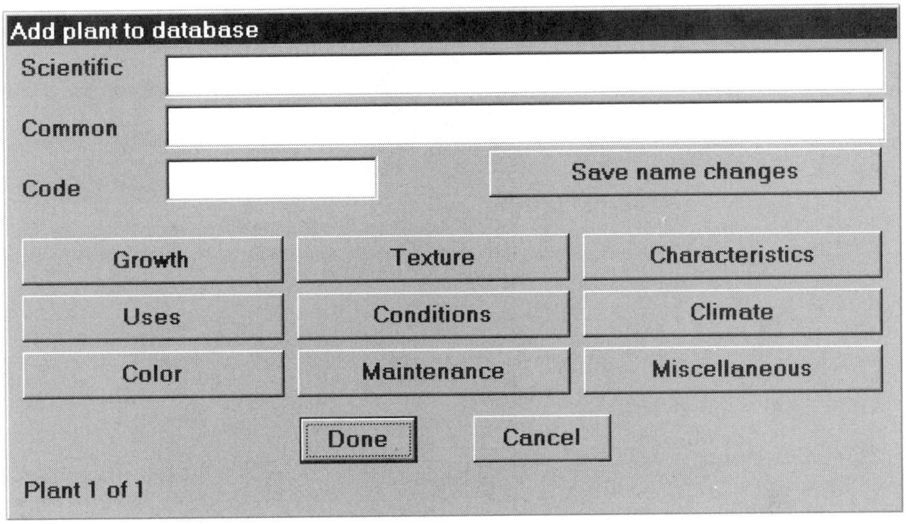

Add plant to database

Scientific

Common

Code

Save name changes

Growth | Texture | Characteristics

Uses | Conditions | Climate

Color | Maintenance | Miscellaneous

Done | Cancel

Plant 1 of 1

Figure 14.27

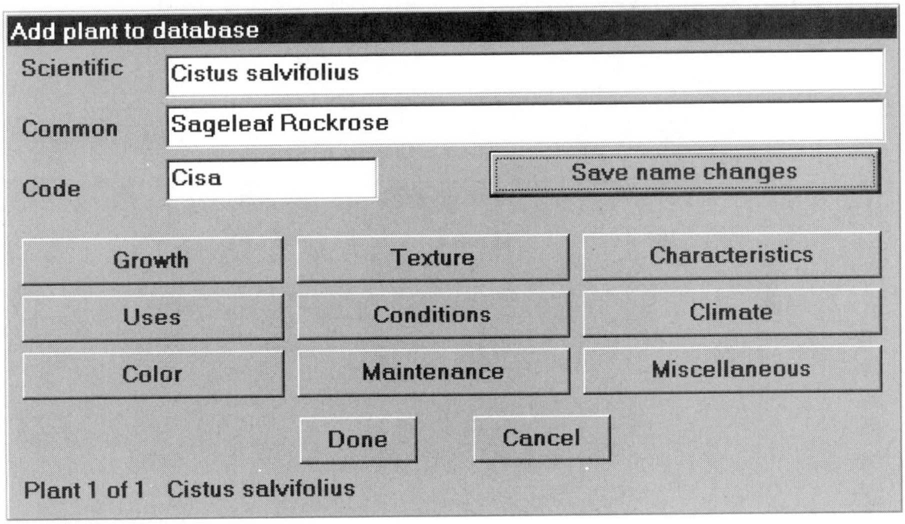

Add plant to database

Scientific: Cistus salvifolius

Common: Sageleaf Rockrose

Code: Cisa

Save name changes

Growth | Texture | Characteristics

Uses | Conditions | Climate

Color | Maintenance | Miscellaneous

Done | Cancel

Plant 1 of 1 Cistus salvifolius

Be sure to press **<Enter>** at the end of each line, then pick **Save name changes** when these entries are complete.

Next, choose **Growth**. This brings up the "Growth parameters" dialogue box (Figure 14.28).

Fill in the **Height** and **Width** boxes, then use the pop-down boxes for **Growth rate**, **Growth form**, and **Plant type**. Click on **Save changes** then on **Texture**. This brings up the "Textures" dialogue box (Figure 14.29).

Figure 14.28

Growth parameters

Height 2'

Width 6'

Growth rate fast

Growth form shrub

Plant type broadleaf evergreen

Uses | Texture | Maintenance | Climate

Color | Conditions | Characteristics | Miscellaneous

|<< | << | >> | >>| | Done | Save changes

Plant 1 of 1 Cistus salvifolius

Figure 14.29

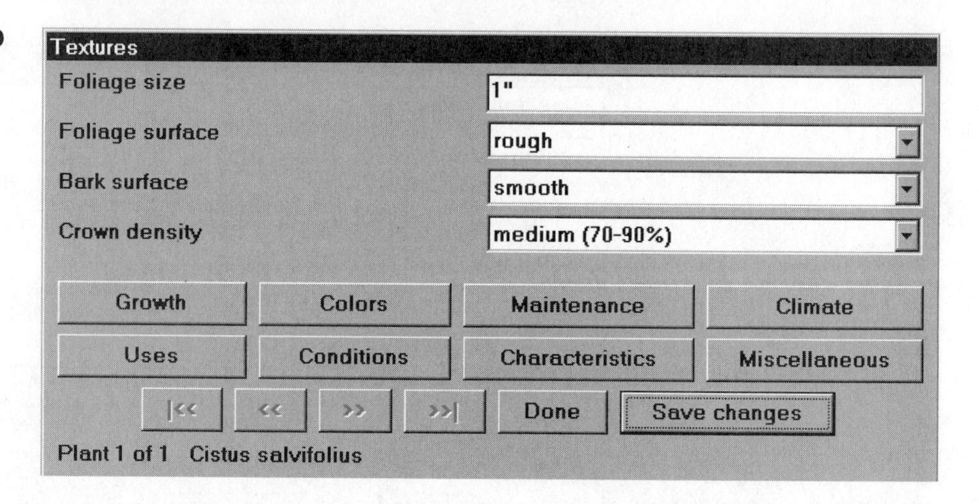

Fill in the **Foliage size** box and use the pop-down boxes for the other entries. Click on **Save changes** then on **Maintenance**. This brings up the "Maintenance parameters" dialogue box (Figure 14.30).

Click on the toggle boxes shown above. Click on **Save changes** then on **Characteristics**. This brings up the "Characteristics" dialogue box (Figure 14.31).

Use the pop-down boxes to complete this dialogue box. Click on **Save changes** then on **Conditions**. This brings up the "Conditions" dialogue box (Figure 14.32).

Figure 14.30

Maintenence parameters

Detriments:

- [ ] Heart rot
- [x] Texas root rot
- [ ] Fireblight
- [ ] Disease susceptible
- [ ] Respiratory irritant
- [ ] Litter
- [ ] Mistletoe
- [ ] Aphids
- [ ] Red spider mites
- [ ] Brittle/wind dmg.
- [ ] Insects
- [ ] Skin irritant
- [ ] Control difficult
- [ ] Cold damage
- [ ] Crown galls
- [ ] Chlorotic
- [ ] Sunburn
- [ ] Poisonous
- [ ] Root suckers
- [ ] Thorns

Tolerances:

- [ ] Requires freezing to fruit
- [ ] Tolerates poor drainage
- [ ] Tap root
- [ ] Shallow root system
- [ ] Difficult to obtain
- [x] Tolerates urban cond.
- [x] Tolerates Pollution
- [ ] Insect Tolerant
- [ ] Disease Tolerant

Growth | Color | Conditions | Climate

Uses | Texture | Characteristics | Miscellaneous

More... | |<< | << | >> | >>| | Done | Save changes

Plant 1 of 1 Cistus salvifolius

Figure 14.31

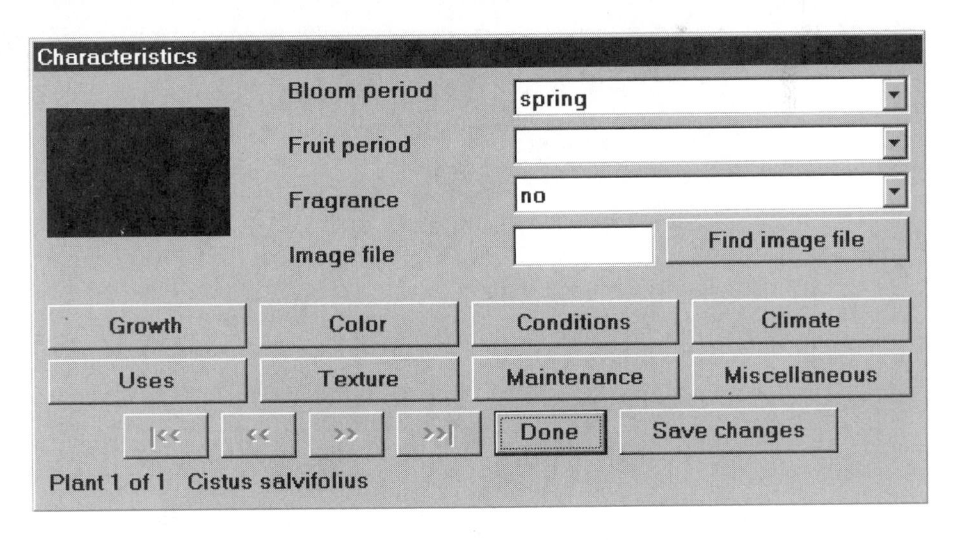

Use the toggle boxes and the pop-down boxes to match the illustration. Click on **Save changes** then on **Uses**. This brings up the "Uses" dialogue box (Figure 14.33).

Figure 14.32

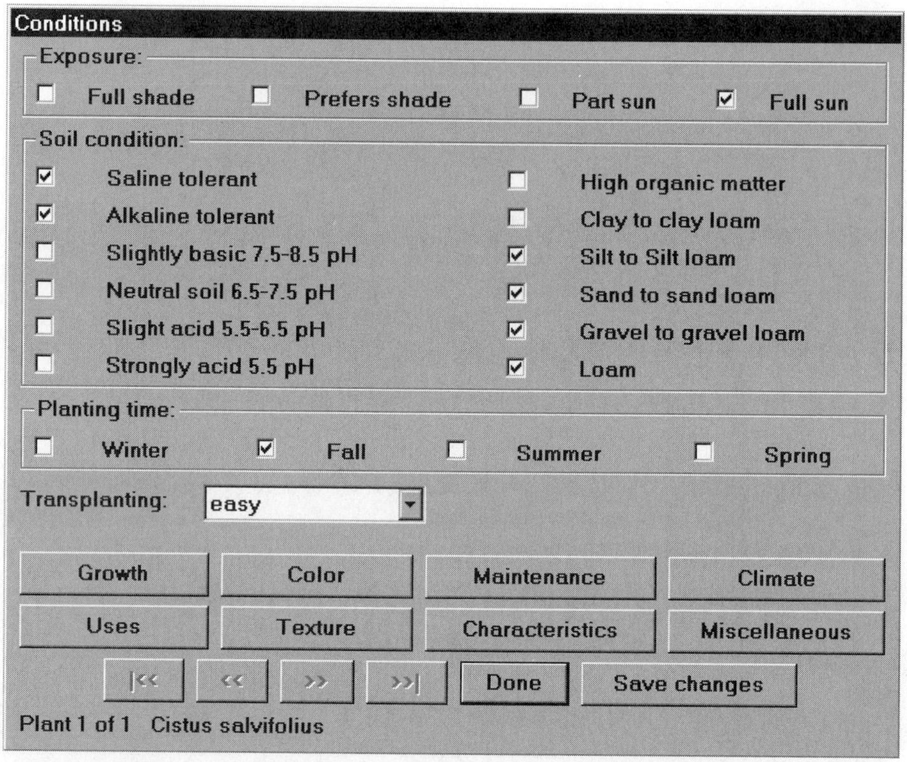

Conditions

Exposure:
- [ ] Full shade
- [ ] Prefers shade
- [ ] Part sun
- [x] Full sun

Soil condition:
- [x] Saline tolerant
- [ ] High organic matter
- [x] Alkaline tolerant
- [ ] Clay to clay loam
- [ ] Slightly basic 7.5-8.5 pH
- [x] Silt to Silt loam
- [ ] Neutral soil 6.5-7.5 pH
- [x] Sand to sand loam
- [ ] Slight acid 5.5-6.5 pH
- [x] Gravel to gravel loam
- [ ] Strongly acid 5.5 pH
- [x] Loam

Planting time:
- [ ] Winter
- [x] Fall
- [ ] Summer
- [ ] Spring

Transplanting: easy

Growth | Color | Maintenance | Climate
Uses | Texture | Characteristics | Miscellaneous
|<< | << | >> | >>| | Done | Save changes

Plant 1 of 1 Cistus salvifolius

Figure 14.33

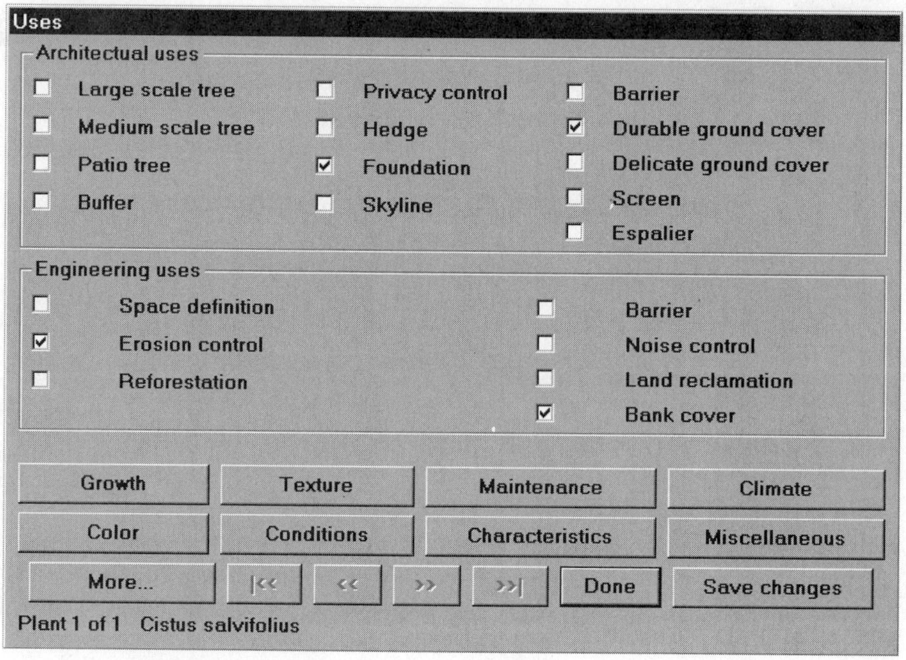

Uses

Architectual uses
- [ ] Large scale tree
- [ ] Privacy control
- [ ] Barrier
- [ ] Medium scale tree
- [ ] Hedge
- [x] Durable ground cover
- [ ] Patio tree
- [x] Foundation
- [ ] Delicate ground cover
- [ ] Buffer
- [ ] Skyline
- [ ] Screen
- [ ] Espalier

Engineering uses
- [ ] Space definition
- [ ] Barrier
- [x] Erosion control
- [ ] Noise control
- [ ] Reforestation
- [ ] Land reclamation
- [x] Bank cover

Growth | Texture | Maintenance | Climate
Color | Conditions | Characteristics | Miscellaneous
More... | |<< | << | >> | >>| | Done | Save changes

Plant 1 of 1 Cistus salvifolius

Use the toggle boxes to match the illustration. Click on **Save changes** then on **More . . .**. This brings up a second "Uses" dialogue box (Figure 14.34).

Use the toggle boxes to complete the entries shown here. Click on **Save changes** then on **Climate**. This brings up the "Climate" dialogue box (Figure 14.35).

Figure 14.34

Figure 14.35

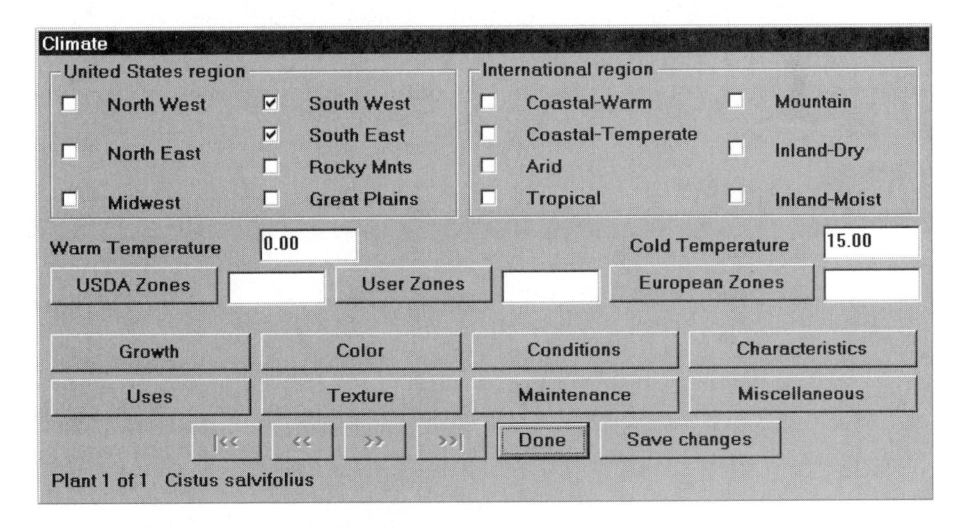

Use the toggle boxes to fill in the **United States region** areas, then click on the **USDA Zones** box to bring up a USDA climate zone map (Figure 14.36).

Use the toggle boxes for **Zones 9** and **10**, then click on **OK**.

Figure 14.36

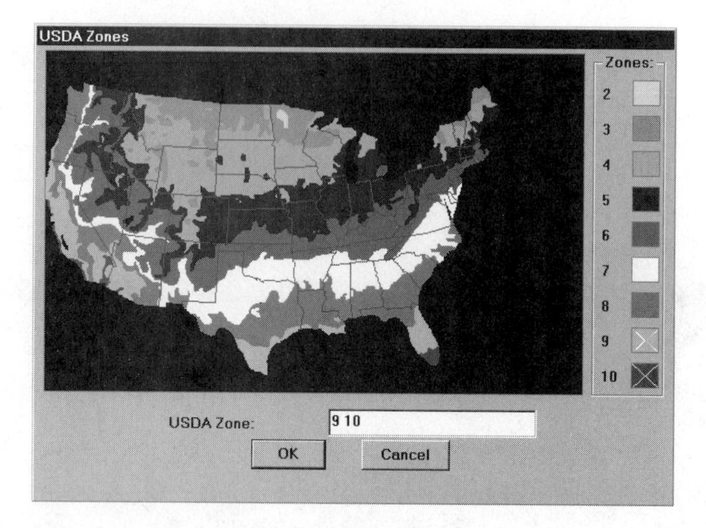

There is also a **European Zones** button, which brings up a climate zone map for western Europe.

Back in the "Climate" dialogue box, click on **User Zones**. This brings up the "User Zones" dialogue box which has 24 climate zones (a purely coincidental parallel to the ***Sunset Western Garden Book*** climate zones) (Figure 14.37).

Use the toggle boxes to indicate the **Sunset** climate zones where **Cistus salvifolius** will grow. Click on **OK** when finished. In the "Climate" dialogue box, click on **Save changes** then on **Done**. After returning to the "Add plant to Database" dialogue box, click on **Done** to complete the entry.

Figure 14.37

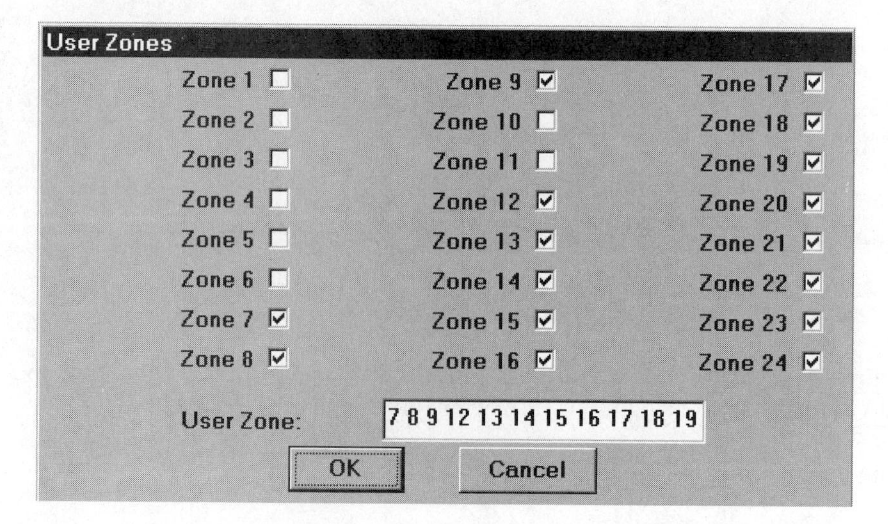

Creating a Plant Name File

After editing, copying and creating plant records, we will create a list of plant names which will be used in the Thomas landscape plan. This plant name list will allow us to quickly choose the name for a plant as we insert and label it. A small amount of work now will save us a considerable amount of work later!

In the **Database** pull-down menu, select **Browse Plants**, then highlight the following 30 plants from the database:

Scientific name	Common name	Code
Acer palmatum "Atropurpurea"	Red Japanese Maple	Acpal
Agapanthus africanus	Blue lily of the Nile	Agaf
Alnus rhombifolia	White alder	Alrh
Camellia japonica	Japanese camellia	Caja
Camellia sasanqua	Appleblossom camellia	Casas
Cassia artemisioides	Feathery cassia	Caart
Cercis canadensis	Eastern redbud	Ceca
Cistus salvifolius	Sageleaf rockrose	Cisa
Festuca ovina glauca	Blue fescue	Feov
Fragaria chiloensis	Wild strawberry	Frch
Gazania rigens "Copper King"	Copper King Gazania	GariC
Grevillea "Noellii"	Noel Grevillea	Grno
Heuchera sanguinea	Coral bells	Hesa
Iris "Louisiana"	Louisiana iris	Irlo
Juniperus conferta	Shore juniper	Jucon
Lagerstroemia indica	Crape myrtle	Lain
Mahonia repens	Creeping mahonia	Maren
Myrtus communis "Compacta"	Compact myrtle	Mycoc
Nandina domestica "Nana"	Dwarf nandina	NadoN

(continued)

Scientific name	Common name	Code
Pittosporum crassifolium "Compactum"	Dwarf Karo	Picrc
Pittosporum tenuifolium	Kohuhu	Pite
Pittosporum tobira "Wheeler's dwarf"	Wheeler's dwarf tobira	PitoW
Podocarpus macrophyllus	Yew podocarpus	Poma
Rhaphiolepis indica	Indian hawthorn	Rain
Rhododendron indicum	Southern Indian hybrids	Rhin
Santolina chamaecyparissus	Lavender cotton	Sach
Stachys lanata	Lambs ear	Stla
Teucrium fruticans	Bush germander	Tefr
Thevetia peruviana	Yellow oleander	Thpe
Viburnum japonicum	Japanese viburnum	Vija

Once you have highlighted all 30 of these plants, pick **Database options**. This will bring up the "Database options" dialogue box (Figure 14.38).

Choose the **Make plant name file** option. This will bring up a dialogue box which allows you to create a plant name file. Name your file **THOMAS.NAM** (Figure 14.39).

Figure 14.38

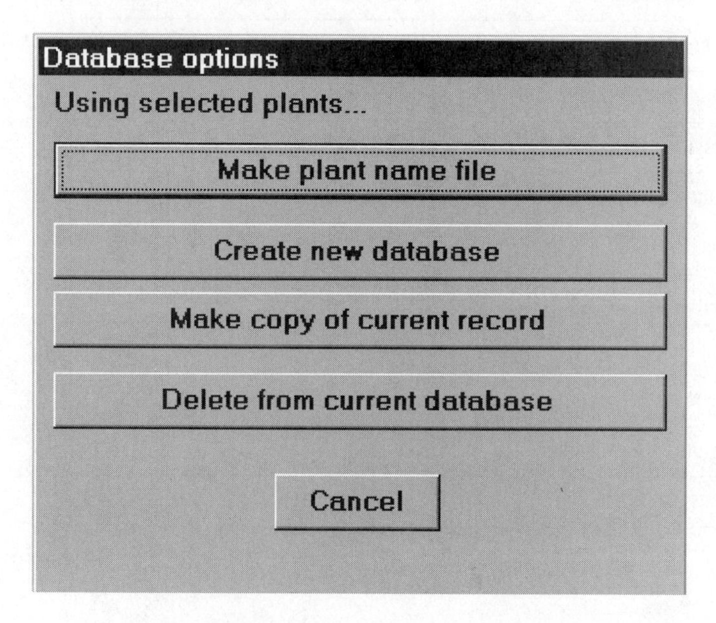

Figure 14.39

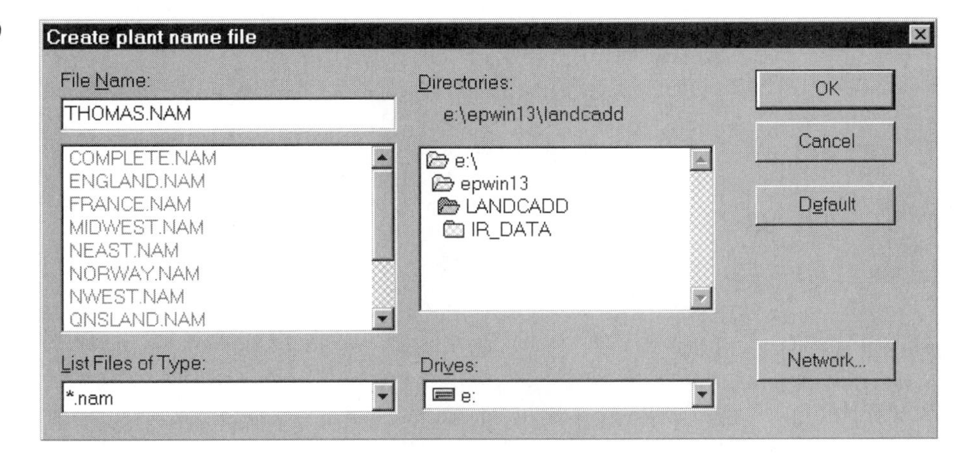

Click on OK when finished.

Once we have completed the plant name file, we are ready to begin inserting plants into the planting plan.

Inserting Plant Symbols

The following section is intended to show you the process of inserting plant symbols. We will follow the selection of a plant symbol, then assigning a plant name, diameter, planting size, elevation, and layer where the plant should be inserted. In addition, we will examine other options for inserting multiple plants. Try these insertion routines on a blank drawing before using them in the Planting Plan drawing. **Do not insert any plant symbols into the Planting Plan until you have finished reading through this section**. This will eliminate confusion when the tree and shrub symbols are inserted later.

To insert a plant symbol, find the LANDCADD Landscape Design pull-down menu Symbols. Different symbol classifications and functions are listed under this pull-down. We will begin with Broadleaf (Figure 14-40).

After you make this selection, you are shown the "Broadleaf Tree Symbols" dialogue box. The pop-down box toward the lower left corner of the screen represents the type of sym-

Figure 14.40

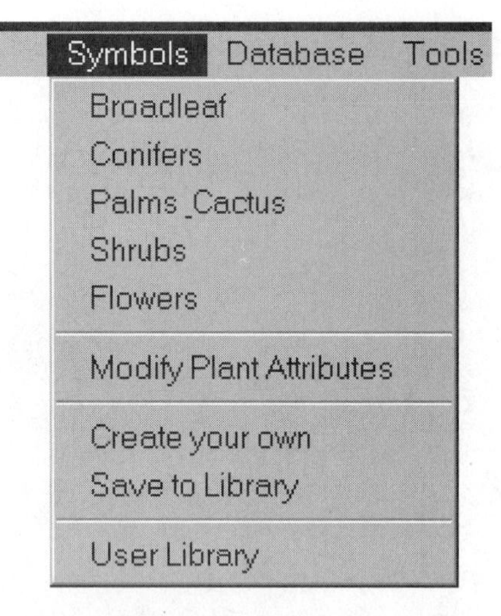

bols being displayed. We can choose to display Simple Plan, Detailed Plan, Section View, Simple 3-D or Detailed 3-D symbols. All of these symbols are provided by LANDCADD. Make your selection the third symbol from the top left (Figure 14.41).

Once the symbol has been selected, you will see the "Plant Symbol Settings" dialogue box (Figure 14.42).

Figure 14.41

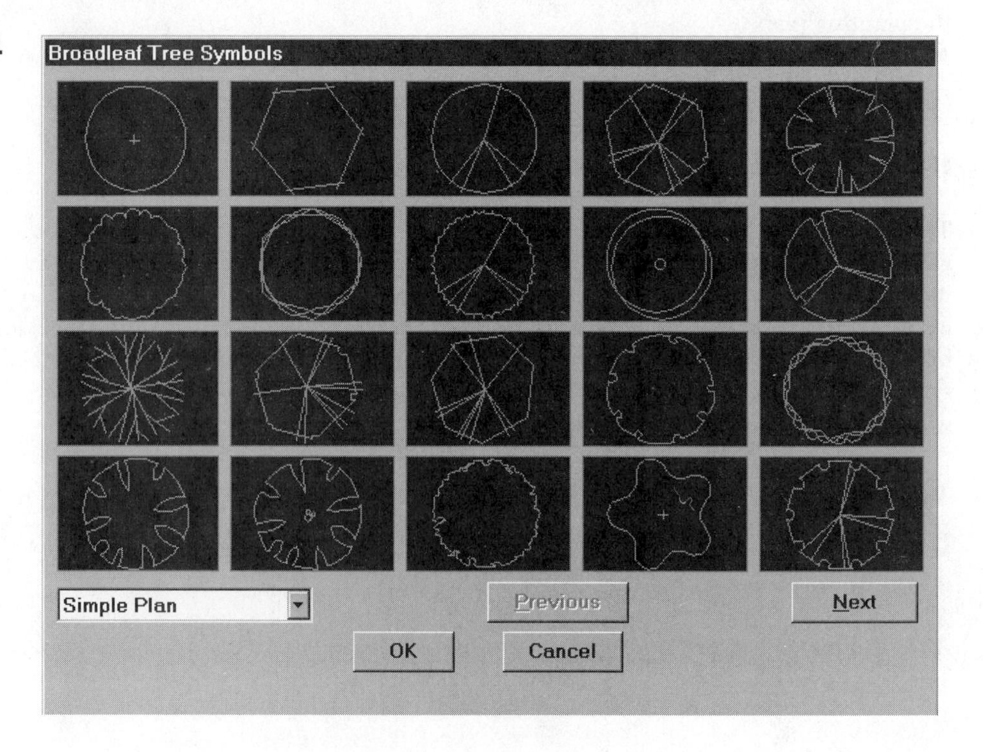

Figure 14.42

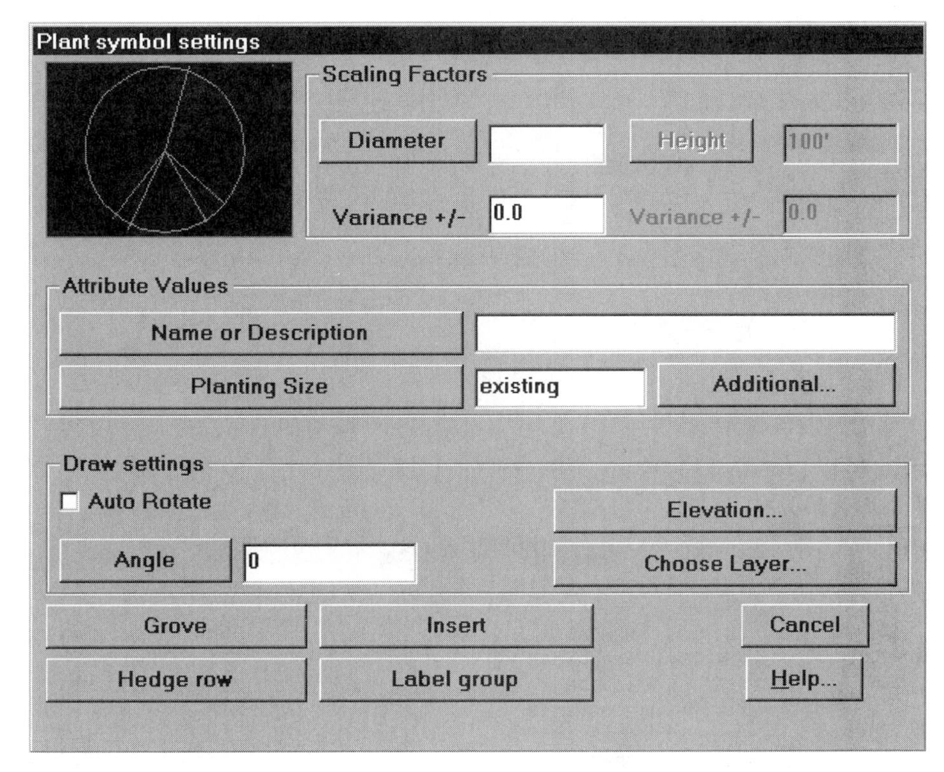

In this dialogue box, you can manipulate the symbol size, accompanying plant name, accompanying planting size, layer, elevation and multiple plant groupings. In addition, you can activate the symbol labeling routine as the symbols are inserted. Note that the settings shown when the dialogue box is opened are left over from the last session working with the plant database. We will modify them before the symbol is actually inserted.

After a symbol is selected, you need to decide how large the symbol should be on the planting plan. To set this, click on **Diameter**. You then see a dialogue box which allows the selection of plant diameters in whatever units are being used. For this plant symbol, select **5'**. Double click on the desired diameter, or highlight the size, then choose **OK** (Figure 14.43).

After the correct plant symbol diameter is chosen, then we can select the appropriate plant name. To make this choice, select the **Name or Description** box. This brings up the "Plant Names" dialogue box (Figure 14.44).

This dialogue box displays all of the plants that are contained in the active plant Name File list. We can select the **THOMAS.NAM** name file that we created before by selecting the **Name File** box. This will bring the "Select Name File" dialogue box which contains

Figure 14.43

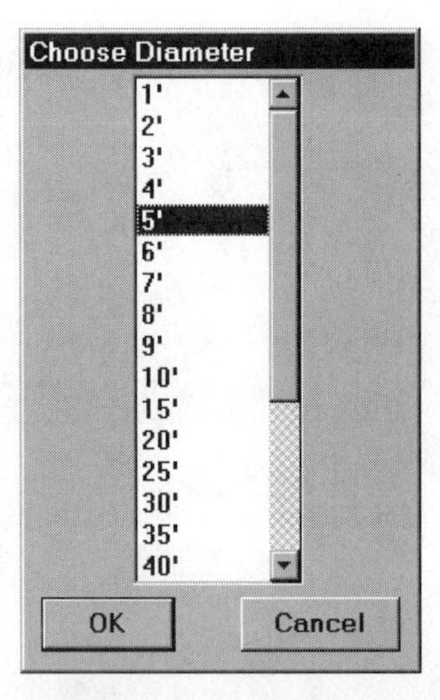

Figure 14.44

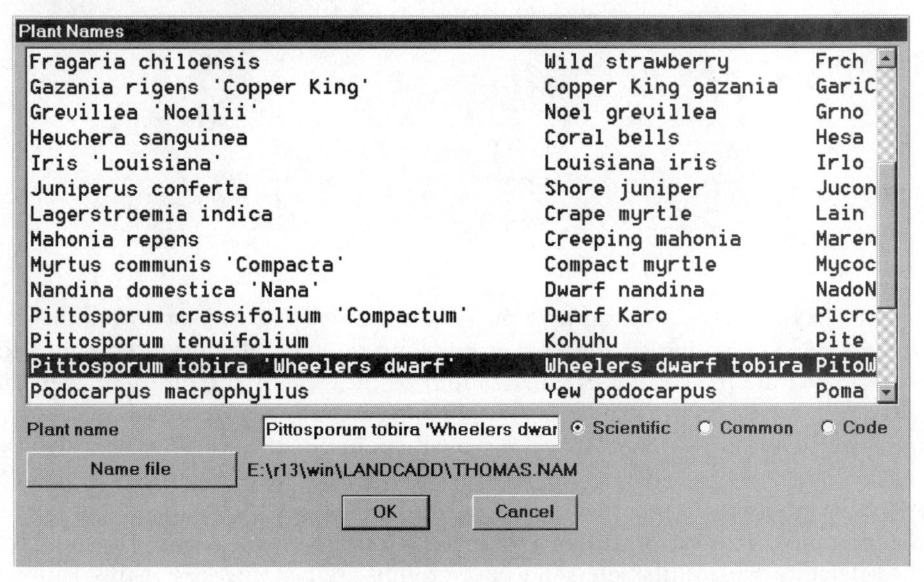

all of the Name Files which are in the EP (or your own) directory. Find the **THOMAS.NAM** file and select it (Figure 14.45).

With the correct name file active, we can then choose the plant name to assign to the symbol. For this plant symbol, select **Pittosporum tobira "Wheeler's Dwarf**."

Next, we need to specify what planting size to use for our plant. Once you click on the **Planting Size** box, the "Planting Size" dialogue box is displayed (Figure 14.46).

Figure 14.45

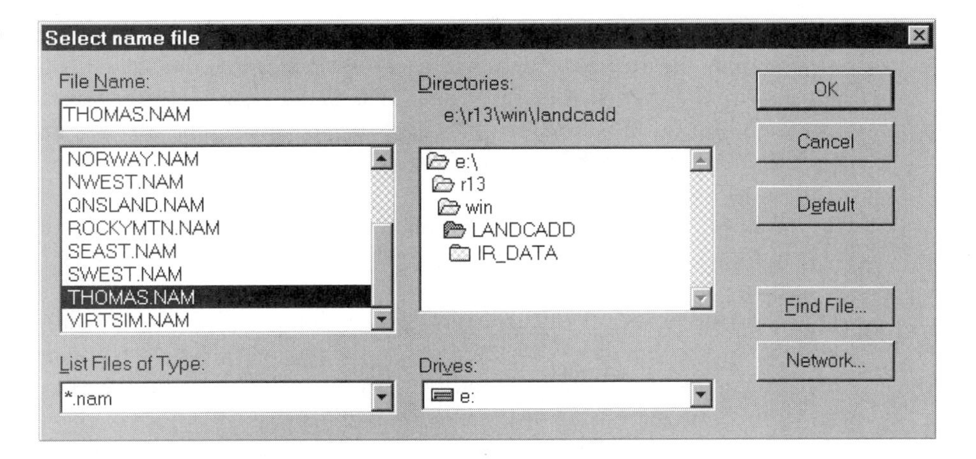

Figure 14.46

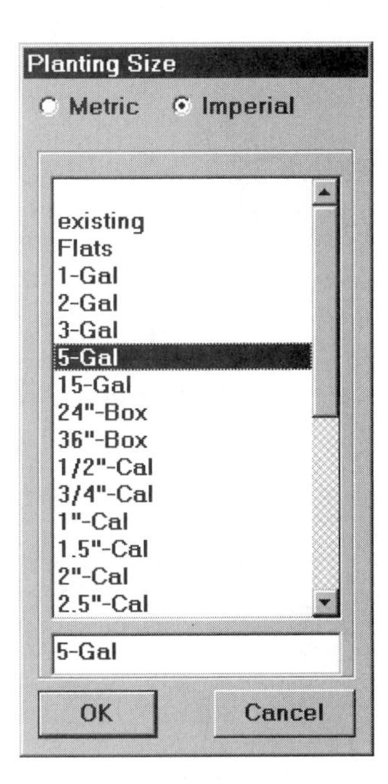

Double click on the desired size **5-Gal** or highlight the desired size and pick **OK**.

At this point we are ready to insert the symbol in the appropriate locations throughout the drawing. Before we do that, however, we should examine some of the other options for plant insertion and labeling.

The **Variance** +/- option allows us to create symbols which vary in size by an amount that we can set. If we set this factor to 2', the plant diameters would range from 3' to 7', instead of retaining a 5' size. The **Auto Rotate** option rotates symbols randomly as they are inserted. **The major drawback of these options is that the "ghost" of the symbol does not display if either of them are used**. This means that a "rubber band" line is used to locate the center of the plants, but their spread will not be obvious until the symbol is placed in the drawing.

Another option under **Attribute Values** is **Additional . . .**. This selection brings up the "Attribute Create" dialogue box (Figure 14.47).

This dialogue box gives us the option of creating more attributes (either visible or not) to store with the plant symbols. An additional attribute could be a descriptive name like "Accent" or "Street Tree", or could be a number like "T-1" for Tree 1.

Other options in the **Draw settings** box include **Elevation . . .** and **Choose Layer . . .**.

Elevation . . . will bring up a dialogue box which allows us to specify an elevation for the symbol, or allows us to pick a contour line which is at the desired elevation. Another option is **Snap to TIN** which allows the symbol to move to the elevation of the triangular region which defines the ground surface in the Surface Modeling module. In this module a Triangulated Irregular Network (TIN) is laid over a site to illustrate its terrain surfaces. A final option is **Prompt at insertion** which allows the user to specify the elevation of each plant as it is inserted (Figure 14.48).

Choose Layer . . . will bring up a dialogue box which allows us to name a layer or pick an object on the layer in which we want to insert the plant symbol. The pop-down box

Figure 14.47

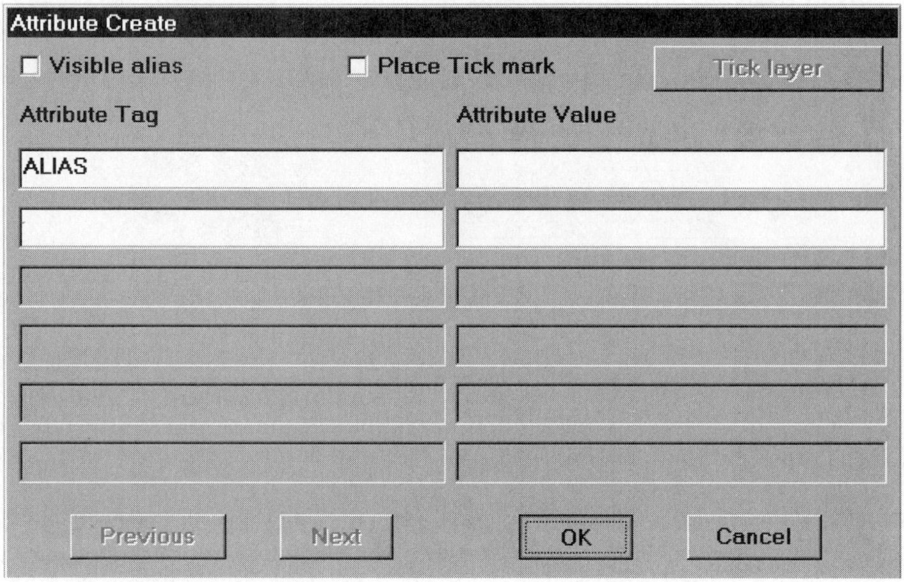

Figure 14.48

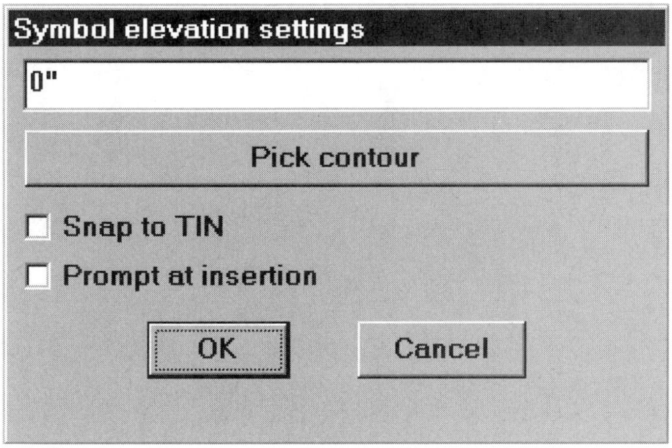

Symbol elevation settings
0"
Pick contour
Snap to TIN
Prompt at insertion
OK
Cancel

below allows us to find the correct layer name without having to remember its exact name (Figure 14.49).

These settings are particularly important for 3-D drawings and those drawings which separate plants into many categories and layers.

The **Grove** option allows you to place a number of plants within an enclosed polyline. The enclosing polyline must exist on the drawing before this command is used, and any islands (areas not to be planted within the polyline-enclosed area) must be surrounded with their own polyline border (Figure 14.50).

Once activated, the command allows plants to be inserted in one of three **Spacing Methods**: **Random**, **Square** or **Triangular**. If Square or Triangular is chosen, a specific **Spacing** distance must be included. If Random is chosen, you must indicate the number

Figure 14.49

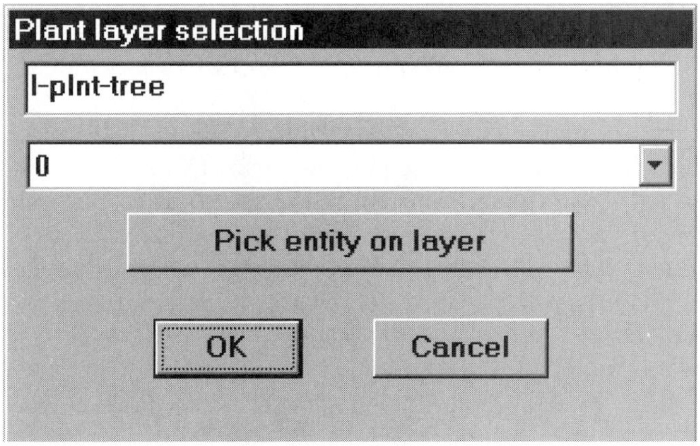

Plant layer selection
l-plnt-tree
0
Pick entity on layer
OK
Cancel

Figure 14.50

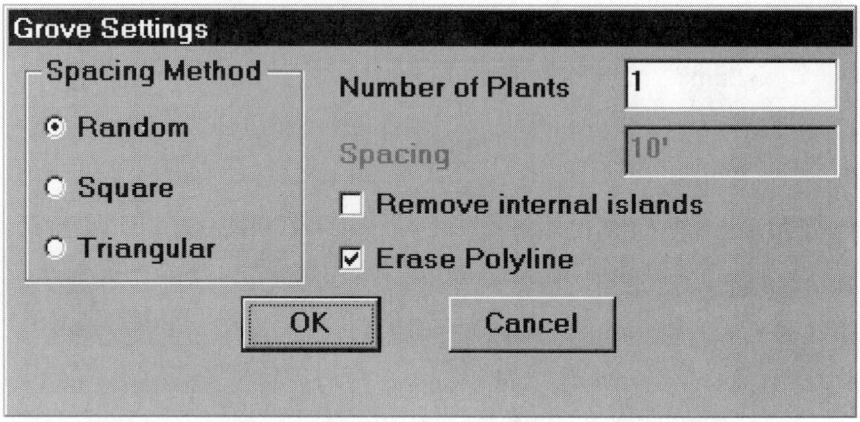

of plants to be inserted, and LANDCADD will place them within the polyline-enclosed area. The surrounding polyline and internal island borders can be removed if their toggle boxes are marked.

The **Hedge Row** button inserts a row of plants along a polyline. The command can be used in one of two ways.

With an existing polyline:

```
Plant spacing distance <5'>:
```
Press **<Enter>** to accept the default spacing (which is the plant diameter) or indicate a desired spacing distance.

```
Follow existing polyline? <Y/N>:
```
Type **Y** then press **<Enter>**.

```
Select defining polyline
Select object:
```
Pick the polyline (either 2-D, or 3-D or polyline arc) that will define the hedge row.

```
Delete defining polyline? <Y/N>:
```
Here, press **Y** if you wish to delete the polyline, or **N** if you do not. The plants are then uniformly inserted along the polyline. An advantage to using the command this way is that the hedge can be inserted along an arc or three-dimensional polyline which may vary in elevation. In addition, the polyline can be edited, splined or straightened before the plants are inserted along its length. The following method does not allow these editing capabilities.

Without an existing polyline:

Plant spacing distance <5'>:
Press **<Enter>** to accept the default spacing (which is the plant diameter) or indicate a desired spacing distance.

Follow existing polyline? <Y/N>:
Type **N** then press **<Enter>**.

Select starting point for hedge:
Click on the desired location to begin the hedge. You then will select points along the polyline which builds the hedge.

Select points to define hedge line:
Pick as many points as needed to define the polyline which the hedge will follow. Press **<Enter>** to leave the command. The defining polyline is deleted as the hedge is finished.

The option **Label group** activates the plant labeling routines immediately after the plant symbols are inserted. This is desirable in some drawings, but not in others. In a fairly complex planting plan, the labels are often inserted after placing all of the plant symbols so that the labels can be more carefully placed and aligned. We will follow this procedure in the Thomas Residence planting plan.

The button begins placing plant symbols wherever you press the **Pick** button. The **Ortho** mode is usually turned on during this command, so that you may need to turn it off to place the symbols into more informal groupings.

To actually insert symbols into the plan, pick **Insert**. The plant symbols are inserted wherever you press the **<Pick>** button. Select the **<Enter>** button to exit the plant insertion command.

Plant Symbols for the Front Yard

The "Rogue's Gallery" of plant symbols and plant names below represents the trees and shrubs used in the front yard. The groundcover plants used (Gazania and Santolina) are represented by a hatch pattern rather than by a symbol. We will create special polyline borders to represent the area covered by the groundcovers (Figures 14.51, 14.52, 14.53, 14.54, 14.55, 14.56, 14.57, 14.58, 14.59, 14.60, 14.61, 14.62, 14.63, 14.64, and 14.65).

Figure 14.51
Acer palmatum "Atropurpureum"

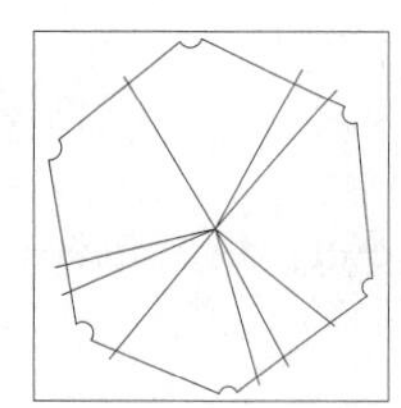

Figure 14.52
Agapanthus africanus

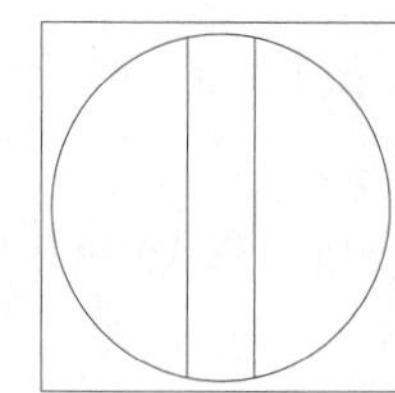

Figure 14.53
Alnus rhombifolia

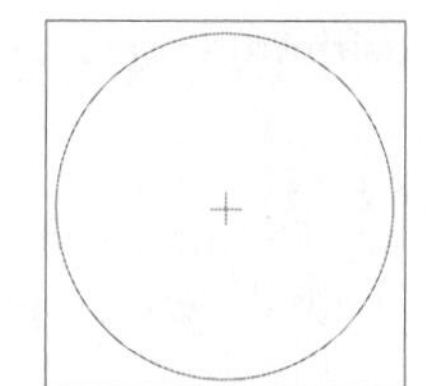

Figure 14.54
Camellia sasanqua

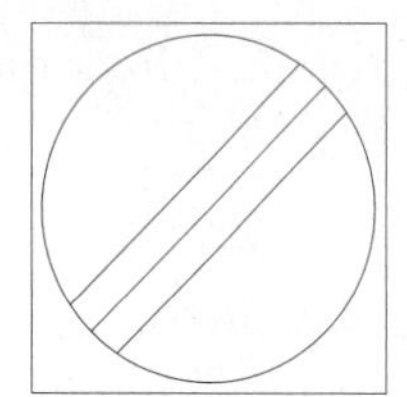

Figure 14.55
Cassia artemisioides

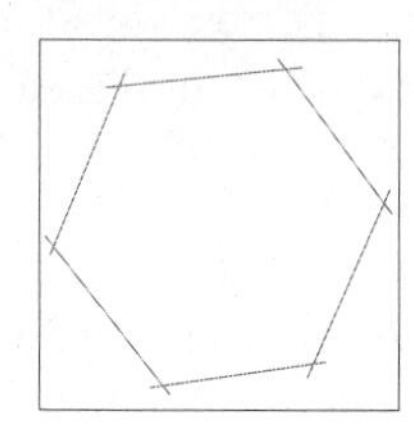

Figure 14.56
Cistus slavifolius

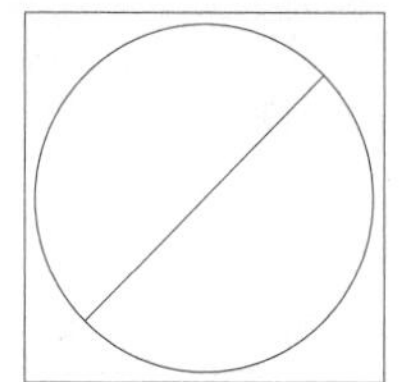

Figure 14.57
Iris "Louisiana"

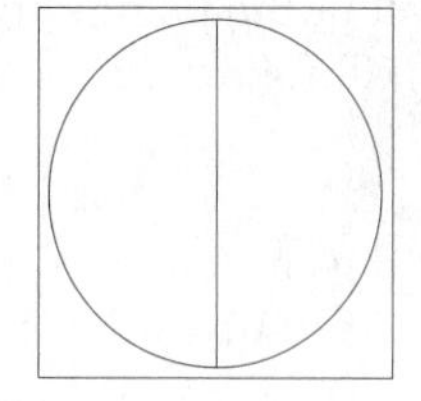

Figure 14.58
Mahonia repens

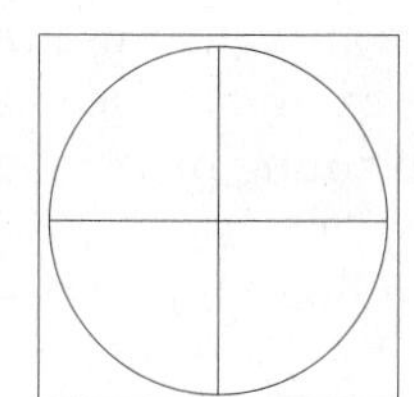

Figure 14.59
Pittosporum crassifolium "Compactum'

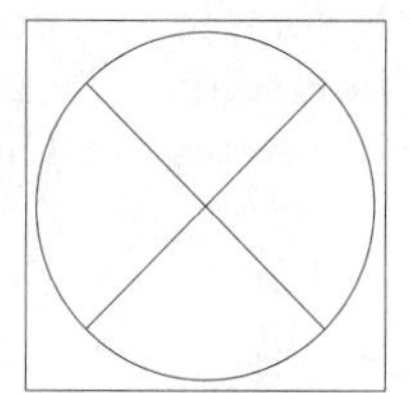

Figure 14.60
Pittosporum tenuifolium

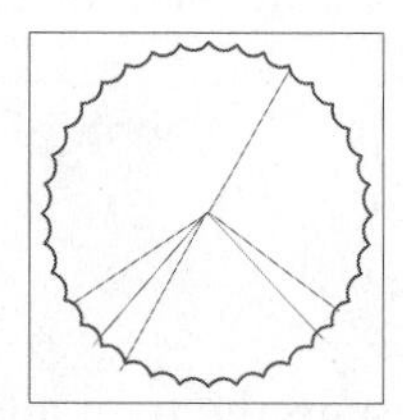

Figure 14.61
Pittosporum tomira "Wheeler's Dwarf"

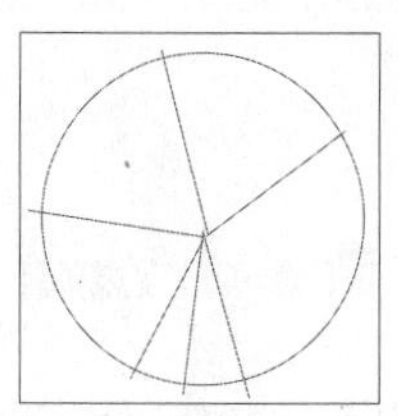

Figure 14.62
Teucrium fruticans

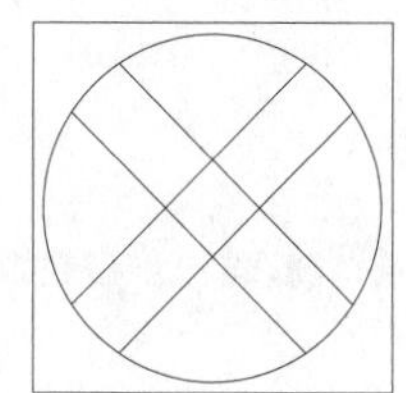

Figure 14.63
Viburnum japonicum

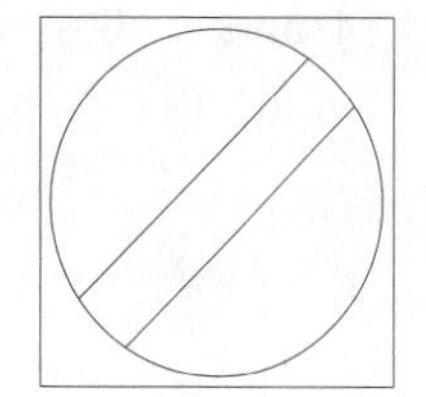

Figure 14.64
Gazania rigens "Copper King"

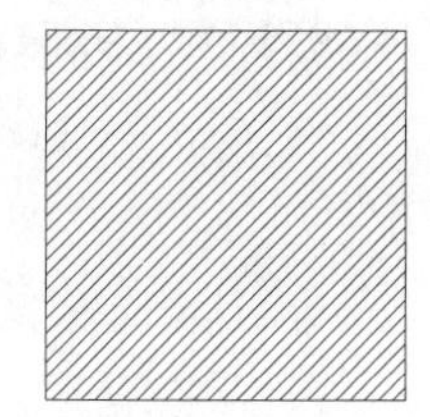

Figure 14.65
Santolina chamaecyparis-sus

Front Yard Planting List

The following chart shows the plant name, quantity, container size, diameter, price and category for all of the trees and shrubs in the front yard. Later in the drawing, we will use the plant name, quantity and size attributes, as well as price and category when a special file — called an **attribute extract file** — is created. We will discuss how this file is created later in this exercise.

Plant Name	Q'ty.	Planting size	Diameter	Price	Category
Acer palmatum "Atropurpureum"	1	15-Gal	10'	$125.00	trees
Agapanthus africanus	14	1-Gal	3'	$7.50	shrubs
Alnus rhombifolia	1	existing	20'	—	—
Camellia sasanqua	1	5-Gal	4'	$20.00	shrubs
Cassia artemisioides	4	5-Gal	5'	$17.50	shrubs
Cistus salvifolius	5	15-Gal	6'	$20.00	shrubs
Gazania rigens "Copper King"*	~12	Flats	6" O.C.	$35.00	groundcover
Iris "Louisiana"	25	1-Gal	3'	$10.00	shrubs
Mahonia repens	9	1-Gal	3'	$10.00	shrubs
Pittosporum crassifolium "Compactum"	24	5-Gal	3'	$17.50	shrubs
Pittosporum tenuifolium	5	15-Gal	10'	$75.00	trees
Pittosporum tobira "Wheeler's dwarf"	14	5-Gal	5'	$15.00	shrubs
Santolina chamaecyparissus*	~45	1-Gal	18" O.C.	$7.50	groundcover
Teucrium fruticans	9	1-gal.	3'	$7.50	shrubs
Viburnum japonicum	3	15-gal.	10'	$75.00	shrubs
* Groundcover quantities will vary with the size of the area within the enclosing polyline.					

The Price and Category are not entered with the plant names, but instead are used in the **Estimating** program included with LANDCADD. You will enter the price and category of each plant later in this exercise.

Beginning with the largest trees first, locate each of the symbols in the approximate locations indicated on the drawing.

Be sure that the quantity of plants used corresponds to the quantity indicated on the chart. As each plant is placed, three pieces of information are stored with each symbol: the plant name, the plant code and the planting size. These pieces of information are attributes — **Item** (the plant name) and **Size** (the planting size) can be tallied and used to develop a planting schedule and cost estimate.

Nearly all blocks in LANDCADD are entered with some sort of attributes. This is very helpful when you want to do cost take-offs on a project. These attributes are stored permanently with the symbol, so that if you copy symbols or erase them, LANDCADD keeps track. When we label the plants later, we don't need to remember the name of the plant. LANDCADD reads the attributes of name and size, and records them in the label. LANDCADD also keeps track of the number of plants of each species (Figures 14.66, 14.67, and 14.68).

Figure 14.66

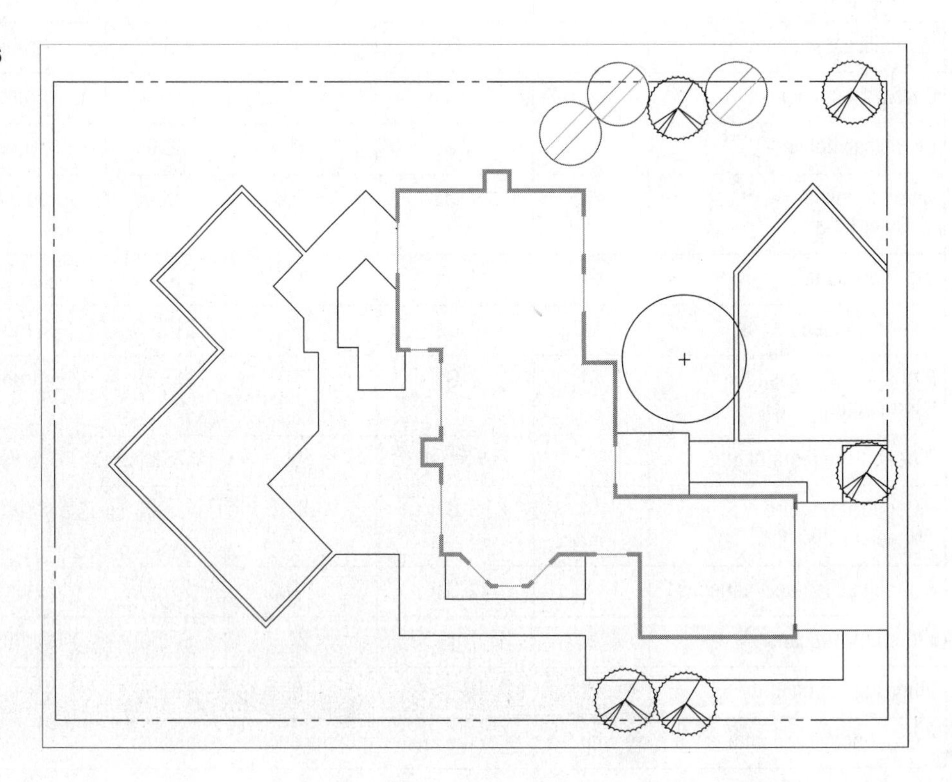

Figure 14.67

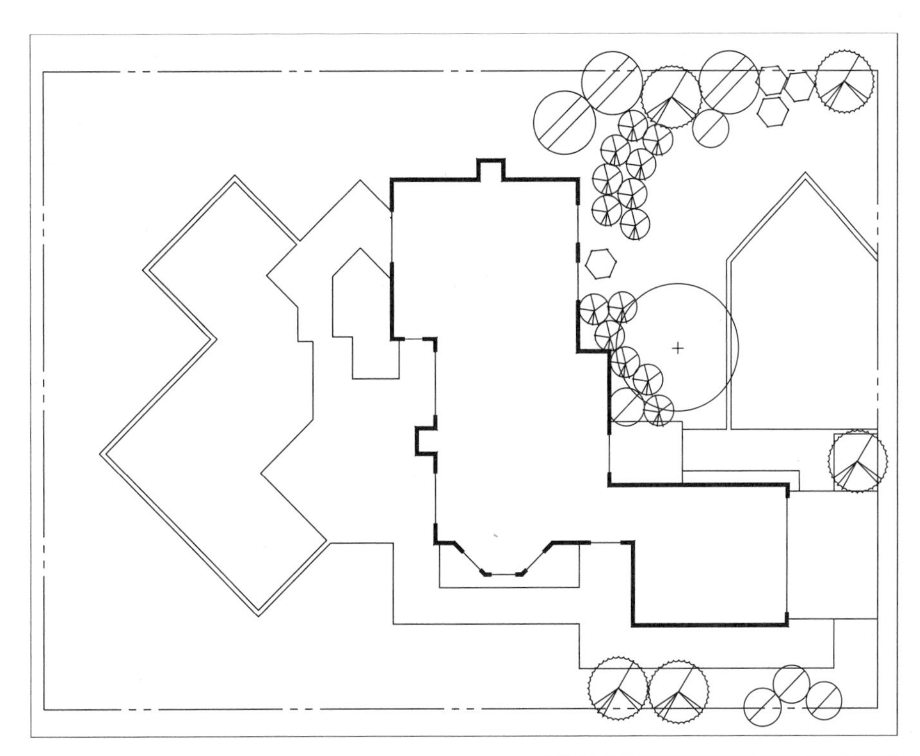

Figure 14.68

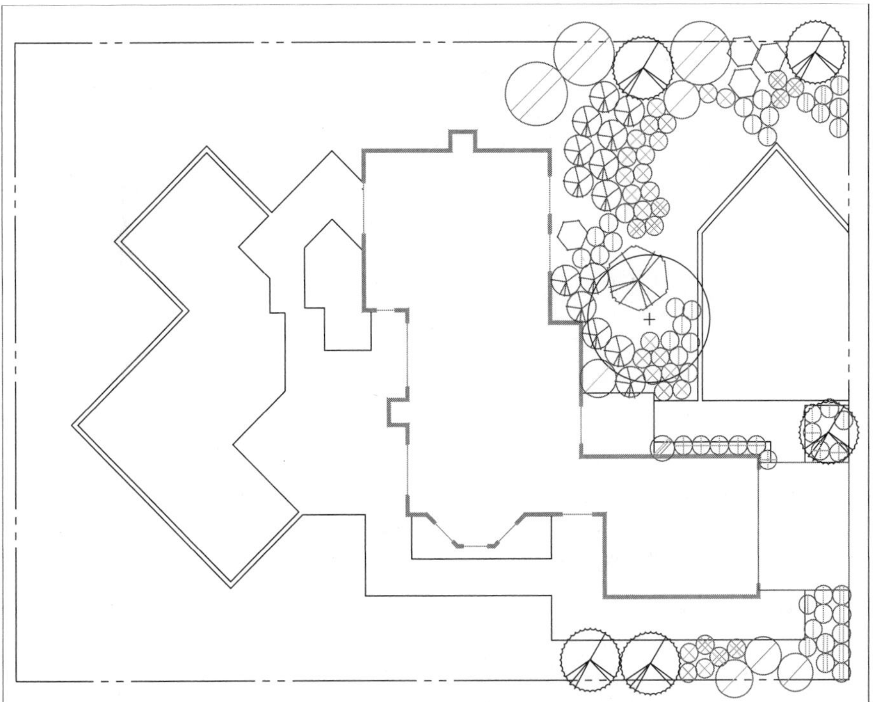

Groundcover Areas

The groundcover areas are drawn using one of the options shown in the **Planting** pull-down menu.

Pick **Ground Cover**, then select **Cloud** (Figure 14.69).

The **Cloud** command creates a polyline made up of arcs. By picking the endpoints of the arcs, a polyline is stretched around the groundcover area. This command gives an attractive edge around the groundcover, but may be a little tricky to use at first. We will examine how to refine your technique with **Cloud** later.

At the prompt:

```
Command: EPLD_cloud
This utility cannot close the pline, and stops only when can-
celed.
First point:
```

Pick the first endpoint of the first arc, then continue building the polyline around the perimeter of the groundcover area (Figures 14.70 and 14.71).

The warning given with this command means that you will have to close the polyline manually (by picking the start point as an ending vertex) and cancel the command rather than pressing **<Enter>** to finish. Although this makes the polyline **look** closed, we must actually close it with the **PEDIT** command.

As you approach the beginning point of the cloud, close the polyline by using the cursor menu to highlight **Endpoint**. Pick the endpoint at the beginning of the cloud. This will close the polyline. Next, press the **Cancel** button or press **<ESC>** (**<Ctrl>-C** for DOS users) to cancel. This stops the command but leaves your polyline cloud intact.

Figure 14.69

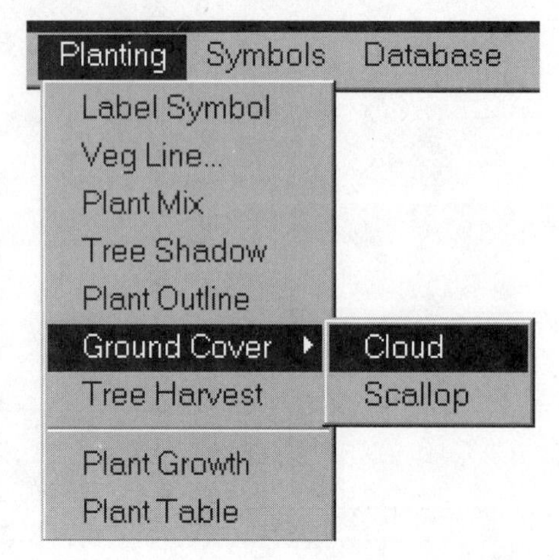

Figure 14.70

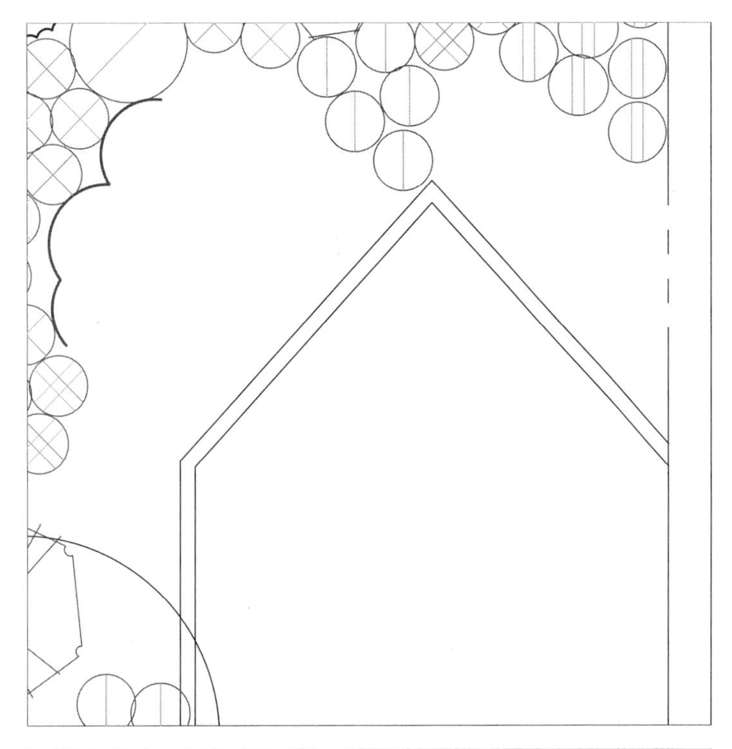

Figure 14.71

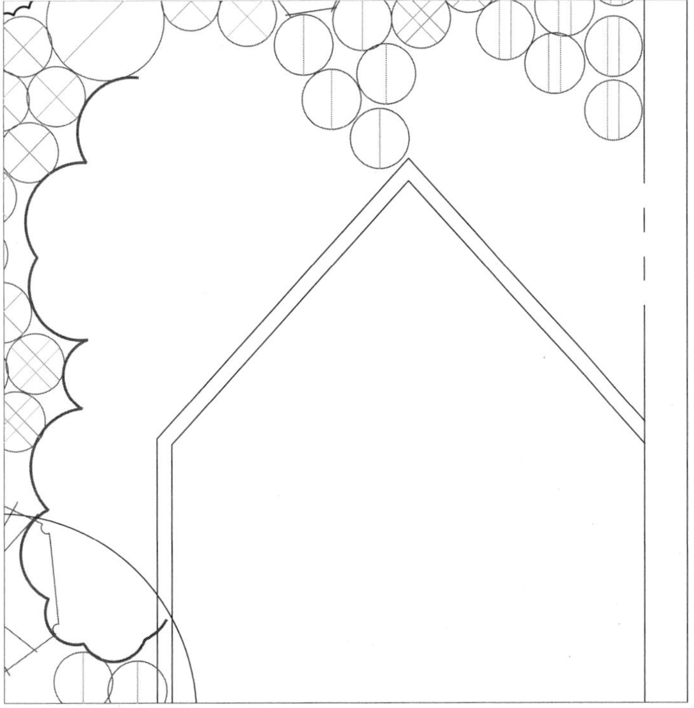

Your groundcover clouds should resemble those below (Figure 14.72).

Figure 14.72

We will edit the polylines for three different purposes. First, we can make the cloud polyline heavier (wider) using this command. Second, we can edit the various vertices (endpoints) if the polyline overlaps too far onto your existing plants or needs other editing. Third, we must formally **Close** the polylines so that AutoCAD will calculate their enclosed areas for later use. The **PEDIT** function can be used to edit, widen and close a polyline.

To edit the polyline, at the prompt:

`Command:`
Type **PEDIT** (or **PE**).

`Select polyline:`
Pick the polyline you wish to edit.

`Close/Join/Width/Edit vertex/Fit/Spline/Decurve/Ltype gen/Undo/eXit <X>:`
Type **E** (for Edit vertex). At this point, a white "X" appears over one of the vertices of the polyline.

`Next/Previous/Break/Insert/Move/Regen/Straighten/Tangent/Width/`
`eXit <N>:`
Type either **N** (for Next) or **P** (for Previous) until the "X" appears over the vertex you wish to move. Once the "X" is properly located, type **M** to move it.

`Enter new location:`
Pick a new location for the vertex and watch the polyline rebuild according to the new location of the point. If the point is still not located properly, simply type **M** again and relocate it. To move to another point, type **N** or **P** as required. Type **X** to leave the **Edit vertex** portion of **PEDIT**.

`Close/Join/Width/Edit vertex/Fit/Spline/Decurve/Ltype`
`gen/Undo/eXit <X>:`
Type **X** to exit the **PEDIT** command.

To widen the polyline, at the prompt:

`Command:`
Type **PEDIT** (or **PE**).

`Select polyline:`
Pick the polyline you wish to edit.

`Close/Join/Width/Edit vertex/Fit/Spline/Decurve/Ltype`
`gen/Undo/eXit <X>:`
Type **W** (for Width) to widen the polyline.

`Enter new width for all segments:`
Type **1** for a 1" polyline.

`Close/Join/Width/Edit vertex/Fit/Spline/Decurve/Ltype`
`gen/Undo/eXit <X>:`
Type **X** to exit the command.

Note that the **Width** option in this portion of the **PEDIT** command allows you to widen all of the segments at one time. The **Width** option in the **Edit vertex** portion of the command allows you to edit the width of each individual segment independently of the other segments.

To close a polyline:

`Command:`
Type **PEDIT** (or **PE**).

```
Select polyline:
```
Pick the polyline you wish to edit.

```
Close/Join/Width/Edit vertex/Fit/Spline/Decurve/Ltype
gen/Undo/eXit <X>:
```
Type **CLOSE**. This will cause the polyline to flicker on the screen, but should not change its appearance. This process simply causes AutoCAD to re-examine the polyline and to calculate the perimeter and area within its borders.

Hatching the Groundcover Areas

We will use the **BHATCH** command to fill in the groundcover areas with a hatch pattern. **BHATCH** stands for "boundary hatch", and provides an easy method for filling in areas defined by a distinct border line.

At the prompt:

```
Command:
```
Type **BHATCH** (or **BH**). This will open an AutoCAD "Boundary Hatch" dialogue box (Figure 14.73).

Figure 14.73

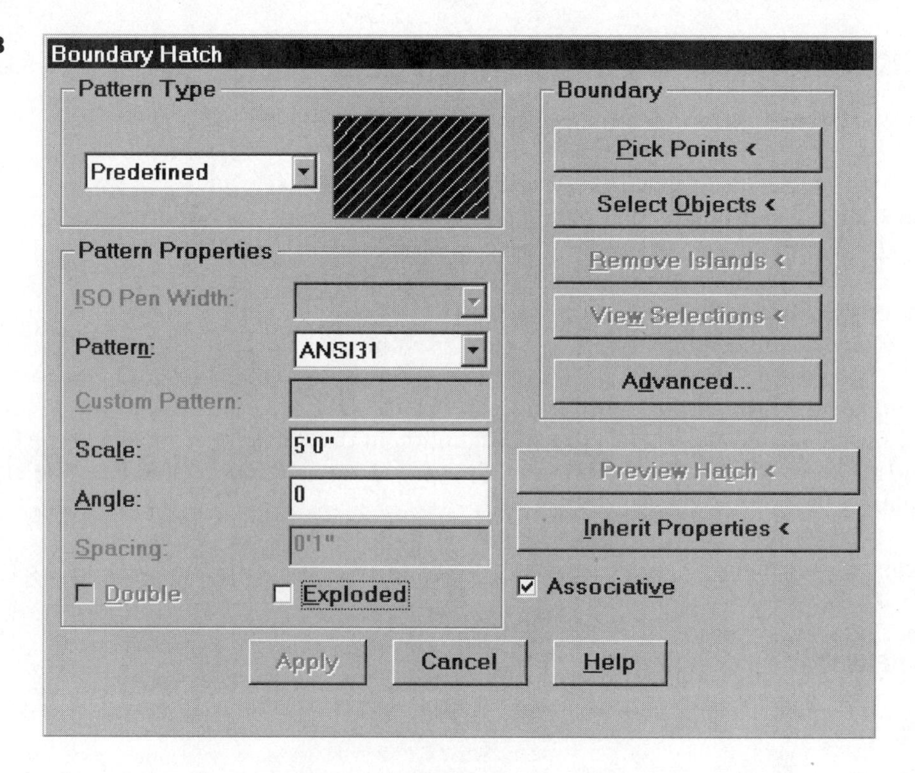

In the **Pattern:** box, accept the default value for **ANSI31** which is a regular diagonal hatch pattern. In the **Scale:** box, type in **5'** to create a pattern which is the correct size.

To select a boundary, pick **Select Objects <**. This will return you to the editing screen. Pick the groundcover cloud on the left side. If this is an enclosed polyline, the "Boundary Hatch" dialogue box returns to the screen. (If the polyline is not enclosed, a warning box will be displayed.)

Once the dialogue box returns, the "grayed out" options become available.

Pick **Preview Hatch <**. This option returns you to the editing screen with the hatch filled in and a "Continue" box in the center (Figure 14.74).

If the hatch looks satisfactory, select continue, then choose **Apply** at the bottom of the dialogue box.

If parts of the hatch require editing, make the necessary changes in the dialogue box. If the hatch pattern is **EXTREMELY DENSE**, **CANCEL IMMEDIATELY**. Using an extremely dense hatch pattern may cause your computer to crash, and result in lost work.

Figure 14.74

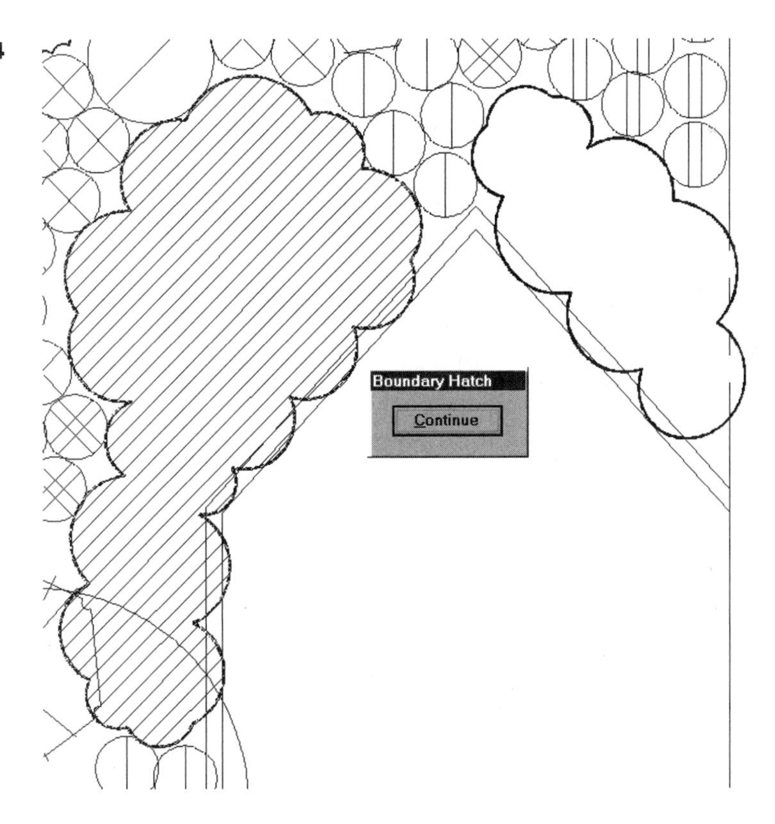

Follow the same procedure to create the hatch pattern in the right side groundcover cloud. Use the hatch pattern **ANSI37**, retaining the scale of **5'**. The results should resemble those below (Figure 14.75).

Labeling Plant Symbols

After inserting the trees, shrubs and groundcover areas in the front yard, we now need to place labels on the plants so that the plant names are properly displayed with the symbols. We will use LANDCADD's automated labeling routines which help us create labels by reading the **attributes** of each symbol we choose to label.

Remember that the attributes you entered are invisible on the screen. (If you actually want to display them, you can type **ATTDISP** then type **On**. This will make **all** of the attributes visible. Type **ATTDISP** then select **Normal** to make the plant attributes invisible again.) (Figures 14.76 and 14.77).

Using the Planting pull-down menu, we find the option Label Symbol. Choose this option to open the dialogue box shown below (Figure 14.78).

Figure 14.75

Figure 14.76

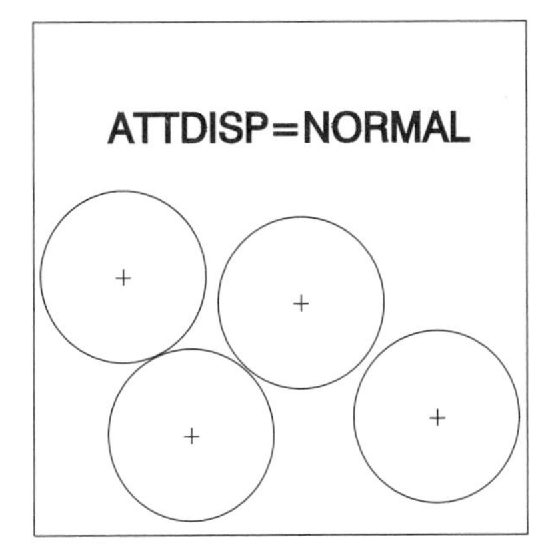

Figure 14.77

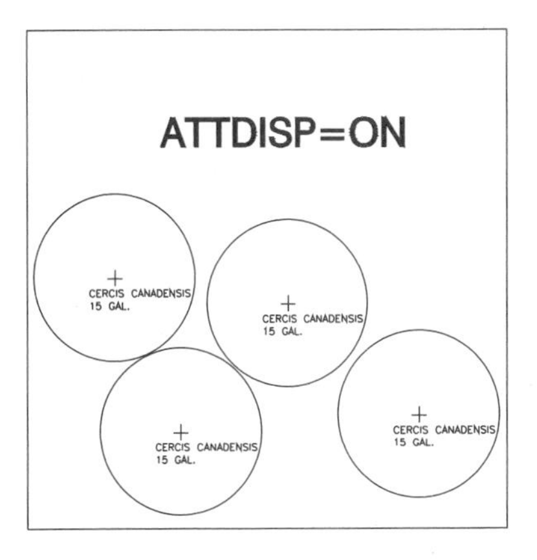

Figure 14.78

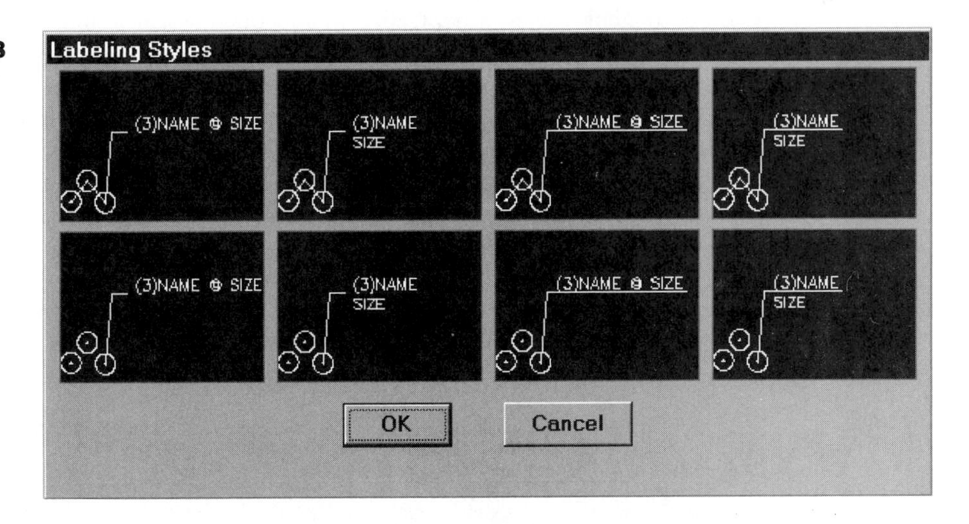

This dialogue box allows us to choose the type of label that will be created. The top row consists of options which draw connecting lines between symbols to indicate the plant groupings. The bottom row of choices omits these connecting lines. The type of style chosen depends on the drawing style of the designer and/or the design office. In addition, the number and complexity of the plant symbols may influence this choice. In our case, we will create labels which omit the connecting lines. This makes the symbols easier to read and simplifies the drawing.

Choose the label style at the lower left corner of the icons.

The prompt tells us:

```
Command: _EPLD_plant_LABEL
Select plants:
Select objects:
```

Pick all five (5) of the symbols corresponding to **Pittosporum tenuifolium**. Press **<Enter>** after the final plant is chosen. This opens the "Apply Plant Label" dialogue box (Figure 14.79).

This dialogue box reports the number of plants that we have chosen, the name of the plant (we specified scientific names earlier), and the planting size of the plants chosen. In the **Output Options** box, we can select whether or not to show the planting size, and how the leader lines will be constructed. In addition, we can choose to use the alias (code) rather than the plant name on the label.

Choose the **Name Only** option, which "grays out" the **Separator** and **Size** boxes above. Use the toggle boxes to **Turn Ortho on** and **Include leader lines**.

We may also to place an "X" in the toggle box marked **Retain this label style as default**. If this box is marked, the "Labeling Styles" dialogue box will no longer appear when we pick **Label Symbols**.

Choose this option.

Figure 14.79

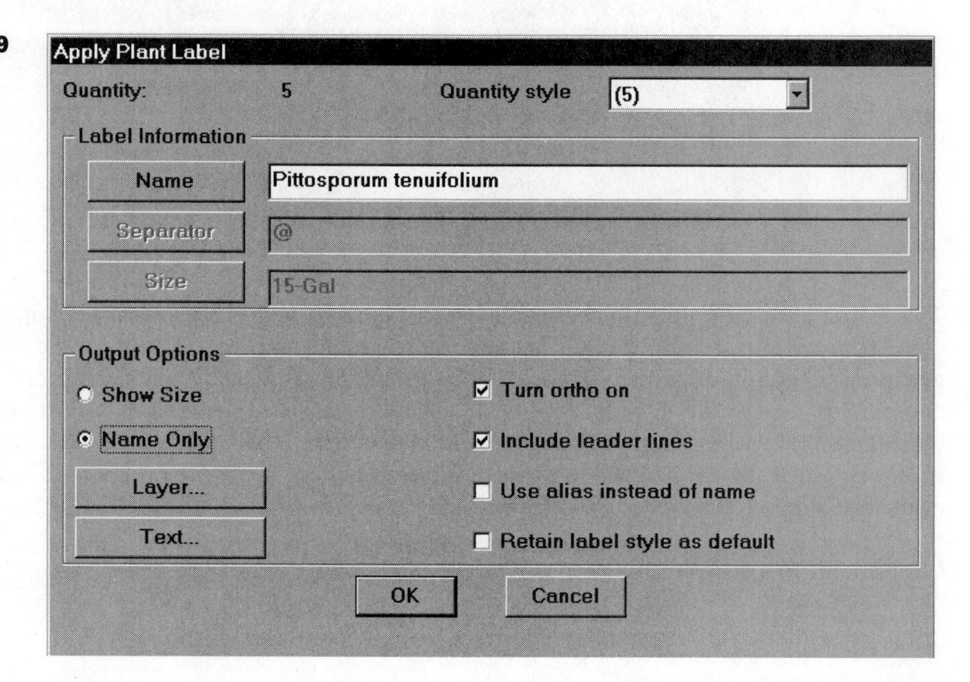

The pop-down box marked **Quantity style** indicates the punctuation used to separate the quantity of plants from the plant name. The second selection (without parentheses or a dash) creates the style used in this drawing. Make this selection. When you finish, the dialogue box should resemble that below (Figure 14.80).

In the **Layer . . .** box, we can select which layer we wish to use for plant labels. The default layer for labels is **PL_PLAN**. The **Text . . .** box brings up a dialogue box for setting the text height (Figure 14.81).

Figure 14.80

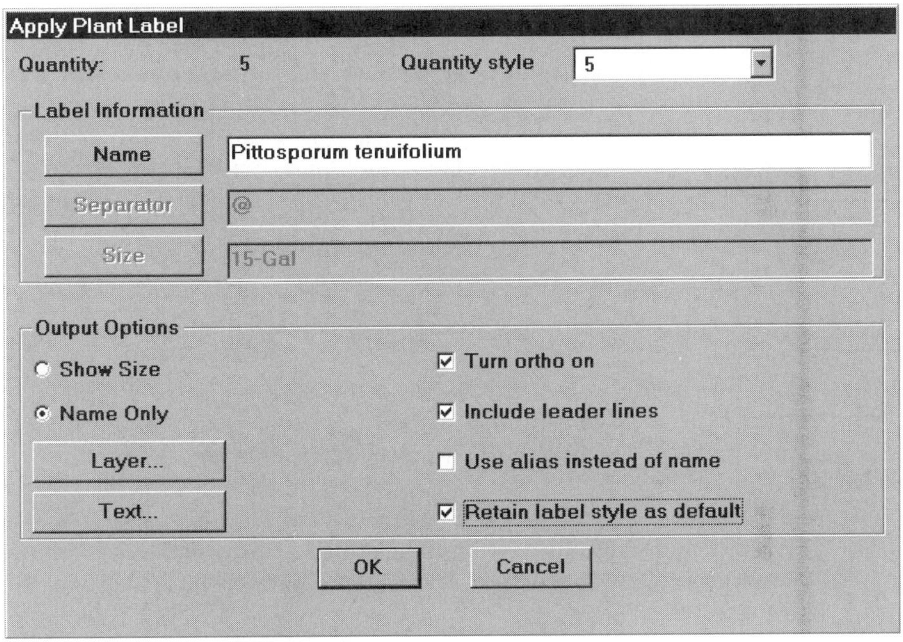

Figure 14.81

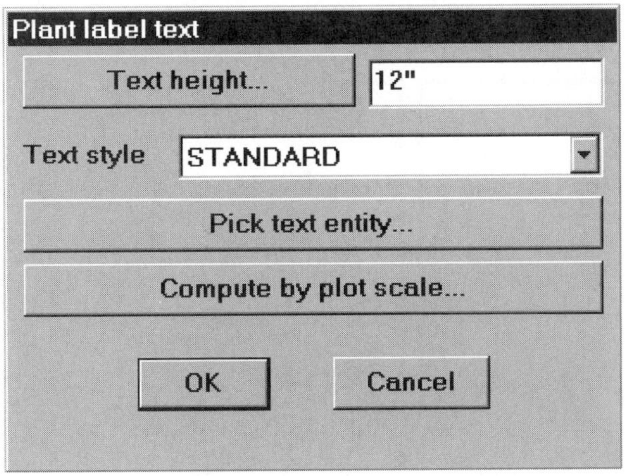

We can either enter the desired height directly in the **Text height . . .** text box or use other methods to set it. The **Pick text entity . . .** option returns us to the drawing screen to select another piece of text with the desired height. The **Compute by plot scale . . .** box permits us to set the height of the text based on the desired height on paper after the drawing is plotted.

If you loaded the proper settings with the Project Manager when the Planting Plan was opened, the default value will always be **12"** (this corresponds to 1/8" lettering on a 1/8" =1.0 ft. scale drawing).

If **12"** is not the default, use the **Text height . . .** text box to enter **12"**. Choose **OK** when finished.

This returns us to the "Apply Plant Label" dialogue box. Pick **OK**.

The prompt then reads:

```
Select start point for leader line:
```

Use the cursor menu to pick the **Insert** point of the tree at the top right corner of the property.

```
Select insertion point label:
```

Move the cursor up and to the right to create an angled leader line.

```
Select point for end of label line:
```

Be sure that **SNAP** is turned on, then stretch a horizontal line slightly to the right, then press **<Pick>**. The label is then placed to the right of the leader line. See example below (Figure 14.82).

Figure 14.82

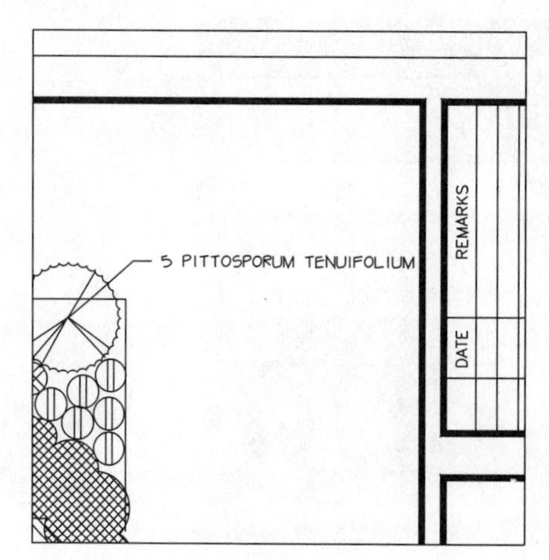

Using **SNAP** will help you keep all of your labels aligned neatly. If your label overlaps onto the title block or is too close to the property line, you can move the label (and horizontal label line) as needed. Remember to reconnect the leader line with the label line using the **STRETCH** or **CHANGE** command. Continue to label the other plants as they are shown in the illustration.

The labels of plants such as **Pittosporum crassifolium "Compactum"** are difficult since they are so long.

In order to break the text into separate lines, **COPY** the label twice directly below the original. Then use the **DDEDIT** (**DD**) command to delete the unwanted text in the dialogue box. When finished, move the words into vertical alignment. Use this procedure to create all of the labels which occupy more than one line (Figures 14.83, 14.84, 14.85, 14.86, 14.87, and 14.88).

Figure 14.83

Figure 14.84

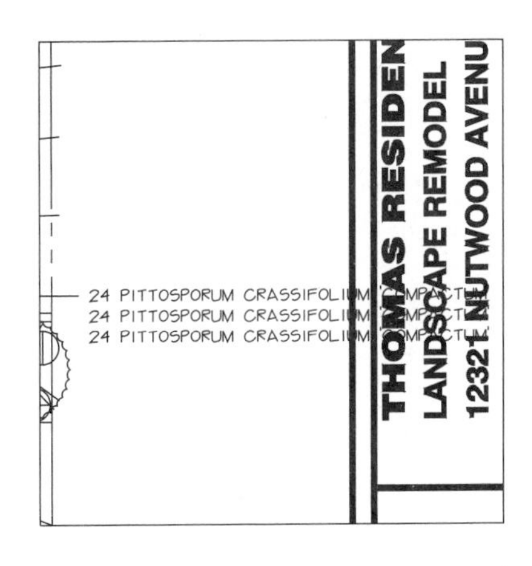

Figure 14.85

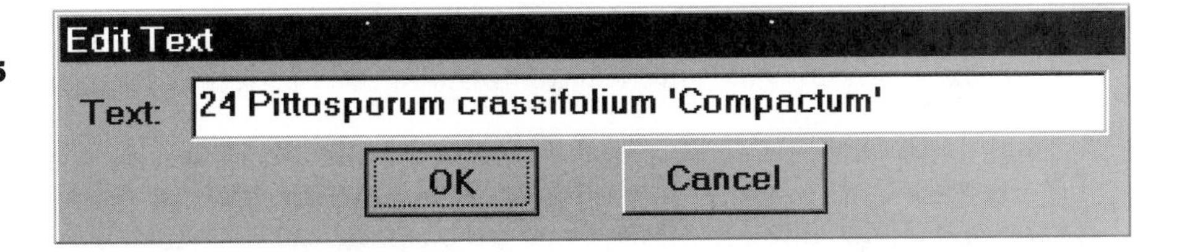

Figure 14.86

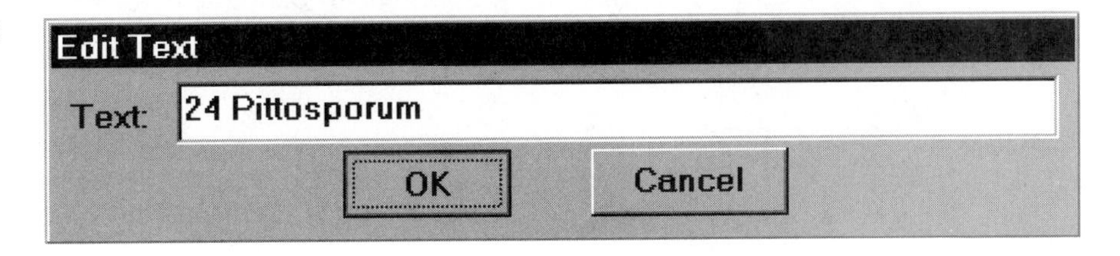

Figure 14.87

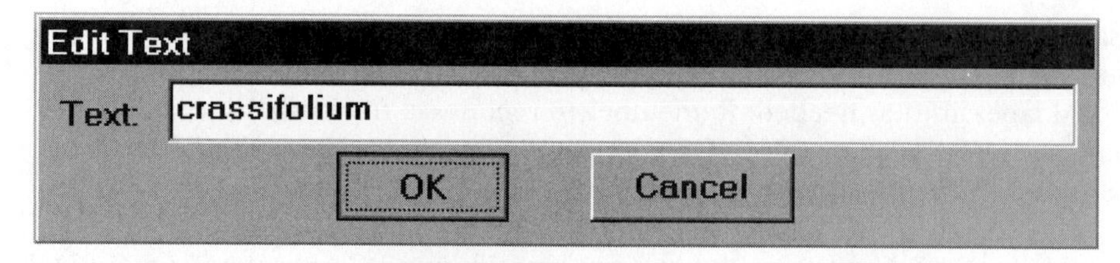

Figure 14.88

For optimum appearance of your drawing, keep the spacing between the labels as uniform as possible. Keep the horizontal label lines at the same length, using **SNAP** as necessary. In addition, try to keep the leader lines from crossing over plant symbols and other objects. Good drafting technique improves the clarity and legibility of your drawing.

Labeling the Existing Alder

The existing Alder tree can be labeled in a similar manner as the new trees and shrubs, with one exception: we will include a planting "size". We will label the Alder tree as "existing" on the plan.

Since we set the previous label style as the default, the "Labeling Styles" dialogue box does not automatically open to permit us to change the style. Instead, we must create a dummy label in order open the "Apply Plant Label" dialogue box. Here, we can turn off the **Retain label style as default** toggle box. This dummy label will have the old style, but can easily be erased later.

Use the **Label Symbol** routine to select the **Alnus rhombifolia** symbol as before.

In the "Apply Plant Label" dialogue box, toggle off the **Retain label style as default** box. Pick OK.

Place the dummy label in the house. Erase it along with the leader and label lines.

Pick Label Symbol again, selecting the **Alnus rhombifolia** again. This will open the "Labeling Styles" dialogue box. Select the second style on the bottom row, where the size is directly under the number and name. Pick OK.

In the "Apply Plant Label" dialogue box, pick the **Show Size** radio button. The Size box should read **EXISTING**. (The Separator box remains "grayed out" since the name and size are on different lines and there is no need to separate the name from the plant size.) Select OK to place the label. The "EXISTING" size should appear below the plant name.

Labeling the Groundcover Areas

In order to label a groundcover area, we use the Planting pull-down menu to gain access to the Plant Mix option.

The Plant Mix command is used to calculate the number of plants required for a specified area after you specify a spacing distance and pattern. After the number of plants is calculated, the command will also generate a plant label to insert into the drawing.

As the name implies, the Plant Mix command can be used to label areas with a single species, or with a mixture of plants, each with its own spacing. These separate mixes can be stored in a file so that they can be retrieved in other drawings.

Hint: The Plant Mix file (called PLANTMIX.MIX) is a simple text file that can contain hundreds of plant mixes in it. The best way to use this feature is to store only a few dozen plant mixes in one file, and create multiple mix files for use in a variety of projects (Figure 14.89).

In the **Plant Mixes** box, we have the opportunity to create, edit, copy and delete mixes from the Plant Mix file. In addition, we can retrieve or save mix files back to the program.

Pick New, then examine the "Planting Mix Editor" dialogue box (Figure 14.90).

This box simply reports the **Plant Mix Name** and the data regarding the **Plant Mix**. Plants listed in the **Plant Mix** box can be edited or deleted.

In your blank dialogue box, move the cursor to the **Plant Mix Name** text box and type **Gazania**.

Figure 14.89

Planting Mix Command
Plant Mixes
New
Copy
Edit
Delete
Mix file
e:\r13\win\LANDCADD\plantmix.mix
Save mix file
Planting Area
New Area
Area Add
Area Sub
Planting Area:
0"
Plant Materials in Current Mix
Label Options
Label
Exit

Figure 14.90

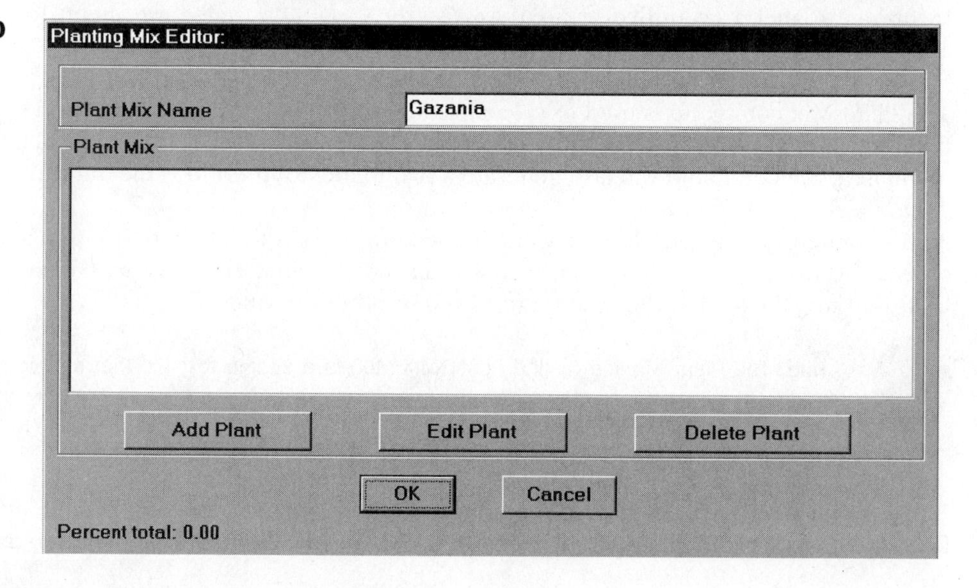

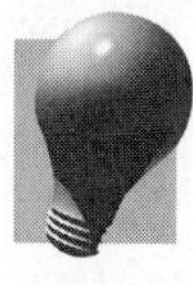

Hint: When naming plant mixes, use the genus name of one of the plants in the mix. If you decide to turn off the visible attributes when labeling the plant mix, the mix name becomes a visible block (the attributes are attached to the word on the screen). The mix name must remain visible on the screen to keep the attributes active in the drawing database. If the mix name is one of the plant names, you can simply build a label around it.

Pick the **Add Plant** button to open the "Plant Editor" dialogue box (Figure 14.91.)

Figure 14.91

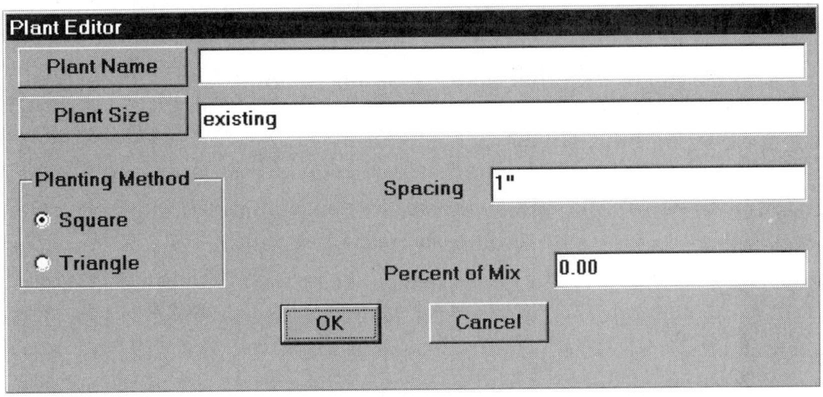

Plant Editor
Plant Name
Plant Size existing
Planting Method
Square
Triangle
Spacing 1"
Percent of Mix 0.00
OK Cancel

To supply the correct plant name (and to avoid having to type it), select the **Plant Name** button. This brings up the listing of plants stored in the **THOMAS.NAM** file (Figure 14.92).

Select the listing for Gazania, then pick **OK**.

This returns us to the "Plant Editor" dialogue box with the correct plant name filled in, but the **Plant Size**, **Planting Method**, **Spacing** and **Percent of Mix** entries must still be completed to place the correct information into the "Planting Mix Editor" dialogue box. Since this groundcover area will be exclusively a clumping style, Gazania, enter the following data:

Figure 14.92

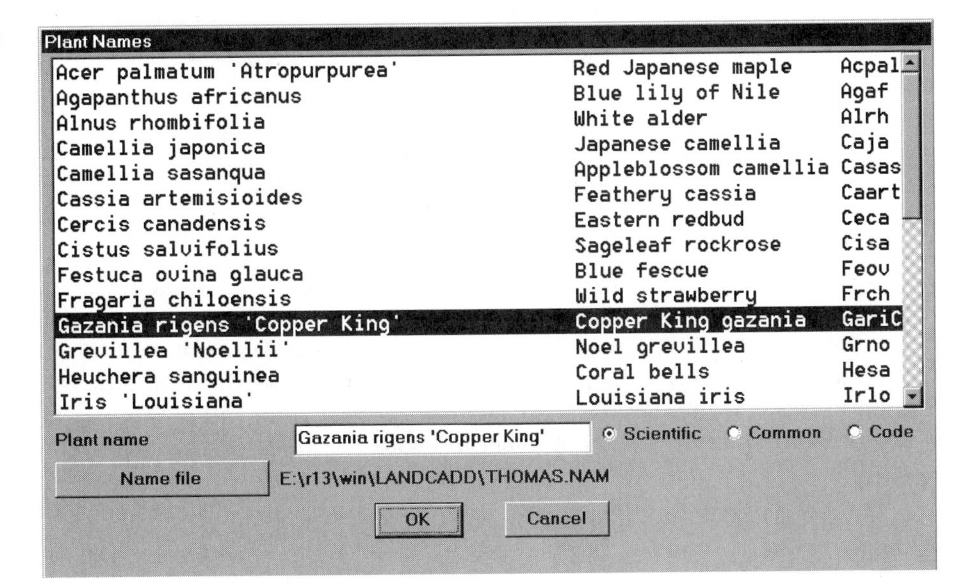

Pick **Plant Name** and find the **Gazania rigens "Copper King"** from the plant name file.

Pick **Plant Size** to retrieve **Flats.**

Pick the **Triangle** radio button for **Planting Method**.

In the **Spacing** box, type **6"**.

In the **Percent of Mix** box, type **100**. (When multiple plants are used in a mix, the percentages must add up to 100 for the mix to be valid.)

Pick **OK**. This is appropriate information for our Gazania area. This returns us to the "Planting Mix Editor" dialogue box with the **Plant Mix** box filled in. If your entries look correct, pick **OK** (Figure 14.93).

Figure 14.93

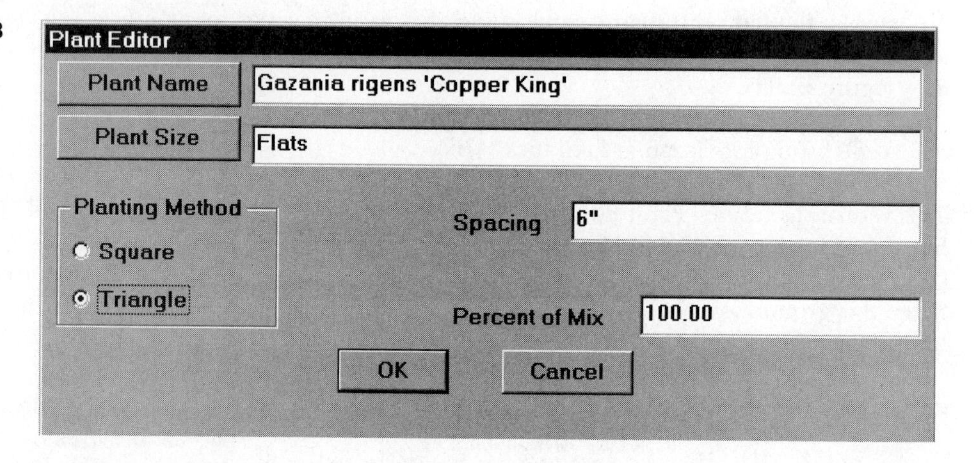

We are then returned to the "Planting Mix Editor" dialogue box, with our information included in the Gazania line (Figure 14.94).

Pick **OK** to return to the "Planting Mix Command" dialogue box (Figure 14.95).

Once the planting mix data has been completed, we can then select the **Planting Area**. In this section, we can select new areas, add other areas or subtract areas from the original area. This feature takes advantage of AutoCAD's treatment of enclosed polylines and boundaries. When AutoCAD creates an enclosed polyline or boundary, it keeps track of the number of square feet (or square inches) within the area. LANDCADD uses this area to generate the number of plants required.

Pick **New Area**. This returns us to the drawing screen, with a choice to make (Figure 14.96).

Figure 14.94

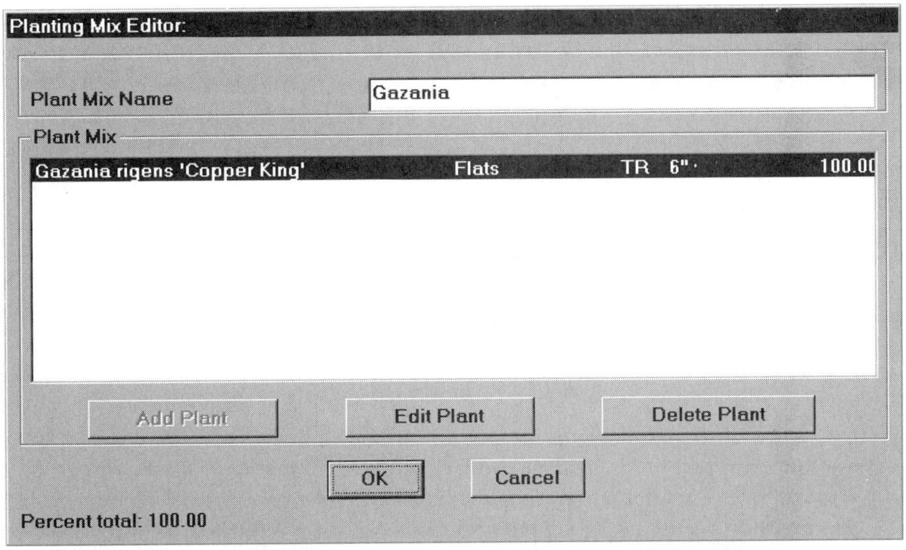

Planting Mix Editor:
Plant Mix Name
Gazania
Plant Mix
Gazania rigens 'Copper King' Flats TR 6" 100.00
Add Plant
Edit Plant
Delete Plant
OK
Cancel
Percent total: 100.00

Figure 14.95

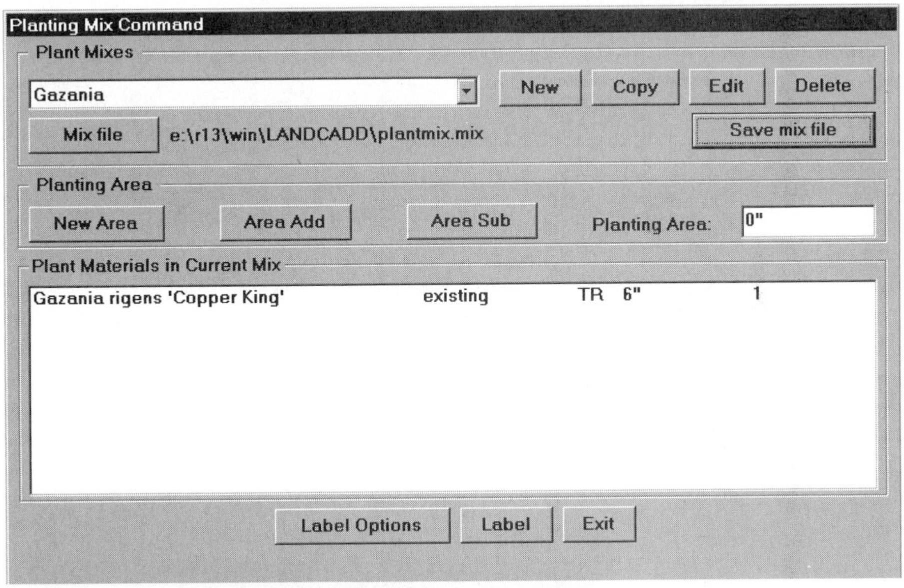

Planting Mix Command
Plant Mixes
Gazania
New
Copy
Edit
Delete
Mix file
e:\r13\win\LANDCADD\plantmix.mix
Save mix file
Planting Area
New Area
Area Add
Area Sub
Planting Area:
0"
Plant Materials in Current Mix
Gazania rigens 'Copper King' existing TR 6" 1
Label Options
Label
Exit

Figure 14.96

Select Area By:
Closed Polyline Method
AutoCAD BPoly Method
Cancel

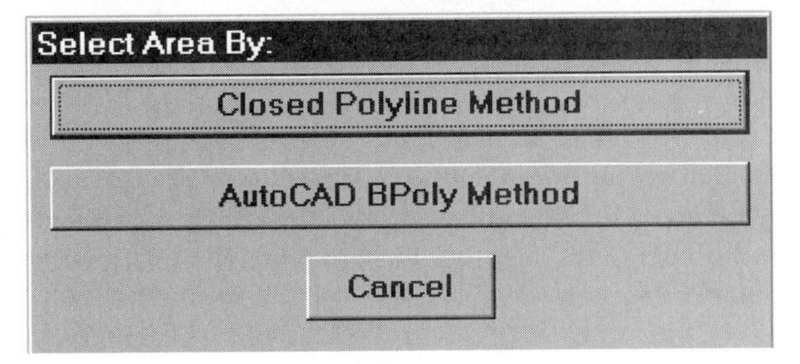

Which option we choose depends on the characteristics of the groundcover area. If the area is fairly complex, but is surrounded by a closed polyline, pick **Closed Polyline Method**. If the area is fairly simple, or is not surrounded by a closed polyline, choose **AutoCAD Bpoly Method**. This method uses the same part of AutoCAD which automatically locates the boundary when placing a hatch inside an area. More complicated areas can cause problems with this option, so be ready to draw a new polyline around an area if necessary.

The groundcover clouds we drew previously can be chosen with the **Closed Polyline Method**. (We know they are closed since we closed them earlier.)

At the prompt:

```
Select POLYLINE:
```

Pick the polyline cloud in the left side of the screen. You may need to zoom in close to the cloud to pick an edge successfully. (The prompt will inform you if you do not pick an enclosed polyline.) This returns us to the "Plant Mix Command" dialogue box with two new entries (Figure 14.97).

Figure 14.97

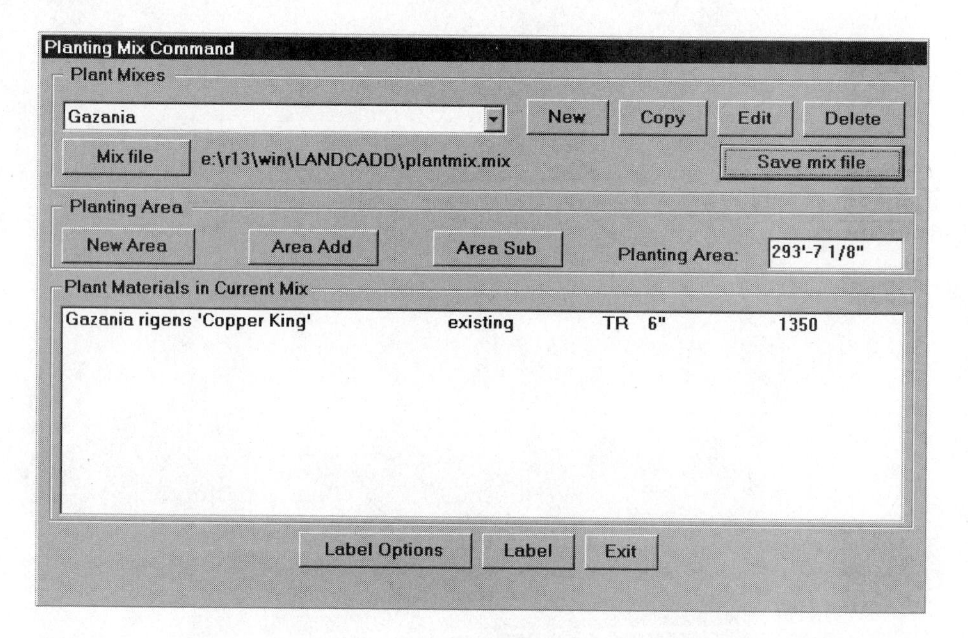

We now see an entry in the **Planting Area** box. This corresponds to the area within the closed polyline. On the far right side of the **Plant Materials in Current Mix**, we see the entry **1350**. This corresponds to the number of plants LANDCADD calculates that we will need to place the groundcover as specified.

Unfortunately, LANDCADD will report that we will need 1350 **flats** of groundcover in the planting label, so we will need to edit this later.

Once the area and number of plants is displayed in the dialogue box, we are then ready to work on the plant label.

Pick **Label Options**. This brings up yet another dialogue box, "Label Options" (Figure 14.98).

Figure 14.98

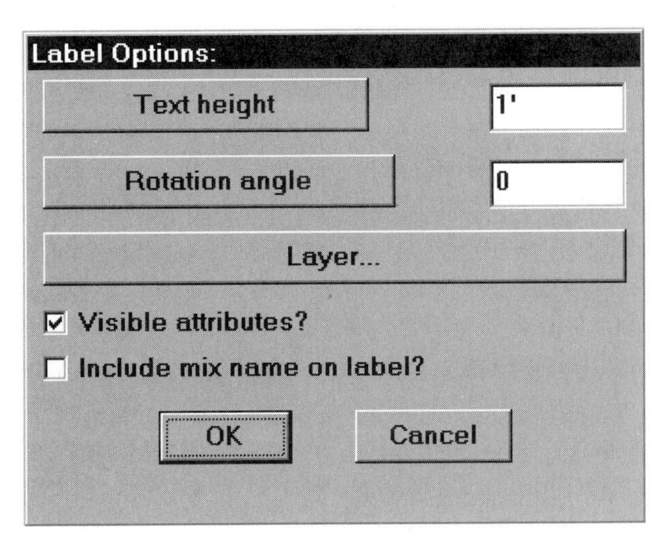

In this box, we can specify what size text to use, what rotation angle it should have, what layer to place it on, and whether the attributes should be visible. The text height, angle and layer are simple enough.

Use **12"** (or **1'**) for the text height, accept the rotation angle of **0**, and accept the default layer for plant labels **PLANT_PL**.

The toggle boxes below require more explanation. Based on the information gathered above, the plant label would look like this:

GAZANIA RIGENS "COPPER KING"

FLATS

1350

This label is actually a series of attributes shown on the screen. The plant name, planting size and quantity are all displayed at once in this stacked arrangement. The problems with this label are factual and aesthetic.

First, we don't need 1350 flats of groundcover. Since Gazania comes in flats with 64 plants each, we really need 1350 ÷ 64 flats. This works out to about 21 flats. We will edit the quantity attribute to reflect this change.

Second, this label does not resemble the other plant labels we have constructed. The other labels display the quantity first, then the plant name. The planting size is not displayed. Placing labels which look different from the others might cause confusion, and will certainly look less pleasing. Because of this, we will choose to make these attributes invisible and create our own label.

Turn off the toggle box marked **Visible attributes?**.

The toggle box marked **Include mix name on label?** is only active when the attributes are visible. If the attributes are not marked as visible, the mix name (a **block**) will automatically be displayed instead. This will allow us to place the mix name on the drawing, and simply create a label around it.

Pick **OK** to continue.

After returning to the "Plant Mix Command" dialogue box, pick **Label**.

The prompt reads:

```
Select placement point of (Gazania) label:
```

Pick a location aligned with the other plant names along the right side of the drawing. We will add other text to complete the label, and add a label and leader line as well. See illustrations below (Figures 14.99, 14.100, 14.101, and 14.102).

Figure 14.99

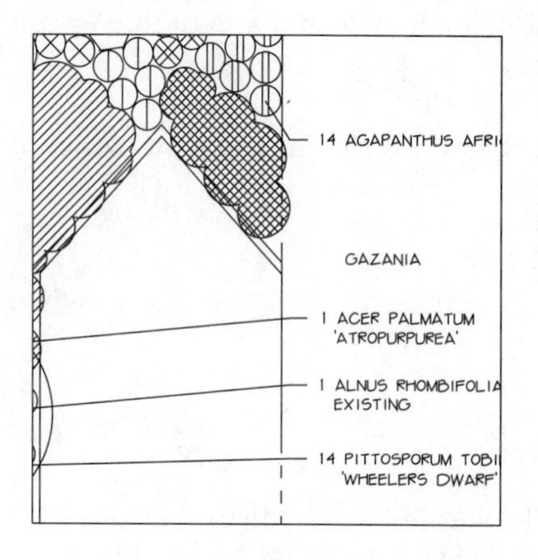

Figure 14.100

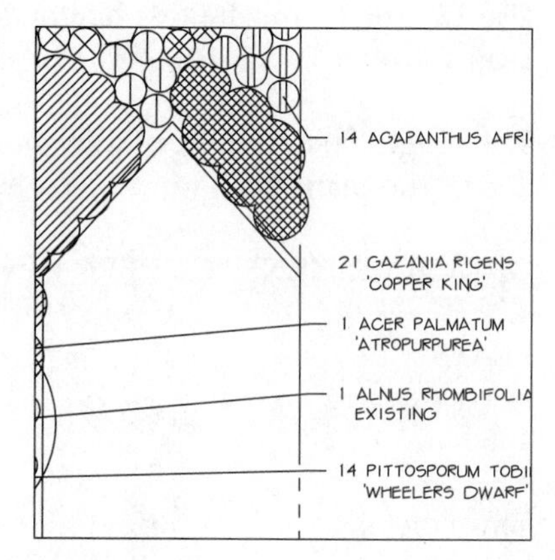

Figure 14.101

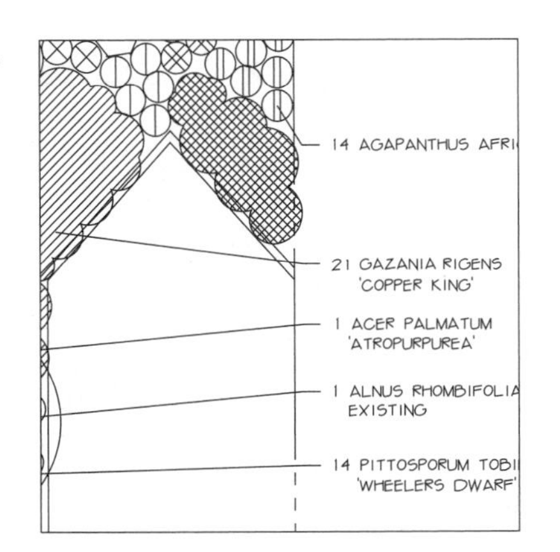

Figure 14.102

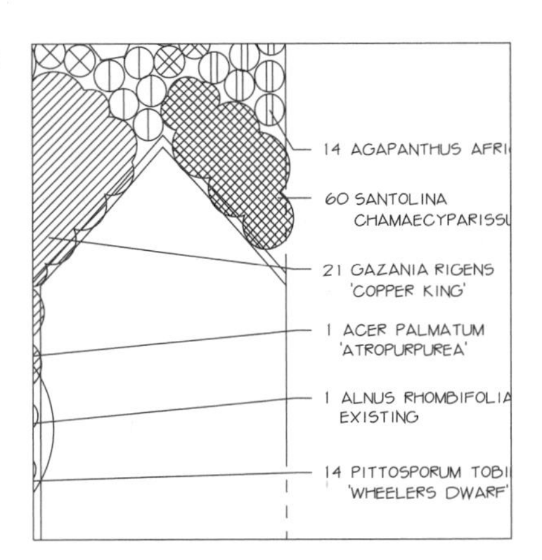

Use the **DTEXT** command to place the number and completed plant names around the **GAZANIA** label. Use **SNAP** to help align the text. Once the text is placed, simply use the **Line** command to create the angled leader line and the straight label line.

To complete the **GAZANIA** label, use the command **ATTEDIT** (**AE**) to edit the quantity attribute.

`Command:`
Type **ATTEDIT** (or **AE**).

`_.DDATTE`
`Select block:`
Pick the original label **GAZANIA**. This opens a "Edit Attributes" dialogue box. Edit the **QTY** attribute as shown (Figure 14.103).

Pick **OK** to complete your editing.

Repeat this procedure for **Santolina chamaecyparissus**. Create a new plant mix in the "Plant Mix Command" dialogue box — call it **Santolina**. Use a spacing of **18"** O.C., making up 100% of the mix. Apply the label to the groundcover cloud on the right side of the front yard. This plant is not planted from flats, so the attributes will not require editing.

Figure 14.103

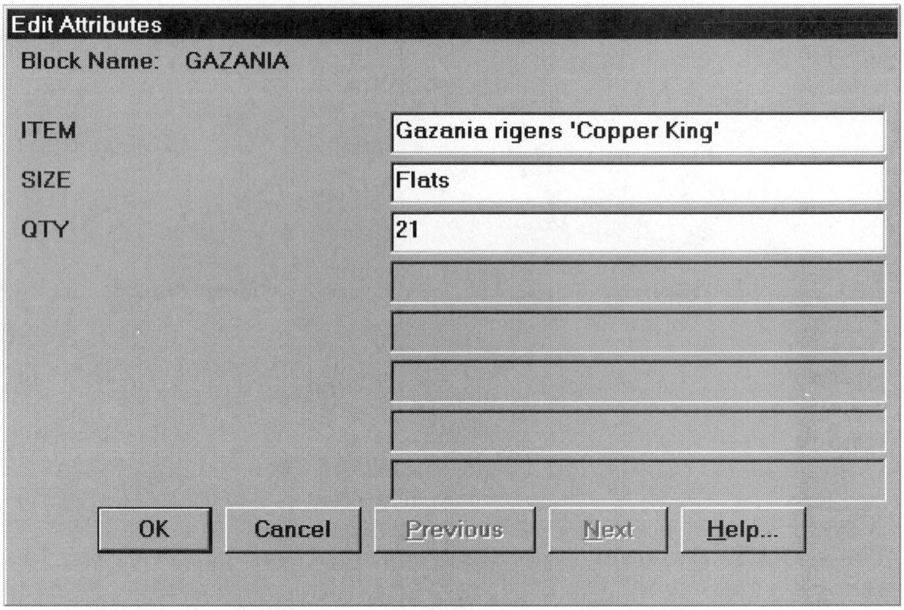

Edit Attributes

Block Name: GAZANIA

ITEM Gazania rigens 'Copper King'

SIZE Flats

QTY 21

OK Cancel Previous Next Help...

The Back Yard Planting Plan

The planting plan for the back yard incorporates many of the same techniques for inserting and labeling symbols that we used in the front yard. In addition to those techniques, we will use some new routines to insert plants in groupings along polylines. We will also define the boundary of groundcover areas with a new type of border, and emphasize planting groups with a heavy outline (Figures 14.104 through 14.118).

The following plant list shows the quantity, planting size and plant diameter, along with the installed price and general plant category. For the sake of this exercise, we will define any plant available in a pot as a tree or shrub, and any plant available in flats as a groundcover. The price and category will be helpful when we look at developing a planting plan cost estimate later in this project.

Figure 14.104 Camellia japonica

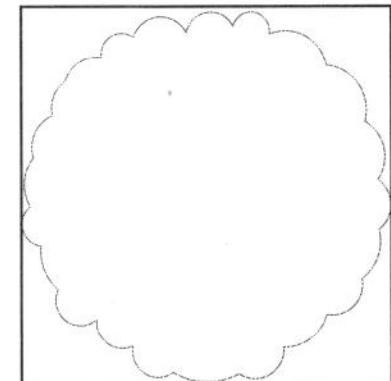

Figure 14.105 Cercis canadensis

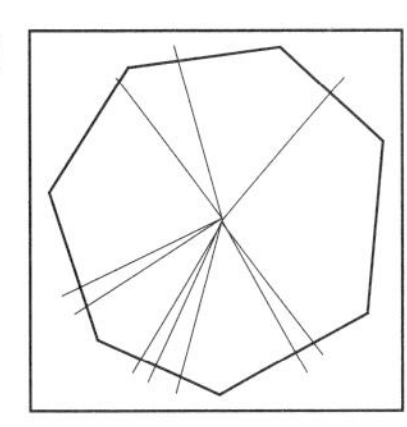

Figure 14.106 Grevillea "Noellii"

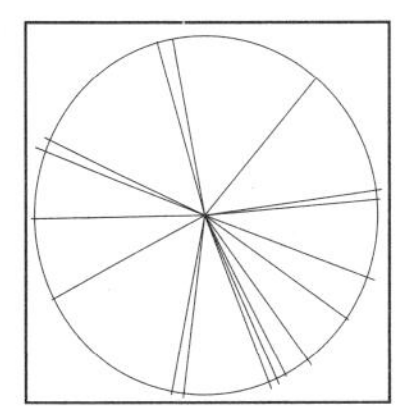

Figure 14.107 Heuchera sanguinea

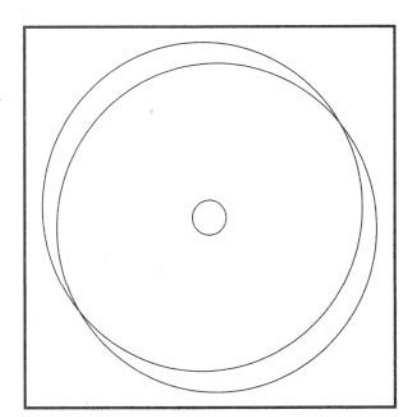

Figure 14.108 Langerstroemia indica

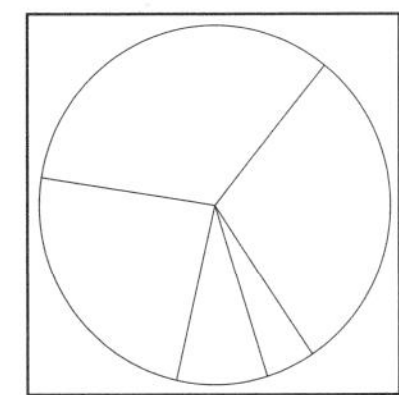

Figure 14.109 Myrtus communis "Compacta"

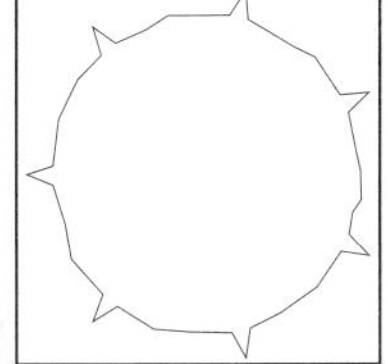

Figure 14.110 Nandia domestica "Nana"

Figure 14.111 Podocarpus macrophyllus

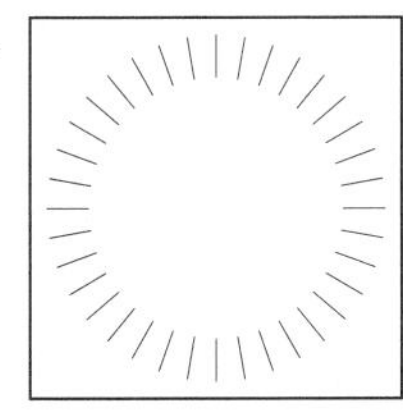

Figure 14.112 Rhaphiolepis indica

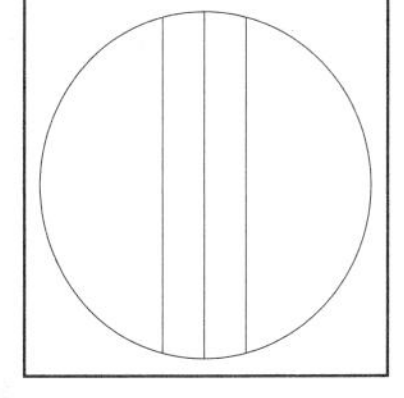

Figure 14.113 Rhododendron indicum

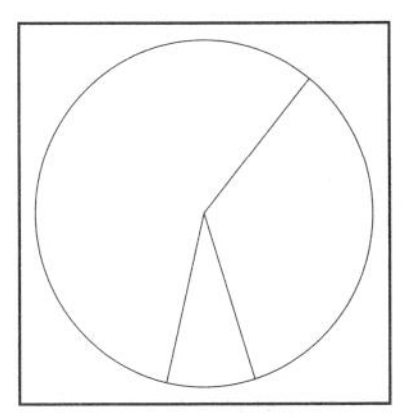

Figure 14.114 Thevetia peruviana

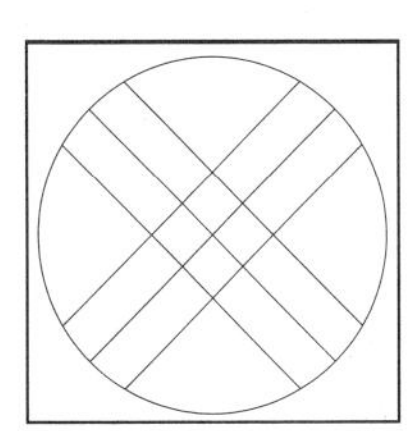

Figure 14.115 Festuca ovina "Glauca"

Figure 14.116 Fragaria chiloensis

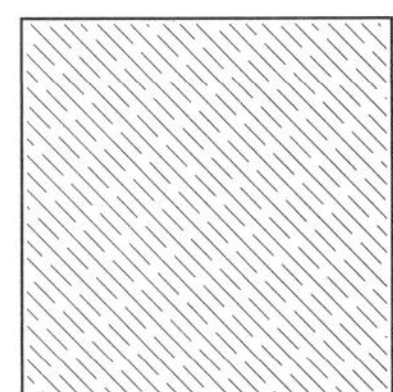

Figure 14.117 Juniperus conferta

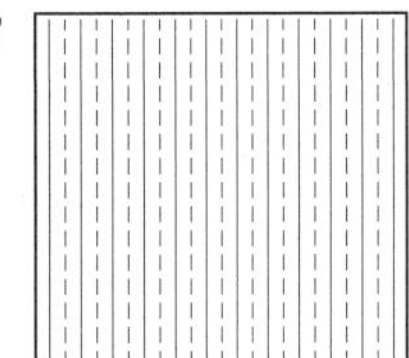

Figure 14.118 Stachys lanata

Plant Name	Qty.	Planting size	Diameter	Price	Category
Camellia japonica	5	5-Gal	6'	$20.00	Shrubs
Cercis canadensis	1	24"-Box	30'	$250.00	Trees
Festuca ovina glauca*	~156	1-Gal	18" O.C.	$6.50	Shrubs
Fragaria chiloensis*	~12	Flats	9" O.C.	$35.00	Groundcovers
Grevillea "Noellii"	13	5-Gal	5'	$17.50	Shrubs
Heuchera sanguinea	61	1-Gal	1'	$7.50	Groundcovers
Juniperus conferta*	~223	1-Gal	2' O.C.	$7.50	Shrubs
Lagerstroemia indica	1	24"-Box	20'	$250.00	Trees
Myrtus communis "Compacta"	12	5-Gal	3'	$17.50	Shrubs
Nandina domestica "Nana"	36	5-Gal	2'	$20.00	Shrubs
Podocarpus macrophyllus	10	15-Gal	8'	$75.00	Trees
Rhaphiolepis indica	11	5-Gal	6'	$20.00	Shrubs
Rhododendron indicum	15	5-Gal	4'	$20.00	Shrubs
Stachys lanata*	~83	1-Gal	18" O.C.	$7.50	Shrubs
Thevetia peruviana	6	15-Gal	10'	$75.00	Shrubs
* Groundcover and plant mix quantities will vary with the size of the area within the enclosing polyline.					

Placing the Plants

As before, begin with the largest trees first, locating each of the symbols in the approximate places shown on the drawing. Be careful that the quantity of plants used is close to the quantity indicated on the chart. This will ensure that your results will be consistent with those indicated in the **Estimating** section (Figure 14.119).

The trees are placed manually as shown here. We will place the Rhaphiolepis plantings along an irregular "hedge row" near the lower left corner of the yard. In order to use this command, we must first draw the polyline for the hedge to follow.

Figure 14.119

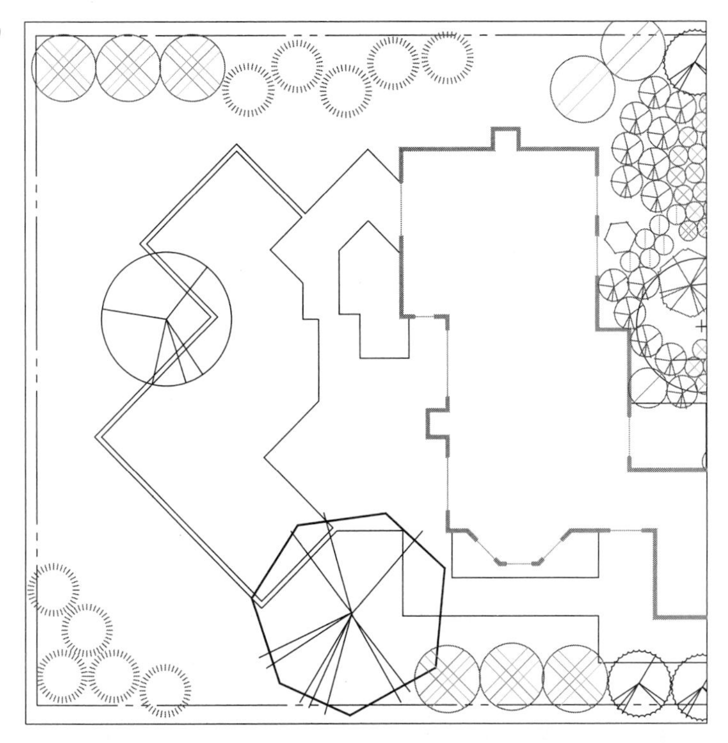

Draw in a polyline (which will be erased during the command) as shown below (Figure 14.120)

Figure 14.120

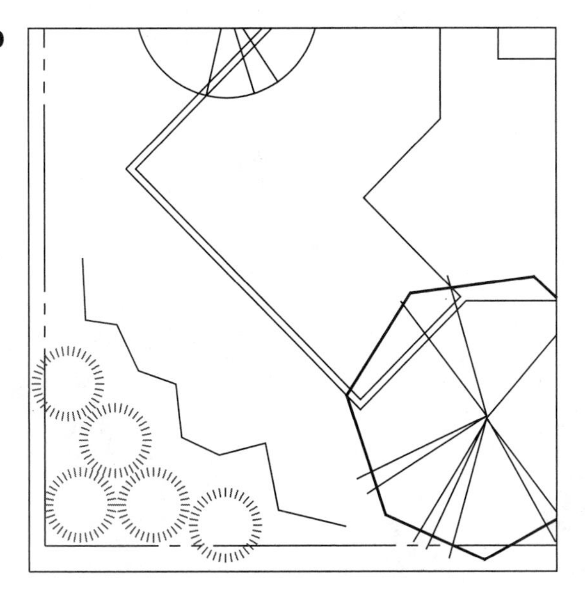

Move to the **Symbols** pull-down menu, then select **User Library**. Pick the symbol used for **Rhaphiolepis** to open the "Plant symbol settings" dialogue box (Figure 14.121).

After entering the correct settings for plant diameter, plant name and planting size, select **Hedge row**.

At the prompt:

```
Command: EPLD_plant_insert
Plant spacing distance <6'>:
```

Press **<Enter>** to accept this distance between plants. LANDCADD will always use the plant diameter as the default spacing. You can specify a larger or smaller distance if you wish.

```
Follow existing polyline <Y/n>
```

Type **Y** for yes. This tells the command to place the plants along an existing polyline rather than along one you create during the command.

```
Select defining polyline
```

Figure 14.121

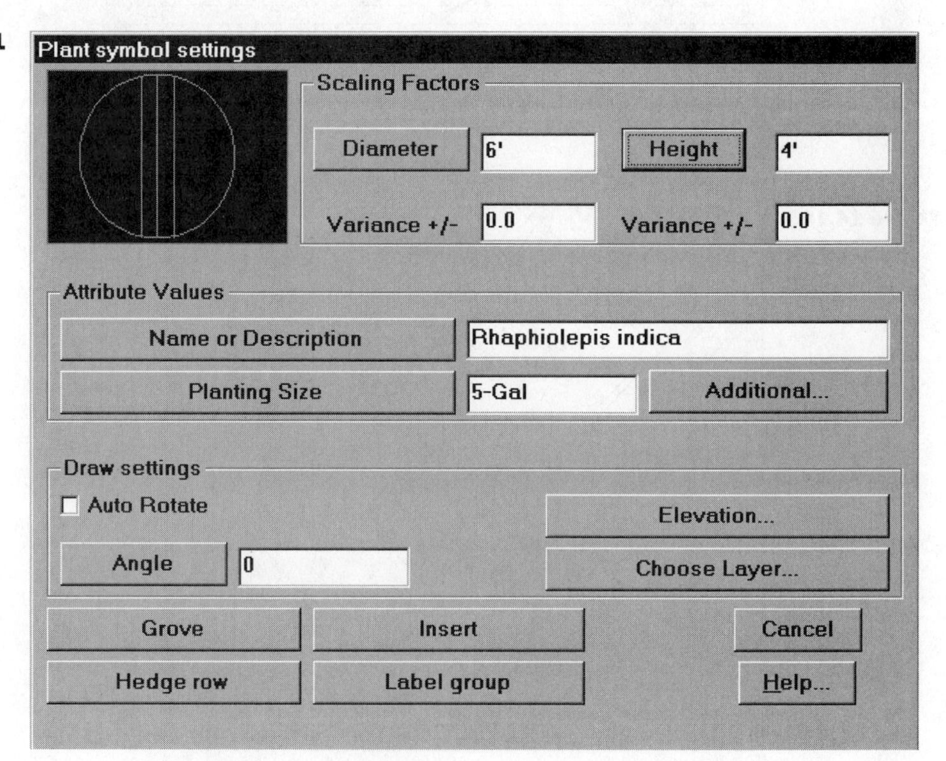

Select object:

Pick the polyline created in the lower left corner of the yard. The plants are then placed along the polyline.

Delete defining polyline? <Y/n>:

Type **Y** for yes. This eliminates the temporary polyline.

Once the plants are placed, they probably will need to be moved slightly to eliminate overlap. See below (Figures 14.122 and 14.123).

The **Hedge row** option makes it easy to place plants in groupings along regular shapes as well. We will place a more formal group of **Myrtus** toward the center of the planter along the back lawn.

Begin by placing a **polyline** arc between the two points of the lawn mow strip (Figure 14.124).

Hint: Although there are many ways to draw an arc (normal or polyline), the easiest one to use here is the three points method. Use the **Intersection** at the upper corner of the mow strip for a start point. For a second point, choose a point along the path of the arc. The end point will be the **Intersection** at the lower corner of the mow strip.

Move through the **Symbols** pull-down menu to the **Shrubs** option. Select the correct symbol for **Myrtus communis "Compacta"** and fill in the correct information (available from the back yard plant list) in the "Plant symbol settings" dialogue box. Select **Hedge row** as before, using the polyline arc as the defining polyline. See the results below (Figure 14.125).

Figure 14.122

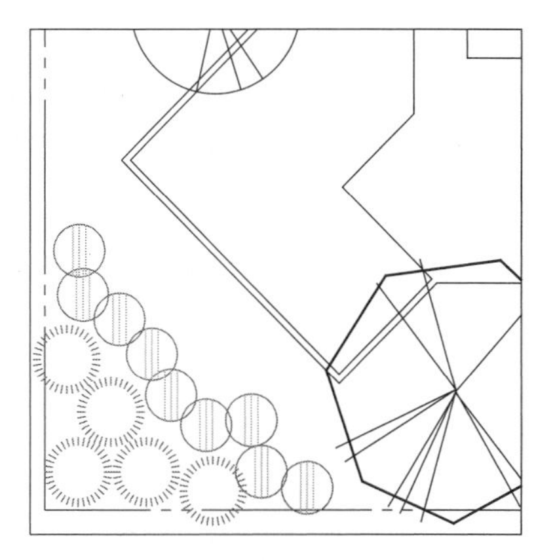

Figure 14.123

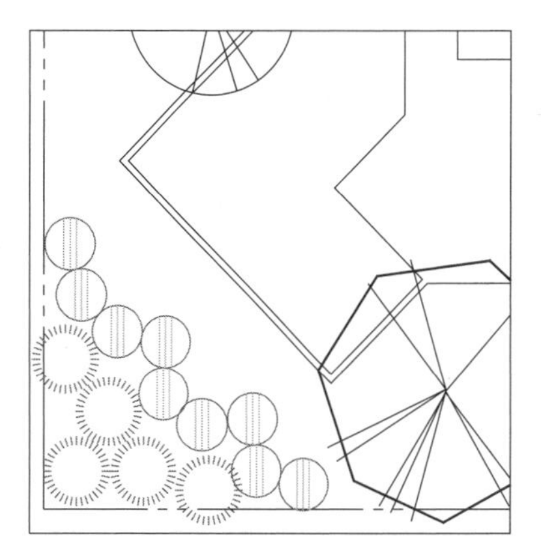

Figure 14.124

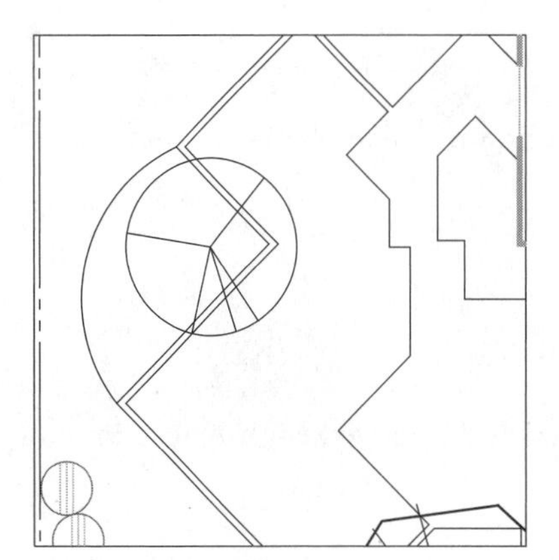

Figure 14.125

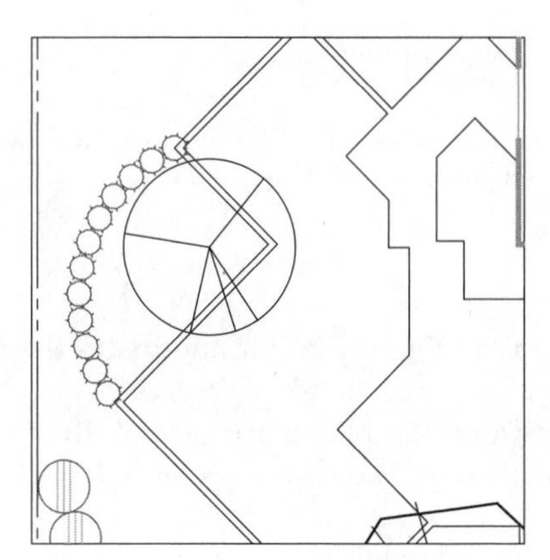

Continue to place the smaller shrubs and plantings around the back yard. Note that most of the plants are placed in groupings, and that the amount of overlap between symbols is minimized. This enhances the clarity of the drawing and makes it easier to select plants when labeling.

With other landscape design graphical styles, the plant symbols may be rotated from one location to the next, or they may vary in size from one plant to the next. LANDCADD allows these variations as options during symbol insertion, but when they are chosen, the plant symbol is not visible as plants are placed, making it much more difficult to prevent plant symbols from overlapping.

When symbols must overlap, as when shrubs are planted beneath tree canopies, try to create a visual difference between the symbols by assigning different line weights to the symbols. Most commonly, trees are given a heavier line weight, while shrubs beneath them are given a lighter one. These line weight differences can be accomplished in several ways. Probably the easiest is to assign the layers containing trees a different color. This color is then assigned a heavier line weight during the plotting process. Shrubs, groundcover and smaller trees might be assigned to a color which has a lighter line weight when plotted. These color-line weight assignments are made during the plotting process, and can be stored to a settings file which can be retrieved at any time. Most architecture offices use a set of standard colors and layers to achieve a consistent look to their drawings. See Chapters 9 and 25 for more information on producing output from your AutoCAD and LANDCADD drawings.

See the enlargement below (Figure 14.126).

Figure 14.126

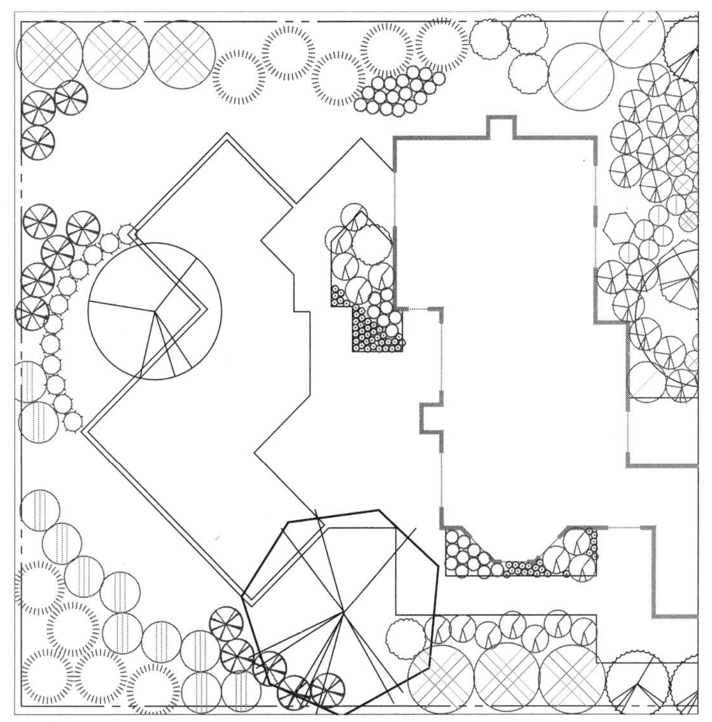

Back yard groundcover areas

The groundcover areas in the back yard are defined using two methods. First, the **Ground Cover** option of **Cloud** is used for both portions of the **Juniperus conferta** area. Use the same procedure to build the polyline border as we used for the groundcover areas in the front yard. The hatch pattern used here is called **Brass**, rotated **90°**, with a scale of **5'**. The second method uses the **Vegline . . .** command (Figure 14.127).

Vegline . . . gives you the opportunity to build a polyline border around your groundcover or hedge areas by: a) following an existing polyline; or b) creating a vegline by picking a succession of endpoints. This second technique is similar to that used by the **Hedge row** routine when a defining polyline is not used. Once the choice whether or not to follow an existing polyline is made, a vegline **option** must be chosen. You can choose **Arc**, **Circle**, **Line**, or **Hedge**, along with a **Width**. In addition to indicating a vegline option, you also specify a segment **length** to establish the scale of the vegline (Figure 14.128).

Figure 14.127

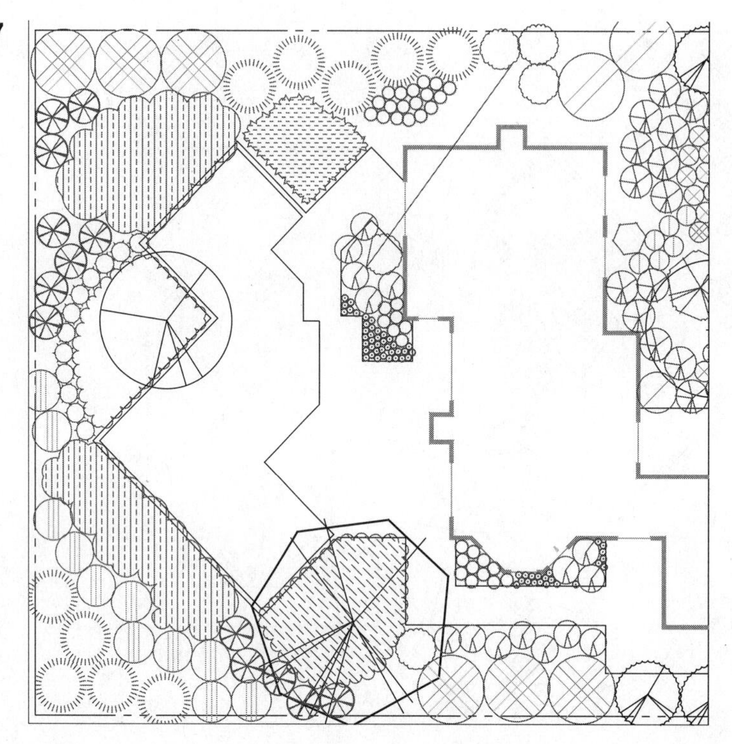

Figure 14.128

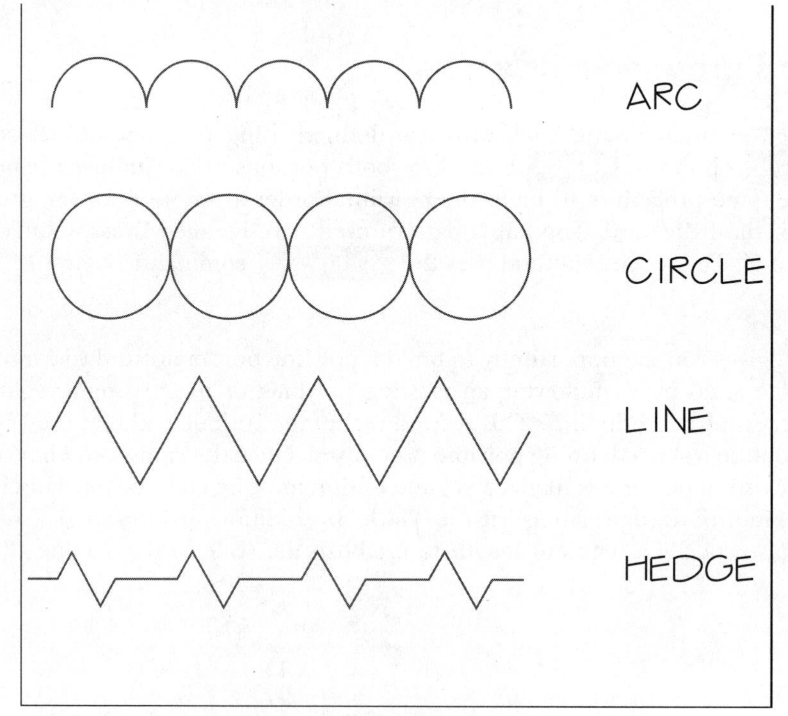

Using Vegline

Begin by using the **PLINE** command to build a polyline border for the **Fragaria** area beneath the **Cercis** tree (Figure 14.129).

Figure 14.129

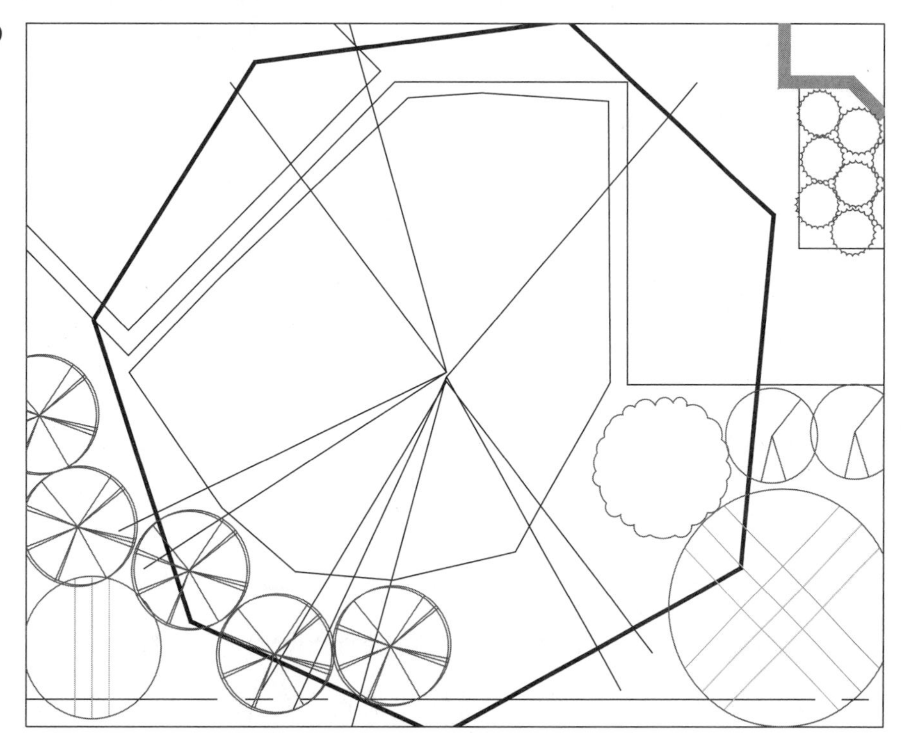

Hint: To make the arcs of the vegline extend outward from the center, be sure to draw this polyline in the **CLOCKWISE** direction. In addition, be sure to close the polyline at the end so that the vegline will close as well.

From the **Planting** pull-down menu, pick **Vegline . . .**.

At the prompt:

```
Command:
EPLD_VEGLINE
Follow existing polyline? <Yes>/No :
```

Type **Y** for yes.

```
Vegline options: Arc/Circle/Line/Hedge/Width <Arc>:
```

Type **Arc** to use this option.

```
Select polyline to turn into a vegline:
```
Pick the polyline border under the **Cercis** tree.

```
Show segment length: <1'>
```
Type **2'** to indicate this segment length. The vegline is then drawn, and the following prompt is displayed:

```
Erase original POLYLINE? Yes/<No>:
```
Type **Y** for yes. The original polyline is eliminated. See below for the results (Figure 14.130).

If the vegline is not closed (it probably will not be), it may be necessary to close it using **GRIPS** or one of the options in the **PEDIT** command.

Use the **PEDIT** command to widen the polyline to **1"**. (It is also possible to assign a width of **1"** to the vegline before it is built.) The hatch used here is **ANSI35**, rotated **90°** and placed with a scale of **5'**.

The border around the **Stachys** area is built without an existing polyline.

```
Command: EPLD_VEGLINE
```

Figure 14.130

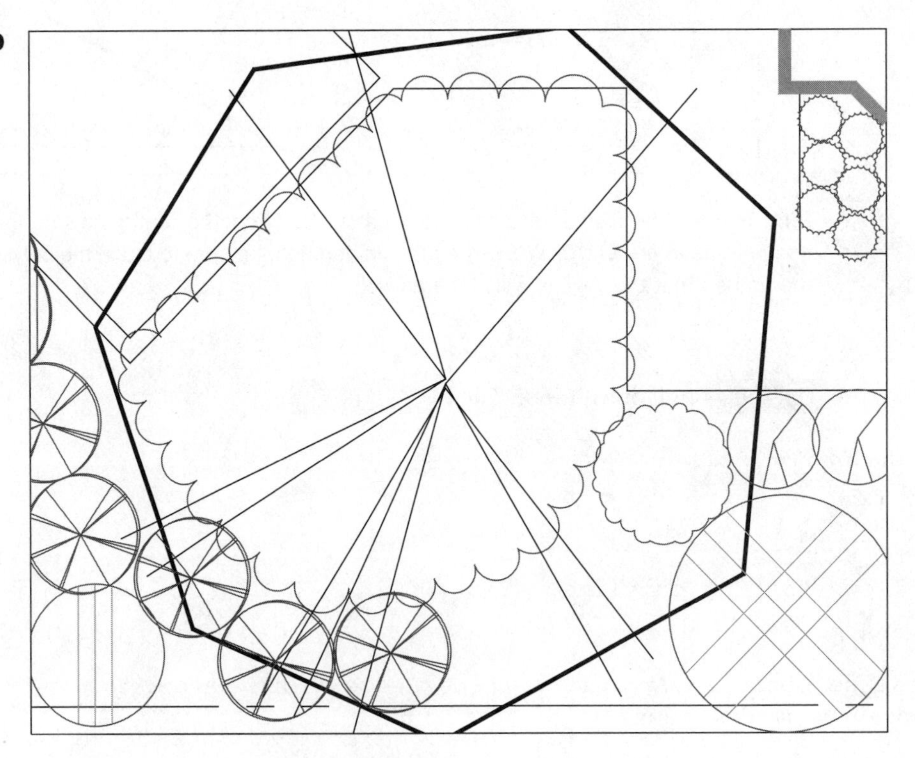

`Follow existing polyline? <Yes>/No:`
Type **N** for no.

`Vegline options: Arc/Circle/Line/Hedge/Width <Arc>:`
Type **Hedge**.

`Show first point:`
Select the first point along the edge you want to border. Choose this point at the far right corner of the groundcover area. Remember that *the edge will be drawn in a clockwise manner* from the first to the second point (Figure 14.131).

`Show second point:`
Choose the second point at the bottom corner (apex) of the groundcover area.

`Show the segment length: <2'>`
Type **<Enter>** to accept this segment length. Continue to build a border around the area shown below. Use the **Close** command to finish the edge line back at the starting point. Use **PEDIT** to change the polyline width to **1"** (Figure 14.132).

In this area, use the hatch pattern **Dash** with a **0°** rotation and a scale of **5'**.

The final area under the **Lagerstroemia** tree is built in a similar manner.

Figure 14.131

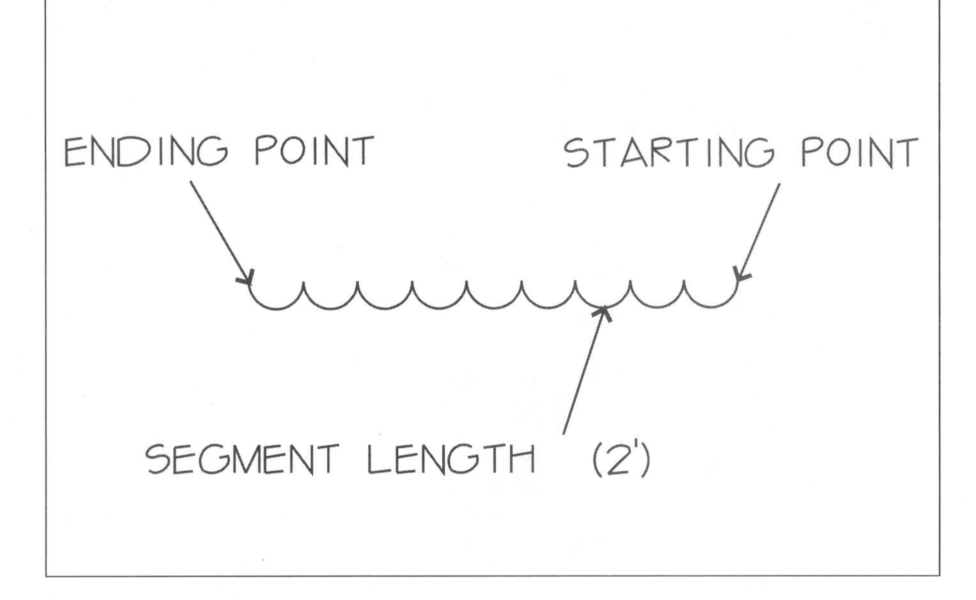

Figure 14.132

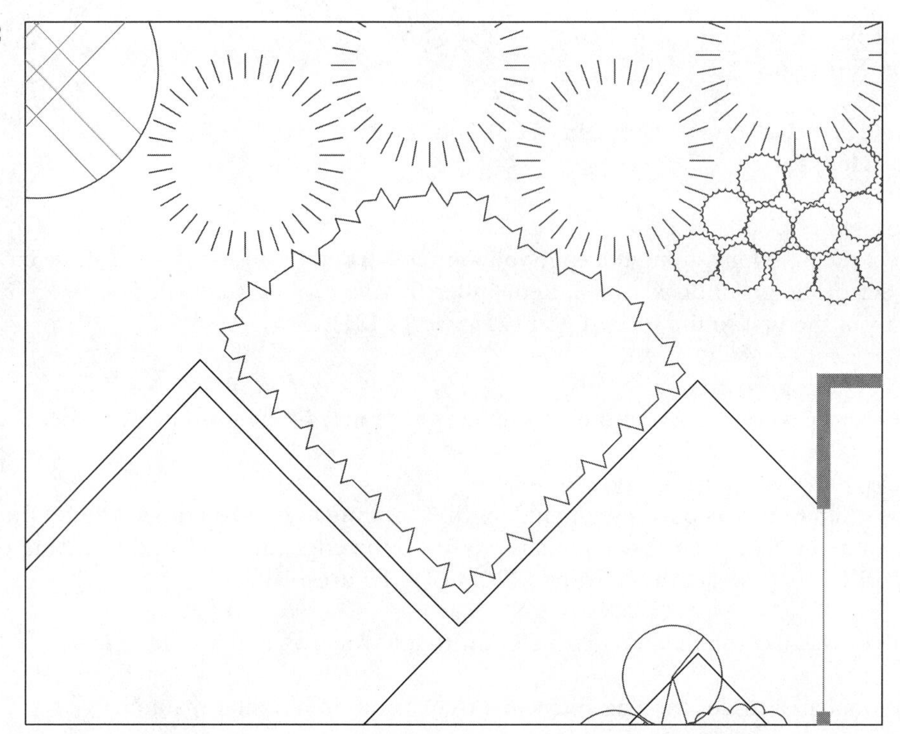

Use the **Arcs** option with a segment length of **2'**. Be sure to build the vegline in a counter-clockwise direction to make the arcs point outward. Be sure to widen and close the polyline with **PEDIT**. The hatch pattern in this area is **Dots**, used with a scale of **5'**. The rotation angle is not significant.

Using Plant Outline

When a grouping of plants requires emphasis or separation from others around them, a plant outline is sometimes used. LANDCADD has a routine to make this process quick and easy. In our example, we will use the plantings of **Nandina** to illustrate the Plant Outline command.

First, zoom in to the area above the top left corner of the house. Use the Planting pull-down menu, then pick Plant Outline.

At the prompt:

```
Command:
Select plants to be outlined:
```

```
Select objects:
```

Pick all of the **Nandina domestica "Nana"** plants in this grouping. Press **<Enter>** when all of the plants are highlighted.

```
Enter polyline width for outline <6″>
```

Type **1"** for the width of the polyline.

```
Enter offset distance of outline <1″>
```

Type **3"** for an offset distance.

```
Should a point be placed at insertion of all plants <Yes>:
```

Type **No** to prevent this process. The program then builds a series of arc polylines surrounding the plants. If there is sufficient space between the plants, polylines may be included in the center of the plant mass. If this occurs, simply erase the unwanted polyline segments. See below for an example of a completed plant outline (Figure 14.133).

Repeat this procedure for the **Nandina** masses near the back door and bay window.

Figure 14.133

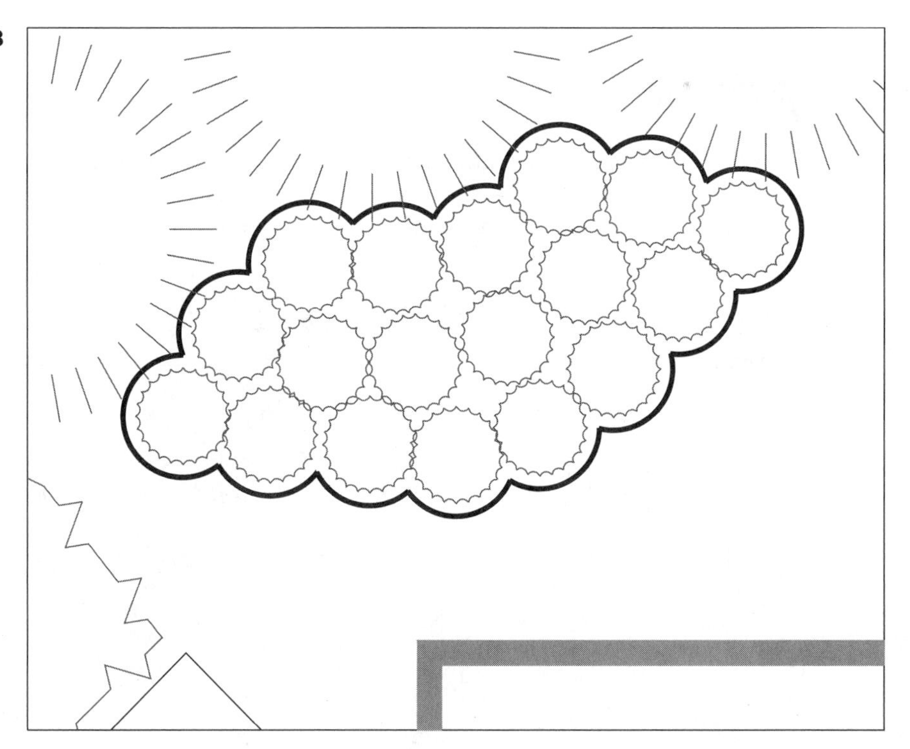

Labeling plants in the back yard

Labeling the back yard plants is performed using the same procedures as with the front yard plantings. Individual plants can be labeled with **Label Symbol** as before, while **Plant Mix** can be employed for all of the groundcover areas.

Be sure to keep the labels in alignment, maintaining two columns of plant labels. The plant labels overlap onto the house area and the patio area, so the lines which define these areas should be broken to make room for the labels.

When a line or polyline segment must be broken, use the **Break** command to eliminate the unwanted section (Figures 14.134 and 14.135).

If two or more lines must be broken, it may be easier to draw two cutting lines, then use the **TRIM** command to remove the unwanted line sections. See below (Figures 14.136 and 14.137).

Be sure to erase the cutting lines to complete the process.

After completing the placement and labeling of all the plants, save your drawing to your hard disk. Be sure to make a floppy disk back-up (Figure 14.138).

LANDCADD Estimating

One of the most convenient features of LANDCADD (and AutoCAD) is its ability to keep track of the number of plants, their sizes and respective prices for a project. With this information, you can develop a bill of materials which helps you provide an estimate to

Figure 14.134

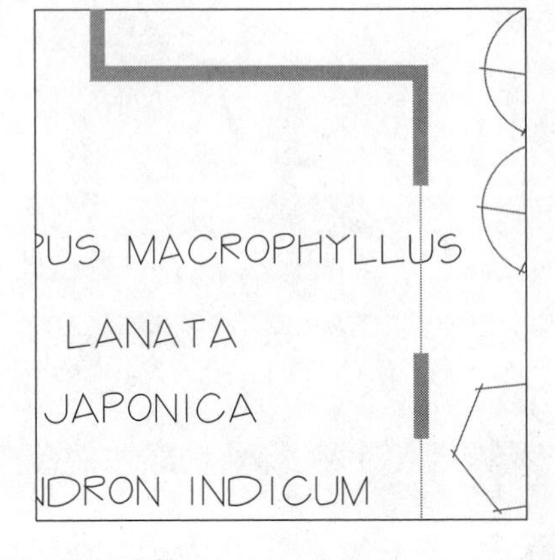

Figure 14.135

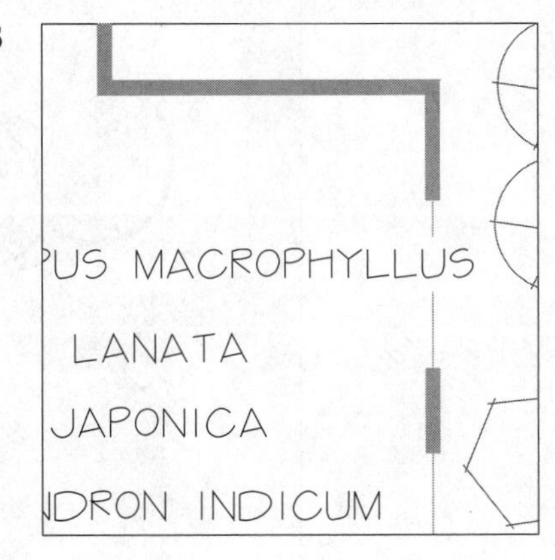

Figure 14.136

Figure 14.137

Figure 14.138

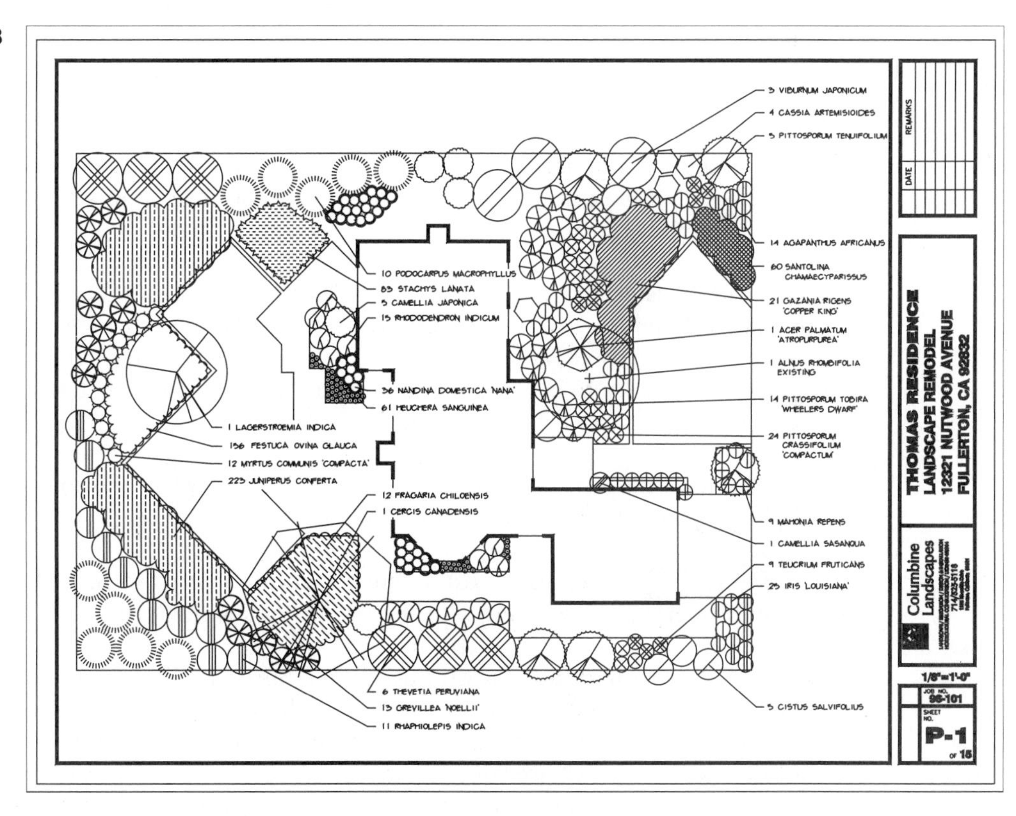

a client as soon as you finish a drawing. The Estimating part of LANDCADD can also be used to determine the amount of each type of pipe used in an irrigation system, along with the number of the various kinds of sprinklers used on a project.

To use the Estimating program effectively, we first need to understand how it does what it does. The first concept to understand is that of **attributes**. Attributes are handles attached to drawing blocks which can convey information about the block. An attribute is named by its **tag**. The text block used in our title block drawing had 8 tags: Job, Name, Address, City, Sheet, Number, Total and Scale.

Each of those attributes had a phrase attached to them which asked us for information. These phrases are **prompts**. Prompts can require that we type in information, or can ask us to choose items from a dialogue box, or to type information in the dialogue box. In any case, we need to enter in a **value** for every attribute assigned to a block. A value is just some sort of entry that relates to the attribute.

With plant symbols, there are three attributes, one tag is **item**, one is **size** and the other is **alias** (or code). The item tag requires that we enter the plant name, and the size tag requires that we enter the planting (container) size. (We will not use the alias attribute in this exercise.) Unlike the **Text** block, when plants are inserted into the drawing, the values of each attribute are not displayed on screen. The attributes are invisible, but are stored with each plant symbol.

The real value of attributes is that they can be tallied, sorted, alphabetized and recorded in a file. This file can then be inserted into a drawing, sent to a word processing program, or combined with price data to create a cost estimate.

Plants can be selected by picking them individually, picking them by layer, or picking them all. For example, we can tally all of the plants placed in the front yard and list them according to their name and size, then record this information in a text file. We can perform the same steps to isolate those plants in the back yard to compare the cost of the front yard to the cost of the back yard.

To begin the estimating program, open your planting plan drawing, then choose the **Estimating** pull-down menu (Figure 14.139).

Pick **Configure** to set up the Estimating program so that it formats the attribute information properly when working with the estimating commands.

For example, we can design the output of the estimate report generated by LANDCADD after the plants are tallied. We can also include added cost factors into the estimate. Added cost factors are additional charges which are added onto the total cost of the job after the individual costs are tallied. If we figured that we should add 10% on to the plant and labor costs of the job for overhead and profits, we can add 10% in the "Configuration" dialogue box (Figure 14.140).

We will enter **Attribute Setup** information when we select the plants to tally, so we can ignore these boxes for now.

Figure 14.139

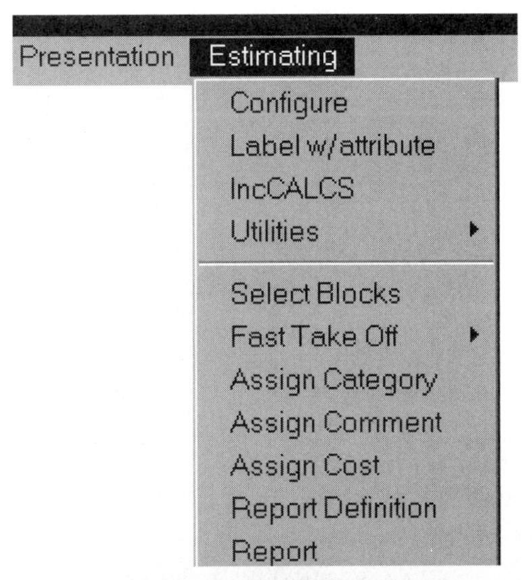

Figure 14.140

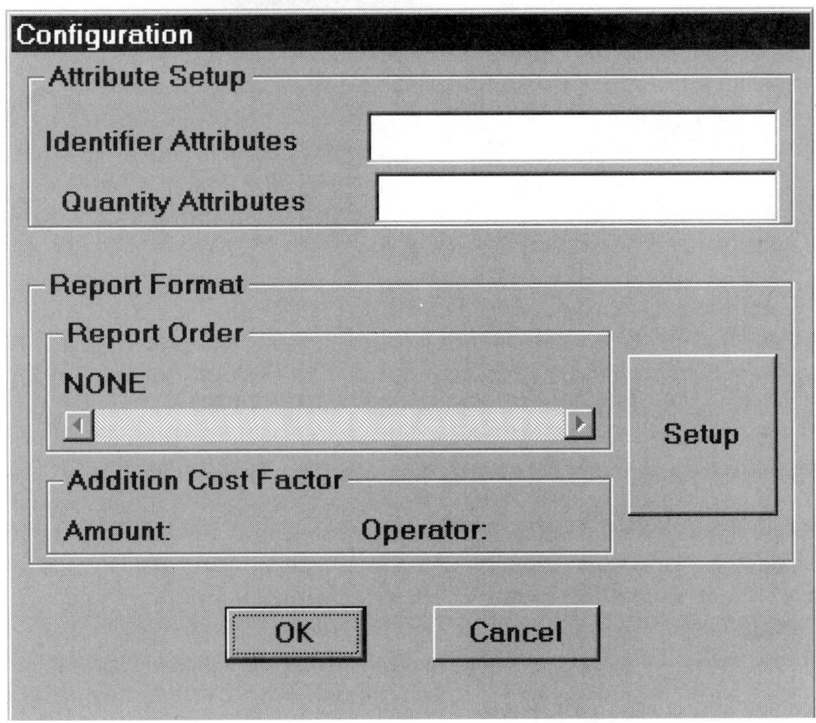

Under **Report Format**, select **Setup**, which moves us into the "Report Format" dialogue box (Figure 14.141).

Figure 14.141

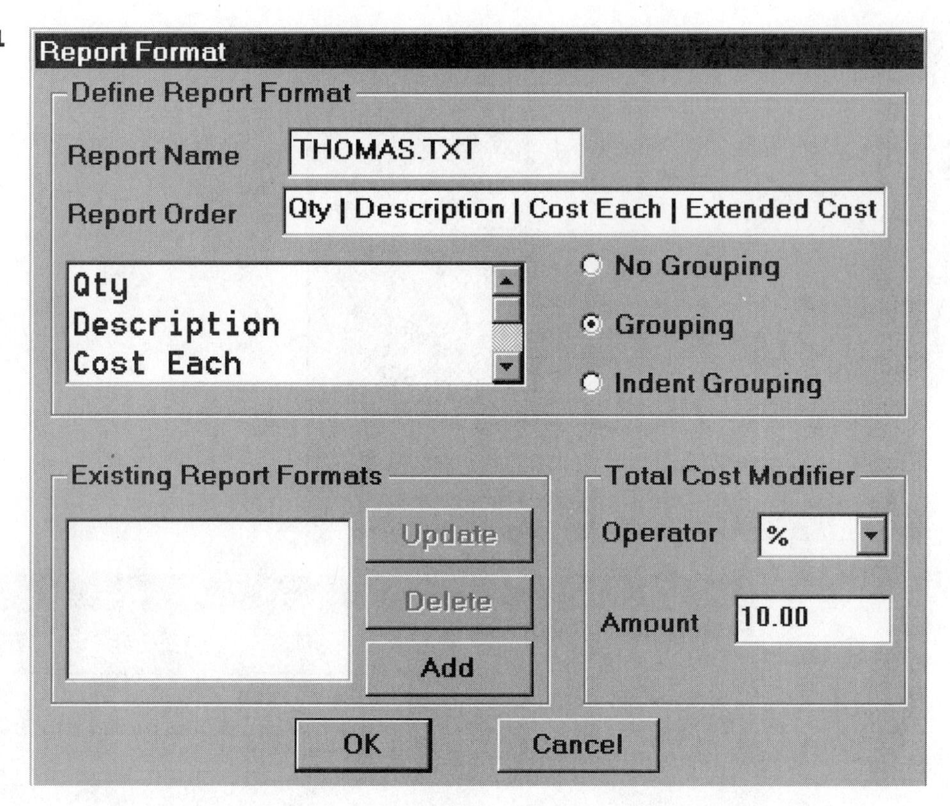

In the **Report Name** box, type **THOMAS.TXT**.

Insert the additional cost factor in the **Total Cost Modifier** box. In the **Operator** pop-down box, select % and in the **Amount** box, type **10.00**.

Create a **Report Order** by moving to the scroll box and selecting each item from the list in order. This places **Qty**, **Description**, **Cost Each**, **Extended Cost** and **Total Cost** in the **Report Order** box. (Extra information not shown in the **Report Order** text box is scrolled to the right.) Be sure that the **Grouping** radio button is selected so that the report will display the plants in their appropriate groupings (i.e., Trees, Shrubs and Groundcovers).

Use the **Add** box to add the **THOMAS.TXT** format into the list of **Existing Report Formats**.

Pick **OK** when finished. This returns us to the "Report Format" dialogue box, where we can again pick **OK**.

Once the Estimating functions have been configured, we then need to choose those items which we wish to count for our estimate. If we attempt to generate reports or estimates without first selecting a group of blocks, we see an error message (Figure 14.142).

To avoid this message, pick **Select Blocks**. This will bring up the "EZ- Estimate Selection" dialogue box which allows you to choose the blocks to include in the estimate (Figure 14.143).

In this dialogue box, we can decide how our target blocks will be selected. The **Select** heading allows us four choices:

All Entities will pick all **blocks** visible in the drawing. Any item which is not a block is ignored.

Layer by Pick allows us to pick an object on the desired layer, after which **all** objects on that layer will be selected.

Figure 14.142

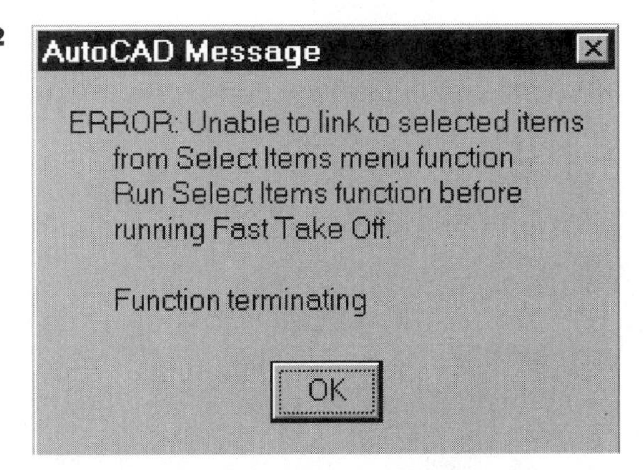

Figure 14.143

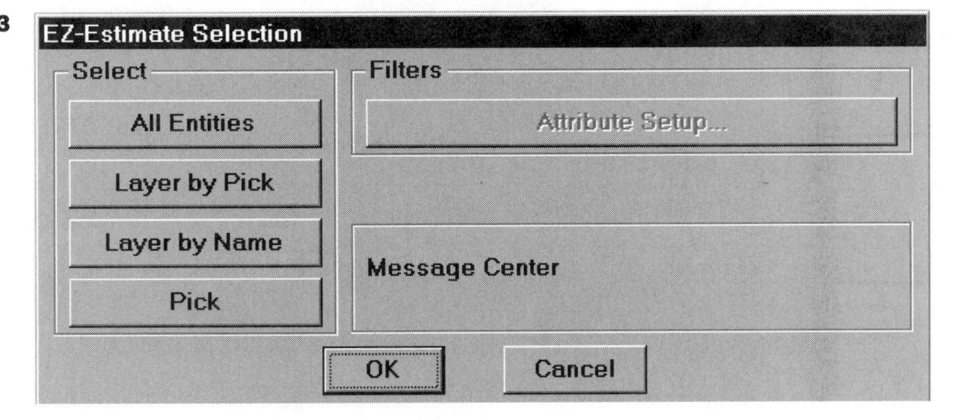

Layer by Name allows us to scroll through the list of layers in the drawing. We can choose the layer which has our selected blocks.

Pick puts us back into the drawing screen where we can pick any of the blocks we wish. As before, only blocks will be selected, so text, lines and polylines will be ignored.

Use the **Pick** option, then select the plants from the front yard plant list. The "EZ-Estimate" dialogue box returns with a Block count (Figure 14.144).

At this point, it is instructive to examine the manner in which the blocks will be examined and reported.

Pick **Attribute Setup . . .** to explore this option (Figure 14.145).

The "Attribute Setup" dialogue box allows us to select which of the attributes LANDCADD will examine when the plant totals are reported. **All** of the attributes on the drawing are available choices, including the attributes that we used when creating the text block in **C18SHEET**. Since these attributes are a part of the Planting Plan, they are considered available.

Figure 14.144

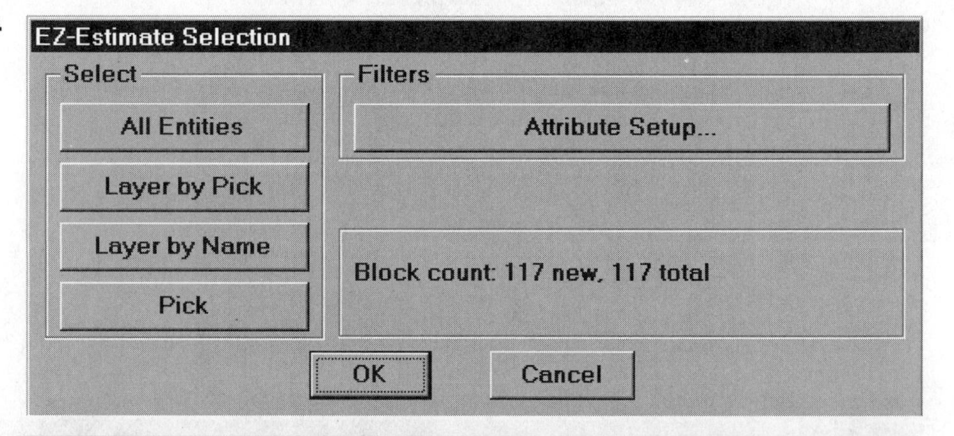

Figure 14.145

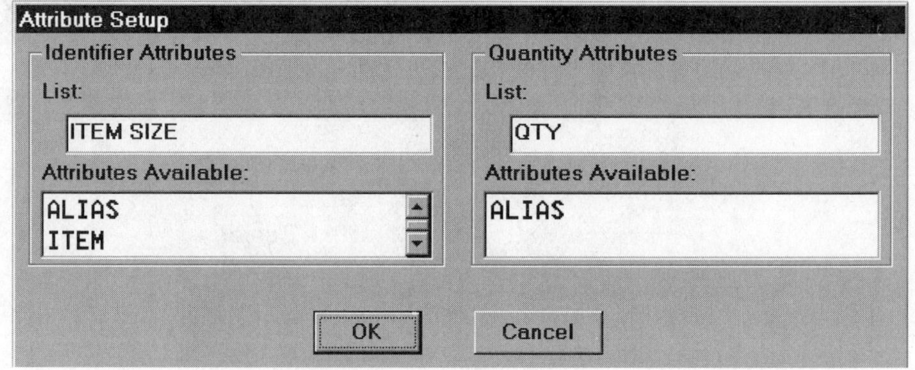

The illustration correctly shows **ITEM** and **SIZE** as the **Identifier Attributes**, while **QTY** is the correct listing for **Quantity Attributes**.

If other attributes are listed here, use the **<Backspace>** key to delete them, then use the **<Pick>** button to add the correct attributes. Pick **OK** when finished.

This returns us to the "EZ-Estimate Selection" dialogue box. Pick **OK**.

Once we have selected all of the objects to include in our estimate, LANDCADD goes to work to compile an **attribute extract file**. This is the list of attributes and the number of times that they occur on the drawing. You will probably see a message at the command prompt:

```
Working.......\
Still working......\
Working...........\
```

After properly creating the attribute extract file, all of LANDCADD's other estimating functions become active. The **Estimating** menu may not appear different on the surface (the appearance does change in the DOS version of LANDCADD), the options listed below the **Select Blocks** line are now available.

The options available allow us to assign categories, comments and costs to each plant in the project. In addition, we can generate a "quick and dirty" take-off list, showing plant names and quantities. For a more formal estimate, we can generate a report. Had we not done so previously, we could define the format of the report.

We will begin by examining the **Fast Take Off** function which lists the names of the blocks (plants) selected and the number of times they are used in a drawing. It is important to note that any of these estimating functions can be performed in any order once the **Select Blocks** functions have been activated. Comments could be created first, costs could be omitted altogether, or the **Fast Take Off** performed last rather than first.

From the **Select Blocks** screen menu, pick **Fast Take Off**.

From the **Fast Take Off** sub menu, pick **w/o Category** (Figure 14.146).

This places the list of selected blocks on the screen, along with their planting sizes and the number of times they occur. The **Fast Take Off** option is an especially good way to check to see that you have the correct number of blocks, that their sizes and spellings are correct, and that the groundcover areas have been successfully labeled (Figure 14.147).

Pick **Write**. Once the **Fast Take Off** has been compiled, there are several choices in the "Export Options" dialogue box (Figure 14.148).

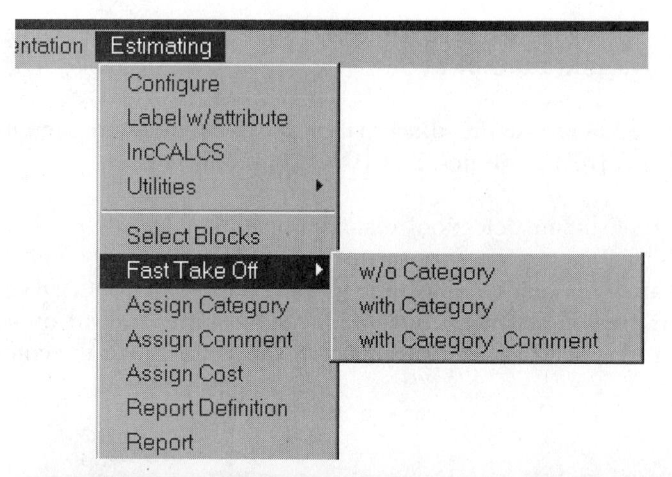

Figure 14.146

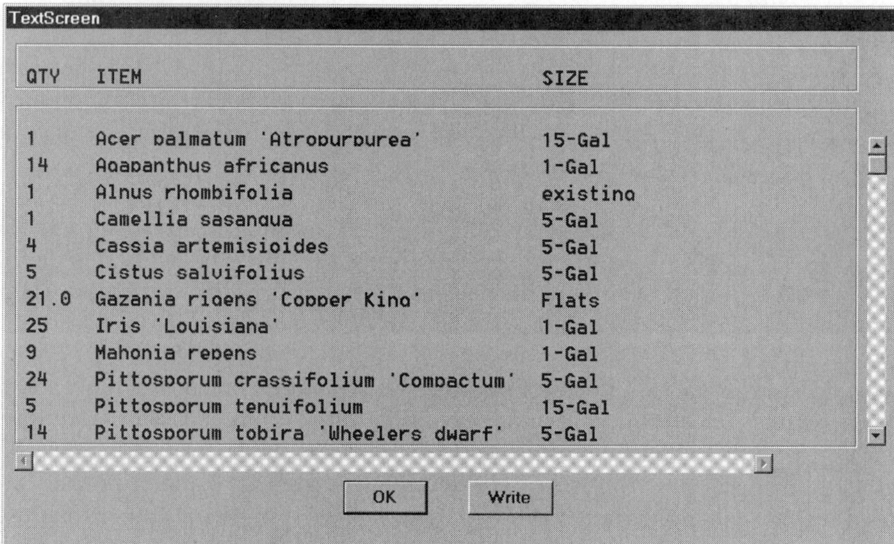

Figure 14.147

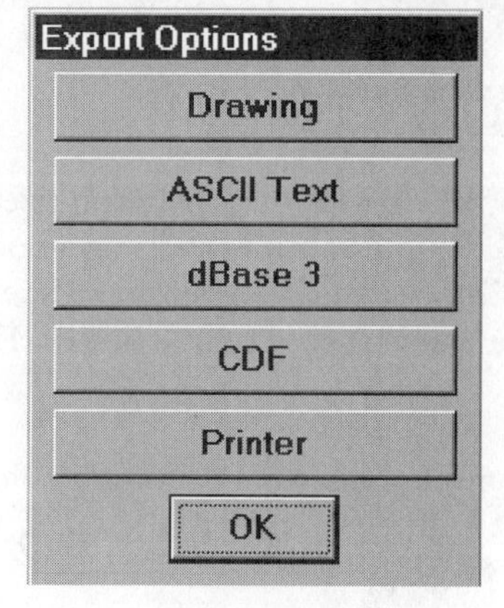

Figure 14.148

The **Fast Take Off** can be placed directly into the drawing as text. It can also be written to an ASCII text file (which could be viewed and edited in a word-processing program), stored as a **dBase III** format file (which can be viewed and edited in most database programs), or stored as a **CDF** (Comma Delimited File)—another format used by certain spreadsheet or database programs. Finally, the **Fast Take Off** can be sent directly to a printer connected to a CAD station.

Pick **Printer** if a printer is connected to your CAD station. Examine the **Fast Take Off** for the front yard

Fast Take Off for Front Yard—Without Category

QTY	ITEM	SIZE
1	Acer palmatum "Atropurpurea"	15-Gal
14	Agapanthus africanus	1-Gal
1	Alnus rhombifolia	existing
1	Camellia sasanqua	5-Gal
4	Cassia artemisioides	5-Gal
5	Cistus salvifolius	5-Gal
21.0	Gazania rigens "Copper King"	Flats
25	Iris "Louisiana"	1-Gal
9	Mahonia repens	1-Gal
24	Pittosporum crassifolium "Compactum"	5-Gal
5	Pittosporum tenuifolium	15-Gal
14	Pittosporum tobira "Wheelers dwarf"	5-Gal
60.0	Santolina chamaecyparissus	1-Gal
9	Teucrium fruticans	1-Gal
3	Viburnum japonicum	15-Gal

(If your entries differ greatly from those values shown above, a section below explains how to edit your drawing to correct any mistakes.)

Next, we will assign the correct category (Trees, Shrubs, Groundcovers) to each plant in the plan.

Select **Assign Category** from the **Select Blocks** screen menu (Figure 14.149).

In the "Category Assignment" dialogue box, you will see a list of plants on your plan, but the category listed will always display **none**. This is because there are no category assignments for the plants yet.

Highlight the first plant on the list, **Acer palmatum "Atropurpurea,"** then select **Modify** (Figure 14.150).

In the "Category Setup" dialogue box, move the cursor to the **Main Category** box, and type **Trees**. Pick **OK**. This returns us to the "Category Assignment" dialogue box with

Figure 14.149

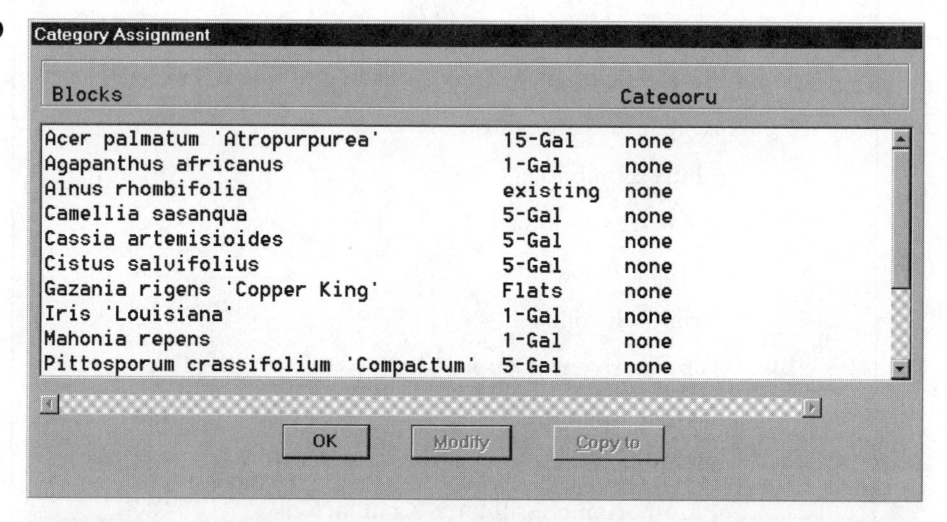

Figure 14.150

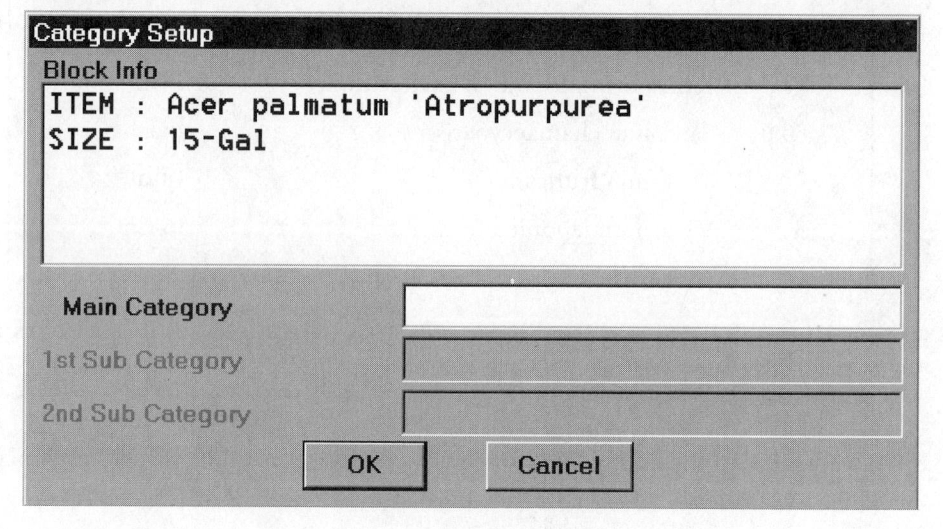

the correct category assigned to **Acer palmatum "Atropurpureum.**"

Repeat this process for the second plant in your list, **Agapanthus africanus**. In the "Category Setup" dialogue box, enter the category **Shrubs**. Return to the "Assign Category" dialogue box (Figure 14.151).

Highlight **Agapanthus africanus** (and its category), then pick **Copy to**. This places a message **Select item to copy to** at the bottom of the dialogue box. Move your highlight to **Camellia sasanqua**, then press **<Pick>**, so that the **Shrubs** category is "pasted" into place. This method saves time and should be used to assign all of the **Trees**, **Shrubs** and **Groundcovers** on the plant list.

Note that a plant can only be assigned to a category once it has been used in the planting plan. Any plants that have been assigned to a category will remain in that category if they are used in drawings created in other drawing sessions. The category listings are saved in a database file **CATEGORY.DBF** (usually found in the LANDCADD subdirectory). This file becomes a permanent part of LANDCADD and means that plants will retain their categories until they are changed.

Hint: If you change workstations, the CATEGORY.DBF and other database files at the new station may be different from the one that you are familiar with. If necessary, you may wish to copy these files from one workstation to others to make sure that they all contain the same databse information.

After all of the plant categories are assigned, we will assign prices to each plant. To assign the correct price to each plant, select **Assign Price** from the **Select Blocks** screen menu (Figure 14.152).

Figure 14.151

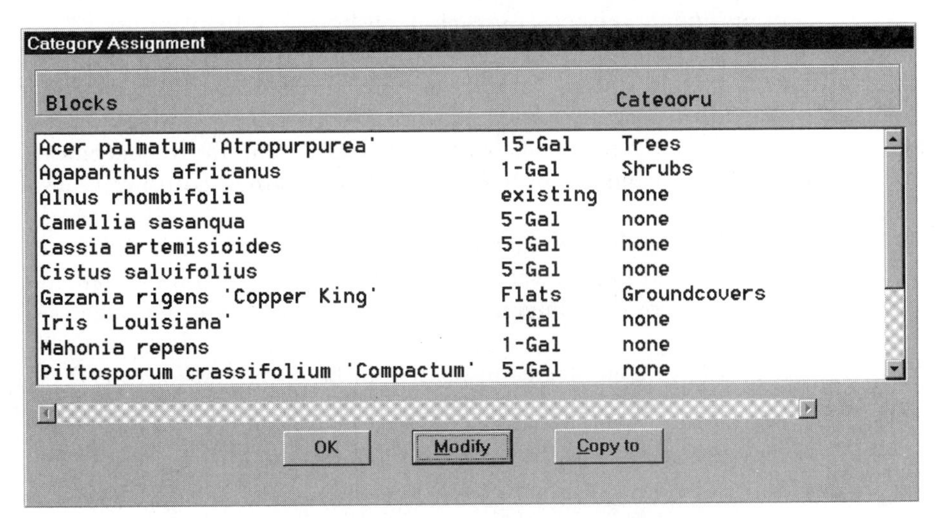

Figure 14.152

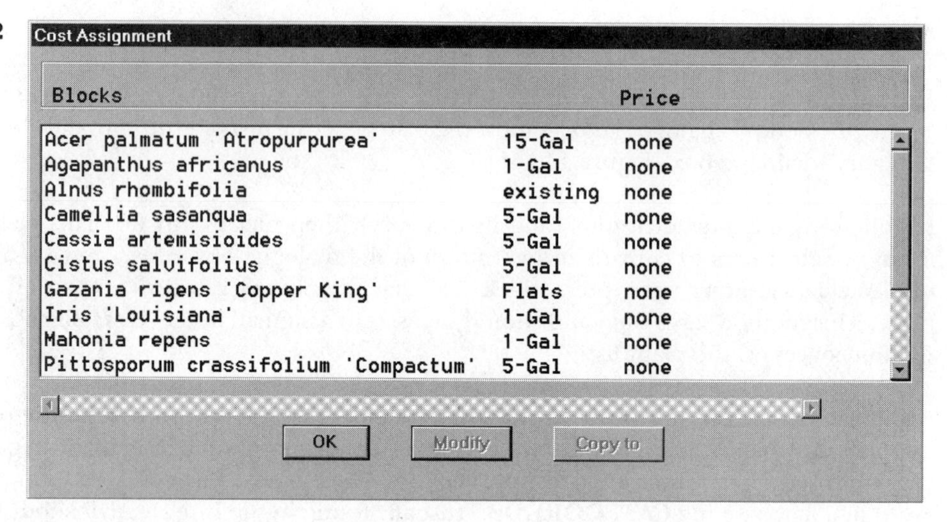

Again, a complete listing of plants used on the drawing is shown, but no prices are yet entered.

Highlight **Acer palmatum "Atropurpureum,"** then pick **Modify** (Figure 14.153).

In the **Cost Each:** box, enter the correct price: **125.00** (the **$** character is not used). After you have entered a price, pick **Accept** to enter that price. Once a cost has been entered and accepted, an **Additional cost** can be applied. Mark-ups, margins or other per-plant costs can be added here. Pick **OK**.

The cost entered is then shown on the plant list (Figure 14.154).

Continue adding the costs for each item shown on the plant list.

Again, note that a plant can only be assigned a cost once it has been used in the planting plan. Plant costs are retained for use in subsequent drawing sessions. The cost assignments are saved in the database file **PRICE.DBF**. This file too becomes a permanent part of LANDCADD and means that plants will retain their costs until they are changed.

In addition to costs and categories, you can also add comments for each plant listing. These comments might relate to sizes, weights, per-flat counts, container notes, etc. The process of entering comments is identical to that of entering categories and costs.

Once you have completed the process of entering categories and costs for all of the plants in the front yard, you can check your work by using **Fast Take Off** once again.

Figure 14.153

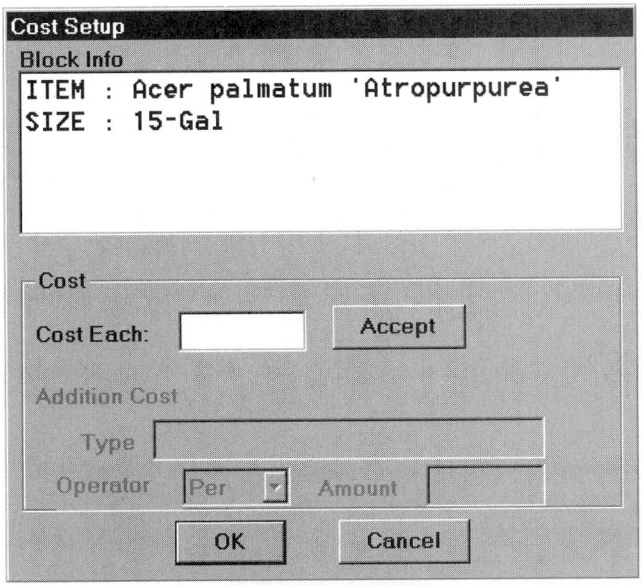

Cost Setup
Block Info
ITEM : Acer palmatum 'Atropurpurea'
SIZE : 15-Gal
Cost
Cost Each:
Accept
Addition Cost
Type
Operator Per Amount
OK Cancel

Figure 14.154

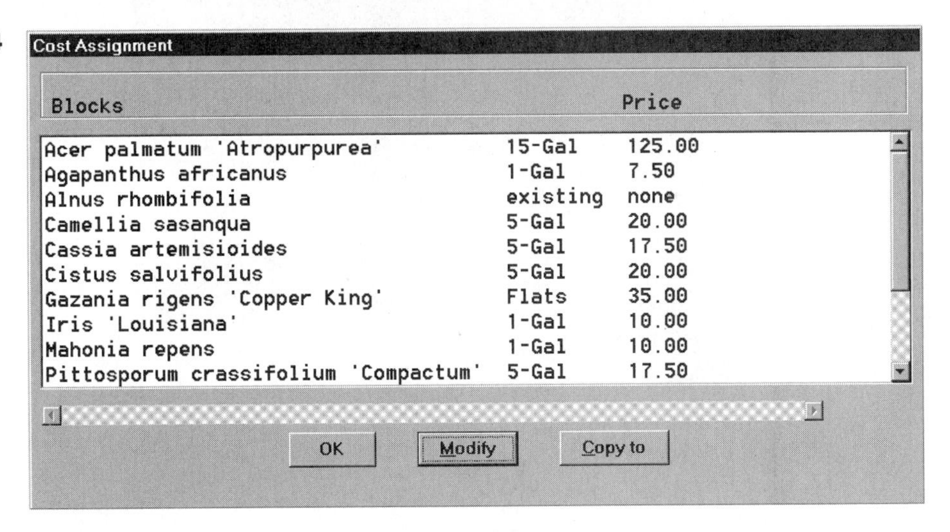

If you select **with Category**, you can see a listing of plants separated by category (Groundcovers, Shrubs and Trees). If a plant is assigned to the wrong category, go back to **Assign Category** to reassign the correct category to the plant. If a plant has an incorrect spelling, or has been assigned an incorrect planting size, more work is required.

There is no way to modify plant names or planting sizes in the Estimating section of LANDCADD. You must return to the drawing screen to modify block attributes.

Once you return to the drawing screen, you have two choices to correct a plant attribute error:

1. change the attribute for each occurrence of the plant, one at a time. Use **DDATTE** (**AE**) to modify the incorrect attribute on the plant, then work your way around the drawing and make the necessary changes to each plant. This method can be slow and tedious when the plant is used many times in a drawing;
2. you can use the global edit function to change all occurrences of the plant in one step.

Select the **Blocks ▸** option from the **Tools** pull-down menu (Figure 14.155).

Figure 14.155

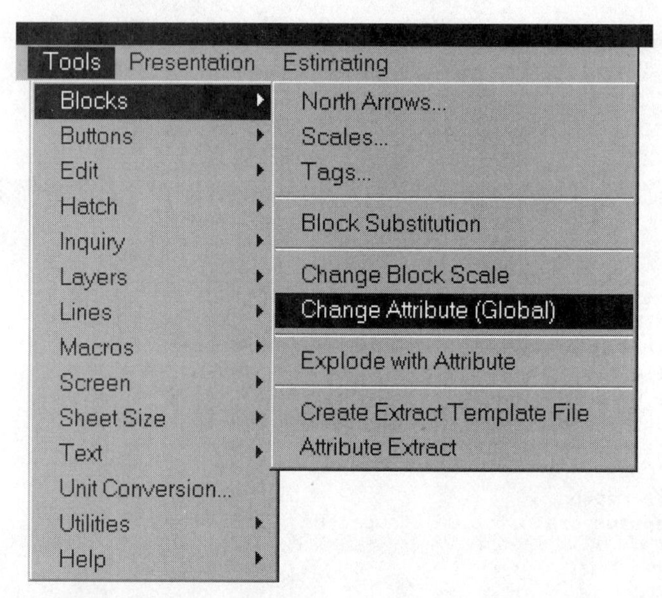

Pick **Change Attribute (Global)**. You then see at the prompt:

```
Select control block
```

Pick the plant symbol whose attribute you need to edit. The "Edit Attributes" dialogue box appears (Figure 14.156).

Figure 14.156

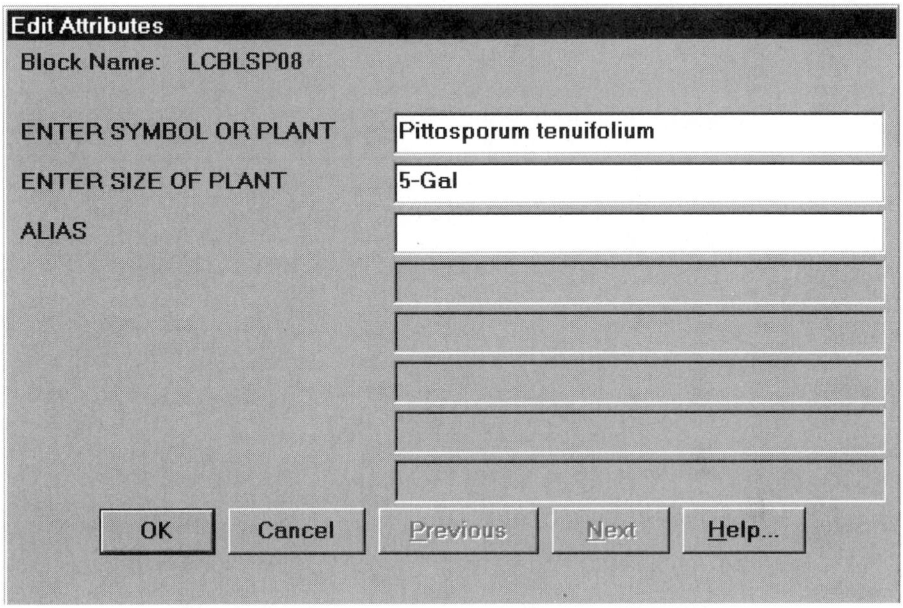

Edit Attributes

Block Name: LCBLSP08

ENTER SYMBOL OR PLANT: Pittosporum tenuifolium

ENTER SIZE OF PLANT: 5-Gal

ALIAS

OK | Cancel | Previous | Next | Help...

Make the appropriate changes to the attributes. (The tree shown here should have a plant size of **15-Gal**.) After your editing is complete, pick **OK**. The prompt then reads:

```
Select blocks to edit or <Enter> for all:
```

Press **<Enter>** to select all of the blocks. At this point all of the matching blocks will flicker briefly as the attributes are changed. You can check the status of one of the blocks by using the **DDATTE** (**AE**) command.

Any changes made during this process will not be reflected in the **Fast Take Off** or **Report** functions until the **Select Blocks** process has been repeated. This is because the blocks must be reselected to reflect the updated attribute information.

Once all of the editing has been completed, reselect the entities and look at the plant list using the **Fast Take Off** option with Category:

Fast Take Off for Front Yard—With Category

DESCRIPTION	SIZE	QTY
Alnus rhombifolia	existing	1
Groundcovers		
Gazania rigens "Copper King"	Flats	21.00
Shrubs		
Agapanthus africanus	1-Gal	14
Camellia sasanqua	5-Gal	1
Cassia artemisioides	5-Gal	4
Cistus salvifolius	5-Gal	5
Iris "Louisiana"	1-Gal	25
Mahonia repens	1-Gal	9
Pittosporum crassifolium "Compactum"	5-Gal	24
Pittosporum tobira "Wheelers dwarf"	5-Gal	14
Santolina chamaecyparissus	1-Gal	60.00
Teucrium fruticans	1-Gal	9
Trees		
Acer palmatum "Atropurpurea"	15-Gal	1
Pittosporum tenuifolium	15-Gal	5
Viburnum japonicum	15-Gal	3

Note that the **Alnus rhombifolia** does not have a category, so it is listed first (as are any other plants without categories). In addition, of course, the **Alnus** tree will show no price since it is an existing tree. Any mistakes in plant category can be changed in the Assign Cost option.

After you examine and/or print your **Fast Take Off** list, you can then print a more complete report using the **Report** function.

As with the **Fast Take Off** option, LANDCADD will then display the complete estimate for the plants in the front yard, using the **Report Format** that we specified earlier. As before, we can review the report on the screen, then write it to the drawing, write it as an ASCII file, or export it in the other file formats.

Although your quantities of groundcover plants may vary from this example, your estimate should total approximately $3500 for the plants in the front yard.

Repeat this same process for the plants in the back yard. The total costs for the plants in the back yard should be approximately $8500. You can also prepare a full report for all the plants by selecting the entire yard. LANDCADD should report a total cost of approximately $12,000 for the entire planting plan.

Front Yard Estimate Report

Qty	Description	Cost Each	Ext. Cost	Add. Cost	Tot. Cost
Groundcovers					
21.00	Gazania rigens "Copper King"	Flats	35.00	735.00	735.00
				735.00 Groundcovers	
Shrubs					
14	Agapanthus africanus	1-Gal	7.50	105.00	105.00
1	Camellia sasanqua	5-Gal	20.00	20.00	20.00
4	Cassia atremisioides	5-Gal	17.50	70.00	70.00
5	Cistus salvifolius	5-Gal	20.00	100.00	100.00
25	Iris "Louisiana"	1-Gal	10.00	250.00	250.00
9	Mahonia repens	1-Gal	10.00	90.00	90.00
24	Pittosporum crassifolium "Compactum"	5-Gal	17.50	420.00	420.00
14	Pittosporum tibra "Wheeler's dwarf"	5-Gal	15.00	210.00	210.00
60.00	Santolina chamaecyparissus	1-Gal	7.50	450.00	450.00
9	Teucrium fruticans	1-Gal	7.50	67.50	67.50
				1782.50 Shrubs	
Trees					
1	Acer palmatum "Atropurpurea"	15-Gal	125.00	125.00	125.00
5	Pittosporum tenuifolium	15-Gal	75.00	375.00	375.00
3	Viburnum japonicum	15-Gal	75.00	225.00	225.00
				725.00 Trees	
				324.25 Total Modifier %10.00	
				3566.75 TOTAL	

Back Yard Estimate Report

Qty	Description	Cost Each	Ext. Cost	Add. Cost	Tot. Cost
Groundcovers					
12.00	Fragaria chiloensis	Flats	35.00	420.00	420.00
				420.00 Groundcovers	
Shrubs					
5	Camellia japonica	5-Gal	20.00	100.00	100.00
156.00	Festuca ovina glauca	1-Gal	6.50	1014.00	1014.00
13	Grevillea "Noellii"	5-Gal	17.50	227.50	227.50
61	Heuchera sanguinea	1-Gal	7.50	457.50	457.50
223.00	Juniperus conferta	1-Gal	7.50	1672.50	1672.50
12	Myrtus communis "Compacta"	5-Gal	17.50	210.00	210.00
36	Nandina domestica "Nana"	5-Gal	20.00	720.00	720.00
11	Rhaphiolepis indica	5-Gal	20.00	220.00	220.00
15	Rhododendron indicum	5-Gal	20.00	300.00	300.00
83.00	Stachys lanata	1-Gal	7.50	622.50	622.50
				5544.00 Shrubs	
Trees					
1	Cercis canadensis	24"-Box	250.00	250.00	250.00
1	Lagerstroemia indica	24"-Box	250.00	250.00	250.00
10	Podocarpus macrophyllus	15-Gal	75.00	750.00	750.00
6	Thevetia peruviana	15-Gal	75.00	450.00	450.00
				1700.00 Trees	
				766.40 Total Modifier %10.00	
				8430.40 TOTAL	

Thomas Residence Planting Plan Estimate Report

Qty	Description	Cost Each	Ext. Cost	Add. Cost	Tot. Cost
Groundcovers					
12.00	Fragaria chiloensis	Flats	35.00	420.00	420.00
21.00	Gazania rigens "Copper King"	Flats	35.00	735.00	735.00
				1155.00 Groundcovers	
Shrubs					
14	Agapanthus africanus	1-Gal	7.50	105.00	105.00
5	Camellia japonica	5-Gal	20.00	100.00	100.00
1	Camellia sasanqua	5-Gal	20.00	20.00	20.00
4	Cassia atremisioides	5-Gal	17.50	70.00	70.00
5	Cistus salvifolius	5-Gal	20.00	100.00	100.00
156.00	Festuca ovina glauca	1-Gal	6.50	1014.00	1014.00
13	Grevillea "Noellii"	5-Gal	17.50	227.50	227.50
61	Heuchera sanguinea	1-Gal	7.50	457.50	457.50
25	Iris "Louisiana"	1-Gal	10.00	250.00	250.00
223.00	Juniperus conferta	1-Gal	7.50	1672.50	1672.50
9	Mahonia repens	1-Gal	10.00	90.00	90.00
12	Myrtus communis "Compacta"	5-Gal	17.50	210.00	210.00
36	Nandina domestica "Nana"	5-Gal	20.00	720.00	720.00
24	Pittosporum crassifolium "Compactum"	5-Gal	17.50	420.00	420.00
14	Pittosporum tibra "Wheeler's dwarf"	5-Gal	15.00	210.00	210.00
11	Rhaphiolepis indica	5-Gal	20.00	220.00	220.00
15	Rhododendron indicum	5-Gal	20.00	300.00	300.00
60.00	Santolina chamaecyparissus	1-Gal	7.50	450.00	450.00
83.00	Stachys lanata	1-Gal	7.50	622.50	622.50
9	Teucrium fruticans	1-Gal	7.50	67.50	67.50
				7326.50 Shrubs	
Trees					
1	Acer palmatum "Atropurpurea"	15-Gal	125.00	125.00	125.00
1	Cercis canadensis	24"-Box	250.00	250.00	250.00
1	Lagerstroemia indica	24"-Box	250.00	250.00	250.00
5	Pittosporum tenuifolium	15-Gal	75.00	375.00	375.00
10	Podocarpus macrophyllus	15-Gal	75.00	750.00	750.00
6	Thevetia peruviana	15-Gal	75.00	450.00	450.00
3	Viburnum japonicum	15-Gal	75.00	225.00	225.00
				2425.00 Trees	
				1090.65 Total Modifier %10.00	
				11997.15 TOTAL	

chapter

Drawings Five and Six—The Irrigation Plan with Materials Schedule and Legend

In the Irrigation Plan, we will develop a design for a sprinkler system to meet the needs of the Thomas residence landscape. This irrigation plan must consider the water requirements of the plants, use appropriate sprinkler equipment, and work within the hydraulic limitations of the Thomas residence water supply. While LANDCADD provides a generous number of tools which will assist us in creating a clear, legible plan, the software cannot provide a substitute for an irrigation designer's knowledge and experience. The judgements inherent in an irrigation plan arise from a substantial knowledge of irrigation components, hydraulics and installation techniques. As with all of the modules, LANDCADD equips us with the tools to illustrate and document a design, but does not make good (or bad) design decisions. Those must come from a trained design professional.

As we build the Irrigation Plan, we will continue to make extensive use of LANDCADD symbols and data files. In the Irrigation Design module, LANDCADD includes extensive **Data files** of sprinkler equipment (heads, nozzles and performance data) produced by different manufacturers to assist us in selecting heads. **Pipe type** files provide information required to calculate friction loss and water velocity through most commonly used pipe materials. LANDCADD uses these data files to place "smart" sprinkler head symbols, pipe segments and control valves which have attached attributes that can be used for several purposes. We can use the flow (gallons per minute, g.p.m.) data attached to irrigation heads to determine the total flow rate in an irrigation zone. We can also use this flow data to size lateral line piping and assign friction loss (p.s.i.) values to each lateral line section. These friction loss values allow us to examine the total piping friction loss to the "critical head" in each irrigation zone. The irrigation designer can even use LANDCADD tools to examine the gross precipitation rate in a zone, and make a quick check to find the weekly station run time for a known weather (ET_O) condition. Although LANDCADD does not make irrigation design decisions, it provides quick access to information which makes those decisions simpler.

LANDCADD's irrigation design tools also help us keep track of the materials used. We will use the Equipment Table function to create a list of materials for the Irrigation Plan, and a legend to key out the use of irrigation symbols. Because each symbol contains attributes which help describe it, the Equipment Table contains a relatively complete (if somewhat disorganized) list of each piece of equipment (including piping) used in the Irrigation Plan. We will edit and organize this table to assist contractors developing bids for the project. The separate Irrigation Legend is intended to explain and provide technical details about the components used in the Irrigation Plan. While it may not be necessary to provide both charts for an irrigation project of this size, the methods shown can be extended to projects of any size and complexity.

While drawing the Irrigation Plan, we will complete the following:

- Reuse the house and hardscape block used in the Planting Plan
- Build closed polylines around eight planting areas
- Examine the different decisions required to select sprinkler heads and the symbols which represent them
- Examine the options for inserting sprinkler heads into the Irrigation Plan
- Use the Autohead routine for inserting sprinkler symbols automatically
- Use the **Distribution Analysis** feature to examine the uniformity of the sprinkler head coverage
- Examine methods to remove the Distribution Analysis templates from the screen
- Use the **Edit Heads** function to move and remove sprinkler heads from the Irrigation Plan
- Manually place bubbler heads in a planter, and edit their attributes
- Build a block of plants from the Planting Plan to assist in the location of drip emitters in the Irrigation Plan
- Use the **Zones** command to determine the flow rate in each planter and turf area
- Draw lateral pipe lines to connect sprinkler heads
- Use the **Pipesize** routine to automatically size and label the lateral pipe segments
- Examine a zone to determine the sprinkler with the lowest available pressure using the **Critical Head** command
- Insert valve labels to call out the valve number, size and flow rate
- Determine the precipitation rate of a sprinkler zone
- Determine the operating time of a sprinkler zone based on a weekly ET_O value

- Edit pipe label styles to clarify the pipe sizing in the Irrigation Plan
- Insert jump lines to indicate locations where pipes cross but do not intersect
- Manually insert irrigation control valves
- Place main lines and pipe sleeves in the Irrigation Plan
- Place drip irrigation piping and control valves
- Place a backflow preventer, point of connection and irrigation controller symbol in the Irrigation Plan
- Size the irrigation main line
- Create an equipment table and irrigation legend for the Irrigation Plan

Beginning the Plan

We can take advantage of the work we have completed previously, since the planter areas and house footprint are stored in our drawing **HOUSE1**. We will use that drawing, along with our **C18SHEET** prototype to begin the Irrigation Plan.

Begin the Irrigation Plan by opening a new project in the Project Manager, using **C18SHEET** as the prototype drawing. Edit the title block text in the usual way, and use the sheet number **I-1**.

Use **DDINSERT** to bring the drawing **HOUSE1** into your drawing. Insert the block with the insertion point **0,0**. **EXPLODE** the block so that you can use parts of the drawing as edges for trimming and extending. Be sure to set the **LTSCALE** to **96**.

Next, create a new layer called **BOUNDARY** and assign it the color **MAGENTA** (color 6). Make this your current layer. Draw **closed polylines** around each planter and lawn area to be irrigated.

These polylines will determine the borders for the irrigation heads to follow. The following illustration shows all of the polylines you will need (Figure 15.1).

When drawing these polylines, follow the existing borders in the hardscape. Use intersections, endpoints and other OSNAP options to build the polylines. Where the planter areas border on the house footprint, be sure to keep the polyline edge outside the edge of the footprint. This prevents LANDCADD from inserting the sprinkler head symbols too close to the house. Where planter edges overlap, be sure that each planter has its own fully enclosed polyline.

Figure 15.1

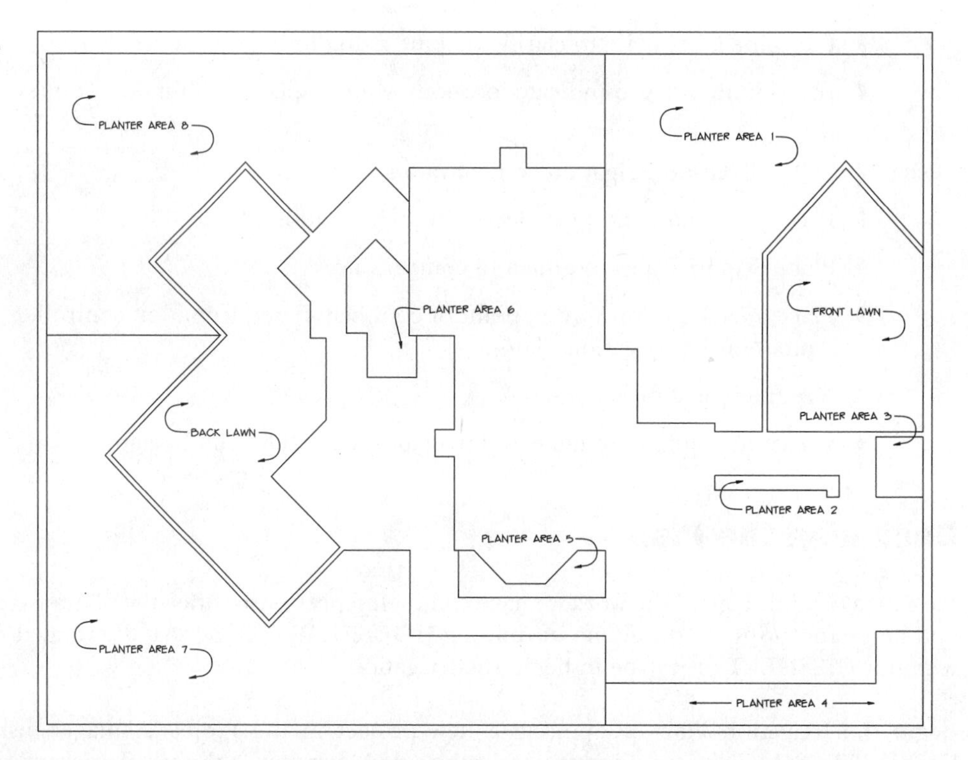

After completing all of the closed polylines, turn off the following layers: **0207-NEW_BLDG**, **CONCRETE**, **EX_HARDSCAPE** and **PROPERTY**. This should leave you with only the boundary polylines and the title block visible on your drawing.

After completing the polylines, we begin the actual process of irrigation design. This involves examining each area to irrigate, and then selecting the proper sprinklers for the plant material, hydraulic conditions and budget constraints. We will begin by selecting heads for the front lawn.

From the **EP** pull-down menu, select **LANDCADD**, then pick **Irrigation Design**. Use the **Heads** pull-down menu to pick **Head Insertion . . .**. After this selection is made, we see the "Sprinkler Head Insertion" dialogue box (Figure 15.2).

In this dialogue box, we define the type of head, nozzle performance data and symbol appearance for the sprinkler type being inserted. To begin this process, pick **Data File**. This brings up the "IRRIGATION HEAD DATA" dialogue box which lists manufacturers and their respective sprinkler head models (Figure 15.3).

In this dialogue box, we select the sprinkler make and model in the **DATA FILES** section. We then specify the type of nozzle (either individual or family grouping) under **NOZZLES**, then select the operating pressure under the **PRESSURE RADIUS** section.

Figure 15.2

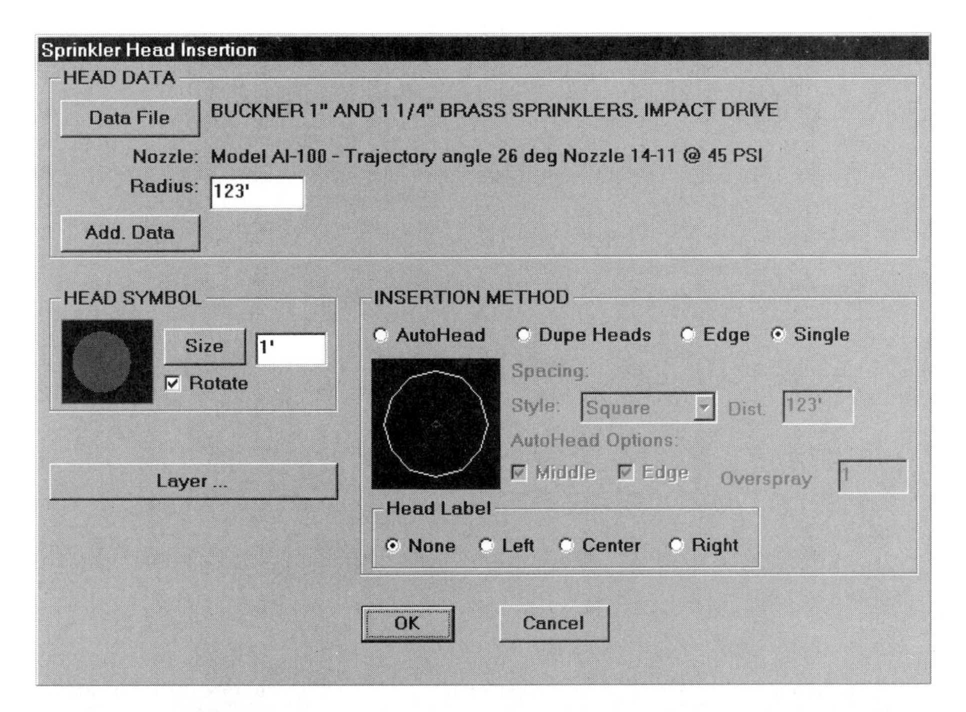

Figure 15.3

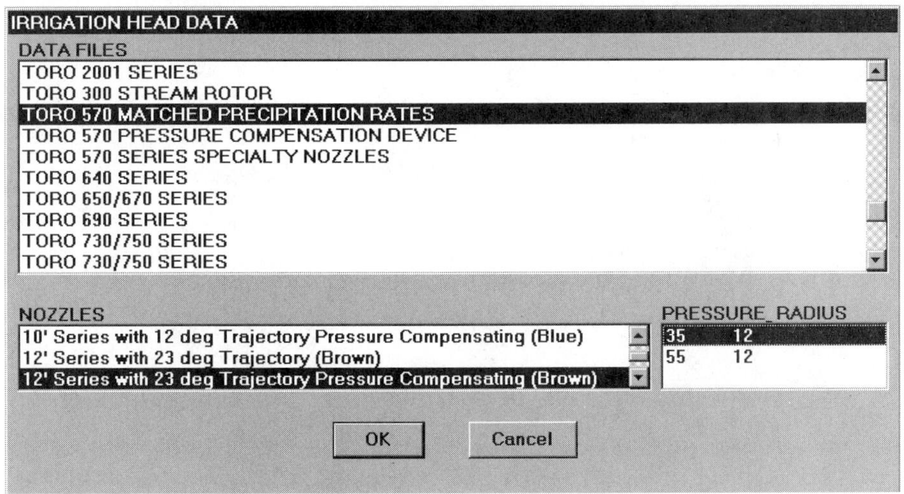

In the **DATA FILES** section, find the **TORO 570 MATCHED PRECIPITATION RATES** file.

In the **NOZZLES** section, select **12' Series with 23 deg Trajectory Pressure Compensating**.

In the **PRESSURE RADIUS** section, select **35 12** (35 PSI, 12' radius).

Pick **OK** when finished.

These selections define a number of parameters which affect the spacing of the heads on the screen, and the attributes assigned to them. The GPM flow, operating PSI, radius and model number information are all designated by choices made in this dialogue box.

This returns us to the "Sprinkler Head Insertion" dialogue box (Figure 15.4).

The **HEAD DATA** section is filled with data pertaining to the choice of heads made previously. The **Add. Data** box provides an opportunity to add extra attributes to the sprinkler symbols such as risers, swing joint fittings, etc. These extra attributes could be assigned categories and costs and included in the Estimating section to give a more accurate cost for the irrigation plan.

In the **HEAD SYMBOL** section, we have a choice of LANDCADD's sprinkler symbol families (Figure 15.5).

Because there are only nine families of heads available, it may be necessary to enter heads at two or three different sizes in order to provide enough symbol variety for a large project. (We will discuss creating customized irrigation head symbols in another exercise.) In this dialogue box, we simply pick the desired symbol style, then pick **OK**.

Figure 15.4

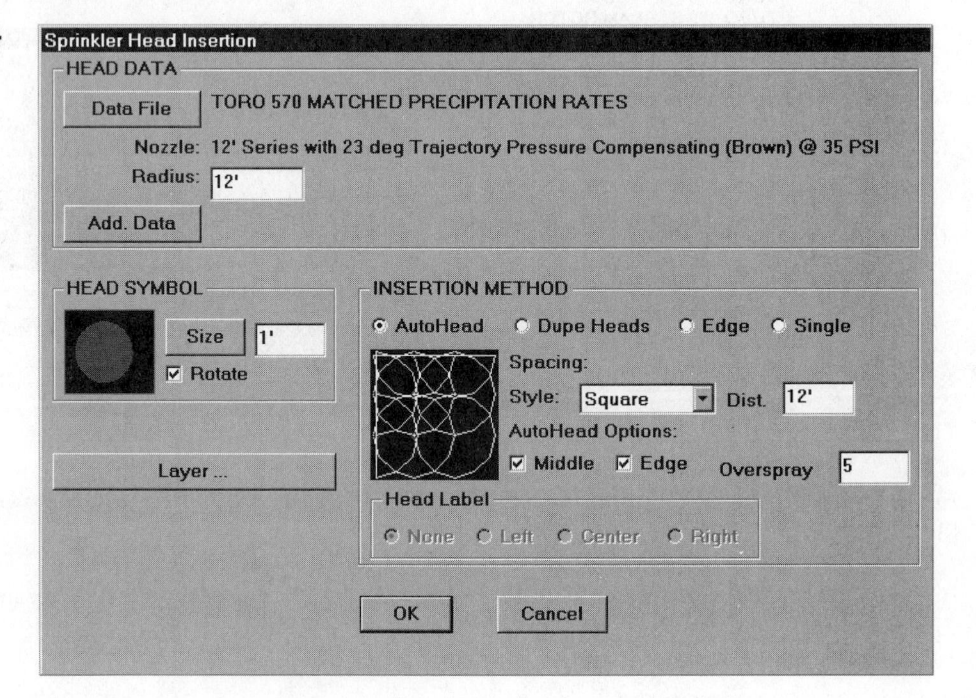

Figure 15.5

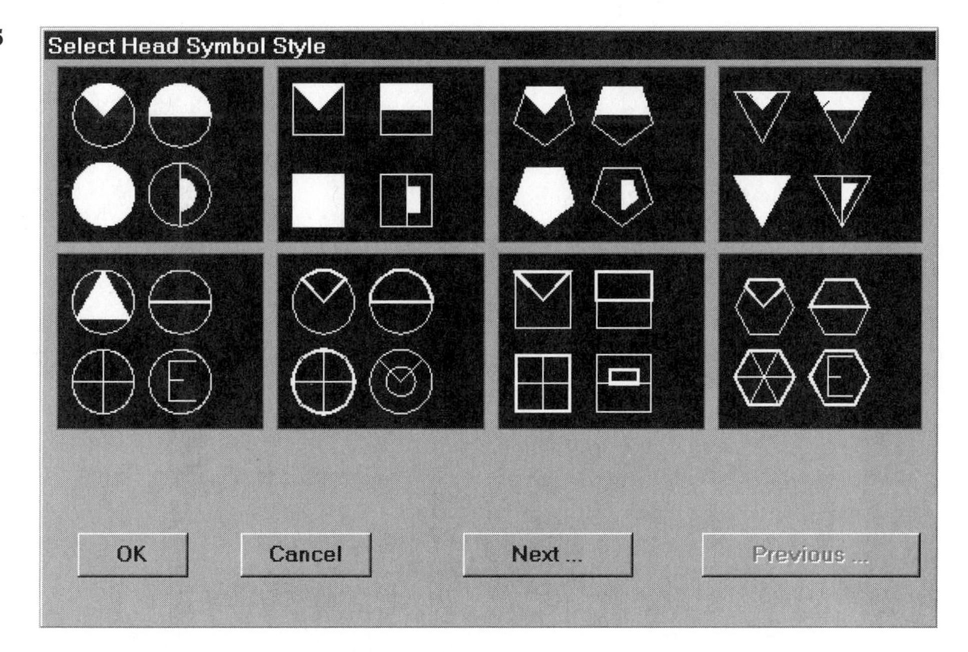

Pick the style shown in the upper left corner, then pick **OK**.

Back in the "Sprinkler Head Insertion" dialogue box, we can see our style selection shown in the **HEAD SYMBOL** portion of the dialogue box. We can also control the size of the symbol, and whether or not the symbol is rotated as it is inserted.

In the **Size** box, enter **1'**. This makes each sprinkler symbol appear at 1/8" diameter in the plotted irrigation plan.

Be sure that the **Rotate** toggle box has an "X" in it to allow the symbols to be rotated as they are inserted.

The **Layer . . .** box will bring up a dialogue box to select the desired layer where sprinkler head symbols will be inserted. We will accept the default layer **SPRINKLERS**.

We have a number of important settings to make in the **INSERTION METHOD** section of the dialogue box (Figure 15.6).

The insertion methods shown here allow us to choose from four options:

- **Autohead** inserts heads automatically to fill an irrigation zone. When this radio button is selected, the options below become active. We can choose whether to place the sprinklers in a **Square** or **Triangular** pattern using a

Figure 15.6

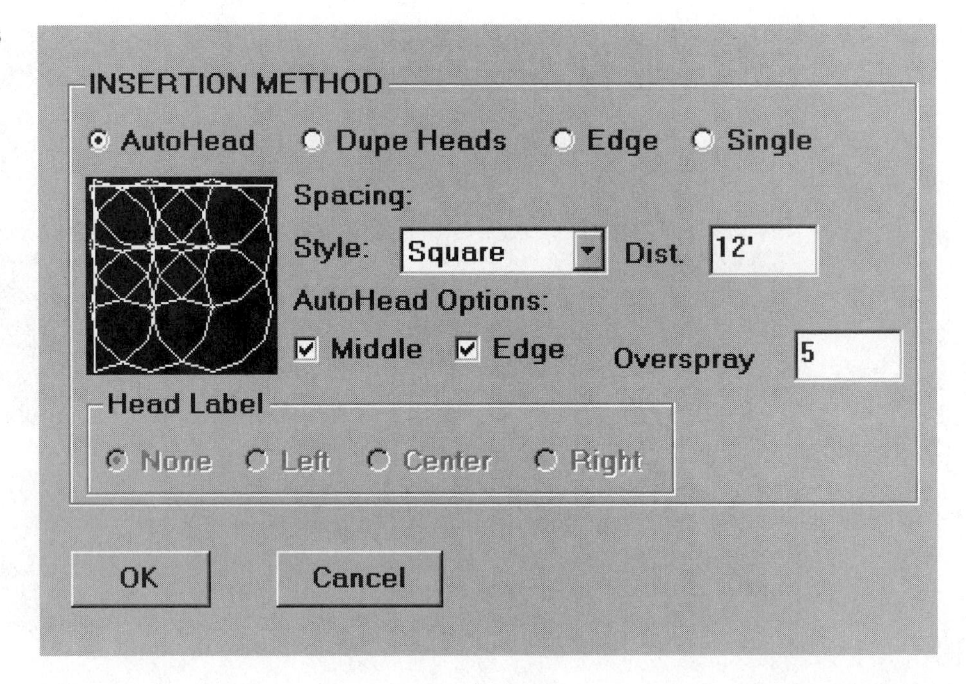

pop-down box. We can set the distance between heads using the **Dist.** box, although this value will be filled in with the radius chosen from the **HEAD DATA** selection. We can choose to allow **Autohead** to place the sprinklers in the middle of the zone, around the edges of the zone, or both.

The **Overspray** box allows us to control which heads are chosen to place along an angled edge. Does the program allow a 120° head to be used where the border makes a 135° angle? If the **Overspray** were set to **15**, the answer would be yes. For most cases, we should accept the value of **5**.

- **Dupe Heads** allows us to insert an array of **full head** sprinklers with a specified number of rows and columns. Once we select this radio button, the **Style** pop-down and the **Dist.** box become active. After setting the pattern and spacing, we are returned to the drawing screen. In the drawing screen, we are then asked to specify a number of (horizonal) rows and (vertical) columns for the sprinklers. We then show the starting point for the array, and specify any angle for the rows and columns to follow. **Dupe Heads** then fills in the sprinklers in the desired pattern.
- **Edge** allows the insertion of **half head** sprinklers along a line. If this radio button is selected, we can then choose the desired spacing between heads. After returning to the drawing screen, we are asked to specify a

starting and ending point for the row of heads. After picking the side where the heads are to cover, the row of sprinklers is placed along the line.

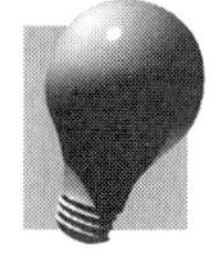

Hint: Note that both the **Dupe Heads** and the **Edge** options do not place **quarter heads** or other types of part-circle sprinklers.

- **Single** allows us to insert single sprinklers of any type which are available from the **HEAD DATA** selections made earlier. For the Toro 570 12' series MPR nozzles, a 120° arc head (1/3 circle) would be available, but a 135° arc head (3/8 circle) would not. This is because Toro does not manufacture a 135° nozzle for the 570 sprinkler. Once **Single** is chosen, the **Head Label** radio boxes become active. The individual heads can be labeled with an attribute as they are inserted. We will not be labeling individual heads, so the **None** option will remain selected. After picking **OK**, a dialogue box displaying available arcs is shown (Figure 15.7).

As shown here, there are six choices: 90° arc, 120° arc, 180° arc, 240° arc, 270° arc and 360° arc. If more choices are available for the head selected, they will be visible here. Pick the head desired and place it on the drawing.

We will take advantage of each of these methods of head insertion as we develop the irrigation plan for the Thomas residence.

Inserting Sprinklers in the Front Yard

We will begin our irrigation design by inserting sprinklers in the front lawn. This planting area has a size and proportion where three rows of 12 foot diameter sprinkler heads

Figure 15.7

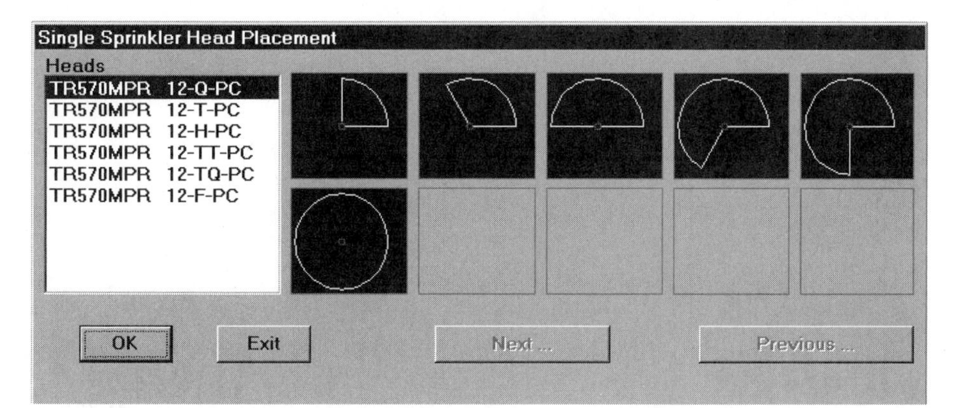

should give us adequate coverage. The unusual shape and angles of the edges of the lawn will probably result in some overspray onto the surrounding planters. With some modifications to the original settings for the Toro heads, we can create a reasonable sprinkler pattern.

In the **INSERTION METHOD** box, select **Autohead**.

In the **Dist.** box, enter a spacing of **10'**.

In the **Spacing Style:** pop-down box, pick **Square**.

Be sure that the **Middle** and **Edge** toggle boxes are filled, and the **Overspray** is set to **5**.

Pick **OK**.

At the prompt:

```
Command:EPLC_runme
Select closed polyline:
```
Pick the polyline which surrounds the front lawn (Figure 15.8).

```
Show lateral angle:
```
Make sure that **ORTHO** is active, then drag a vertical line toward the top of the screen. This assures that the rows of heads will be arranged vertically.

```
Select any areas to avoid head placement:
Select objects:
```
If there were any island areas within the lawn where we did not want sprinkler heads, we would pick them here. Press **<Enter>** to continue.

Figure 15.8

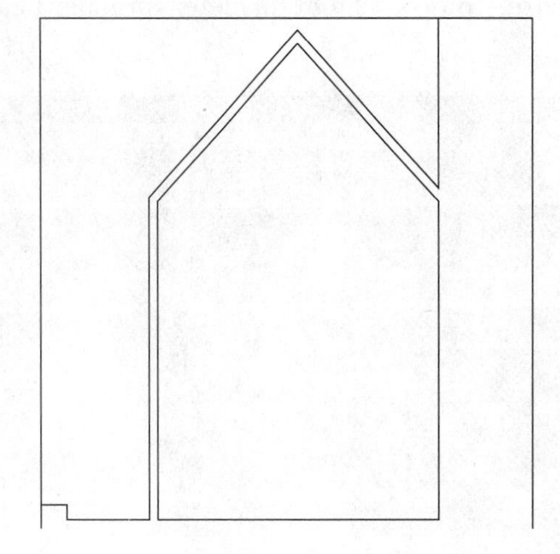

```
Auto locate 1st full circle head or pick location (Auto/Pick)
<Auto>:
```

Type **A** or **<Enter>** if **Auto** is the default. We will allow the **Autohead** routine to place the first full circle head.

```
Processing polyline please wait ...
```

The heads are inserted into the lawn area as shown below (Figure 15.9.)

Distribution Analysis

To check on the uniformity of coverage in the lawn area, we can use the **Distribution Analysis** function. This command will place screened coverage templates over each head so that we can see whether we have dry spots, excessive overspray, or uniform coverage. The **Distribution Analysis** function places these templates on a special layer that can be turned off when not in use.

To perform a distribution analysis, pick **Distrib. Analysis** under the **Heads** pull-down menu (Figure 15.10).

Figure 15.9

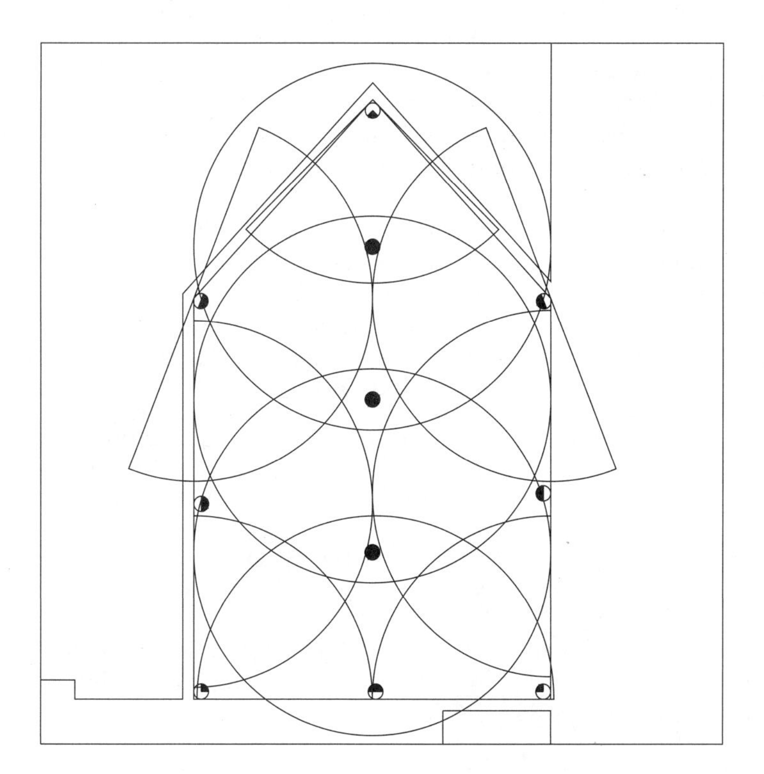

Figure 15.10

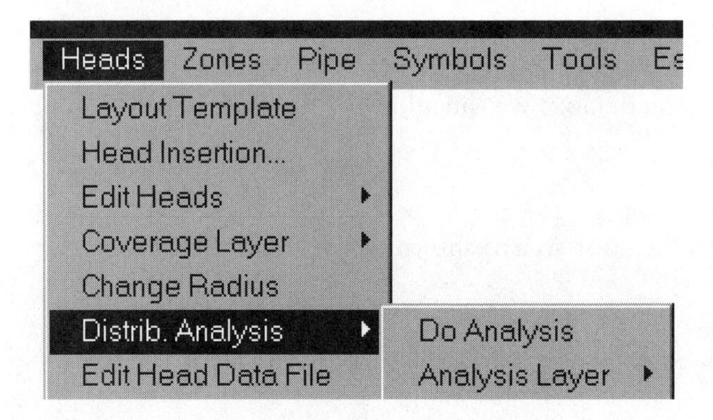

Pick **Do Analysis**, then select all the heads in the front lawn, to see the "densogram" for the front lawn sprinklers (Figure 15.11).

The "densogram" indicates reasonable uniformity without excessive overspray onto surrounding areas. To remove the Distribution Analysis templates from the screen, pick the **Analysis Layer** function, then select **OFF**.

Remember that the distribution analysis templates are not erased each time the **Analysis Layer** routine is used. Instead, the **L_IRRIG_DIST** (or equivalent) layer is turned off. This means that if an analysis were done in an area, and new heads were added later to eliminate dry spots, the old analysis templates would still be there. Your new distribution analysis would be written over the top of the old one, making it impossible to read. **You need to erase the old templates each time you perform a new Distribution Analysis**. There are two easy ways to accomplish this:

1. Immediately after performing the Distribution Analysis, simply type **UNDO**, then choose **BACK**. This will undo all of the commands until a special marker is encountered. Many routines place a marker before they start, so that you can undo them easily. Distribution Analysis works this way. When you choose **UNDO/BACK** all of the Distribution Analysis templates that were just drawn will disappear, but no other work will be lost.

2. Isolate the **L_IRRIG_DIST** (or equivalent) layer. This means that all other layers are turned off (or frozen) so that the templates can be erased easily without affecting objects on other layers.

   To isolate a layer, choose **Layers ▸** in the cursor menu, then select **By OFF** under **Isolate Entity Layer:**. See below (Figure 15.12).

After you pick one of the templates, all other layers are turned off. Erase the Distribution Analysis templates. Return to the cursor menu to turn on all the layers which were turned off (use the **Layer Control . . .** option).

Figure 15.11

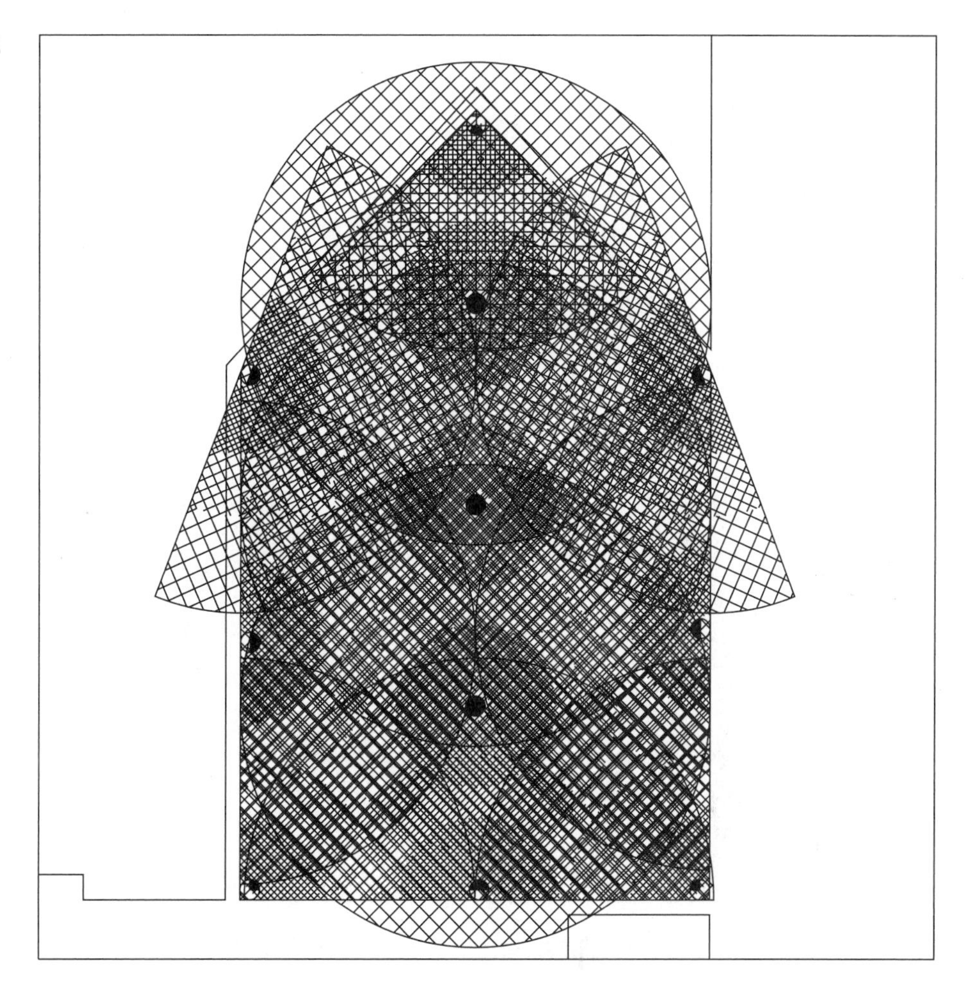

If the coverage pattern did not appear to be adequate, we would add sprinklers, or even change nozzles (and head diameters) to try to improve the uniformity. This process is very common in irrigation design, and is made easy in LANDCADD.

Should you make an error in the type of head you have selected, and/or the coverage pattern is not what you had in mind, getting rid of all of the heads and coverage arcs is easy. Either use the **UNDO/BACK** option, or use the Edit Heads option under the **Heads** pull-down menu. When Edit Heads is selected, choose Erase. Pick the heads you wish to erase. The heads (and their coverage arcs) will disappear. All other objects on the screen will be filtered out and not affected by this command.

Figure 15.12

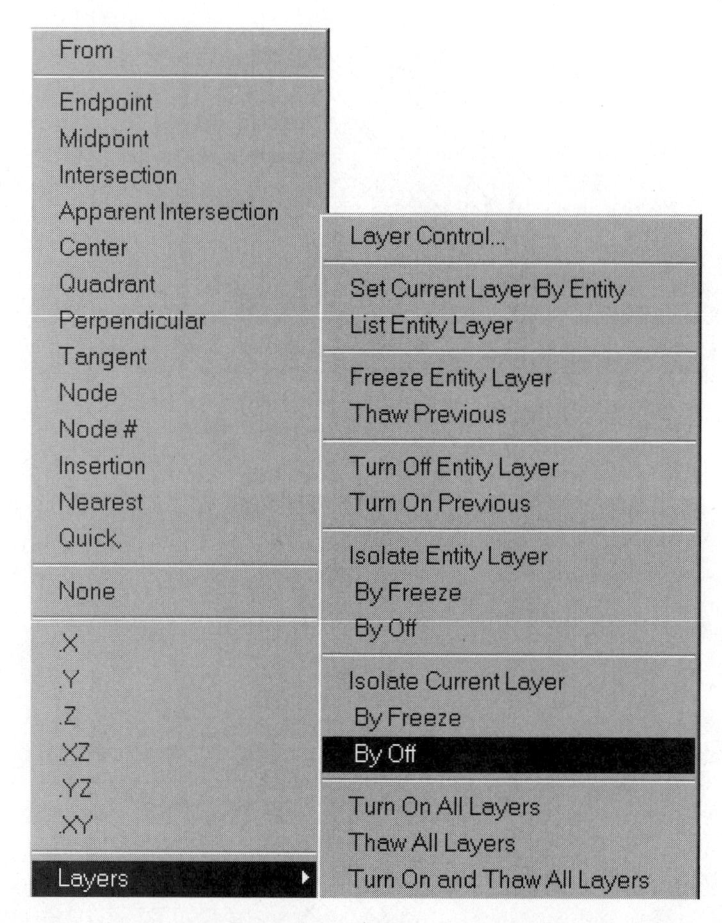

Inserting Sprinklers in Planter Area 1

Planter area 1 has a very irregular shape which features both wide and narrow angles around its perimeter. Generally, for a planter of this type, a smaller radius head is used to help conform to these unusual angles. For this reason, we will switch to the Toro 570 10' Series MPR nozzles.

Pick **Head Insertion . . .** under the **Heads** **Heads** pull-down menu.

Pick **Data File**, then accept the **DATA FILE** used previously: **TORO 570 MATCHED PRECIPITATION RATES.**

In the **NOZZLES** section, pick **10' Series with 12 deg. Trajectory Pressure Compensating**.

In the **PRESSURE RADIUS** section, select **35 10** (35 PSI, 10' radius).

Pick **OK** when finished.

We have defined all the parameters which affect the spacing and performance of the sprinkler heads.

In the "Sprinkler Head Insertion" dialogue box, move to the **HEAD SYMBOL** box, and select a family of symbols which are different than those used in the front lawn. Even though the sprinkler itself is the same, the nozzle is different, so a different symbol must be used to denote this. Accept the **Size** standard of **1'**.

Accept the **Autohead** options for inserting both **Middle** and **Edge** sprinklers, and leave the **Spacing Style:** set at **Square.**

Pick **OK** when finished.

At the prompt:

```
Command:EPLC_runme
Select closed polyline:
```
Pick the polyline which surrounds planter area 1.

```
Show lateral angle:
```
Make sure that **ORTHO** is active, then drag a vertical line toward the top of the screen. This assures that the rows of heads will be arranged vertically.

```
Select any areas to avoid head placement:
Select objects:
```
If there were any island areas within the lawn where we did not want sprinkler heads, we would pick them here. Press **<Enter>** to continue.

```
Auto locate 1st full circle head or pick location (Auto/Pick)
<Auto>:
```
Type **P** for **PICK.**

```
Pick location of 1st full circle head:
```
Pick on the point **117',112'**. This initial point causes **Autohead** to place the full heads in rows that fit well within Planter area 1.

```
Processing polyline please wait ...
```
The heads are inserted into the planter as shown below (Figure 15.13).

After **Autohead** is completed, you will find some extra heads and some full heads out of position in the area near the front door.

Figure 15.13

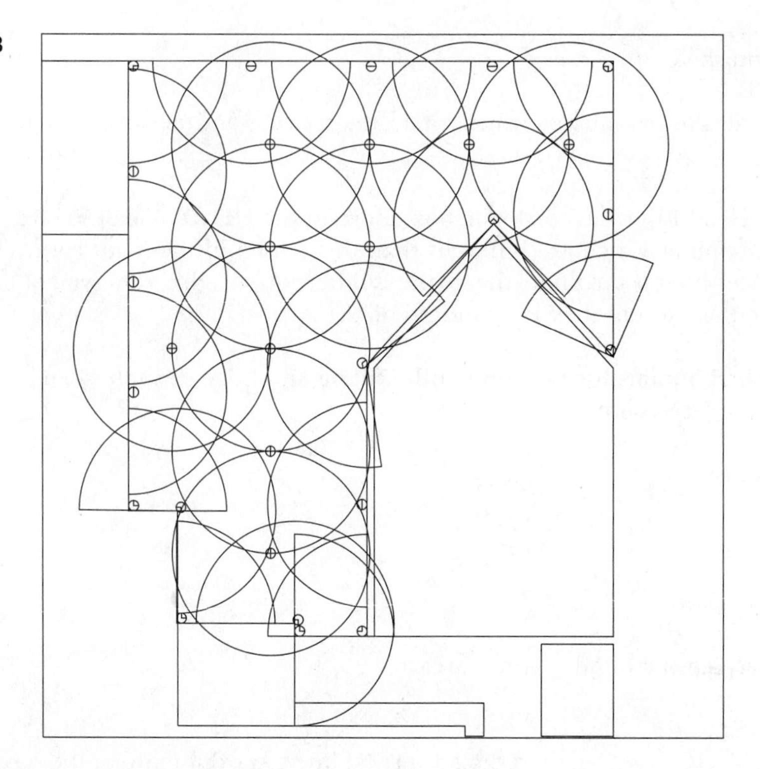

Use **Edit Heads ▸** to remove the extra 1/4 circle (90°) head, 3/4 circle (270°) heads and full (360%) heads from the area near the front door. See the illustrations for details. (Figures 15.14, 15.15, and 15.16).

Use the **Single** option under **Head Insertion . . .** to insert the additional 1/2 (180°) heads along the perimeter of the planter near the front door.

The only remaining problem in Planter area 1 relates to the area near the point of the front lawn. We need to add two 1/2 (180°) heads along the edge of the planter to assure proper coverage (Figures 15.17 and 15.18).

Figure 15.14

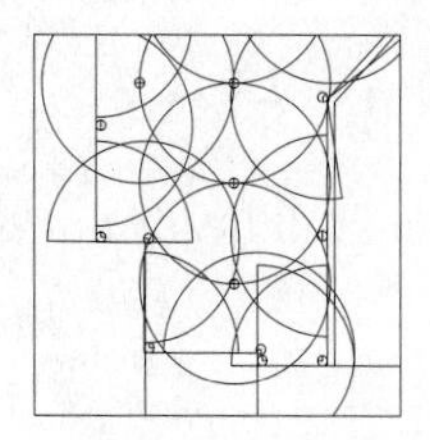

Figure 15.15

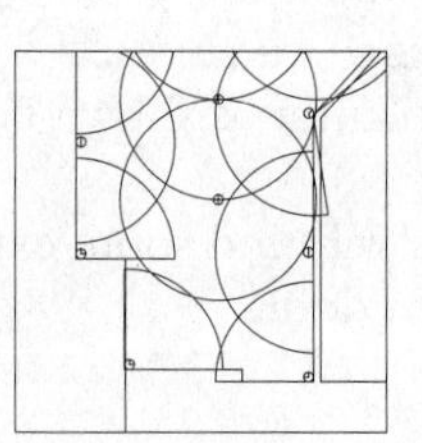

Figure 15.16

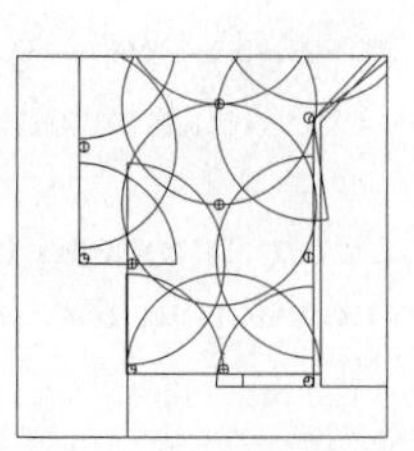

Figure 15.17

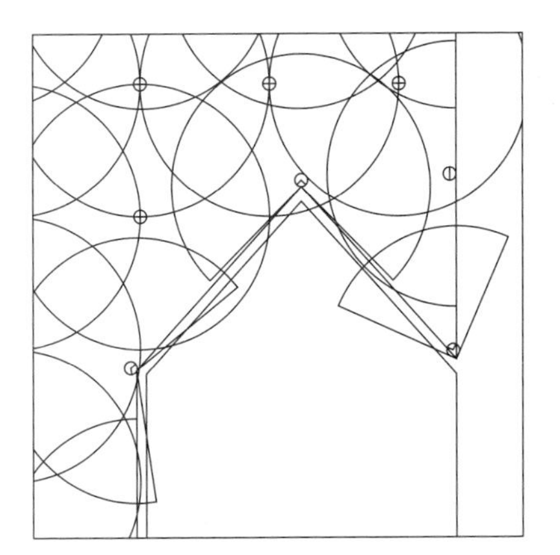

Figure 15.18

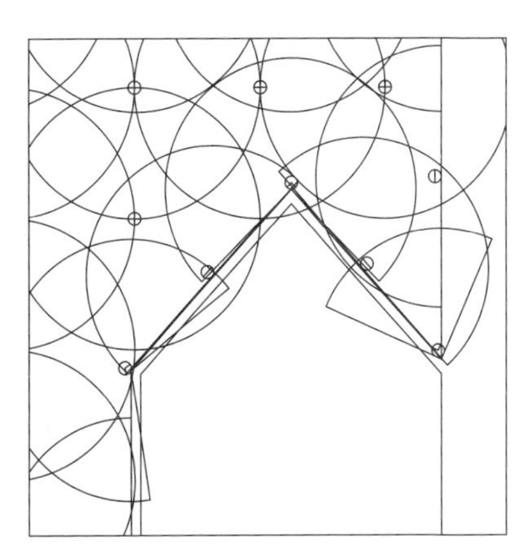

Planter Areas 2 and 3

With the head placement in the Front Lawn and Planter area 1 completed, we can proceed to the smaller planters in the front yard. We need to use small radius sprinklers and bubblers in the small planter areas shown below (Figure 15.19).

Planter area 3 on the right is about 7'-3" across and about 10' long. Because this area is planted with shrubs—and not turf—we can locate traditional spray heads or microspray heads at the corners. Although this coverage is not ideal, it will be adequate for the plants involved.

In the **DATA FILES** section, select **TORO 570 MATCHED PRECIPITATION RATES**.

In the **NOZZLES** section, select **8' Series with 5 deg Trajectory Pressure Compensating**.

Figure 15.19

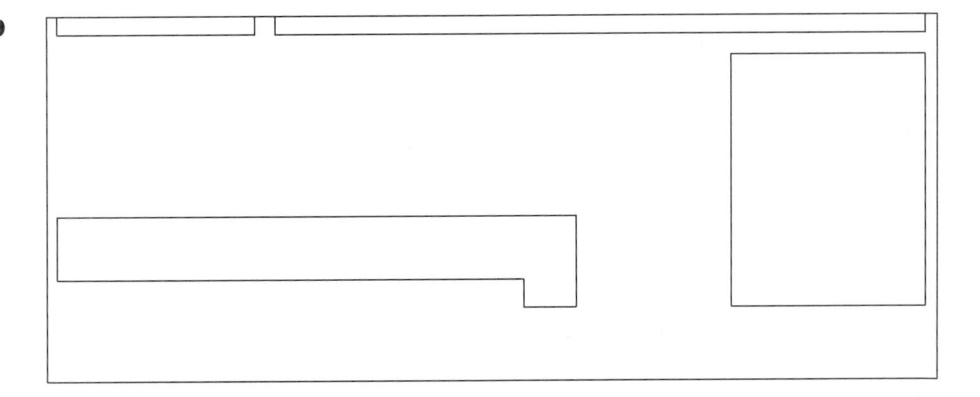

In the **PRESSURE RADIUS** section, select **35 8** (35 PSI, 8' radius).

Back in the "Sprinkler Head Insertion" dialogue box, adjust the **Radius:** box to **7'**. We will specify the use of the radius adjustment screw to reduce the radius of these heads.

Change the **HEAD SYMBOL** to another symbol family, retaining the **1'** symbol size.

Use **Autohead** to insert the sprinklers into Planter area 3. The heads will only appear at the edges of the planter, regardless of whether or not the **Middle** toggle box is activated, and regardless of which **Spacing Style:** is chosen. See the results below (Figure 15.20).

Planter area 2 has a series of spreading groundcover plants which will have fairly extensive root systems. This means that bubbler heads would probably be appropriate here, so long as the heads are close enough together to give adequate coverage. We will assume that the heads should be no more than 2' apart.

To place bubbler heads, select the **Equipment** option from the Symbols pull-down menu. This brings up the "Irrigation Peripheral Equipment" icon menu (Figure 15.21).

Select the bubbler symbol, then pick **OK**.

At the prompt:

`Locate symbol:`
Locate the bubbler toward the left end of the planter, centering it between the edges.

`Scale symbol:`
Type **5** to scale the bubbler to the correct size.

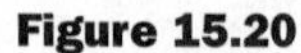

Figure 15.20

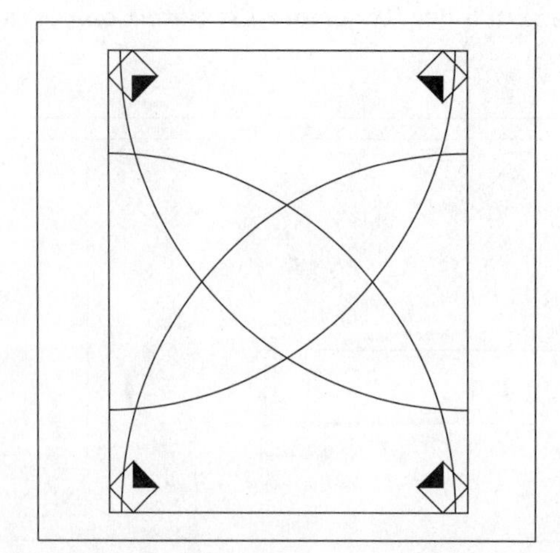

Figure 15.21

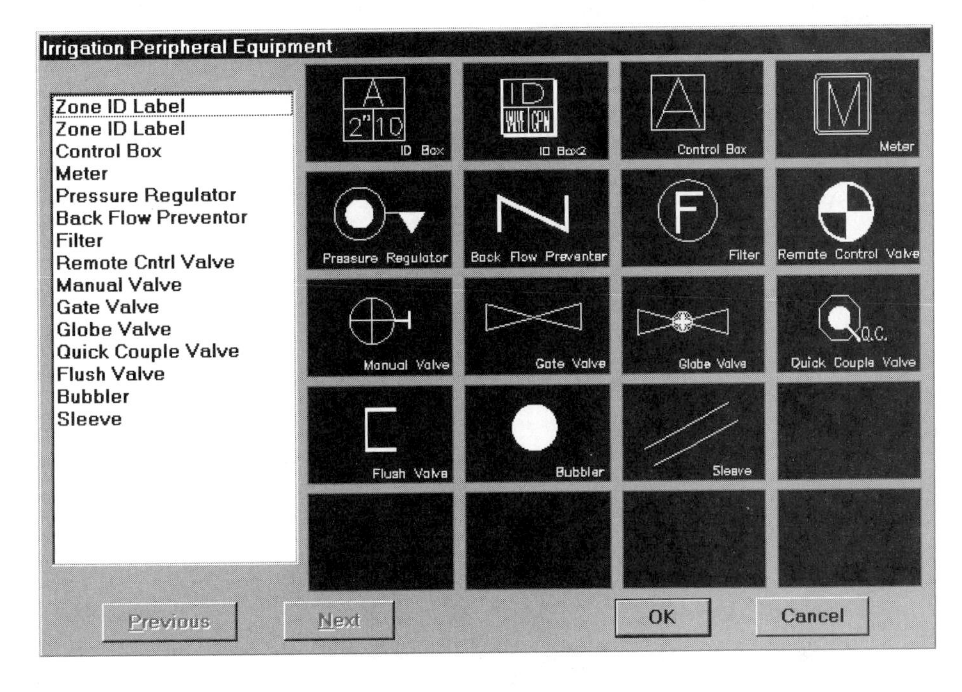

`Rotate symbol:`
Pick any rotation angle (since the symbol is round, this does not matter).

Before the symbol is inserted, an "Edit Attributes" dialogue box appears (Figure 15.22).

Figure 15.22

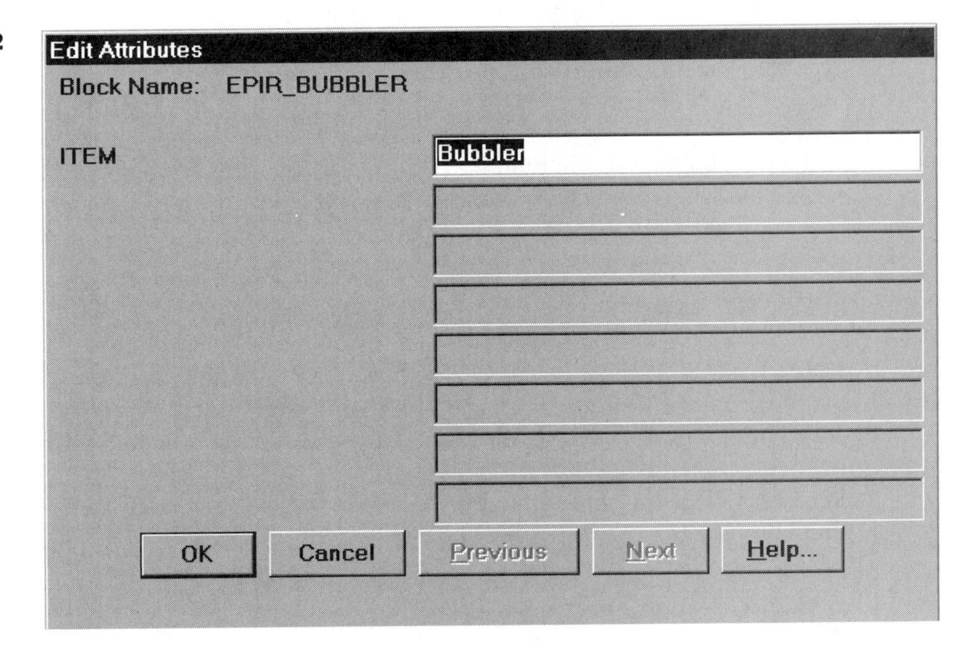

Change the attribute to **FB-100-PC**, the model number of a 1 GPM Toro pressure compensating flood bubbler (Figure 15.23).

Create multiple copies of the bubbler using the **Array** command.

At the prompt:

`Command:`
Type **ARRAY** (or **AR**).

`_.ARRAY`
`Select objects:`
Pick the bubbler you just inserted.

`Select objects: 1 found`
Press **<Enter>** to complete the selection process.

`Rectangular or Polar array (R/P) <R>:`
Press **<Enter>** to accept a **Rectangular** array.

`Number of rows (--)<1>:`
Press **<Enter>** to accept 1 row.

`Number of columns (|||)<1>:`
Type **10** to create a total of 10 bubblers across the planter.

`Distance between columns (|||):`
Type **23**, which gives a uniform spacing of 23 inches between bubblers. The distance between bubblers may vary slightly depending on where the first bubbler was inserted.

Add a final bubbler to the right end of the planter using the **Copy** command. Place this symbol at the approximate spot shown below (Figure 15.24).

Planter Area 4

To place heads in Planter area 4, we will continue to use the **Toro 570 MPR** nozzles with the **8'** radius—reduced to **7'**. Because the same equipment is used with the same operating conditions, we can use the same symbols as we used in Planter area 3.

Retaining the same settings as before, use **Autohead** to place the heads. The results should look similar to those shown below (Figure 15.25).

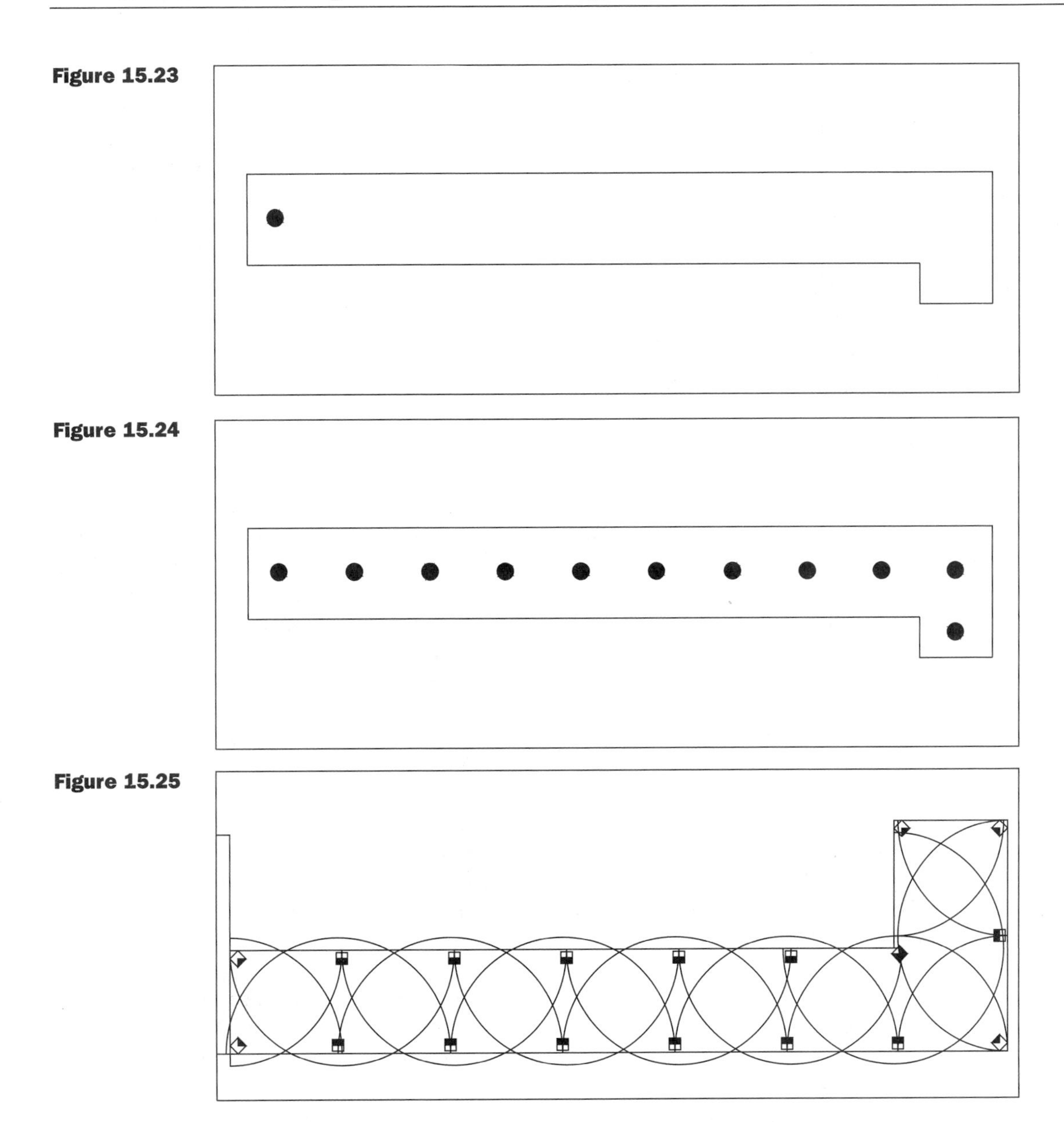

Figure 15.23

Figure 15.24

Figure 15.25

Planter Areas 5 and 6

Planter areas 5 and 6 are small, densely planted beds positioned close to the house. These beds are ideal candidates for drip irrigation. LANDCADD has several methods for building drip irrigation stations.

One of the methods for developing a drip system involves placing drip emitters at each plant in a given planter area. Since we developed our irrigation plan without including the plant symbols, we will need to retrieve them using the **WBLOCK** command. We will use this command to build a planting drawing—containing only the plants contained in Planter area 5 and Planter area 6. We will the use the **DDINSERT** command to place these plants into the sprinkler plan. After we finish locating the drip emitters, we can freeze the planting layers to prevent them from being visible in the irrigation plan.

Some designers will include all of the landscape information on a single plan, placing information on separate layers. Theoretically, a base plan, hardscape plan, construction plant, planting plan, and irrigation plan could be condensed into a single drawing file. The designer would then turn layers off and on in order to control the information visible on the screen. There are advantages and disadvantages to this technique. The advantages include fewer drawing files, access to **all** plan information at all times, and the ability to plot multiple drawings from a single drawing file. The disadvantages include large drawing file size (drawings larger than 1 megabyte are common) which can slow system performance, and large numbers of drawing layers which must be controlled by the user. Controlling multiple-layer display can be time consuming, especially if there are thirty or more layers in a drawing. Individual office standards, drawing size and complexity, and layering techniques must all be considered when choosing methods for controlling the different drawings for a particular project.

To build the planting block, save and close your current drawing, and reopen the Planting Plan. Use **WBLOCK** to create a drawing block of the plants in Planter areas 5 and 6. See below (Figures 15.26 and 15.27).

Select all of the plants in Planter areas 5 and 6, and save them as the drawing **DRIP**. Use special care to include all of the plants in the block, but avoid selecting any of the concrete border lines, plant outline arcs or other unwanted entities. Use the lower left corner of Planter area 5 as the insertion point for the block. This makes it easy to locate the block in the irrigation plan.

Use **DDINSERT** to insert the **DRIP** block into the irrigation plan. Use the corner point from Planter area 5 to locate the block in the corresponding point in the plan.

See the results below (Figure 15.28).

This new block is visible in the irrigation plan, but the plants cannot be recognized until the block has been exploded. Use **EXPLODE** to separate the **DRIP** block into individual plants which the drip irrigation routines can recognize.

Next, we will use the drip irrigation routine to locate drip emitters in the plan. To begin the command, use the **Pipe** pull-down menu, and select **Drip . . .**. This opens the "Drip Irrigation" dialogue box (Figure 15.29).

Figure 15.26

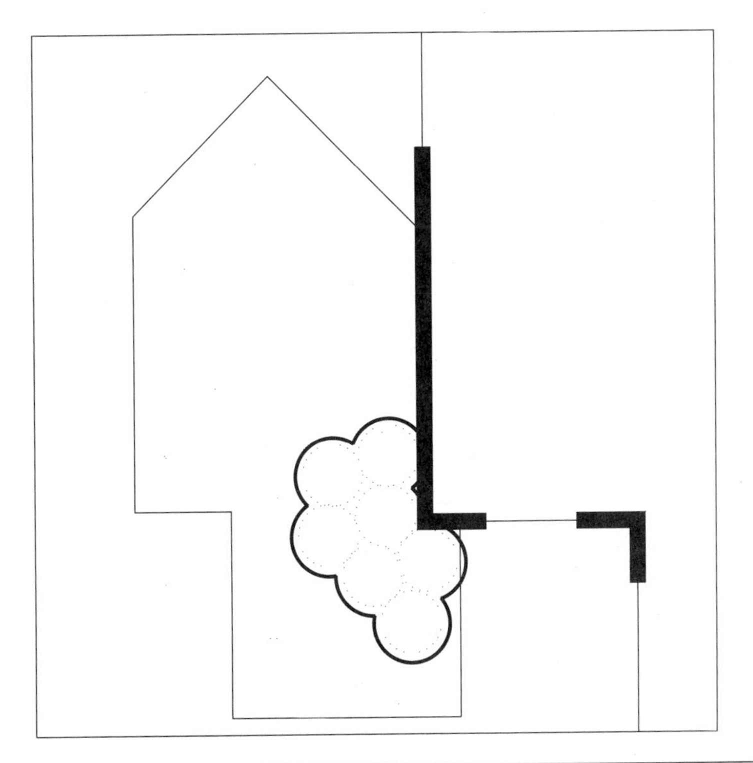

Figure 15.27

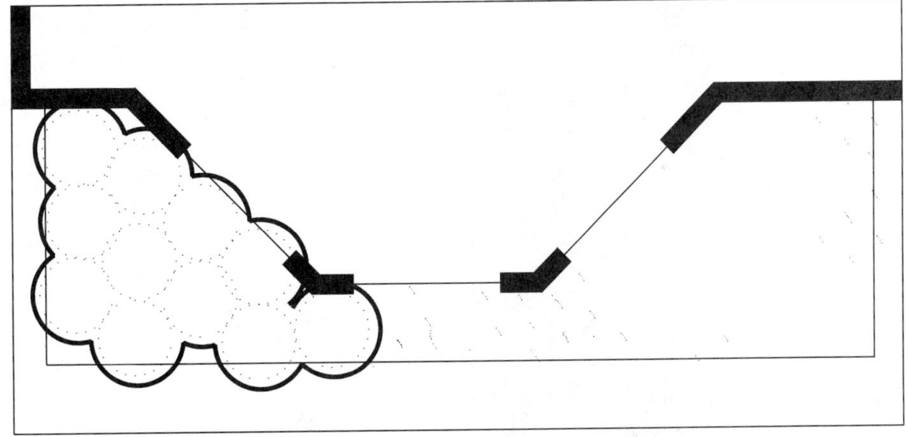

This dialogue box allows access to three primary functions for designing drip irrigation. In the **Emitters** section, we can choose plants which will be provided with drip emitters. We can specify the flow rate of the emitter, the size of the emitter symbol, and the number of emitters per plant. After the selection of plants is made, the total number of emitters is displayed, along with the total flow rate in gallons per minute.

Figure 15.28

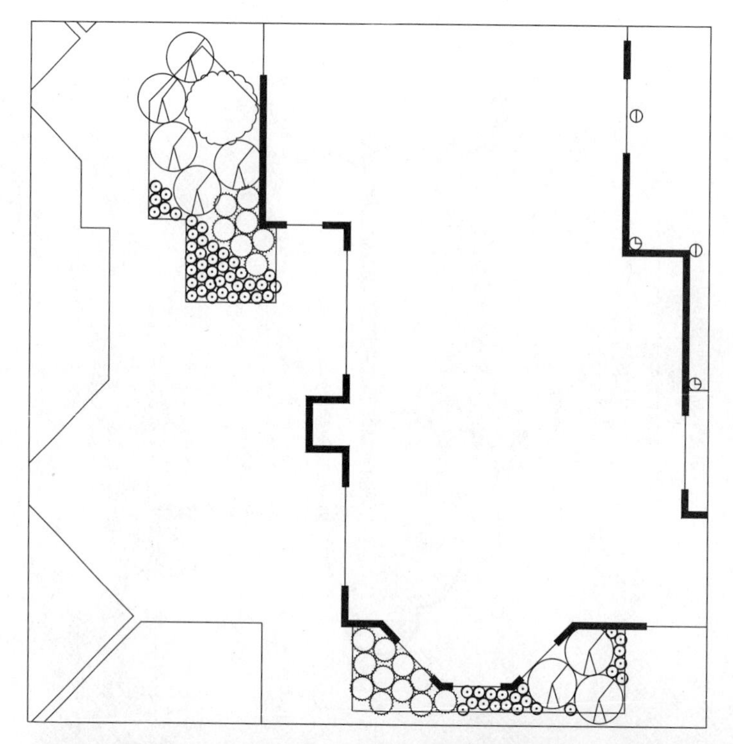

Figure 15.29

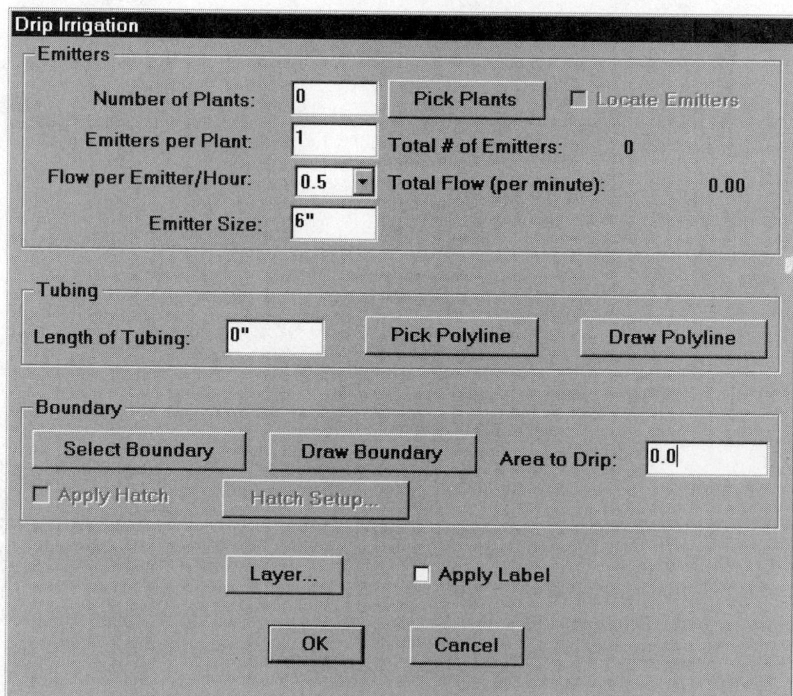

In many drip systems, the designer specifies different emitter sizes for different plants. This way, larger, more thirsty plants can be run on the same station as smaller, more drought-tolerant plants. In addition, a designer may choose to provide more than one emitter per plant in order to meet that plant's water requirements. A good rule of thumb is to have the wetted soil area under the plant be 50% or more of the total plant canopy area. There are a number of excellent drip irrigation publications available for further reading.

In the **Tubing** section, the designer can draw a polyline to represent drip tubing between emitters. We will return to this function when we add piping to the irrigation plan.

The **Boundary** section allows the designer to hatch an area to be provided with drip equipment. We will not use this function in our example.

We will begin by providing 2 GPH emitters for the larger plants.

First, pick **Pick Plants**. This routine then returns you to the drawing screen, allowing you to select any plants you wish to provide with emitters. Pick the medium (2 ft. and 3 ft.) diameter plants in both planter areas. Do not select the small (1 foot) diameter or the largest (5 ft.) diameter plants at this time. Press **<Enter>** when finished selecting these plants.

In the **Number of Plants:** box, be sure the number is **26**.

In the **Emitters per Plant:** box, accept the default value of **1**.

In the **Flow per Emitter/Hour:** pop-down box, select the entry **2**.

For **Emitter Size:** enter **6"**.

Be sure to place an "X" in the **Locate Emitters** toggle box.

Pick **OK** when finished. LANDCADD places a predefined emitter symbol at the center of each plant.

Repeat this process for the small diameter plants, providing them with a single **.5 GPH** emitter. See below (Figure 15.30).

Finally, we will place **three 1 GPH** emitters around the large Camellia plant in Planter area 6.

See the illustrations below for drip emitter placement (Figures 15.31, 15.32, 15.33, and 15.34).

Figure 15.30

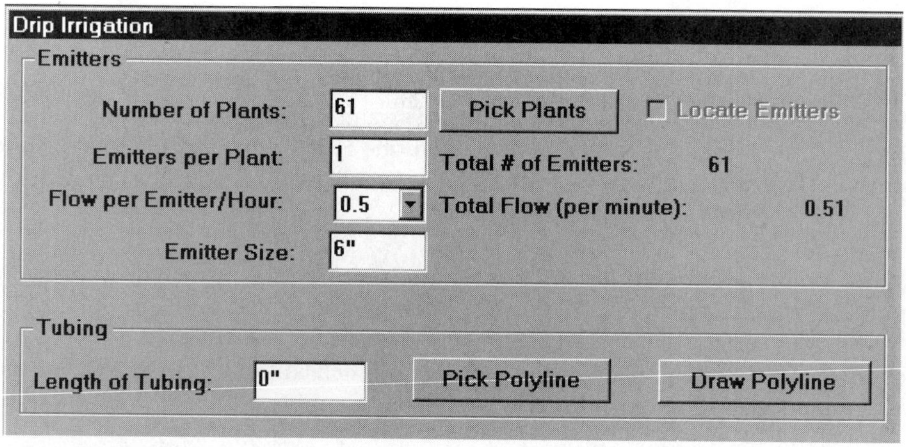

Drip Irrigation

Emitters

Number of Plants: 61 | Pick Plants | ☐ Locate Emitters

Emitters per Plant: 1 | Total # of Emitters: 61

Flow per Emitter/Hour: 0.5 | Total Flow (per minute): 0.51

Emitter Size: 6"

Tubing

Length of Tubing: 0" | Pick Polyline | Draw Polyline

Figure 15.31

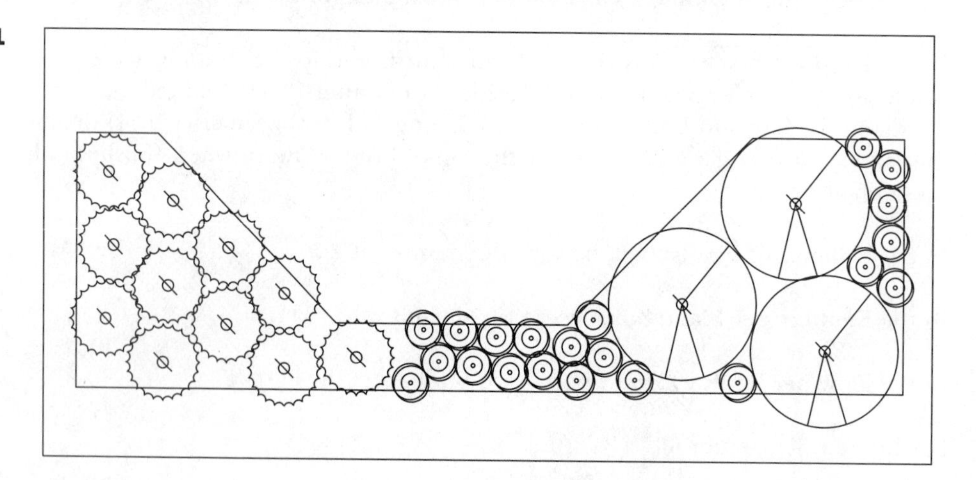

Figure 15.32

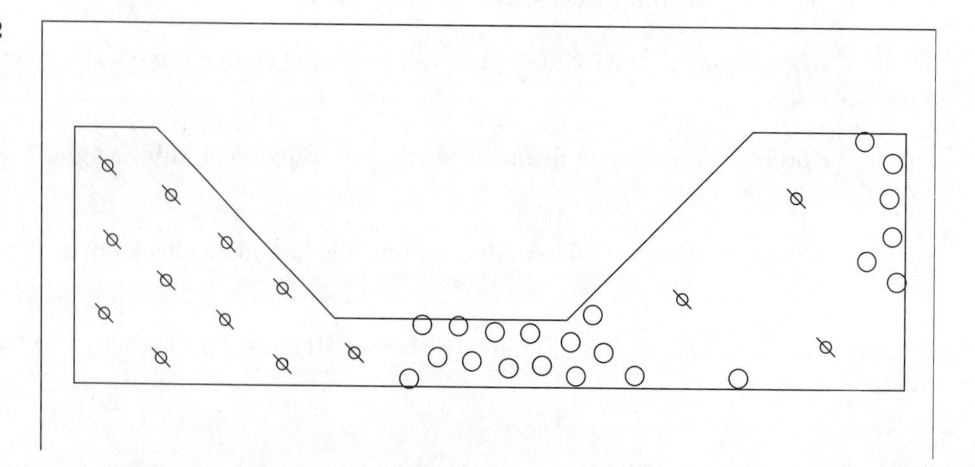

Figure 15.33

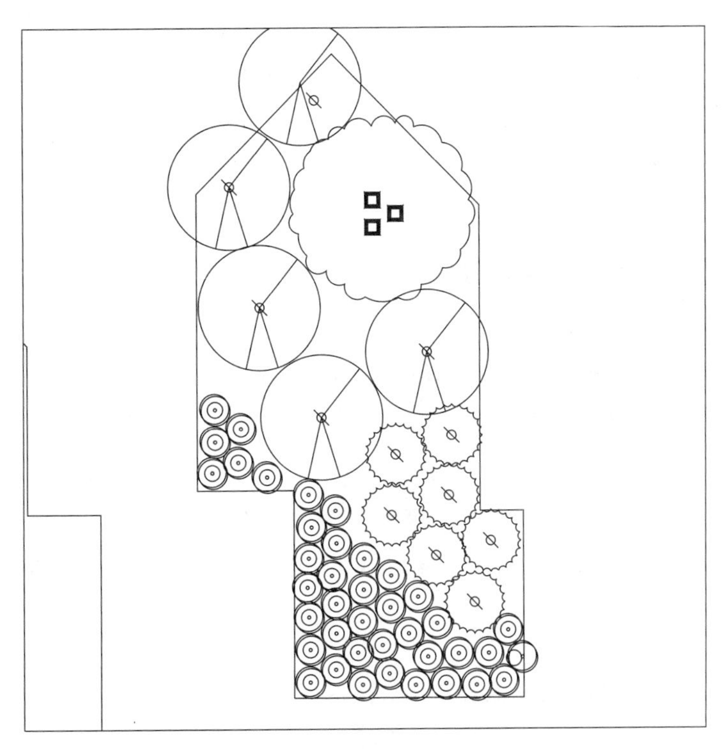

Figure 15.34

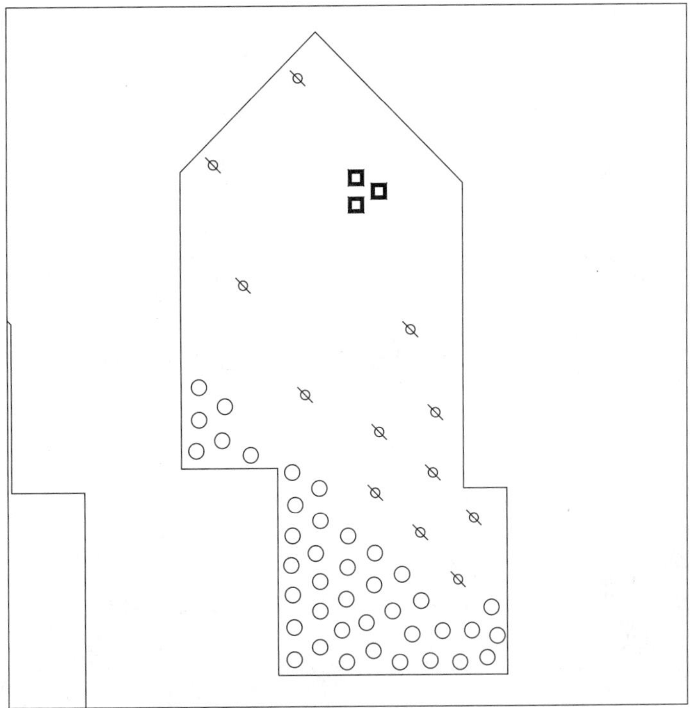

Use the cursor menu to freeze the layers which contain the plant symbols. Select **Layers ▸** then **Freeze Entity Layer**, then pick any of the plant symbols. They should all disappear from the screen. If some plants do not disappear immediately, the symbols remaining are on a different layer. Repeat the command as necessary to make all of the plants disappear.

Planter Areas 7 and 8

The irrigation system in Planter area 7 makes use of the **Rainbird 1800 Series** pop-up spray heads. We will use the **15' Series** nozzles. The **Autohead** command can be used to establish the basic pattern, while a few heads must be erased, relocated and added to make the coverage more uniform. See below for details (Figure 15.35).

Pick **Head Insertion . . .** under the **Heads** pull-down menu.

Pick **Data File**, then pick the **DATA FILE** for **RAINBIRD 1800 SERIES.**

Pick **15' Series trajectory 30 deg.**

In the **PRESSURE RADIUS** section, select **30 15** (30 PSI, 15' radius).

Pick **OK** when finished.

In the “Sprinkler Head Insertion” dialogue box, move to the **HEAD SYMBOL** box, and select a new family of symbols. Accept the **Size** standard of **1'**.

Accept the **Autohead** options for inserting both **Middle** and **Edge** sprinklers, and leave the **Spacing Style:** set at **Square**.

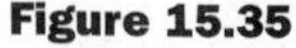
Figure 15.35

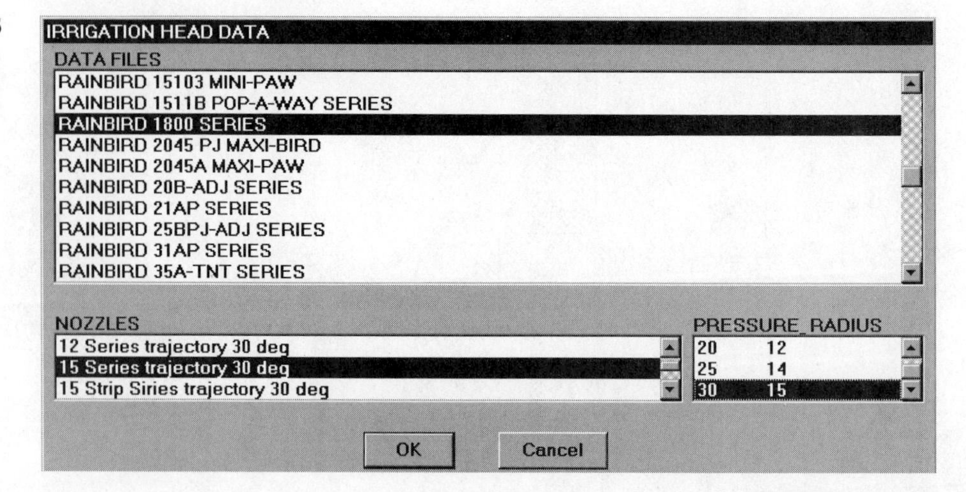

Pick **OK** when finished.

At the prompt:

```
Command:EPLC_runme
Select closed polyline:
```
Pick the polyline which surrounds Planter area 7.

```
Show lateral angle:
```
Make sure that **ORTHO** is active, then drag a horizontal line toward the back property line.

```
Select any areas to avoid head placement:
Select objects:
```
Press **<Enter>** to continue.

```
Auto locate 1st full circle head or pick location (Auto/Pick)
<Auto>:
```
Type **A** for **Auto**.

```
Processing polyline please wait ...
```
The heads are inserted in normal fashion. The only problem area centers near the narrow point of the back lawn (Figure 15.36).

Note that the full-circle and the 3/4 (270°) heads cause unnecessary overspray. In addition, the 1/4 (90°) head is rotated incorrectly.

Use the **Edit Heads** function to erase the full and 3/4 (270°) heads, then insert half heads along the perimeter to take their place.

Rotate the 1/4 (90°) head into position. See below for the correct head placement (Figure 15.37).

Use the same procedure (and sprinkler heads) to place heads in Planter area 8. There are a number of problem areas in this planter. We will begin in the area above the bedroom patio (Figure 15.38).

There are several extra 3/4 (270°) heads along the perimeter, and full-circle head is obviously misplaced, so remove those first (Figure 15.39).

Insert a 1/2 (180°) head and a 2/3 (240°) head to correct the coverage deficiencies (Figure 15.40).

Figure 15.36

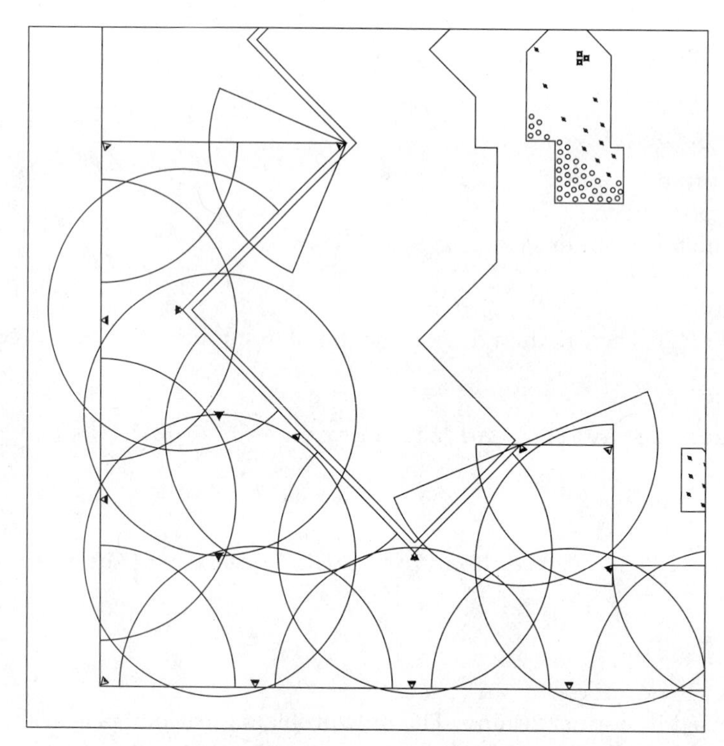

Figure 15.37

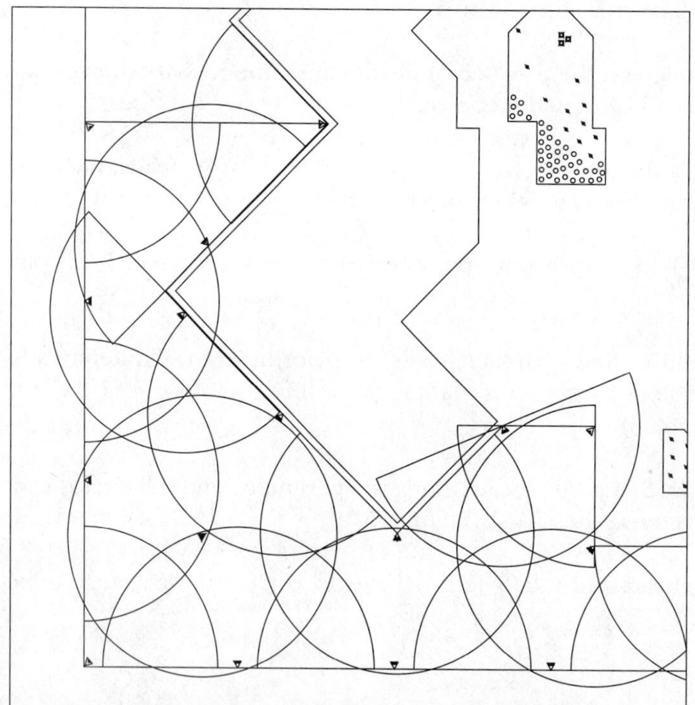

Figure 15.38

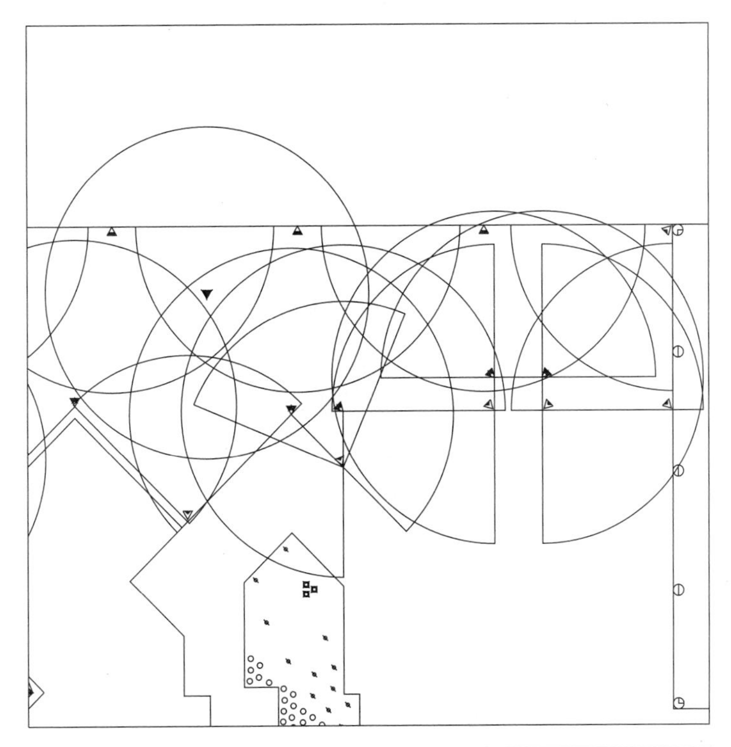

Figure 15.39

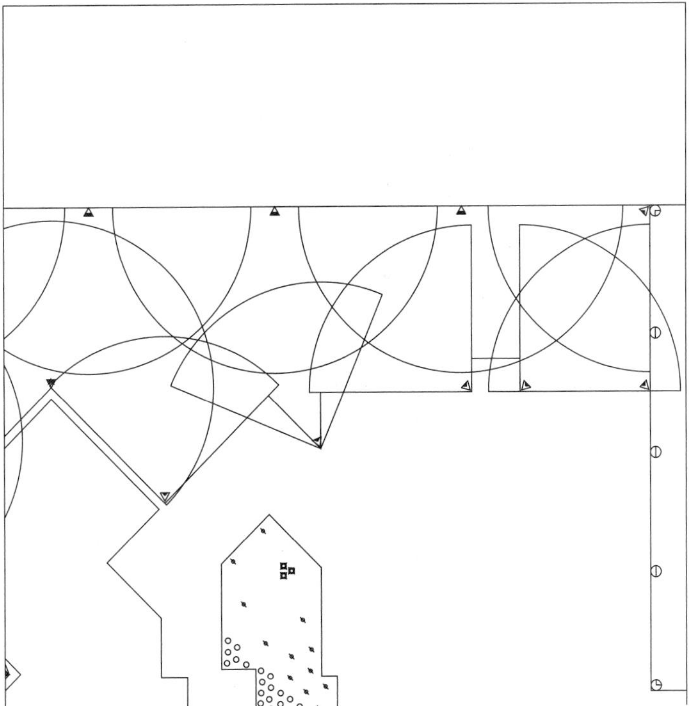

Figure 15.40

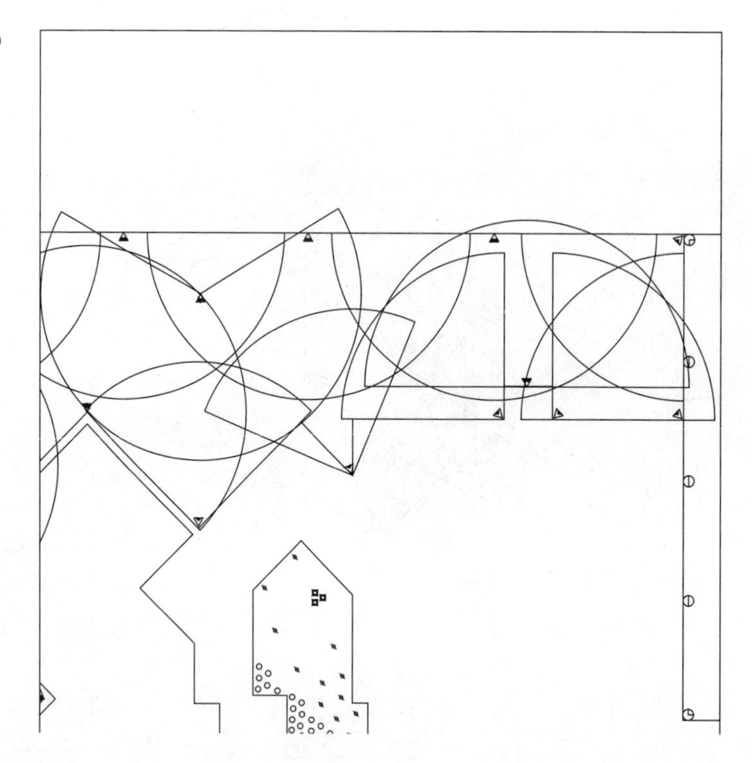

The next area requiring our attention lies near the narrow point in the back lawn (again) (Figure 15.41).

Remove the extra 1/4 (90°) head at the corner point, and replace the two 1/4 (90°) heads with a single 1/2 (180°) head. Finally, move the full-circle head near the upper left corner of the back yard into a better position (Figure 15.42).

Dividing the large back planter into Planter areas 7 and 8 caused some duplication of heads, and forced some extra editing. However, the advantage of this method lies in the ability to more carefully control the placement of a smaller number of heads. In addition, the division of this area into two planter areas might roughly correspond to our sprinkler stations. We will examine the division of the yard into irrigation zones when we add piping to the irrigation plan.

Figure 15.41

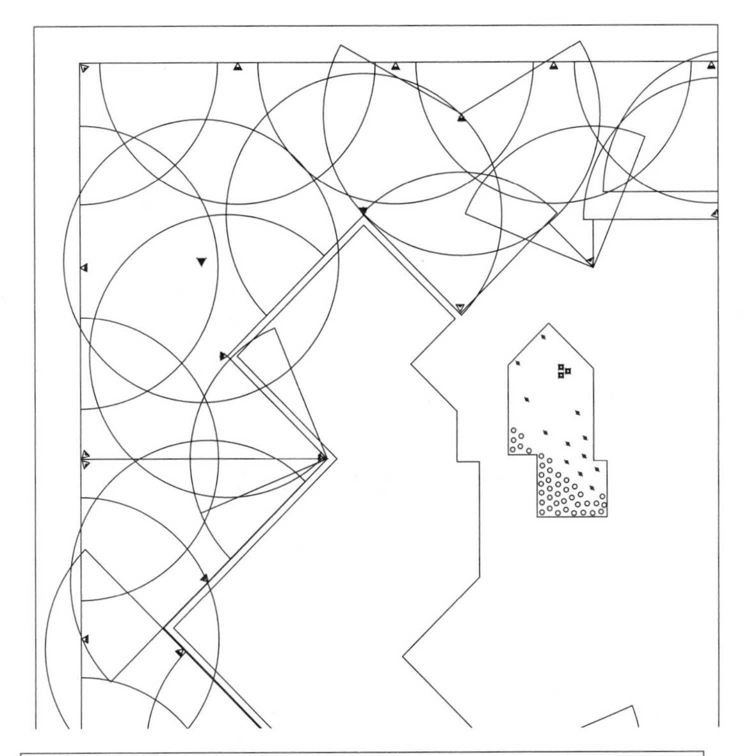

Figure 15.42

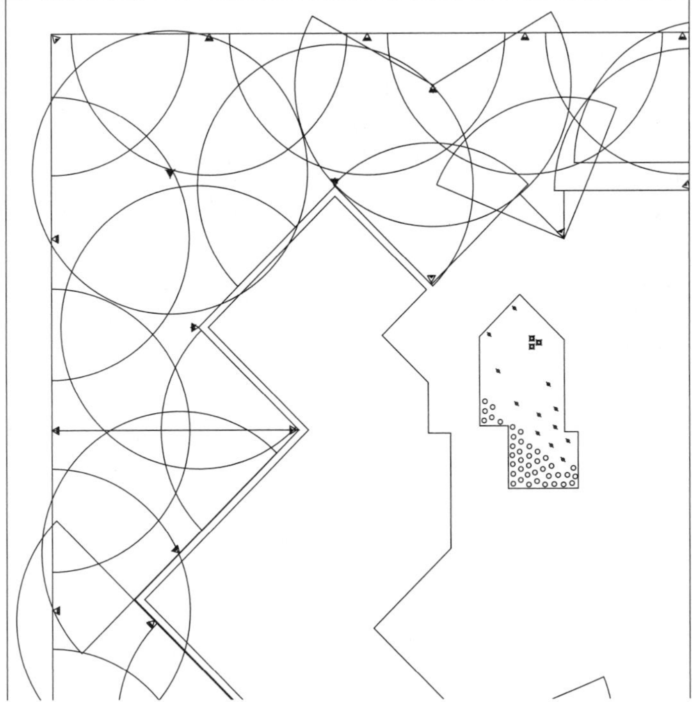

The Back Lawn

In the back lawn, we will continue to use the **RAINBIRD 1800 SERIES** equipped with **15'** radius nozzles. We will, however, be using a different type of sprinkler body in the turf area. We need to select a different symbol for these heads, even though the **DATA FILE**, **NOZZLES** and **PRESSURE RADIUS** information remain the same.

Pick **Head Insertion . . .** under the **Heads** pull-down menu.

Accept the setting used previously in **Data File**.

Move to the **HEAD SYMBOL** box, and select a new family of symbols. Accept the **Size** standard of **1'**.

Accept the **Autohead** options for inserting both **Middle** and **Edge** sprinklers.

Pick **OK** when finished.

At the prompt:

```
Command:EPLC_runme
Select closed polyline:
```
Pick the polyline which surrounds the Back Lawn.

```
Show lateral angle:
```
Drag a diagonal line toward the upper left corner of your screen.

```
Select any areas to avoid head placement:
Select objects:
```
Press **<Enter>** to continue.

```
Auto locate 1st full circle head or pick location (Auto/Pick)
<Auto>:
```
Press **<Enter>** to accept the default.

```
Processing polyline please wait ...
```
The heads are inserted into the back lawn as expected. Because of the unusual shape of the lawn, there are a few heads to remove and a few to add to provide better coverage.

First, remove the 3/4 (270°) head which causes the overspray onto the patio (Figures 15.43 and 51.44).

Figure 15.43

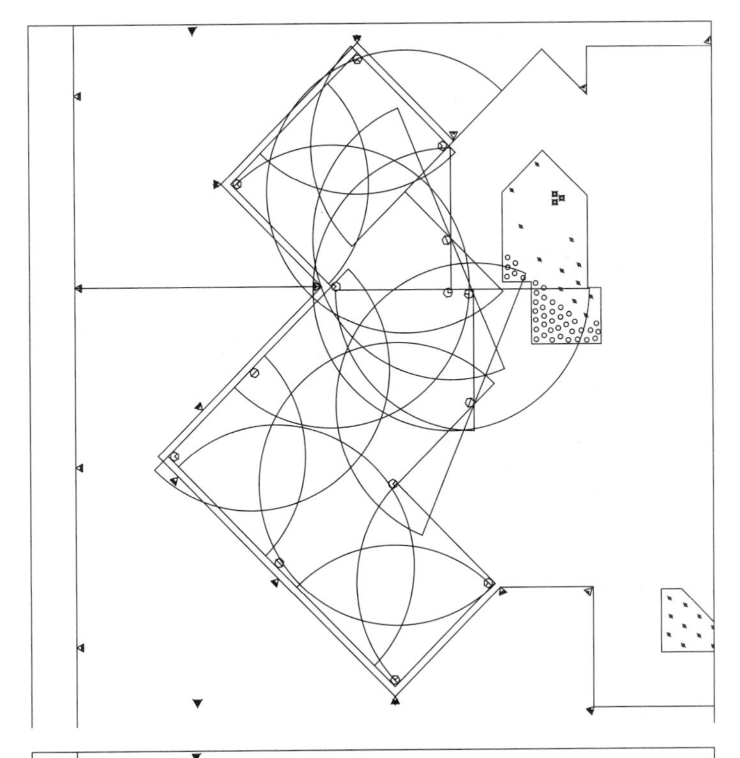

Figure 15.44

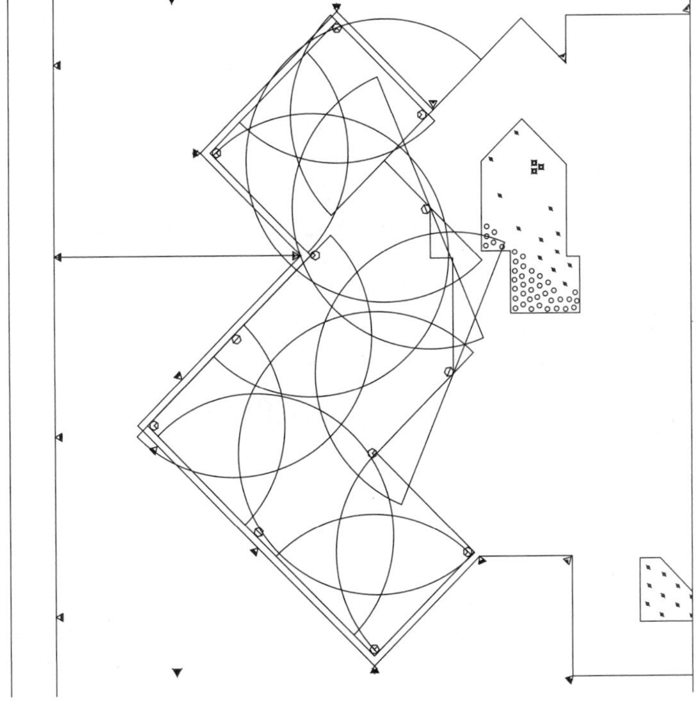

Next, use the **Edit Heads ▸** function to **Move** and **Copy** the 1/2 (180°) head at the lower left edge of the lawn. This will compensate for the "stretched" spacing of heads along that edge. Finally, add a 2/3 (240°)"fill-in" head toward the center of the lower part of the lawn (Figures 15.45 and 15.46).

Piping the System

After the sprinkler heads and drip emitters have all been located, we can begin to concern ourselves with piping. Before we begin to lay out the piping in the individual stations, we must first establish some important hydraulic data:

Thomas Residence Hydraulic information

1. The Thomas residence is equipped with a 1" water meter.
2. The static pressure is 60 PSI at the water meter.
3. The main irrigation supply line is 1 1/2" Schedule 40 PVC.
4. The maximum available flow rate is 30 gallons per minute

This hydraulic data will assist us in determining how many sprinklers can be operated simultaneously on a single control valve. The maximum safe flow of 30 GPM represents an upper limit. We can only operate as many sprinklers as will keep the accumulated flow under the limit.

Zones

Before we can begin piping a sprinkler station, we need to know how many heads will operate and how many GPMs have accumulated. This function is easily accomplished with the **Sum Flow** function under the **Zones** pull-down menu. We simply select all of the sprinkler heads we wish to operate on a single control valve, and LANDCADD will supply us with the total accumulated flow for all of the heads.

Open the **Heads** pull-down menu, then select **Coverage: OFF** to turn off the sprinkler coverage layer. This greatly simplifies the drawing on the screen.

Open the **Zones** pull-down menu, then select **Sum Flow**.

At the prompt:

```
EPLC_runme
Select irrigation heads:
```

Select all of the irrigation heads in the front lawn. Any lines, polylines, or other objects (other than sprinkler heads) will be filtered out and ignored.

Figure 15.45

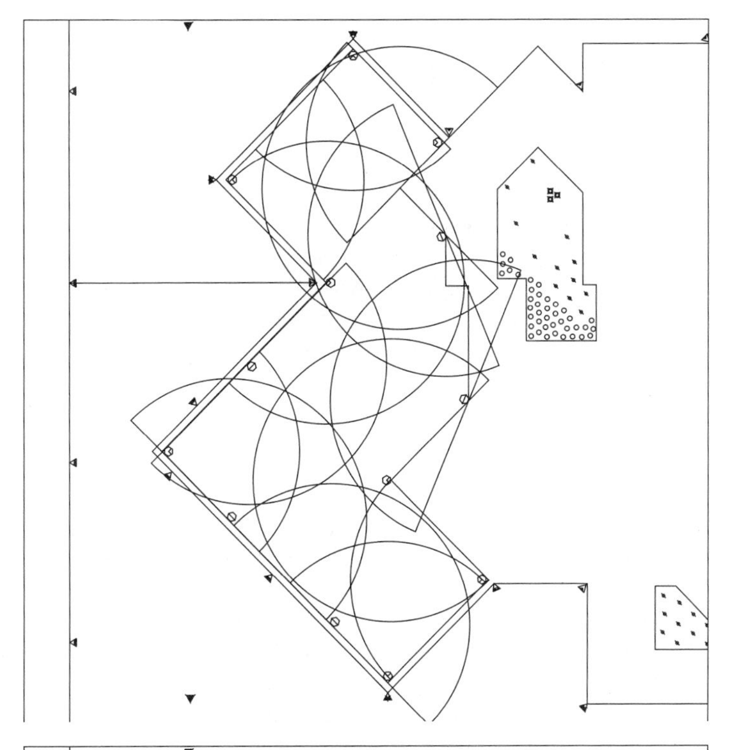

Figure 15.46

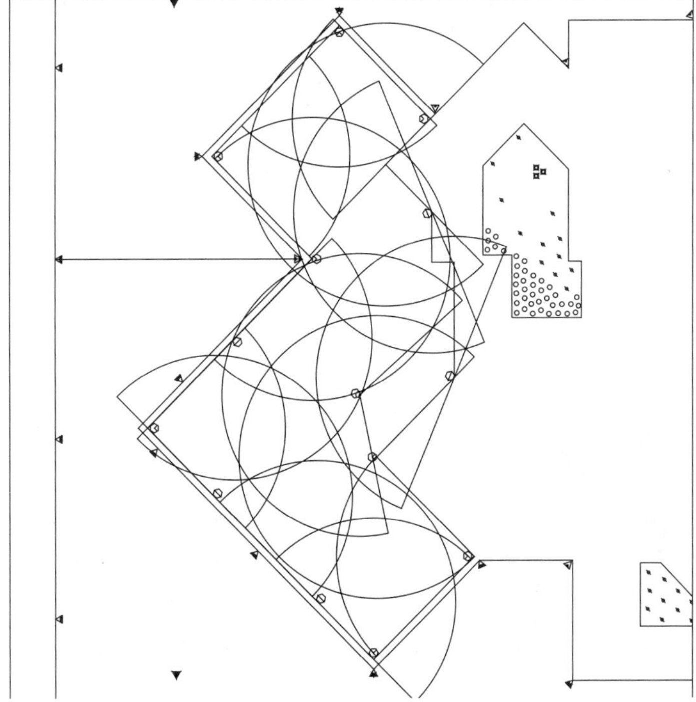

```
Select objects: 11 found
```

After all the sprinklers are chosen, press **<Enter>** to continue.

```
11 new heads selected, 11 total heads selected, discharging
12.00 GPM.
Select additional heads or press <Enter> to quit:
```

Press **<Enter>** to end the command.

This command informs us that the lawn sprinklers total 12 GPM, well below our 30 GPM limit. If the GPM flow were over 30 GPM, we would be forced to split the zone into two smaller ones, each with a total flow of less than 30 GPM.

We will use this command every time we plan to add piping to a group of sprinklers, so that we know the accumulated flow for the heads before we begin to place the piping.

Piping the Front Lawn

The piping process in LANDCADD is done in two separate steps:

- First, generic lateral lines are placed to connect all of the heads into a logical circuit. The lines placed here should show the actual location of the piping, but can extend beyond sprinklers and crossing pipes. The pipe sizing routines used later will clean up and trim extending lines back to the appropriate heads or pipe. These lateral lines do not yet possess attributes of any kind. They are simply lines on the screen.
- Second, the lateral piping is sized. The sizing routines assign pipe material and type, size, length, flow and friction loss attributes. These attributes are used to keep track of total pipe required for a job, and the total friction loss accumulated to the critical head in a circuit. In addition, pipe size labels can be placed on appropriate pipe segments.

Irrigation Pipe Location

Pipe placement should make sense from an installer's viewpoint. Long, straight trenches are always preferred to short, choppy trenches. (Ask anyone who has operated a trencher if you want to know why....) Piping should be connected at right angles, (or at an occasional 45° angle) and should run out toward the distant sprinkler heads, not double back toward the valves.

Pipe placement should make sense hydraulically. That is, the pipe should be sized correctly and should join together in such a way as to minimize the amount of large diameter pipe required. This means that most stations should be fed from the center of the station rather than from one end or the other. While most of these factors are simple enough, considering them all at once is complicated. An irrigation design class is helpful.

We will begin placing pipe lines in the front lawn.

Open the **Pipe** pull-down menu, then select **Draw Lateral**.

At the prompt:

```
EPLC_runme_.line From point:
```
Use the cursor menu to select **Insert**, then pick the sprinkler at the top right corner of the front lawn.

```
To point:
```
Use the cursor menu to select **Insert**, then pick the sprinkler at the bottom right corner of the front lawn. The line drawn extends through the head in the center. Press **<Enter>** to complete the line command.

Press **<Enter>** again to restart the **Draw Lateral** routine.

Draw a lateral pipe for the center row of heads, then for the left side row of heads.

Hint: The LANDCADD pipe sizing routine will trim and clean up pipe lines which overhang symbols and which do not intersect cleanly. The most important consideration is to make pipe lines cross at least halfway through each sprinkler symbol. If the pipes do not cross through (or touch) the insertion point of each sprinkler symbol, the pipe sizing routine will not sense the connection between the head and the pipe. This will make the pipe sizes and other calculated values incorrect.

Restart the **Draw Lateral** command for a final header line which will connect the heads to the valve. Be sure to extend the header far enough into Planter area 1 so that the valve does not interfere with the sprinklers there. Your results should resemble those below (Figure 15.47).

After the heads and pipe lines are located, we can size the pipe.

Open the **Pipe** pull-down menu, then select **Pipe Size ▸** and pick **Auto Size** (Figure 15.48).

This opens the "Auto Pipe Sizing" dialogue box (Figure 15.49).

In this dialogue box, we supply important information relating to the desired characteristics of the lateral piping, the hydraulic data for the project, the size and type of pipe label, and the size of the control valve symbol.

In the **Pipe Type:** pop-down box, select the pipe type **200 PSI SDR 21 PVC PIPE ENGLISH UNITS** (Figure 15.50).

Figure 15.47

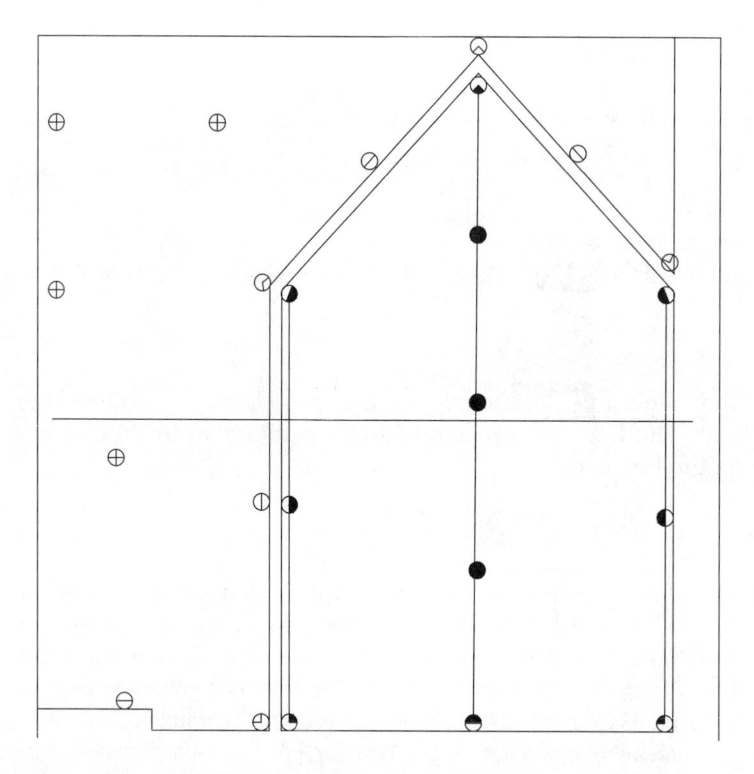

This is the appropriate type of pipe to use in a residential irrigation system where the ground does not freeze. Other regional conditions may dictate the use of other types of pipe. (In the author's opinion, 125 PSI 32.5 SDR PVC PIPE is **never** an appropriate choice.)

In the **Minimum Size:** pop-down box, select **1/2"** (Figure 15.51).

Figure 15.48

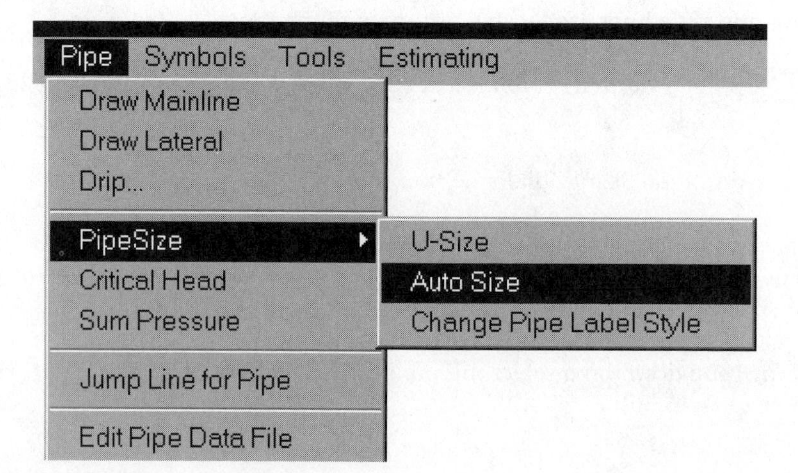

Figure 15.49

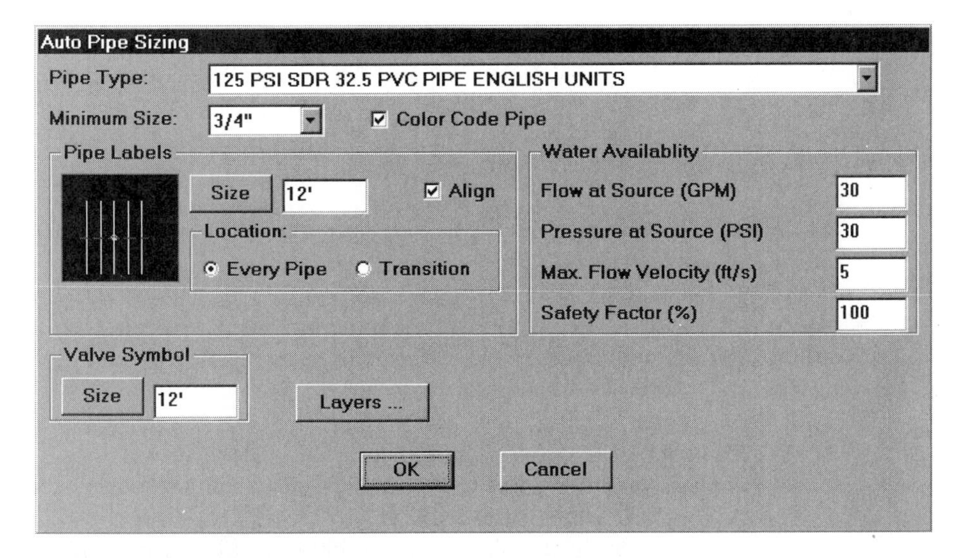

Figure 15.50

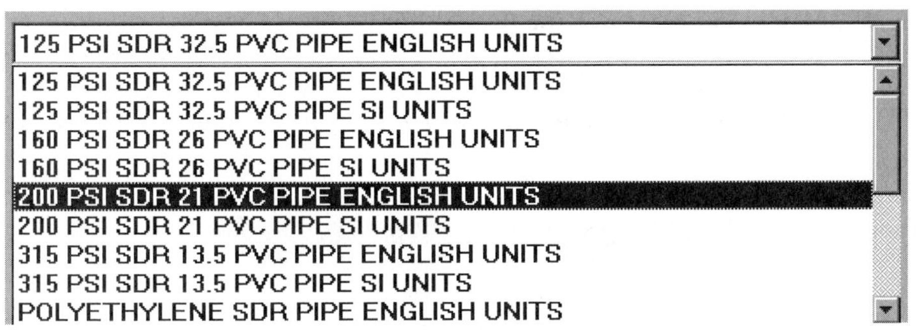

Figure 15.51

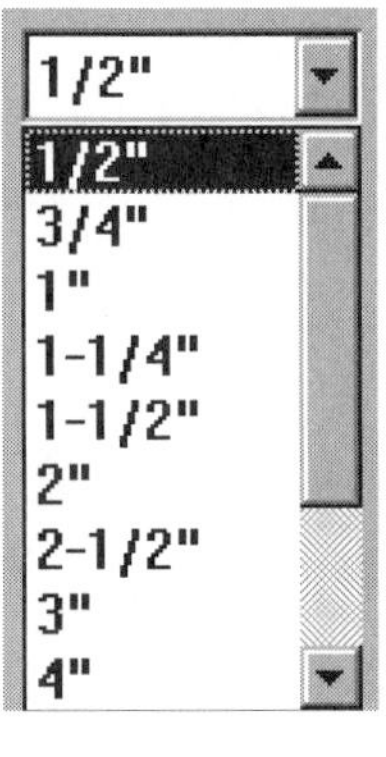

This allows the sizing routine to use a pipe diameter as small as 1/2". This is also an appropriate choice in a residential irrigation system.

In the **Water Availability** boxes, fill in the information as follows:

Flow at Source (GPM) — enter **30**.

Pressure at Source (PSI) — enter **60**.

Max. Flow Velocity (ft/s) — accept the default of **5**. This is a standard maximum for most piping systems.

Safety Factor (%) — accept the default value of **100**. We will keep track of the amount of flow as we select the heads to use in a sprinkler zone. If we entered another value here, LANDCADD would alert us if we exceeded a predefined percentage of the maximum flow.

In the **Valve Symbol** entry, we enter the desired size for the remote control valve symbol.

Enter a value of **2'-6"** in the **Size** box. This creates a valve symbol in proportion to the other symbols used in the Irrigation Plan.

In the **Layers . . .** box, we can specify the layer, color and linetype of the components used in the pipe sizing routines. This opens the "Auto Size Pipe Layers" dialogue box (Figure 15.52).

Figure 15.52

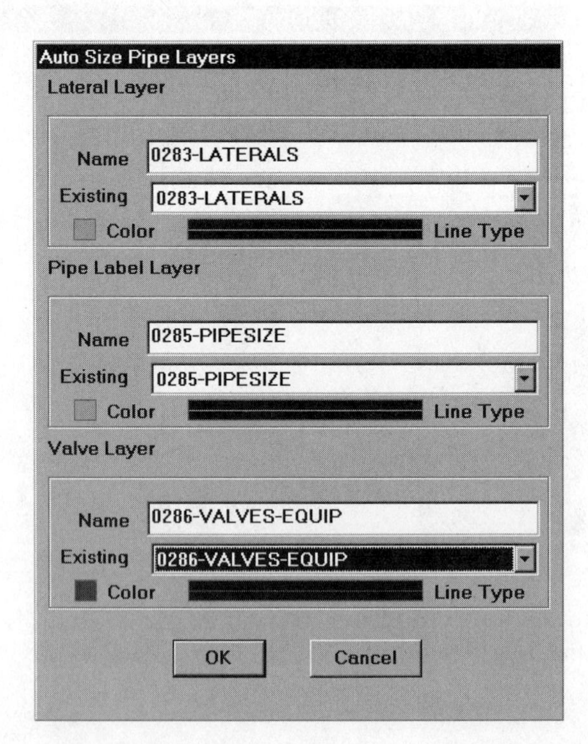

In this dialogue box, we can specify the layer, color and linetype of the lateral piping, the label and the remote control valve symbol. While the defaults are acceptable here, we can modify the properties of these layers at any time. This becomes especially valuable when we plot the drawings.

In the last portion of the "Auto Pipe Sizing" dialogue box, we can control the features of the pipe labels themselves.

To control the type of label, pick on the illustration box. This opens the "Select Pipe Label Style" icon menu (Figure 15.53).

The first style (in the upper left corner) places a slash across the pipe to designate its size. The more slashes, the larger the pipe size. The other choices show a label placed either above or below the pipe — with a loop, or without. Once a label style has been chosen, that style will be used until the style is changed.

Individual labels can have their style changed very easily. Simply pick **Change Pipe Label Style** under the **PipeSize** option of the **Pipe** pull-down menu. Pick the label(s) to change, then pick the new style for that (those) label(s). LAND-CADD changes the label style immediately.

Select the second style, with the label placed above — using a loop to identify the pipe.

The final choices relate to the label size and the frequency with which labels are placed on piping.

Figure 15.53

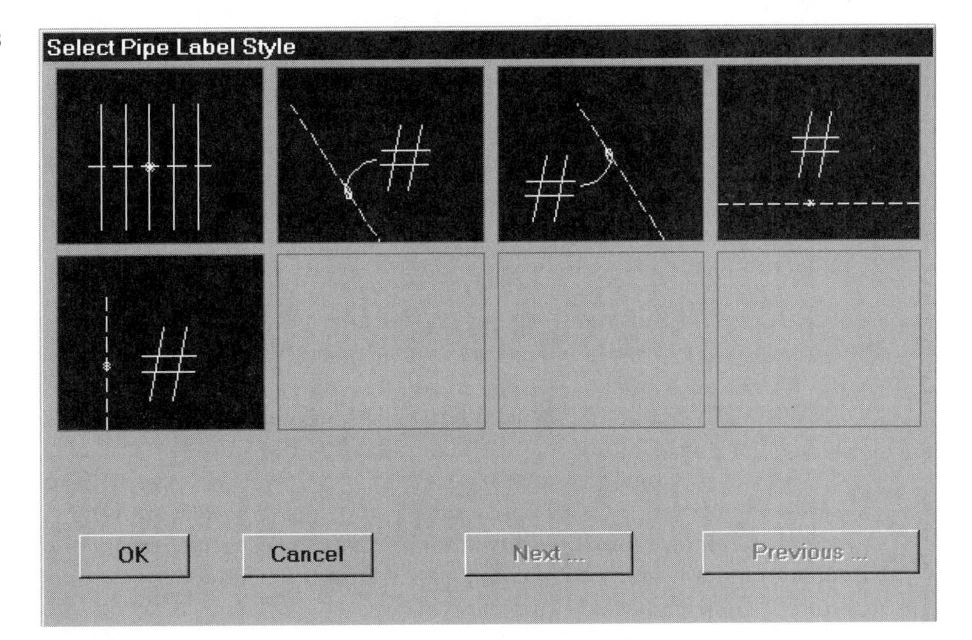

In the **Size** box, enter **20'**.

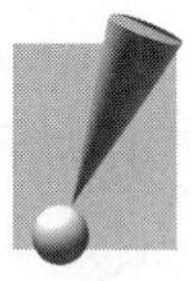

Warning: The "Auto Pipe Sizing" dialogue box does not accept text units properly. Any entries placed in the **Size** text box must be multiplied by **12** for text to be sized correctly. If you need **20"** tall text, enter **20'**.

This unlikely entry is a result of two problems: First, a LANDCADD bug does not handle text size properly in this dialogue box (see LANDCADD Warning above). This forces the user to multiply the size of the text value by 12 to create the proper size labels. Second, LANDCADD uses stacked fractions and a cramped style that makes the labels difficult to read at 1/8" plotted size. To compensate for the text size problems, we will use a text height of **20'**.

In the **Location:** entry, pick the radio button **Transition**.

This causes LANDCADD to place the labels at pipes where the size makes a transition. Once a pipe label of **3/4"** is shown, the reader (specifier, contractor, etc.) should assume that all piping along that lateral is this size until the next size label, **1"**, is shown. This simplifies the drawing and makes it much easier to read.

Once we have completed entries in the "Auto Pipe Sizing" dialogue box, we are ready to actually select the piping and heads.

At the prompt:

```
EPLC_runme
Select lines & heads
Select objects:
```

Pick all of the heads and pipes in the front lawn. Any objects which are not heads or pipe lines will be filtered out and not selected. Press **<Enter>** when completed.

```
WORKING ....\
Pick valve location:
```

Pick the end of the lateral pipe in Planter area 1.

```
WORKING ..../
```

LANDCADD then converts each lateral line into pipe blocks. Each block (pipe segment) is then assigned a pipe size, pipe material, GPM flow, and friction loss entry. Each transition from one pipe size to the next is labeled. The beginning of each lateral section (toward the outer heads) is sized at 1/2", while the central header pipe increases to 3/4" and 1". A remote control valve symbol is placed at the end of the 1" header. The control valve symbol is assigned a size and a GPM flow (Figure 15.54).

Figure 15.54

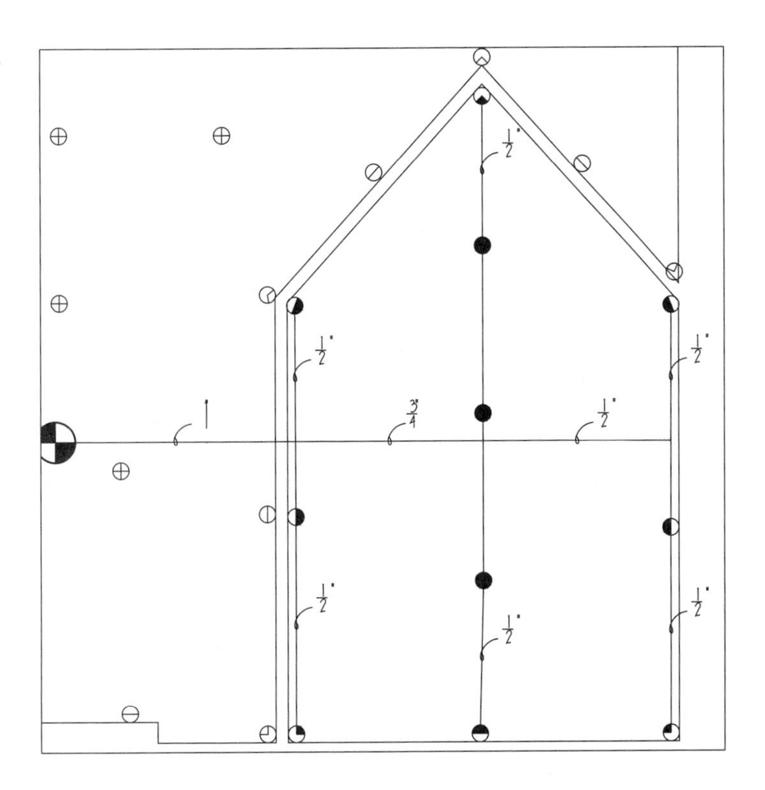

Note: LANDCADD builds a new text style for pipe labels called **EPIR_PIPELABEL**. If you want these labels to appear with the same style as the rest of your architectural lettering, you can modify the style to change the font from **SIMPLEX.SHX** to **HLTR.SHX** or another font. In addition, you might wish to modify the width factor of the lettering. Use the **DDSTYLE** command to experiment with the appearance of this text. (The **1"** labels sometimes show up with the " overlapping the **1**, so you may have to explode those labels to correct their appearance.)

After the pipe sizing is complete, each pipe segment is assigned its own attributes so that a full estimate can be developed which will list the amount of each type of pipe required to complete the job. In addition, LANDCADD can calculate the amount of friction loss occurring in any piping path. The selection **Critical Head** under the **Pipe** pull-down menu will identify the most distant sprinkler head from the control valve—where the greatest friction loss occurs. This head is then identified, and the total friction loss is reported.

Pick **Critical Head** under the **Pipe** pull-down menu.

```
Command:
EPLC_runme
Pick valve:
```

Pick the remote control valve symbol. LANDCADD then identifies the critical head and reports the friction loss and pipe length.

```
Critical Head Pressure Loss: 0.859 PSI
Critical Head Pipe Length: 44'-10 9/16" feet
```

(Figure 15.55)

Note: Depending on where you place your central header pipe, the critical head location might change. If you move the header further toward the top of the screen, the critical head may show as the one in the lower right corner. LANDCADD's mathematical calculations are very precise!

The boxes around the critical head (actually a slide) can be removed by zooming or panning to any location on the screen.

We can now return to the **Zones** pull-down menu to assign a label to the zone, and to examine other important characteristics for the front lawn irrigation zone (Figure 15.56).

The **Define** function simply allows you to draw a surrounding polyline to lay out different areas for irrigation stations. These zone polylines can be turned on or off, or assigned to any desired layer.

Figure 15.55

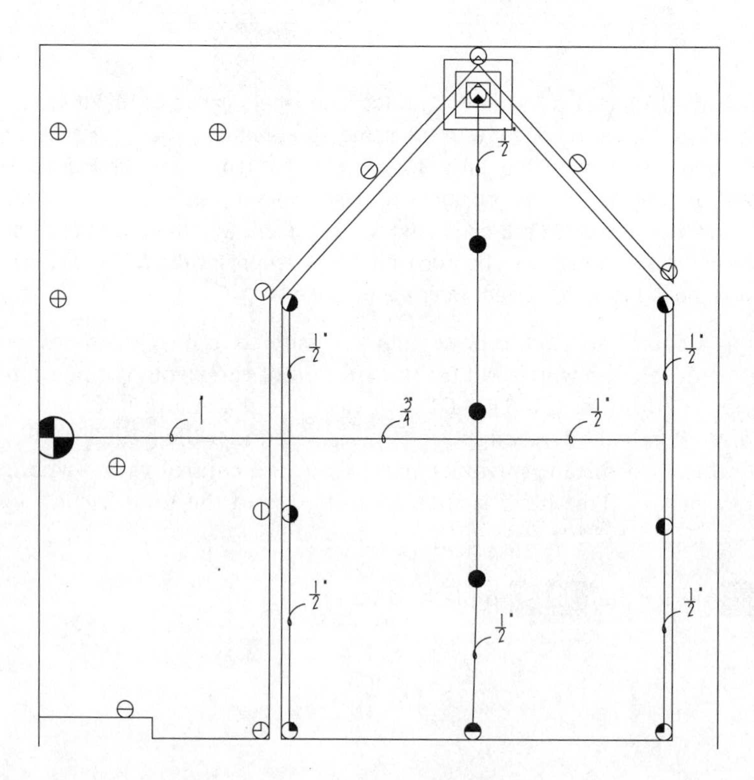

Figure 15.56

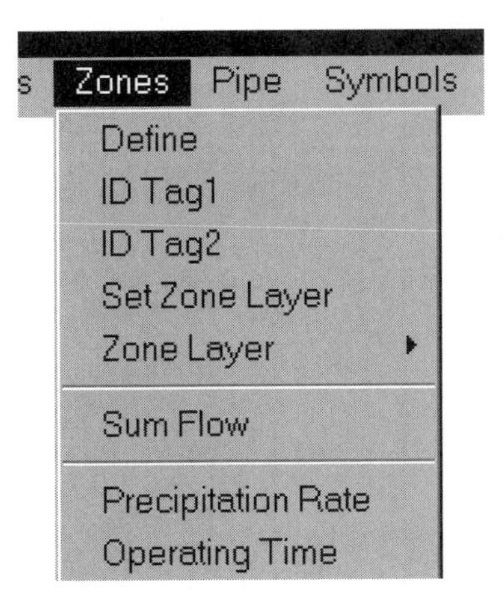

The **ID TAG** functions allow us to insert a zone call-out tag to identify the irrigation zone, the valve size and the GPM flow for the station. LANDCADD automatically inserts the zone tag with the appropriate information.

Select **ID Tag2** under the **Zones** pull-down menu.

```
Command: EPLC_runme
Pick valve for zone:
```
Pick the control valve symbol inserted for the front lawn irrigation station.

```
Enter zone ID:
```
Type **1**.

```
Locate Zone Tag:
```
Pick a location directly across from the valve, inside the house.

```
Scale Zone Tag:
```
Type **40**. This creates a call-out about 3/8" tall.

```
Rotate Zone Tag:
```
Be sure that **ORTHO** is turned on, and rotate the call-out so that it is horizontal (Figure 15.57).

We will continue to use **ID TAG2** as out call-out style throughout the irrigation plan.

The **Precipitation Rate** function will calculate the gross precipitation rate of the sprinkler station and report it.

```
Command: EPLC-runme
Select irrigation heads:
```

Figure 15.57

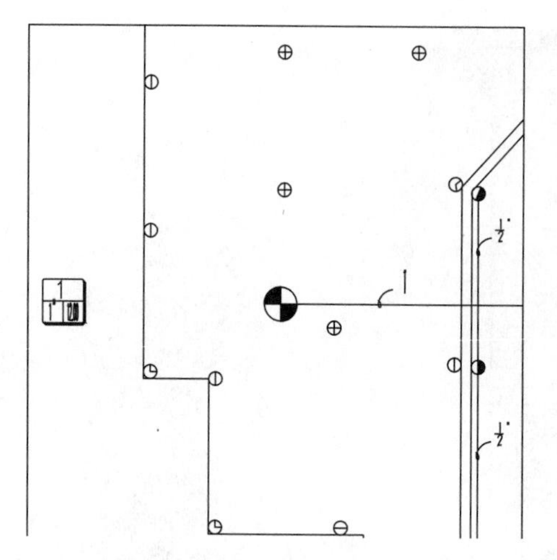

```
Select objects:
```
Pick all of the heads in the front lawn. Any objects other than heads will be filtered out and discarded.

```
Select objects:
```
Press **<Enter>** when finished.

```
11 loops extracted.
```
LANDCADD reports how many sprinkler heads are being examined. Be sure that **11** loops are being extracted which corresponds to 11 sprinkler heads in the front lawn.

```
11 Regions created
Area = 145967.76 square in. (1013.6650 square feet.), Length =
131'-33-3/4"
11 heads selected
Average precipitation rate for selected heads: 1.14" per hour
```
LANDCADD makes this calculation based upon the area contained within the boundary polyline and the number of GPM flowing into this area. The calculation is then made for the zone.

The **Operating Time** function will report the number of minutes of operating time that the station must run when you enter the total water loss per week.

```
Command: EPLC_runme
Irrigation requirement in inches per week?
```
Type **1.5** for a moderate (for California) summer ET_O water use.

```
Days available for irrigation per week?
```
Type **5** for number of available days.

```
Precipitation rate in inches or C to calculate:
```
Type **C** to allow LANDCADD to calculate the precipitation rate.

```
Select irrigation heads:
Select objects:
```
Pick all of the sprinklers in the front lawn.

```
Select objects:
```
Press **<Enter>** when finished.

```
11 loops extracted.
11 Regions created
Area = 145967.76 square in. (1013.6650 square feet.), Length =
131'-33-3/4"
11 heads selected
Average precipitation rate for selected heads: 1.14" per hour
Operating time = 15.79 minutes per day, 5 days a week.
```
LANDCADD takes the irrigation requirement, the available days per week and the precipitation rate to calculate the number of watering minutes. Note that the calculation of precipitation rate is duplicated here. If you remember the precipitation rate, you can simply enter it rather than waiting for LANDCADD to make the calculation.

Planter Area 1 Piping

After completing work on the front lawn irrigation circuit (station 1), we are ready to proceed to Planter area 1, which is adjacent to the front lawn.

Use the **Draw Lateral** command, under the **Pipe** pull-down menu, to draw the laterals as shown below (Figure 15.58).

Note that the pipe occasionally crosses and extends beyond the intersecting pipe. This is acceptable because LANDCADD will clean up these intersections when we use the pipe sizing routine. Pipe lines are normally run from the **Insertion** point (use the **Insert** option in the cursor menu) of one sprinkler to the **Insertion** point of another. Most of the pipe meets at a 90° angle, or an occasional 45° angle. This is important for the installer, who only has 90° and 45° pipe fittings available. Terminate the last lateral pipe section directly above the remote control valve symbol from station 1. This will cause the next valve to line up with the first one.

Use the **Auto Size** pipe sizing routine to size the piping for this circuit. You should select **28** sprinkler heads. The **Auto Size** routine should yield the results shown below (Figure 15.59).

Although the sizing shown here is technically correct, there are several size extraneous labels some of which can be erased, and some which are incorrectly located. We will use two methods for changing pipe label locations.

Figure 15.58

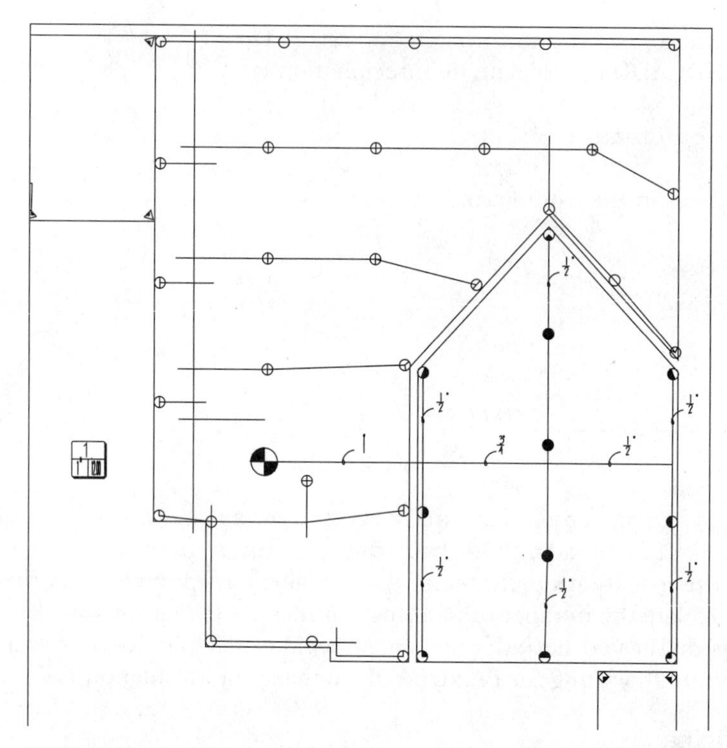

Figure 15.59

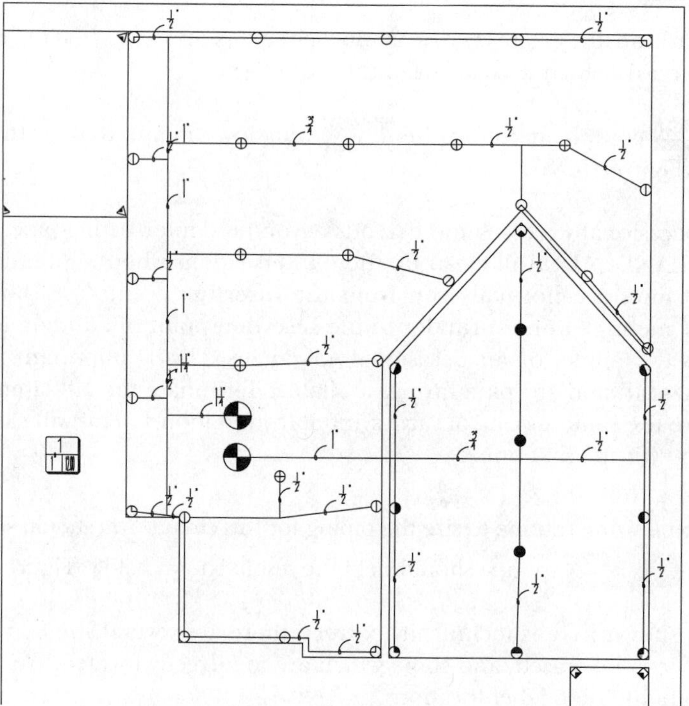

The first method takes advantage of LANDCADD's **Change Pipe Label Style** command (under the **PipeSize ▸** selection of the **Pipe** pull-down menu). This command allows us to select specific pipe labels, and change their style to another style.

Pick **Change Pipe Label Style** from the **PipeSize ▸** sub-menu.

```
Command: EPLC_runme
Select pipe labels to change:
```

Pick the top two **1/2"** labels and the three **1/2"** labels on the left side of the station which overlap onto the header pipe. (We will modify the other **1/2"** label, the **1"** label and the **1 1/4"** label later.)

```
Select objects:
```

Press **<Enter>** to indicate that you have finished selecting objects.

You then will be presented with the "Change Pipe Label Style" dialogue box. Pick the large box in the center to view the "Select Pipe Label Style" icon menu. Pick the style with the label which appears **below** the pipe, using a loop to identify the pipe. This is the third style from the left (Figure 15.60).

The "Change Pipe Label Style" box reappears with the new style appearing in the large box. Press **OK** to complete the command. This changes the **1/2"** labels to the new style. See below (Figure 15.61).

Figure 15.60

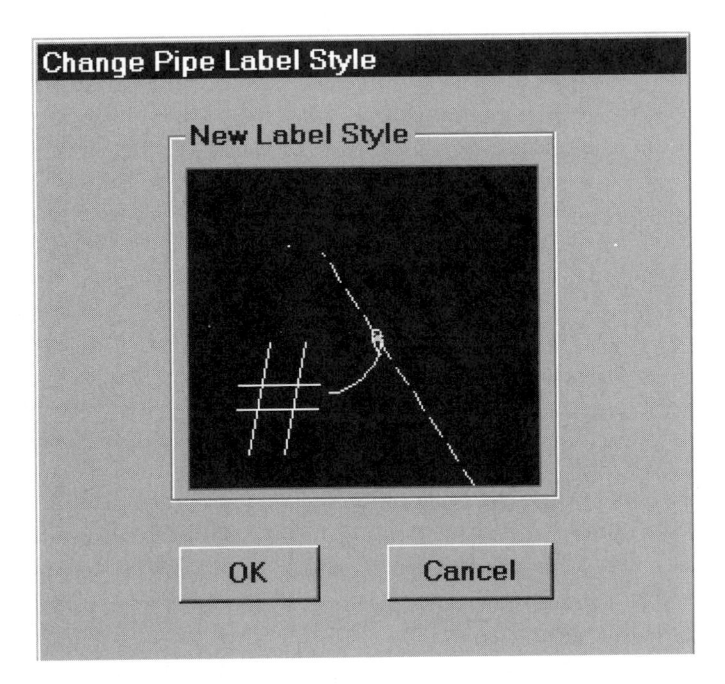

Figure 15.61

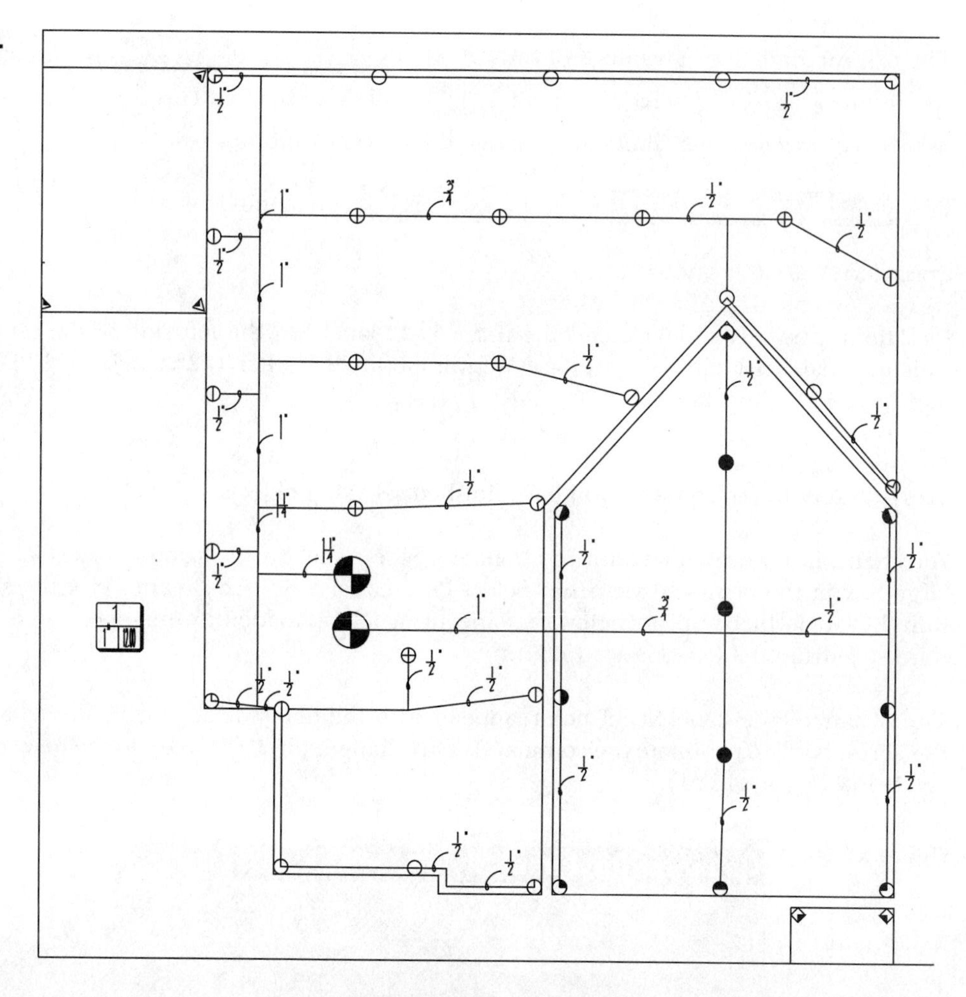

The remaining labels overlapping other pipes cannot easily be switched with another style if we wish to keep the identifying loops. We will use different AutoCAD editing commands to modify these labels. See below (Figures 15.62 through 15.70).

Since the labels do not contain any information which is vital to the drawing, the labels can be exploded and edited without harming the piping information. Remember that exploding any object causes it to be moved to the **O** layer. Move all of the exploded blocks back to the **0285–PIPESIZE** layer when finished.

Use **ID Tag2**, under the **Zones** pull-down menu, to insert a zone call-out for Planter area 1. Use the zone ID number **2**, and place the tag directly over the call-out for station 1.

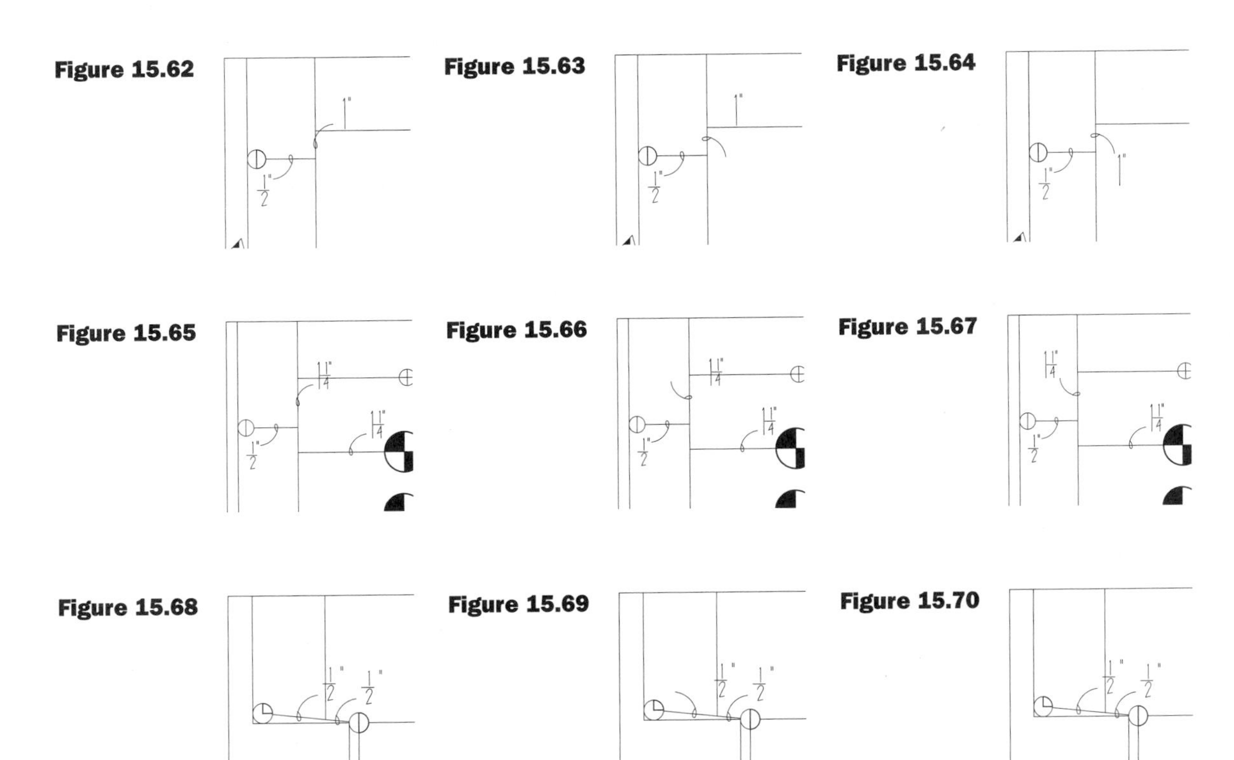

Figure 15.62 Figure 15.63 Figure 15.64

Figure 15.65 Figure 15.66 Figure 15.67

Figure 15.68 Figure 15.69 Figure 15.70

Planter Area 2 Piping

Now that we have completed the piping for the first two stations, we are ready to tackle Planter area 2. The piping which serves this planter will run through the lawn area, then down to the planter.

Use the **Draw Lateral** command under the **Pipe** pull-down menu. Draw lateral lines through the sprinkler heads in Planter area 3. Draw the header line back into Planter area 1 as shown. Notice that the header pipe line passes through a pipe label (Figure 15.71).

Use the **MOVE** command to move the pipe label down lower on the lateral so that it is clear of the new header line. This enhances the clarity of the drawing and does not sacrifice its accuracy.

Use the **Auto Size** selection under the **PipeSize ▸** sub-menu. Accept the values entered in the "Auto Pipe Sizing" dialogue box and pick **OK**.

Figure 15.71

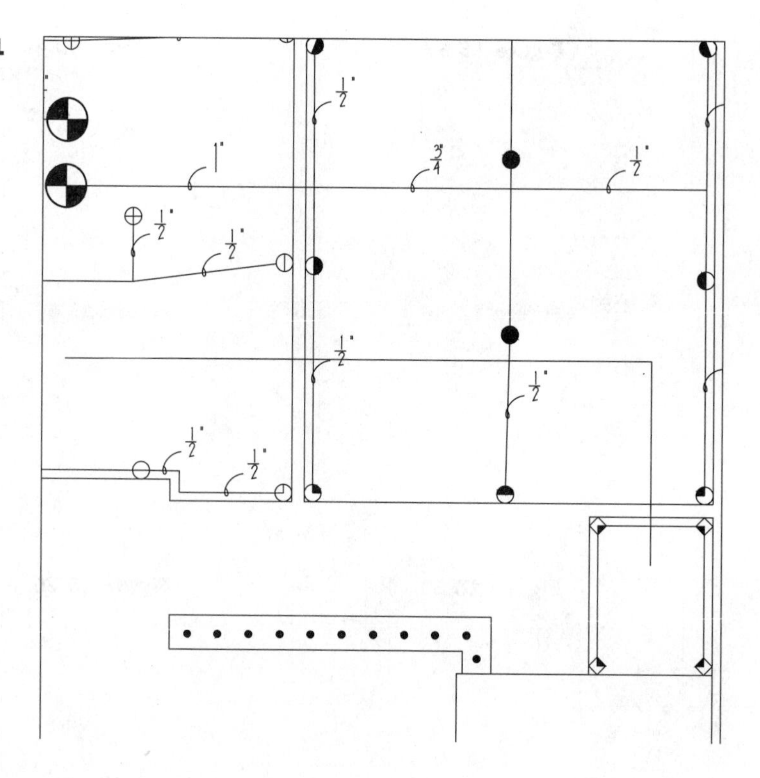

At the prompt:

```
EPLC_runme
Select lines & heads
Select objects:
```

Pick all of the heads and pipes in Planter area 3, along with the header pipes extending into Planter area 1. Press **<Enter>** when completed.

```
WORKING ....\
Pick valve location:
```

Pick the end of the header pipe in Planter area 1.

```
WORKING ..../
```

LANDCADD sizes the pipes and inserts a remote control valve symbol (Figure 15.72).

The piping is inserted correctly, but the header line crosses the piping in the front lawn area. To indicate that these pipes cross, but are not connected, we use the **Jump Line for Pipe** command (Figure 15.73).

Figure 15.72

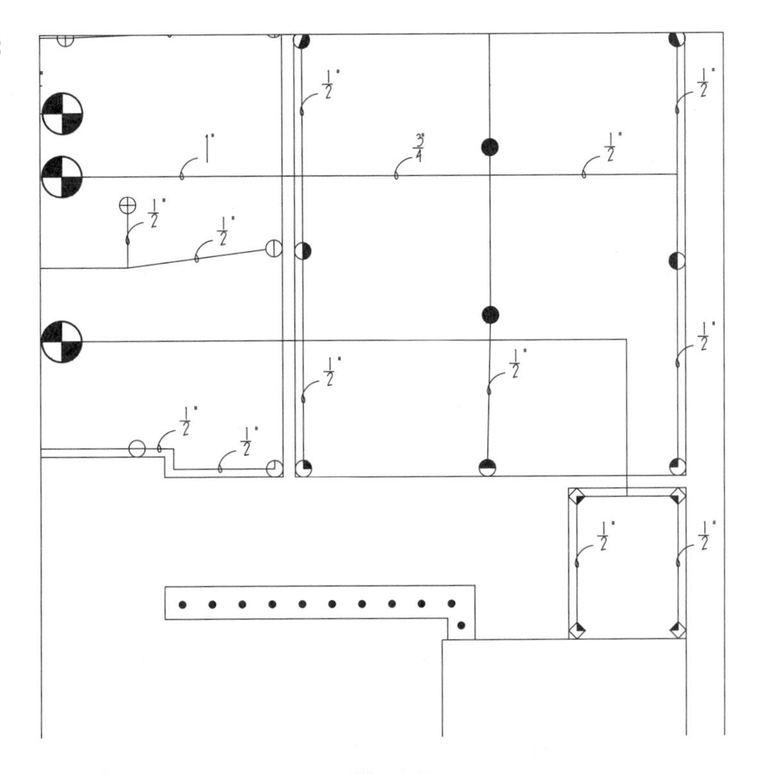

Figure 15.73

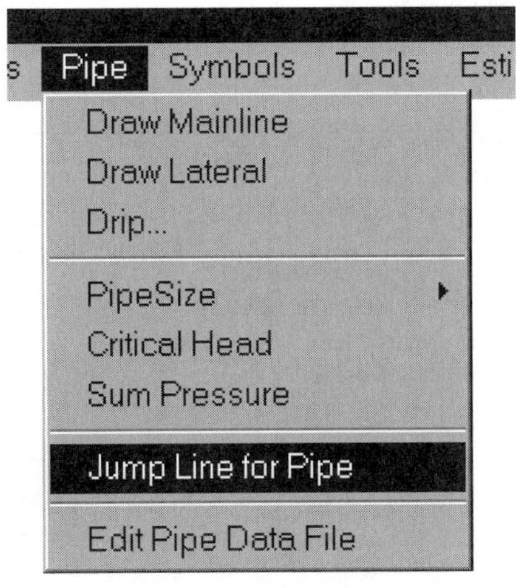

This command places breaks the pipe line, then places an arc to indicate the crossing location.

Jump Line for Pipe must be used when working with pipe lines. The **Jump Line** command under the **Tools** pull-down menu will not work with piping.

At the prompt:

```
Command: EPLC_runme
Pick jumpline START point on pipe:
```

Pick one end of the desired jump line. Remember that the arc created will be drawn in a counter-clockwise direction.

```
Pick jumpline END point on pipe:
```

Pick the opposite end of the desired jump line. Try to center the crossing line between the jump line endpoints.

```
Mirror jumpline? Yes/No <No>:
```

LANDCADD inserts the jump line arc. If the jump line appears as you wish it to, type **N**. If the jump line would look better in mirror image, type **Y**.

```
Redefining block EPIR_PIPE_0
```

LANDCADD breaks the line at the arc endpoints, then redefines the pipe as a block (Figures 15.74 and 15.75).

Insert a jump line at both locations where the header pipe crosses the front lawn piping (Figure 15.76).

Use **ID Tag2** to insert a zone call-out for Planter area 2. Use the zone ID number **3**. Place the tag opposite the valve placed for Planter area 2.

Planter Area 3 Piping

The piping for Planter area 3 presents a special set of challenges. Because we selected an equipment symbol — the bubbler— instead of a sprinkler symbol, there are no GPM attributes for the **Auto Size** routine to use in pipe sizing. Instead of using this routine, we will use the **U-Size** routine under the **PipeSize ▸** sub-menu.

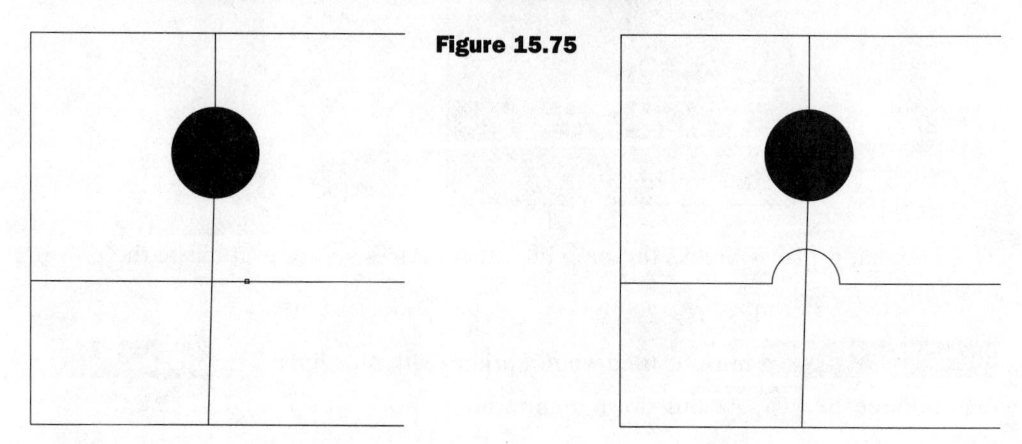

Figure 15.74

Figure 15.75

Figure 15.76

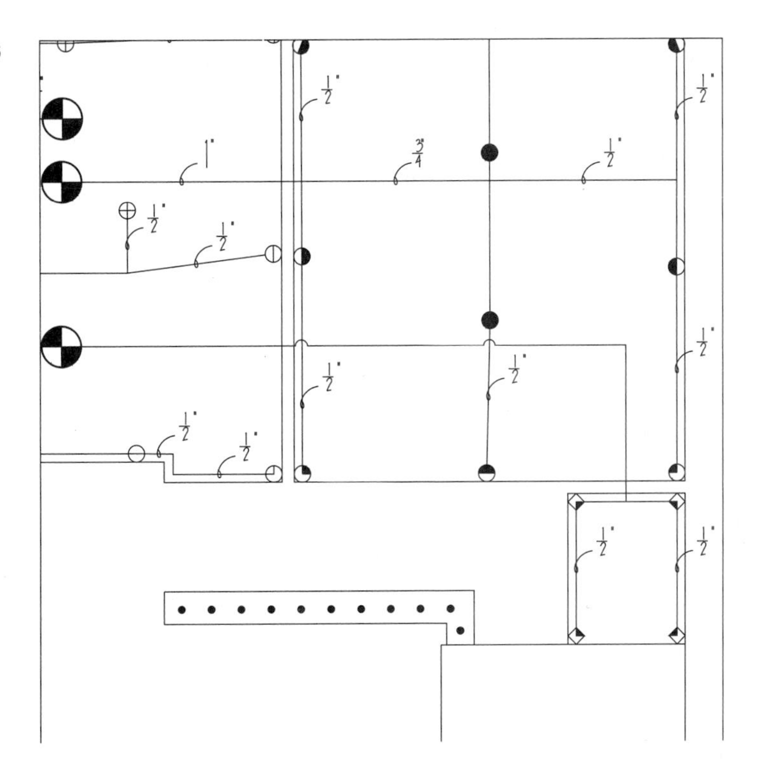

This routine allows us to convert pipe lines into pipe blocks (the same as the **Auto Size** routine), but we must select the size of the pipe manually. The resulting pipe block has pipe type, size and length attributes, but does not possess any flow, velocity or friction loss attributes. In addition, there is no automatic valve insertion at the completion of the command. We must insert a remote control valve manually.

Before we begin to place pipe lines in the drawing, we will take the precaution of turning on a few layers which may affect the placement of the pipe lines and labels.

Use the layer control box to turn on the layers **0207-NEW_BLDG**, **EX_HARDSCAPE**, **CONCRETE** and **PROPERTY**. We will turn these off again when we work on other planter areas.

To begin, we will place our pipe line in the station. The first pipe segments will correspond to the **1/2"** pipe. We will enter those pipe line segments first, then convert them to pipe blocks before proceeding on to the **3/4"** pipe lines.

Pick **Draw Lateral** from the **Pipe** pull-down menu.

Draw the pipe line as shown below. Be sure to keep the pipe lines aligned to run through the center of the bubbler symbols (Figure 15.77).

Next, pick the **U-Size** routine under the **PipeSize ▸** sub-menu. This brings up the "Usize Pipe" dialogue box (Figure 15.78).

Figure 15.77

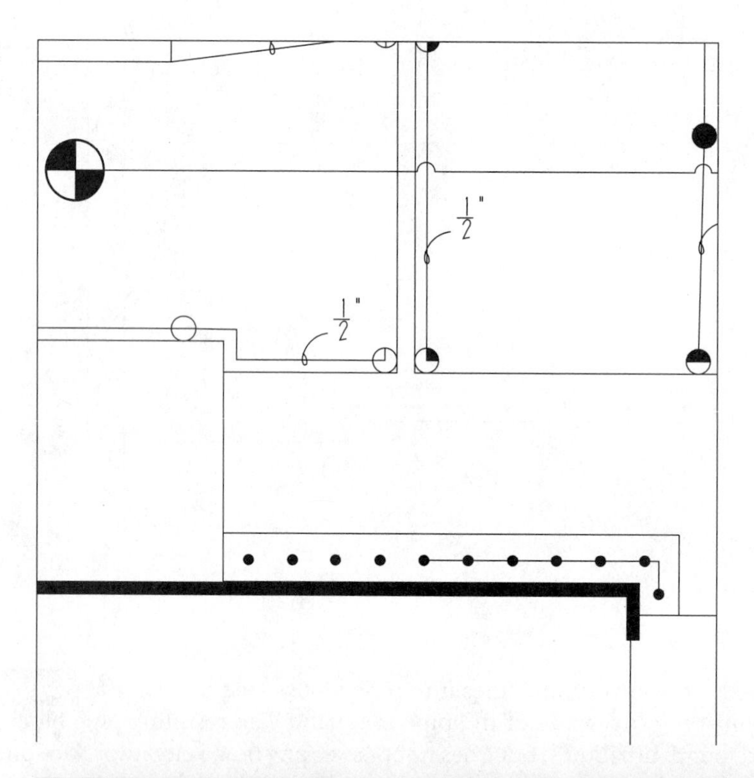

Figure 15.78

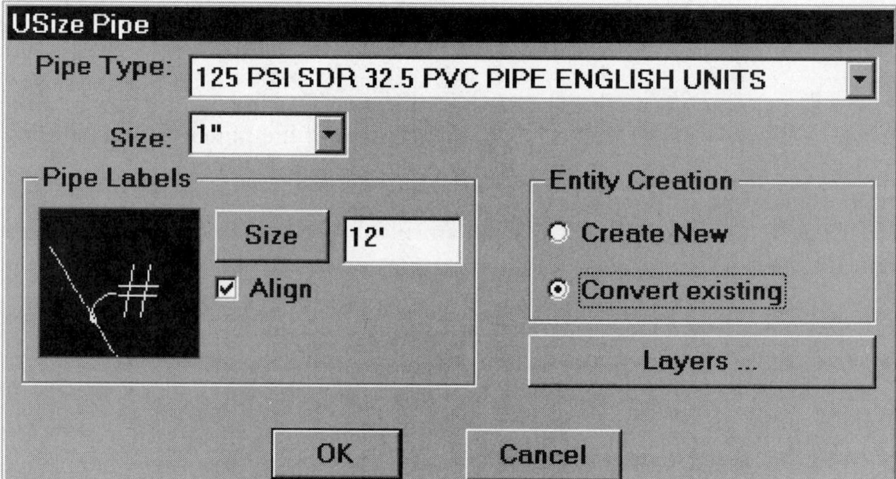

For **Pipe Type:** select **200 PSI SDR 21 PVC PIPE ENGLISH UNITS** from the pop-down box. We will continue to use this pipe for all of our lateral lines.

For **Size:** select **1/2"** from the pop-down box. This corresponds to the pipe size we will use first.

In the **Pipe Labels** box, select the loop style with the pipe label above the pipe. Set the **Size** at **20'**, and be sure that the **Align** toggle box is empty.

In the **Entity Creation** box, pick the **Convert existing** radio button to choose it. We will convert existing lines into blocks, rather than create pipe blocks directly.

Pick **OK** when finished.

At the prompt:

```
Command: EPLC_runme
Select objects:
```

Pick the two pipe line segments drawn through the bubblers. Press **<Enter>**.

```
Select objects: 2 found
```

The pipe segments are converted into blocks and labeled. Since the 1/2" pipe consists of two segments, both are labeled. Since we are only labeling the pipe size transitions, we will erase the pipe size on the long pipe segment when we are finished converting and labeling all of the pipe used in Planter area 3.

We placed the first pipe segments to correspond to the 1/2" pipe. The final pipe segment carries 6 GPM, from the accumulated flow of the bubblers downstream. The next pipe segments carry more flow, which require a larger pipe size. We will first place the pipe lines, then convert them to blocks for labeling.

Pick **Draw Lateral** from the **Pipe** pull-down menu.

Draw the remaining pipe lines as shown below. Terminate the pipe line so that the end point will be aligned with the end point of the pipe above (Figure 15.79).

Pick the **U-Size** routine under the **PipeSize ▸** sub-menu. In the “Usize Pipe” dialogue box, change the **Size:** to **3/4"** using the pop-down box. Accept the other values.

Pick **OK** when finished.

Pick the three pipe segments to convert to blocks and label. See the results below (Figure 15.80).

Figure 15.79

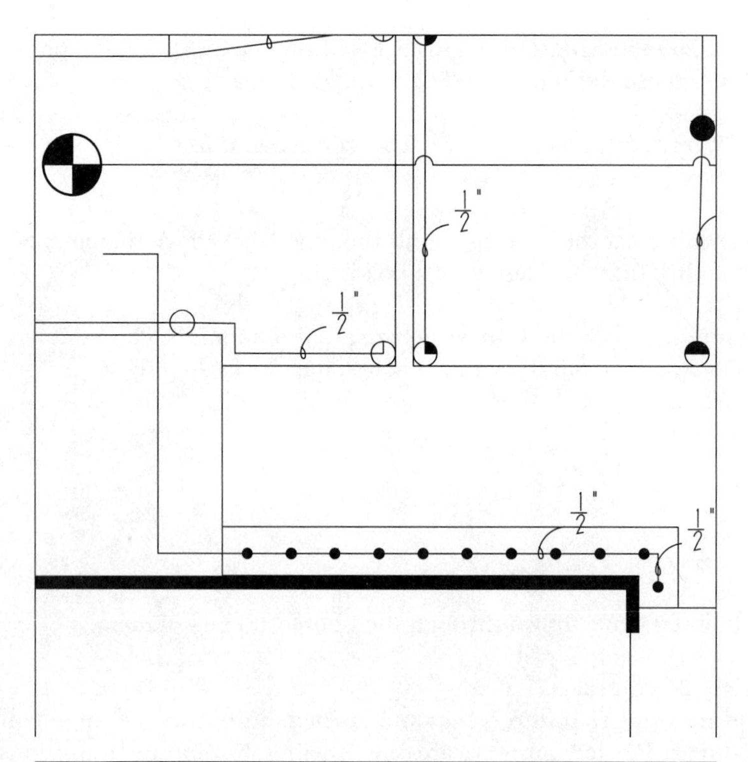

Figure 15.80

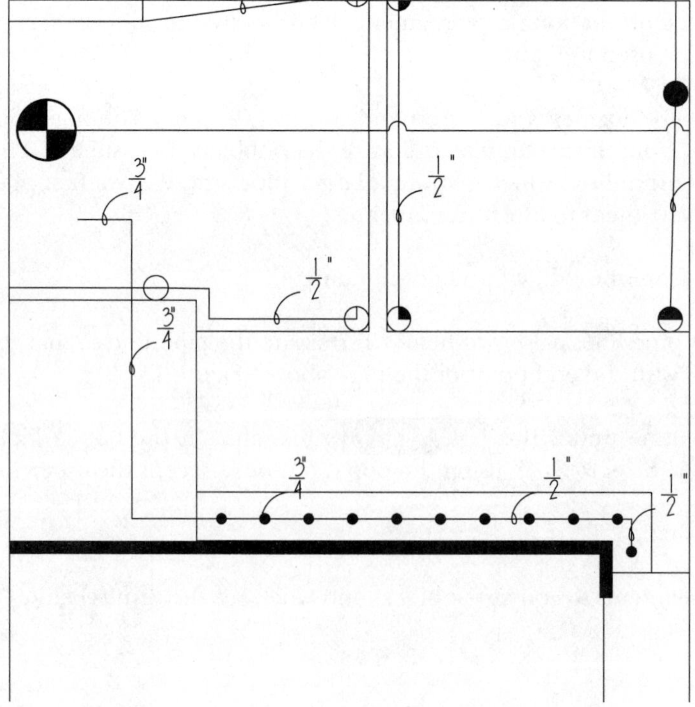

Erase the two extra **3/4"** labels and the one extra **1/2"** label. Move the labels so that they mark only the transitions between pipe sizes. In the case of the **1/2"** pipe, this label can be left at the first pipe segment, at the end of the lateral line.

The **3/4"** remaining label must be moved to the correct pipe segment to mark the size transition, and should be modified to prevent the obstruction of the crossing planter border (Figures 15.81, 15.82, and 15.83).

First use **EXPLODE** to separate the label block into its components.

MOVE the text above the crossing line.

Use **PEDIT** or **GRIPS** to edit the polyline to change the ending vertex up toward the relocated text.

Use **DDMODIFY** to move these objects back to the **0285-PIPESIZE** layer.

Once the piping is properly blocked and labeled, we are ready to insert the remote control valve symbol and the ID tag call-out.

From the **Symbols** pull-down menu, select **Equipment . . .**. This opens the "Irrigation Peripheral Equipment" icon menu (Figure 15.84).

Select the **Remote Control Valve** symbol from the icon menu.

```
Command:
Locate symbol:
```
Pick a point directly below the center of the remote control valve above, aligned with the endpoint of the short 3/4" header pipe.

```
Scale symbol:
```
Type **30**. This scales the symbol to match the existing remote control valve symbols.

Figure 15.81

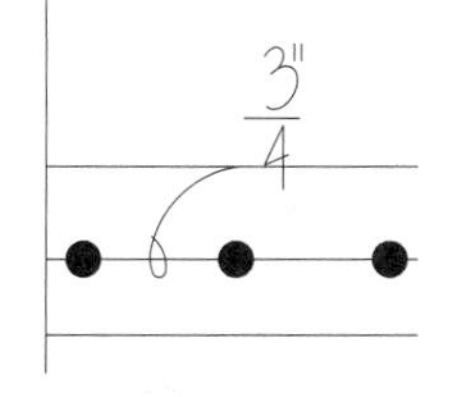

Figure 15.82

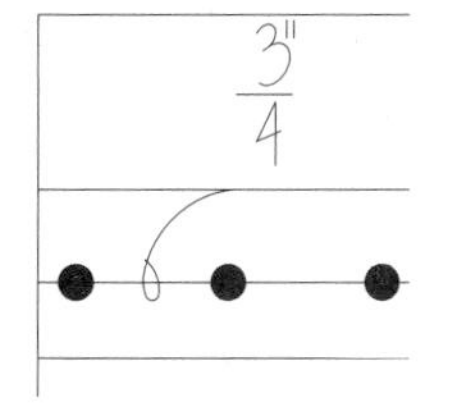

Figure 15.83

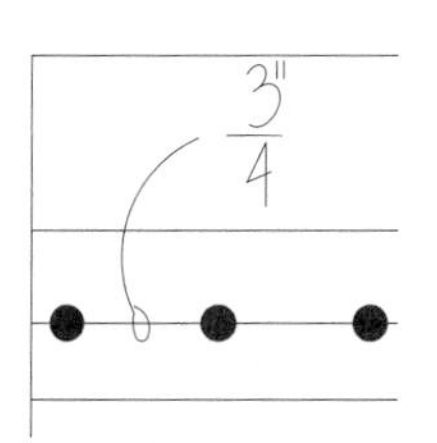

Figure 15.84

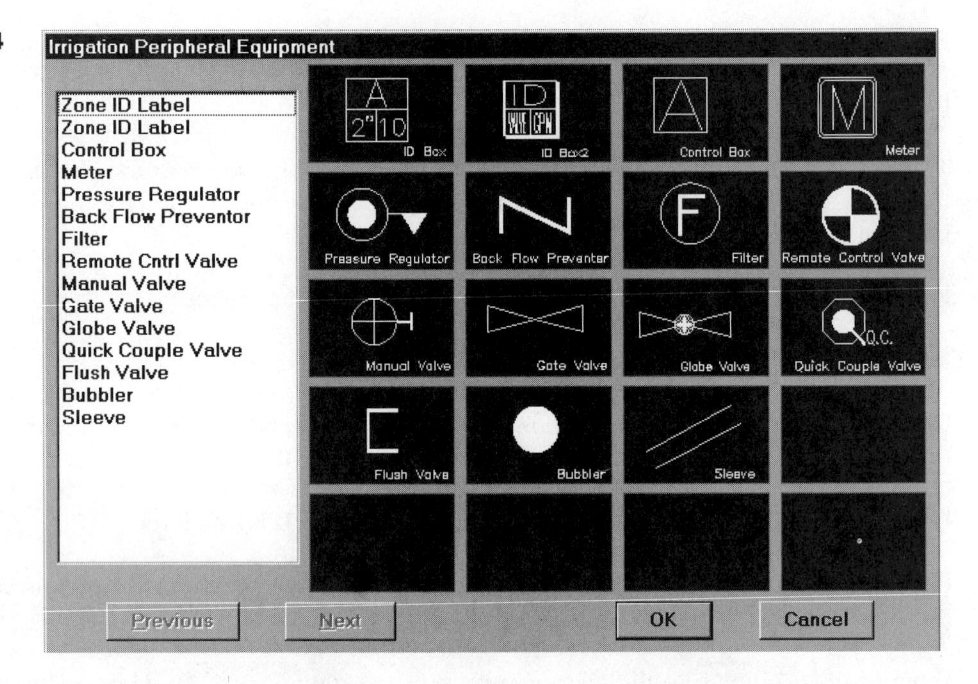


```
Rotate symbol:
```

With **ORTHO** active, pick a valve rotation so that it matches the other remote control valve symbols.

After the symbol is inserted, we see the "Edit Attributes" dialogue box, allowing us to insert attribute information about the control valve (Figure 15.85).

In the **ITEM** box, change the entry to **RC Valve**.

In the **SIZE** box, enter **3/4"**.

These attributes are then attached (invisibly) to the remote control valve symbol.

After the control valve symbol is inserted, we need to place a zone call-out for Planter area 3. Because this irrigation station was created manually, we will need to create its call-out manually as well. Instead of reading the attributes from the remote control valve symbol, this routine presents a dialogue box, permitting us to create our own entries.

Under the **Symbols** pull-down menu, select **Equipment . . .**. Pick the **ID Box2** from the "Irrigation Peripheral Equipment" icon menu.

Figure 15.85

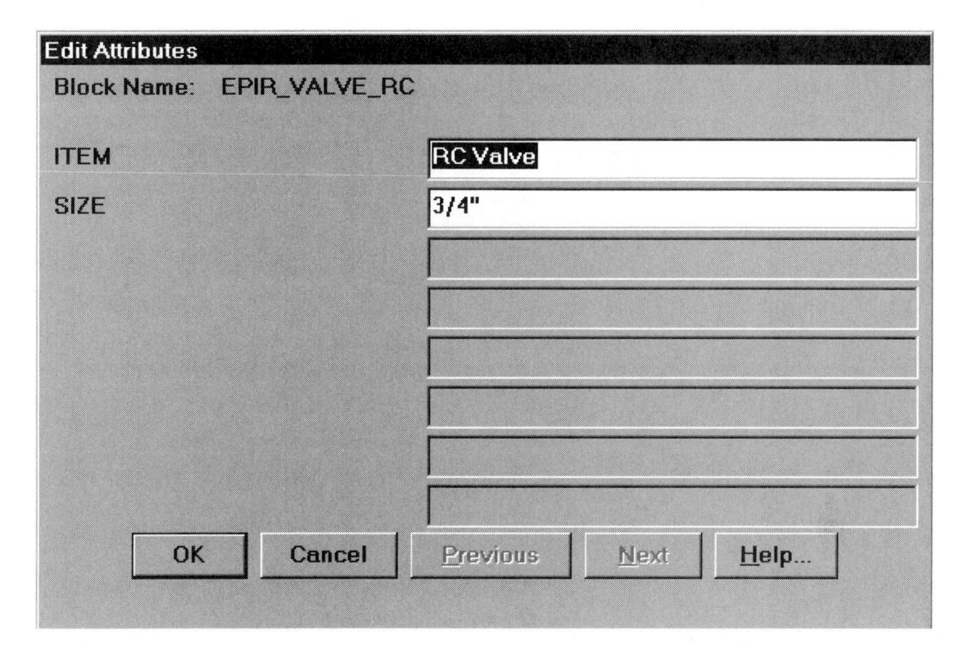

Warning: The current version (13.2) of LANDCADD ships with an error in the Landscape Irrigation menu file so that the "Irrigation Peripheral Equipment" icon menu does not properly retrieve **ID Box2**, but substitutes **ID Box1** in its place. If this happens, simply copy an existing **ID Box** from another valve, and edit the attributes. **If you have experience editing AutoCAD menu files**, you can fix the error by editing the IRAWMENU.MNU file. If you search for the string **zone id** you can find the **EPIR_ZONETAG_1** improperly listed for the second Zone ID tag. You must reload the menu file for the changes to take effect.

`Command:`

`Locate symbol:`

Pick a point aligned with the left edge of the existing zone ID tags, directly opposite the new remote control valve symbol.

`Scale symbol:`

Type **40** to make the size match the existing zone call-outs.

`Rotate symbol:`

With **ORTHO** active, rotate the symbol into position.

After the symbol is located, an "Edit Attributes" dialogue box is ready to receive information about our new remote control valve (Figure 15.86).

In the **EPIR_ZONETAG_VALVE** box, type **AVB Valve.**

Figure 15.86

Edit Attributes
Block Name: EPIR_ZONETAG_2

EPIR_ZONETAG_VALVE — AVB Valve
Enter Valve Size: — 1"
Enter Zone ID: — 4
Enter Flow Rate: — 11.00

OK | Cancel | Previous | Next | Help...

In the **Enter Valve Size:** box, type **3/4".**

In the **Enter Zone ID:** box, type **4**.

In the **Enter Flow Rate:** box, type **11.00**. (The zeros are place holders, making the call-out more closely resemble the other zone call-outs.)

The zone call-out is inserted as before.

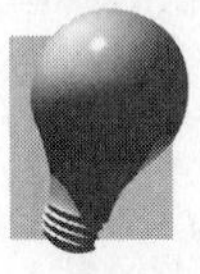

Hint: The text in the **Zone ID** (upper) box in manually placed Zone Tags is usually distorted. To remove this distortion, add one or two spaces before and after the **Zone ID** number to correct the distortion and center the number in the Zone Tag.

Locating Main Line

Before we begin piping the back yard sprinklers, we will place an irrigation main line across the front yard and continue it around into the back. The main lines are drawn as polylines, which can be assigned a width. Main lines are traditionally shown as heavier lines than lateral lines. LANDCADD normally draws the main lines as continuous, but we will change the linetype of the main line layer to create a dashed main line.

Placement of Main Lines on Irrigation Plans

The placement of irrigation main lines and control valves on irrigation plans is often a source of confusion. It is important to remember that **main line placement is diagrammatic**. This means that, as far as main line placement is concerned, accuracy is sacrificed for clarity. Main lines are often shown running through houses, through hardscaped areas, outside property lines, etc. The designer knows that the lines cannot possibly be installed there, but places the main lines outside the planting areas so that the lateral lines and the sprinkler heads are clear and uncluttered. The location of laterals and sprinklers is intended to be very accurate.

This same standard is usually applied to remote control valves. Valve symbols are normally placed adjacent to the main line, somewhere close to the area that they will irrigate. It is considered acceptable to place valve symbols outside the property, in sidewalks or even inside the residence if it makes the head symbols and lateral line piping clearer and easier to read.

The irrigation designer usually allows the contractor to place main line piping in the most efficient manner possible. The irrigation general notes and specifications often specify that the contractor must provide the client with a set of "as-built" plans which indicate the actual location of the remote control valves and main line. The contractor will often choose to locate the valves into groups (or manifolds) to allow for more efficient installation and wiring. Although the plan may show valves spread out, they are often grouped together when installed.

To place main line, use the **Draw Mainline** option under the **Pipe** pull-down menu.

At the prompt:

```
Command: EPLC_runme
Enter polyline width <0">
```

Enter **3"** for the mainline width.

```
_.pline
From point:
```

Pick a point near the remote control valve symbol closest to the top of the drawing in the front yard.

```
Current line-width is 0'-3"
Arc/Close/Halfwidth/Length/Undo/Width/<Endpoint of line:
```

Pick points to create the main line shown below (Figure 15.87).

Continue to build the main line down the front walkway, across the driveway and front yard, outside the lower part of the property line, up outside the back property line, and across to Planter area 6. Add the "stub" main line which serves Planter area 5.

Add a jump line where the pipe from Zone 4 crosses the main line.

Use the AutoCAD layer control box to change the linetype of the **0287-MAINLINE** layer to **EPLC_SHTDASH**. This gives a different appearance to the main line as shown below (Figure 15.88).

Figure 15.87

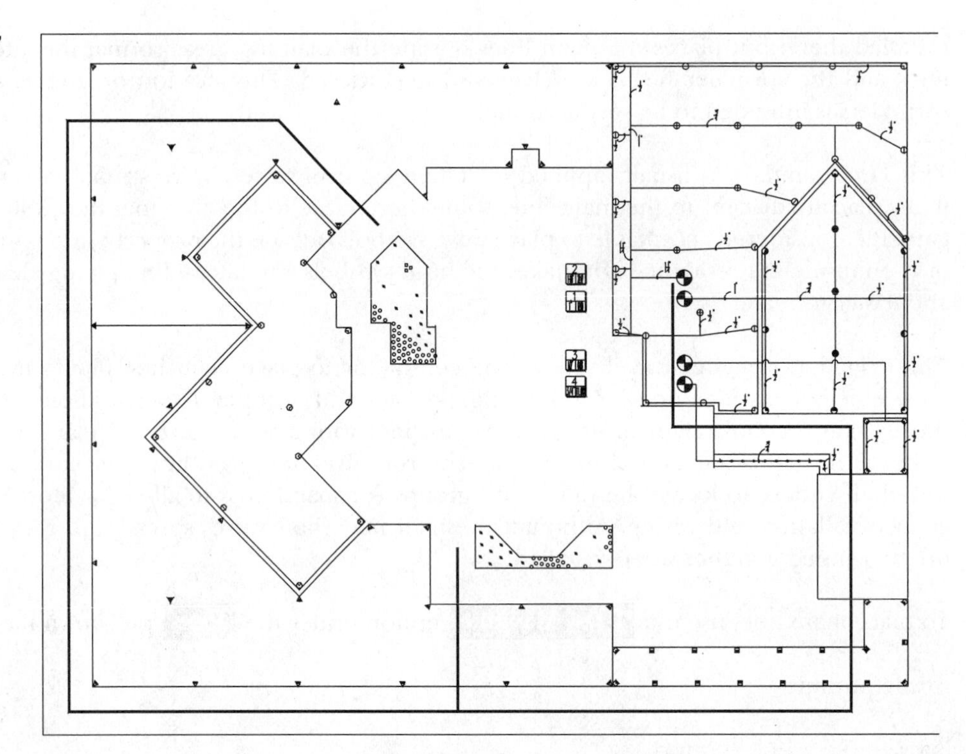

Figure 15.88

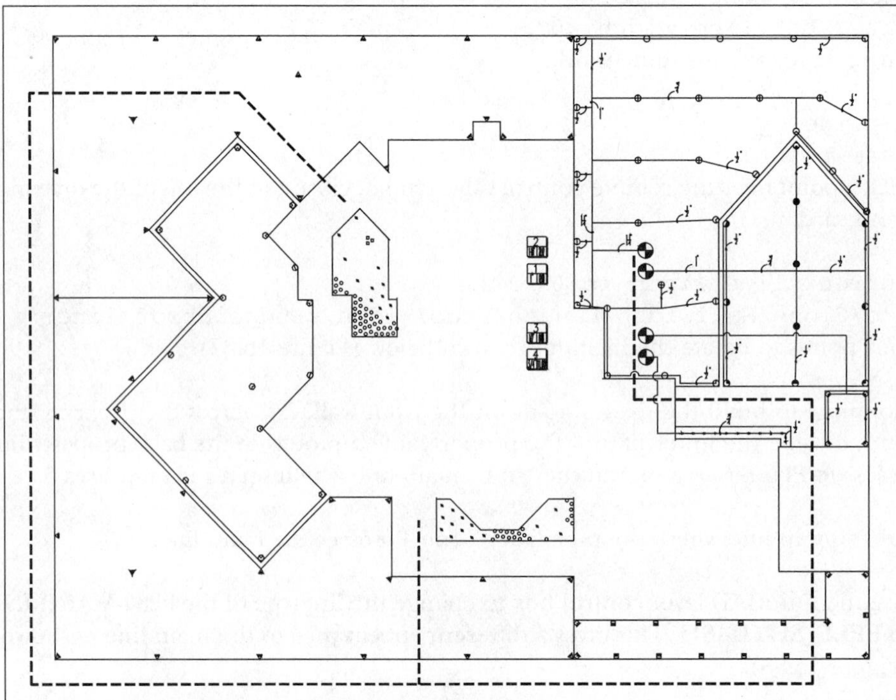

Sleeves

In the locations where piping will run beneath concrete, we will indicate the presence of a sleeve pipe. The sleeve pipe is usually shown by a double line surrounding the original pipe (main or lateral) line.

Use the **Symbols** pull-down menu to access the **Equipment . . .** icon menu.

Pick the icon marked **Sleeve.**

At the prompt:

```
Command:
Starting point of sleeve:
```
Pick the intersection of the main line and the top of the driveway.

```
Ending point of sleeve:
```
Pick the intersection of the main line and the bottom of the driveway.

```
Offset width of sleeve <0'-2">
```
Type **4"**. This creates two lines offset 4" on each side of the center. The two lines are 8" apart.

Add sleeves to the following locations (Figures 15.89, 15.90, and 15.91).

Each sleeve which appears at the end of the main line will have a valve placed at its terminal point, which will improve the appearance of the drawing. The only sleeves on this drawing will be for main lines. Although it may be necessary to place a sleeve for the header pipe in Zone 4, we really are not sure where the contractor will install the control valves, so we can not specify where the sleeve should go. A general note would inform the contractor to use sleeves whenever any type of pipe is run under concrete.

While it is possible to indicate the location of the main line at this point, it is not possible to size it. This is because we do not yet know the greatest flow demand for all of the control valves in the plan—the determining factor in sizing the main line. We will only be able to size the main line when all of the control valves have been located. We will proceed on to the next valve.

Figure 15.89

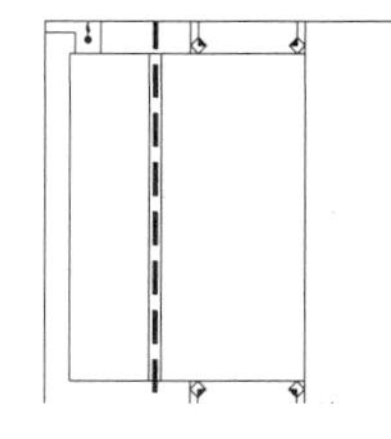

Figure 15.90

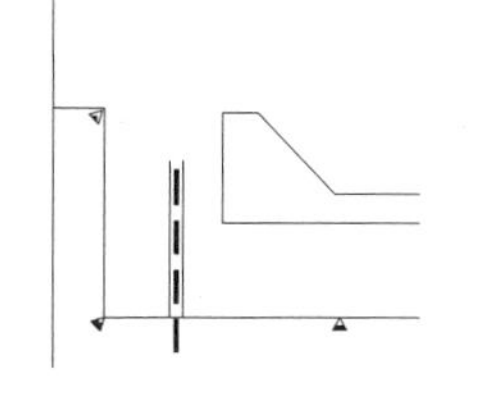

Figure 15.91

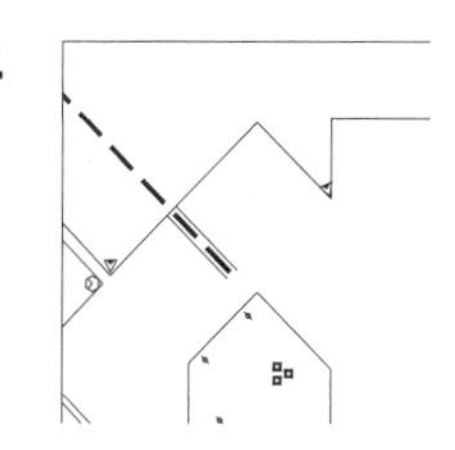

Planter Area 4

Planter area 4 is a station which consists of two parallel legs of pipe, each running along the outside edge of the planter. For maximum efficiency, it is best to place the valve as close to the center of the station as possible. All of our piping will run toward this center valve.

Use the **Draw Lateral** command under the **Pipe** pull-down menu. Draw lateral lines through the sprinkler heads in Planter area 4 (Figure 15.92).

Figure 15.92

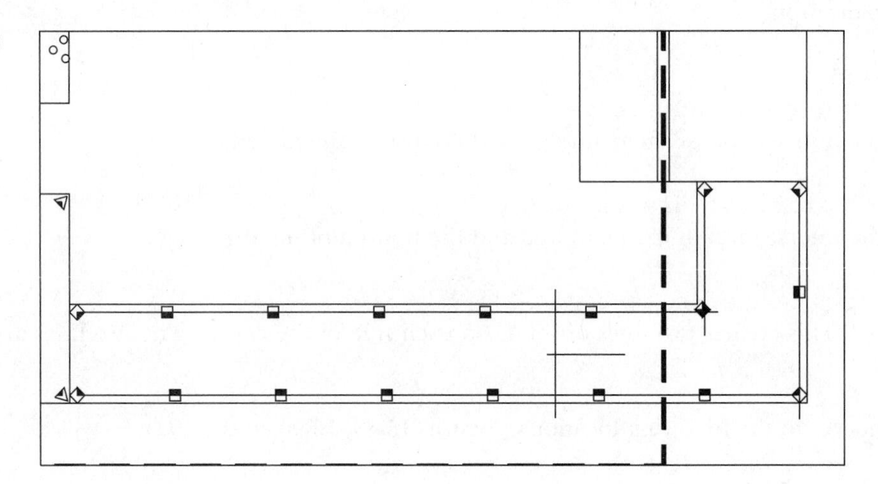

Note that the final header pipe ends near to, but not touching, the main line. There is only enough room remaining to place the remote control valve symbol. Remember that the pipe lines should pass completely through the head symbols, and need not intersect each other neatly. The **Auto Size** routine will clean up the pipe intersections.

Use the **Auto Size** selection under the **PipeSize ▸** sub-menu. Accept the values entered in the "Auto Pipe Sizing" dialogue box and pick **OK**.

At the prompt:

```
EPLC_runme
Select lines & heads
Select objects:
```
Pick all of the heads and pipes in Planter area 4.

```
WORKING ....\
Pick valve location:
```

Pick the end of the header pipe near the main line.

```
WORKING ..../
```

LANDCADD sizes the pipes and inserts a remote control valve symbol (Figure 15.93).

Figure 15.93

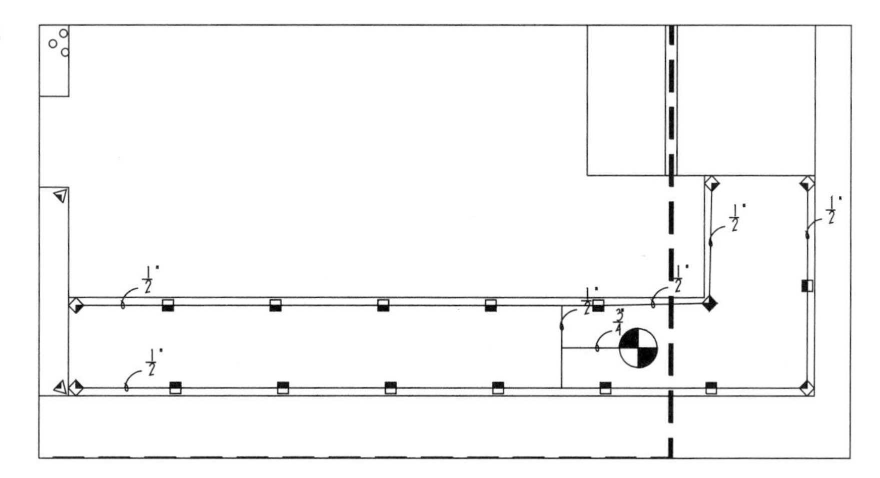

Be sure to use the **Jump Line for Pipe** to add jump lines where the lateral pipe lines cross the main line. Move or modify any pipe labels which are unclear or obstructed by other objects in the drawing. In addition, use **ID Tag** to label the remote control valve symbol (Figure 15.94).

Figure 15.94

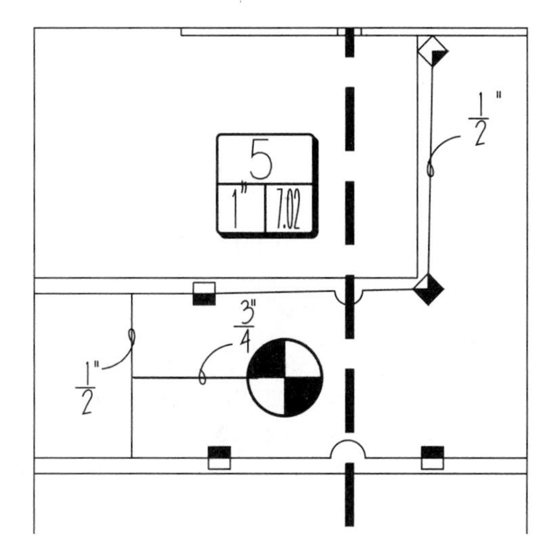

Planter Areas 7 and 8

Earlier in the development of the irrigation plan, we drew boundary lines which acted to define the areas for head insertion. At this point, we need to determine whether or not these areas contain the correct number of heads to create irrigation zones. If the areas contain too many heads, the maximum available flow will be exceeded. If the areas contain too few heads, the irrigation system will not be efficient because too many valves will be required. To test the total flow for Planter areas 7 and 8, we can use the **Sum Flow** routine.

Under the **Zones** pull-down menu, choose **Sum Flow**.

At the prompt:

```
Command: EPLC_runme
Select irrigation heads:
Select objects:
```

Pick all of the irrigation heads in the Planter areas 7 and 8. Press **<Enter>** when finished selecting all of the heads. There should be **38** heads.

```
Select objects: 38 found
3 were filtered out
38 New heads selected, 38 total heads selected, discharging
66.39 GPM
Select additional heads or press <Enter> to quit:
Select objects:
```

Press **<Enter>** to complete the command.

From this information, we see that the sprinkler heads in Planter areas 7 and 8 total **66.39** GPM. This represents too many gallons per minute to run on two control valves. Each would need to carry over 30 GPM, our maximum available flow. Because of this, we should divide the areas into three zones rather than two. Each zone should carry about 22 GPM for a balanced flow. We need to divide the area according to reasonable piping and head locations, along with flow considerations.

One of the easiest ways to divide up an area into irrigation zones is to use the **Define** function under the **Zones** pull-down menu.

This function draws a polyline around the heads that we wish to include in a single zone. These enclosing polylines are only for marking purposes, and perform no other function. They simply allow us to keep track of which heads are grouped together. See the zones defined below (Figure 15.95).

These zones divide the back area into approximate thirds. By using the **Sum Flow** routine, we find that the bottom zone has 13 heads discharging 24.08 GPM. The middle zone has 11 heads discharging 21.29 GPM, while the top zone has 14 heads discharging 21.02 GPM. This represents acceptable flow balance, and places the heads into logical circuits.

Figure 15.95

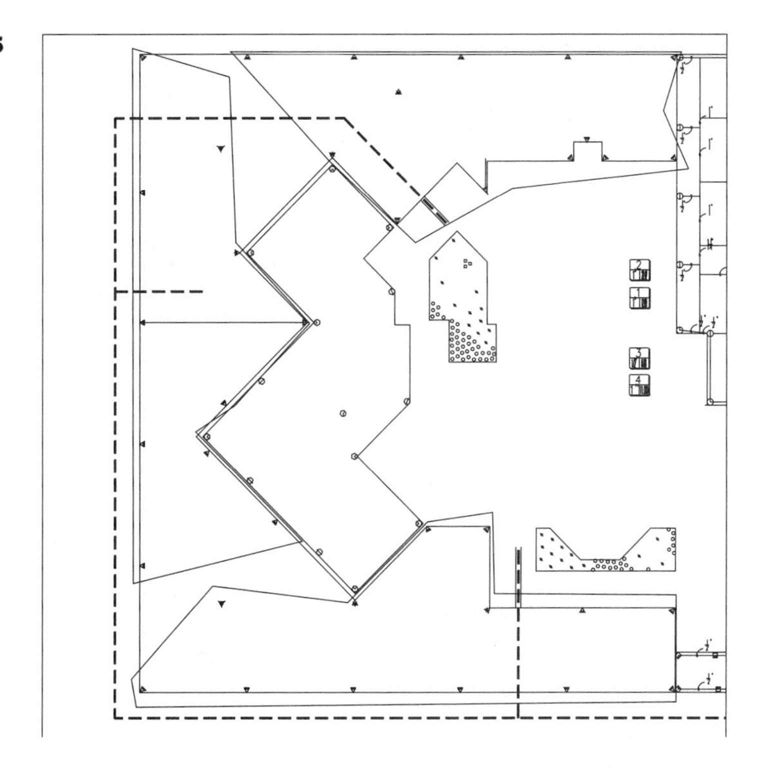

Next, lay out the piping for the bottom zone, which we will call Zone 6.

Under the **Pipe** pull-down menu, use the **Draw Lateral** command. Draw lateral lines through the sprinkler heads in Zone 6 (Figure 15.96).

This arrangement of piping places the control valve near the center of the zone, next to the main line. All of the piping runs away from the control valve and does not double back toward the valve.

Figure 15.96

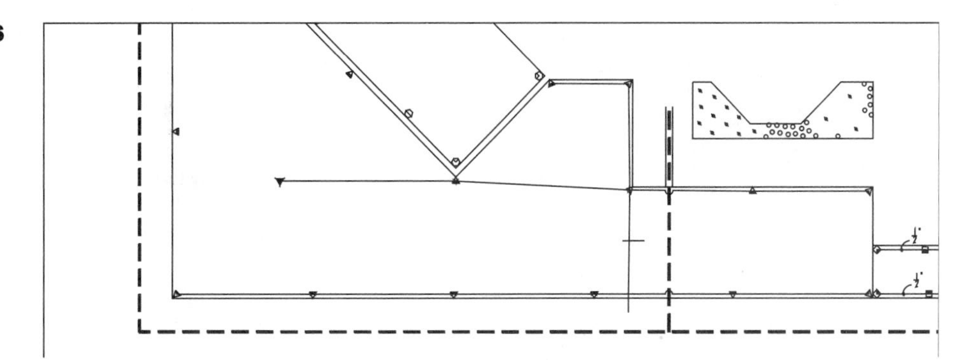

Use the **Auto Size** selection under the **PipeSize ▸** sub-menu. Accept the values entered in the "Auto Pipe Sizing" dialogue box and pick **OK**.

At the prompt:

```
EPLC_runme
Select lines & heads
Select objects:
```

Pick all of the heads and pipes in Zone 6.

```
WORKING ....\
Pick valve location:
```

Pick the end of the header pipe near the main line.

```
WORKING ..../
```

LANDCADD sizes the pipes and inserts a remote control valve symbol (Figure 15.97.)

To eliminate the congestion surrounding the remote control valve, change the style, move and modify the labels to keep them from overlapping onto pipe lines and the control valve symbol (Figures 15.98 and 15.99).

Use the **Jump Line for Pipe** to add jump lines where the lateral pipe lines cross the main line then use **ID Tag 2** to label the remote control valve symbol (Figure 15.100).

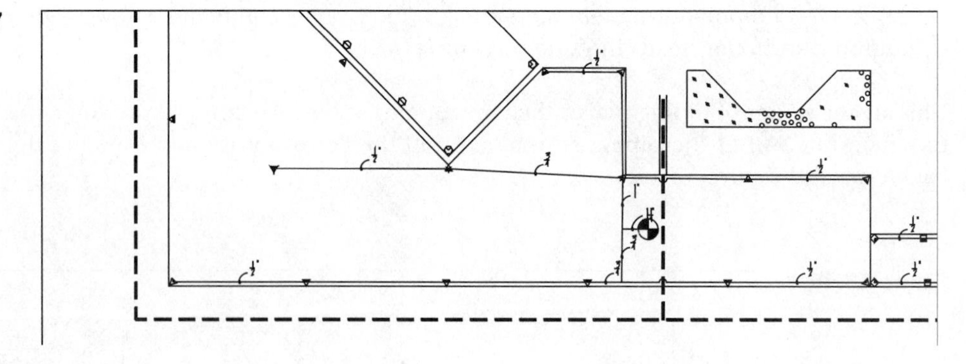

Figure 15.97

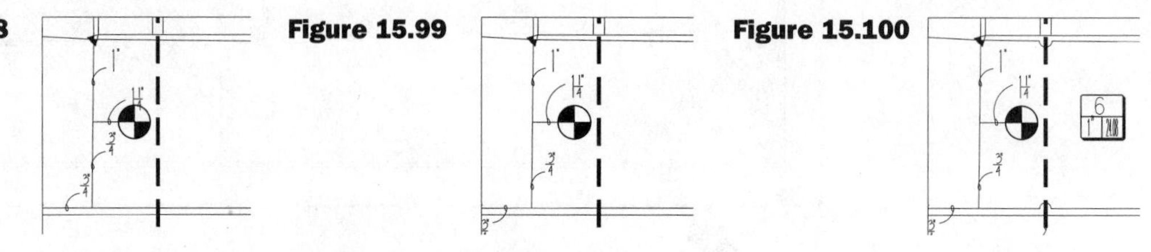

Figure 15.98

Figure 15.99

Figure 15.100

After completing the piping for the bottom zone, we are ready to lay out the pipe for the middle zone, Zone 7.

Use the **Draw Lateral** command under the **Pipe** pull-down menu. Draw lateral lines through the sprinkler heads in Zone 7. Use the **Auto Size** selection under the **PipeSize ▸** sub-menu (Figures 15.101 and 15.102).

Accept the values entered in the "Auto Pipe Sizing" dialogue box and pick **OK**.

At the prompt:

```
EPLC_runme
Select lines & heads
Select objects:
```
Pick all of the heads and pipes in Zone 7.

```
WORKING ....\
Pick valve location:
```
Pick the end of the header pipe in the center of the zone.

Figure 15.101

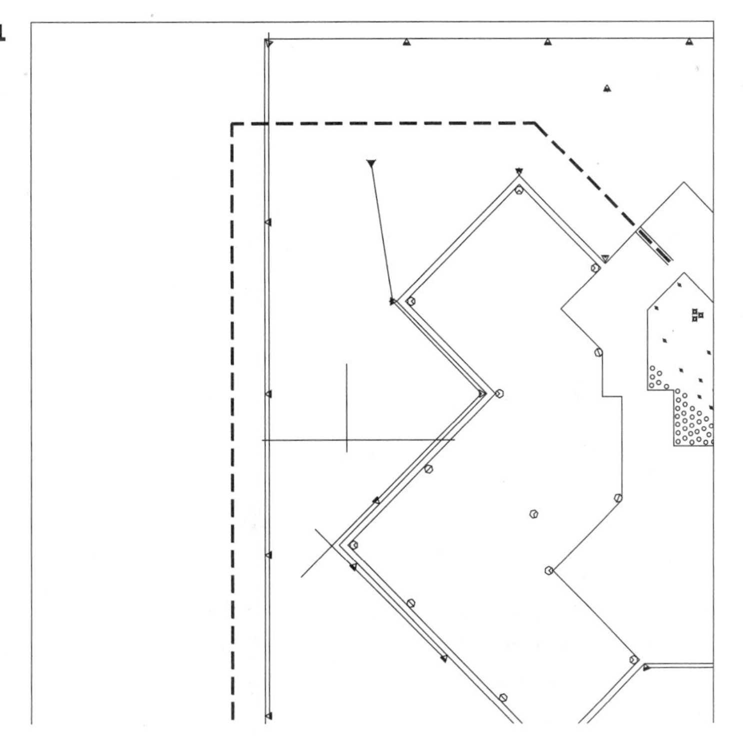

Figure 15.102

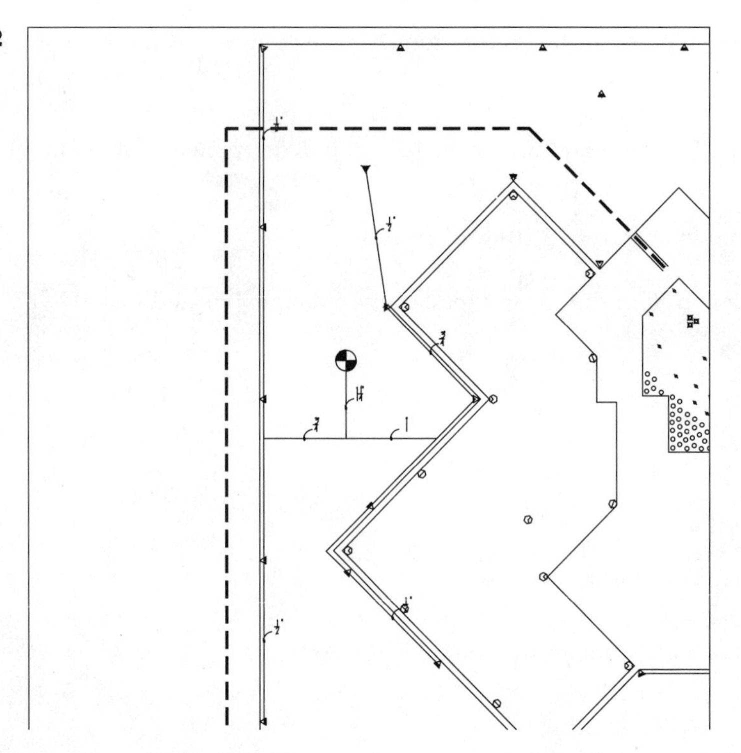

```
WORKING ..../
```

LANDCADD sizes the pipes and inserts a remote control valve symbol. The location of the control valve will require that we add a main line to serve it (Figure 15.103).

Complete work on Zone 7 by following these steps:

Use the **Draw Main Line** command (under the **Pipe** pull-down menu) to add a main line segment to the control valve for Zone 7.

Move any pipe labels which interfere with other pipe lines.

Use **Jump Line for Pipe** to place jump lines where lateral lines cross the main line.

Add the **ID Tag 2** to label the remote control valve symbol.

Begin the piping for Zone 8 by using **Draw Lateral** (under the **Pipe** pull-down menu) to place the lateral lines as shown (Figure 15.104).

This pipe arrangement is somewhat awkward, but still represents an approximation of the way a contractor would install the lateral lines. Placing most of the pipe toward the outer edges of a zone minimizes the chances of damaging the pipe when installing trees and shrubs.

Figure 15.103

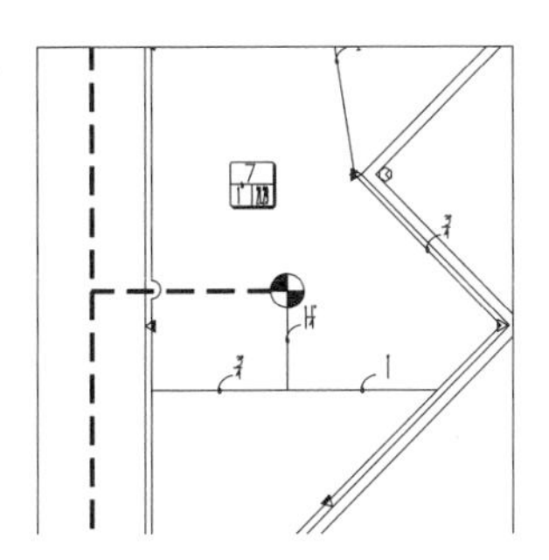

Figure 15.104

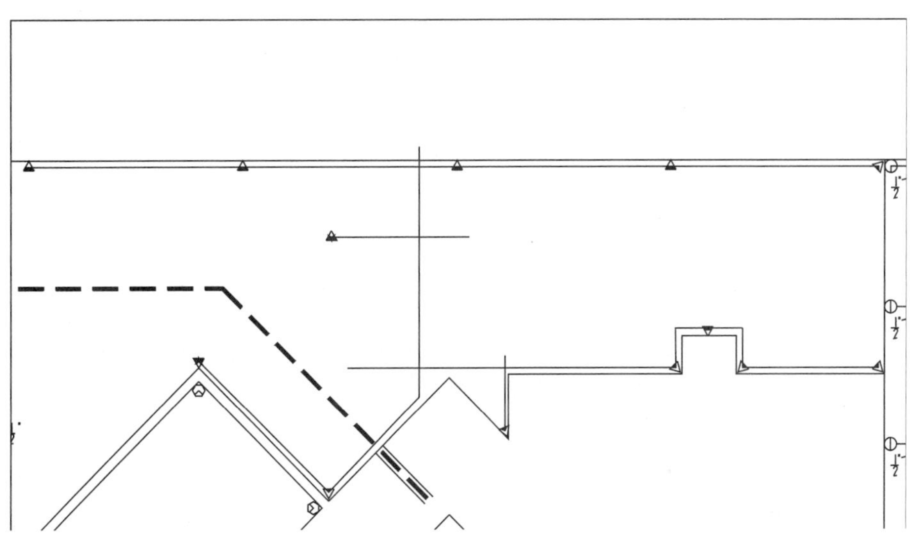

Use the **Auto Size** command under the **PipeSize ▸** sub-menu. Accept the values entered in the "Auto Pipe Sizing" dialogue box and pick **OK**.

At the prompt:

```
EPLC_runme
Select lines & heads
Select objects:
```

Pick all of the heads and pipes in Zone 8.

```
WORKING ....\
Pick valve location:
```

Pick the end of the header pipe ending near the main line.

```
WORKING ..../
```

LANDCADD sizes the pipes and inserts a remote control valve symbol (Figure 15.105).

Figure 15.105

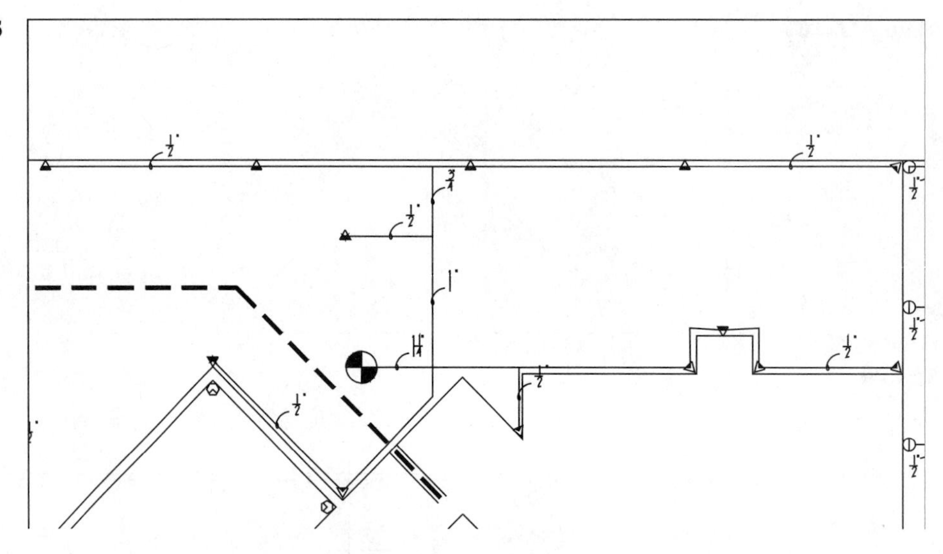

Complete work on Zone 8 by following these steps:

Change the pipe label styles and move any pipe labels which interfere with other objects in the drawing. Look at the 1/2" line near the house, and the 1/2" labels at the top of the screen.

Use **Jump Line for Pipe** to place a jump line where the lateral line crosses the main line.

Add the **ID Tag 2** to label the remote control valve symbol (Figure 15.106).

Figure 15.106

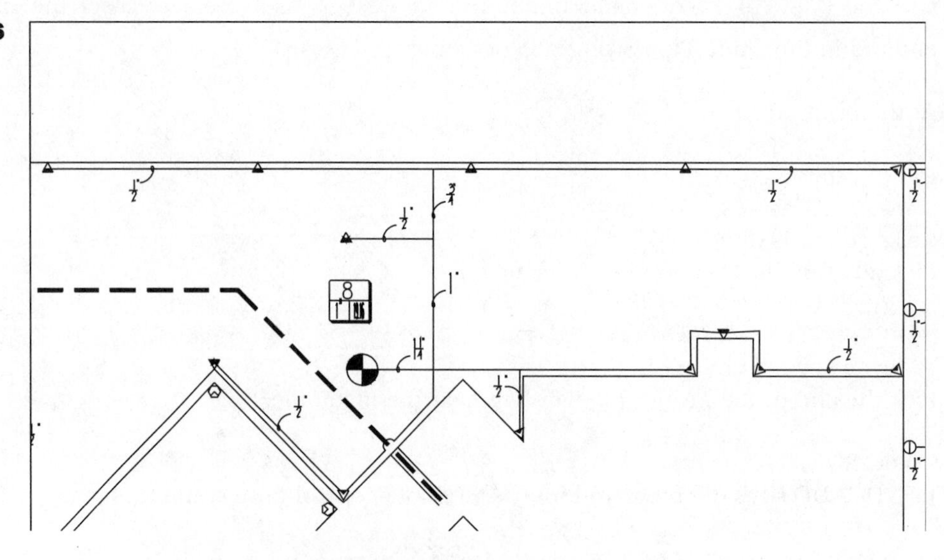

Piping for the Back Lawn

When designing the irrigation system for the back lawn, we are faced with a similar situation as that when we worked with Zones 6, 7 and 8. That is, we do not yet know exactly how many GPM are used when running all of the heads in the back lawn. To obtain this information, we use the **Sum Flow** function under the **Zones** pull-down menu.

At the prompt:

```
Command: EPLC_runme
Select irrigation heads:
Select objects:
```
Pick all of the sprinkler heads in the back lawn. There should be **16** heads.

```
Select objects:
16 found
```
Press **<Enter>** when all of the heads are highlighted.

```
16 new heads selected, 16 total heads selected, discharging
26.58 GPM
Select additional heads or press <Enter> to quit:
Select objects:
```
Press **<Enter>** to finish the command.

```
16 heads selected, discharging 26.58 GPM
```
The flow is close to, but does not exceed 30 GPM. Therefore, we can connect all of these heads to one control valve and still be confident that the system will operate properly. There should be sufficient pressure to make all of the heads cover the lawn area as designed.

Use the **Draw Lateral** routine to draw the piping for the back lawn. Place the pipes as shown (Figure 15.107).

This arrangement places the pipe at the perimeter of the lawn, and minimizes the amount of cross-trenching. The pipe labels should be easy to read and should not interfere with the other pipe labels already on the plan.

Use the **Auto Size** routine to size the pipe and place the remote control valve symbol (Figure 15.108).

Change the label styles where necessary, and move and modify the labels to make them easier to read. Avoid conflicting with the existing pipe labels.

Figure 15.107

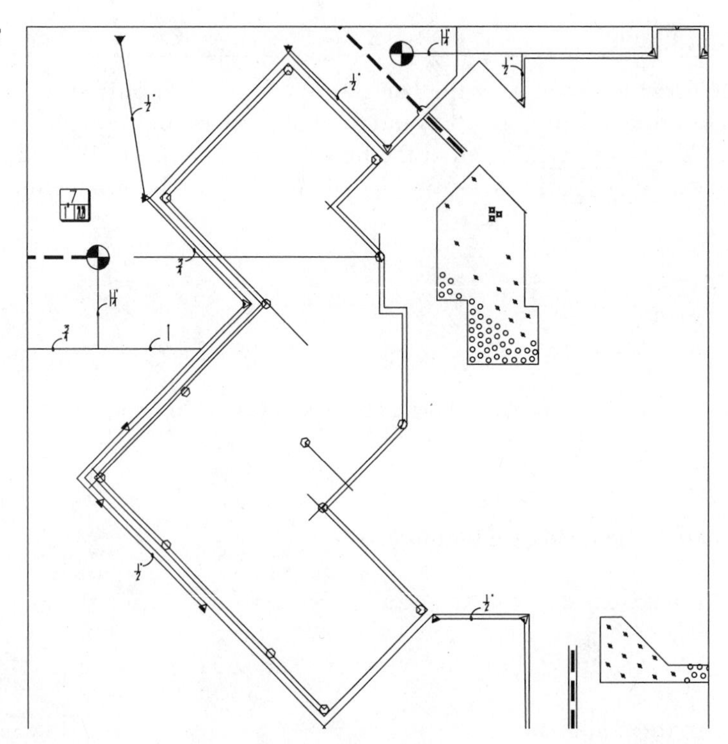

Figure 15.108

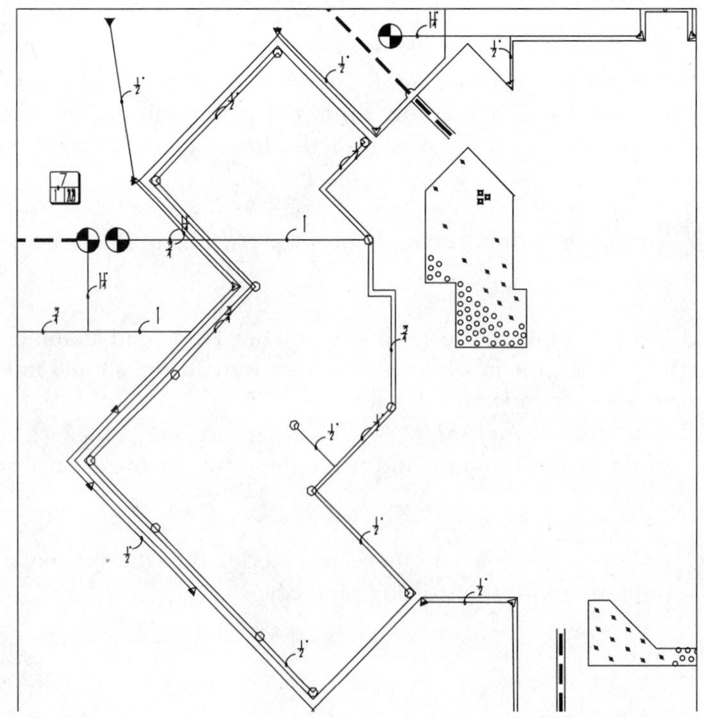

Use the **Jump Line for Pipe** routine to place a jump line where the piping crosses the pipe from Zone 7. In addition, insert the **ID Tag 2** symbol on the drawing to label this control valve as Zone 9.

Note that where possible, the pipe labels extend into the area of the irrigation zone. This makes the labels more obvious and the drawing easier to understand (Figure 15.109).

Figure 15.109

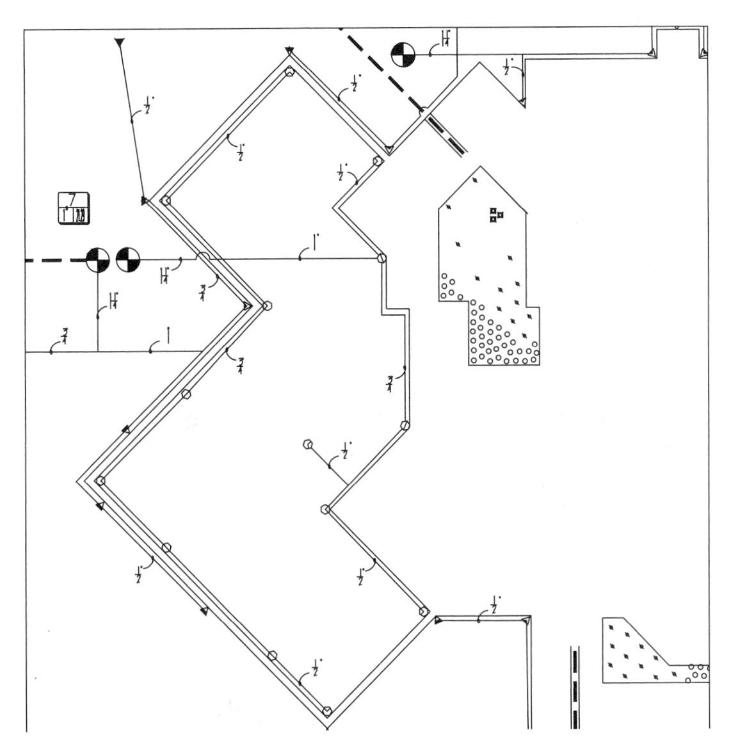

Piping the Drip Irrigation Stations

Placing pipe in the drip irrigation stations requires the use of LANDCADD routines we have used previously, but in new ways.

We will use the **Drip . . .** option under the **Pipe** pull-down menu to create polylines that connect the emitters to the valve, then place a valve and ID tag to identify the valve. Finally, we will use the **U size** option under the **PipeSize ▸** sub-menu to assign the correct pipe size, material and label to the drip tubing.

Begin by moving some of the emitters in Planter area 6 back towards the center of the planter. Their initial placement was directly at the center of the plant symbols, so a few must be moved toward the center of the area (Figure 15.110).

Figure 15.110

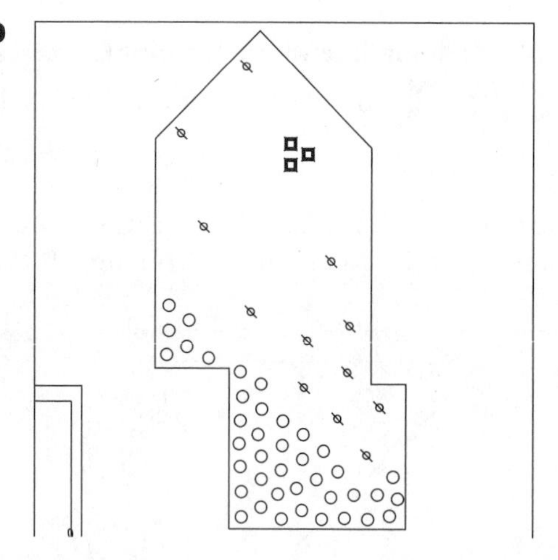

Pick the Drip . . . option, found under the Pipe pull-down menu, to open the "Drip Irrigation" dialogue box (Figure 15.111).

Figure 15.111

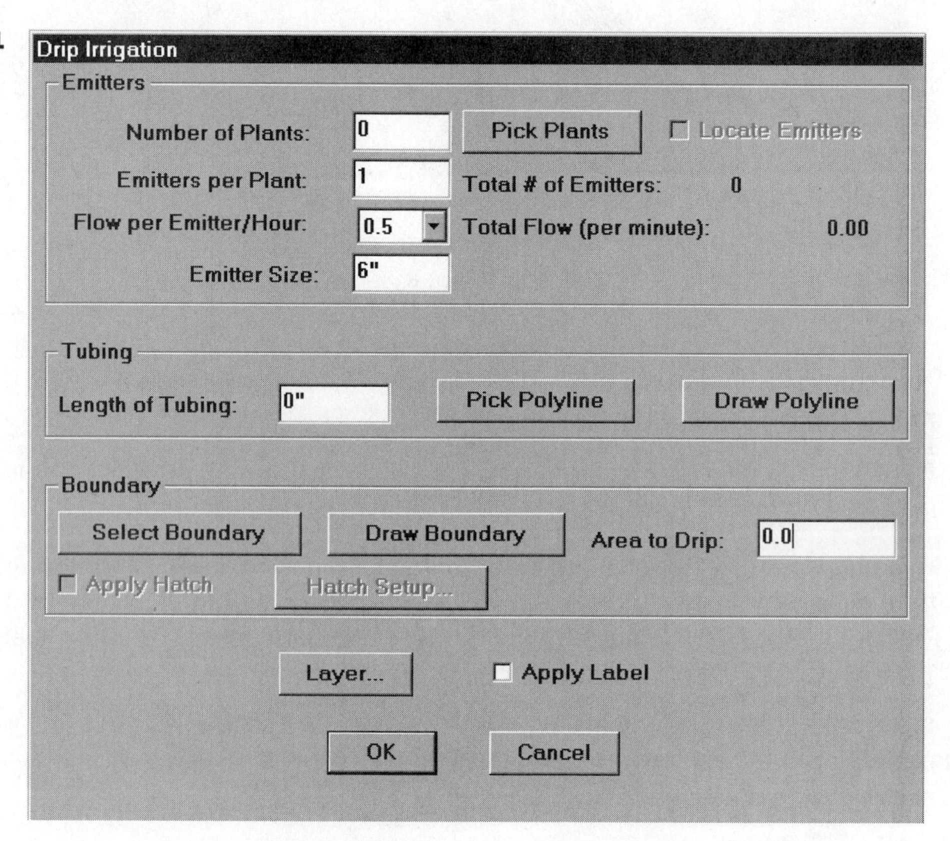

Pick the **Draw Polyline** option to begin to draw the drip pipe polylines. See below for a suggested sequence for piping (Figures 15.112 and 15.113).

Figure 15.112

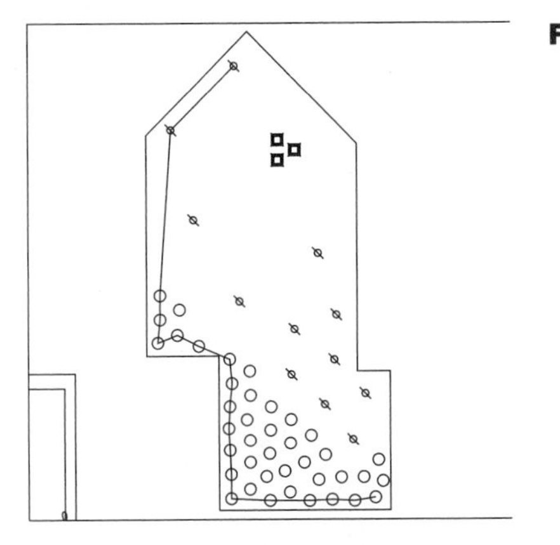

Figure 15.113

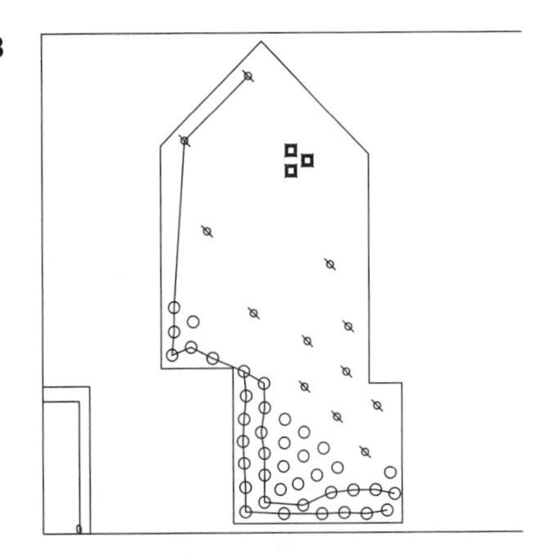

Be sure to pick the insertion points for the emitter symbols to act as the vertices in the pipe polylines. Continue placing pipe polylines which connect back to the original pipe.

When completed, your piping should look like the following illustration (Figure 15.114).

Figure 15.114

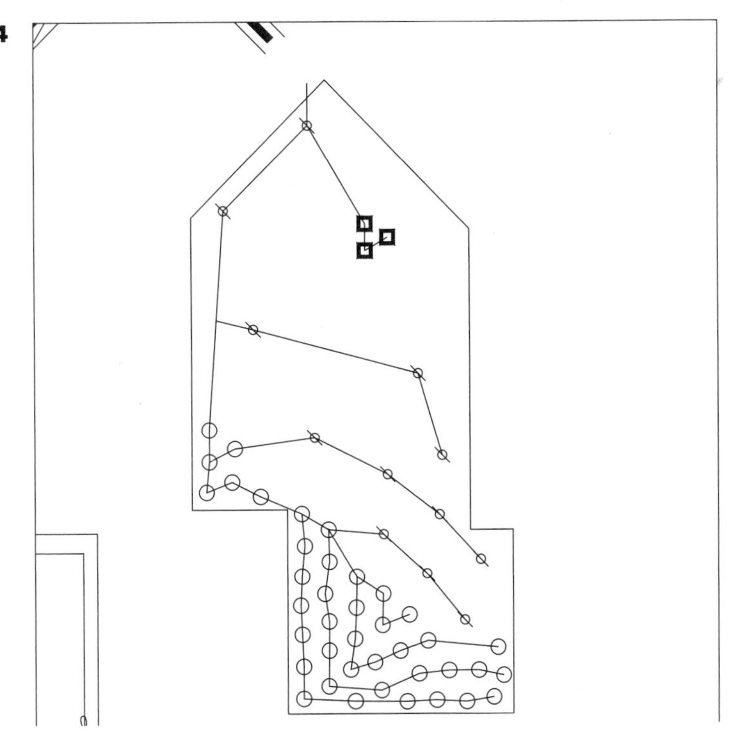

Once the piping is complete, we can insert the remote control valve symbol and the zone call-out.

Open the Symbols pull-down menu, then pick Equipment . . . . In the "Irrigation Peripheral Equipment" icon menu, pick **Remote Control Valve**.

```
Command:
Locate symbol:
```

Locate the symbol at the end of the main line, adjacent to the end of the lateral line extending from the last drip emitter.

```
Scale symbol:
```

Type **22.** This makes the symbol for the drip valves appear different from the regular remote control valves.

```
Rotate symbol:
```

With **ORTHO** active, rotate the valve so that it matches the other remote control valves.

When the "Edit Attributes" dialogue opens, fill in the boxes:

In the **ITEM** box, change the entry to **RC Valve**.

In the **SIZE** box, enter **3/4"**. This is the smallest size automatic remote control valve available.

To place the zone call-out, reopen the **Equipment . . .** selection, bringing up the "Irrigation Peripheral Equipment" icon menu. Pick **ID Box2**. (Remember the icon menu may not behave properly, so you may have to copy an existing ID Box and edit its attributes.)

```
Command:
Locate symbol:
```

Place the zone call-out to the left of the control valve symbol.

```
Scale symbol:
```

Type **40** to make the size match the existing zone call-outs.

```
Rotate symbol:
```

With **ORTHO** active, rotate the symbol into position.

In the "Edit Attributes" dialogue box, fill out the information as follows (Figure 15.115).

In the **EPIR_ZONETAG_VALVE** box, type **Drip Valve**.

Figure 15.115

In the **Enter Valve Size** box, type **3/4"**.

In the **Enter Zone ID:** box type **10**. Remember to insert spaces as necessary into this text box. This will assure that the Zone ID number will not appear distorted in the finished zone call-out.

In the **Enter Flow Rate:** box type **0.78**. This corresponds to the emitter flow for this station. (There are forty 0.5 GPH emitters, twelve 2.0 GPH emitters and three 1.0 GPH emitters. Since 60 GPH = 1 GPM, this yields 47 GPH or 0.78 GPM.)

The zone call-out is inserted with its appropriate information.

After the remote control valve and zone call-out are inserted, we are ready to size and label the pipe. This process is important for two reasons. First, the pipe will be properly sized so that the installer will know what size pipe to use. Second, the pipe polylines will be turned into blocks which can store attribute information. The information includes pipe material (which will be polyethylene drip tubing), pipe length and pipe size. Labeling and creating pipe blocks will assist us in developing an irrigation legend and schedule.

To size the drip pipe, we will use the **U size** routine.

Open the **PipeSize ▸** sub-menu under the **Pipe** pull-down menu. Pick **U size** (Figure 15.116).

Figure 15.116

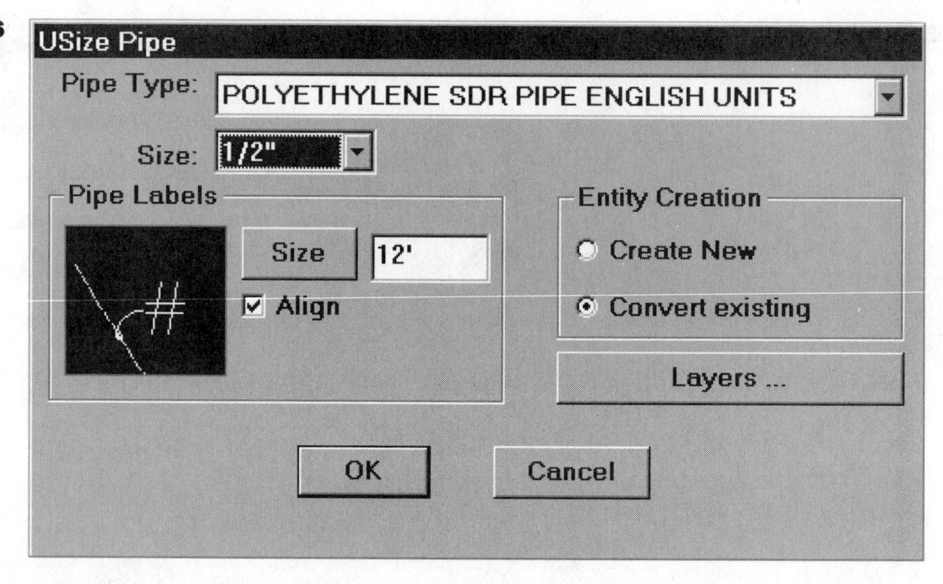

In the "Usize Pipe" dialogue box, set the **Pipe Type:** pop-down box to read **POLYETHYLENE SDR PIPE ENGLISH UNITS**.

Set the **Size:** pop-down box to read **1/2"**.

To make certain that the drip piping remains on its correct layer, pick the **Layers . . .** button to open the "Usize Pipe Layers" dialogue box (Figure 15.117.)

Figure 15.117

USize Pipe Layers
Lateral Layer
Name 0283-LATERALS
Existing 0283-LATERALS
Color Line Type
Label Layer
Name 0285-PIPESIZE
Existing 0285-PIPESIZE
Color Line Type
OK Cancel

Change the **Lateral Layer** box to keep the drip lateral lines in the layer **0288-DRIP_**. The **Label Layer** can be left as **0285-PIPESIZE**.

If we were to leave the layers as originally set, the drip piping would move to the same layer as the hard pipe lateral line. If we wish to assign a different color or linetype to the drip tubing, it will be very helpful to keep it on its own layer.

The **Pipe Labels** box can be left as it was set for the previous labels.

Leave the **Entity Creation** radio button set to **Convert Existing**, since we will be converting existing polylines into pipe blocks.

Pick **OK** when finished.

At the prompt:

```
Command: EPLC_runme
Select objects:
```

Pick all of the drip tube polylines connecting the emitters. The example shows 9 polyline segments. Press **<Enter>** when finished choosing the pipe.

```
Select objects: 9 found
```

The pipe polylines are converted into blocks and labeled. The results, while accurate, are unclear (Figure 15.118).

Use **EXPLODE**, **MOVE** and **Change Pipe Label Style** to modify the labels so that they are easier to read. Move the labels so that each reads more clearly. See the results below (Figure 15.119).

The final drip station is Zone 11, located below the bay window at the bottom edge of the house.

Use the **Drip . . .** option (under the **Pipe** pull-down menu) to open the "Drip Irrigation" dialogue box.

Pick the **Draw Polyline** to draw the drip pipe polylines. Your results should resemble those below (Figure 15.120).

Note that there are only three branches connected together at the emitter closest to the remote control valve location.

Open the **Symbols** pull-down menu, then pick **Equipment . . .**. In the "Irrigation Peripheral Equipment" icon menu, pick **Remote Control Valve**.

Figure 15.118

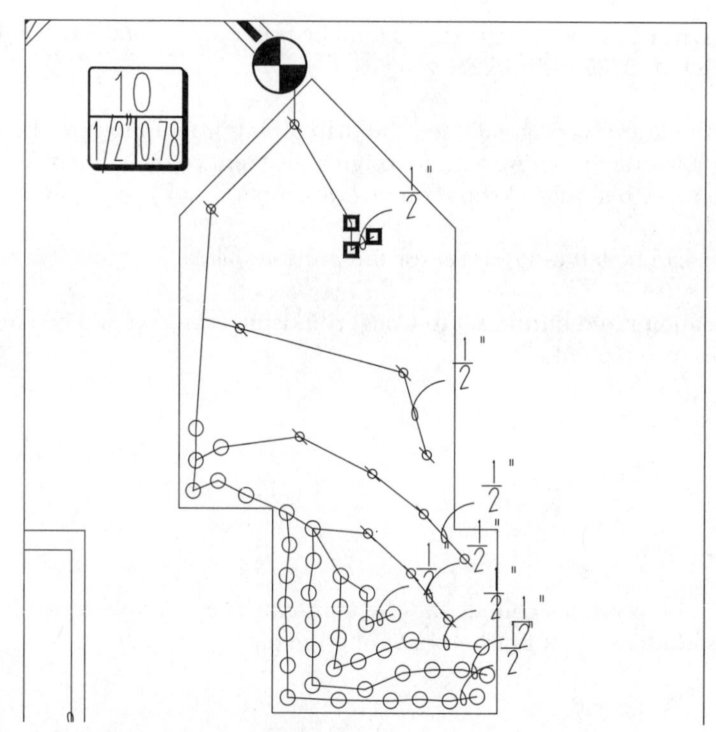

Figure 15.119

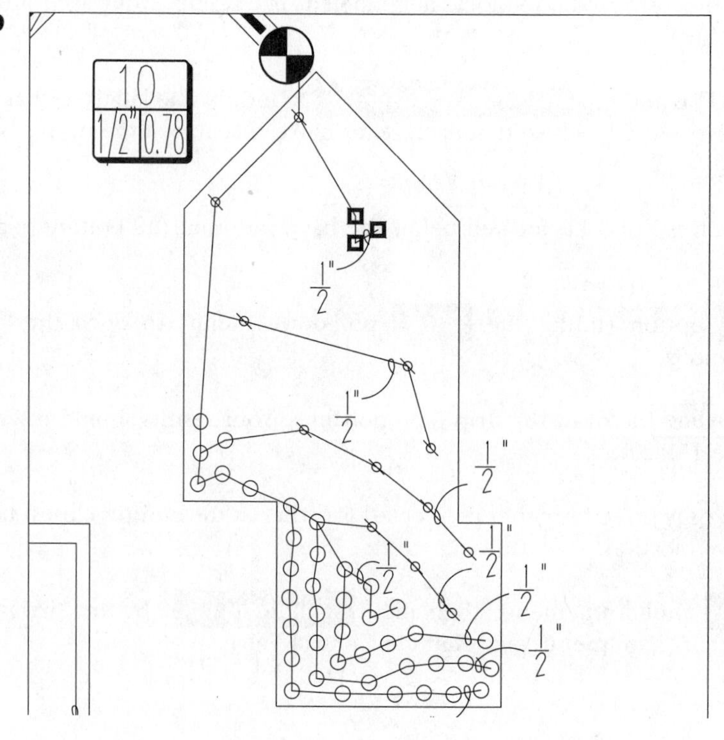

Figure 15.120

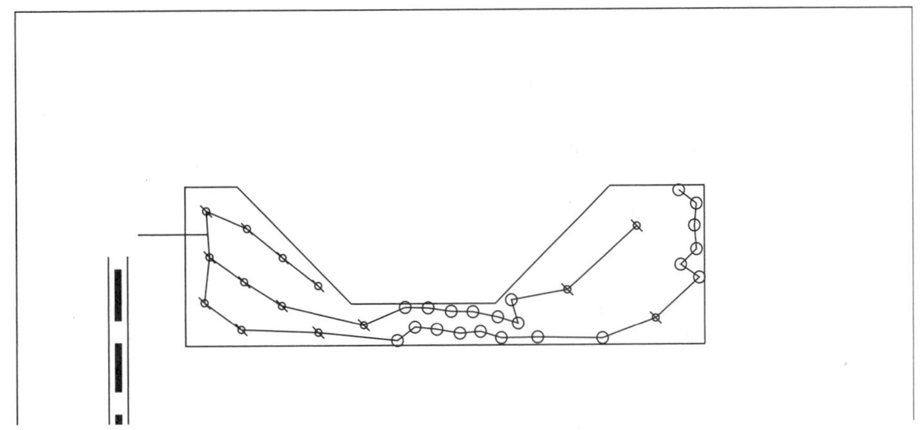

`Command:`
`Locate symbol:`
Locate the symbol at the end of the main line, adjacent to the end of the lateral line extending from the last drip emitter.

`Scale symbol:`
Type **22**.

`Rotate symbol:`
With **ORTHO** active, rotate the valve so that it matches the other remote control valves.

When the "Edit Attributes" dialogue opens, fill in the boxes:

In the **ITEM** box, change the entry to **RC Valve**.

In the **SIZE** box, enter **3/4"**.

To place the zone call-out, reopen the **Equipment . . .** selection, bringing up the "Irrigation Peripheral Equipment" icon menu. Pick **ID Box2**. (Remember the LAND-CADD bug.)

`Command:`
`Locate symbol:`
Place the zone call-out to the left of the control valve symbol.

`Scale symbol:`
Type **40** to make the size match the existing zone call-outs.

`Rotate symbol:`
With **ORTHO** active, rotate the symbol into position.

In the "Edit Attributes" dialogue box, fill out the information as follows:

In the **EPIR_ZONETAG_VALVE** box, type **Drip Valve.**

In the **Enter Valve Size** box, type **1/2".**

In the **Enter Zone ID:** box type **11**. Use spaces as needed.

In the **Enter Flow Rate:** box type **0.64**. The flow for this station is created by twenty-one 0.5 GPH emitters and fourteen 2.0 GPH emitters. This yields 38.5 GPH or 0.64 GPM.

The zone call-out is inserted with its appropriate information.

Size the drip pipe polylines as before.

Open the **PipeSize ▸** sub-menu under the **Pipe** pull-down menu. Pick **U size**.

Pick **OK** to accept the current values in the "Usize Pipe" dialogue box.

At the prompt:

```
Command: EPLC_runme
Select objects:
```

Pick the drip pipe polylines connecting the emitters. Press **<Enter>** when finished choosing the pipe.

```
Select objects: 3 found
```

The pipe polylines are converted into blocks and labeled. The results, while not as bad as the previous zone, still require some work (Figure 15.121).

As before, use **EXPLODE**, **MOVE** and **Change Pipe Label Style** to modify the labels so that they can be read easily. Again, place the labels toward the end of each branch of drip tubing. See the results below (Figure 15.122).

This completes the process of piping all of the irrigation zones. Once this process is complete, we have enough information to properly size the main line piping. Until we knew how much flow demand would be created by each remote control valve, we could not assign the proper size to the main line pipe. With all of the irrigation zones set up, we can make decisions about main line pipe sizes, the backflow prevention device size and the number of stations required for the irrigation controller.

Figure 15.121

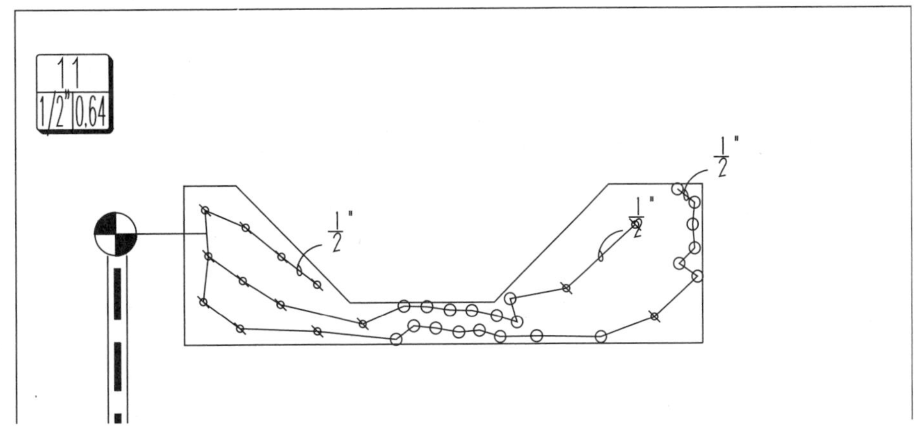

Figure 15.122

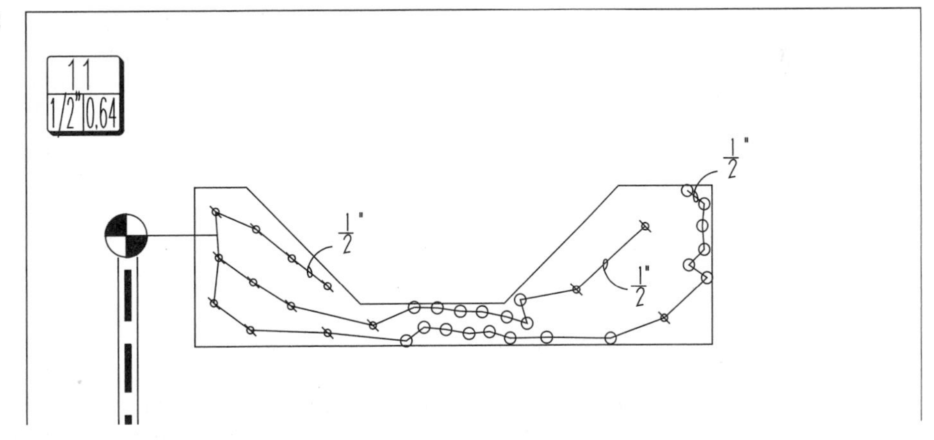

Placing the Backflow Preventer and Point of Connection

We placed the main line across the front yard without regard for the backflow device and the point of connection to the existing water service line. We will add a symbol for a backflow preventer and designate a point of connection now.

Open the **Symbols** pull-down menu to gain access to the **Equipment . . .** icon menu.

Pick the **Backflow Preventer** symbol.

```
Command:
Locate symbol:
```
Pick a point at the end of the main line near Zone 2.

```
Scale symbol:
```
Type **20**.

```
Rotate symbol:
```
With **ORTHO** active, rotate the symbol into the position shown in Figure 15.123.

After the symbol is inserted, we see the "Edit Attributes" dialogue box, allowing us to insert attribute information about the backflow preventer.

For **Item**, change the attribute to **R.P. Backflow Preventer**.

For **Size**, enter the attribute **1-1/4"**.

Pick **OK** when finished.

Next, we will extend the main line and place a symbol for the point of connection to the existing water service line.

Pick the **Draw Mainline** option under the **Pipe** pull-down menu. Using a polyline width of 3", place more main line to extend back to the approximate crossing point of the irrigation service line drawn in the Base Plan.

While working on the **0287-MAINLINE** layer, add a circle (with a **16"** radius) at the end of the main line, then use the **OFFSET** command (distance **4"**) to create a second circle inside the original. This creates a distinctive dashed double circle to symbolize a point of connection.

Add **Jumplines** to any lateral pipes which cross this new main line extension (Figure 15.123).

After the backflow preventer and point of connection are indicated, we need to label the point of connection symbol, and place a symbol for the irrigation controller.

To quickly move to a layer more suitable for placing labels and leaders, open the cursor menu. Pick **Layers ▸**, then pick **Set Current Layer By Entity**.

At the prompt:

```
Command:
Select object:
```
Pick any pipe label. The current layer will automatically switch to the **0283-PIPESIZE** layer. This is a good layer choice since this layer has "Continuous" as the linetype, and has other labels on it. We will add two new labels to this layer (Figure 15.124).

Because we inserted a block from our hardscape plan (**HOUSE1.DWG**) into the irrigation plan, the dimension styles created in the hardscape plan are present in this drawing as well. This means that we can build a dimension immediately, and the previously established style will be active.

Figure 15.123

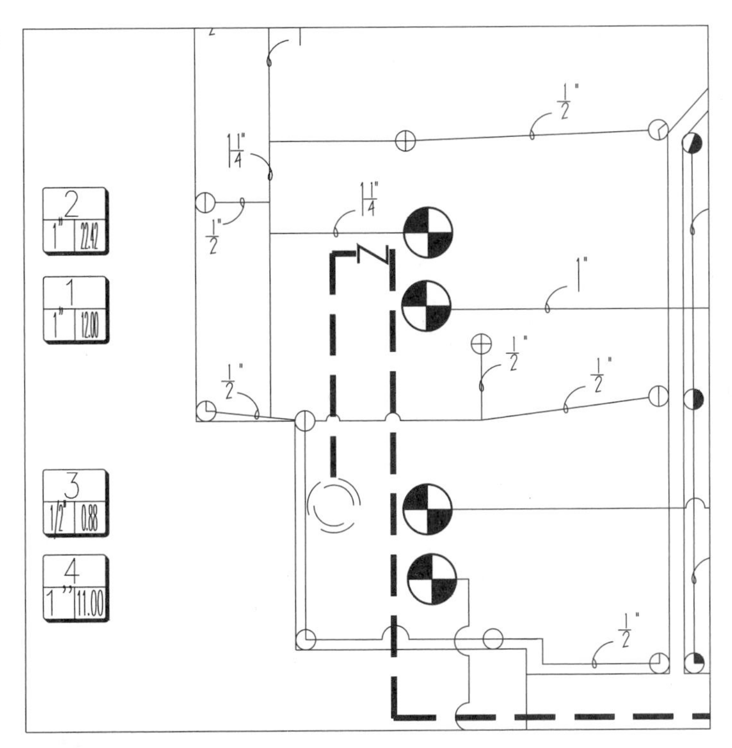

Figure 15.124

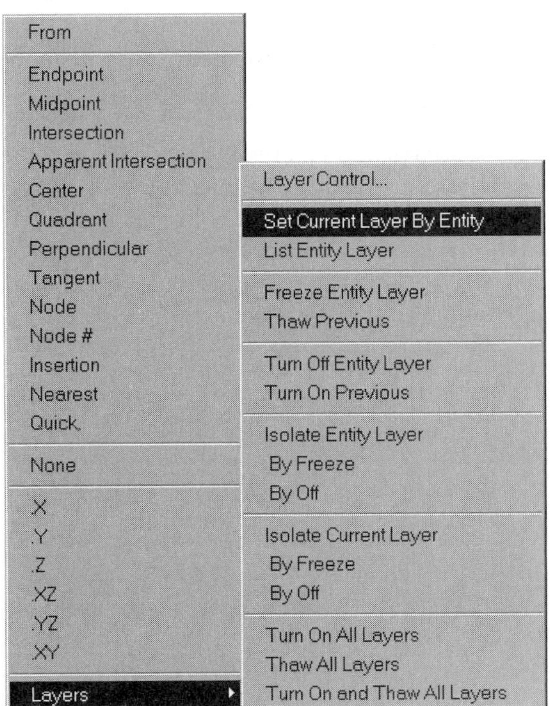

At the prompt:

`Command:`
Type **LE** to build a leader.

`From point:`
Pick a point on the double circle at the end of the main line.

`To point (Format/Annotation/Undo)<Annotation>:`
Drag a point below and to the left of the double circle.

`To point (Format/Annotation/Undo)<Annotation>:`
Turn **ORTHO** on, then stretch the leader to the left. Pick a point to create a short horizontal line.

`To point:`
Type **A** for Annotation.

`Annotation:`
Type **P.O.C.**

`Mtext:`
Press **<Enter>** to complete the command.

This label alerts the contractor where to attach the new main line to the existing service line.

We will build a new symbol for an irrigation controller. To make sure that the symbol will be available to use in the irrigation legend, we will use the **WBLOCK** command to save the symbol to the hard drive. We will also make it a "smart" symbol which will contain attributes for the brand and model of the controller.

Begin by zooming into a blank space in the garage area.

At the prompt:

`Command:`
Type **POLYGON**.

`Number of sides <4>:`
Type **3**.

`Edge/<Center of polygon>:`
Pick a point in the garage.

`Inscribed in circle/Circumscribed about circle (I/C) <I>:`
Type **I**.

`Radius of circle:`
Type **18**.

Use **DTEXT** to place the letter **C** inside the triangle. ("C" stands for controller.) (Figure 15.125).

After the symbol is finished, we can add two attributes to it which will remain invisible. Those attributes will specify the brand name of the controller and the model number.

To add attributes to the controller symbol, move to the **AutoCAD and EP** menu, then pull down the **Construct** option. Select **Attribute** to open the "Attribute Definition " dialogue box (Figure 15.126).

Figure 15.125

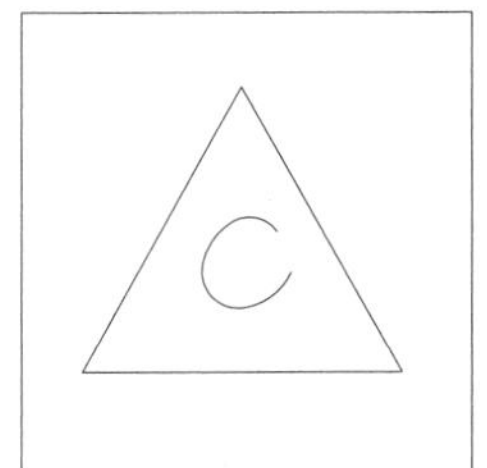

Figure 15.126

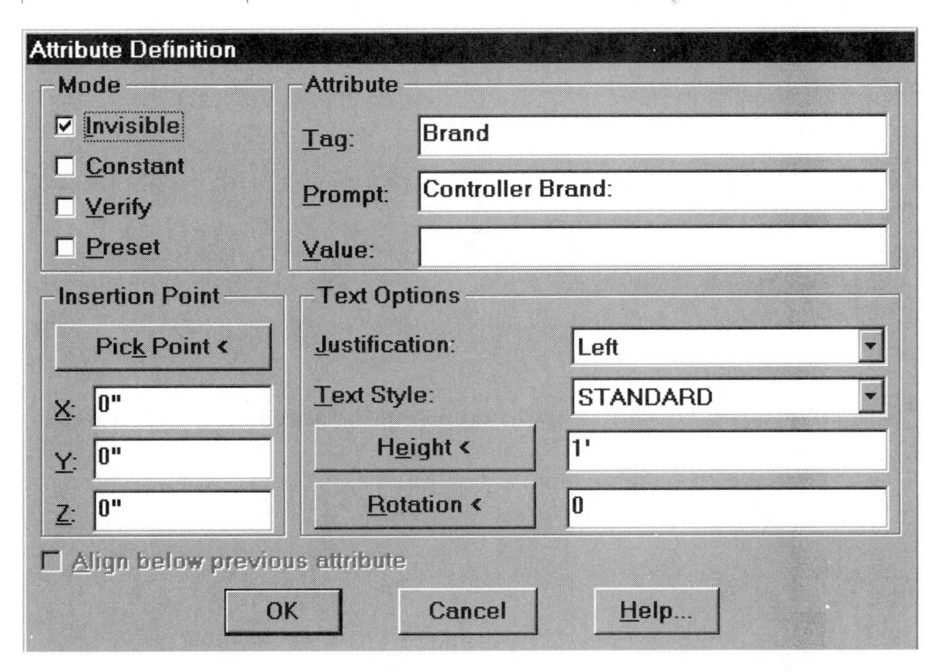

Be sure that the **Invisible** toggle box is marked, and create an attribute called **Brand**. Use the prompt **Controller Brand:**. Place it somewhere next to the controller symbol. Since this attribute will remain invisible, its precise location and size are not important.

Repeat this process with another attribute with the tag **Model**. Make the prompt **Controller Model:**. Place this attribute below the **Brand** tag.

Your finished symbol should resemble the one shown below (Figure 15.127).

Figure 15.127

Use the **WBLOCK** command to make a separate drawing out of the symbol. Name it **CONTRL**.

`Block name:`
Press **<Enter>**.

`Insertion base point:`
Pick the lower left corner of the triangle.

`Select objects:`
Pick the triangle and the letter "C", then the **Brand** tag, and finally the **Model** tag. The symbol (naturally) will disappear.

Insert the **CONTRL** symbol, and fill out the attributes as shown (Figure 15.128).

Place the controller symbol at the upper edge inside the garage (Figure 15.129).

Figure 15.128

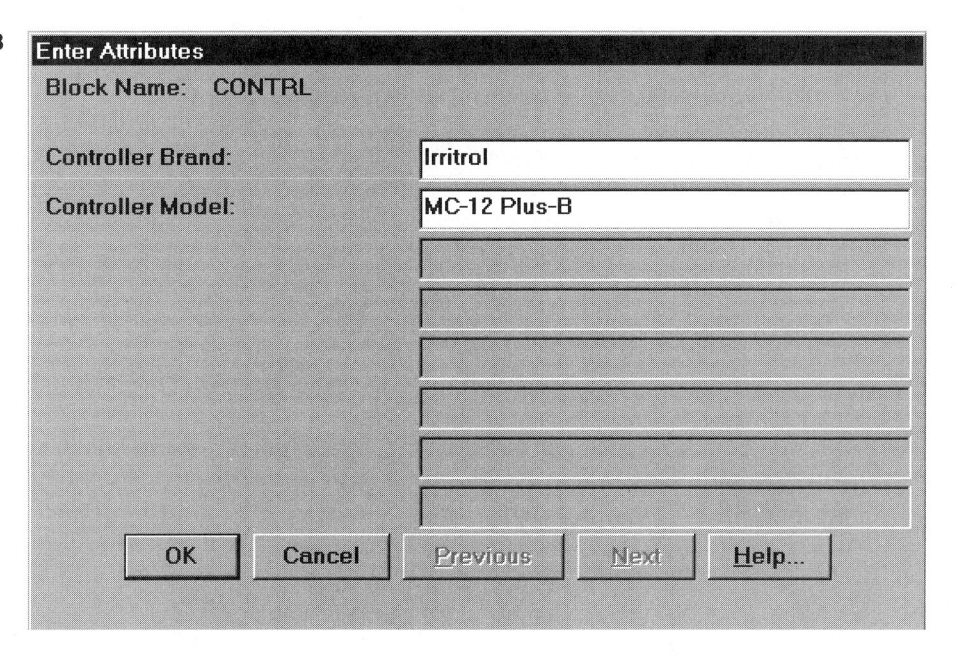

Figure 15.129

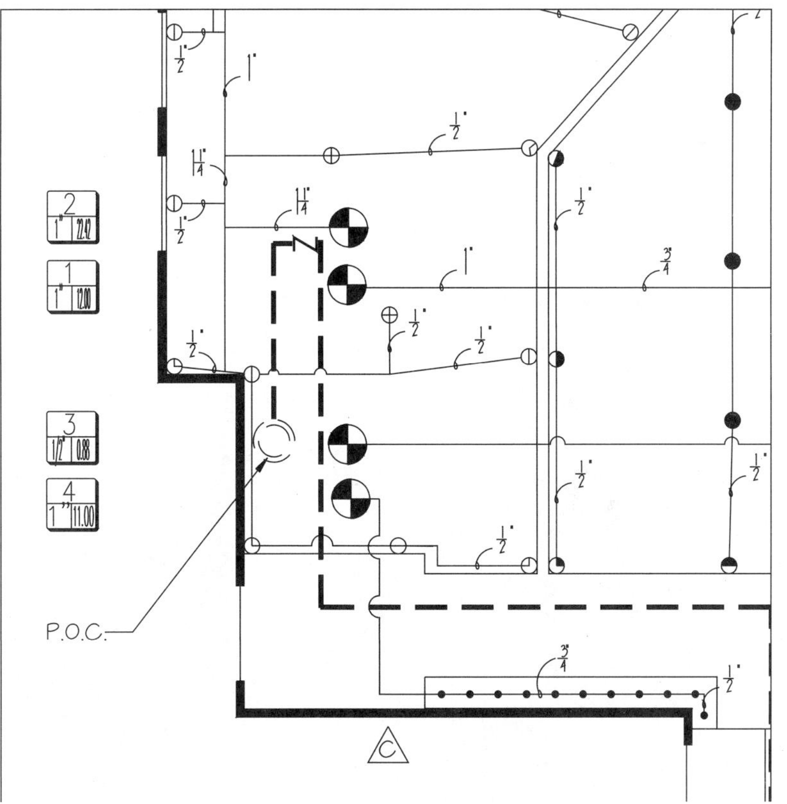

Sizing the Main Line

After the irrigation zones are established, we can select the appropriate size main line. Although there are several branches on the main line which could be sized differently, the flow demand at each branch is near or over 20 Gallons Per Minute. According to normal practice, the velocity for these main line pipes should not exceed 5 Feet Per Second, so the appropriate choice for main line, using Schedule 40 PVC pipe is 1-1/2". To simplify matters, we will make all of the main line this size. To change the characteristics of the main line polyline into a block which contains attributes, we again will use the U size routine found under the PipeSize ▸ sub-menu (Figure 15.130).

Open the PipeSize ▸ sub-menu under the Pipe pull-down menu. Pick U size.

In the "Usize Pipe" dialogue box, set the **Pipe Type:** pop-down box to read **SCHEDULE 40 PVC PIPE ENGLISH UNITS**. This is an appropriate choice for a main line in a residential site where the ground does not freeze.

Set the **Size:** pop-down menu to read **1-1/2"**.

To keep the main line on the correct layer, pick the **Layers . . .** button to open the "Usize Pipe Layers" dialogue box (Figure 15.131).

Even though this is a main line, change the setting in the **Lateral Layer** box. It should read **0284-MAINLINE**. Leave the **Label Layer** set to **0285-PIPESIZE**.

Pick **OK** when finished.

Figure 15.130

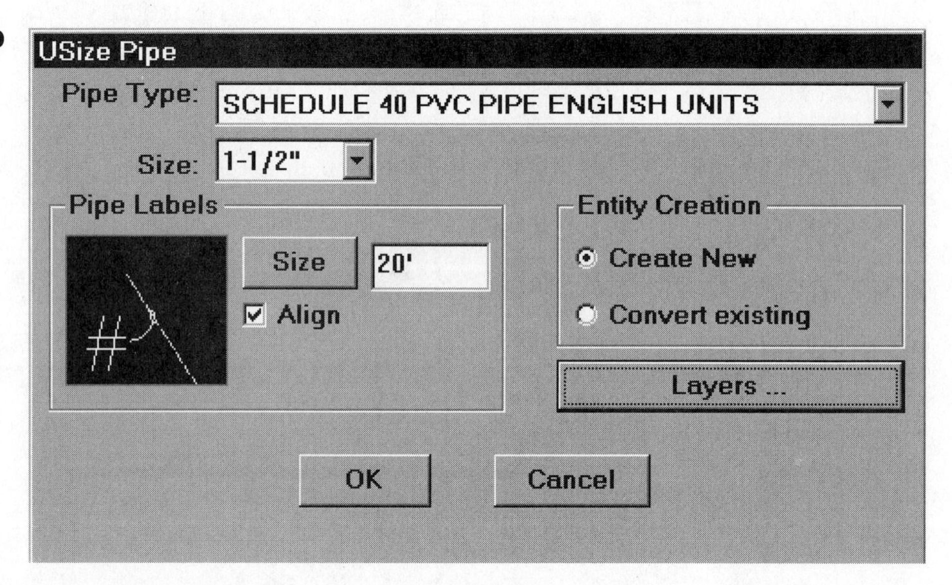

Figure 15.131

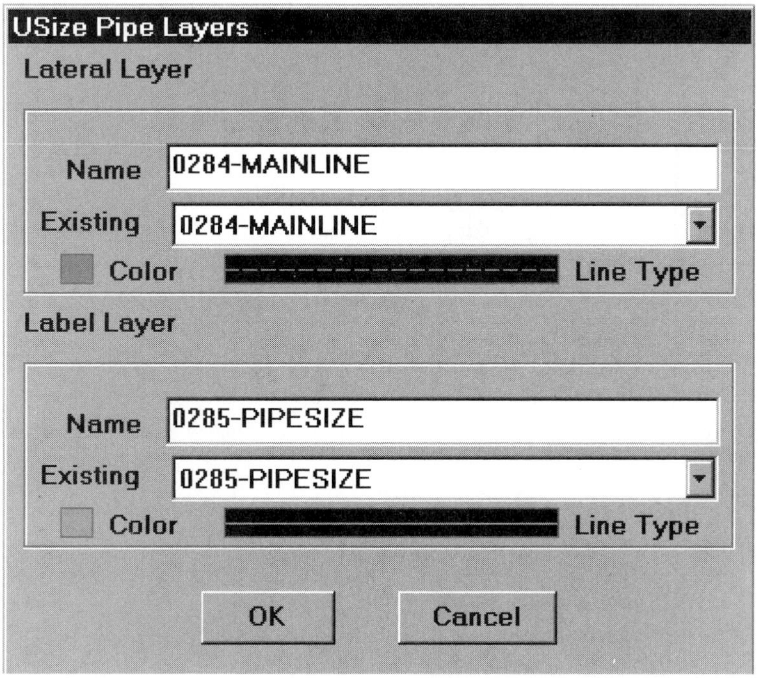

USize Pipe Layers
Lateral Layer
Name 0284-MAINLINE
Existing 0284-MAINLINE
Color Line Type
Label Layer
Name 0285-PIPESIZE
Existing 0285-PIPESIZE
Color Line Type
OK Cancel

After returning to the "Usize Pipe" box, leave the **Pipe Labels** box set as it was for previous labels.

Leave the **Entity Creation** radio button set to **Convert Existing**, since we wish to convert the existing main line polylines into pipe blocks.

Pick **OK** when finished.

At the prompt:

```
Command: EPLC_runme
Select objects:
```

Pick **all** of the sections of main line which appear on the drawing. There may be as many as four or five sections of main line.

```
Select objects: 4 found
```

The main line piping is converted into blocks and the labels are placed on the drawing. These labels need to be moved to make them more clear and obvious, but do not need serious modification as the drip piping required.

See below for some suggested locations for the main line pipe labels (Figures 15.132, 15.133, 15.134, and 15.135).

Figure 15.132

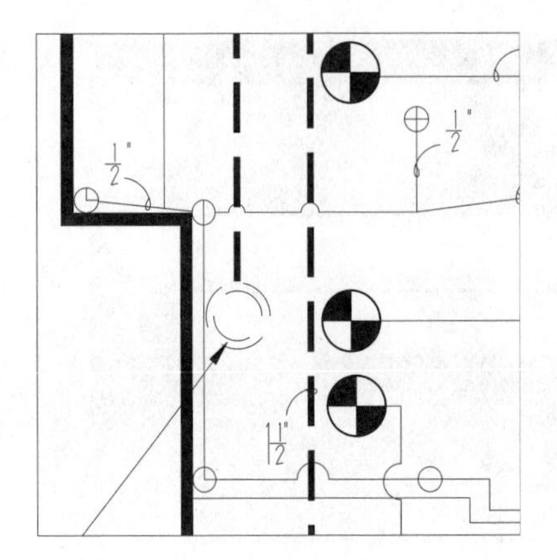

Figure 15.133

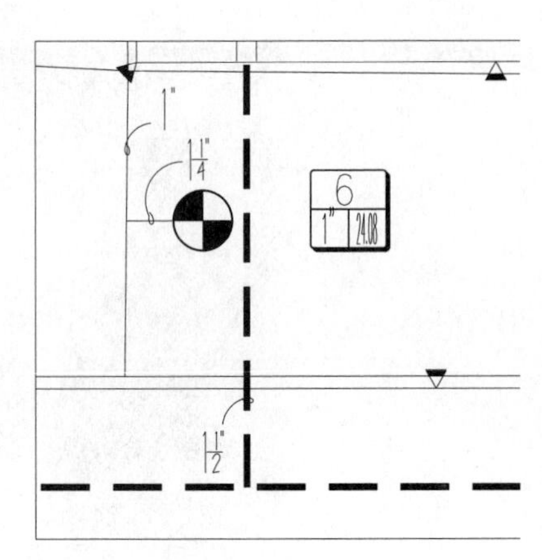

Figure 15.134

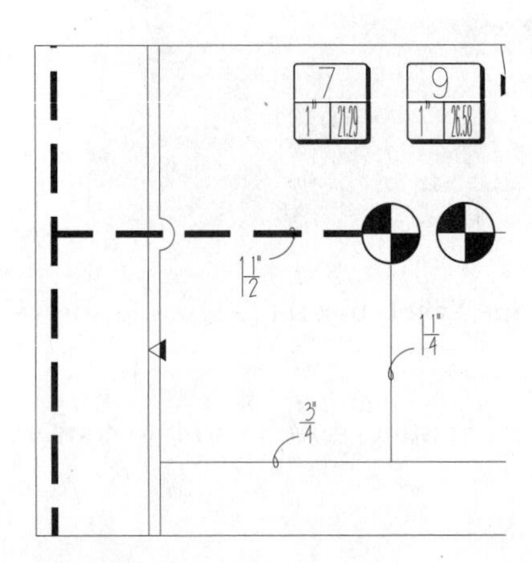

Figure 15.135

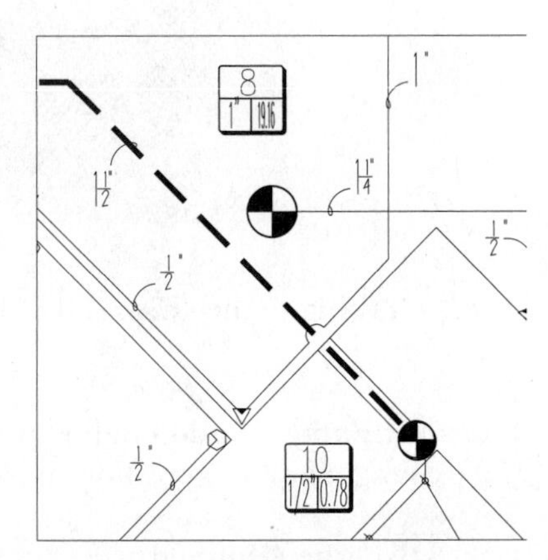

At this point, all of the piping has been labeled and the irrigation design is completed.

BE SURE TO SAVE YOUR DRAWING NOW (Figure 15.136).

An important part of the plan, however, is still missing. We need to build an Irrigation Legend. This legend should contain an example of each symbol used on the drawing, and a key to the meaning of each one. We can also provide an Irrigation Equipment Schedule to list how many (or how much) of each item is required. Developing a full legend and schedule could be a complicated and time-consuming process, but LANDCADD provides some tools to make this process much easier and faster.

Figure 15.136

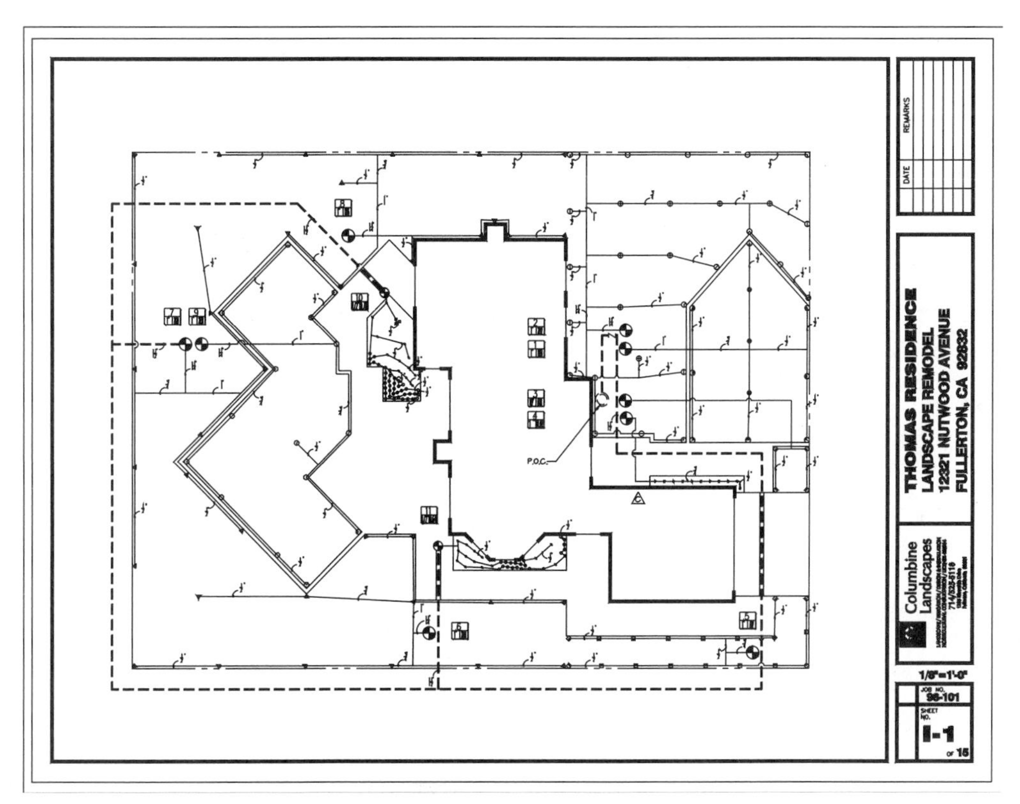

Building the Irrigation Materials Schedule

To build a legend and materials schedule, we will take advantage of the **Equipment Table** function found under the **Symbols** pull-down menu. This routine works much as the **Estimating** functions do in the planting plan. But instead of presenting the attribute and block information in a text report, the **Equipment Table** reports data in a tabular form that is inserted into a drawing. This table provides us with a valuable starting point to develop a functional materials schedule and legend.

To begin the **Equipment Table** function, it is helpful to zoom out past the edges of the drawing, so that the table can be inserted outside the title block. Were we working on a larger size paper, or in a smaller drawing scale, we might be able to fit the table and the drawing onto the same page.

At the prompt:

```
Command:
```

Type **Zoom** then **Dynamic** (or **ZD**) then zoom out so that the drawing is placed on the left side of the screen, with blank space on the right.

Pick **Equipment Table** from the **Symbols** pull-down menu.

```
Command:
Select blocks or search for All (S/A) <S>
```
Type **A** for All.

```
Select objects: 620 found
209 were filtered out
```

```
Select objects:
```
Press **<Enter>** to build the table.

```
Table textsize: <1'>
```
If the default does not read **1'** or **12"**, type in that size. If it does, press **<Enter>**.

```
Working /
```
LANDCADD then places a table of pipe, heads, valves, etc. at point **0,0**, and prompts:

```
Select table location:
```
Turn **ORTHO** off, then drag the table over to the blank part of the screen to the right of the drawing (Figure 15.137).

Although the size and content of the equipment table may be satisfactory for certain uses, it cannot be used in the existing drawing. We will use the **WBLOCK** command to

Figure 15.137

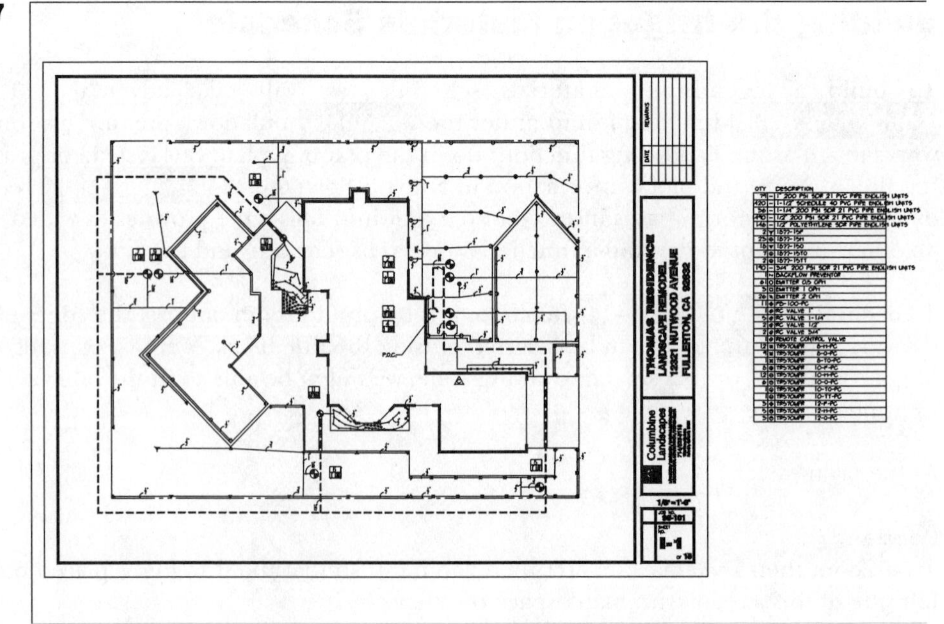

make the table into a drawing block, then save it to the hard disk. We will then insert it into a new drawing, using the **C18SHEET** drawing as a prototype.

`Command:`
Type **WBLOCK**. This opens the "Create Drawing File" dialogue box. Name the drawing **SCHEDULE**, then pick **OK**.

`Block name:`
Press **<Enter>**. Do not enter a name at the prompt.

`Insertion base point:`
Pick the lower left corner of the equipment table.

`Select objects:`
Select the entire equipment table. Be sure to include all of the symbols, lines and text in the block. The entire equipment table disappears from the drawing. As before, this is supposed to happen.

At this point, we will open another drawing. When the prompt asks us if we wish to save changes to the current drawing, it is best use the **Discard changes** option. This prevents adding the equipment table block to our existing irrigation plan, keeping the drawing file size as small as possible.

Open the **EP** pull-down menu, then select **Project Manager**.

Pick **New ...**. In the "New Project" dialogue box, complete the information as follows:

For **Description:** enter **Irrigation Legend**.

For **Drawing Name . . .**, enter a new drawing name, being sure to specify the correct path where the drawing should be located.

Set the **Prototype Drawing . . .** to the **C18SHEET** drawing used as the title block for all of the drawings.

Once these entries are complete, pick **Open**, then use **AE** (**DDATTEDIT**) to edit the title block.

For **SHEET NUMBER:** enter **I-2**.

For **DRAWING SCALE:** enter **NONE**, since this drawing will only contain an Irrigation Legend and Irrigation Equipment Schedule. See below (Figure 15.138).

Figure 15.138

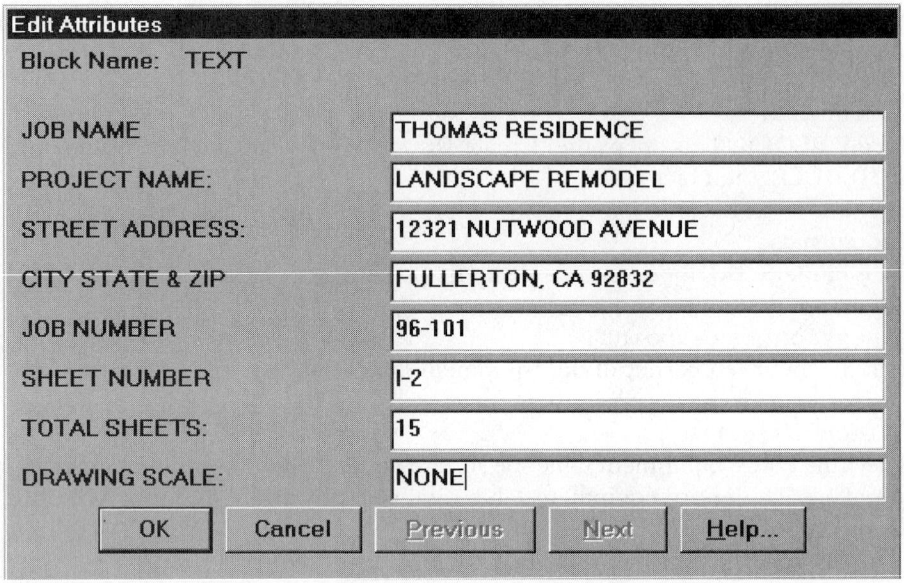

Edit Attributes

Block Name: TEXT

JOB NAME	THOMAS RESIDENCE
PROJECT NAME:	LANDSCAPE REMODEL
STREET ADDRESS:	12321 NUTWOOD AVENUE
CITY STATE & ZIP	FULLERTON, CA 92832
JOB NUMBER	96-101
SHEET NUMBER	I-2
TOTAL SHEETS:	15
DRAWING SCALE:	NONE

OK | Cancel | Previous | Next | Help...

Now we will use the **DDINSERT** command to insert the **SCHEDULE** block into the new drawing.

`Command:`
Type **DDINSERT**. This opens the "Insert" dialogue box.

Pick **File . . .** then find the **SCHEDULE** drawing that was created with the **WBLOCK** command. Pick **OK** when finished. Be sure that **Snap** is turned **ON**.

At the prompt:

`Insertion point:`
Pick a point so that the equipment table is placed toward the left side of the drawing page.

`X scale factor <1>/Corner/XYZ:`
Press **<Enter>** to accept the default value of **1**.

`Y scale factor (default=1):`
Press **<Enter>** to accept the default value.

`Rotation angle <0>:`
Press **<Enter>** to accept the default value. This places the equipment table on the drawing page.

`Command:`

Use **EXPLODE** (**EX**) to explode the drawing block to be able to edit its components (Figure 15.139).

Figure 15.139

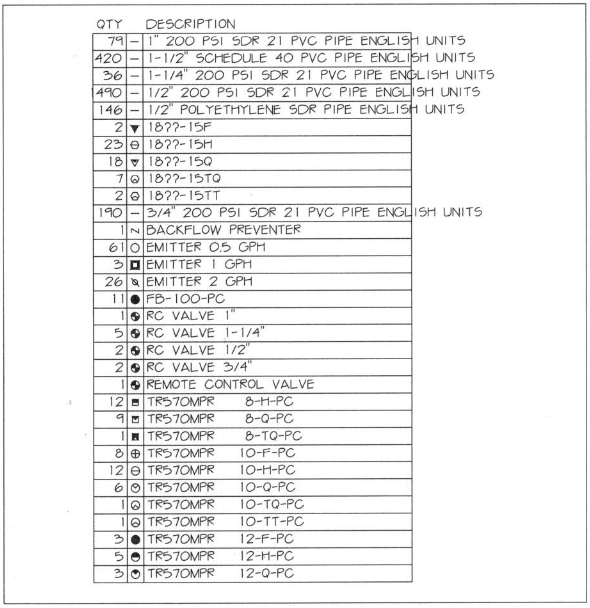

QTY		DESCRIPTION
79	–	1" 200 PSI SDR 21 PVC PIPE ENGLISH UNITS
420	–	1-1/2" SCHEDULE 40 PVC PIPE ENGLISH UNITS
36	–	1-1/4" 200 PSI SDR 21 PVC PIPE ENGLISH UNITS
490	–	1/2" 200 PSI SDR 21 PVC PIPE ENGLISH UNITS
146	–	1/2" POLYETHYLENE SDR PIPE ENGLISH UNITS
2	▼	18??-15F
23	⊖	18??-15H
18	▽	18??-15Q
7	⊘	18??-15TQ
2	⊘	18??-15TT
190	–	3/4" 200 PSI SDR 21 PVC PIPE ENGLISH UNITS
1	~	BACKFLOW PREVENTER
61	○	EMITTER 0.5 GPH
3	■	EMITTER 1 GPH
26	⊘	EMITTER 2 GPH
11	●	FB-100-PC
1	●	RC VALVE 1"
5	●	RC VALVE 1-1/4"
2	●	RC VALVE 1/2"
2	●	RC VALVE 3/4"
1	●	REMOTE CONTROL VALVE
12	■	TR570MPR 8-H-PC
9	■	TR570MPR 8-Q-PC
1	■	TR570MPR 8-TQ-PC
8	⊕	TR570MPR 10-F-PC
12	⊖	TR570MPR 10-H-PC
6	⊘	TR570MPR 10-Q-PC
1	⊘	TR570MPR 10-TQ-PC
1	⊘	TR570MPR 10-TT-PC
3	●	TR570MPR 12-F-PC
5	●	TR570MPR 12-H-PC
3	●	TR570MPR 12-Q-PC

The equipment table is, in its current form, disorganized and primitive. Aside from the obvious overhang of the text past its borders, items are listed in no particular sequence, some item descriptions are incomplete, and others include unneeded information. At this point we will edit the information with these specific goals:

1. organize the items by classification: pipe, valves, heads and controller
2. specify the exact brand and model for all irrigation equipment
3. edit the pipe descriptions to eliminate "English Units"
4. edit and add symbols to accurately reflect the contents of the drawing

It is useful to consider the following information about the equipment table as it is currently drawn:

- All of the text is 12" high.
- The lines of text are spaced 24" apart.
- The perimeter of the table and the two vertical lines in the table are polylines.
- The horizontal lines are a **POLYFACE**, which cannot be edited as drawn.

All of the editing commands we perform should consider the construction of the table. As we use **OFFSET**, **MOVE**, **COPY** and other editing commands, we will take advantage of the construction of the table to make the changes to the table undetectable.

We will begin by resizing the table to fit the text and descriptions we will be using. First, explode the perimeter polyline and the horizontal polyface. This will create separate lines which can be edited independently. Then create three extra lines below the bottom of the table, spaced 24" apart. Add vertical lines (or use **GRIPS** to extend the existing vertical polylines) to match the columns above. This will provide extra spaces for symbols which were not included in the table. Finally, move the left edge of the table outward by **12"** and the right edge of the table outward by **6'**. We will use a specialized version of the **EXTEND** command to extend all of the horizontal lines to the outside borders in one step:

At the command prompt:

```
Command:
```
Type **EXTEND** (**X**).

```
Select boundary edges:(Projmode=UCS,Edgemode=No extend)
Select objects:
```
Pick the right outer vertical line, then press **<Enter>**.

```
<Select object to extend>/Project/Edge/Undo:
```
Type **F** (for **FENCE**), then press **<Enter>**.

```
First fence point:
```
Pick a point above all of the horizontal lines to extend.

```
Undo/<Endpoint of line>:
```
Pick a point below the horizontal lines to extend. The highlight line should extend through all of the horizontal lines.

```
Undo/<Endpoint of line>:
```
Press **<Enter>**. The command extends all of the lines simultaneously.

```
Undo/<Endpoint of line>:
```
Press **Enter>** to finish the command.

Repeat this technique to extend the horizontal lines to the left outer vertical line.

Your results should resemble those below (Figure 15.140).

Figure 15.140

7		18??-15TQ
2		18??-15TT
190	–	3/4" 200 PSI SDR 21 PVC PIPE ENGLISH UNITS
1	N	BACKFLOW PREVENTOR
61	○	EMITTER 0.5 GPH
3	◘	EMITTER 1 GPH
26		EMITTER 2 GPH
11	●	FB-100-PC
1		RC VALVE 1"
5		RC VALVE 1-1/4"
2		RC VALVE 1/2"
2		RC VALVE 3/4"
1		REMOTE CONTROL VALVE
12		TR570MPR 8-H-PC
9		TR570MPR 8-Q-PC
1		TR570MPR 8-TQ-PC
8	⊕	TR570MPR 10-F-PC
12	⊖	TR570MPR 10-H-PC
6		TR570MPR 10-Q-PC
1		TR570MPR 10-TQ-PC
1		TR570MPR 10-TT-PC
3	●	TR570MPR 12-F-PC
5		TR570MPR 12-H-PC
3		TR570MPR 12-Q-PC

Next, remove the phrases **ENGLISH UNITS** and **PSI** from each of the pipe listings. This is unnecessary (and confusing) information. Use **DDEDIT** (**DD**) to edit each line of text. In addition, add the word **CLASS** and listing for each Class 200 pipe listing. The line for 1/2" lateral line should read:

1/2" CLASS 200 SDR 21 PVC PIPE

After the spaces are added, and the pipe text edited, reorder the pipe symbols from smallest to largest lateral (Class 200) pipe, main line (Schedule 40) pipe, then drip (polyethylene) pipe. Note that the line segment for the main line does not reflect the width of the polyline used for the main line in the Irrigation Plan. Use **PEDIT** to make the line into a polyline, then assign it a width of **3"**. To complete the pipe list, add a line to show **35** feet of **3" CLASS 315 SDR 13.5 SLEEVE**. This line can be added easily by copying the line above, placing it in the proper location, then using **DDEDIT** (**DD**) to edit the text. Make sure that the symbol shows a double line to correspond with the sleeve symbol used in the Irrigation Plan. See below (Figure 15.141).

As we look to the valve symbols, we need to revise the equipment table to reflect the actual type of valves that an irrigation designer would specify for a job. The original table shows listings for RC (remote control) valves ranging from 1/2" up to 1-1/4", based on the flow to the final header pipe connecting to the control valve. These sizes do not reflect two important realities: first, remote control valves are often sized at least one size below the pipe line size; second, irrigation equipment manufacturers do not build 1/2" remote control valves. With these facts in mind, we will make some important changes to the valve symbols.

Figure 15.141

QTY		DESCRIPTION
1490	—	1/2" 200 PSI SDR 21 PVC PIPE
190	—	3/4" 200 PSI SDR 21 PVC PIPE
79	—	1" 200 PSI SDR 21 PVC PIPE
36	—	1-1/4" 200 PSI SDR 21 PVC PIPE
420	▬	1-1/2" SCHEDULE 40 PVC PIPE
146	—	1/2" POLYETHYLENE SDR PIPE
35	=	3" 315 PSI SDR 13.5 PVC SLEEVE
2	▼	18??-15F
23	⊖	18??-15H
18	▽	18??-15Q
7	⊗	18??-15TQ
2	⊗	18??-15TT

First, assume that all of the larger (1" and 1-1/4") remote control valves will be sized at 1". This places up to 24 GPM through the remote control valves, but a well-selected valve will be able to withstand this flow condition. Second, assume that the smaller (1/2" and 3/4") remote control valves will be sized at 3/4". Without endorsing any manufacturer or product, typical selections might be:

1" Remote Control Valve: Hunter HPV-101-A (8 required)

3/4" Remote Control Valve: Hardie 700-.75 (3 required)

Erase the unneeded valve symbols, and edit the entries to reflect these changes. Edit the backflow preventer entry to specify the following RP backflow device (without endorsement or recommendation):

1-1/4" RP Backflow Preventer Febco 825Y-125

Skip a line under the pipe section, then place the control valves and backflow preventer symbols (Figure 15.142).

The final section will list the types of sprinkler heads, emitters and the irrigation controller. Edit the descriptions to match those found in the manufacturers' catalogs. We have specified Rainbird 1800 sprinklers to be used in the areas requiring 15' nozzles, but have not yet selected a pop-up height. We will use a 6" pop-up head in the shrub and turf areas in the plan. To reflect the selection of a pop-up height, and the name of the manufacturer, edit the line:

Figure 15.142

QTY		DESCRIPTION
1490	—	1/2" 200 PSI SDR 21 PVC PIPE
190	—	3/4" 200 PSI SDR 21 PVC PIPE
79	—	1" 200 PSI SDR 21 PVC PIPE
36	—	1-1/4" 200 PSI SDR 21 PVC PIPE
420	▬	1-1/2" SCHEDULE 40 PVC PIPE
146	—	1/2" POLYETHYLENE SDR PIPE
35	—	3" 315 PSI SDR 13.5 PVC SLEEVE
8	◕	HUNTER HPV-101-1 R.C. VALVE 1"
3	◕	HARDIE 700-.75 R.C. VALVE 3/4"
1	N	FEBCO 825Y-125 R.P. BACKFLOW PREVENTER
2	▼	18??-15F
23	⊖	18??-15H
18	▽	18??-15Q
7	⊘	18??-15TQ
2	⊘	18??-15TT

18??-15F to read **RAINBIRD (10 spaces) 1806-15F**

Repeat this procedure for all of the 1800 series heads.

The Toro sprinkler descriptions also require editing. We will also specify a 6" pop-up head for these symbols. To list the manufacturer and head correctly, edit the line:

TR570MPR 8-H-PC to read **TORO 570Z 6P (5 spaces) 8-H-PC**

Repeat this for all of the Toro 570 heads.

The drip emitters specified for this project might (without endorsement or recommendation) be:

0.5 GPH Emitter: HARDIE DKL02

1.0 GPH Emitter: HARDIE DKL04

2.0 GPH Emitter: HARDIE DKL08

Edit the lines in the equipment table to match these specified items (Figure 15.143).

Figure 15.143

Qty	Symbol	Equipment	Model
1	N	FEBCO 825Y-125 R.P. BACKFLOW PREVENTER	
2	▼	RAINBIRD	1806-15F
23	⊖	RAINBIRD	1806-15H
18	▽	RAINBIRD	1806-15Q
7	⊗	RAINBIRD	1806-15TQ
2	⊗	RAINBIRD	1806-15TT
11	•	TORO 570Z 6P	FB-100-PC
12	■	TORO 570Z 6P	8-H-PC
9	■	TORO 570Z 6P	8-Q-PC
1	■	TORO 570Z 6P	8-TQ-PC
8	⊕	TORO 570Z 6P	10-F-PC
12	⊖	TORO 570Z 6P	10-H-PC
6	⊗	TORO 570Z 6P	10-Q-PC
1	⊗	TORO 570Z 6P	10-TQ-PC
1	⊗	TORO 570Z 6P	10-TT-PC
3	●	TORO 570Z 6P	12-F-PC
5	●	TORO 570Z 6P	12-H-PC
3	●	TORO 570Z 6P	12-Q-PC
61	O	HARDIE DKL02 EMITTER 0.5 GPH	
3	◘	HARDIE DKL04 EMITTER 1 GPH	
26	⊗	HARDIE DKL08 EMITTER 2 GPH	

Notice that the symbol for the Toro 12-F-PC is the same size as the Toro 12 FB-100-PC in this equipment table, even though they are different sizes in the drawing. To correct this, **Scale** the Toro FB-100-PC symbol by **.5** to make it the same size as it is in the Irrigation Plan. To complete the equipment table, use **DDINSERT** to add the **CONTRL** symbol to the drawing. Insert the symbol using the scale factor of **.5**. This reduces the symbol to fit into the size of the column in the table. Label the irrigation controller symbol according to the brand and model we selected earlier:

IRRITROL MC-12 PLUS-B CONTROLLER

Move the controller symbol on the bottom line of the table, then add the text. Finally, place a title at the top of the table. Since this is now an accurate listing of the materials and equipment required to build the system, we can call it an Irrigation Materials Schedule (Figure 15.144).

Building the Irrigation Legend

The Irrigation Materials Schedule is meant as a listing of the components to build the irrigation system. It is not particularly useful as an Irrigation Legend, since the pipe listings are repetitive and several of the symbols are incorrectly scaled. In addition, the format of the Irrigation Materials Schedule does not include enough columns to convey all

Figure 15.144

IRRIGATION MATERIALS SCHEDULE

QTY		DESCRIPTION
1490	–	1/2" 200 PSI SDR 21 PVC PIPE
190	–	3/4" 200 PSI SDR 21 PVC PIPE
79	–	1" 200 PSI SDR 21 PVC PIPE
36	–	1-1/4" 200 PSI SDR 21 PVC PIPE
420	▬	1-1/2" SCHEDULE 40 PVC PIPE
146	–	1/2" POLYETHYLENE SDR PIPE
35	=	3" 315 PSI SDR 13.5 PVC SLEEVE
8	●	HUNTER HPV-101-1 R.C. VALVE 1"
3	●	HARDIE 700-.75 R.C. VALVE 3/4"
1	~	FEBCO 825Y-125 R.P. BACKFLOW PREVENTER
2	▼	RAINBIRD 1806-15F
23	⊖	RAINBIRD 1806-15H
18	▽	RAINBIRD 1806-15Q
7	◎	RAINBIRD 1806-15TQ
2	◎	RAINBIRD 1806-15TT
11	•	TORO 570Z 6P FB-100-PC
12	■	TORO 570Z 6P 8-H-PC
9	■	TORO 570Z 6P 8-Q-PC
1	■	TORO 570Z 6P 8-TQ-PC
8	⊕	TORO 570Z 6P 10-F-PC
12	⊖	TORO 570Z 6P 10-H-PC
6	◎	TORO 570Z 6P 10-Q-PC
1	◎	TORO 570Z 6P 10-TQ-PC
1	◎	TORO 570Z 6P 10-TT-PC
3	●	TORO 570Z 6P 12-F-PC
5	◓	TORO 570Z 6P 12-H-PC
3	●	TORO 570Z 6P 12-Q-PC
61	○	HARDIE DKL02 EMITTER 0.5 GPH
3	◘	HARDIE DKL04 EMITTER 1 GPH
26	⊗	HARDIE DKL08 EMITTER 2 GPH
1	△C	IRRITROL MC-12 PLUS-B CONTROLLER

of the information required in an Irrigation Legend. Although we may not use the materials schedule in our legend, we can borrow most of its construction techniques.

First, copy the left boundary line and the horizontal lines from the materials schedule to an area on the right side of the drawing page.

At the prompt:

`Command:`
Type **OFFSET** (**OF**).

`Offset distance or Through <Through>:`
Type **76'**.

`Select object to offset:`
Pick the vertical line at the left border of the legend.

`Side to offset?`
Move the cursor to the right and press the **<Pick>** button.

Extend the horizontal lines over to the right vertical line using the same **EXTEND/FENCE** technique as before.

Use the **DTEXT** command to place headings over the columns of the Irrigation Legend.

The headings are:

- SYM. (SYMBOL)
- MANUFACTURER
- HEAD
- NOZZLE
- PRESSURE
- RADIUS
- G.P.M. (GALLONS PER MINUTE)

Arrange these headings so that they are spread out to resemble the illustration below (Figure 15.145).

Once the Irrigation Legend box has been constructed, we are ready to insert the various head and emitter symbols which appear in the drawing. To make this simple, copy the

Figure 15.145

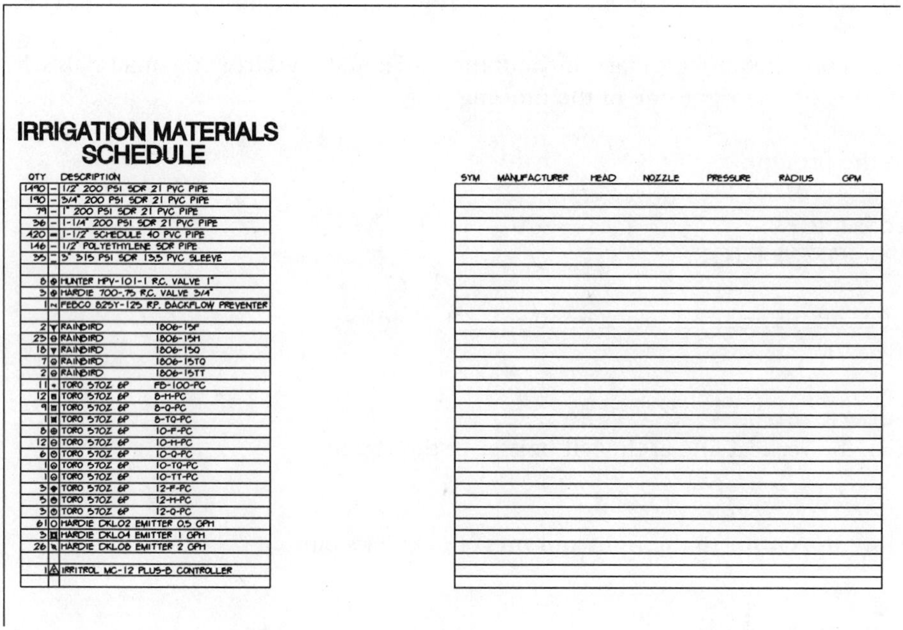

IRRIGATION MATERIALS SCHEDULE

QTY		DESCRIPTION	
1490	–	1/2" 200 PSI SDR 21 PVC PIPE	
190	–	3/4" 200 PSI SDR 21 PVC PIPE	
79	–	1" 200 PSI SDR 21 PVC PIPE	
36	–	1-1/4" 200 PSI SDR 21 PVC PIPE	
420	–	1-1/2" SCHEDULE 40 PVC PIPE	
146	–	1/2" POLYETHYLENE SDR PIPE	
35		3" 315 PSI SDR 13.5 PVC SLEEVE	
8		HUNTER HPV-101-1 R.C. VALVE 1"	
3		HARDIE 700-.75 R.C. VALVE 3/4"	
1		FEBCO 825Y-125 R.P. BACKFLOW PREVENTER	
2		RAINBIRD	1806-15F
23		RAINBIRD	1806-15H
18		RAINBIRD	1806-15Q
7		RAINBIRD	1806-15TQ
2		RAINBIRD	1806-15TT
11		TORO 570Z 6P	FB-100-PC
12		TORO 570Z 6P	8-H-PC
9		TORO 570Z 6P	8-Q-PC
1		TORO 570Z 6P	8-TQ-PC
8		TORO 570Z 6P	10-F-PC
12		TORO 570Z 6P	10-H-PC
6		TORO 570Z 6P	10-Q-PC
1		TORO 570Z 6P	10-TQ-PC
1		TORO 570Z 6P	10-TT-PC
3		TORO 570Z 6P	12-F-PC
5		TORO 570Z 6P	12-H-PC
3		TORO 570Z 6P	12-Q-PC
61		HARDIE DKL02 EMITTER 0.5 GPH	
3		HARDIE DKL04 EMITTER 1 GPH	
26		HARDIE DKL08 EMITTER 2 GPH	
1		IRRITROL MC-12 PLUS-B CONTROLLER	

SYM	MANUFACTURER	HEAD	NOZZLE	PRESSURE	RADIUS	GPM

column of head and emitter symbols from the Irrigation Materials Schedule and place them in the Irrigation Legend. Using **Snap** will help keep the symbols in alignment with other objects in the legend (Figure 15-146).

The first five sprinkler symbols are Rainbird heads, while the next 12 are Toro heads. We will use the **ARRAY** command to copy the manufacturer names in an aligned column.

Use **DTEXT** to place the name **RAINBIRD** in the top line of the **MANUFACTURER** column. Be sure that the word is centered in the column and centered between the lines.

At the prompt:

```
Command:
```

Type **ARRAY** (**AR**).

```
Select objects:
```

Pick the text **RAINBIRD**, then press **<Enter>**.

```
Rectangle or Polar Array (R/P)<R>:
```

Press **<Enter>** to accept the default (rectangular) array.

```
Number of rows (--) <1>:
```

Type **6**.

Figure 15.146

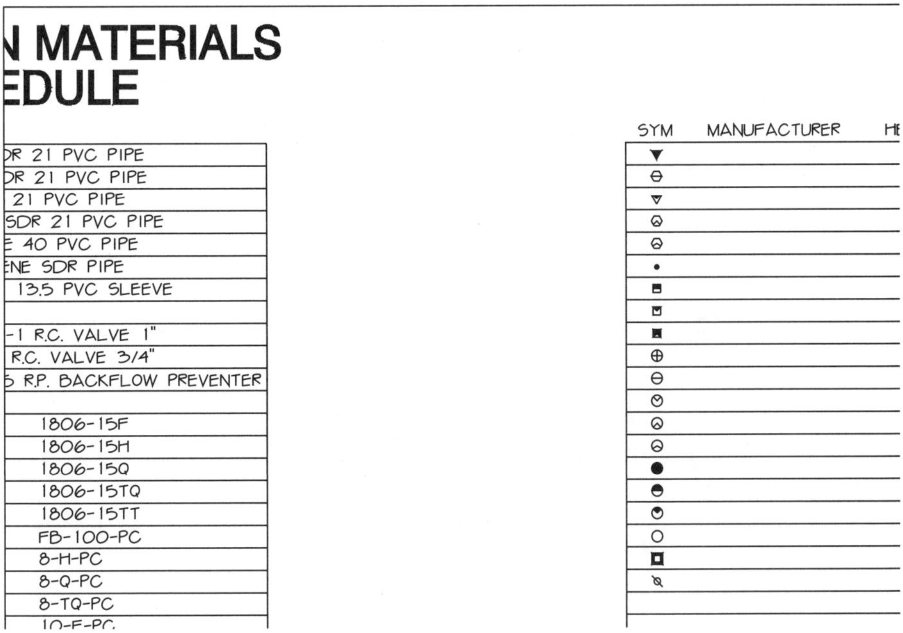

N MATERIALS
EDULE

DR 21 PVC PIPE
DR 21 PVC PIPE
21 PVC PIPE
SDR 21 PVC PIPE
E 40 PVC PIPE
ENE SDR PIPE
13.5 PVC SLEEVE
-1 R.C. VALVE 1"
R.C. VALVE 3/4"
5 R.P. BACKFLOW PREVENTER
1806-15F
1806-15H
1806-15Q
1806-15TQ
1806-15TT
FB-100-PC
8-H-PC
8-Q-PC
8-TQ-PC
10-F-PC

SYM MANUFACTURER HE

Number of columns (|||) <1>:
Type **1**.

Unit cell or distance between rows (--):
Type **-24**. This creates a total of six words in a column. The spacing of **-24** creates the listings 24" apart, moving down the screen. Use **DDEDIT** (**DD**) to change the last **RAINBIRD** listing into **TORO** (Figure 15.147).

Figure 15.147

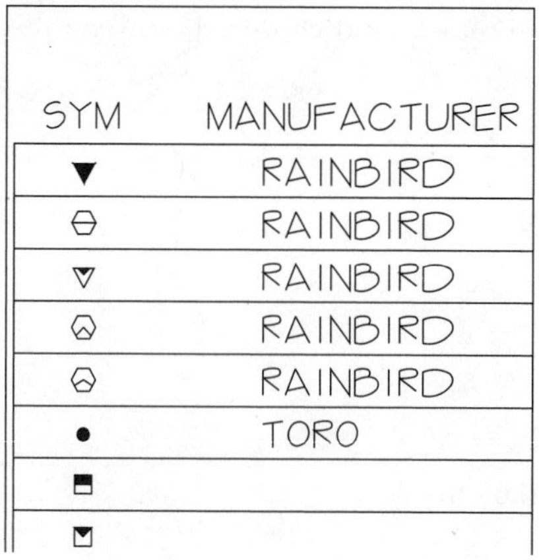

SYM	MANUFACTURER
▼	RAINBIRD
	RAINBIRD
	RAINBIRD
	RAINBIRD
	RAINBIRD
•	TORO

Use the **ARRAY** command again to make **13** listings of **TORO**, then use **DDEDIT** to change the last one to **HARDIE**. Use **ARRAY** a final time to create three listings of **HARDIE**. This technique is very useful in building all kinds of tables and graphs (Figure 15.148).

Build the entries in the **HEAD** column using the same techniques. The model of head listed first is the **1806**. Place the **1806** text entry in the column, centered under the heading. (Either use the **DTEXT** command to create the text, or simply **COPY** the word **RAINBIRD**, then edit it to read **1806**.)

After the first entry is placed (and carefully centered) in the **HEAD** column, use the **ARRAY** command to make a total of six listings.

Use **DDEDIT** to change the last listing to **570Z-6P**, then use **ARRAY** to make **13** listings. Edit the last listing to convert it to **DKL02**.

Use **Array** to make three listings, then edit the last two to read **DKL04** and **DKL08** respectively (Figure 15.149).

Adding the entries in the **NOZZLE** column is somewhat more time-consuming, since each entry is different. One way to create the entries is to use **ARRAY** to fill the spaces with text, then edit each line separately.

Figure 15.148

SYM	MANUFACTURER	HEAD	NOZZLE
▼	RAINBIRD		
⊖	RAINBIRD		
▽	RAINBIRD		
⊗	RAINBIRD		
⊗	RAINBIRD		
•	TORO		
⬒	TORO		
⊠	TORO		
◙	TORO		
⊕	TORO		
⊖	TORO		
⊗	TORO		
⊗	TORO		
⊗	TORO		
●	TORO		
◓	TORO		
◔	TORO		
○	HARDIE		
◘	HARDIE		
⊘	HARDIE		

Figure 15.149

SYM	MANUFACTURER	HEAD	NOZZLE
▼	RAINBIRD	1806	
⊖	RAINBIRD	1806	
▽	RAINBIRD	1806	
⊗	RAINBIRD	1806	
⊗	RAINBIRD	1806	
•	TORO	570Z-6P	
⬒	TORO	570Z-6P	
⊠	TORO	570Z-6P	
◙	TORO	570Z-6P	
⊕	TORO	570Z-6P	
⊖	TORO	570Z-6P	
⊗	TORO	570Z-6P	
⊗	TORO	570Z-6P	
⊗	TORO	570Z-6P	
●	TORO	570Z-6P	
◓	TORO	570Z-6P	
◔	TORO	570Z-6P	
○	HARDIE	DKL02	
◘	HARDIE	DKL04	
⊘	HARDIE	DKL08	

The listing at the top of the **Nozzle** column is **15F**. Add this text into the first line, then use **ARRAY** to make **17** listings (Figure 15.150).

Figure 15.150

MANUFACTURER	HEAD	NOZZLE	PRESSU
RAINBIRD	1806	15F	
RAINBIRD	1806	15F	
RAINBIRD	1806	15F	
RAINBIRD	1806	15F	
RAINBIRD	1806	15F	
TORO	570Z-6P	15F	
TORO	570Z-6P	15F	
TORO	570Z-6P	15F	
TORO	570Z-6P	15F	
TORO	570Z-6P	15F	
TORO	570Z-6P	15F	
TORO	570Z-6P	15F	
TORO	570Z-6P	15F	
TORO	570Z-6P	15F	
TORO	570Z-6P	15F	
TORO	570Z-6P	15F	
TORO	570Z-6P	15F	
HARDIE	DKL02		
HARDIE	DKL04		
HARDIE	DKL08		

The listings in the **NOZZLE** column match the nozzles shown in the Irrigation Materials Schedule. Use the **DDEDIT** (**DD**) command to edit each line (Figure 15.151).

Add the word **EMITTER** to substitute for a nozzle listing for the three types of drip emitters.

Figure 15.151

MANUFACTURER	HEAD	NOZZLE	PRESSURE
RAINBIRD	1806	15F	
RAINBIRD	1806	15H	
RAINBIRD	1806	15Q	
RAINBIRD	1806	15TQ	
RAINBIRD	1806	15TT	
TORO	570Z-6P	FB-100-PC	
TORO	570Z-6P	8-H-PC	
TORO	570Z-6P	8-Q-PC	
TORO	570Z-6P	8-TQ-PC	
TORO	570Z-6P	10-F-PC	
TORO	570Z-6P	10-H-PC	
TORO	570Z-6P	10-Q-PC	
TORO	570Z-6P	10-TQ-PC	
TORO	570Z-6P	10-TT-PC	
TORO	570Z-6P	12-F-PC	
TORO	570Z-6P	12-H-PC	
TORO	570Z-6P	12-Q-PC	
HARDIE	DKL02		
HARDIE	DKL04		
HARDIE	DKL08		

The **PRESSURE** column is very easy to complete, since all of the sprinkler heads were specified with a **30 PSI** design pressure. The drip emitters were specified with a **20 PSI** design pressure.

Add the text **30 PSI** to the top line, centering it carefully. Use **ARRAY** to make **20** listings, then edit the last three to read **20 PSI**.

The **RADIUS** column features four listings, corresponding to the number at the beginning of the **NOZZLE** column. The Toro flood bubbler and the Hardie drip emitters do not have a radius, so the appropriate entry for them is **N/A**.

The final column to complete is **G.P.M.**. The gallons per minute column is important for the designer and installer. It is a reminder that the system was designed for heads with specific flow rates, and that these flow rates are listed in the manufacturers' catalogs. The designer would typically examine these flow rates when selecting a sprinkler to use in a project. The listings below reflect the manufacturers' specifications for the sprinkler heads called out in the irrigation plan (Figure 15.152).

Note that the **G.P.M.** listings for the drip emitters include the units **G.P.H** since these emitters generate flow in Gallons Per Hour rather than Gallons Per Minute. This completes the sprinkler head and drip emitter listings.

Figure 15.152

SYM	MANUFACTURER	HEAD	NOZZLE	PRESSURE	RADIUS	GPM
▼	RAINBIRD	1806	15F	30 PSI	15 FT.	3.70
⊖	RAINBIRD	1806	15H	30 PSI	15 FT.	1.85
▽	RAINBIRD	1806	15Q	30 PSI	15 FT.	0.93
	RAINBIRD	1806	15TQ	30 PSI	15 FT.	2.78
	RAINBIRD	1806	15TT	30 PSI	15 FT.	2.48
•	TORO	570Z-6P	FB-100-PC	30 PSI	N/A	1.00
	TORO	570Z-6P	8-H-PC	30 PSI	8 FT.	0.44
	TORO	570Z-6P	8-Q-PC	30 PSI	8 FT.	0.22
	TORO	570Z-6P	8-TQ-PC	30 PSI	8 FT.	0.64
⊕	TORO	570Z-6P	10-F-PC	30 PSI	10 FT.	1.33
⊖	TORO	570Z-6P	10-H-PC	30 PSI	10 FT.	0.66
	TORO	570Z-6P	10-Q-PC	30 PSI	10 FT.	0.33
	TORO	570Z-6P	10-TQ-PC	30 PSI	10 FT.	0.99
	TORO	570Z-6P	10-TT-PC	30 PSI	10 FT.	0.89
●	TORO	570Z-6P	12-F-PC	30 PSI	12 FT.	1.29
	TORO	570Z-6P	12-H-PC	30 PSI	12 FT.	0.96
	TORO	570Z-6P	12-Q-PC	30 PSI	12 FT.	0.48
○	HARDIE	DKL02	EMITTER	20 PSI	N/A	0.5 G.P.H.
◘	HARDIE	DKL04	EMITTER	20 PSI	N/A	1.0 G.P.H.
	HARDIE	DKL08	EMITTER	20 PSI	N/A	2.0 G.P.H.

The symbols for the valves, backflow preventer and irrigation controller are larger than shown in the Irrigation Materials Schedule, and will require more than one line to illustrate at the same scale as the irrigation plan. To accommodate these larger symbols, use the **ERASE** command to eliminate every other line to create four large rows. See below (Figure 15.153).

Use the **COPY** command to add the symbols and listings for the control valves, backflow preventer and irrigation controller into the legend. Be sure to keep **SNAP** active to assist in the alignment of the text and symbols.

Use the **SCALE** command to enlarge the Hunter control valve symbol, the backflow preventer symbol, and the irrigation controller symbol. Use a **Scale Factor** of **2**. This makes them the same size as they appear on the irrigation plan.

Move the text over to align with the **MANUFACTURER** column, then edit the text for the backflow preventer to read:

FEBCO 825Y-125 1-1/4" R.P. BACKFLOW PREVENTER

The Irrigation Materials Schedule was too narrow to hold this complete listing (Figure 15.154).

The final listings in the Irrigation Legend are those for piping. Since the sizes for the lateral piping are called out on the Irrigation Plan, showing a separate symbol for each size is unnecessary. In addition, we should make sure that the symbol for the drip tubing

Figure 15.153

SYM	MANUFACTURER	HEAD	NOZZLE	PRESSURE	RADIUS	GPM
▼	RAINBIRD	1806	15F	30 PSI	15 FT.	3.70
⊖	RAINBIRD	1806	15H	30 PSI	15 FT.	1.85
▽	RAINBIRD	1806	15Q	30 PSI	15 FT.	0.93
	RAINBIRD	1806	15TQ	30 PSI	15 FT.	2.78
	RAINBIRD	1806	15TT	30 PSI	15 FT.	2.48
•	TORO	570Z-6P	FB-100-PC	30 PSI	N/A	1.00
	TORO	570Z-6P	8-H-PC	30 PSI	8 FT.	0.44
	TORO	570Z-6P	8-Q-PC	30 PSI	8 FT.	0.22
	TORO	570Z-6P	8-TQ-PC	30 PSI	8 FT.	0.64
⊕	TORO	570Z-6P	10-F-PC	30 PSI	10 FT.	1.33
⊖	TORO	570Z-6P	10-H-PC	30 PSI	10 FT.	0.66
	TORO	570Z-6P	10-Q-PC	30 PSI	10 FT.	0.33
	TORO	570Z-6P	10-TQ-PC	30 PSI	10 FT.	0.99
	TORO	570Z-6P	10-TT-PC	30 PSI	10 FT.	0.89
●	TORO	570Z-6P	12-F-PC	30 PSI	12 FT.	1.29
	TORO	570Z-6P	12-H-PC	30 PSI	12 FT.	0.96
	TORO	570Z-6P	12-Q-PC	30 PSI	12 FT.	0.48
○	HARDIE	DKL02	EMITTER	20 PSI	N/A	0.5 G.P.H.
	HARDIE	DKL04	EMITTER	20 PSI	N/A	1.0 G.P.H.
	HARDIE	DKL08	EMITTER	20 PSI	N/A	2.0 G.P.H.

appears different from the lateral piping on our plotted drawing. Because the drip equipment was drawn on its own layer, we will be able to control the color (and line width) of each type of piping on our plotted drawings.

To place the appropriate piping symbols onto the Irrigation Legend, copy the lowest four listings of pipe in the Irrigation Materials Schedule. These include:

1-1/4" CLASS 200 PSI SDR 21 PVC PIPE

1-1/2" SCHEDULE 40 PVC PIPE

1/2" POLYETHYLENE SDR PIPE

3" CLASS 315 PSI SDR 13.5 PVC SLEEVE

Move these into position below the listing for the irrigation controller (Figure 15.155).

Move the text into alignment below the other listings, then edit the top text line to read:

CLASS 200 SDR 21 PVC PIPE (SIZE AS NOTED)

Extend the pipe symbols to fill most of the space between the left edge line and the text, then erase the lowest horizontal lines in the irrigation legend. Trim the vertical lines to the new bottom edge of the box. To complete the legend, add the title **IRRIGATION LEGEND** above the legend box (Figure 15.156).

Return to the Irrigation Materials Schedule, and remove extra lower lines and trim the vertical lines to fit. Save your completed Irrigation Materials Schedule and Legend to your hard disk (Figure 15.157).

Figure 15.154

TORO	570Z-6P	8-Q-PC	30 PSI
TORO	570Z-6P	8-TQ-PC	30 PSI
TORO	570Z-6P	10-F-PC	30 PSI
TORO	570Z-6P	10-H-PC	30 PSI
TORO	570Z-6P	10-Q-PC	30 PSI
TORO	570Z-6P	10-TQ-PC	30 PSI
TORO	570Z-6P	10-TT-PC	30 PSI
TORO	570Z-6P	12-F-PC	30 PSI
TORO	570Z-6P	12-H-PC	30 PSI
TORO	570Z-6P	12-Q-PC	30 PSI
HARDIE	DKL02	EMITTER	20 PSI
HARDIE	DKL04	EMITTER	20 PSI
HARDIE	DKL08	EMITTER	20 PSI

HUNTER HPV-101-1 R.C. VALVE 1"

HARDIE 700-.75 R.C. VALVE 3/4"

FEBCO 825Y-125 R.P. 1-1/4" BACKFLOW PREVENTER

IRRITROL MC-12 PLUS-B CONTROLLER

Figure 15.155

HUNTER HPV-101-1 R.C. V

HARDIE 700-.75 R.C. VAL

FEBCO 825Y-125 R.P. 1-1

IRRITROL MC-12 PLUS-B C

1-1/4" 200 PSI SDR 21 PVC PIF

1-1/2" SCHEDULE 40 PVC PIPE

1/2" POLYETHYLENE SDR PIPE

3" 315 PSI SDR 13.5 PVC SLEE

Figure 15.156

IRRIGATION LEGEND

SYM	MANUFACTURER	HEAD	NOZZLE	PRESSURE	RADIUS	GPM
▼	RAINBIRD	1806	15F	30 PSI	15 FT.	3.70
⊖	RAINBIRD	1806	15H	30 PSI	15 FT.	1.85
▽	RAINBIRD	1806	15Q	30 PSI	15 FT.	0.93
⊗	RAINBIRD	1806	15TQ	30 PSI	15 FT.	2.78
⊖	RAINBIRD	1806	15TT	30 PSI	15 FT.	2.48
•	TORO	570Z-6P	FB-100-PC	30 PSI	N/A	1.00
■	TORO	570Z-6P	8-H-PC	30 PSI	8 FT.	0.44
■	TORO	570Z-6P	8-Q-PC	30 PSI	8 FT.	0.22
■	TORO	570Z-6P	8-TQ-PC	30 PSI	8 FT.	0.64
⊕	TORO	570Z-6P	10-F-PC	30 PSI	10 FT.	1.33
⊖	TORO	570Z-6P	10-H-PC	30 PSI	10 FT.	0.66
⊙	TORO	570Z-6P	10-Q-PC	30 PSI	10 FT.	0.33
⊗	TORO	570Z-6P	10-TQ-PC	30 PSI	10 FT.	0.99
⊖	TORO	570Z-6P	10-TT-PC	30 PSI	10 FT.	0.89
●	TORO	570Z-6P	12-F-PC	30 PSI	12 FT.	1.29
◓	TORO	570Z-6P	12-H-PC	30 PSI	12 FT.	0.96
⊙	TORO	570Z-6P	12-Q-PC	30 PSI	12 FT.	0.48
○	HARDIE	DKL02	EMITTER	20 PSI	N/A	0.5 G.P.H.
◘	HARDIE	DKL04	EMITTER	20 PSI	N/A	1.0 G.P.H.
⊘	HARDIE	DKL08	EMITTER	20 PSI	N/A	2.0 G.P.H.

SYM	DESCRIPTION
◕	HUNTER HPV-101-1 R.C. VALVE 1"
●	HARDIE 700-.75 R.C. VALVE 3/4"
N	FEBCO 825Y-125 R.P. 1-1/4" BACKFLOW PREVENTER
△C	IRRITROL MC-12 PLUS-B CONTROLLER
———	200 PSI SDR 21 PVC PIPE (SIZE AS NOTED)
— —	1-1/2" SCHEDULE 40 PVC PIPE
———	1/2" POLYETHYLENE SDR PIPE
═══	3" 315 PSI SDR 13.5 PVC SLEEVE

Figure 15.157

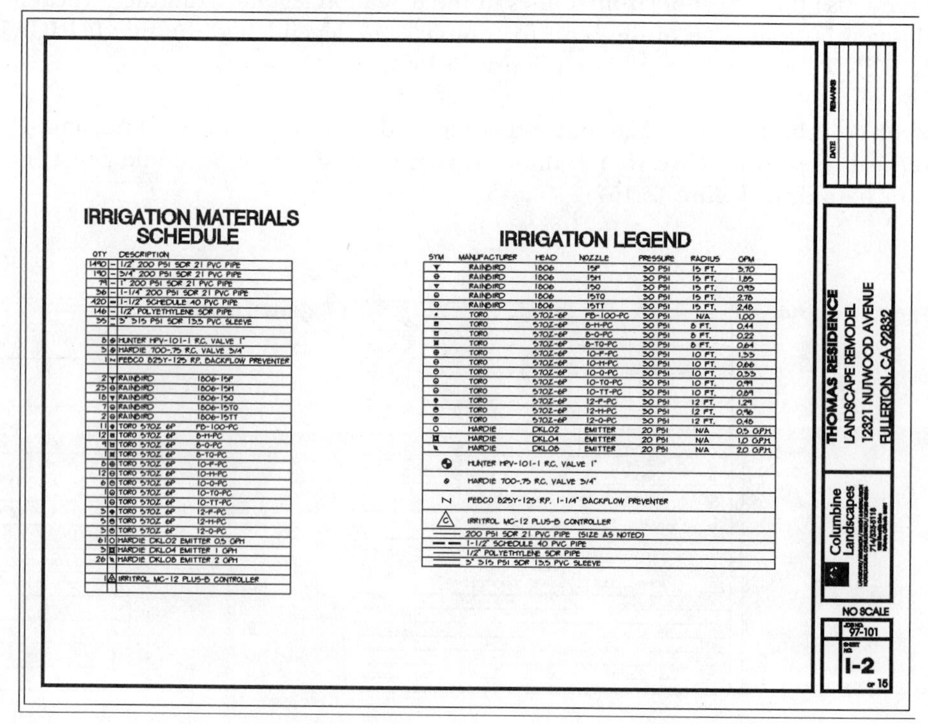

IRRIGATION MATERIALS SCHEDULE

QTY		DESCRIPTION	
1490	—	1/2" 200 PSI SDR 21 PVC PIPE	
190	—	3/4" 200 PSI SDR 21 PVC PIPE	
79	—	1" 200 PSI SDR 21 PVC PIPE	
36	—	1-1/4" 200 PSI SDR 21 PVC PIPE	
420	—	1-1/2" SCHEDULE 40 PVC PIPE	
146	—	1/2" POLYETHYLENE SDR PIPE	
35	═	3" 315 PSI SDR 13.5 PVC SLEEVE	
8	◕	HUNTER HPV-101-1 R.C. VALVE 1"	
3	●	HARDIE 700-.75 R.C. VALVE 3/4"	
1	N	FEBCO 825Y-125 RP. BACKFLOW PREVENTER	
2	▼	RAINBIRD	1806-15F
23	⊖	RAINBIRD	1806-15H
18	▽	RAINBIRD	1806-15Q
7	⊗	RAINBIRD	1806-15TQ
2	⊖	RAINBIRD	1806-15TT
11	•	TORO 570Z 6P	FB-100-PC
12	■	TORO 570Z 6P	8-H-PC
9	■	TORO 570Z 6P	8-Q-PC
1	■	TORO 570Z 6P	8-TQ-PC
8	⊕	TORO 570Z 6P	10-F-PC
12	⊖	TORO 570Z 6P	10-H-PC
6	⊙	TORO 570Z 6P	10-Q-PC
1	⊗	TORO 570Z 6P	10-TQ-PC
1	⊖	TORO 570Z 6P	10-TT-PC
3	●	TORO 570Z 6P	12-F-PC
5	◓	TORO 570Z 6P	12-H-PC
3	⊙	TORO 570Z 6P	12-Q-PC
61	○	HARDIE DKL02 EMITTER 0.5 GPH	
3	◘	HARDIE DKL04 EMITTER 1 GPH	
26	⊘	HARDIE DKL08 EMITTER 2 GPH	
1	△C	IRRITROL MC-12 PLUS-B CONTROLLER	

IRRIGATION LEGEND

SYM	MANUFACTURER	HEAD	NOZZLE	PRESSURE	RADIUS	GPM
▼	RAINBIRD	1806	15F	30 PSI	15 FT.	3.70
⊖	RAINBIRD	1806	15H	30 PSI	15 FT.	1.85
▽	RAINBIRD	1806	15Q	30 PSI	15 FT.	0.93
⊗	RAINBIRD	1806	15TQ	30 PSI	15 FT.	2.78
⊖	RAINBIRD	1806	15TT	30 PSI	15 FT.	2.48
•	TORO	570Z-6P	FB-100-PC	30 PSI	N/A	1.00
■	TORO	570Z-6P	8-H-PC	30 PSI	8 FT.	0.44
■	TORO	570Z-6P	8-Q-PC	30 PSI	8 FT.	0.22
■	TORO	570Z-6P	8-TQ-PC	30 PSI	8 FT.	0.64
⊕	TORO	570Z-6P	10-F-PC	30 PSI	10 FT.	1.33
⊖	TORO	570Z-6P	10-H-PC	30 PSI	10 FT.	0.66
⊙	TORO	570Z-6P	10-Q-PC	30 PSI	10 FT.	0.33
⊗	TORO	570Z-6P	10-TQ-PC	30 PSI	10 FT.	0.99
⊖	TORO	570Z-6P	10-TT-PC	30 PSI	10 FT.	0.89
●	TORO	570Z-6P	12-F-PC	30 PSI	12 FT.	1.29
◓	TORO	570Z-6P	12-H-PC	30 PSI	12 FT.	0.96
⊙	TORO	570Z-6P	12-Q-PC	30 PSI	12 FT.	0.48
○	HARDIE	DKL02	EMITTER	20 PSI	N/A	0.5 G.P.H.
◘	HARDIE	DKL04	EMITTER	20 PSI	N/A	1.0 G.P.H.
⊘	HARDIE	DKL08	EMITTER	20 PSI	N/A	2.0 G.P.H.

SYM	DESCRIPTION
◕	HUNTER HPV-101-1 R.C. VALVE 1"
●	HARDIE 700-.75 R.C. VALVE 3/4"
N	FEBCO 825Y-125 R.P. 1-1/4" BACKFLOW PREVENTER
△C	IRRITROL MC-12 PLUS-B CONTROLLER
———	200 PSI SDR 21 PVC PIPE (SIZE AS NOTED)
— —	1-1/2" SCHEDULE 40 PVC PIPE
———	1/2" POLYETHYLENE SDR PIPE
═══	3" 315 PSI SDR 13.5 PVC SLEEVE

c h a p t e r

Drawing Seven—The Detail Sheet

In this exercise, we will build a Detail Sheet with detail drawings for the Planting Plan and Irrigation Plan. Detail Sheets are usually provided to illustrate specific instructions and explanations for the contractor to aid in the correct installation of the job. Because local conditions and building regulations determine so much of the content of good installation details, using "packaged" details in a landscape architecture practice is a mixed blessing. Although the general idea of the detail may be correct, specifics may not. You should be very careful when using details which have been prepared by others. Neither LANDCADD nor any manufacturer or illustrator will be held liable for your decision to include unsuitable details in your Detail Sheets. Nonetheless, you may find that the detail drawings included in this module will save you a considerable amount of time if you use them for a basis of your own detail collection.

As is often the case with detail drawings, most of the details we use in this sheet will require modification of one sort or another. There are several reasons for this. First, because LANDCADD's details have been assembled over several years (and versions), they are not as uniform as we might wish. Detail drawings should show a common style and "look" to give a professional appearance. Second, the LANDCADD details use a square 12" x 12" box to frame the details, which works well when you build a 24 x 36 (D-size) detail sheet. Since we have used a C-size title block in all of our previous exercises, we will build our Detail Sheet to conform to the package of drawings we have already prepared. With a C-size page, the basic module for our details will be a rectangle about 5" x 7". This means that we will need to scale the details as we insert them, and rearrange the label and drawing information to make it proportional to the space available. Finally, the arrangement of the title text, scale and drawing elements does not agree with the basic style of our detail page. This means that we need to move the title text and scale information to make the details fit with the Detail Sheet we are building. Under ideal circumstances, you would have all of the detail drawings correctly drawn so that they fit perfectly into your detail sheets without any modification whatsoever.

Another issue with details is the standardization of size and text styles. Since any of your details could be used in combination, it is important that they use the same style text and are of the same proportions. The actual size of your details will depend on the title block you use and the amount of space you want to devote to each detail. Ideally, all of the details should be the same size and use the same text font, size, and style of leader. Using a consistent text style name scheme is important since different text styles with the same name will conflict, giving unexpected results. In addition, the layering of all of your details should be consistent so that the drawing objects, text and title will be on appropriate layers. Some details you build will not fit into a single module. Large details, or those which might be very long or tall might actually span two or more module spaces. We will see an example of this with our tree installation detail later.

As you build this detail sheet, you will see the value in saving your individual details in a library that can be used again and again. The details you build can be stored in your own separate library (this is the best idea), or they can be stored with LANDCADD's detail drawings. If you choose to store your details in a separate library, be sure to keep accurate records of drawing names and keep a binder of hard copy plots of each drawing. This will aid in retrieving your details. If you choose to place your details in the LANDCADD system, you can set up the dialogue boxes so that you can retrieve them along with LANDCADD's details.

For the source of our detail drawings, we will take advantage of those provided in the LANDCADD Construction Details module, and import others from the *Using LANDCADD* CD-ROM. While the details provided by LANDCADD will require modification for use in our project, the ones provided on the disk should fit correctly with the Detail Sheet we are building. This should illustrate the benefit of creating details that fit into the format without further editing.

We will begin this exercise with the **C18SHEET** drawing, and use it as a basis for a new type of drawing sheet specifically intended to display detail drawings. After opening the **BLANK** drawing, we will insert the **C18SHEET** drawing as a block, and scale it to a 1:1 page. Since the original scale factor for this sheet is **96**, we can scale it by a factor of **1/96** to shrink it to a 1:1 sheet on C-size paper. Once the drawing has been scaled, we will explode it, then modify it to create nine spaces for our details. Since this page can be used again for more details, we will save it under the name **CDTSHEET**—for C-size Detail Sheet. We can also place this drawing in the Project Manager if we want to retrieve it easily.

After the basic detail sheet is established, we can begin to insert details. Since each drawing will require some modification as it is inserted, we will need to scale the drawing as it is inserted, then explode it. Because the Construction Details user interface does not allow us to explode the drawing as it is inserted (unlike the **DDINSERT** command), we must perform this task in two separate steps. After each drawing is inserted, we will modify it to fit the space allocated, and rearrange the text elements until the drawing fits our scheme. We will repeat this process for eight details on the page. Once all of the drawings are inserted and edited, we will add final labeling then save the drawing.

Remember that the detail drawings that you build should always be saved for later use. If the details provided here look as though you might be able to use them in your work, by all means save them so that you will not need to draw them again later.

While building the DETAILS drawing, we will complete the following:

- Create a 1:1 scale version of **C18SHEET**
- Use **DDPTYPE** to change the display of points
- Use **DIVIDE** to place equally spaced points along a line
- Add lines to the title block to create detail drawing frames
- Change the lines to polylines to complete the blank Detail Sheet
- Save the Detail Sheet as **CDTSHEET**
- Use **CDTSHEET** as a prototype drawing for the Thomas Residence Detail Sheet
- Use the LANDCADD Construction Details menu to retrieve detail drawings
- Explode and modify details to fit in the **CDTSHEET** format
- Use **WBLOCK** to save a detail into the LANDCADD detail drawing folder
- Modify the **CD_DETLS.US** file to display the modified detail in LANDCADD's details list
- Insert detail drawings from the *Using LANDCADD* CD-ROM
- Add labeling information to the Detail Sheet
- Complete the Detail Sheet drawing by completing the title block information

Creating the 1:1 Detail Sheet

We will begin the process of building the Detail Sheet by creating a 1:1 version of the **C18SHEET** drawing. This process is similar to that used when creating the Paper Space title block, except that the drawing is left in Model Space. We will modify the 1:1 title block until it is suitable for use as a Detail Sheet.

Use the Project Manager to open the **BLANK** drawing, which only contains a few settings.

Next, use **DDINSERT** to insert **C18SHEET**, being sure to select the **Explode** toggle box, and remove the mark from the **Specify Parameters on Screen** toggle box. Set the **X Scale** factor at **1/96**.

Command:

Type **DDINSERT** (Figure 16.1).

This inserts, scales and explodes the **C18SHEET** drawing. In order to edit objects and navigate in the reduced drawing size, we need to reset the **Limits**, **Grid** and **Snap**.

Command:

Under the AutoCAD **Options** pull-down menu, select **Drawing Aids...**, then modify the **Grid** and **Snap** settings there (Figure 16.2).

Set the **Snap** to **1/8"** and your **Grid** to **1"**. Notice that AutoCAD takes your entry in the **X Spacing** and copies it to the **Y Spacing** box. Pick **OK**.

`Command:`
Type **LIMITS**.

`Reset Model space limits:`
`ON/OFF/<Lower left corner> <0'-0",0'-0">:`
Press **<Enter>**.

Figure 16.1

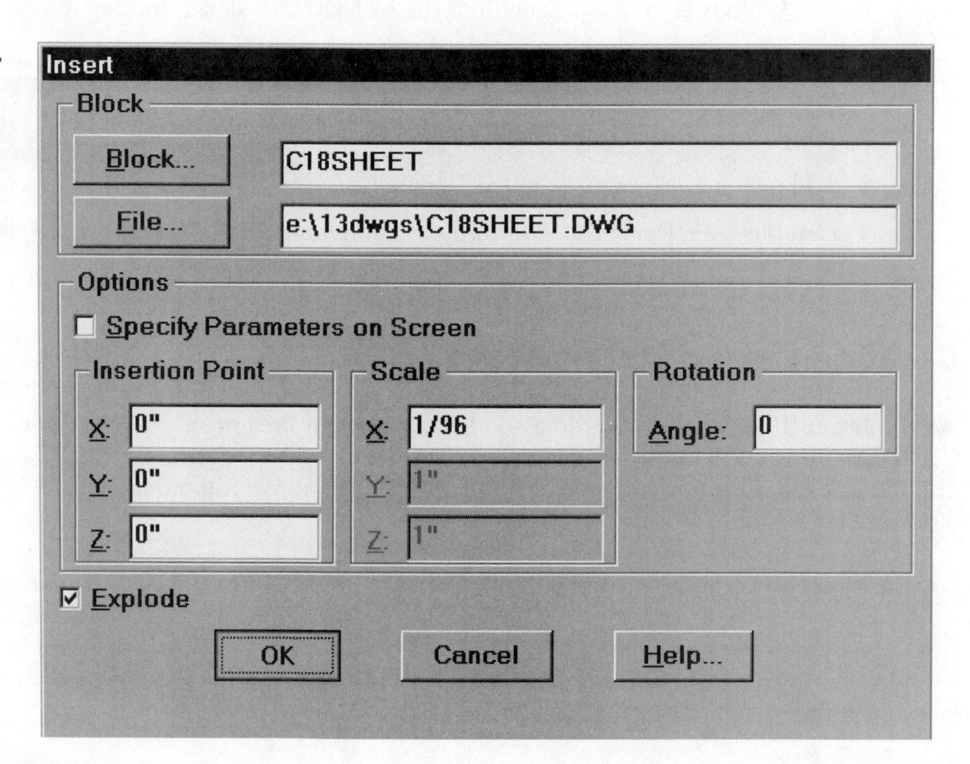

Figure 16.2

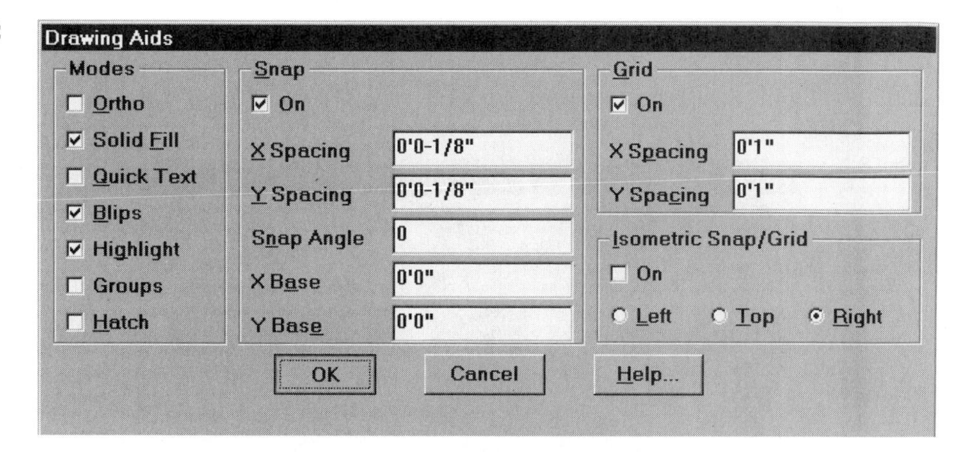

`Upper right corner <250'-0",185'-0">:`
Type **25,19** to set the limits to just outside the 24" x 18" title block.

Use one of the options available to make **SHEET** your current layer.

After the drawing area has been properly configured, we can begin to work on the title block to convert it into a Detail Sheet. We will begin by exploding the polyline rectangle which surrounds the drawing area at the left side of the title block. When the polyline is exploded, it is converted to lines. We will use these lines as the basis for the boxes which define the borders of our individual detail drawings.

`Command:`
Type **EX** (or **EXPLODE**).

`EXPLODE`
`Select objects:`
Pick the large rectangle which encloses the drawing area of the title block.

`Select objects:`
Press **<Enter>** to complete the selection set and return to the **EXPLODE** command.

`Exploding this polyline has lost width information.`
`The UNDO command will restore it.`
This warning is always issued when you explode a polyline.

After the rectangular polyline is converted to lines, we can use the **DIVIDE** command to help us create equal-sized rectangles for the details themselves. **DIVIDE** places equally spaced points along a line that we select. Before we can take advantage of these points, we need to alter the way that they are displayed on the screen.

```
Command:
```

Type **DDPTYPE** to modify the way that points are displayed on the screen. This opens the “Point Style” dialogue box. Pick the style shown in the illustration (Figure 16.3).

Figure 16.3

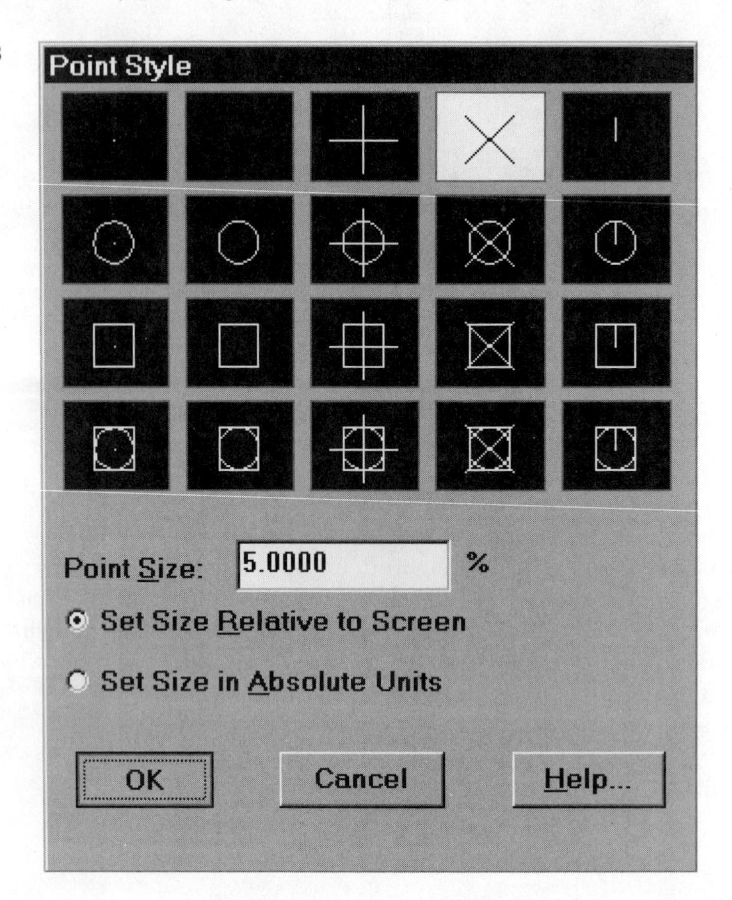

This temporary setting will allow us to see the points used by the **DIVIDE** command when they are placed along the rectangle lines. Notice that the original setting for “Point Style” would make the points impossible to see when placed along a line. Now we can use the **DIVIDE** command.

```
Command:
```

Type **DIVIDE**.

```
Select object to divide:
```

Pick the line at the top of the rectangle you just exploded.

```
<Number of segments>/Block:
```

Type **3**. This places the point marks on the line, dividing it into thirds. Repeat this procedure for the line at the left side of the rectangle. Your results should resemble those shown below (Figure 16.4).

Figure 16.4

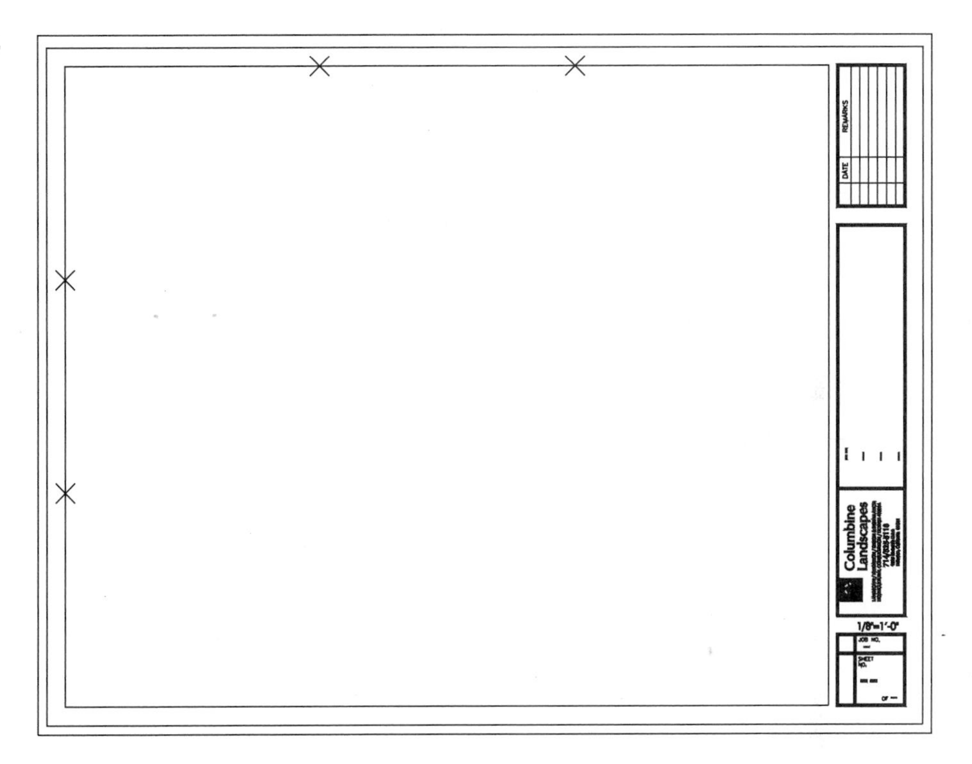

Since these points are located at precisely the correct location to divide the lines into equal segments, we can use them to create equally sized boxes. We will draw two horizontal lines and two vertical lines, using the dividing points as the starting points for all of the lines. This is an excellent use of the **OSNAPS** available in AutoCAD.

`Command:`
Type **L** (or **LINE**).

`LINE From point:`
Type (or pick) the **OSNAP** mode **NODE**. This will select the nearest point object.

`_nod of`
Pick the left point of the horizontal line at the top of the rectangle.

`To point:`
Type or pick the **OSNAP** mode **PER** to build the line to terminate perpendicular to the chosen line.

`_per to`
Pick the line at the bottom of the rectangle.

`To point:`
Press **<Enter>** to end the **LINE** command.

Repeat this procedure for building two horizontal and two vertical lines through the drawing area (Figure 16.5).

Figure 16.5

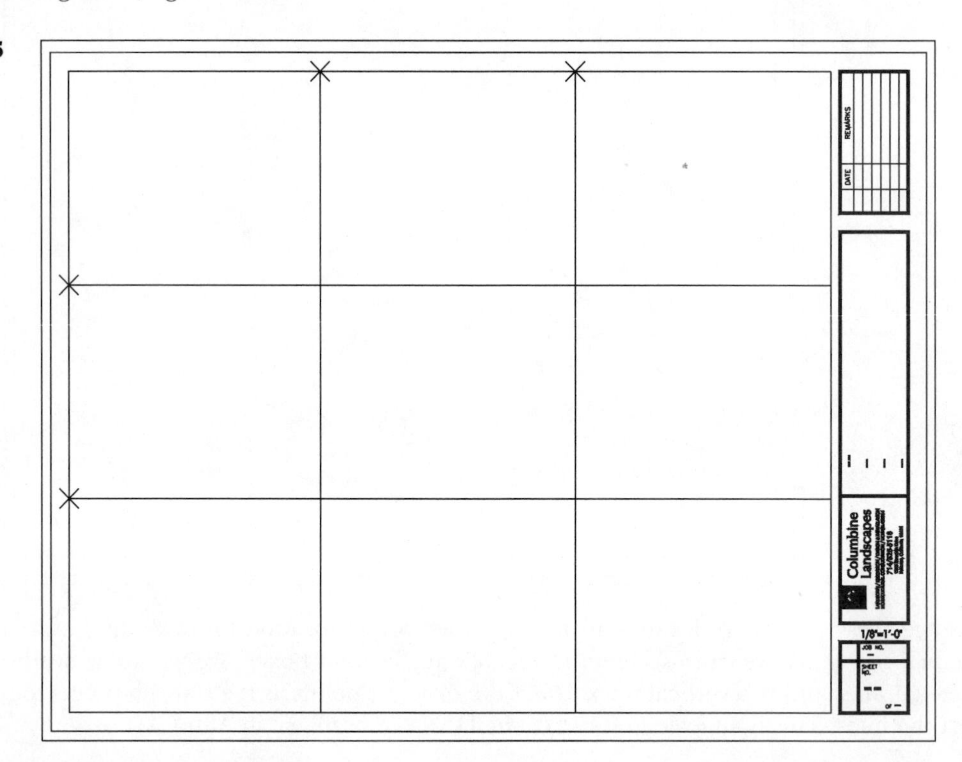

Since we are finished with the points, we can change their appearance so that they are no longer visible on the screen.

`Command:`
Type **DDPTYPE**. Use the "Point Style" to change the appearance of the points so that the points are displayed as dots (the upper left box). Pick **OK**. The points will disappear if you type **REGEN** to regenerate the screen.

Next, we need to build the label areas below each detail box. To do this, use the **OFFSET** command to create copies of the three lower horizontal lines. Each copy should be **1/2"** above the original line. We do not copy the top line, since we do not need a label area above the top row of details.

```
Command:
```
Type **OF** (or **OFFSET**).

```
Offset distance or Through <Through>:
```
Type **1/2**.

```
Select object to offset:
```
Pick one of the horizontal lines.

```
Side to offset?
```
Pick the side **above** the original. This makes each rectangle the same size. After you offset each of the lower three horizontal lines, you should see a label area below each rectangular box.

We should now transform all of these lines into polylines so that they can display the correct width. In addition, we can reconstruct the original outer polyline rectangle. To do this, we use the **PEDIT** command.

```
Command:
```
Type **PE** (or **PEDIT**).

```
Select polyline:
```
Pick the top line of the rectangle.

```
Entity selected is not a polyline.
Do you want to turn it into one? <Y>:
```
Press **<Enter>** to accept this choice.

```
Close/Join/Width/Edit vertex/Fit/Spline/Decurve/Ltype
gen/Undo/eXit <X>:
```
Type **J** to **JOIN** other objects to the first polyline.

```
Select objects:
```
Pick the remaining three sides of the original outer rectangle.

```
Select objects:
```
Press **<Enter>** to finish the selection set and return to the **PEDIT** command.

```
3 segments added to polyline
Open/Join/Width/Edit vertex/Fit/Spline/Decurve/Ltype
gen/Undo/eXit <X>:
```
Type **W** to change the width of the entire polyline.

```
Enter new width for all segments:
```

Type **1/16** to make the polyline the correct width.

Use the **PEDIT** command to convert each of the remaining eight lines into polylines, and assign them the correct width of **1/16"**. Your results should resemble the illustration below (Figure 16.6).

Figure 16.6

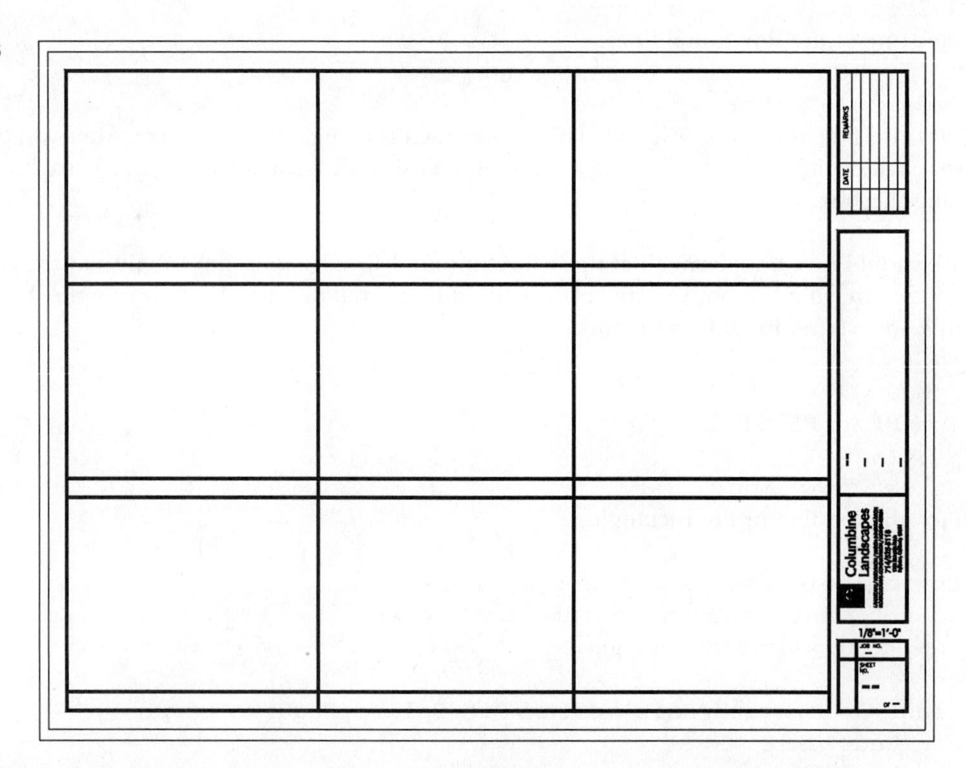

To complete the Detail Sheet prototype, we should add a layer specifically for the insertion of new detail drawings. Using any of the options available, add a new layer **DETAILS** to the layers in this drawing. Keep the color **WHITE** and the linetype **CONTINUOUS**.

Use the **SAVEAS** command to save this drawing under a new name. Name it **CDT-SHEET**. This creates a new drawing, and makes it the current drawing. Since we might want to use this Detail Sheet as a template for other detail sheets in other projects, we should not use it as our Thomas Residence Detail Sheet. Instead, we should use it as a prototype drawing for our Detail Sheet.

Use the Project Manager to open a new project, and use the **CDTSHEET** as the prototype drawing (Figure 16.7).

Figure 16.7

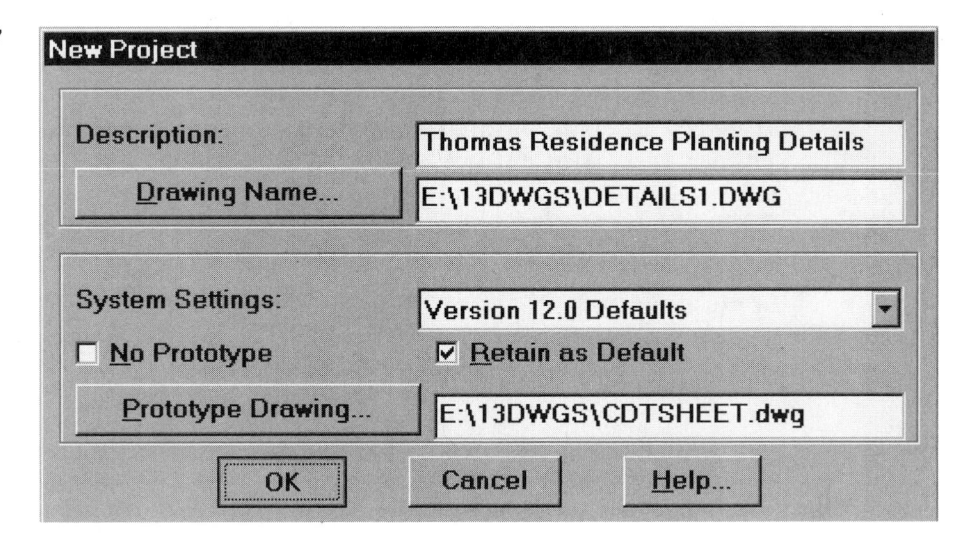

Inserting Construction Detail Drawings

Once the new project is opened, we can load the **Construction Details** module. This module makes the various details available and displays different categories of details which can be used in landscape-related detail sheets.

Command:

From the AutoCAD and EP menu, Pick EP, then select LANDCADD. From the LANDCADD menu, select Construction Details.

Assuming that you have all of the Construction Details available, this displays the pull-down options for Accessories, Civil, SiteDesign, Irrigation, Metric, as well as the normal options for EP and Tools.

Once we have this menu loaded, we should switch to the layer upon which we want to have the details inserted. Using any of the options available, make **DETAILS** the current layer.

We will select our first three details from the SiteDesign pull-down menu.

Command:

From the SiteDesign pull-down menu, pick Planting. This opens the "Select Detail to Insert" dialogue box. From this dialogue box, we can see the planting details that are available. The list displayed here can be edited to include more drawings if we wish to add them later (Figure 16.8).

Figure 16.8

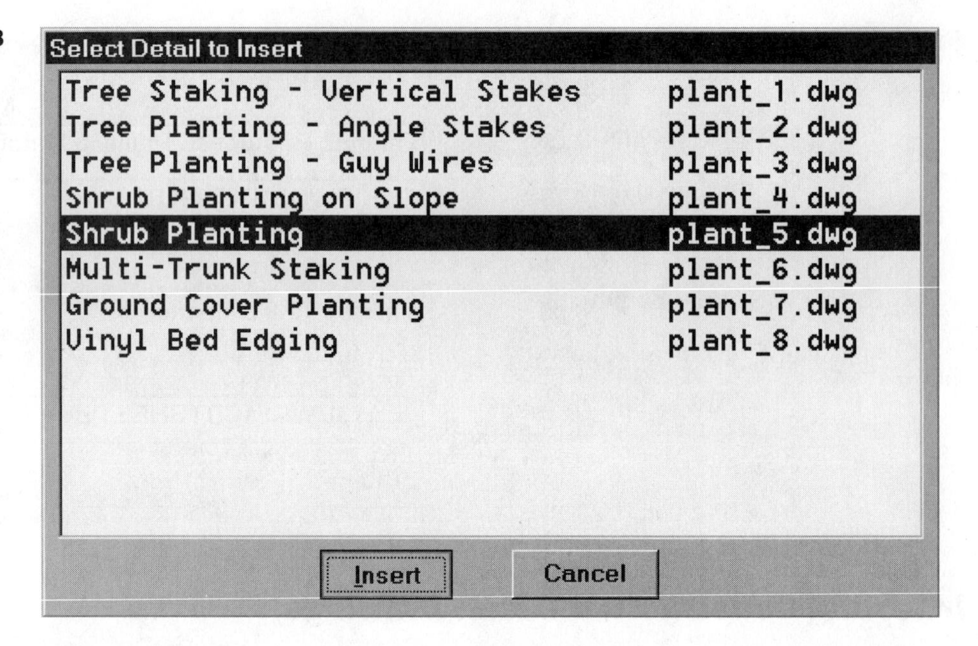

Pick the **Shrub Planting** detail (plant_5.dwg) as the first drawing to insert. We will insert this drawing in the first (upper left) detail box. For the insertion point for this detail, use the intersection of the two polylines at the lower right corner of the label box below the detail. Use this location (the lower right corner of the label box) as the insertion point for all of the subsequent details. See below (Figure 16.9).

Pick **Insert**.

At the command prompt:

```
Locating detail..._.insert Block name (or ?)<>:
E:\EPWIN13\DETAILS\plant_5.dwg Insertion point:
```

Pick the insertion point shown in the illustration. Be sure to use the **INT OSNAP** selection.

```
X scale factor <1> / Corner / XYZ:
```

Type **.75**.

```
Y scale factor (default=X):
```

Press **<Enter>** to accept the default value.

```
Rotation angle <0>:
```

Press **<Enter>** to accept the default value.

Figure 16.9

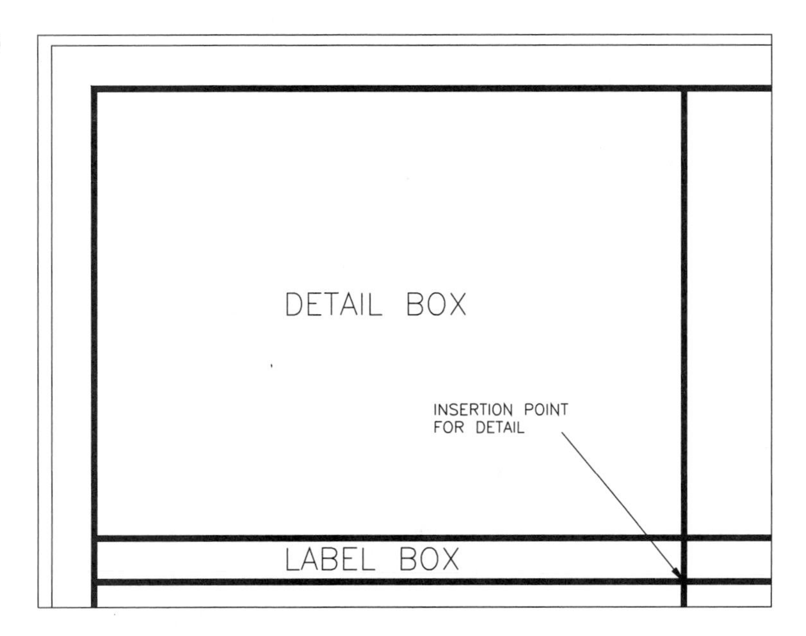

The detail is inserted in the box as shown (Figure 16.10).

The detail looks good initially, but if we look closely, there are a few problems to resolve. First, the **SCALE: NOT TO SCALE** tag is below our detail box. Second, the **SHRUB PLANTING** tag is not centered in the label box. Finally, there is a polyline below the

Figure 16.10

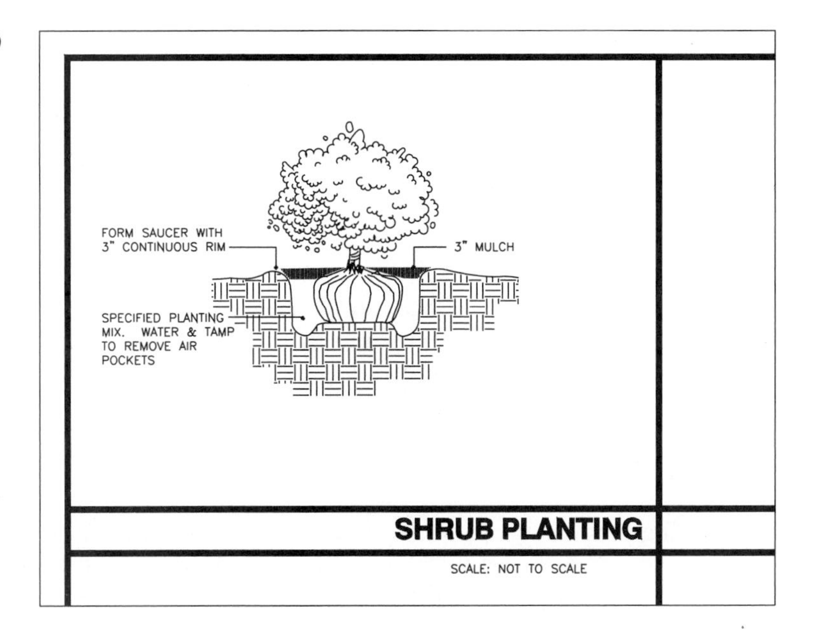

SHRUB PLANTING tag that is superimposed over our **CDTSHEET** polylines. Since the detail is a block, we must explode it before we can edit it in any way.

`Command:`

Type **EX** (or **EXPLODE**), then pick the detail. Erase the polyline below the **SHRUB PLANTING** label (you may find it easiest to use a window-type box for your selection set), then move the labels into the locations shown below (Figure 16.11).

Figure 16.11

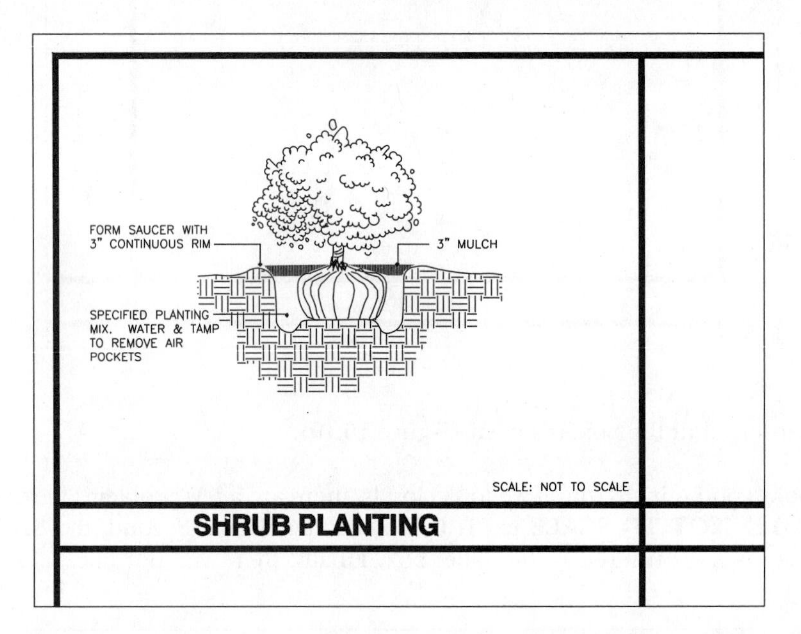

This should finish the work necessary on our first detail. For the second detail, repeat the steps above, but pick the detail **Multi-Trunk Staking** (plant_6.dwg). Place it in the box directly below the previous detail. Remember to use the insertion point of the intersection at the lower right corner of the label box, and use the scale factor of **.75** (Figure 16.12).

As before, the labeling needs work, but the image itself will also require some modification. Explode the drawing, then relocate the labels and erase the extra polyline at the lower part of the drawing. Finally, relocate and rearrange the drawing itself to center the drawing in the box and place the labeling information within the limits of our detail box. See below for the correct result (Figure 16.13).

For the sake of alignment, it may be easier to erase the **SCALE: NO SCALE** tag from the new drawing, then simply copy it to the detail box below, using the **ORTHO** option.

The third drawing to use is the detail **Ground Cover Planting** (plant_7.dwg). Using the same procedure as before, insert this drawing directly below the previous two details.

Figure 16.12

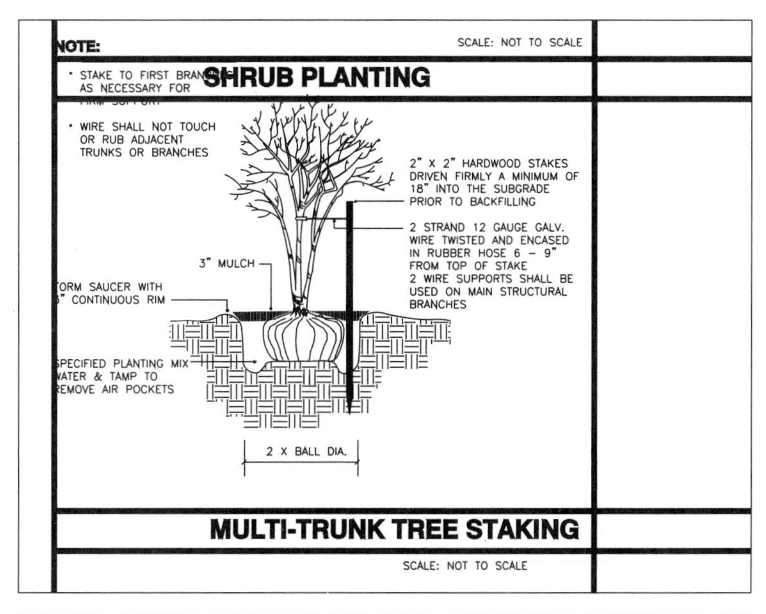

Figure 16.13

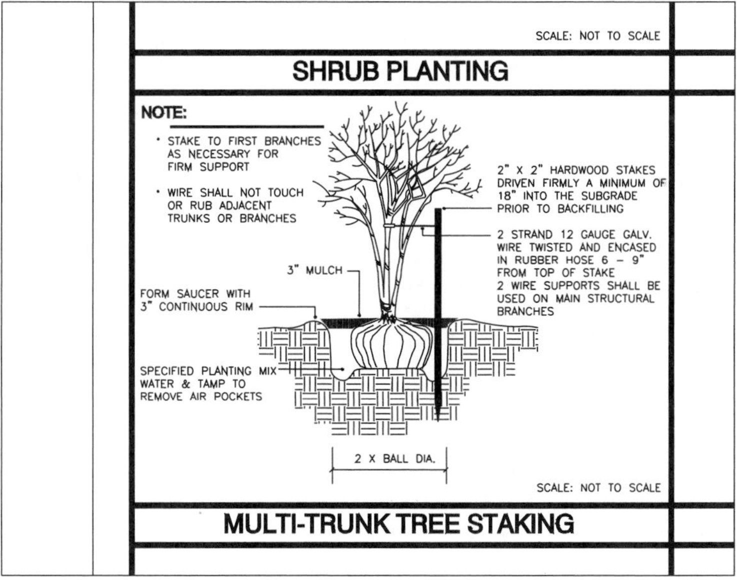

This drawing needs only to be centered and have its labels modified as before. See below for an illustration (Figure 16.14).

The final planting detail will present more of a challenge. To make this drawing fit properly into our format, we will need to create an extra-tall detail box. We will use the **TRIM** command to create this space, then insert the drawing.

Figure 16.14

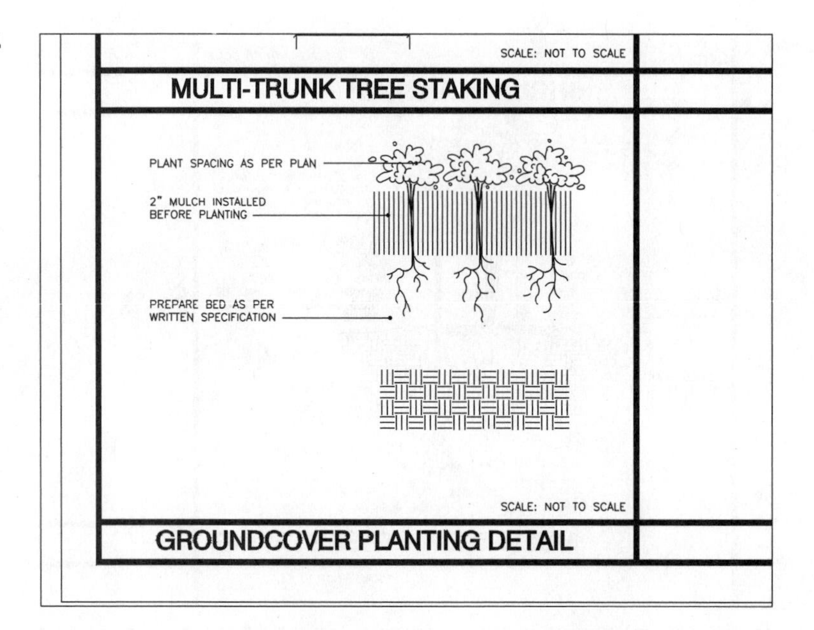

In the center column of drawing boxes, trim away the polylines separating the top detail and label box from the detail box below.

`Command:`
Type **TR** (or **TRIM**), the trim the polylines as shown (Figure 16.15).

Figure 16.15

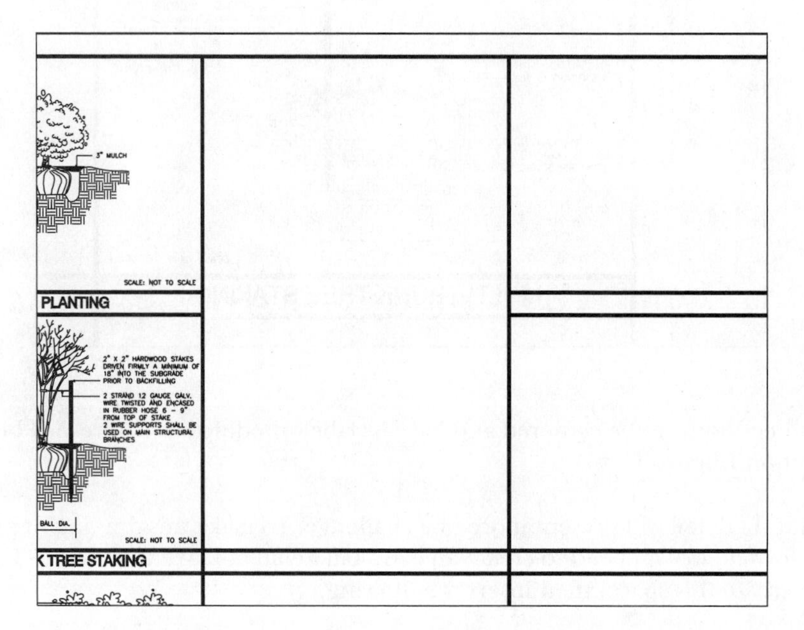

Next, insert the final planting detail **Tree Staking — Vertical Stakes** (plant_1.dwg) at the same relative location as before (Figure 16.16).

To make this detail look more proportional and balanced, we need to center the drawing, then stretch the tree. Since the tree itself is a nested block within the **PLANT_1.DWG** block, we must explode it to edit it. Finally, we will need to rearrange the drawing labeling so that it is evenly spaced up the sides of the tree. See below for a finished result (Figure 16.17).

This results in a more proportional tree, and a better distribution of text in the detail.

The next details should be loaded from the *Using LANDCADD* CD-ROM. These details have been prepared to fit within the format of the **CDTSHEET** and should not require any modifications after insertion.

`Command:`
Type **DDINSERT**, then pick **FILE**. Navigate to your *Using LANDCADD* CD-ROM, then load the first drawing **WLMTCTLR.DWG**. As before, use the lower right corner of the label box as the insertion point. Insert it directly below the previous (Tree Planting) detail (Figure 16.18).

Notice that like the details provided by LANDCADD, this drawing has its objects drawn on layer **0**, while the text is on a layer called **LABELS**. The dimension style is called **TEMPLATE**, while the text styles are named **1**, **2** and **3**. This is intended to make the details behave in a predictable manner.

Figure 16.16

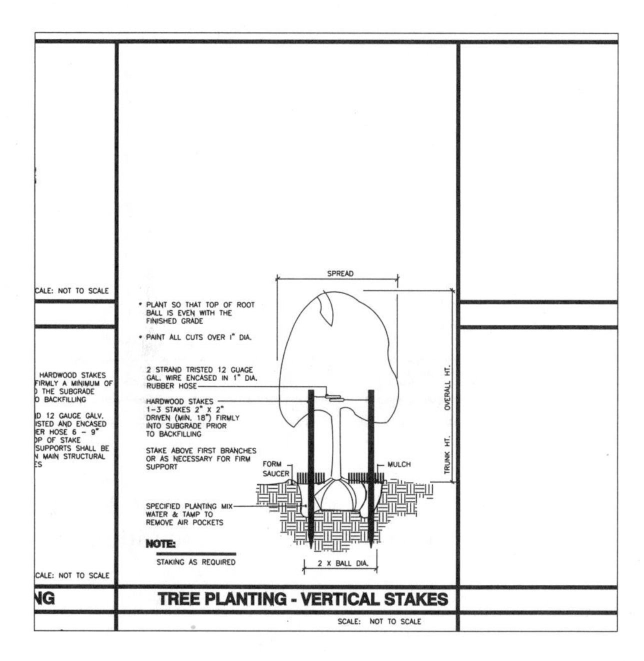

Figure 16.17

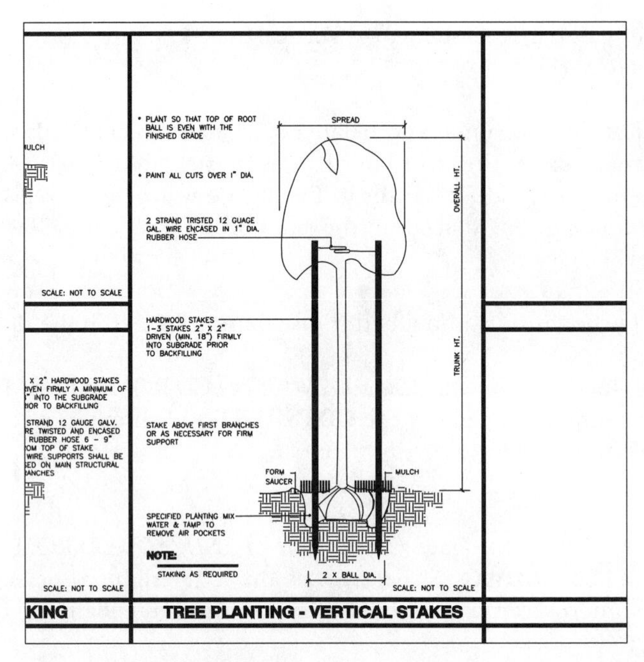

Figure 16.18

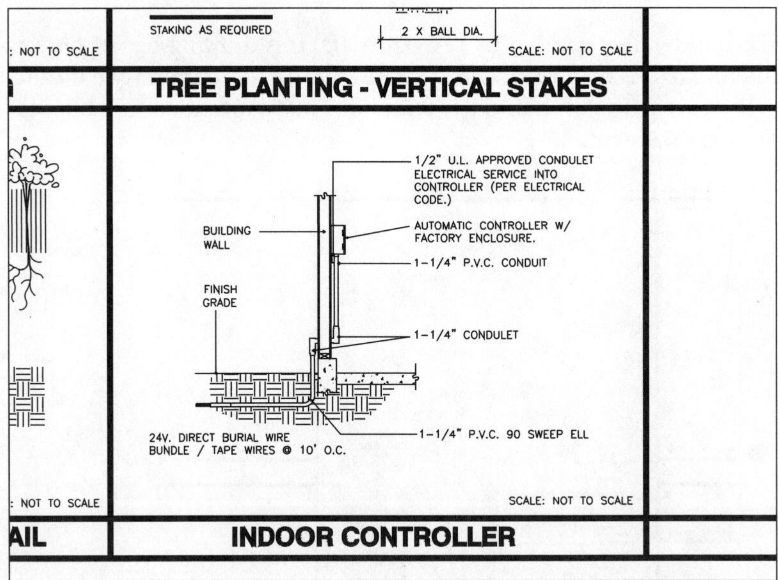

Insert the second detail from your *Using LANDCADD* CD-ROM, **RCVALVE.DWG** in the top right corner. As before, this drawing should be inserted without need for modification (Figure 16.19).

Figure 16.19

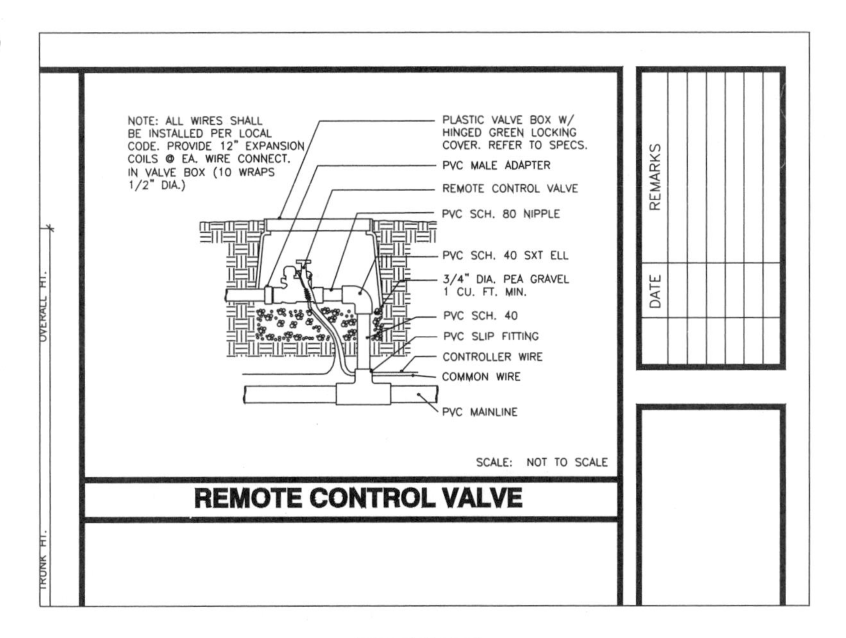

The final two details will be taken from the **Irrigation** pull-down menu under the category **TORO**. Choose the first detail from the **Sprinklers** category (Figure 16.20).

Select the drawing **6" pop-up 570-HP on swing joint** (toro115.dwg). This detail does not match the other drawings in its insertion point, labeling or other features. Because of this, we will need to modify it after we insert it.

Figure 16.20

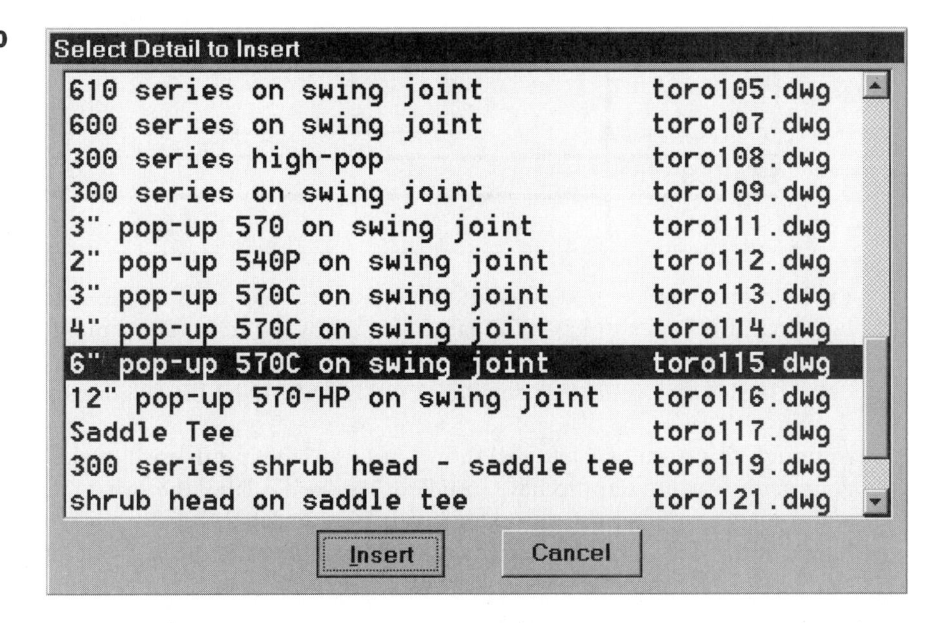

At the prompt:

```
Command:
Locating detail..._.insert Block name (or ?)<>:
E:\EPWIN13\DETAILS\toro115.dwg Insertion point:
```

Pick the **lower left** corner of the label box as the insertion point for the detail.

```
X scale factor <1> / Corner / XYZ:
```

Type **.5** as the correct scale factor.

```
Y scale factor (default=X):
```

Press **<Enter>** to accept this value.

```
Rotation angle <0>:
```

Press **<Enter>** to accept the default (Figure 16.21).

Figure 16.21

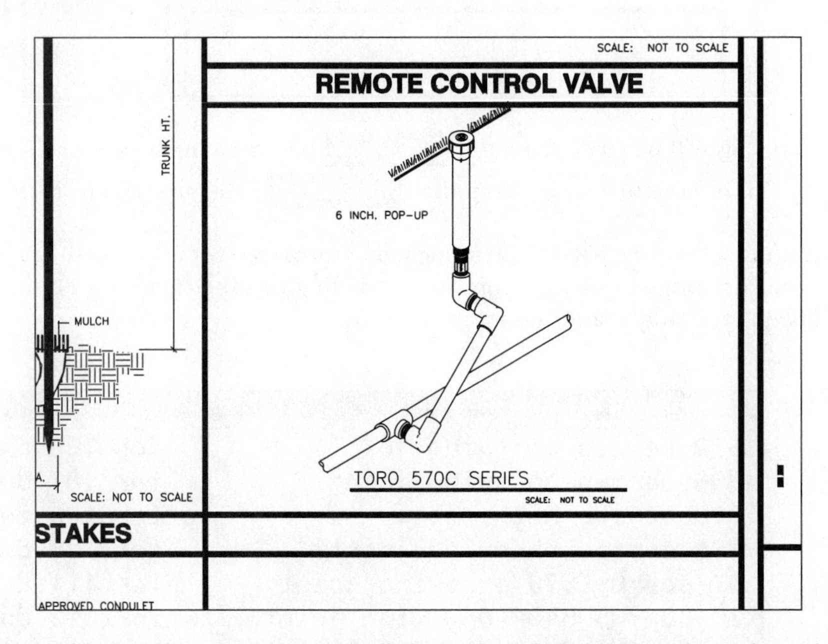

Explode this detail so that you can edit it. Rather than convert the style and size of the lettering, it is easier to erase the label text and scale text, then simply copy it from an adjoining detail. Once the text is in place, you can modify the lettering to agree with the detail. In this case, label this detail **6" POP-UP SPRINKLER** (Figure 16.22).

Although this detail is fairly well drawn, it lacks labeling that will make it clearer to the contractor what we want to have installed. We should label the swing joint assembly, and add a leader to label the sprinkler head. We can label these easily because a suitable dimension style is already present in the Detail Sheet drawing — we imported it when

Figure 16.22

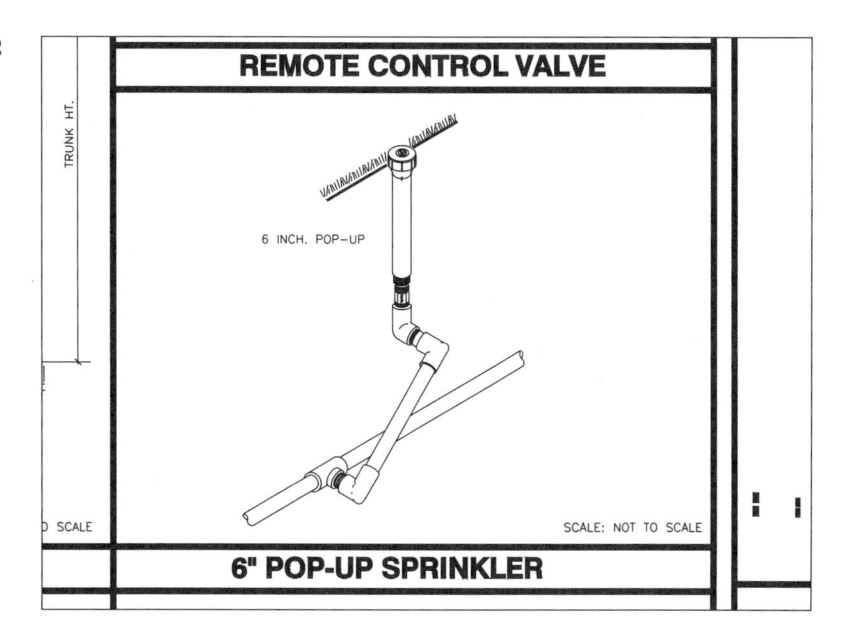

we brought in the two drawings from the *Using LANDCADD* disk. This dimension style is called TEMPLATE, and features very small (1/32") dots at the end of leader lines.

`Command:`
Type **DDIM** to open the "Dimension Styles" dialogue box (Figure 16.23).

Figure 16.23

Dimension Styles
Dimension Style
Current: TEMPLATE
Name: TEMPLATE
Save
Rename
Family
Parent
Linear
Diameter
Radial
Ordinate
Angular
Leader
Geometry...
Format...
Annotation...
OK
Cancel
Help...

Use the pop-down box to find the **TEMPLATE** dimension style, and select it.

Using one of the choices available, make **LABELS** the current layer.

```
Command:
```
Type **LE** (or **LEADER**) to build a leader. Add the text and leader as shown below (Figure 16.24).

For the final detail, select the **Irrigation** pull-down menu, then pick **TORO**. From the list of TORO options, select **Drip**. This opens the "Select Detail to Insert" dialogue box (Figure 16.25).

Select the detail **Emitter w/Lime cap on hose** (toro21.dwg).

At the prompt:

```
Command:
Locating detail..._.insert Block name (or ?)<>:
E:\EPWIN13\DETAILS\toro21.dwg Insertion point:
```
Pick the **lower left** corner of the label box as the insertion point for the detail.

```
X scale factor <1> / Corner / XYZ:
```
Type **.5** as the correct scale factor.

Figure 16.24

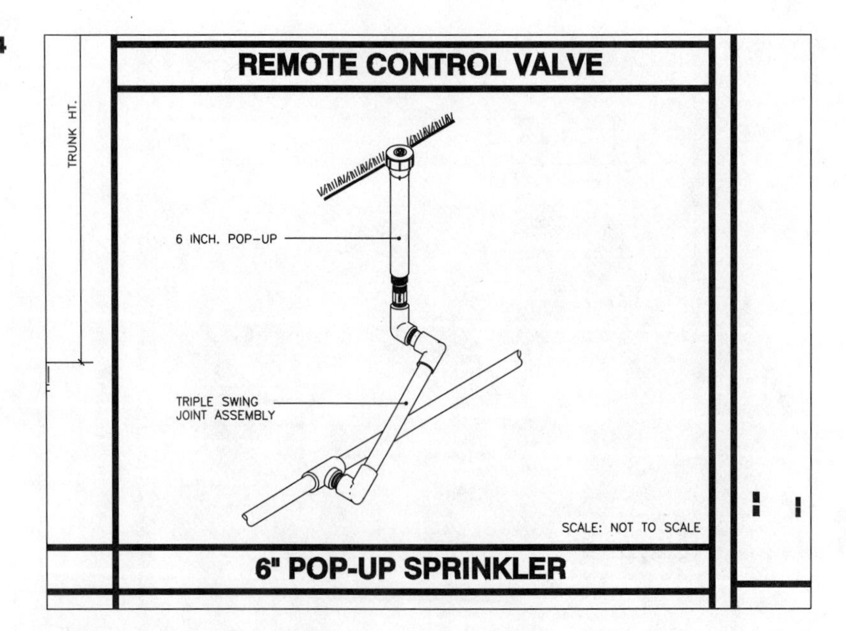

Figure 16.25

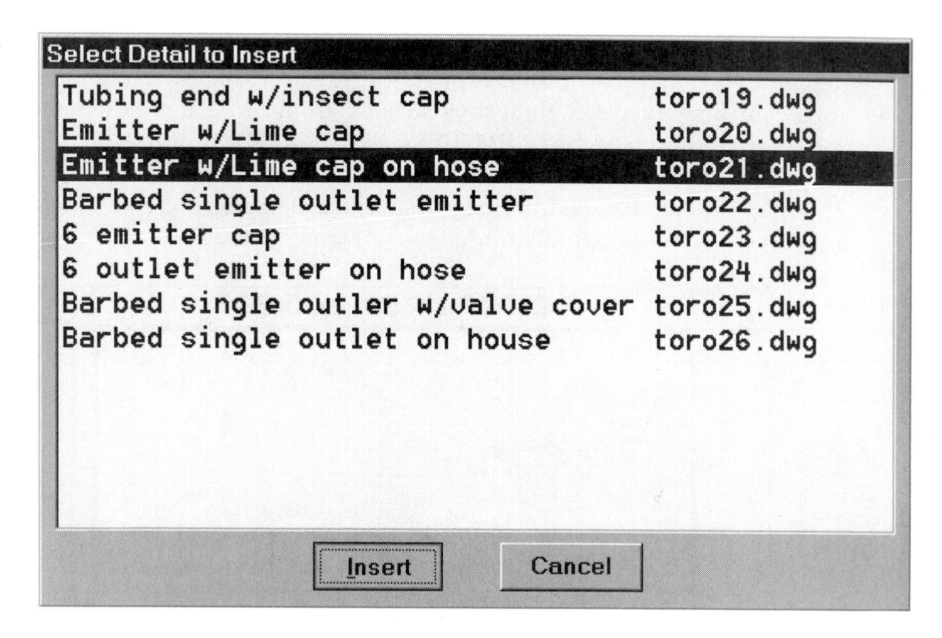

`Y scale factor (default=X):`
Press **<Enter>** to accept this value.

`Rotation angle <0>:`
Press **<Enter>** to accept the default (Figure 16.26).

Figure 16.26

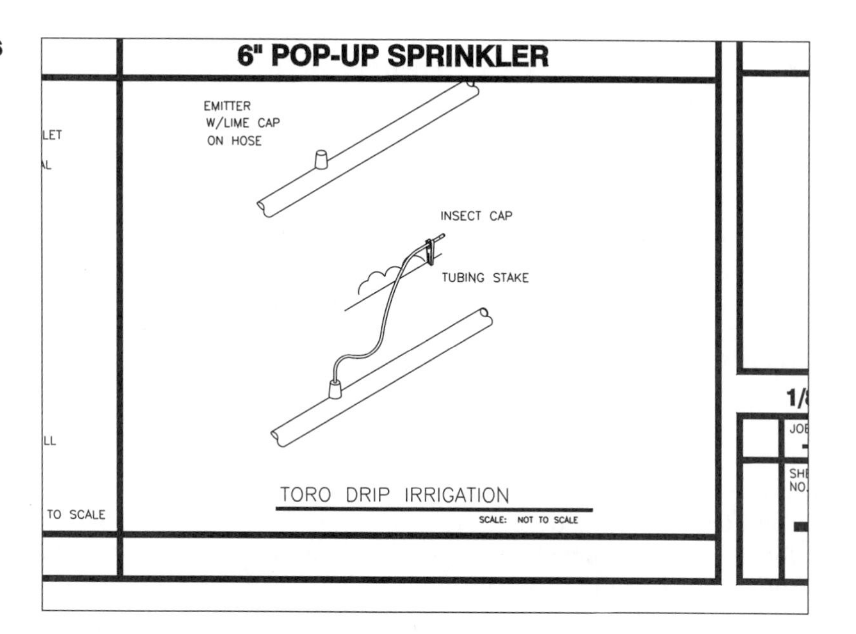

Command:
Type **EX** (or **EXPLODE**), then explode the detail. Using the technique from the previous detail, erase the label, then copy the label from above and move it into position. Edit the detail label to read **DRIP EMITTER PLACEMENT**.

Move the drawing objects into the center of the detail box, then add leaders to the text. Your results should resemble those below (Figure 16.27).

Figure 16.27

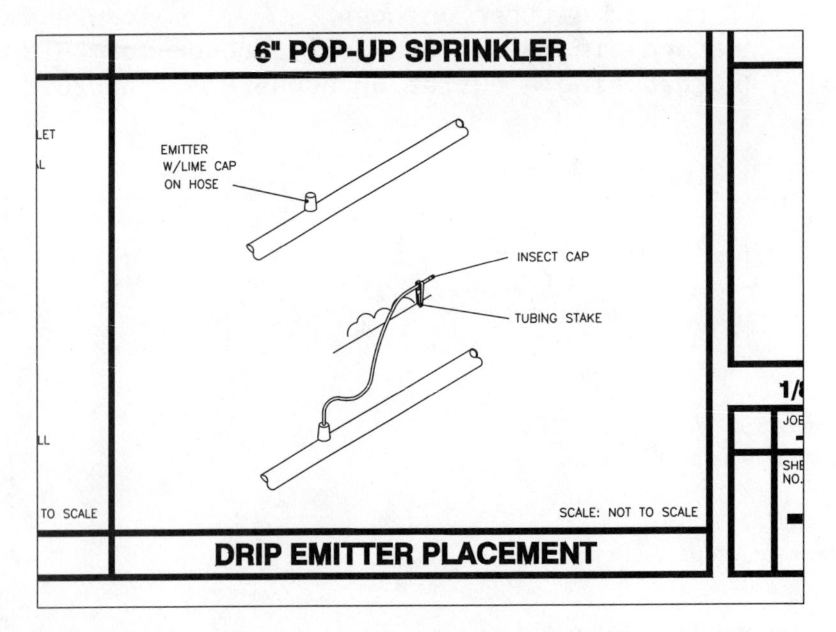

Adding Details to the Construction Details Module

If you wish to add details to the collection contained in LANDCADD's Construction Details module, this technique will help you.

After you have finished editing a detail, use **WBLOCK** to create a separate copy of the detail in the **C:\EPWIN13\DETAILS** folder (or the counterpart found on your system). Once the drawing is copied into the folder (directory), you then must add the drawing title and drawing name found in the detail list. This list is found in the file called **CD_DETLS.US** for the English units details, and **CD_DETLS.MET** for the metric details.

As an example, move to the box containing the Remote Control Valve detail drawing.

Command:
Type **WBLOCK**. This opens the "Create Drawing File" dialogue box. Navigate to the correct folder where the detail drawings are stored, then use the name **RCVALVE** for the copy of the block (Figure 16.28).

Figure 16.28

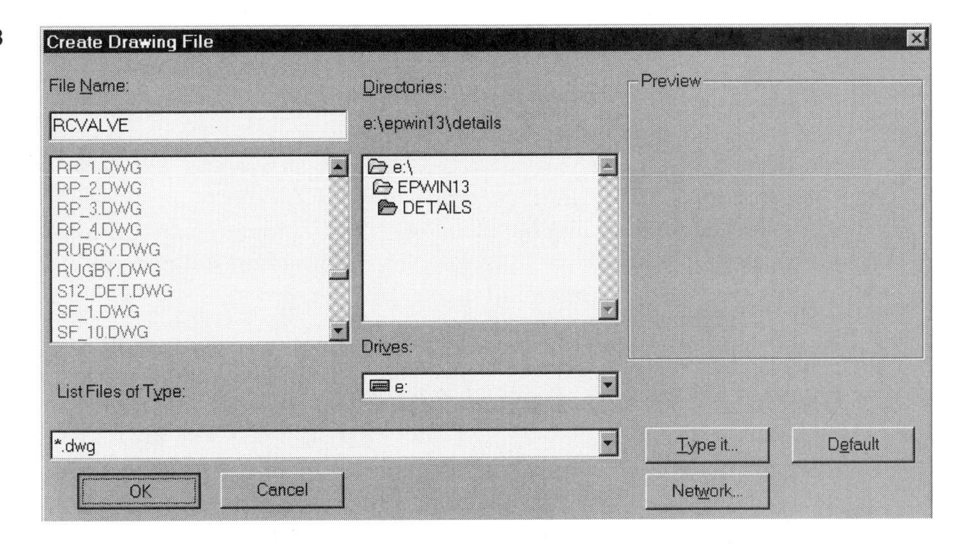

`Block name:`

Type **RCVALVE**. This creates a copy of the block in the correct directory. To add the listing to the LANDCADD Construction Details listing, open the file **CD_DETLS.US** with a text editor. The illustration shows the Windows text editor **WordPad** (Figure 16.29).

Figure 16.29

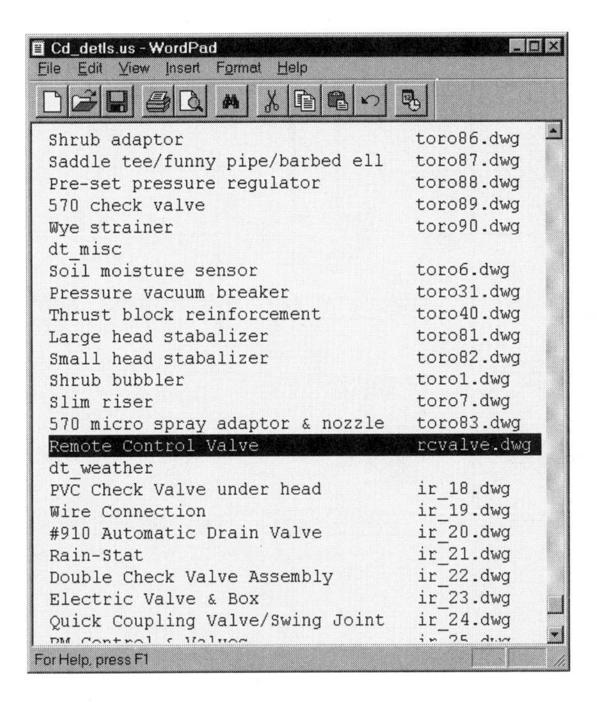

Under the listing dt_misc , add the line **Remote Control Valve rcvalve.dwg**. The first listing is the title, while the second listing is the drawing name that will be retrieved when you select the detail. Remember **not** to use the **<Tab>** key to create the spaces. Instead, simply use the **<Space Bar>** to create your spaces.

You can add as many drawings in as many different locations as you wish. Just be sure that you accurately record the name of each drawing, then add it carefully to the list of the drawing names shown. As with most information entered on a computer, typing accuracy is extremely important if you want the drawing to be retrieved properly.

Completing the Detail Sheet

After the details are inserted, we are ready to modify the attribute text block to reflect the contents of the drawing sheet, and then assign letter designations to each detail. Using the text style **3**, begin with the detail at the upper left, and insert a letter for each label. Save your drawing as **DETAIL1.DWG** (Figure 16.30).

Figure 16.30

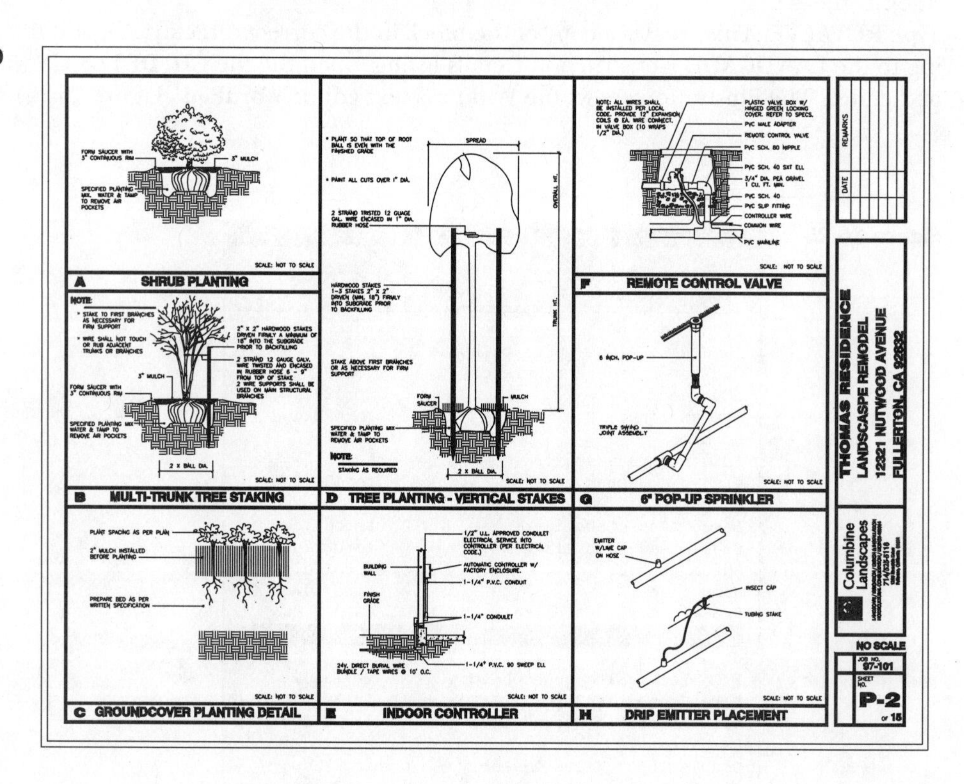

chapter

Drawing Eight—2-D to 3-D Drawing Conversion

For this exercise in *Using LANDCADD*, we will use LANDCADD's tools for substituting blocks in a drawing. More specifically, we will convert the 2-D plan view plant symbols into 3-D symbols in a landscape plan. Once we perform this transformation, we will create several different forms of drawing output to illustrate the potential use of LANDCADD for producing presentation drawings and concept plans for different landscape designs.

The basis of this exercise is a drawing provided on the *Using LANDCADD* CD-ROM. We will use the drawing **3DPLAN** as a prototype drawing for a new drawing that you can save on your hard disk. This is an alternative to copying a file from your floppy disk to your hard disk before opening a file. As the name implies, the **3DPLAN** drawing is a three-dimensional representation of a residence, complete with house, raised planters, concrete patio, porch and driveway, grass and fences.

The house drawing was taken from the LANDCADD **Site Planning** module, from the Bldgs pull-down menu under Shelters:.

The fences were built from the commands in the Circul pull-down menu under Fences.

The raised planters, patio slabs, grass area and driveway were drawn as **regions**. A **region** is a closed set of lines or arcs or a polyline which forms an enclosing edge. Once a region has been created, it can be assigned a depth (or thickness) by using the **EXTRUDE** command. The concrete surfaces have a stipple hatch on their upper surfaces to help give the appearance of thickness, and to add texture when creating different views. There is also a hatch surface on the lawn area. The plant symbols are inserted at the appropriate elevations, so that the plants in the raised planters are at their correct levels. A final feature of the **3DPLAN** drawing is its saved views. There are four views saved for this drawing, each from a different location. These views give a prospective client a number of different viewpoints from which to view the drawing of their residence.

In working with 2-D to 3-D block conversions, we will first observe the effects of a block substitution file which swaps one set of LANDCADD symbols for another. This **block sub-**

stitution file provides a "script" for which 3-D blocks will replace which 2-D blocks. This file is a simple ASCII text file which can be edited by any text editing program. Once the 3-D blocks are inserted, we will examine the drawing from the different views which have been saved in the drawing. We will also try out the **SHADE** and **RENDER** commands to see the effects of the 3-D symbols. After we finish viewing the results of this block substitution file, we will reverse the process, restoring the original 2-D blocks and removing the 3-D blocks. We will then produce new block substitution files which give different results than the substitution file provided on your *Using LANDCADD* disk. Finally, we will use **SHADE** and **RENDER** again to see the results of the new block substitution.

While completing the exercises with the **3DPLAN** drawing we will complete the following:

- Open a new drawing using the **3DPLAN** drawing on the *Using LANDCADD* disk as a prototype
- Examine the **3DPLAN** drawing from four different views saved with the drawing
- Use **HIDE** to see a clarified wireframe drawing
- Use the **Block Substitution** commands to exchange 3-D plant symbols for the existing 2-D plant symbols
- Use the **Substitution file (reverse)** option to restore the original blocks to a drawing
- Use the **Select blocks graphically** option to create a new block substitution file to exchange 3-D tree symbols for existing 2-D tree symbols
- Use the **Type block names manually** option to create a block substitution file to exchange 3-D shrub symbols for existing 2-D shrub symbols
- Examine the four different views with the 3-D plant symbols inserted
- Use **SHADE** to examine different views of the drawing with 3-D surfaces painted with solid colors
- Use **RENDER** to examine the different views with shaded and lighted 3-D surfaces
- Save the results of a **RENDER** to a bitmap file which can be printed or edited in a paint program

Opening the New Drawing

We will begin this exercise by opening a drawing in a new way. We start with a new drawing and use a drawing directly from the *Using LANDCADD* CD-ROM as a prototype. When working with drawing files stored on a CD-ROM, we must make allowance for the fact that they are always "read only" files. A "read only" file can (as the name implies) only be read by a drive—it cannot be altered and work cannot be saved to it. In order to save work done on a "read only" file, it must be saved under a new name. This problem is avoided when we use the "read only" file as a prototype for the new file. Another "work around" for this problem is to copy the file from the CD-ROM to your local hard disk, remove the "read only" file attribute, then open the drawing directly. In either case, the original drawing on the CD-ROM is unaltered.

`Command:`

Type **NEW**. This opens the "Open New Drawing" dialogue box. Pick the **Prototype...** box. This opens the "Prototype Drawing File" box. Select **3DPLAN** from your CD-ROM drive (Figure 17.1).

After you return to the "Create New Drawing" dialogue box, type in the complete name (including path) of your new drawing — here shown as **C:\13DWGS\NEW3DPLN.DWG** (Figure 17.2).

Click on **OK** to open your new drawing. Your current drawing will be the one that you named in the "Create New Drawing" dialogue box.

Figure 17.1

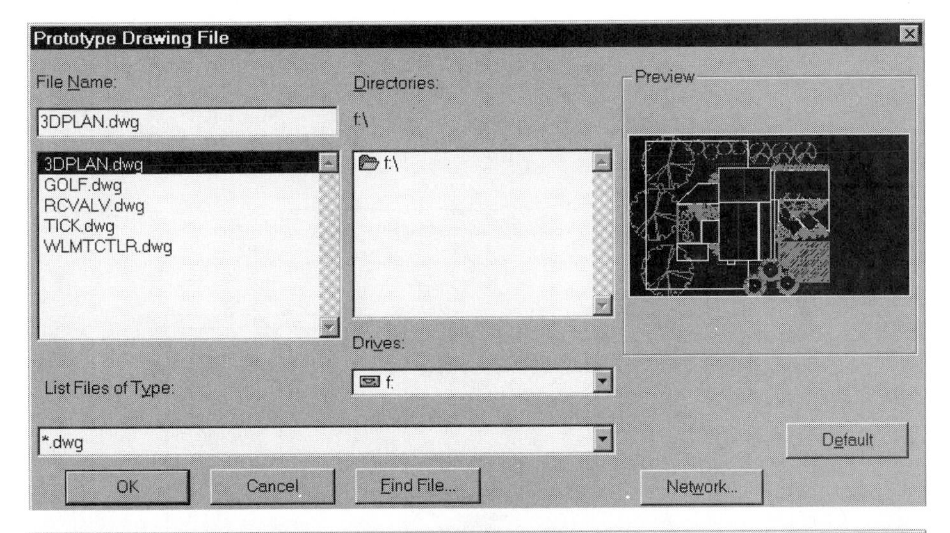

Figure 17.2

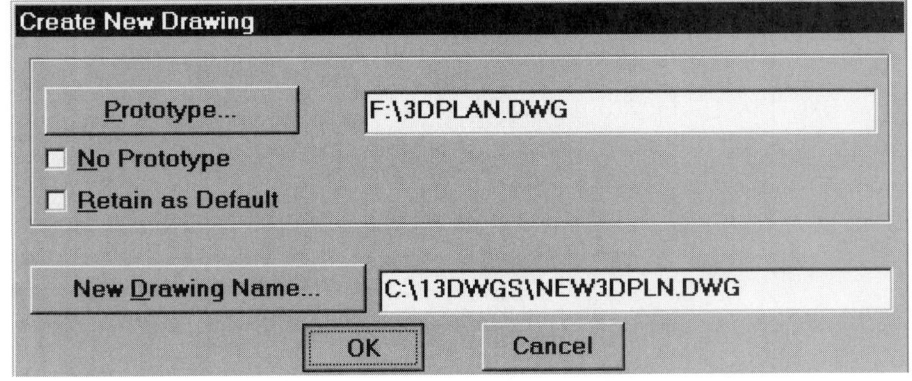

Once the new drawing is on the screen, you can see that it consists of a plan view drawing of a residence in a small yard (Figure 17.3).

This plan view landscape features a number of three-dimensional elements in its construction. The house has a detailed set of doors and windows, a porch and two-story roof line. The yard features several concrete pads and two raised planters, and a three-dimensional fence which encloses the back portion of the property. Because this plan contains three-dimensional elements, it can (and should) be viewed from several different locations to see the relationships between the elements. There are four separate views which are stored with the drawing. To retrieve a view, you can use the **VIEW** command:

`Command:`
Type **VIEW**.

`?/Delete/Restore/Save/Window:`
Type **RESTORE** (or **R**).

`View name to restore:`
Type **1**.

Figure 17.3

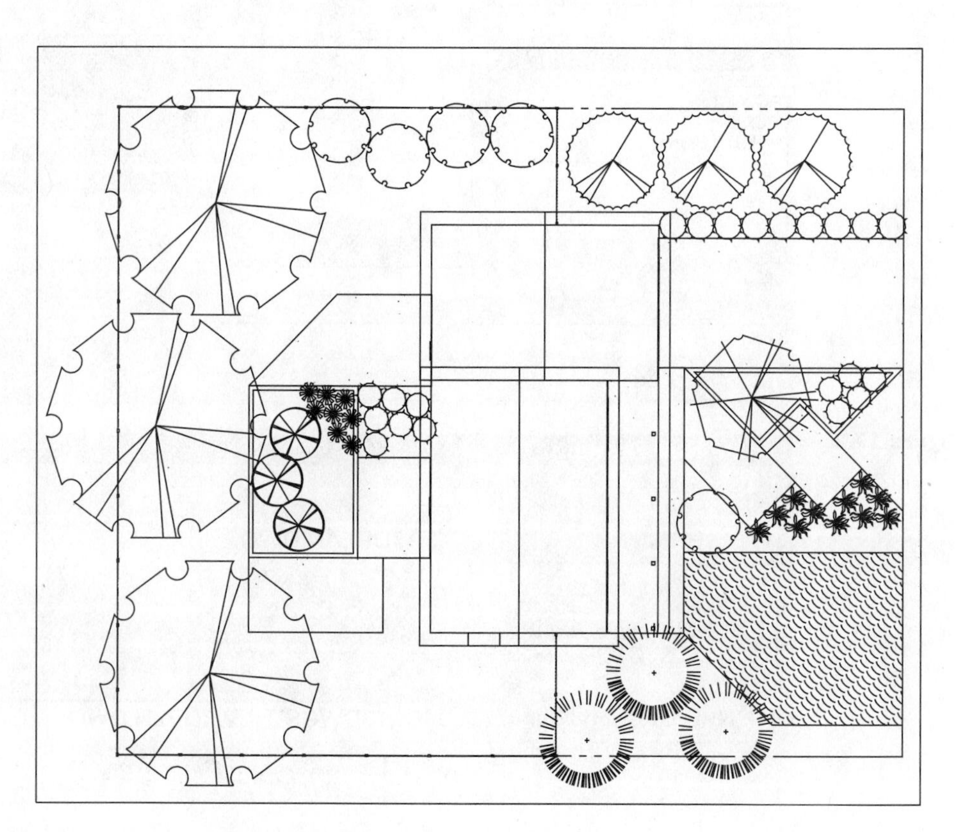

`Regenerating drawing.`
Type **HIDE**. This will make the planar surfaces appear solid, and hide the lines which occur behind them. This makes the drawing appear much clearer and less cluttered.

`Regenerating drawing.`
(Figure 17-4).

View 1 provides a "bird's eye" view of the property, and shows the flat "crime-scene" aspect to the plant symbols. We will modify those later.

`Command:`
Type **VIEW**.

`?/Delete/Restore/Save/Window:`
Type **RESTORE** (or **R**).

`View name to restore:`
Type **2**.

`Regenerating drawing.`
Type **HIDE**.

`Regenerating drawing` (Figure 17.5).
View 2 provides a closer view of the back of the property. As the trees and shrubs are converted into 3-D objects, the privacy and shade they supply will be more evident. Note that

Figure 17.4

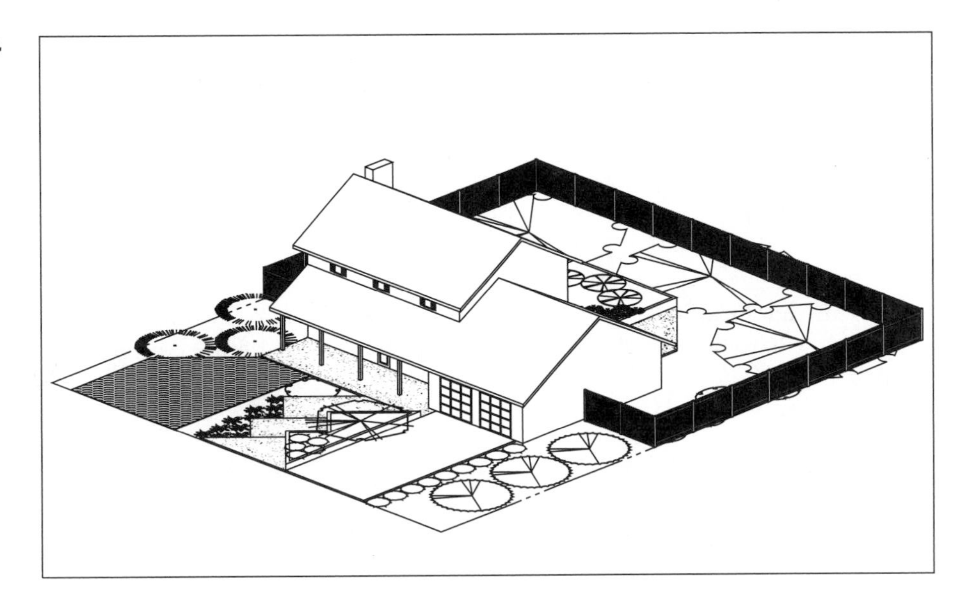

Figure 17.5

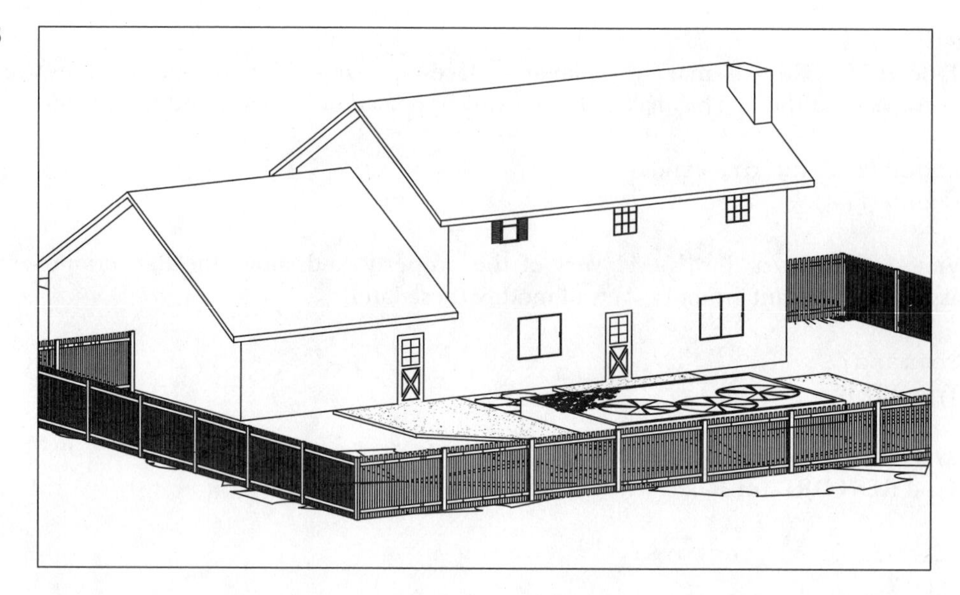

the shrubs in the raised planter are placed at their proper elevation. In addition, the hatch pattern on the patio slabs are placed on top. This is important when we use the **SHADE** and **RENDER** commands.

`Command:`
Type **VIEW**.

`?/Delete/Restore/Save/Window:`
Type **RESTORE** (or **R**).

`View name to restore:`
Type **3**.

`Regenerating drawing.`
Type **HIDE**.

`Regenerating drawing` (Figure 17.6).
This view examines the entry of the residence from a wide-angle perspective view. This view will become much more interesting when the plants are converted into 3-D symbols.

`Command:`
Type **VIEW**.

`?/Delete/Restore/Save/Window:`
Type **RESTORE** (or **R**).

Figure 17.6

`View name to restore:`
Type **4**.

`Regenerating drawing.`
Type **HIDE**.

`Regenerating drawing` (Figure 17.7).

Figure 17.7

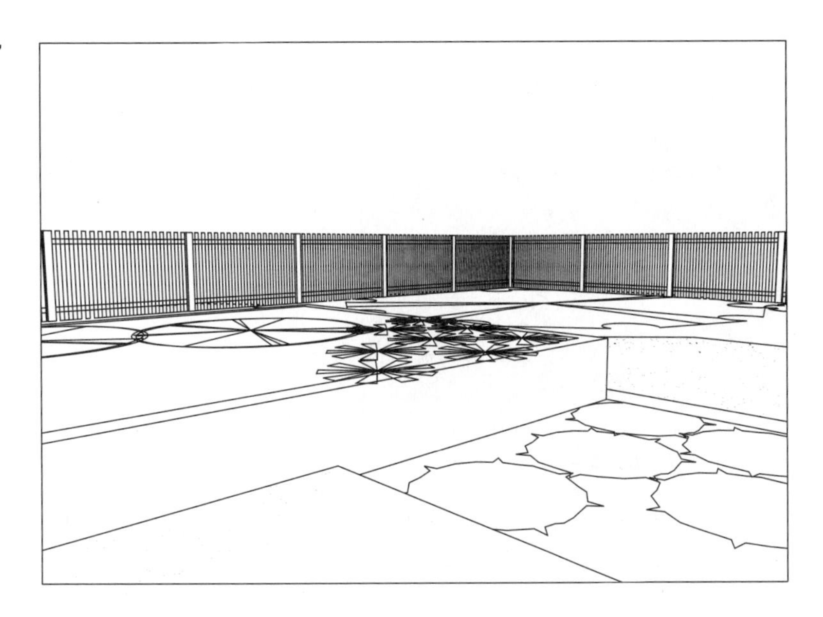

This final view is a perspective view into the back yard from a vantage point near the back door of the residence, looking across the raised planter toward the upper fence corner.

After you have examined all of the available views, try building a few more. You can use the **VIEW** command to save, restore and list views that you create. Remember that views can be built in **Perspective** mode using the **Perspective: Create** option found under the **Presentation** pull-down menu.

Another useful tool in creating different views is the **DDVPOINT** command. This tool allows precise control over the viewpoint and direction of view.

`Command:`
Type **DDVPOINT** to open the "Viewpoint Presets" command (Figure 17.8).

As you change the value in the **From: X Axis** box, your viewing point around the center of the drawing changes. As you change the angle in the **XY Plane:** box, your elevation relative to the flat UCS plan changes with it. The higher the value here, the higher above (or further below) you are compared to the original X-Y plane. Experiment with this option until you have created at least three more views. After you have created more views, you are ready to convert the 2-D plant symbols to 3-D plant symbols.

Running a Substitution File

To see an immediate transformation of the 2-D plant symbols into 3-D symbols, you can use LANDCADD's **Block Substitution** options. To gain access to these options, choose

Figure 17.8

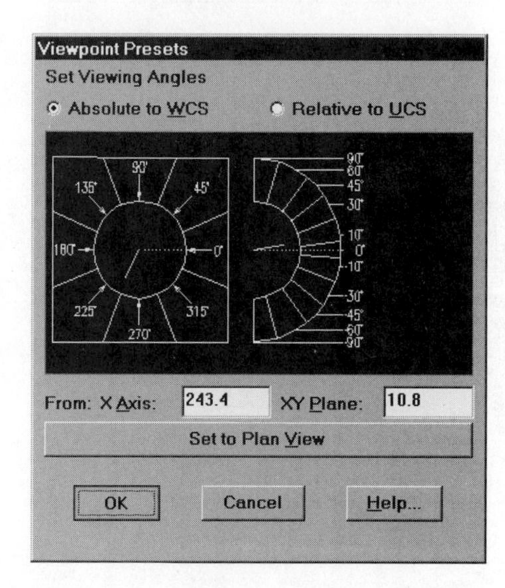

Tools, then pick Blocks ▸. Once you pick Block Substitution, you can see the substitution options at your disposal (Figure 17.9).

For our first procedure, we will use the Substitution file (Normal) option. This option uses a simple ASCII text file to name the original blocks and then the substitution blocks. The file can be used to convert simple blocks into complex ones, 2-D blocks into 3-D blocks, or any sort of transformation that you want. The substitution file included in the *Using LANDCADD* disk is called **NEW3D.SUB**. We will use this file to perform our first block substitution.

`Command:`
Pick Substitution file (Normal). This opens the "Select substitution file to use" dialogue box (Figure 17.10).

Figure 17.9

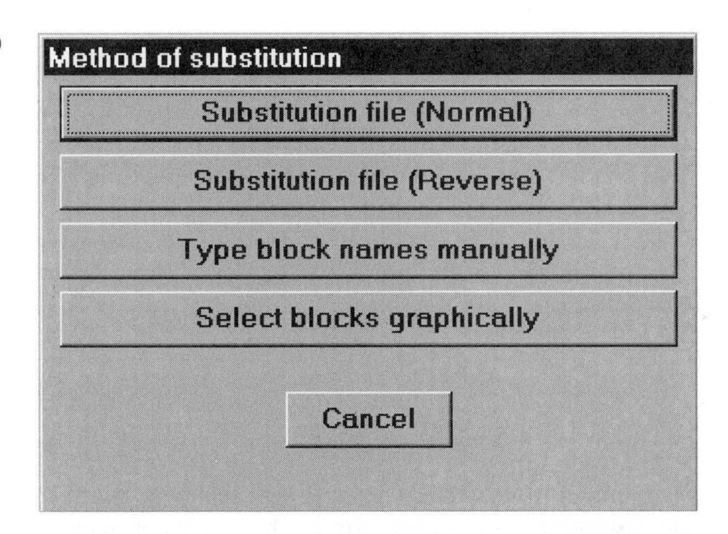

Figure 17.10

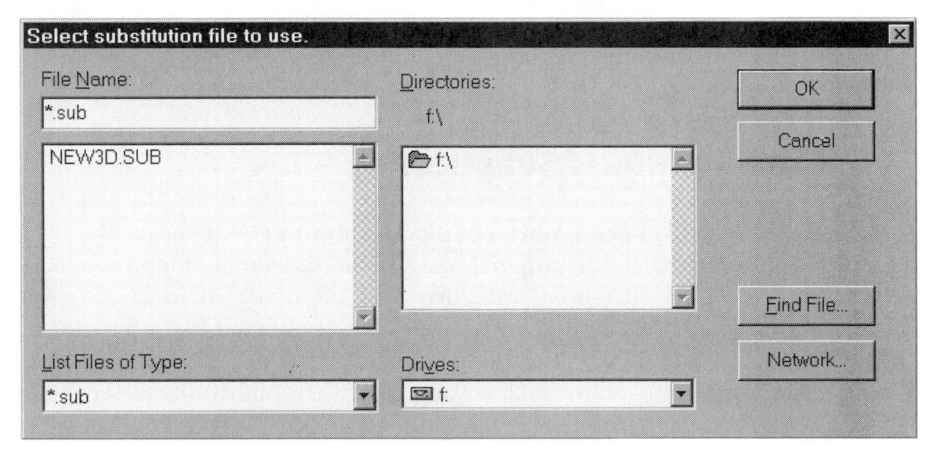

Navigate to your floppy disk drive, then pick **NEW3D.SUB**, then pick **OK**.

```
Command: EPLC_blocksub
Reading substitution file
```

After your computer digests the information, the new 3-D plant symbols are inserted in place of the original 2-D symbols.

You will notice the change both on the screen and in the response of your computer. The 3-D plant symbols used are quite large and greatly expand the size of the original drawing. If your computer has a limited amount of memory, a slow processor or a slow video card, you may find performance to be unacceptably slow with such a large drawing file loaded on your machine. Because of this drag on performance, you should perform block substitutions just before you are ready to print, plot or prepare presentation images for your drawings. Trying to perform drawing and editing functions with 3-D symbols loaded into your drawing is like driving your car with the emergency brake on.

If you decide that you want to return to a normal editing mode after performing a block substitution, you can use the **Substitution file (reverse)** to replace the substituted blocks with the original ones. This returns your drawing to its original status. Switching back and forth between 2-D and 3-D symbols allows you to fine-tune your drawing with 2-D symbols in place, then examine the results of your work with the 3-D symbols on screen, perhaps taking advantage of one of the views that you have saved. As a general rule, even with fast computers, it is best not to try to perform complex functions with 3-D symbols on screen. Running out of available memory not only slows performance, it may cause AutoCAD to crash, which can result in wasted time and lost work.

Your substituted blocks should make your drawings look much more life-like. See below (Figures 17.11, 17.12, 17.13, and 17.14).

After you have examined the views to see how the symbols look at different angles, perform a reverse block substitution to restore the 2-D blocks. When selecting the substitution file, remember to choose **NEW3D.SUB**. This will reverse the block substitution and restore the drawing to its original condition.

Creating Block Substitution Files

The process of creating your own block substitution scheme can be performed two ways. First, you can select the original blocks from the screen, then pick each substitute block from the selected symbol category. This method is used when you pick the choice **Select blocks graphically**. After you have completed the process of selecting the original blocks and their substitute counterparts, you are asked to save your substitutions to a file. You should always take advantage of the opportunity to create a substitution file,

Figure 17.11

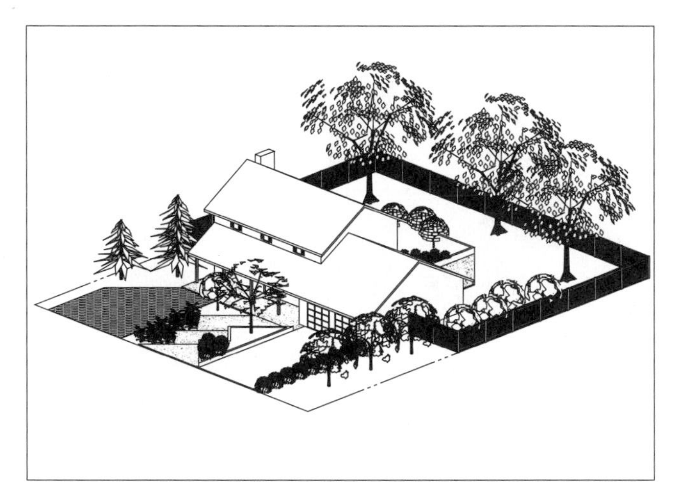

Figure 17.12

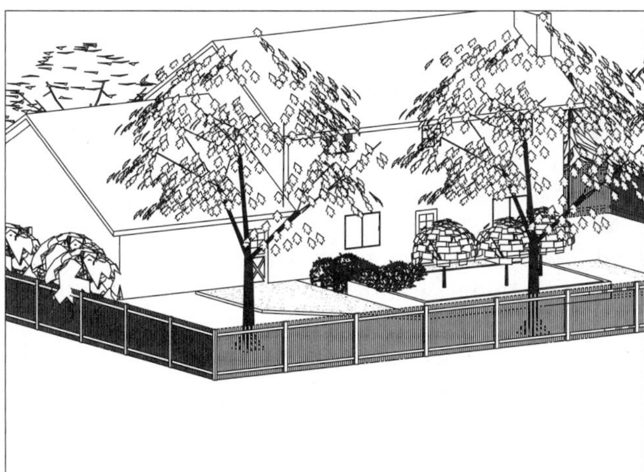

Figure 17.13

Figure 17.14

since it allows you to reverse the effects of your substitution scheme. This might be necessary if your results were not satisfactory, or if you want to return to your original blocks for general editing. We will perform a graphical block substitution later in this exercise.

The second method for block substitution is performed manually. In this case, you type in the name of the original block and the name of the corresponding substitution block. This method is used if you pick **Type block names manually**. As before, the substitution list created here can also be saved to a standard substitution file. This type of substitution is probably most useful if you use blocks that you have created, and whose names are easy to remember, since there is no way to access the names of the blocks when you are typing them into the dialogue box designed for this purpose. For the sake of experience, we will perform a manual block substitution later in this exercise, even though the LANDCADD symbol names can be challenging to remember.

When we create our block substitution files, we will build one for trees and another for shrubs. This reduces the number of steps required for each substitution, and simplifies keeping track of LANDCADD's block file names. The block file names follow a standardized format:

LCBLSP12	indicates	LC	(LANDCADD)
		BL	(BroadLeaf)
		SP	(Simple Plan)
		12	(Number 12)

OR

LCCN3D03	indicates	LC	(LANDCADD)
		CN	(Conifer)
		3D	(3-Dimensional)
		03	(Number 3)

Performing a Graphical Substitution

We will first build a graphical substitution file graphically, using the tree symbols on the plan. There are five trees to work with (Figures 17.15, 17.16, 17.17, 17.18, and 17.19).

Figure 17.15

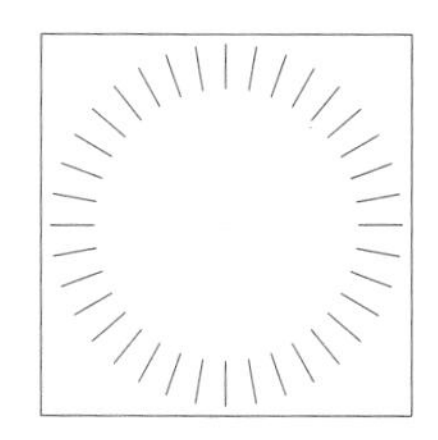

Figure 17.16

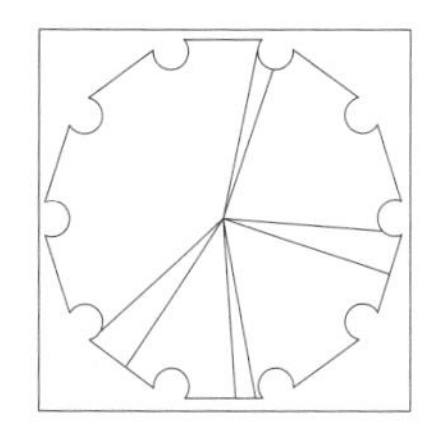

Figure 17.17

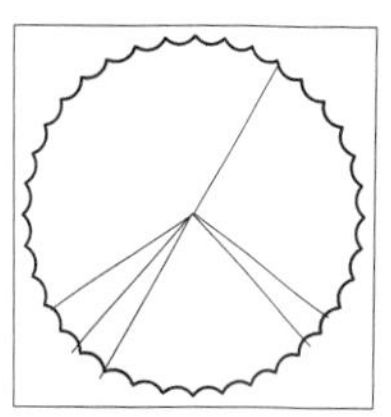

Figure 17.18

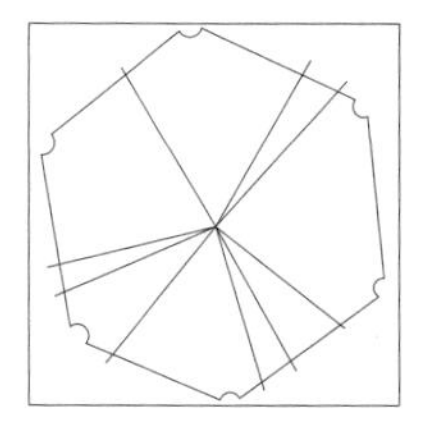

Figure 17.19

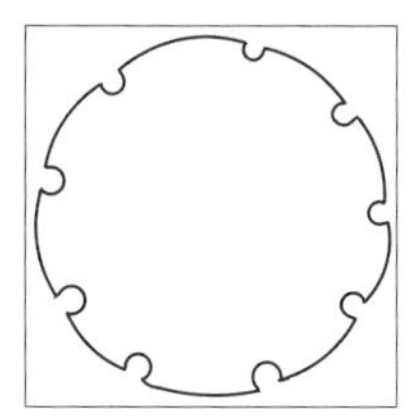

To make this process as easy as possible, we will select these blocks **in the order they appear here**. Keeping this sequence will make it easier to assign the 3-D blocks to the correct 2-D block.

`Command:`
Open the **Block Substitution** options by selecting **Blocks ▸** from the **Tools** pull-down menu.

Pick the **Select blocks graphically** option.

```
Command: EPLC_blocksub
Select one of each block to substitute.
```
Pick each of the trees shown **in the order they appear on the page**.

```
Select objects:
```
Press **<Enter>** when finished selecting the five trees. This opens the "Substitute what kind of symbol" dialogue box (Figure 17.20).

The **Original block name** is displayed under the heading. If you are in doubt which block this is, you can select **Show this block**. This will cause the original block to highlight in the drawing. After highlighting the category of plant you want to use (**Conifer plants** in this case), select **Pick new block**. This opens the "Conifer Tree Symbols" icon menu. From this menu, you can select **Detailed 3D** from the pop-down box to examine the correct 3-D tree symbols (Figure 17.21).

Select the center tree symbol in the second row. If you were to choose to leave the original symbol intact, you could select **Skip this block**.

The remainder of the trees are broadleaf plants, so all of these choices are made in the "Broadleaf Tree Symbols" icon menu. See below for the substitute blocks (Figures 17.22, 17.23, 17.24, 17.25, and 17.26).

Figure 17.20

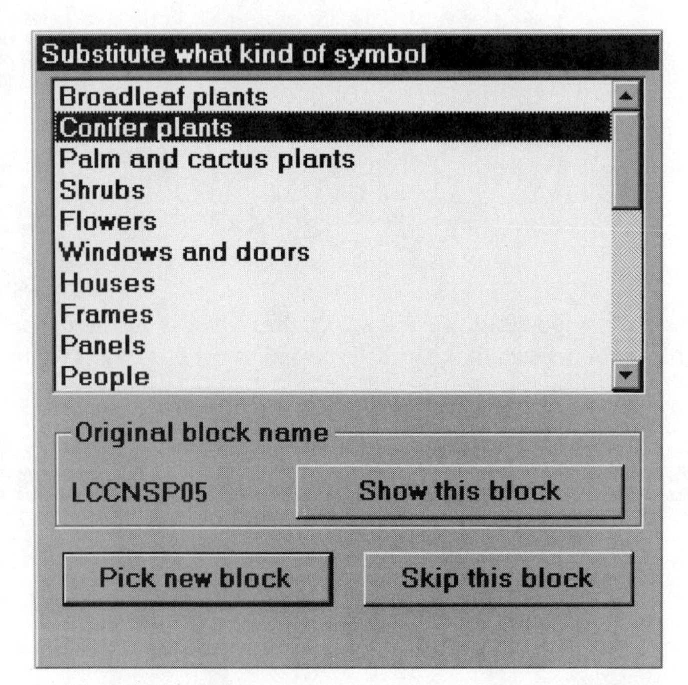

Figure 17.21

Figure 17.22

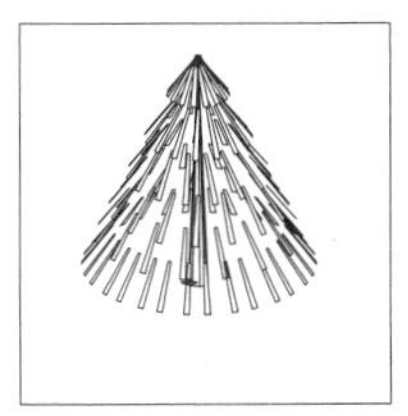

Figure 17.23

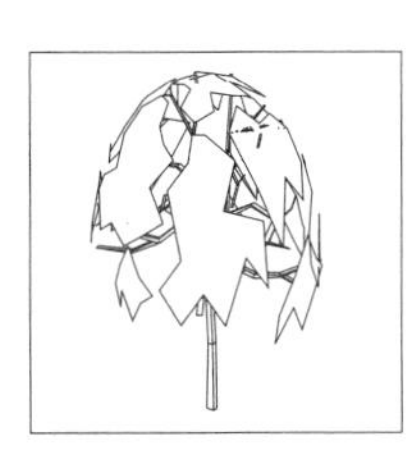

Figure 17.24

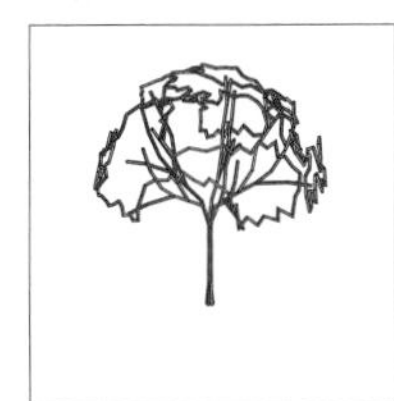

Figure 17.25

Figure 17.26

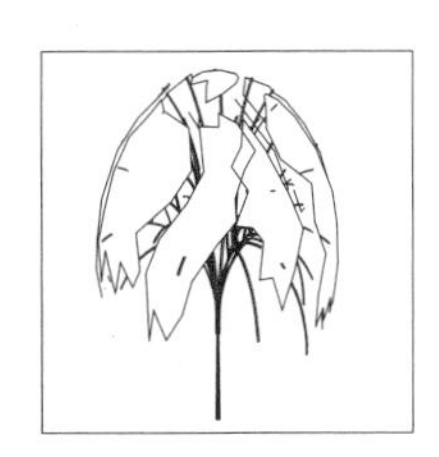

The most difficult aspect of this task lies in the fact that once you begin a substitution routine, you cannot correct mistakes or stop in the middle of the routine. If you make a mistake, you must start over. This is not a problem when very few plants are involved, but if you are trying to substitute twenty plants, it can be a source of frustration. To prevent this frustration, it is best to break up large groups of plant substitutions into smaller groups, and write a substitution file for each group. When you are finished with your substitutions, you are shown the "Output results to substitution file" dialogue box (Figure 17.27).

Here you can name the substitution file, and select the location on your hard disk. Name your substitution file **TREES.SUB**, then pick **OK**. This will save the file to your hard disk, and allow you to reverse your substitutions when you wish.

Performing a Manual Substitution

Performing a manual block substitution is a simple procedure **if you already know the names of the original and substitute blocks**. If you do not already know the names of these blocks, finding out their names can be a pain. For the original blocks, you can simply use the **LIST** command to determine the name of the block (and its attributes, if you need them). But to determine the names of the substitute blocks, you need to examine the naming system used by LANDCADD, then count the blocks in the icon menu (starting from the upper left), to arrive at the full block name. While not particularly difficult, this task can be tedious and irritating if you have many blocks to substitute. This is especially true because you cannot leave the "Type block substitutions" dialogue box to check on the names while you are typing them. When you are performing a manual block substitution, it is best to have the lists of blocks recorded ahead of time to prevent false starts and errors.

Figure 17.27

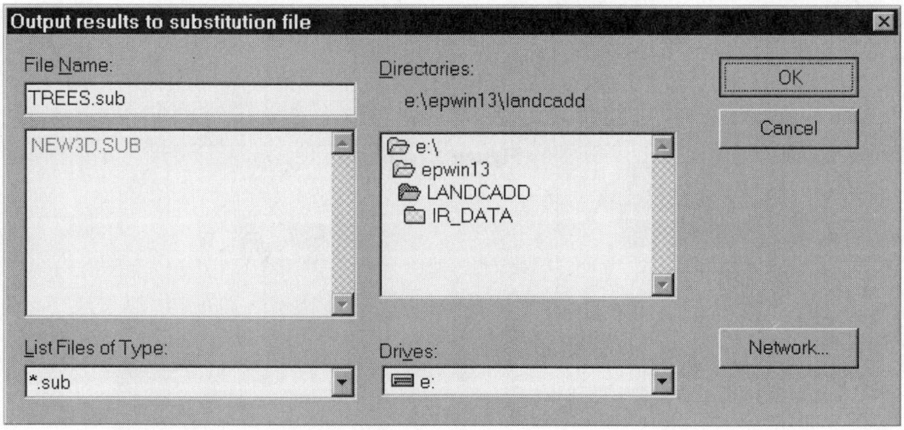

Output results to substitution file

File Name: TREES.sub

NEW3D.SUB

Directories: e:\epwin13\landcadd

e:\
epwin13
LANDCADD
IR_DATA

OK

Cancel

Network...

List Files of Type: *.sub

Drives: e:

```
Command:
```

Following the same sequence as before, select the **Block Substitution** options, then pick **Type block names manually**. This opens the "Type block substitutions" dialogue box (Figure 17.28).

Use the block names shown in the illustration to create your substitutions. After you pick **OK**, the symbols will be substituted in the drawing. As soon as this process is completed, the "Output results to substitution file" dialogue box opens, prompting you to save the list to a file. Use the file name **SHRUBS.SUB**. This allows you to keep the substitution of trees and shrubs separated into individual files.

Examining a Substitution File

As you read earlier, LANDCADD's substitution files are simple ASCII text files. If you followed the directions above, the **SHRUBS.SUB** file will be the same as that below:

SHRUBS.SUB

```
LCSHSP03
LCSH3D05

LCSHSP12
LCSH3D04
```

Figure 17.28

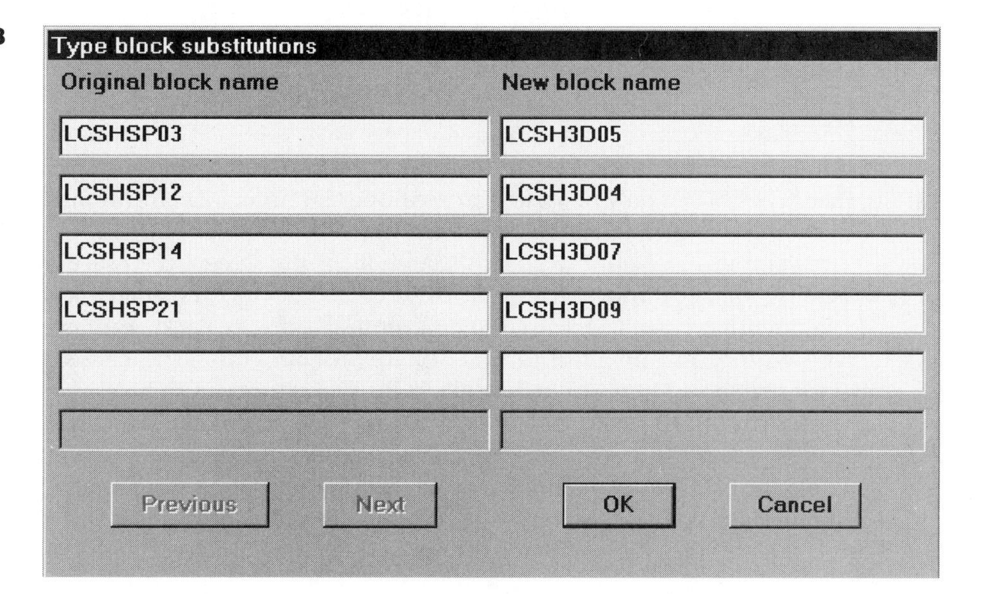

```
LCSHSP14
LCSH3D07

LCSHSP21
LCSH3D09
```

The format is extremely simple. The top name in the pair is the original block name, while the bottom name of the pair is the substitute block name. If you want to change the substitute block name, simply **reverse the substitution first**, then modify the substitution file to try a new 3-D block. You should always reverse the substitution first, so that all of the original blocks are restored. If you change the substitution file before reversing, the reversing process will not find your new (renamed) 3-D block, and the old 3-D block will be "stranded" in the drawing. Because the reversing process does not recognize the old 3-D block (you changed its name in the file) it cannot be restored to its original block. You will either have to erase the block (and reinsert it), or construct a substitution file especially for a single plant. In either case, the process wastes time.

Using the SHADE and RENDER Commands

The advanced features of today's CAD programs include the ability to create 3-D models and color them to make them look more realistic. The **SHADE** command performs this coloring task in a simple and direct way. The command simply assigns the color of the edge of an object to the entire object. If the edges of a leaf are green in a 3-D block, **SHADE** will turn the entire leaf (the plane and edge that defines it) green. AutoCAD can color the drawing with up to 256 colors. The resulting image is just that: an image. It cannot be edited or plotted. In fact, a shaded image only lasts until you change the display. As soon as you edit any object on the screen, zoom, pan or regenerate the drawing, the shaded image is lost.

There are two ways to save a shaded image. First, you can use **MSLIDE** to create a slide of the image. This slide file can be viewed any time you wish by using the **VSLIDE** command. Slides can only be viewed from within the AutoCAD environment (or from within slide library file editors), so they are not very versatile. The other option is to use a screen-capture program to take a "snapshot" of the screen, then save the file in one of the raster-image formats (BMP, TIF, GIF, PCX, etc.). Once a shaded image has been saved to a raster file, it can be edited using a "painting" software package, or even imported into a word-processing program. Screen-capture programs can be purchased for any platform (DOS, Windows 3.x and Windows 95/NT). The Windows Clipboard feature can also be used to "grab" screen areas, and copy them to other programs (Figure 17.29).

While the **SHADE** command is a useful feature to produce more realistic images, it lacks several things. First, there are very few options and controls over the image produced. You can vary the edging color, the number of colors displayed (16 or 256) and the com-

Figure 17.29

position of the light source (ambient or diffuse). There are no controls for light position, surface textures and shading, which creates a cartoonish look to the images.

If you want to exert greater control in the display of your three-dimensional models ("drawings" become "models" in 3-D), you can use the **RENDER** command. Although it is beyond the scope of this text to discuss the details of AutoCAD rendering (there are several books and articles which treat the subject in detail), we can discuss some of the general controls and concepts in rendering. We will also examine the output options for rendered images.

When you create a rendered image in AutoCAD, you can control the source and properties of the lights which shine on your model. In addition, you can control the materials which define the surfaces of your model. That is, you can assign texture, color, and reflectance characteristics to different surfaces in the image. The combination of light and materials can give a much more realistic image than is possible to create with the **SHADE** command (Figure 17.30).

The **SHADE** and **RENDER** commands give a hint as to the capabilities that 3-D models have to allow you and your clients to better visualize your projects. There are many programs available with advanced rendering capabilities to make your AutoCAD 3-D models extremely lifelike. Among those available, be sure to look at AutoVision by Autodesk and AccelVIEW by AccelGraphics for starters.

Figure 17.30

Using RENDER and Saving Rendered Views

To use the **RENDER** command, restore one of your stored views, then type **RENDER**. This opens the "Render" dialogue box (Figure 17.31).

One of the important options in this dialogue box relates to the **Destination** of the rendered image. You can choose to place the image directly on the screen (**Viewport**), into a screen window which can be saved or captured and used in another program (**Render Window**) or sent out to a file (**File**). If you select the last choice, you can pick the **More Options...** to establish the parameters of the file you wish to save. This opens the "File Output Configuration" dialogue box (Figure 17.32).

Figure 17.31

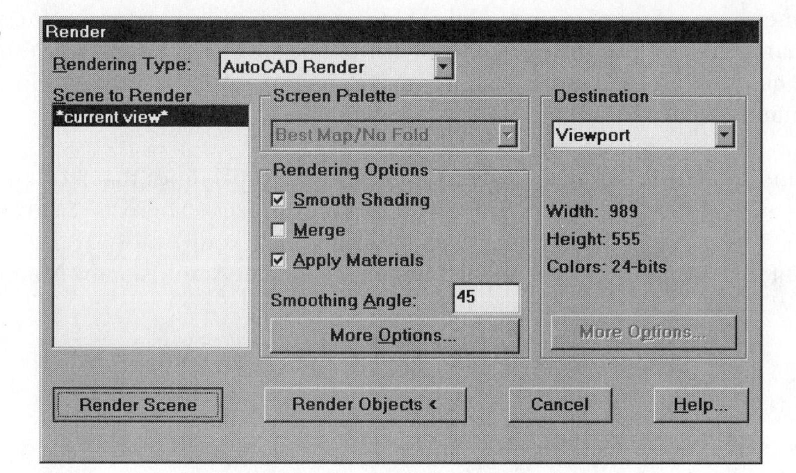

Figure 17.32

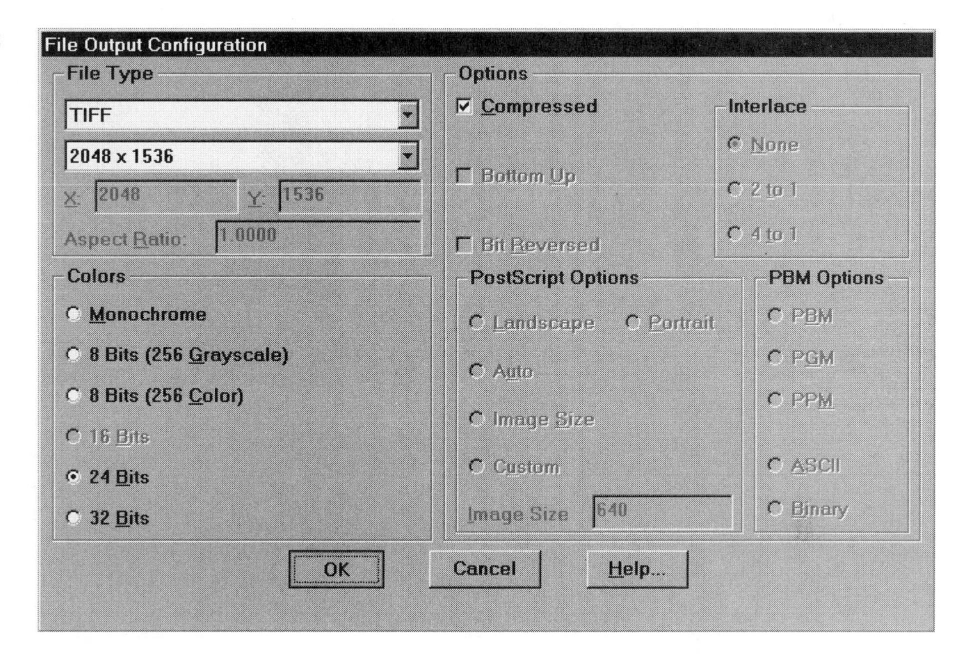

You can select from amongst the popular raster image file formats (BMP, TGA, PCX, GIF, TIF, etc.), and also select the size, resolution and number of colors in the file you want to create. The file will then be written to a name and destination of your choice (Figure 17.33).

Try the **RENDER** command on several of your views to examine the effects of this command (Figure 17.34).

To create a bitmap image that can be used by most Windows programs, set **RENDER** to plot to a file, then specify the file type **BMP**, and use the resolution **1024 x 768**. Use as many colors as your video card will support, then put the file in your drawing directory.

You can open the file in Windows Paint program, or any other paint program which supports the BMP file format (almost all do).

Figure 17.33

Figure 17.34

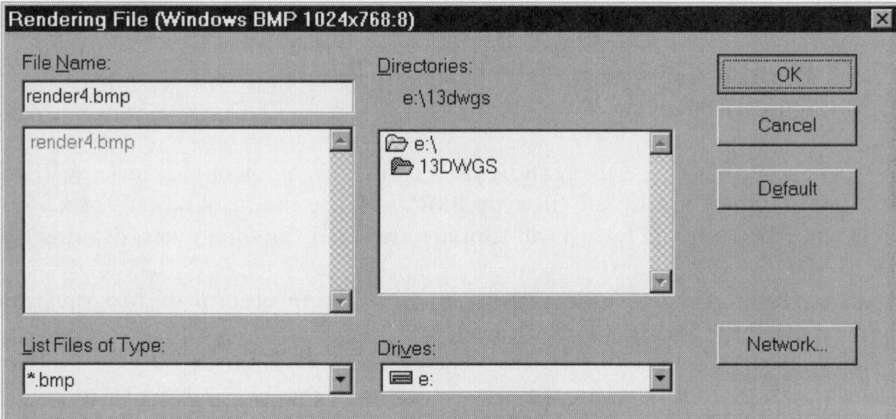

chapter

Drawing Nine—Paper Space Drawing Exercise

In this exercise, we will make use of **Paper Space** to present three different displays of the same **Model Space** drawing. When we work in Paper Space, we can create different viewing areas called **viewports**. These viewports can be set to different drawing scales, and each can have different layers displayed. This ability, along with the convenience of plotting at 1:1 scale makes Paper Space an important portion of the AutoCAD program. Although few of the commands used in this drawing are specific to LANDCADD, it is important that CAD landscape design professionals understand how to use Paper Space, and why it might make certain tasks much easier.

We will begin the exercise by using the **C18SHEET** drawing as the basis for a new title block drawing in Paper Space. Since we created this title block drawing specifically for 1/8" scale drawings using C-size paper, we will need to scale this title block down to 1:1 (or 1" = 1") scale when we insert it into our Paper Space drawing environment. Paper Space should always be used at full 1:1 scale, so that the plotting scale is always the same. We use title blocks at 1:1 scale, then create Model Space viewports within them to actually display Model Space. Remember that Model Space and Paper Space are mutually exclusive. If you are working in Model Space, you cannot edit Paper Space objects and vice versa. In fact, the best use of Paper Space is to **display** Model Space drawings, not edit them.

Typically, a title block is created in Paper Space first, then Model Space viewports are created. If you have a configuration that will be used repeatedly, save the Paper Space title block as a separate drawing. Once the viewports are created, enter Model Space, then make one of the viewports active. Then (not before!) insert a drawing into Model Space. Although the drawing may not be visible in any of the viewports, it can be made visible by using the **ZOOM/Extents** command in each viewport. The same drawing is displayed in each viewport. You cannot display different drawings in each viewport. If you want to display multiple drawings in a Paper Space title block, you will need to combine several drawings into one drawing using normal Model Space functions such as **INSERT** or **XREF**.

We can control the scale of the Model Space drawing within each viewport using a **zoom scale factor**. This factor is actually the inverse of the **scale factor** used when calculating plotting sizes, text heights and other important display calculations. The **zoom scale factor** is established after the drawing is visible in the viewport since it makes it much easier to **PAN** into the correct position. Once the **zoom scale factor** is set, do not use the **ZOOM** command again, since it will affect the scale of the drawing display in the viewport.

Once the **zoom scale factor** is established in each viewport, and the displays have been properly centered and aligned, we can then control the display of layers and linetypes in each viewport. This control makes Paper Space an especially valuable tool, since it allows you to create very detailed drawings which can be displayed selectively. When large scale displays are used to create an overall view, the detailed portions (which might be completely illegible) can be removed by freezing the detail layers. When enlargement viewports are used to display detail areas, these layers can be displayed to provide a clearer view.

After the viewport displays have been completed, we can return to Paper Space to complete the title block, label each viewport and plot the drawing. All Paper Space objects will be created at 1:1 scale, so certain modifications may be needed to make LANDCADD routines function properly.

While building the PSGOLF drawing we will complete the following:

- Control the **TILEMODE** setting to allow access to Paper Space
- Toggle between Paper Space and Model Space
- Set LIMITS, GRID and SNAP for a Paper Space title block
- Insert the C18SHEET drawing at a reduced scale to create a Paper Space title block
- Build a separate layer for Model Space viewports
- Create three Model Space viewports on the title block
- Use **XREF** to insert the *image* of the drawing **GOLF.DWG** which is included on your *Using LANDCADD* CD-ROM
- Control the display of linetypes in different viewports
- Change the display of point objects in the drawing
- Set the **zoom scale factor** and center the display in each viewport
- Control the display of layers and linetype scale in different viewports
- Complete the title block by adding polylines, inserting scale symbols and labeling text for each viewport.

Creating the Paper Space Title Block

To begin, we will open the BLANK drawing, which has the 1/8" scale, C-size limits set up. Then we will set our drawing to view Paper Space using the TILEMODE setting. The TILEMODE setting allows us to control how viewports are arranged on the screen. When TILEMODE is set to **1** (one), a single viewport fills the screen. When TILEMODE is set to **0** (zero) we can establish multiple, independent viewports each allowing us to look into Model Space. When TILEMODE is set to **0** the Paper Space icon appears in the lower left corner of the screen (Figure 18.1).

Using the Project Manager, open the BLANK drawing, which only contains a few settings for snap, grid and limits.

Next, reset the TILEMODE variable to **0**.

`Command:`
`Type` **`TILEMODE.`**
New value for TILEMODE <1>:

Type **0**.

`Regenerating drawing.`
Windows users can create the same effect by double-clicking on the box at the bottom of the screen marked **TILE**. When this box contains darkened letters, the TILEMODE setting is 1 and the box to the left will say **MODEL**. After you double-click on the **TILE** box, the letters are “grayed out,” the TILEMODE setting changes to **0**, and the box to the left says **PAPER** (Figure 18.2, and 18.3).

Figure 18.1

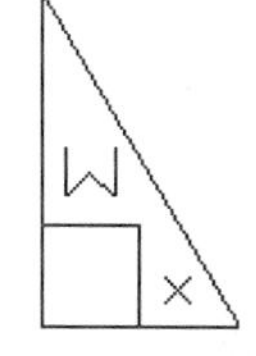

Figure 18.2

MODEL TILE 10:57 AM

Figure 18.3

PAPER TILE 10:57 AM

After resetting the TILEMODE value, notice that there is a dramatic change in the way the cursor moves on the screen (if it does at all). Turn off **SNAP**, then move your cursor around in the new drawing space. The drawing limits of your initial Paper Space area are set to 12" X 9". We will need to reset the drawing limits, snap and grid to values more appropriate for our real title block size.

From the AutoCAD and EP menu, pick **Drawing Aids...** from the **Options** pull-down menu.

Set the **Snap** to **1/8"**.

Set the **Grid** to **1"**.

We need to set the drawing limits to reflect the paper we will be using, C-size. We will allow a 1/2" margin around the paper, so we will set the limits to **25" X 19"** for our 24" x 18" title block.

```
Command:
```
Type **Limits**.

```
Reset Paper space limits:
ON/OFF/<Lower left corner> <0'-0",0'-0">:
```
Press **<Enter>** to accept this value.

```
Upper right corner <1'-0",0'-9">:
```
Type **25",19"**.

We are now ready to insert the title block created for a 1/8" C-size sheet. We can insert this drawing at a reduced scale to make it a 1:1 title block. The easiest way to do this is to use a scaling factor at insertion. Since we built the original title block for a 1/8" scale drawing, the initial scale factor we used for text, plotting, etc. was **96**. The C18SHEET title block is actually 96 times 1:1 size. Our scaling factor at insertion will need to be **1/96** to reduce this to the correct size. We will use the DDINSERT command to insert the drawing, explode the block and scale it to the correct size.

```
Command:
```
Type **DDINSERT** or pick **Insert ▶** from the **Draw** pull-down menu found in the AutoCAD and Eagle Point menu. Choose **Block...** to open the "Insert" dialogue box.

We will use this dialogue box to select the **file** to insert (**C18SHEET.DWG**), specify the insertion point (**0, 0**), the scale for the block (**1/96**), and allow the drawing to be exploded as we insert it. This is required since the drawing has a text block that requires editing after it is inserted.

After the drawing is inserted, we see the title block on our screen at **actual size**. When we plot this drawing, our plotting scale will be set to 1" = 1". Paper Space title blocks are always created at this 1:1 drawing scale.

We need to perform four functions with layers before we build viewports for our drawing. We need to turn **OFF** the layer **SHEET**, build a new layer called **VPORTS**, assign it the layer color **YELLOW**, then make it the current layer.

These tasks can be performed using the "Layer Control" dialogue box accessible from the AutoCAD **Data** pull-down menu, the LANDCADD **Tools** pull-down menu, the cursor menu, the Layer icon (for Windows users), or using the command **DDLMODES**. Another way to control layers is from the command line, using the **LAYER** command. This method is fast and sometimes very efficient. The instructions below follow the manipulation of the layers using the **LAYER** command.

`Command:`
Type **Layer** (or **LA**).

`LAYER`
`?/Make/Set/New/ON/OFF/Color/Ltype/Freeze/Thaw/LOck/Unlock:`
Type **OFF** to turn off a layer.

`Layer name(s) to turn Off:`
Type **SHEET**. This will turn off the layer and allow us to continue to manipulate layers.

`?/Make/Set/New/ON/OFF/Color/Ltype/Freeze/Thaw/LOck/Unlock:`
Type **Make** (or **M**). This will build a new layer and make it the current layer at the same time.

`New current layer <0>:`
Type **VPORTS** for the new layer name.

`?/Make/Set/New/ON/OFF/Color/Ltype/Freeze/Thaw/LOck/Unlock:`
Type **Color** (or **CO**) to set the color for the new layer.

`Color:`
Type **Yellow** or **2** to set the new color.

`Layer name(s) for color 2 (yellow) <VPORTS>:`
Press **<Enter>** to accept the assignment of the color to **Vports** layer only.

`?/Make/Set/New/ON/OFF/Color/Ltype/Freeze/Thaw/LOck/Unlock:`
Press **<Enter>** to end the command.

We now have a screen which displays the edges of the paper, and we are working on the layer which will contain the viewports we will build. This layer is important to control, since we need the viewports, but do not want their edge lines to display when we plot the drawing. We will shut off this layer when we plot. We will build the viewports next.

Building the Model Space Viewports

Building viewports is done with the **MVIEW** command. This command allows us to create viewports and control some of their features. The various options include:

On—turns on the display of the model space objects in a viewport. Each viewport display can be controlled independently of the others.

Off—turns off the display of the model space objects in a viewport.

Hideplot—causes 3-D hidden lines to be removed from the viewport when the drawing is plotted.

Fit—creates a viewport as large as the current display.

2/3/4—creates this number of viewports in different configurations. These viewports are immediately adjacent to each other with no space between them.

Restore—restores a named configuration of tiled viewports created with the **Vports** command.

We will not use any of these options for our example, but simply place three viewports on the screen for our model space drawing.

`Command:`

Type **MVIEW** or pick **Floating Viewports ▸** from AutoCAD's **View** pull-down menu, then select the option **1 Viewport**.

`ON/OFF/Hideplot/Fit/2/3/4/Restore/<First Point>:`
Type or pick the point **1/2,1/2** (or **.5,.5**)—remember that our base unit is **inches** when we work in architectural scale, so we can omit the inch marks if we choose.

`Other corner:`
Type or pick the point **21-3/8,12**.

`Regenerating drawing.`
This creates Viewport 1 in our title block (Figure 18.4).

The viewports created with the **Mview** have very interesting properties. Although they are very flexible, powerful tools for examining Model Space objects at different scales, AutoCAD treats viewports as ordinary objects. You can use ordinary editing commands to copy, move, stretch, erase, rotate, or array a viewport. This means that we can create Viewport 2, and simply copy it in a new location to create Viewport 3.

Figure 18.4

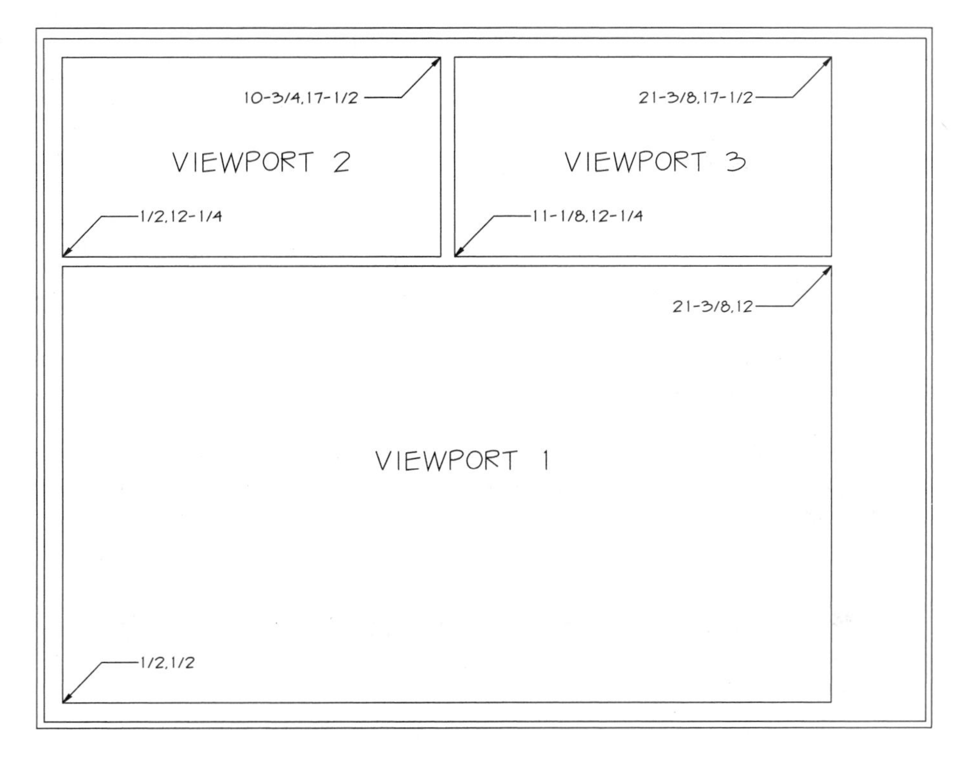

`Command:`
Type or pick **MVIEW**.

`ON/OFF/Hideplot/Fit/2/3/4/Restore/<First Point>:`
Type or pick the point **1/2,12-1/4** (or **.5,12.25**).

`Other corner:`
Type or pick the point **10-3/4,17-1/2** (or **10.75,17.5**). This creates Viewport 2. We will simply copy it to create Viewport 3.

`Command:`
Type **COPY** (or **C**).

`Select objects:`
Pick Viewport 2.

`Select objects: 1 found`
`Select objects:`
Press **<Enter>** to complete selecting objects to copy.

```
<Base point or displacement>/Multiple:
```
Pick any point on the screen.

```
Second point of displacement:
```
Type **@10-5/8<0** (or **@10.625<0**).

```
Regenerating drawing.
```
Viewport 3 is created in the correct location. Use **SAVEAS** to save this drawing as **CPS1-1** or a name to indicate a C title block in Paper Space at 1:1 scale.

We will use these three viewports to illustrate the use of some important Paper Space functions and **XREF** functions.

Now that you have some Model Space viewports on the screen, you can enter Model Space to observe some of the procedures for moving between viewports, and for moving between Model Space and Paper Space.

Maneuvering in Paper Space and Model Space

Model Space and Paper Space

In Paper Space, you will see a single Paper Space icon at the lower left corner of the screen. When in Paper Space, you can perform functions relating to the title block, the Paper Space linetype scaling and other matters relating to the plotting of your drawing. You cannot edit anything in Model Space when you are working in Paper Space.

In Model Space, you can see a UCS icon at the bottom of each viewport. The active viewport will be highlighted at the outer edge, while the cursor will be restricted to movement within the viewport. When you move the cursor outside the viewport, it becomes an arrow to use with the pull-down menus, or other functions.

1. Windows AutoCAD users can toggle back and forth between Paper Space and Model Space using the boxes at the bottom of the screen.
2. Use the keyboard macros **PS** and **MS** to move between Paper Space and Model Space
3. Under the View menu, toggle between Floating Model Space and Paper Space settings.

Viewport to Viewport

When in Model Space, move your cursor from one viewport to another, then press **<Pick>**. The border around the viewport will highlight, and the cursor will be restricted to that viewport. To move to another viewport, repeat the procedure. You cannot access information within the viewports when you are working in Paper Space.

Using XREF to Attach a Drawing

We will begin by placing an image of the GOLF drawing in our title block using the **XREF** command. Be sure that this drawing is available on your hard drive in a folder (or directory) where you can find it. If it is not located in the directory where you want it, move it **before** you perform the XREF. If you move the drawing after the XREF command is used, you may need to identify a new path to specify where the file is located.

The **XREF** command is very useful in placing the image of one drawing into another drawing. Because the actual drawing is not inserted (only its image) the result is that plotting versions of drawings can be stored separately from original drawings without taking up large amounts of disk space. Additionally, XREF drawings are updated as the original drawing is changed. This permits changing base information one time, then having the changes transmitted to several other drawings which make use of this material in the form of an XREF.

`Command:`
Using any preferred method, enter Model Space. Make Viewport 1 active.

`Command:`
Type **XREF**.

?/Bind/Detach/Path/Reload/Overlay/<Attach>:
Type **Attach** to attach a new XREF to the drawing. This will open the "Select file to attach" dialogue box. Pick the GOLF drawing from the directory where it has been stored (Figure 18.5).

Figure 18.5

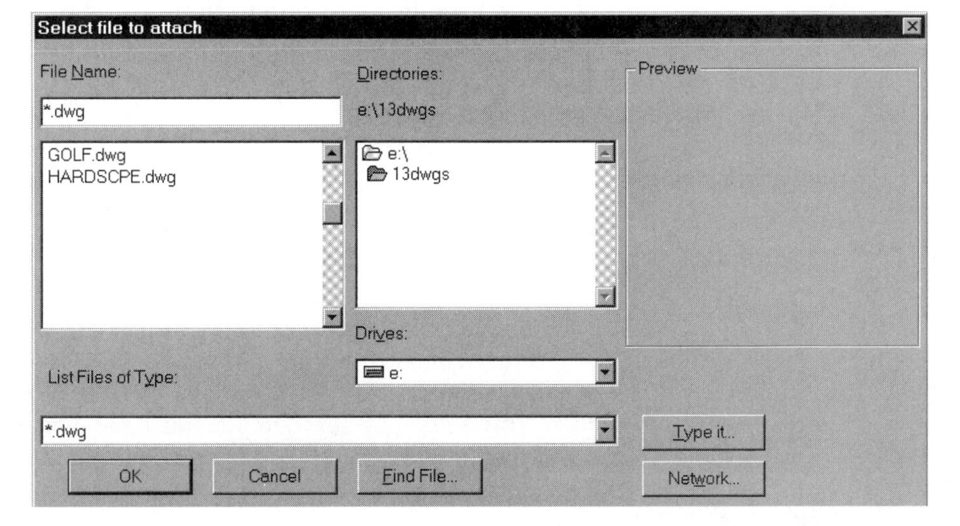

After you pick **OK**, you will be returned to the command line for instructions about the insertion point, scale and rotation angle.

`Attach Xref GOLF: GOLF.dwg`
`GOLF loaded.`
`Insertion point:`
Type **0,0**.

`X scale factor <1> / Corner / XYZ:`
Press **<Enter>** to accept this value.

`Y scale factor (default=X):`
Press **<Enter>** to accept this value.

`Rotation angle <0>:`
Press **<Enter>** to accept this value.

This completes the attachment of the XREF, but nothing is visible in any of the viewports. We will need to zoom in each of these viewports in order see the drawing at the proper scale.

Setting Paper Space Display Options

One of the important features that we can control from Paper Space is the generation of linetypes in the Model Space viewports. We can control the linetype scale with a single global adjustment in Paper Space, or we can opt to control the generation independently in each viewport. We will use the second method. To create the independent linetype scale setting, we need to change the Paper Space setting **PSLTSCALE**. If this is set to **0**, the LTSCALE setting is globally adjusted from Paper Space. If the **PSLTSCALE** setting is **1**, we can set the LTSCALE separately in each viewport.

`Command:`
Type **PSLTSCALE**.

`New value for PSLTSCALE <0>:`
Type **1**. This resets the value to the desired setting.

Another setting that will require adjustment relates to the display of **POINT** objects in the drawing. The GOLF drawing contains several spot elevations that will not be displayed correctly unless the correct AutoCAD system variable **PDMODE** is set properly. The easiest way to change this setting is through the dialogue box found in the **AutoCAD and Eagle Point** menu under the **Options** pull-down menu.

`Command:`
Pick **Display ▸** from the **Options** pull-down menu, then pick **Point Style**. This opens the "Point style" dialogue box. Be sure to highlight the style shown in the illustration. Set the **Point Size:** to **5.0** as indicated in the illustration. This will force the points to be displayed at a set proportion of the screen size regardless of the **zoom scale factor** used in the viewports. Pick **OK** to continue (Figure 18-6).

Figure 18.6

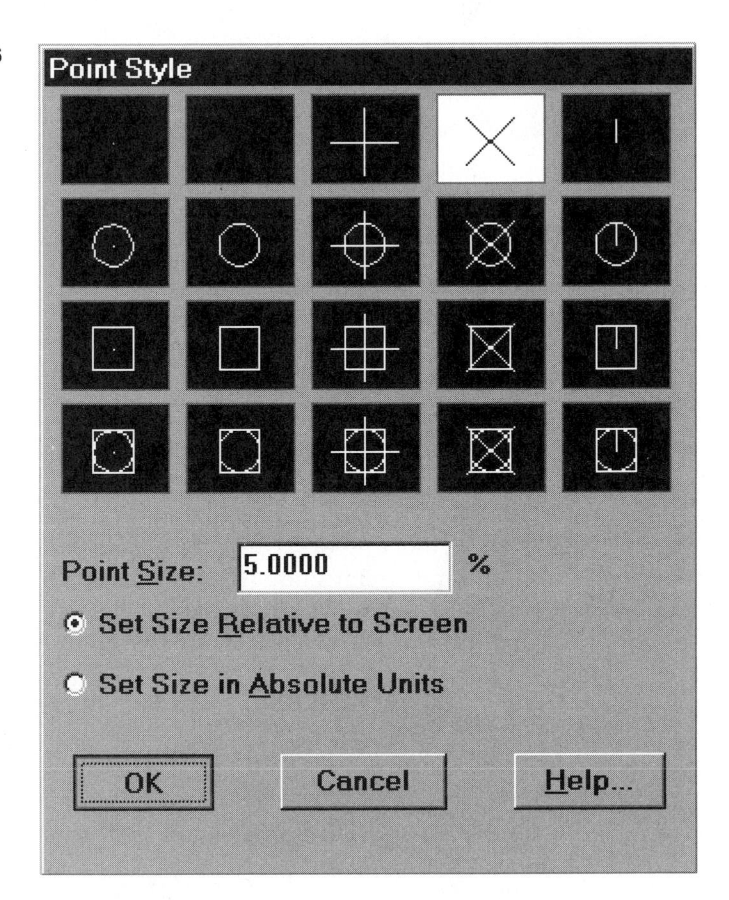

Setting Model Space Zoom Scale Factors

Now we are ready to work in each Model Space viewport. When we are working within a particular viewport, we can control the scale at which the drawing is displayed, the layers which are on or frozen, and the linetype scale for that viewport.

Setting the scale of the viewport is accomplished by the using the **Zoom** command. By controlling the **zoom scale factor**, we can create a precise scalar relationship between the

Paper Space view and the Model Space view. The **zoom scale factor** is a relationship defined using the option **XP**. This **zoom scale factor** is the reciprocal of the **drawing scale factor**. Using a 1/4" architectural drawing as an example:

Original drawing	1:1
Plotting Scale:	1/4" = 1'-0"
Equivalent Scale:	1" = 4'-0"
Equivalent Scale:	1" = 48"
Scale Ratio:	1:48
Scale Factor:	48
Zoom Scale Factor:	**1/48XP**

This method is useful whether the drawing is created with architectural or engineering units. Engineering drawings (and Engineering Units in AutoCAD) use one **foot** as a base unit where Architectural drawings (and Architectural Units in AutoCAD) use one **inch** as a base unit. This means that the engineering plotting relationship is actually 12X smaller than for architectural plotting. An example for an 30 scale engineering drawing:

Original drawing:	1:1
Plotting Scale:	1 = 30 (one plotted unit foot equals 30 plotted units)
Scale Ratio:	1:30
Scale Factor:	30
Zoom Scale Factor:	**1/30XP**

Using Paper Space is useful whether the Model Space drawing was built using Engineering Units or Architectural Units. As long as you know which unit system was used, the drawing can be correctly scaled for Paper Space plotting. The GOLF drawing is an **engineering drawing** (using Engineering Units in AutoCAD) of the 7th hole at a public golf course in Southern California. We will set our **zoom scale factor** for our viewports with this in mind.

Before we can establish the **zoom scale factor**, we need to see the drawing in the viewport. After the drawing is visible, we can zoom and pan appropriately. First, move to Viewport 1, then use **ZOOM/Extents**.

`Command:`
Move to make Viewport 1 active.

`Command:`
Type **ZOOM**.

`All/Center/Dynamic/Extents/Left/Previous/Vmax/Window/<Scale(X/X P)>:`
Type **Extents**.

`Regenerating drawing.`
The entire viewport is filled with the GOLF drawing.

Hint: If the GOLF drawing does not appear after the **ZOOM/Extents** command, you have probably attached your XREF to the Paper Space view of the drawing. If this is true, use the **XREF** command to DETACH the XREF, then enter a Model Space viewport (it does not matter which viewport). Use **XREF** again to attach the GOLF drawing, then try **Zoom/Extents** again. This should make it possible to view the entire drawing.

Now we can set the **zoom scale factor** to make the drawing appear at **40** scale:

`Command:`
Type **ZOOM**.

`All/Center/Dynamic/Extents/Left/Previous/Vmax/Window/<Scale(X/X P)>:`
Type **1/40XP**. This changes the scale of the drawing so that it is somewhat smaller in the viewport. Use the **PAN** command to center the drawing in the viewport.

Do not use the **ZOOM** command again, since it will alter the scale of the drawing display. If you make the mistake of using **ZOOM** again, you can always reset the **zoom scale factor** with the **1/40XP** factor again. Your drawing should resemble the one below (Figure 18.7).

Figure 18.7

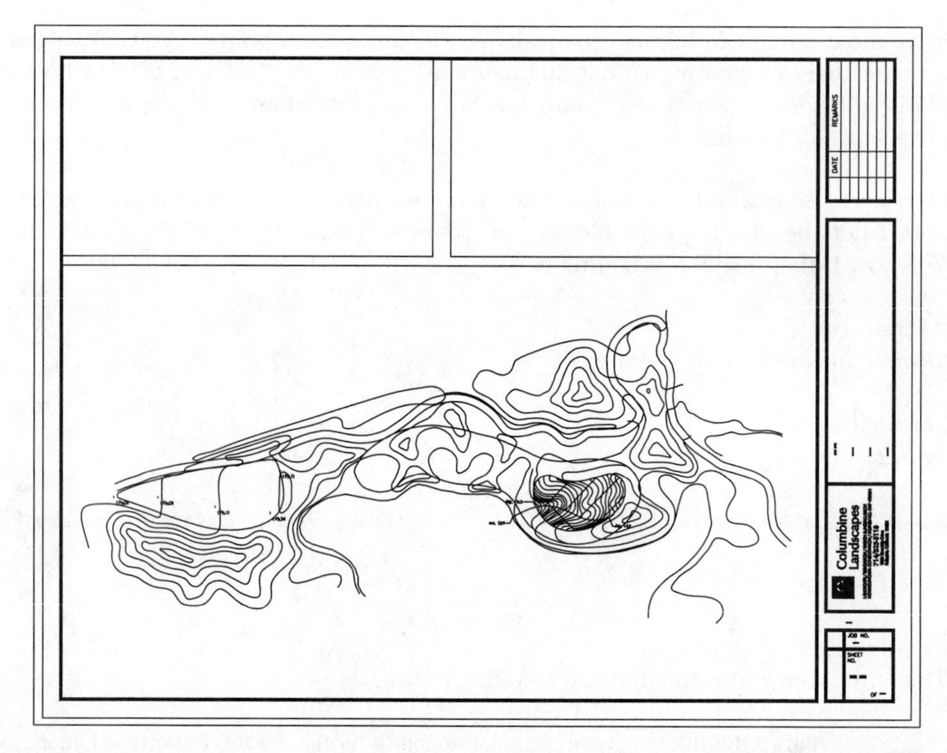

Once we have set Viewport 1 to the proper scale, move to Viewports 2 and 3, and use the **ZOOM/Extents** command to make the GOLF drawing fill each viewport. Viewports 2 and 3 are intended to show enlargements of the tee area and the green area, respectively. With increased magnification, we can see the detail contained in these areas that is not obvious in Viewport 1.

`Command:`
Move to Viewport 2, and make it the active viewport.

`Command:`
Type **ZOOM**.

`All/Center/Dynamic/Extents/Left/Previous/Vmax/Window/<Scale(X/XP)>:`
Type Extents.

`Regenerating drawing.`
The GOLF drawing fills the viewport.

`Command:`
Type **ZOOM**.

```
All/Center/Dynamic/Extents/Left/Previous/Vmax/Window/<Scale(X/X
P)>:
```

Type **1/20XP**.

```
Regenerating drawing.
```

Using the **PAN** command, pan across the drawing until the tee area fills this viewport. Be sure that the spot elevations are visible at both ends of the tee. The portion of the drawing visible in Viewport 2 is shown below (Figure 18.8).

Figure 18.8

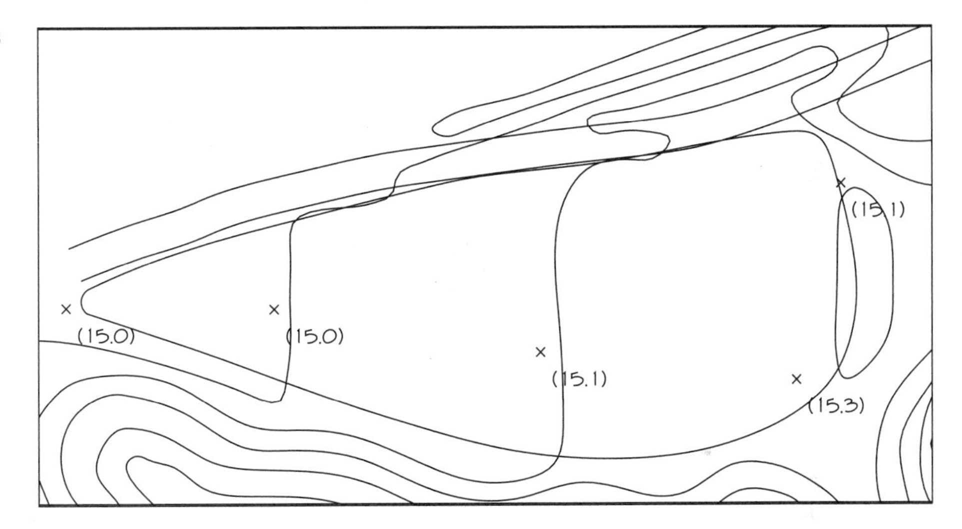

Viewport 3 will contain an enlargement of the information shown at the green. We will locate the drawing, then create the correct **zoom scale factor** to display the enlarged area at **20 scale.**

```
Command:
```

Move to Viewport 3, and make it the active viewport.

```
Command:
```

Type **ZOOM**.

```
All/Center/Dynamic/Extents/Left/Previous/Vmax/Window/<Scale(X/X
P)>:
```

Type **Extents**.

```
Regenerating drawing.
```

The GOLF drawing fills the viewport.

```
Command:
```
Type **ZOOM**.

```
All/Center/Dynamic/Extents/Left/Previous/Vmax/Window/<Scale(X/X
P)>:
```
Type **1/20XP**.

```
Regenerating drawing.
```
Use the **PAN** command to center the green area in the viewport. The drawing display should resemble that below (Figure 18.9).

Figure 18.9

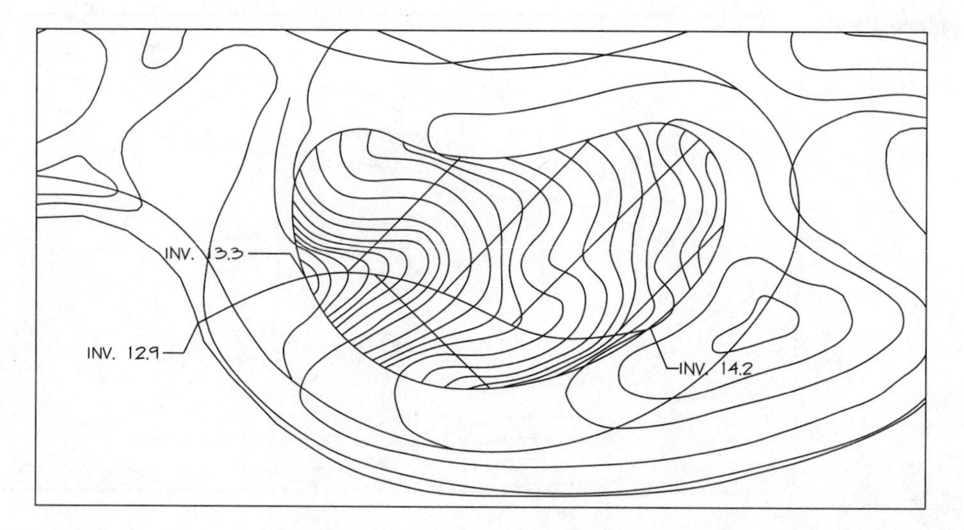

In this area, there are very fine (1") contour lines and a drainage system that are displayed. These are clear and legible at 20 scale, but are too small to be very useful at 40 scale. We will allow the layers that contain this information to display in Viewport 3, but not in Viewport 1. The **linetype scale** of this viewport should allow the drain lines to display as dashed lines. To create this effect, we need to reset the **linetype scale** to a smaller value.

```
Command:
```
Type **LTSCALE**.
LTSCALE New scale factor <1.0000>:

Type **.25**.

```
Regenerating drawing.
```
This should force the drain lines to be generated as dashed lines.

To control the display of layers in Viewport 1, we will take advantage of the independent layer control available to us when Paper Space is used.

Controlling Layers in Model Space Viewports

`Command:`
Move to Viewport 1 and make it the active viewport.

`Command:`
Using any preferred method, open the "Layer Control" dialogue box (Figure 18.10).

Figure 18.10

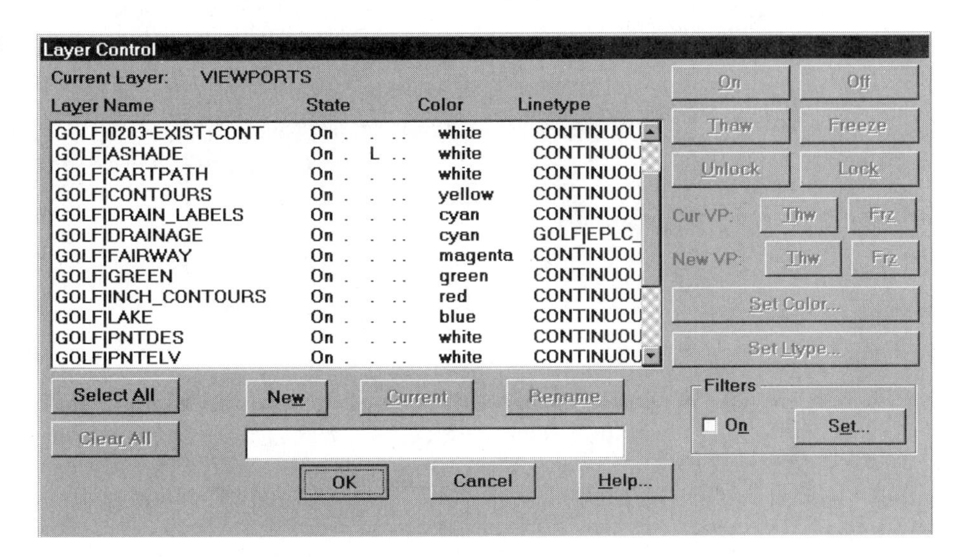

The XREF layers appear with the prefix of the XREF drawing name. The layer containing the green boundary has the name **GOLF|GREEN**. Highlight the following layers to freeze in Viewport 1.

GOLF|INCH_CONTOURS

GOLF|DRAINAGE

GOLF|DRAIN_LABELS

GOLF|PNTDES

GOLF|PNTELV

GOLF|PNTNO

After the layers are highlighted, you will notice that there are options available that were not available when we used Model Space exclusively. The **Cur VP:** area allows you to freeze and thaw layers in the current viewport. The **New VP:** controls the display of any viewports created after the settings are created here (Figure 18.11).

Figure 18.11

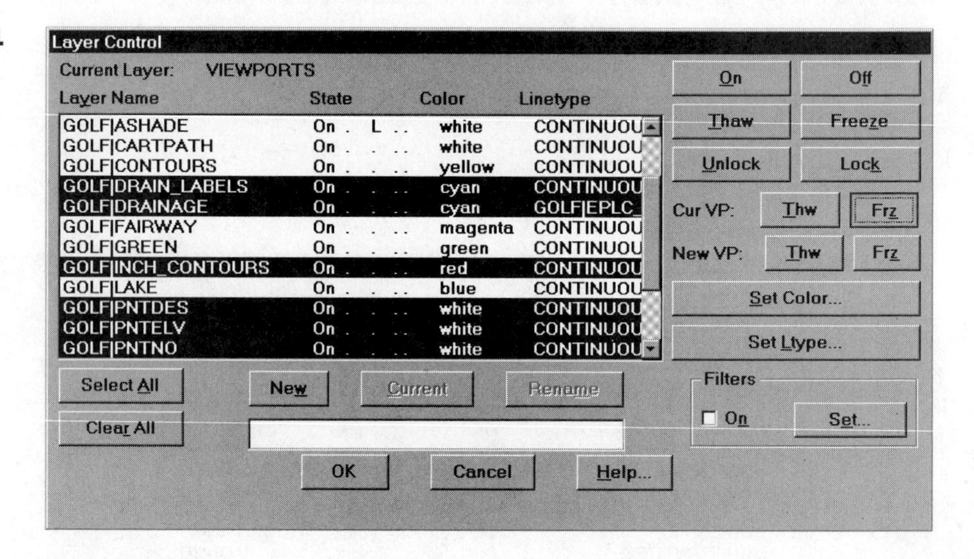

Use the **Cur VP:** option to freeze these layers in Viewport 1.Viewport 1 should resemble the following illustration (Figure 18.12).

Figure 18.12

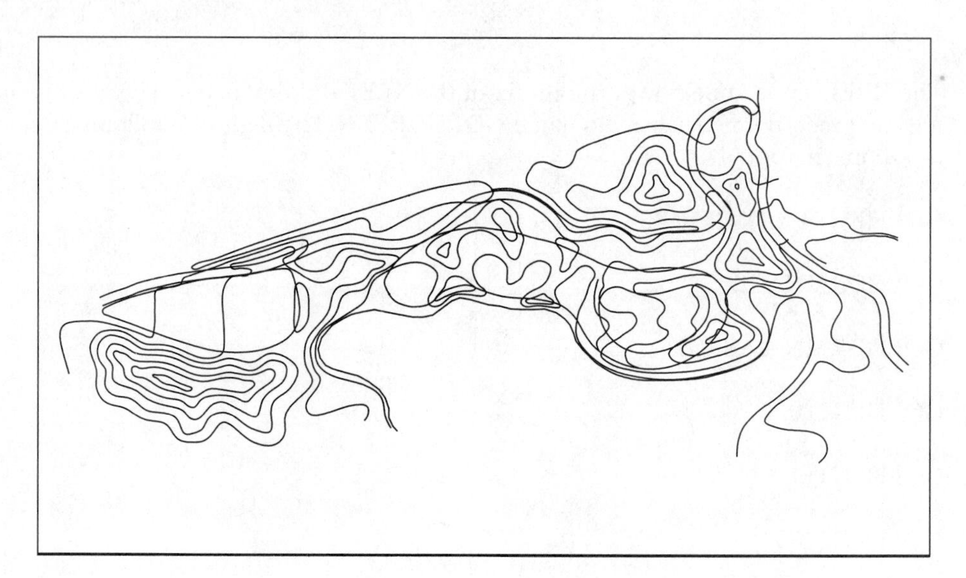

Completing the Title Block

Return to Paper Space to turn on the **SHEET** layer, then fill out the title block text with the following information (Figure 18.13).

Because the individual viewports were created to display at different drawing scales, we need to indicate the drawing scale in each viewport. We can insert scale symbols in our Paper Space drawing, using a modified version of a command found in LANDCADD's **Blocks ▸** option under the **Tools** pull-down menu. We will insert these blocks and labeling text objects on the layer **TEXT**, since we will be turning off the **VPORTS** layer when we plot.

`Command:`
Using your preferred method, open the "Layer Control" dialogue box, then turn the **VPORTS** layer **OFF**, then make **TEXT** the current layer.

`Command:`
Move to the LANDCADD menu, pick **Landscape Design** or another module that includes the **Tools** pull-down menu.

Pick **Blocks ▸**, then select **Scales....** This will open the "Symbol Insertion" icon menu (Figure 18.14).

Pick the first symbol on the left side of the icon menu.

Figure 18.13

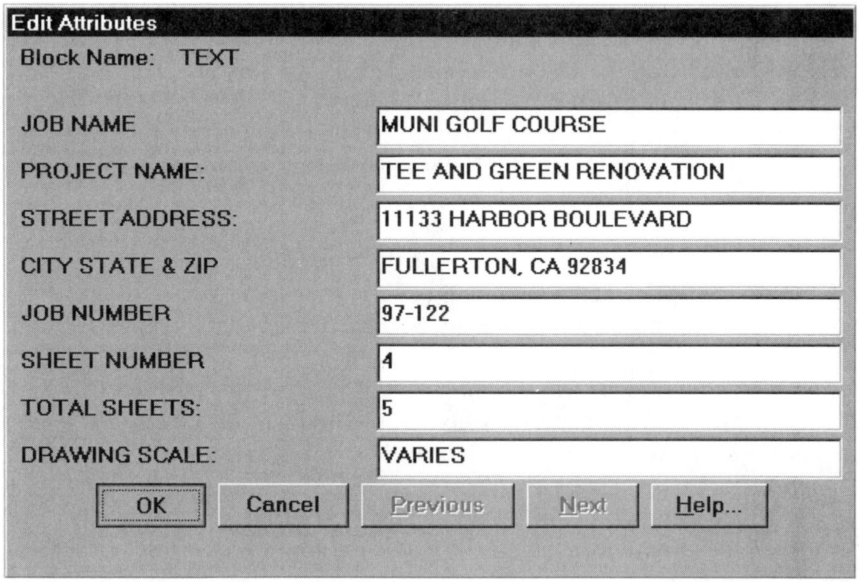

Edit Attributes

Block Name: TEXT

JOB NAME	MUNI GOLF COURSE
PROJECT NAME:	TEE AND GREEN RENOVATION
STREET ADDRESS:	11133 HARBOR BOULEVARD
CITY STATE & ZIP	FULLERTON, CA 92834
JOB NUMBER	97-122
SHEET NUMBER	4
TOTAL SHEETS:	5
DRAWING SCALE:	VARIES

OK | Cancel | Previous | Next | Help...

Figure 18.14

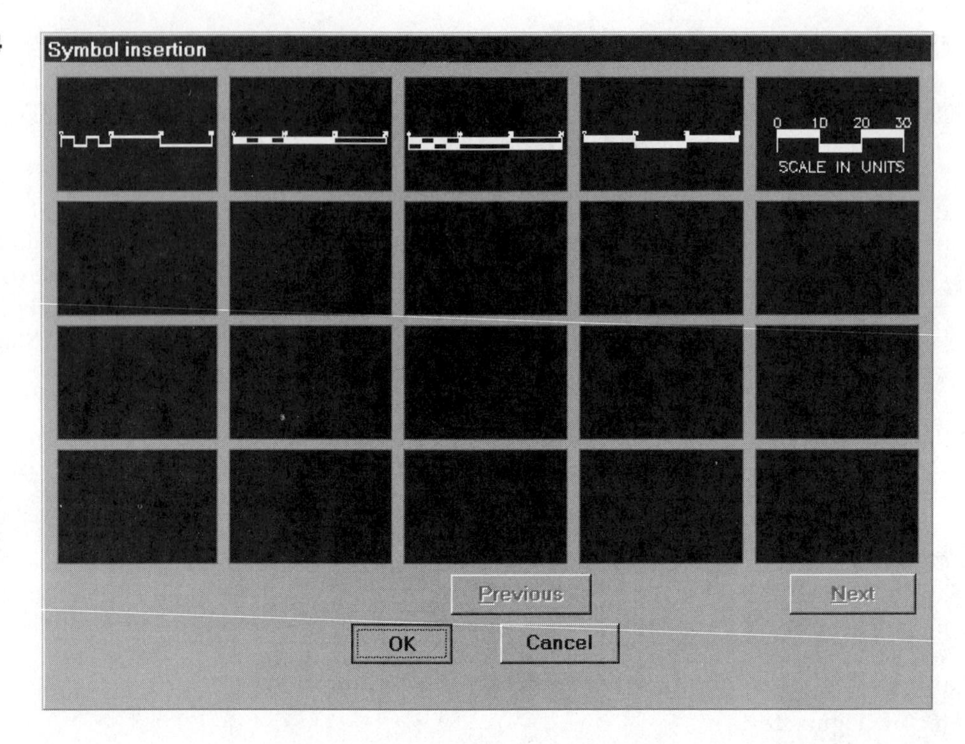


```
Command: EPLC_symbol_insert
Enter scale - 1 = :
```
Type **1**.

Since we will place this scale symbol in Paper Space, which is always used at 1:1, this is the correct scale for the symbol. After the symbol is inserted onto the Paper Space title block, we can edit the attributes of the block to reflect the correct drawing scale.

Place the scale symbol at the lower right corner of Viewport 1.

```
Command:
```
Type **DDATTEDIT** (or **AE**) to edit the attributes of this symbol. The attributes correspond to the text entry at the top of each scale inch. We should reset these to reflect the **40** scale drawing we have created. See the dialogue box below (Figure 18.15).

Repeat this process for Viewport 2, but change the attributes to **0'**, **20'**, **40'**, and **60'** respectively. Move the scale symbol to the lower right corner of Viewport 2. Make a copy of the scale symbol and place it at the lower right corner of Viewport 3.

The final step in preparing the drawing is to create labeling text for each viewport, then create dividing polylines matching our existing sheet border to separate the viewports.

Figure 18.15

Place the text on the **TEXT** layer, but create the polylines on the **SHEET** layer.

Create the text in Style 2:

`Command:`
Type **STYLE** (or **St**).

`_.STYLE Text style name (or ?) <STANDARD>:`
`Text style name (or ?) <STANDARD>:`
Type **2**.

`Existing style.`
This opens the "Select Font File" dialogue box, allowing you to redefine your font. Press **<Enter>** to accept your entry.

`Height <0'-0">:`
Press **<Enter>**.

`Width factor <1.0000>:`
Press **<Enter>**.

`Obliquing angle <0>:`
Press **<Enter>**.

Backwards? <N>
Press **<Enter>**.

Upside-down? <N>
Press **<Enter>**.

Vertical? <N>
Press **<Enter>**.

2 is now the current text style.
Command:
Type **DTEXT** (or **Dt**).

_.DTEXT Justify/Style/<Start point>:
Pick a point below Viewport 2.

Height <1'-0">:
Set the height to **1/4"**.

Rotation angle <0>:
Press **<Enter>** to accept this rotation angle.

Text:
Type **TEE DETAIL**.

Text:
Press **<Enter>** twice to leave the command.

Place the title **GREEN DETAIL** under Viewport 3 and **7th HOLE** under Viewport 1.

Make **SHEET** the current layer, then use the **PLINE** command to build the final polyline borders.

Command:
Type **PLINE** (or **PL**).

From point:
Type or pick **1/2,11-1/2**.

Current line-width is 0'-0"
Arc/Close/Halfwidth/Length/Undo/Width/<Endpoint of line>:
Type **WIDTH** (or **W**).

`Starting width <0'-0">:`
Type **1/16**.

`Ending width <0'-0 1/16">:`
Press **<Enter>** to accept this value.

`Arc/Close/Halfwidth/Length/Undo/Width/<Endpoint of line>:`
Type or pick **21-3/8,11-1/2**. This creates the polyline at the bottom of Viewports 2 and 3. Create a second polyline from the **MIDPOINT** of this polyline to the top of the sheet boundary. After the polylines divide the title block properly, you can turn off the **VIEWPORTS** layer for plotting or other editing.

Save this drawing as **PSGOLF** to make it distinct from the original **GOLF** drawing. Your drawing should resemble the one below (Figure 18.16).

Figure 18.16

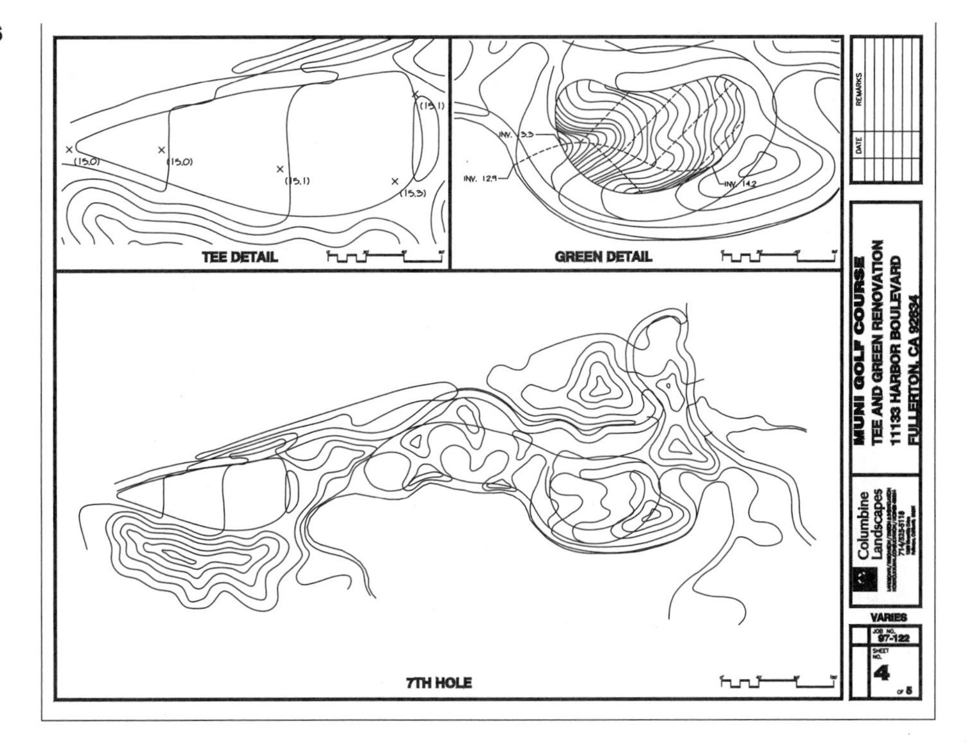

chapter

Adding Customized Irrigation Head Symbols

One of the first questions that new users of LANDCADD often ask is "How can I use MY irrigation symbols in LANDCADD?" The process for building customized symbols is not difficult, but it is exacting. Your custom drawn symbols must meet certain requirements in order for the LANDCADD Irrigation Design module to recognize and use them. They must be stored in the **BLOCKS** sub-directory (under LANDCADD's normal working directory) on the hard disk, and must adhere to certain naming and drawing standards (Figure 19.1).

Drawing and File Name Structure

LANDCADD's irrigation symbols are organized using a "family" structure. A single LANDCADD irrigation head drawing file contains symbol blocks for 45°, 90°, 120°, 135°,

Figure 19.1

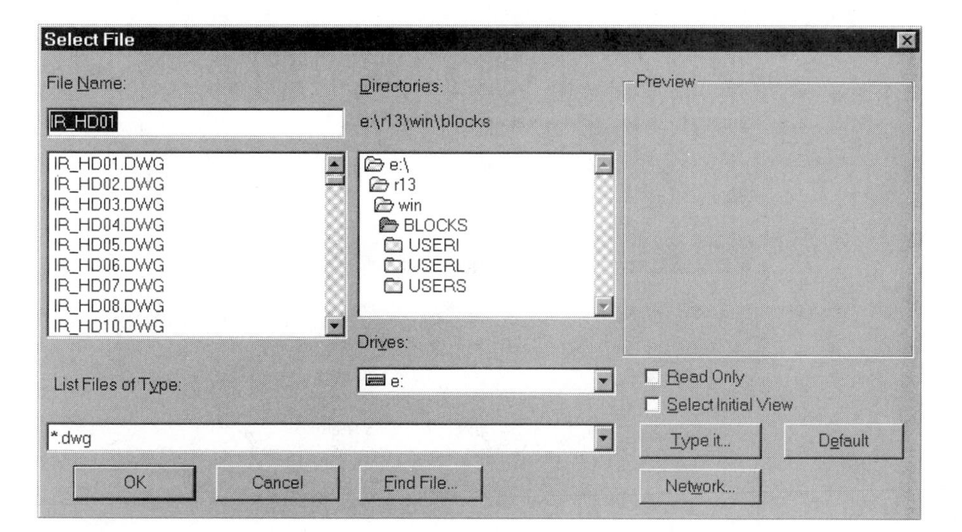

180°, 225°, 240°, 270°, 360°, EST (end strip), CST (center strip), SST (side strip), ADJ (adjustable arc) and ADM (adjustable arc with matched precipitation rate) heads. These fourteen blocks are stored invisibly within a drawing file with a file name format:

IR_HD??.DWG

where the ?? represents a two-digit number which defines the head symbol family. You can examine the files in the BLOCKS directory to see which numbers are already defined in the LANDCADD Irrigation Design module. Since these drawings each contain fourteen symbol block definitions, but no visible objects, we will call them **block definition files**.

The symbol blocks individually stored within the block definition file are named using the syntax:

EPIR_HEAD??_ □□□

where ?? represents the number of the block definition file (from above) while the □□□ represents the arc designation (Figure 19.2).

The arc designation can be any of the arc angles, or any of the three-letter abbreviations shown below. A block definition file must have a minimum of two symbol blocks (and a maximum of fourteen), to be recognized by LANDCADD.

Figure 19.2

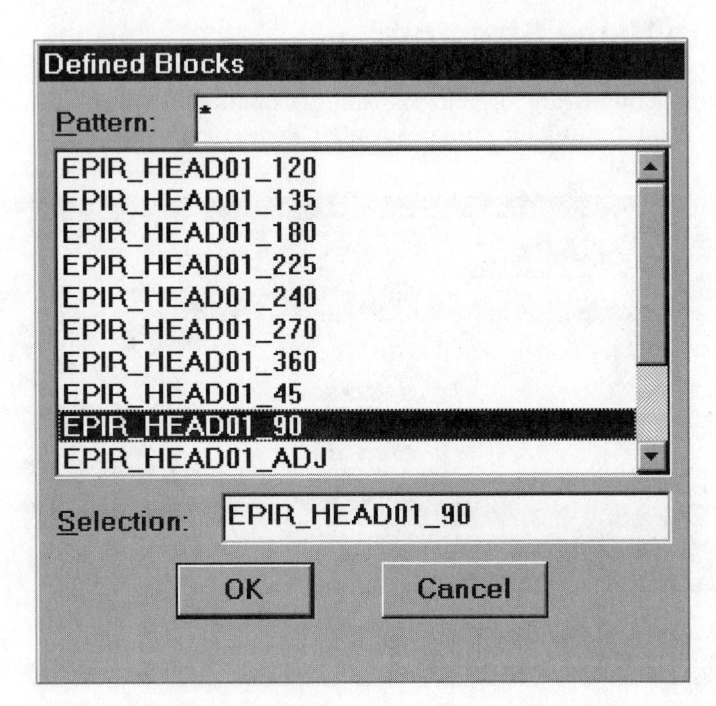

EPIR_HEAD??_120

EPIR_HEAD??_135

EPIR_HEAD??_180

EPIR_HEAD??_225

EPIR_HEAD??_240

EPIR_HEAD??_270

EPIR_HEAD??_360

EPIR_HEAD??_45

EPIR_HEAD??_90

EPIR_HEAD??_ADJ

EPIR_HEAD??_ADM

EPIR_HEAD??_CST

EPIR_HEAD??_EST

EPIR_HEAD??_SST

LANDCADD's Irrigation Design module will recognize any of these arc numbers or abbreviations when inserting symbols in a drawing. The numbers and/or abbreviations at the end of these block names are used to match up with the corresponding arcs listed in the sprinkler data file you select when you begin to insert heads. If a particular arc is not shown in the data file for that head, the symbol is ignored and will not be used (Figure 19.3).

Symbol Blocks

Each symbol block must be drawn to a specific size and must contain a single attribute to be used by LANDCADD's Irrigation Design module. All blocks must be scaled to fit within a **.5 radius** circle, and must contain an attribute:

EPIR_HDSYM_SIDES

This attribute text is typically **.2** units in height, and should be placed directly on top of the symbol. In addition, the attribute should be defined to be invisible, and possess a constant value. This value will correspond to the number of sides of the symbol. (A circle has zero sides.) Finally, the insertion point of each symbol block must be at the center of the circle which defines the outer edge of the symbol (Figure 19.4).

Symbol Slides

To be able to display the family of head symbols in the "Select Head Symbol Style" dialogue box in LANDCADD, we need to create two slide files. The first slide is a "family portrait" which displays several of the symbols to show the general appearance of the symbols. The file name of this slide is always:

IR_HD??A.SLD (Figure 19.5).

The second slide shows a single example of the head family. This is typically the 360° head symbol. This slide is displayed in the "Sprinkler Head Insertion" dialogue box, and indicates which family is active when the heads are inserted. This single head slide is always named:

IR_HD??S.SLD (Figure 19.6).

How Many Symbols?

It is important to recognize that we do not need to create all fourteen possible symbols to make a useable symbol block family. In fact, it may not be necessary or practical to do so.

Figure 19.3

Figure 19.4

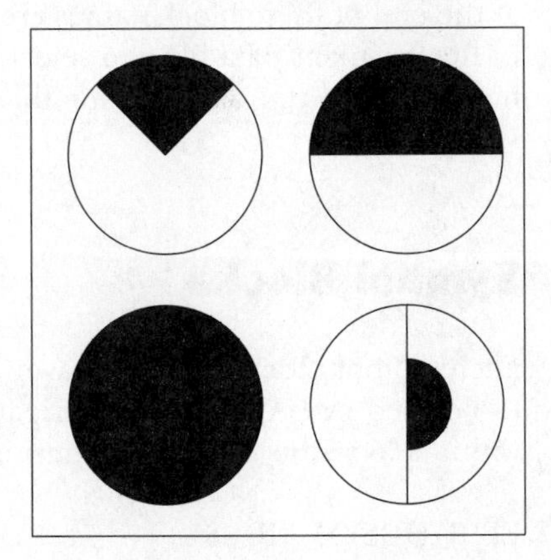

Figure 19.5

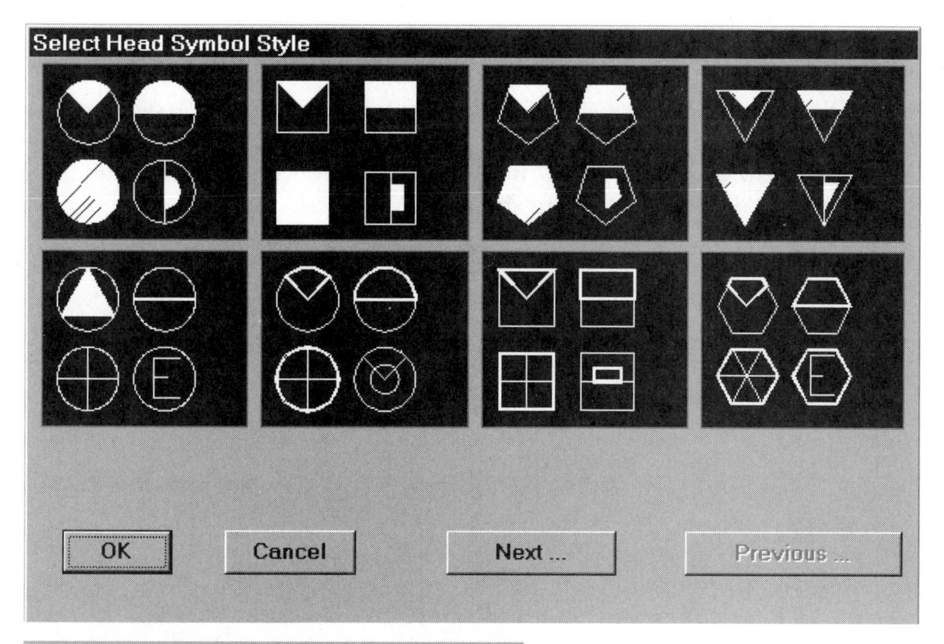

Figure 19.6

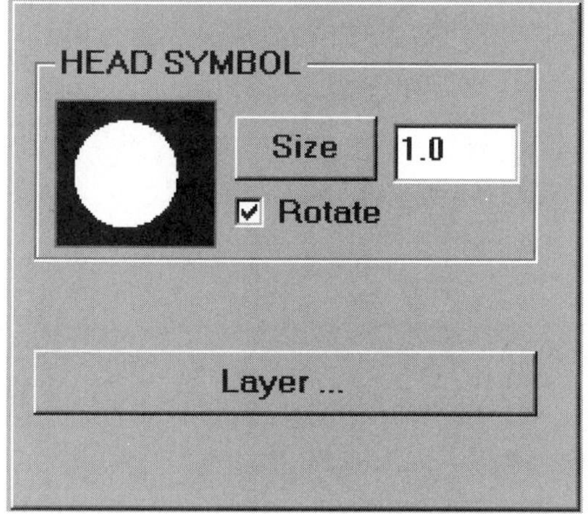

Many designers use only six or eight symbols in any one head family. In the case of rotor heads, there may be only two models available: full-circle and part-circle. A designer might define only two symbols for that family. Many designers use specific symbols to indicate pop-up spray heads, others to indicate shrub spray heads, and still others to indicate rotors, bubblers, micro-spray heads and drip emitters. If this is the case, it makes sense to create only those symbols appropriate for the type of head being used. In the case of rotor heads with only two models available, a designer might define several similar families with only two heads each to show the different nozzles available for the rotor in question. The standard practices of the irrigation designer will usually dictate the best organization of head symbols.

Drawing Symbols

We will build a set of six irrigation block symbols for the Irrigation Module of LANDCADD. Begin by opening a new project in the Project Manager. Name the drawing **IR_HD15.DWG**. Place this drawing in the **BLOCKS** sub-directory under the standard LANDCADD directory. This drawing number should not yet be defined as a drawing block. If it is, choose another number (Figure 19.7).

Be sure that the **No Prototype** toggle box is checked, and that the **System Settings:** pop-down box is set to **Version 12.0 Defaults**. This will place the text at the correct size and scale. The drawing itself will use decimal values for drawing units, although it is not necessary to switch to engineering units while working on this drawing.

Once the drawing has opened, draw a circle with a radius of **.5**.

At the prompt:

`Command:`
Type **0** (or **CIRCLE**).

`CIRCLE 3P/2P/TTR/<Center point>:`
Pick a point toward the center of the screen.

`Diameter/<Radius>: .5`
The circle is placed in the drawing (Figure 19.8).

Figure 19.7

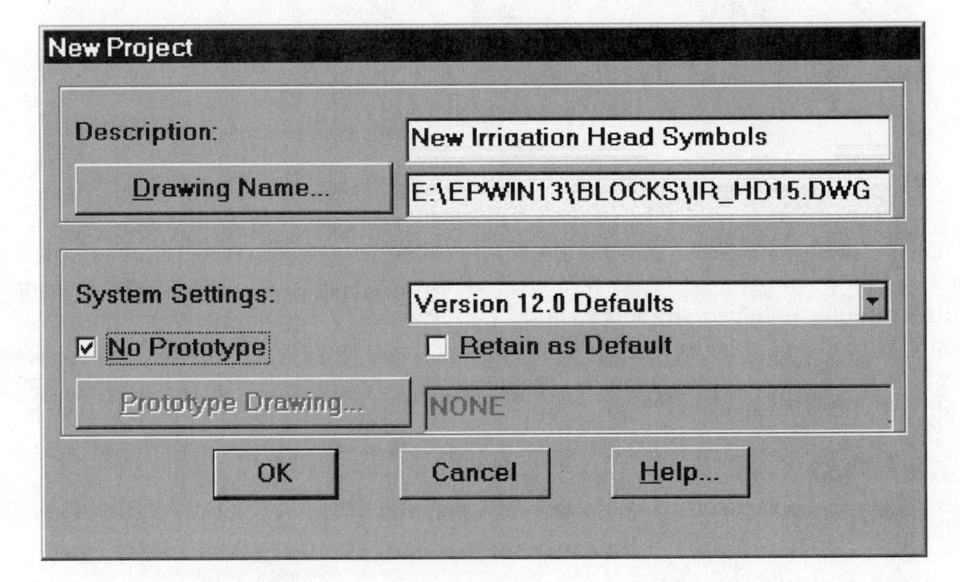

After the circle is placed, we will place some lines within it to define the head symbols. We will begin by inserting a triangle within the circle.

At the prompt:

`Command:`
Type **polygon**.

`Number of sides <4>:`
Type **3** to define the triangle.

`Edge/<Center of polygon>:`
Use the cursor menu to pick the center of the circle.

`Inscribed in circle/Circumscribed about circle (I/C) <I>:`
Press **<Enter>** to accept the placement of the triangle inside the circle.

`Radius of circle:`
Type **.5**.

This places the triangle inside the circle, but rotated upside-down (Figure 19.9.)

Rotate the triangle so that the flat side appears at the top.

`Command:`
Type **RO** (or **ROTATE**).

```
ROTATE
Select objects:
```
Pick the triangle.

```
Select objects: 1 found
Select objects:
```
Press **<Enter>** to finish selecting objects.

`Base point:`
Use the cursor menu to pick the center of the circle.

`<Rotation angle>/Reference:`
At this point, the triangle highlights and rotates as you move your cursor. With **ORTHO** active, rotate the triangle so that the flat side is located on top (Figure 19.10).

Explode the triangle so that we can edit the sides of the triangle independently of each other.

Figure 19.8

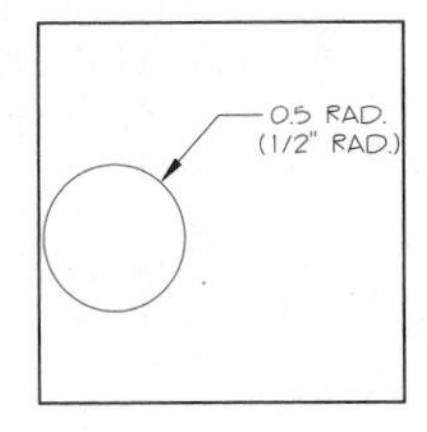

Figure 19.9

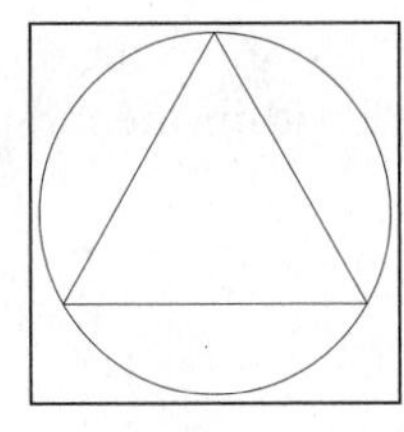

Figure 19.10

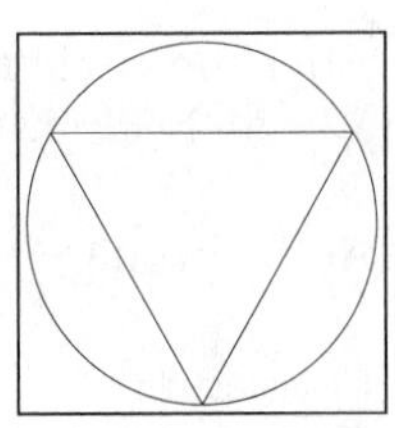

```
Command:
```
Type **EX** (or **EXPLODE**).

```
EXPLODE
Select objects:
```
Pick the triangle.

```
Select objects: 1 found
Select objects:
```
Press **<Enter>** to finish selecting objects to explode. The triangle is then ready to edit as independent lines.

The next step in the symbol creation process is to insert the properly defined attribute into the drawing.

Change menus to use the **AutoCAD and EP** commands. Pull down the **Construct** menu item, then select **Attribute...**. Use the "Attribute Definition" dialogue box to create the attribute shown (Figure 19.11).

Be **sure** to mark the **Invisible** and **Constant** toggle boxes. Type the tag name **EPIR_HEADSYM_SIDES** in the **Tag** text box. Use a standard **Value** of **3**. This constant value informs LANDCADD that the symbols are based on a triangular (3-sided) form. Accept the text size defaults.

For the text **Insertion Point**, choose the **Pick Point <** option, then insert the attribute over the top of the symbol (Figure 19.12.)

This drawing will be the basis for all of the symbol blocks stored in the **IR_HD15** drawing. We will simply use the **BLOCK** command to store six different versions of this basic symbol.

At the prompt:

```
Command:
```
Type **B** (or **BLOCK**).

Figure 19.11

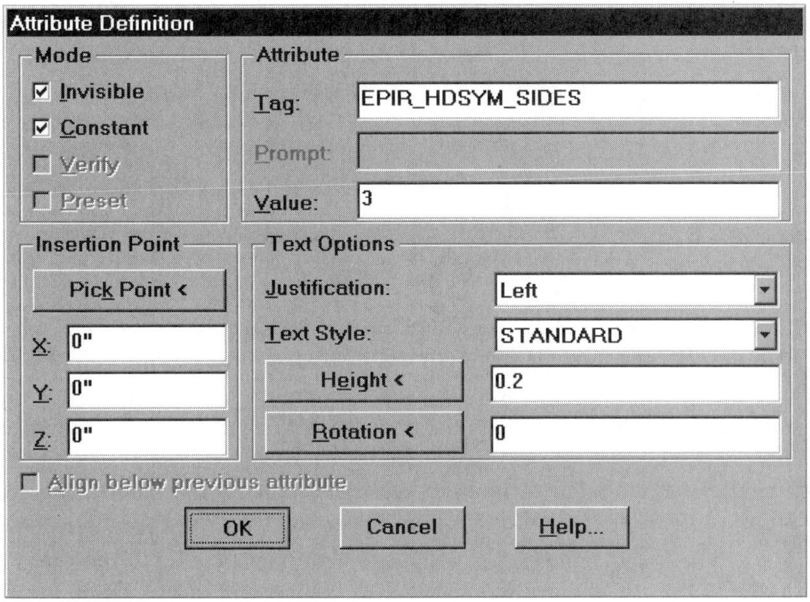

```
BLOCK
Block name (or ?):
```
Type the block name **EPIR_HEAD15_360**.

```
Insertion base point:
```
Use the cursor menu to pick the **center** of the circle around the triangle.

```
Select objects:
```
Pick all of the sides of the triangle, **and** the attribute text. Do **not** pick the circle itself.

```
Select objects: 4 found
Select objects:
```
Press **<Enter>** to finish selecting objects for the block. The objects you chose will disappear from the screen. You have just completed your first block.

Type **OOPS**. This returns the erased objects to the screen. As a result of the **OOPS** command, you can easily build the next block without having to redraw the objects or the attribute.

Pick the objects shown below for the symbol **EPIR_HEAD15_180** (Figure 19.13).

Figure 19.12

Figure 19.13

Examine the illustrations below to see how to build each of the six symbols. Be sure to leave the circle on the drawing so that you can use its center for the base point of each of the symbols (Figures 19.14, 19.15, 19.16, 19.17, 19.18, and 19.19).

After the last symbol is blocked, erase the circle and save the drawing **IR_HD15.DWG**. This completes the actual drawing process. To check your work, perform this test:

`Command:`

Type **DDINSERT**, then pick **Block...**. This opens the "Define Blocks" dialogue box. You should see these symbol blocks defined in your drawing (Figure 19.20).

If all of these symbols are listed, you have stored the blocks correctly. Try inserting each of them to see if the blocks look like the examples. Be sure to erase them all before proceeding to make the slides.

Figure 19.14

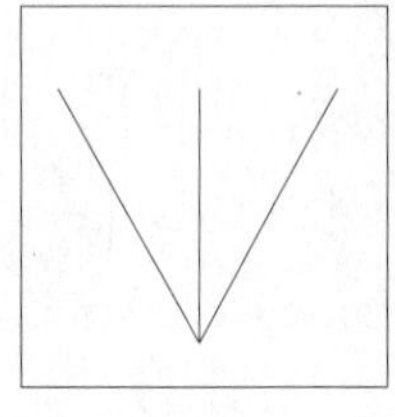

Figure 19.15

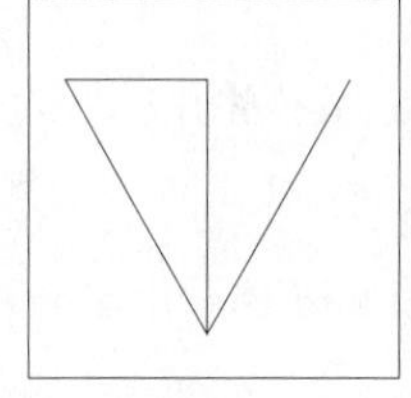

Figure 19.16

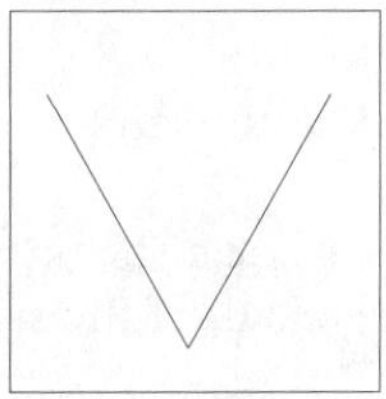

Figure 19.17

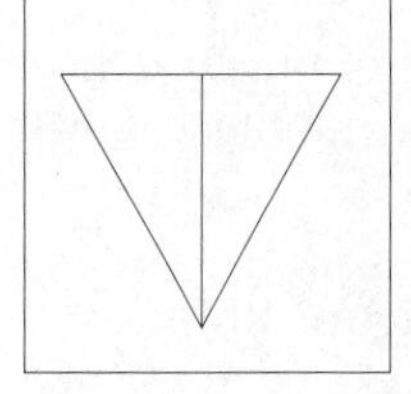

Figure 19.18

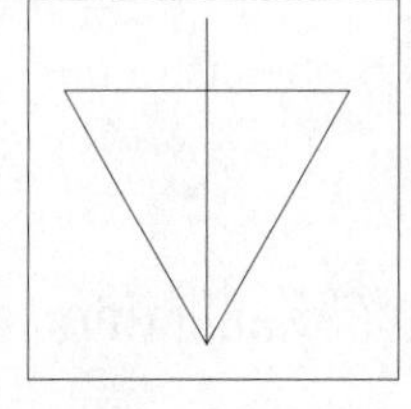

Figure 19.19

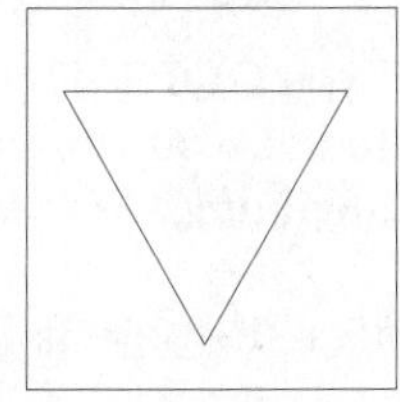

Figure 19.20

Defined Blocks

Pattern: *

EPIR_HEAD15_120
EPIR_HEAD15_180
EPIR_HEAD15_240
EPIR_HEAD15_270
EPIR_HEAD15_360
EPIR_HEAD15_90

Selection:

OK Cancel

Creating the Slides

We will use the **MSLIDE** command to create the slides that LANDCADD uses to display the head symbol choices in the "Select Symbol Style" and "Sprinkler Head Insertion" dialogue boxes. The slide files are easy to compose and create, but may require some trial-and-error work before you are satisfied with their appearance in the dialogue box.

Begin by inserting the symbols for a 90°, 180°, 270° and 360° head in a square pattern (Figure 19.21).

Use **ZOOM - EXTENTS** to make the symbols fill the screen.

Hint: Windows users (Windows 3.x, Windows 95 and Windows NT) may want to turn off scrollbars, toolbars and minimize the text window to maximize the drawing screen while making slides. Even so, your slide will usually be skewed to the left side of the screen, with the objects usually in the lower left part of the slide. Experiment with the **MSLIDE** command until you are satisfied with the slide you have created.

Center the symbols across the screen, then make the slide:

`Command:`
Type **MSLIDE**. This opens the "Create Slide File" dialogue box (Figure 19.22).

Figure 19.21

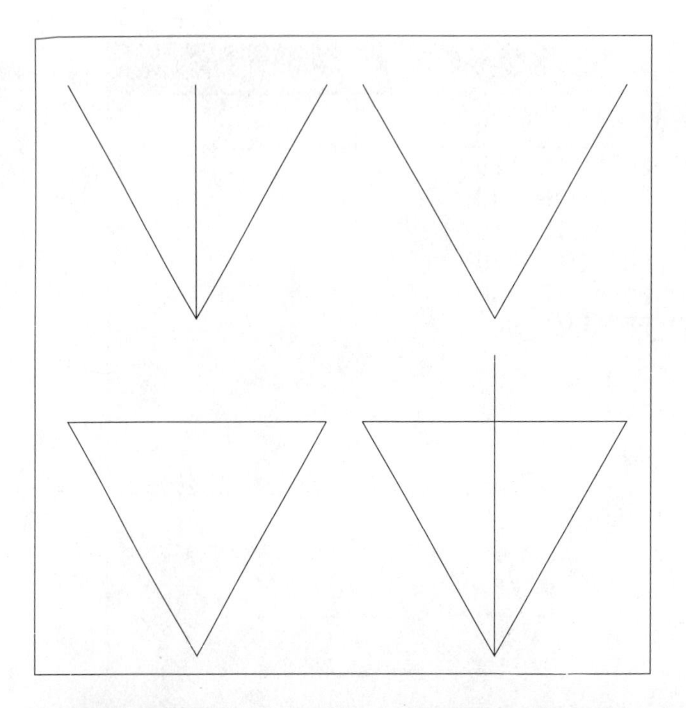

Figure 19.22

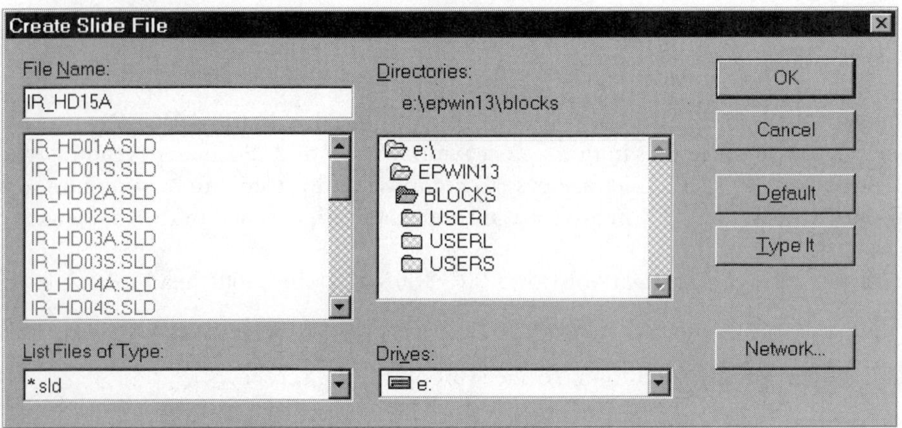

Name the slide **IR_HD15A**. This is the "family portrait" which is displayed in the "Select Head Symbol Style" dialogue box.

Next, erase all but the 360° symbol. Use **ZOOM-EXTENTS** to zoom in on the symbol, then center it on the screen. Use **MSLIDE** to create another slide. Name it **IR_HD15S** (Figure 19.23).

Figure 19.23

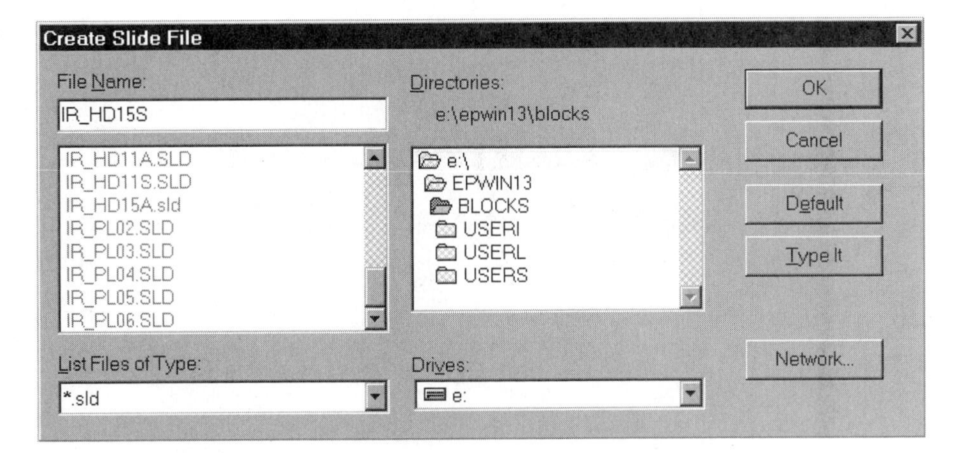

This is the single slide that is displayed in the "Sprinkler Head Insertion" dialogue box to indicate which symbol set is active.

Hint: Both the "**S**" and "**A**" slides must be complete before the symbol set will appear in the dialogue boxes.

To check your work, open the **Heads** menu in the Irrigation Design module of LANDCADD. Select the **Head Insertion...** option. To check the appearance of the slide, pick the **Next...** box. Your slide should appear next to the last LANDCADD symbol slide (Figure 19.24).

Pick your symbol set, then examine the "Sprinkler Head Insertion" dialogue box to check the appearance of the other slide (Figure 19.25).

If you are not satisfied with the appearance of either slide, do not hesitate to create new ones. Slides are easy to create, but cannot be edited, so you may need several attempts to get the look you want.

Hint: Leave the heads white, or choose another color (besides red) for the symbols on the slide. This will help you distinguish between your symbols and LANDCADD symbols.

Figure 19.24

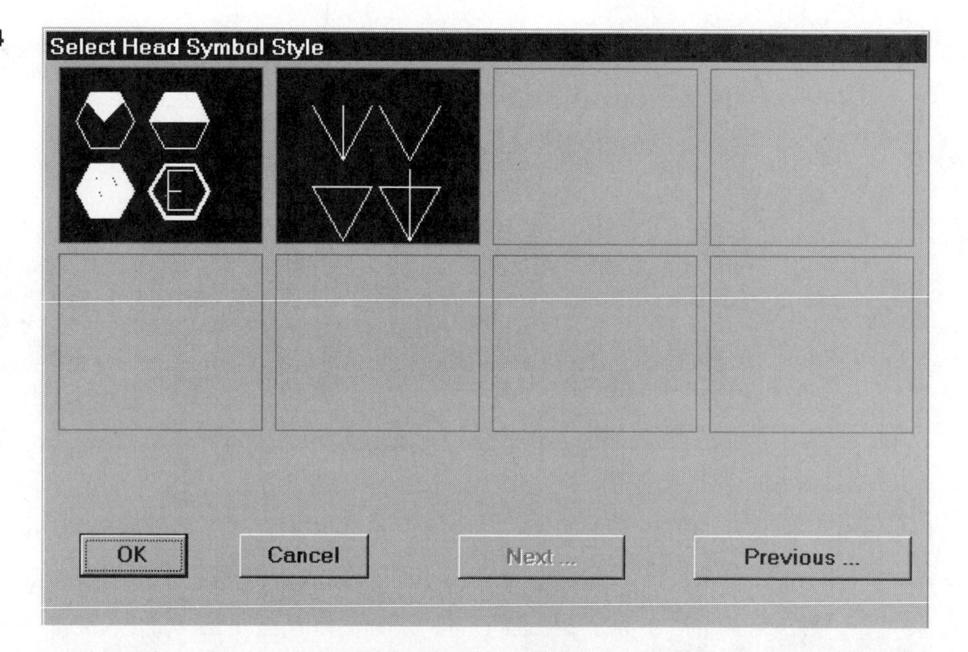

Figure 19.25

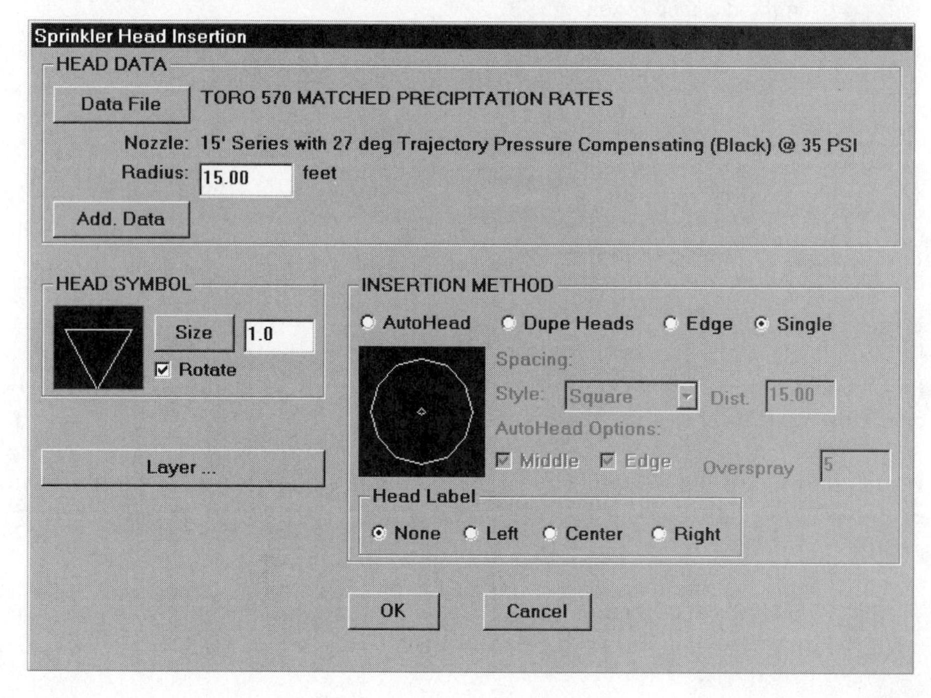

Your symbols are now ready to use in the LANDCADD Irrigation Design module. They will be inserted as expected for any sprinklers chosen except that there will be no ADJ, ADM, SST, EST and CST nozzles available. In addition, the LANDCADD piping routines will have slight difficulty trimming pipe lines at the edges of those symbols which do not have a solid horizontal line (or completed circle) across the top. See below for an example (Figure 19.26).

Some quick editing will correct these errors, but LANDCADD recommends using polygons with a horizontal line at the top or side of the symbol.

Figure 19.26

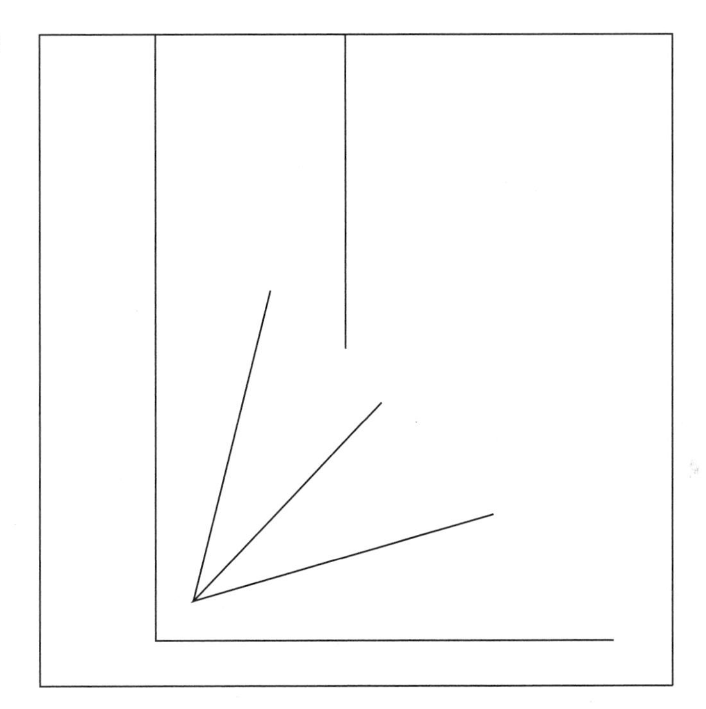

chapter

The Relationship Between AutoCAD and LANDCADD

AutoCAD is a very complex, complete desktop Computer Assisted Drafting program. It has an extensive collection of *generic* tools to perform a huge number of design and drafting tasks. AutoCAD was written to be useful in any type of design and drafting work, but was not written to serve the particular needs of any specific discipline. The success of AutoCAD lies in the fact that it is designed with "open architecture." The way that AutoCAD program code is written has invited a very diverse group of third party developers to create "application programs" to exploit the features of AutoCAD with commands and routines that reflect specific design disciplines. LANDCADD uses AutoCAD as a platform to run customized commands which parallel the kind of work that a landscape design professional would be doing on his/her projects. By taking advantage of the powerful time-saving and coordination benefits of CAD drafting and the customization that LANDCADD provides, a landscape design professional can be more productive and offer more and better services to clients.

AutoCAD is a program with enormous strengths and power, but also one with glaring weaknesses for landscape designers and land planners. AutoCAD gives us generalized commands and menus that relate to manipulating objects based on lines, shapes, volumes, sizes and forms, not specific commands related to plant symbols, irrigation coverage or parking lot layout. Although AutoCAD gives us block and attribute functions that make blocks "intelligent", AutoCAD also requires us to define and attach attributes to every block we create and use. An unpleasant side effect of the great power of AutoCAD is the nested menus where access to a command may require three to four steps into a sub-sub-sub-submenu. Memorizing the steps required to find a particular function is time-consuming and tedious. In short, AutoCAD gives us extremely powerful drafting tools but without any customized features which make landscape-oriented drawing easier and faster.

LANDCADD improves on AutoCAD's power because it was written with landscape architectural work in mind. Specific features in LANDCADD include:

Customized blocks and hatch patterns:

The LANDCADD program includes over 1000 drawing blocks already prepared; everything from sprinkler symbols to semi-trucks. There are more than 65 hatch patterns included specifically for use in landscape architecture and design work

Drawing tools:

LANDCADD provides the ability to manipulate objects, layers, attributes and blocks in different (and sometimes more convenient) ways than AutoCAD.

Modular functions:

The LANDCADD program allows access to specific tools during specific phases of drawing—planting tools are not needed when laying out a site or developing the irrigation plan. This way the designer has a more manageable number of options at any one time during drawing work. As a result of the modular approach, commands are not nested as deeply, allowing for more immediate access to them.

Better interface:

LANDCADD includes many dialogue boxes and icon menus to make it easier to make visual choices when performing different design functions. LANDCADD command routines often combine functions from several different AutoCAD commands into a logical (and often invisible) sequence for landscape design professionals. Included short-cuts provide methods for doing common operations in one or two steps that may require six or eight steps in AutoCAD. LANDCADD's creators have tried to provide intuitive sequences that allow the user to select from the right options at the right times when performing common landscape-related design tasks.

Attribute manipulation:

The LANDCADD program furnishes us with routines which automatically assign names, models, numbers and other values to blocks as they are inserted into the drawing. We have access to command routines which can gather attributes assigned to blocks and use them to provide quantity information for decision-making and bill of materials functions.

Customized Databases:

Although LANDCADD acts to "customize" AutoCAD for landscape design professionals, LANDCADD can be customized itself. The database containing plant names and usage information can be appended and edited, and new plant species can be added. Sprinkler manufacturer and model information is contained in a file which can also be easily edited. New models and brands of sprinklers can be added at any time. The irrigation pipe specifications and flow characteristics are include in another file which can also be edited by the user. When using the estimating functions in LANDCADD, the total cost of the project and the format of the estimate report depends on customized cost, category and comment databases prepared by the user.

LANDCADD is a valuable tool for the landscape design professional when used "out of the box." But LANDCADD becomes even more powerful when it is customized to suit the needs of your specific office practices and standards.

The LANDCADD program code includes three specific kinds of program files which drive the landscape oriented functions. These program file types include:

ARX—AutoCAD Runtime Extension computer programming interface. This programming environment can communicate directly with AutoCAD's program core and executes functions very quickly as a result. ARX programs are used in LANDCADD Release 13 when loaded with AutoCAD Release 13. LANDCADD ARX programs are not used with AutoCAD 12.

ADS—AutoCAD Development System. This is a C language programming connection to AutoCAD's program core. ADS programs are used in LANDCADD Release 13 when loaded with AutoCAD Release 12. LANDCADD ADS programs are not used with AutoCAD 13.

AutoLISP—smaller programs built in AutoCAD's native programming language. AutoLISP is used to build tools to perform AutoCAD tasks automatically. AutoLISP programming is often used by more sophisticated AutoCAD users to build customized routines and programs.

Loading AutoCAD and LANDCADD

When you pick the Eagle Point icon in Windows, or start the Eagle Point batch file in DOS, you are really opening a specialized session of AutoCAD. AutoCAD program functions are loaded before and concurrently with LANDCADD program functions. LANDCADD functions shape the AutoCAD drawing environment by loading some programs into memory, calling for specific menus, and making other files available if they are required. The entire sequence is designed to occur automatically; although you see the Eagle Point logo screen, you do not need to make any changes or adjustments as the programs begin. To open LANDCADD and AutoCAD together, you begin by picking the Eagle Point icon (in Windows) or starting the Eagle Point batch file (in DOS). This causes your computer to load EPWIN.EXE (or its DOS counterpart). The sequence to start LANDCADD continues in this order:

1. **EPWIN.EXE**—the program that starts all Eagle Point software, including LANDCADD. This program displays the Eagle Point logo, calls up an initialization file EP.INI, then begins AutoCAD by calling for ACAD.EXE—AutoCAD's actual program file.

2 **EP.INI** establishes the Eagle Point (LANDCADD) and AutoCAD operating environment. This operating environment includes SET statements which tell LANDCADD functions where (in what directory) to find the files that are needed when a particular key term (like ACAD, EP, EPDWG, EPSUPPORT, etc.) is used in the programs.

   Eagle Point SET statements

startupdwg=epwin13\dwg\epstart.dwg	the name of the Eagle Point's start-up drawing
ep=epwin13	where to find the Eagle Point files
epdwg=epwin13\dwg	where to find Eagle Point's drawing files
epsupport=epwin13\support	where to find Eagle Point's support files

3. **ACAD.EXE** is AutoCAD's *executable* program file, the real core AutoCAD program. ACAD.EXE begins by looking for its own initialization file: ACAD.INI.

4. **ACAD.INI** is AutoCAD's own initialization file that adds SET statements to AutoCAD's operating environment.

   AutoCAD SET statements

acad= epwin13\program	where to find the basic LANDCADD functions
epwin13\dwg	where to find the LANDCADD/Eagle Point drawing files
r13\com\support	where to find AutoCAD's support files

r13\com\fonts	where to find AutoCAD's fonts
r13\support\acad.exe	where to find the main AutoCAD executable file
r13\win\acad.exe	if not at the other location, where to find ACAD.EXE

After the AutoCAD environment is completely configured, the AutoCAD program begins an automatic search of start-up files that will automatically run if they are present. When third party applications are not installed, these programs are very short or absent altogether.

AutoCAD Release 13 begins by loading any ARX programs into memory. These programs load entirely into memory, as long as AutoCAD is running, and operate extremely fast when requested by AutoCAD. The file accessed first is called ACAD.RX

5. **ACAD.RX**—initiates Eagle Point ARX functions; this file calls for two Eagle Point ARX files:
   - **a.** EPAWCMMD.ARX—Opens Eagle Point functions, loads Eagle Point additions to AutoCAD menu
   - **b.** EPAWCALL.ARX—Starts other Eagle Point functions and loads Project Manager

   In AutoCAD for DOS and for all versions of AutoCAD Release 12, this step is skipped, and the functions are loaded in ***8A** below.

6. **ACAD.MNS**, **ACAD.MNR** and **ACAD.MNC**—these files load AutoCAD's Windows menus. The ACAD.MNR and ACAD.MNC files are *compiled* and cannot be edited. These files are derived from a simple (though long) text file called ACAD.MNU. In DOS, the ACAD.MNU file creates a compiled version called ACAD.MNX. There are also equivalent menu files for all of Eagle Point and LANDCADD modules. Whatever your operating system, the menu files are loaded before the program is actually visible on the screen.
7. **ACAD.MNL**—this file adds AutoLISP command functions into the AutoCAD menus and is loaded each time the AutoCAD menu is loaded. There is also an equivalent Eagle Point file EPASWAP.MNL. Each time a menu loads, its **MNL** (if it has one) file is loaded with it. Although ARX and ADS functions are functional throughout an AutoCAD session, menu files and their corresponding commands must be reloaded each time a new drawing is opened.

8. **ACAD.LSP**—this is an AutoLISP file that AutoCAD automatically loads without a command. This file is created by the Eagle Point installation program if one does not already exist in your system. You can append your own custom AutoLISP routines onto the end of this file.

*8A. **ACAD.ADS**—this file is the AutoCAD Release 12 (and Release 13 DOS) version of ACAD.ARX. This file calls other files which correspond to the ARX files:

EPADCALL.EXP

EPADCMMD.EXP

EPADNODE.EXP

ACADAPP.EXP

9. **LANDCADD.INI** contains settings that were in force when LANDCADD was last closed. This file is usually overwritten each time LANDCADD is run.

   EPSTART.DWG or **EPA_SEED.DWG** or **EPSTRT13.DWG**—prototype drawings that will be used as AutoCAD's **initial drawing**. This drawing is in the background when the Project Manager opens. If there are no projects available in the Project Manager, you will be editing this drawing when you pick **Close**.

10. **Eagle Point Project Manager**—Eagle Point builds in this feature to assist in loading files (both drawings and companion files which may contain important settings) and organizing projects. The Project Manager *must* be started for LANDCADD to load its menus properly. If you pick **Close** at the bottom of the Project Manager dialogue box instead of loading a file, you will not see the normal AutoCAD/Eagle Point menu. Your abbreviated menu will only list **File** and **Help**. You cannot load LANDCADD menus from this opening menu. To restore the AutoCAD/Eagle Point menu, follow this procedure:

    `Command:`
    Type **SETACMENUS**. This should restore the AutoCAD/Eagle Point menu. You should be able to load all of your LANDCADD modules from this menu.

chapter

Setting Up and Configuring LANDCADD

LANDCADD and AutoCAD Operating Platforms

The operating system you use for running AutoCAD and LANDCADD can have a great deal to do with these programs' performance and reliability.

DOS (5.0, 6.0, 6.2, etc.)

DOS is the original platform for IBM-based PCs. DOS issues instructions in 16-bit pieces, which makes it a 16-bit operating system. This operating system only runs programs on a one-at-a-time basis. On modern computers, DOS is very dependent on memory management techniques to make newer, larger programs operate properly. These programs often require the use of complex memory manager programs because 16-bit programs must operate in a very confined range (640 KB) of memory addresses. Because memory addresses can only be occupied by one program at a time, this limited space must be jealously guarded for large programs (like AutoCAD). Memory managers move smaller, more flexible DOS function and TSR (**t**erminate, **s**tay **r**esident) programs (which need not occupy this "lower" memory space) into "upper" memory address space. If this sounds complicated, it often is. Individual systems with different hardware installed can require specific "tuning" techniques to make them run AutoCAD properly.

DOS is one of the fastest platforms for AutoCAD performance. With sufficient system memory, aggressive lower memory management techniques and a willingness to experiment, you can make AutoCAD perform well in the DOS environment. Remember that in DOS, and all other operating systems, you should always optimize and configure AutoCAD completely, so that it performs well **before** you attempt to install LANDCADD.

Windows 3.x (Windows 3.1, Windows for Workgroups 3.11, etc.)

Windows 3.x is really an add-on to DOS. The primary attraction of Windows is the ability to operate several programs at one time, flipping between one program and another. This is called *multi-tasking*. In Windows 3.x the multi-tasking is *cooperative*, meaning that all of the applications (programs) must agree to participate. If a program chooses not to participate in multi-tasking, it can hog all of the processor's attention, and slow down or crash the entire operating system. In reality, the Windows 3.x operating system really only adds a graphical user interface onto the existing DOS functions. Computer memory management is still dependent on memory manager programs and techniques which free up the lower 640K of memory for programs. This inefficient scheme often results in error messages reporting insufficient memory when AutoCAD runs in Windows 3.x.

AutoCAD running in Windows 3.x is very dependent on a **permanent swap file** for optimal performance. This swap file is used as an overflow area for those portions of programs that will not fit into system memory. By designating a permanent swap file, you are dedicating a fixed portion of your hard disk for Windows to use for this purpose. Because AutoCAD is a large program which routinely writes many temporary files to the hard disk during a drawing session, it is essential to establish a permanent swap file on your hard disk for AutoCAD to run properly. See your Windows documentation on **Virtual Memory** for details.

Windows 3.x is the slowest and least efficient platform on which to run an AutoCAD system. AutoCAD Release 12 will operate satisfactorily (if somewhat slowly) in Windows 3.x, but Release 13 is another matter. The AutoCAD program code in Release 13 is written in **32-bit code**. This takes advantage of Windows 95 and Windows NT 32-bit operating systems which use longer 32-bit instructions. These computer operations must be filtered through the Windows 3.x "interpreter" **WIN32S** which converts 32-bit instructions to 16-bit instructions that are understood by Windows 3.x. This conversion is time-consuming and extracts a serious toll on operating speed. Most users find AutoCAD Release 13 in Windows 3.x unacceptably slow.

Autodesk will support neither Windows 3.x nor DOS in future releases of AutoCAD. Instead, it will devote its efforts to supporting only 32-bit operating systems found in Windows 95 and Windows NT.

Windows 95

Windows 95 is a *preemptive* multi-tasking operating system. Preemptive multi-tasking *forces* all programs to share processing time in the computer. This helps programs run more efficiently and reduces the number of system crashes. In addition, Windows 95 is 32-bit operating system that has the built-in flexibility to operate 16-bit and 32-bit programs. The roots of Windows 95 still extend into DOS however, so that memory management programs can still benefit 16-bit programs which operate in the lower 640K memory area. The fact that Windows 95 can operate 32-bit programs is a real bonus for AutoCAD users. The performance gains created by Windows 95 make AutoCAD Release 13 operate nearly as fast in Windows 95 as it does in DOS. Other features supplied with all Windows versions of AutoCAD make them attractive compared to the DOS version. These include an improved user interface, tool bars, spell-checking, paragraph text editing, dialogue box text style creation, and cut-and-paste options linking AutoCAD and other programs.

When these features are coupled with the performance benefits of a 32-bit operating system, multiple simultaneous AutoCAD sessions, and preemptive multi-tasking abilities built into the Windows 95 platform, there are compelling reasons to consider Windows 95 as an operating system for CAD users.

Because Windows 95 has been written to support 16-bit programs, Microsoft has taken the approach of supporting many of the computers that run these applications as well. Microsoft has been very thorough in supplying drivers for an enormous number of older and less well-known hardware devices. Because of this, Windows 95 can run on older 486 "legacy" systems reasonably well. Windows 95 does require a somewhat larger amount of memory than do DOS and Windows 3.x. To make AutoCAD and LANDCADD perform well, a Windows 95 system should be equipped with a minimum of 16 MB of RAM. To get optimal performance, 32 MB may be preferable. With RAM available at less than $8 per megabyte, this represents a sound investment for any computer system.

Windows NT

Windows NT is a true preemptive multi-tasking, 32-bit operating system which reliably operates only 32-bit programs. Most DOS and Windows 3.x programs will not run on Windows NT. In exchange for this lack of flexibility, NT users are rewarded with a very stable operating system which is extraordinarily resistant to system crashes. This is because the operating system divides the system's programs into separate *threads* so that each is kept separated from the others. This is called a *multi-threading* environment. The operating system itself is well insulated from the applications, so that if a single application crashes, it is very unlikely to crash the entire operating system. Windows NT is also equipped to provide different levels of security to different users, so that password protection can be employed to protect sensitive directories and files. Like Windows 95, Windows NT can be used to operate peer-to-peer networks, allowing up to 10 user nodes within the legal limitations of the software license. Windows NT also can take advantage of the performance benefits found in recent hardware introductions like Pentium Pro processors and dual-processor systems, where Windows 95 cannot.

On the down side, Windows NT does not operate many 16-bit programs effectively, and requires a large amount of hard disk space to install. Another drawback is that Windows NT requires much more memory that does Windows 95. An AutoCAD installation operating in Windows NT really requires 32 MB of memory to operate well, while users with large drawings may find 64 MB a more realistic number. In addition, Windows NT supports about one third fewer peripheral devices than does Windows 95, and is not as easy to install and configure. Windows NT is truly an industrial strength operating system which is probably most suitable on larger networks with more modern equipment.

LANDCADD Program Organization

When you install LANDCADD, program files are installed in several different locations. Certain portions of the program refer to the Eagle Point group of programs, while others refer specifically to LANDCADD. The file categories and their directories include:

File Type	Directory*	Function
LANDCADD program files	C:\EPXXX\PROGRAM	LANDCADD commands, menus, slides, icon menus
LANDCADD support files	C:\EPXXX\SUPPORT	Eagle Point settings and configuration files
Eagle Point drawing files	C:\EPXXX\DWG	Drawing blocks used in Eagle Point civil engineering modules
LANDCADD symbols	C:\EPXXX\BLOCKS	Drawing blocks used in all LANDCADD modules
LANDCADD plant texture bitmaps	C:\EPXXX\TEXTURES	Textures used in LANDCADD's Virtual Simulator and Virtual Image programs
LANDCADD detail drawings	C:\EPXXX\DETAILS	Detail drawings used in LANDCADD Construction Details module
LANDCADD tutorial files	C:\EPXXX\TUTORIAL	Tutorial drawings used in LANDCADD tutorial exercises.

* The drive letter (C:\) will depend on your installation. The name of the main Eagle Point directory (EPXXX) will depend on whether you are installing the DOS version or the Windows versions of LANDCADD, and whether you are using LANDCADD Release 12 or Release 13.

The programs installation will follow this pattern whether you install LANDCADD on the DOS, Windows 3.x, Windows 95 or Windows NT platforms.

Installing LANDCADD

Installing LANDCADD is a more exacting process than installing other programs because of its intimate connection with AutoCAD. The type of installation procedure you use will depend on the operating system on your computer, the version of AutoCAD you use, and the type of media you selected when you purchased LANDCADD. By far, the easiest procedure is with a CD-ROM. If you have not yet invested in one of these devices, you should consider it. Ignoring the multi-media programs and games available on CD-ROM, CD-ROM drives make the installation of programs **much** faster. The disk version of DOS LANDCADD requires more than 20 separate disks. Installation can easily consume 45 minutes or more. A complete version of every LANDCADD module and every electronic documentation file fits on a single CD-ROM, and installation can easily be accomplished in 15 minutes or less.

When you begin **any** installation of LANDCADD, you should have the following information available:

1. The version of AutoCAD installed on your computer.
2. The name of the directory (or folder) containing ACAD.EXE. For AutoCAD Release 13 users, this will be either **C:\R13\WIN** or **C:\R13\DOS**. For Release 12 users, this might be either **C:\ACADR12** or **C:\ACADWIN**.
3. The name of the directory (or folder) containing the AutoCAD common files. For AutoCAD Release 13 users, this will be **C:\R13\COM**. AutoCAD Release 12 users are not prompted for the common directory.
4. The drive from which you will be installing LANDCADD. This will be a floppy disk drive (A:\ or B:\) or a CD-ROM drive (probably D:\ or E:\ or another letter).
5. The desired target drive and directory (or folder) where you want the Eagle Point software to be installed. You have the option of installing LANDCADD to any directory you specify, but the default directory is typically **C:\EPWIN13** or **C:\EP**. Note that you can change this directory, and that it need not reside on the same drive as the AutoCAD program.

DOS Installations

Installing LANDCADD

LANDCADD is typically installed from the DOS prompt. Insert disk #1 into your floppy disk drive, then type the command:

A:\INSTALL

(Use the correct drive letter which corresponds to your floppy disk drive.) After you specify the required information above, LANDCADD will be installed on the drive and in the directory you have chosen. If you are installing LANDCADD for the first time, it will be configured as it is installed.

Upgrading LANDCADD or AutoCAD

If you are upgrading your version of LANDCADD or AutoCAD, you may need to reconfigure your menus and your AutoCAD environment. This is accomplished by running the configuration command **EPCFG.EXE**. To run this command, change directories to the Eagle Point directory, (probably **C:\EP**), then type:

EPCFG

This program configures your LANDCADD menus and creates a batch file that will run LANDCADD. We will examine this batch file next.

LANDCADD Batch File

Once all of the LANDCADD files are installed and configured, the batch file LANDCADD.BAT is created to properly start AutoCAD and LANDCADD in DOS. The batch file causes several modifications in the AutoCAD environment when LANDCADD and AutoCAD begin together. (The changes in this batch file are comparable to the initialization changes made in the EP.INI file run in Windows.) The **set** statements in the LANDCADD.BAT file modify several of AutoCAD's environment variables.

The set statement **ACAD=** normally contains a list of directories which tells AutoCAD where to find program files, menus, fonts, and drawing files. LANDCADD.BAT adds two directories to the normal list — C:\EP\PROGRAM and C:\EP\DWG. This allows the program to find extra Eagle Point and LANDCADD functions when AutoCAD is running.

The set statement **ACADCFG=** normally contains the directory where the AutoCAD configuration file (ACAD.CFG) is stored. Advanced users may wish to configure AutoCAD differently when LANDCADD is running than when AutoCAD is running by itself. This can be accomplished if two separate configuration files are created and stored in different directories. Use **ACADCFG=** in the LANDCADD batch file to point to the directory where the desired version of the configuration files is located.

The set statement for **ACADDRV=** is used to point to the directory which contains the AutoCAD device drivers. This environment variable is normally left with the same setting as was used when AutoCAD was installed. The default value for AutoCAD Release 13 for DOS is **C:\R13\DOS\DRV.**

The next three set statements created relate to Eagle Point environment variables. These variables tell the Eagle Point programs (of which LANDCADD is one), where to find certain files. The set statement **EP=** tells the Eagle Point functions where the main Eagle Point directory is located. This is normally **C:\EP**. The set statement for **EPDWG=** is used to tell Eagle Point programs where to find prototype and Eagle Point drawing blocks. This is normally **C:\EP\DWG**. The final set statement used is **EPSUPPORT=,** which is used to tell Eagle Point programs where to find (what else) the support file directory. This is normally **C:\EP\SUPPORT**.

Another very important set statement used in the LANDCADD batch file is **path=**. The **path=** is normally defined in the **AUTOEXEC.BAT** file on your computer, and is used to set the *search path* for any executable files. Usually, the **path=** statement is set at boot-up, then left alone until the computer is shut down. Sometimes, however, a program may require the ability to gain access to a directory not listed in the original **path=** statement. This is the case with LANDCADD. The LANDCADD batch file adds the Eagle Point directory to the existing path to tell the Eagle Point Programs (and AutoCAD) where to find any of the executable commands it might need to operate. Typically, the LANDCADD batch file names the **existing** path with a variable name **$path** this way:

set $path=%path%

where "%path%" represents the existing path.

Then the new path is defined:

set path=C:\EP\PROGRAM;%path%

This has the effect of appending the Eagle Point program directory onto the front of the existing path when LANDCADD is running. After you exit from LANDCADD the path is then redefined back to its original state by the set statement:

set path=%$path%

This is a "bookend" to the first statement because it resets the path to the same directories that were contained in **$path**, the original condition.

Hint: Be careful with the length of LANDCADD batch file **set** statements. If these statements are more than 127 characters in length, the remaining characters are ignored. This DOS limitation may require that you substitute logical drive names for directory names or even (under difficult circumstances) rename directories to shorten the path or other **set** statements. See your DOS manual for details.

Troubleshooting Tip: If the slides in LANDCADD's icon menus fail to display properly, the most likely culprit is the **path=** setting in the LANDCADD batch file. Be sure that the proper path is indicated and that the entire path statement is less than 127 characters.

The remainder of the LANDCADD batch file is devoted to loading a special driver to enable the use of the up- and down-arrows in dialogue boxes, and then to reset all of the **set=** statements back to their original condition before the batch file ran.

Using the LANDCADD Batch File

Often, you may choose to work on your drawings in a different directory than the C:\EP\DWGS directory specified in the LANDCADD batch file. You may also wish to return to the root directory or another directory when you exit LANDCADD. If you want to alter the behavior of the system before and after the LANDCADD batch file operates, you can do this by creating your own batch file, then using the **CALL** command to invoke the LANDCADD batch file. Your custom batch file can specify a working directory, "call" the LANDCADD batch file, then return you to the root directory when you quit LANDCADD. See the example LAND13.BAT below:

LAND13.BAT

@ECHO OFF	turns off display of commands
CLS	clears the screen
CD\CADDWGS	changes to the working directory CADDWGS
CALL C:\EP\LANDCADD	calls the LANDCADD.BAT file
CD\	returns to the root directory when you exit LANDCADD
CLS	clears the screen

This file could be placed in your batch file directory or in your root directory. See your DOS manual for more information regarding the creation and editing of batch files.

AutoCAD ADS

The file ACAD.ADS (if it exists) is modified when Eagle Point/LANDCADD is installed. If it does not exist, LANDCADD creates it. The ACAD.ADS file causes AutoCAD to automatically load any ADS files listed here. The ADS files listed here typically include **ACADAPP**, **EPADCALL**, **EPADCMMD** and **EPADNODE**. These files are very closely paralleled by the loading of ARX files when LANDCADD for Windows (3.x, 95 and NT) loads.

DOS Environment

Often, the addition of Eagle Point software will create the need for extra environment space for the command interpreter **COMMAND.COM**. The extra **set** statements in the LANDCADD batch file (and any other **set** statements contained in the **AUTOEXEC.BAT** file) occupy a small amount of memory. If enough **set** statements accumulate, the size of the DOS environment available to interpret them is exceeded. This causes the error message:

```
Out of environment space
```

If this occurs, you will need to increase the size of the command environment in your **CONFIG.SYS** file. Use any text editor to add this line to the file (if this line already exists, increase the value to the one shown here):

```
shell=c:\command.com /e:1024 /p
```

This increases the value for the command environment from its original size (256 bytes) to 1024 bytes, and makes **COMMAND.COM** the permanent command interpreter. After you add this line, save the file and reboot the computer.

If you continue to see the same error message, you can add to the environment space in 16 byte increments until the error message no longer appears. Use larger increments (any multiple of 16) if the error messages do not disappear quickly.

Windows Installations

Installing LANDCADD in Windows 95 and Windows NT 4.0

LANDCADD can be installed using several different methods. Windows 95 or Windows NT 4.0 users can use the **Start** menu on the task bar, then select **Run**. In the **Open:** text box, type **A:\SETUP.EXE** or substitute the correct drive letter for **A:** (Figure 21.1).

Figure 21.1

You can also use the **Add/Remove Programs** function in the Control Panel found under the "My Computer" function (Figure 21.2).

When newer applications are installed using this option, Windows 95 and NT create a record to allow you to **uninstall** the program automatically. Since it can be very difficult to remove programs in Windows 95 and NT without this record, The **Add/Remove Programs** is a good option.

The newest CD-ROM releases of some software packages contain an AUTORUN file which causes them to display an installation icon immediately after the disk is inserted and the drive tray closed. This allows you to begin the installation process without doing anything more difficult than inserting the CD-ROM. When this function is included, the program will normally create the same type of uninstall record mentioned above.

Finally, you can use the Windows Explorer to find the **SETUP.EXE** command on the floppy drive or CD-ROM (Figure 21.3).

Once you find and highlight this file, press the right mouse button to open the file (Figure 21.4).

Figure 21.2

Figure 21.3

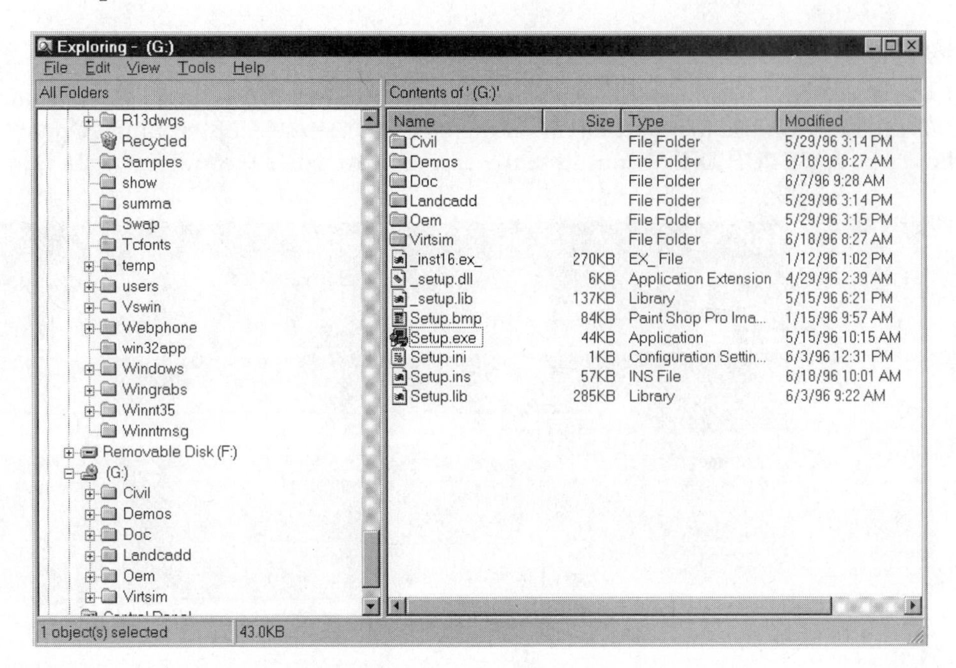

After you specify the location of the **ACAD.EXE** file (the default is **C:\R13\WIN**) and the location of the common files (the default is **C:\R13\COM**), and the other information above, LANDCADD will be installed on the drive and in the directory you have chosen. If you are installing LANDCADD for the first time, your installation will be configured as it is installed.

Installing LANDCADD in Windows 3.x and Windows NT 3.5

Windows 3.x and Windows NT 3.5 users can select **Run...** from the **File** pull-down menu (Figure 21.5), then type **A:\SETUP.EXE** at the command line (Figure 21.6).

Another option is to use the File Manager to locate and open the correct drive, then double click on the **SETUP.EXE** command (Figure 21.7).

Figure 21.4

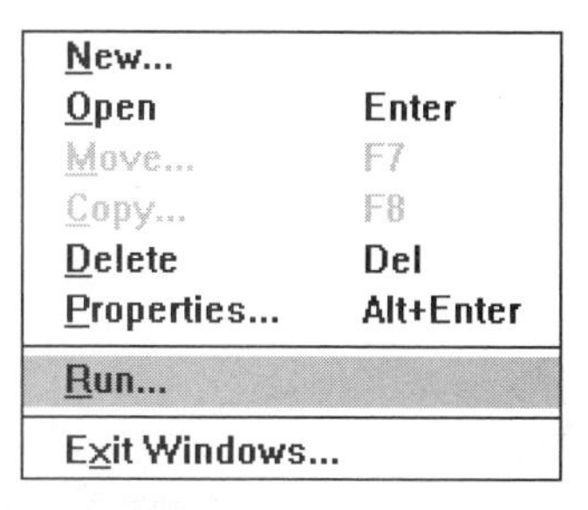

Figure 21.5

Figure 21.6

Run

Command Line:

a:\setup.exe

Run Minimized

OK

Cancel

Browse...

Help

Figure 21.7

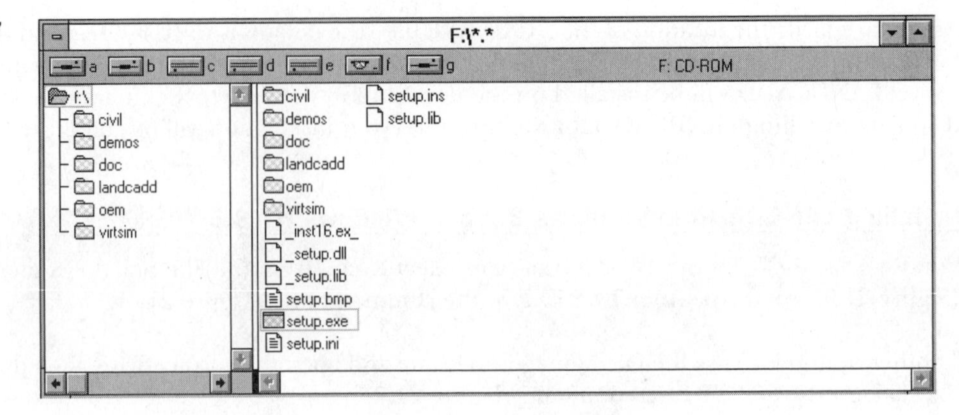

Starting LANDCADD in Windows Platforms

Once LANDCADD is installed on your computer, you can use a few different methods to open the program. In Windows 95 and Windows NT 4.0, once the program is installed, it will normally be added to the Programs section of the Start menu. You can begin the program by clicking on the program option of your choice.

If you use LANDCADD every day, you can customize the Start menu so that LANDCADD appears as one of the primary options on the original menu. See your Windows documentation or one of the excellent "after-market" books available on Windows 95 or Windows NT(Figure 21.8).

Figure 21.8

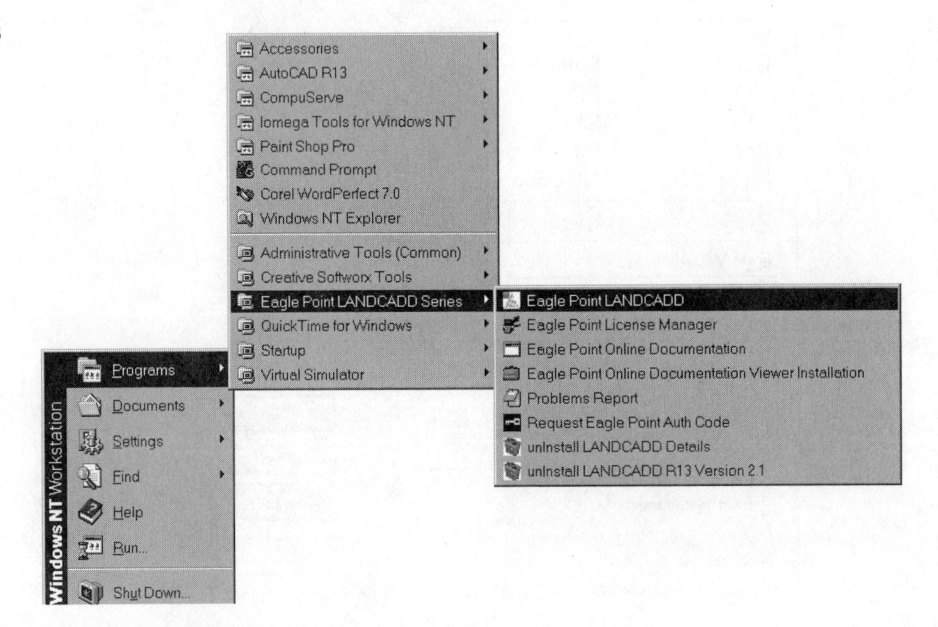

Another method of starting LANDCADD which appeals to many users is the **short-cut**. In Windows 95 and Windows NT 4.0, you can create a short-cut which enables you to access LANDCADD without using the Start menu at all.

To build a short-cut to LANDCADD, begin by using the Windows Explorer to examine the contents of the folder **EPWIN13** (Figure 21.9).

Locate the **Program** folder, then find the program file called **Epwin.exe** (Figure 21.10).

Highlight this file, then click the right mouse button (Figure 21.11).

Figure 21.9

Contents of 'Epwin13'

Name	Size	Type	Modified
Blocks		File Folder	1/31/97 5:07 PM
Details		File Folder	1/31/97 5:07 PM
Dwg		File Folder	1/31/97 5:07 PM
Landcadd		File Folder	1/31/97 5:07 PM
Program		File Folder	1/31/97 5:07 PM
Support		File Folder	1/31/97 5:07 PM
Textures		File Folder	1/31/97 5:07 PM
Tutorial		File Folder	1/31/97 5:07 PM

Figure 21.10

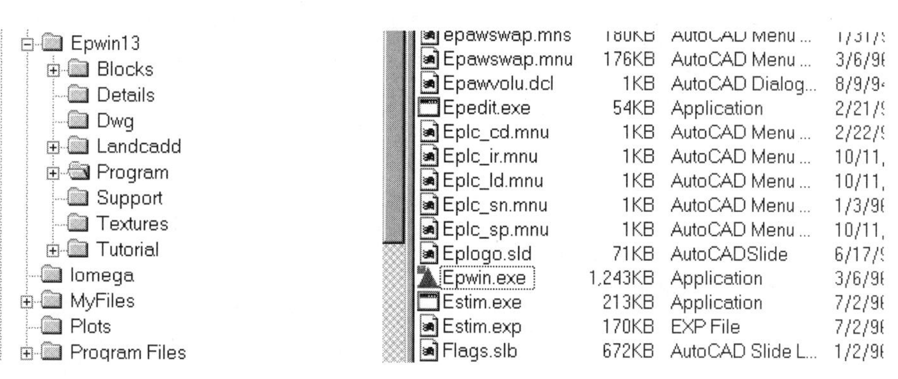

Figure 21.11

Pick the **Create Shortcut** option. This places a short-cut at the bottom of the file list (Figure 21.12).

Highlight this short-cut, then drag it out to the desktop. The short-cut becomes a pointer which then can start LANDCADD directly from the desktop. You can use the original icon (it will match the icon for the program file), substitute your own, or edit the original as desired. Rename the short-cut to a brief name that will allow you to recognize the icon at a glance (Figure 21.13).

The properties of a short-cut can be modified. If you pick the short-cut to highlight it, then use the right mouse button, you can edit the **Properties** of the short-cut (Figures 21.14 and 21.15).

The **General** tab describes the properties of the short-cut and the date it was created. The **Shortcut** tab describes the location of the target file (the program that opens when the icon is picked) and the **Start In** location. This becomes the default location for files created in LANDCADD (Figure 21.16).

Figure 21.12

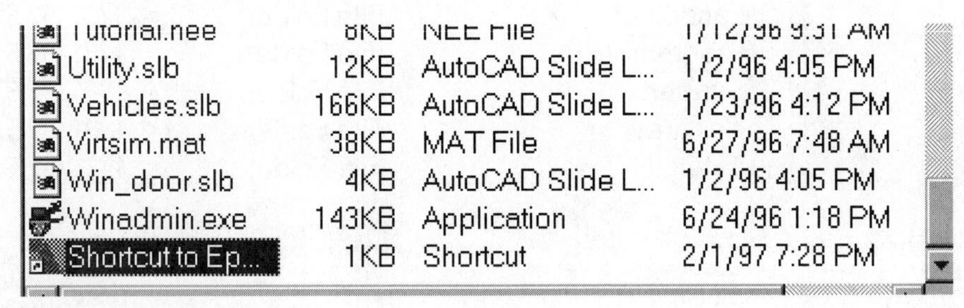

Utility.slb 12KB AutoCAD Slide L... 1/2/96 4:05 PM
Vehicles.slb 166KB AutoCAD Slide L... 1/23/96 4:12 PM
Virtsim.mat 38KB MAT File 6/27/96 7:48 AM
Win_door.slb 4KB AutoCAD Slide L... 1/2/96 4:05 PM
Winadmin.exe 143KB Application 6/24/96 1:18 PM
Shortcut to Ep... 1KB Shortcut 2/1/97 7:28 PM

Figure 21.13

Figure 21.14

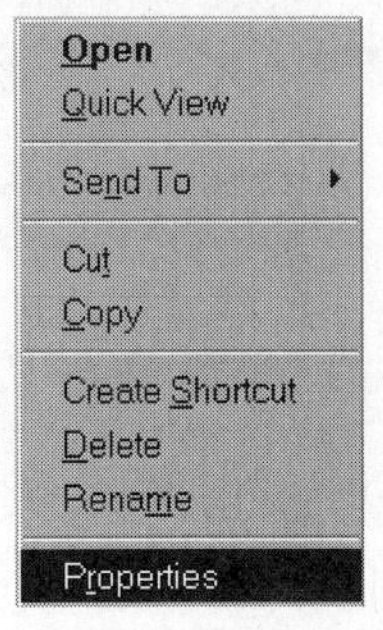

Figure 21.15

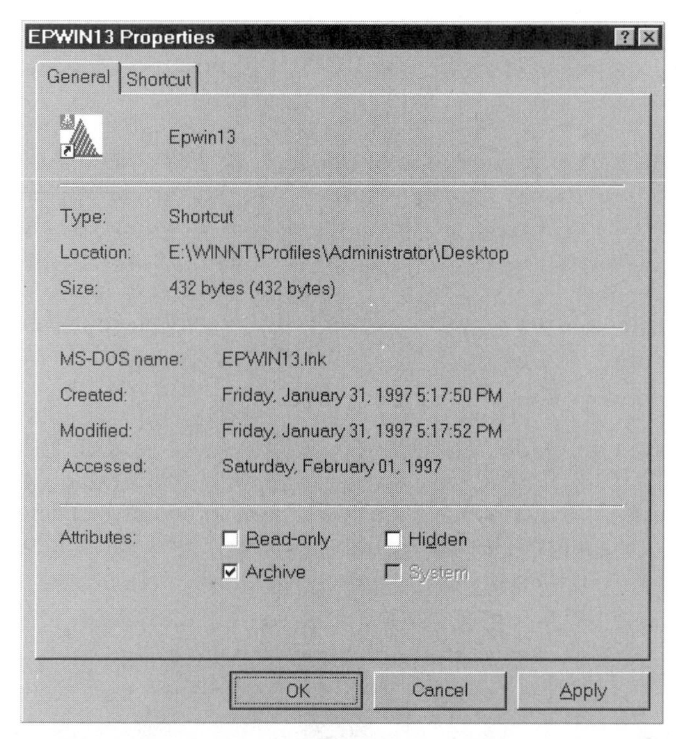

Figure 21.16

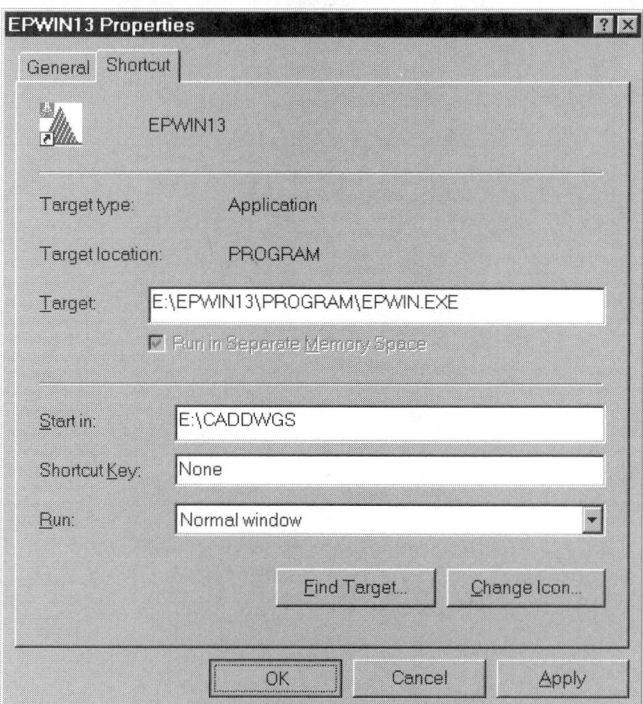

After LANDCADD is installed in Windows 3.x or Windows NT 3.51, users will find a new program group, **Eagle Point LANDCADD Series**, available on their desktops (Figure 21.17).

To begin LANDCADD, you can double-click on the **EPWIN13** icon. Although short-cuts are not available in Windows 3.x and Windows NT 3.51, you can still customize the way that LANDCADD opens. The properties of the EPWIN13 icon can be edited to allow you to begin your application from a **Working Directory** that you specify. This working directory performs the same function as the **Start In** option found in Windows 95/NT 4.0. The Working Directory becomes the default location of all drawing files created in LANDCADD.

Another strategy employed by some AutoCAD users is to build several different starting icons, but specify a different Working Directory for each one. This allows you to begin AutoCAD (or LANDCADD) in different drawing directory locations throughout your computer or network system. The Project Manager function in Eagle Point software makes this less important for LANDCADD users, since access to different files and directories can be built into the projects you define there.

Figure 21.17

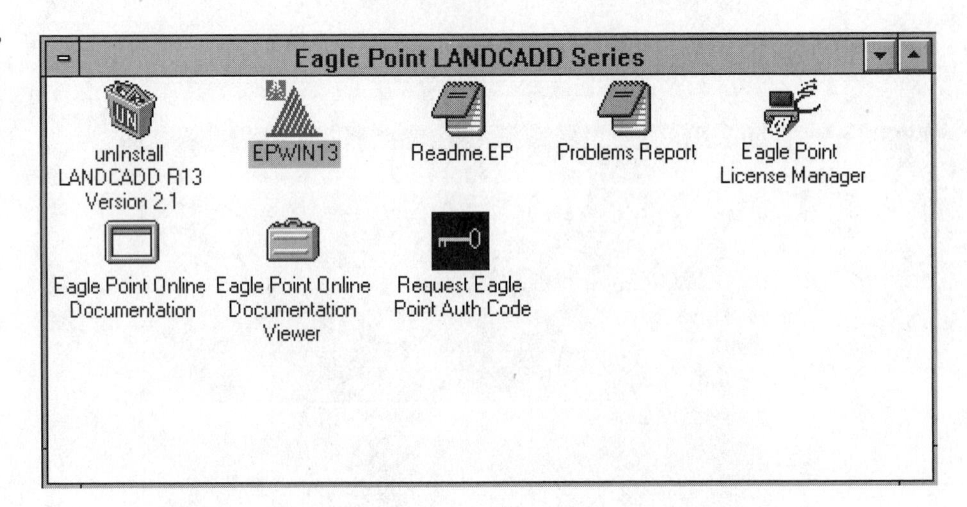

chapter

Customizing LANDCADD and AutoCAD Functions

The most challenging aspect of AutoCAD and LANDCADD installation is getting everything **working**. Setting up AutoCAD in your operating system, optimizing AutoCAD performance, and being able to create output using the available devices all represent a substantial amount of time and effort. Add to it the effort spent in setting up LANDCADD properly, and you have a substantial amount of time and money invested in CAD before a single hour of billable time is generated. But just because your LANDCADD installation is completed doesn't mean that it can't evolve and change to meet your needs. In fact, there are many ways to continue to customize and improve LANDCADD and AutoCAD to suit your own working style and the needs of your particular projects. There are several methods to employ to customize LANDCADD and AutoCAD to suit your purposes.

Note: The customizing techniques mentioned here are amplified and more fully explained in several excellent AutoCAD customizing books. Refer to Appendix 2: References and Resources section for book publishers and titles.

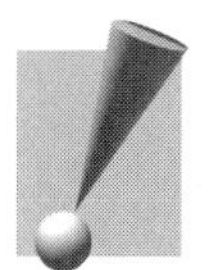

Warning: Always make copies of your original files before attempting to customize existing LANDCADD or AutoCAD files. If you change or damage these files so that AutoCAD cannot read them, you may see error messages, or even prevent AutoCAD from loading properly. This type of customization should be undertaken by experienced users **ONLY**.

Keyboard Customization

Many AutoCAD and LANDCADD users find that in spite of the increased use of dialogue boxes, pull-down menus and toolbars, the fastest way to enter many commands is with the keyboard. Most programs allow you to abbreviate simple commands with keyboard macros. For simplicity's sake, these programs usually offer only a single method to do this. AutoCAD and LANDCADD allow several different ways to create customized keyboard abbreviations to automate your most commonly-used commands. The method you use to automate and abbreviate commands will depend on the type of command you want to automate and the type of operating system you use.

Editing the ACAD.PGP file

The easiest way to add customized keyboard functions to AutoCAD and LANDCADD is through the AutoCAD file ACAD.PGP. This file was originally intended as the **P**rogram **P**arameter file because it originally was used to establish a method for calling external programs during an AutoCAD session. Although the ACAD.PGP file can still be used for this purpose, it has limited use in Windows programs which already allow multi-tasking. You do not really need to invoke another program in AutoCAD if you can easily open it in Windows. The most important use of the ACAD.PGP file for Windows (and DOS) users of AutoCAD is for the definition of command aliases.

The ACAD.PGP is found in the SUPPORT directory found under the COM directory of R13 in AutoCAD Release 13. AutoCAD Release 12 users can find this file under the normal AutoCAD directory in the SUPPORT directory.

This is a section of the ACAD.PGP which contains the common command aliases contained in AutoCAD when it comes "right out of the box".

```
; Command alias format:
; <Alias>,*<Full command name>
; Sample aliases for AutoCAD Commands
; These examples reflect the most frequently used commands.
; Each alias uses a small amount of memory, so don't go
; overboard on systems with tight memory.

A,          *ARC
C,          *CIRCLE
CP,         *COPY
DV,         *DVIEW
E,          *ERASE
L,          *LINE
LA,         *LAYER
M,          *MOVE
MS,         *MSPACE
P,          *PAN
PL,         *PLINE
PS,         *PSPACE
```

```
R,          *REDRAW
T,          *MTEXT
Z,          *ZOOM

3DLINE,     *LINE

; Give Windows an AV command like DOS has
AV,         *DSVIEWER
; Menu says "Exit", so make an alias
EXIT,       *QUIT

; easy access to _PKSER (serial number) system variable
SERIAL, *_PKSER
; Dimensioning Commands.

DIMALI,     *DIMALIGNED
DIMANG,     *DIMANGULAR
DIMBASE,    *DIMBASELINE
DIMCONT,    *DIMCONTINUE
DIMDIA,     *DIMDIAMETER
DIMED,      *DIMEDIT
DIMTED,     *DIMTEDIT
DIMLIN,     *DIMLINEAR
DIMORD,     *DIMORDINATE
DIMRAD,     *DIMRADIUS
DIMSTY,     *DIMSTYLE
DIMOVER,    *DIMOVERRIDE
LEAD,       *LEADER
TOL,        *TOLERANCE
```

The syntax for adding commands to this file is very simple.

```
[abbreviation], *[command name]
```

The **abbreviation** used can be any combination of letters or numbers. The **command name** can be any single AutoCAD command. The number of spaces between them is unimportant, although they are used to make the file more readable. In addition, the ACAD.PGP file is not case sensitive, so capital letters and lower case letters can be used interchangeably. Remember that the command aliases defined here can only be used for "single-step" commands. That is, commands like "EXTEND" or "QSAVE" are appropriate, but not "Zoom All" or "Fillet Radius 0". Multiple-step commands can be placed elsewhere. Use names for your command aliases that make sense to you, and that you can remember easily. There is no limit to the number of command aliases that you can place in this file. With the amount of memory usually required for today's programs, the warning about memory size contained in this file is somewhat out-of-date.

An expanded ACAD.PGP example file is included in your *Using LANDCADD* CD-ROM.

One of the best reasons for using this method for adding macros to AutoCAD is its simplicity. You do not need to know any AutoLISP programming techniques to add to this file, and you can make immediate improvements in your own productivity if you can type in a simple letter combination to trim, extend, edit text, save, stretch, etc. Another good reason to use the ACAD.PGP file relates to the fact that it is automatically loaded each time AutoCAD opens a drawing. No extra programming or commands are needed to load this file. If you change the contents of the ACAD.PGP file during a drawing session, you can force AutoCAD to reinitialize, which can make the ACAD.PGP file reload. To reinitialize the ACAD.PGP file, use the following sequence:

`Command:`
Type **REINIT**. This opens the "Re-initialization" dialogue box. Mark the **PGP File** toggle, then pick **OK**. (Figure 22.1).

Be careful to test each abbreviation to make sure that it works properly before you load and work on drawings. If you enter aliases improperly, AutoCAD will report an error message.

The largest drawback of this method of defining keyboard macros is, of course, the single command limitation. This allows us to build in simple short-cuts, but deprives us of some really useful commands such as Zoom-Window, or Copy-Multiple. AutoCAD users normally build these "two-step" macros in a file called ACAD.LSP. This file is not normally included in AutoCAD when it is installed "out of the box". Third-party applications (like LANDCADD) routinely build a version of ACAD.LSP to load menus, add functions and define certain AutoCAD system variables. More sophisticated AutoCAD users often customize this file to include commonly used keyboard macros which require more than one step. LANDCADD users have this option and another as well.

Figure 22.1

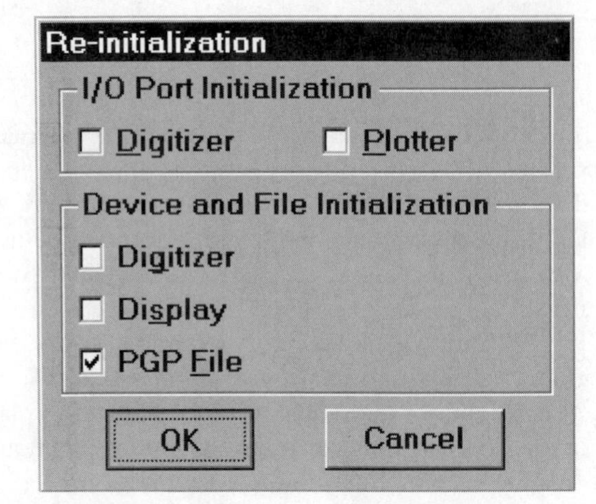

ACAD.LSP

LANDCADD builds a version of ACAD.LSP which is found in the EPWIN13\PROGRAM directory or (for DOS users) the EP\PROGRAM directory. The file reads this way after LANDCADD is installed:

```
;; AutoCad loads the first <acad.lsp> that it finds . . .

(if (not s::startup)
   ;; If the menu didn't fail then there is a S::STARTUP from
   the MNL
   (defun s::startup ()
      (setvar "filedia" 0)
      (setvar "cmdecho" 0)
      (command "_.menu" "epawswap") ;; Load up correct menu
      (command "_.menuload" "epawmenu")
      (menucmd "P1=+EP.POP1")
      (menucmd "P2=+EP.POP4")
      (menucmd "P3=+EP.POP5")
      (menucmd "P4=+EP.POP6")
      (menucmd "P5=+EP.POP7")
      (menucmd "P6=+EP.POP8")
      (menucmd "P7=+EP.POP9")
      (menucmd "P8=+EP.POP10")
      (menucmd "P9=+EP.POP11")
      (menucmd "P10=+EP.POP12")
      (menucmd "P11=+EP.POP13")
      (menucmd "GEP.EP_EAGLA=~")
      (setvar "filedia" 1)
      (setvar "cmdecho" 1)
      ;; AutoCad only loads MNL file on success
      (load "epawswap.mnl") ;; Load MNL for menu
      (s::startup) ;; Call S::STARTUP from MNL
      (princ)
   )
)
```

You can add keyboard macro functions onto the end of this file, or (preferably) edit another file which LANDCADD provides to load its own keyboard macros. This file is called LC_KB.LSP.

LC_KB.LSP

LANDCADD's keyboard macros are found in the file **LC_KB.LSP** which is found in the **EPWIN\PROGRAM\LISP** directory or (for DOS users) the **EP\PROGRAM\LISP** directory. The **LC_KB.LSP** file loads whenever a LANDCADD menu is called from the **EP** pull-down menu. This file is not necessary to make LANDCADD run, but is provided "for ease of use only". This is a partial listing of the **LC_KB.LSP** file, showing the syntax of the entries shown there.

```
;;; 1 and 2 character macros for quick keyboard entry
;;; These functions are not required but are provided
;;; ease of use only.
;;; When using with ACAD 11, these should be deleted
;;; and put in the ACAD.PGP file!
(defun C:$ ()
   (command "_.SKETCH")
   (princ)
)

(defun C:0 ()
   (command "_.CIRCLE")
   (princ)
)

(defun C:AD ()
   (command "_.ATTDEF")
   (princ)
)

(defun C:AE ()
   (command "_.DDATTE")
   (princ)
)

(defun C:ZD ()
   (command "_.ZOOM" "_D")
   (princ)
)

(defun C:ZE ()
   (command "_.ZOOM" "_E")
   (princ)
)
```

```
(defun C:ZP ()
    (command "_.ZOOM" "_P")
    (princ)
)

(defun C:ZW ()
    (command "_.ZOOM" "_W")
    (princ)
)
```

The following breakdown explains the various parts of the **ZW** command

`(defun`
defun tells AutoLISP that this command will **de**fine a **fun**ction. The opening parenthesis will match the close parenthesis in the final line.

`C:ZP`
C: tells AutoLISP that the function will be an AutoCAD-type command that you can type at the command prompt (like any other AutoCAD command) rather than a more obscure programming function. **ZP** is the name of the newly defined command.

`()`
This empty pair of parentheses can hold local variables or symbols for other AutoLISP functions. In the case of simple command definitions, the parentheses are normally empty.

`(command`
This tells AutoLISP that the next statement will be a command sent directly to the AutoCAD command prompt. The open parenthesis matches the close parenthesis at the end of the line.

`"_.ZOOM"`
"_.ZOOM" starts AutoLISP **ZOOM** command. The quotations tell AutoLISP to treat **_.ZOOM** as a "string" just as though you had typed it at the command line. The underscore and period before the command name are optional. The underscore indicates to AutoCAD that the command name is in English (important for foreign-language versions of the program), while the period tells AutoCAD to use its own **ZOOM** command, even if someone (or some program) had redefined the ZOOM command.

`"_W"`
This selection is used to activate a specific portion of the ZOOM command. Once the **ZOOM** command is invoked, "_W" starts the **Window** option. Again, the quotation marks are required but the underscore is optional. Periods are not used with command options.

```
)
```

This close parenthesis matches the open parenthesis before command, and tells AutoLISP that the command portion of the definition is completed.

```
(princ)
```

(princ)is a term that tells AutoLISP not to display a **nil** on the command line after **ZP** is run. This is an example of "tidy" AutoLISP programming since it remains mostly invisible to the user.

```
)
```

The final close parenthesis matches the open parenthesis before defun, and tells AutoLISP that the function definition is complete.

This material can act as a template to create other simple commands in AutoLISP. A new listing which erases the previous selection set might be listed this way:

```
(defun C:EP ()
    (command "_.ERASE" "_PREVIOUS")
    (princ)
)
```

As before, the command begins with a parenthesis, then defines the function called EP. The empty parentheses indicate the absence of local variables. The command ERASE is issued, then the selection set is predefined as PREVIOUS. The princ line prevents **nil** from appearing on the command line, and is followed by a closing parenthesis to complete the command.

AutoLISP, like any other programming language, is very particular about the placement of punctuation in general and parentheses in particular. Be careful to double check your typing when adding to the **LC_KB.LSP** . As a precaution, be sure to save the original version of **LC_KB.LSP** as **LC_KB.OLD** or some other identifiable file name. This insures that you can retrieve the original file and use it again if you create problems in the revised file.

You have the option of adding your customized commands to the **ACAD.LSP** file or the **LC_KB.LSP**. The primary difference is that the **ACAD.LSP** file will be loaded even if you decide not to load any LANDCADD menus during your drawing session. The functions created in **LC_KB.LSP** will only be loaded after you choose the LANDCADD option under the EP pull-down menu. Another option to consider is to cull out the single-function macros from the **LC_KB.LSP** file and place their equivalents in the **ACAD.PGP** file. The command aliases defined in the **ACAD.PGP** file are available any time AutoCAD is started.

Although the topic of AutoLISP programming is beyond the scope of this book, there are many references on the subject. Several books are available on AutoLISP programming, and specialized AutoCAD magazines like *Cadence,* and *Cadalyst* feature regular columns on customized AutoLISP files. Other excellent sources of information about AutoLISP are found on the Internet and in CompuServe® and America On Line® Autodesk forums. As a final note, if you have a keen interest in learning AutoLISP, consider enrolling in a local community college or technical school course in the subject. You will be able to learn more information more quickly in a structured course than you can learn from any text, including this one.

Windows Accelerator Keys

AutoCAD Release 13 for Windows has a built-in facility for adding keyboard macros into menu files. AutoCAD defines these as "accelerator keys," but they are really the same as the keyboard macros we have already examined. The only difference is that the Accelerator Keys defined here can include the **CTRL**, **SHIFT**, **INSERT**, **DELETE** and other keys as well as function keys. You can create combinations of keystrokes to call up any AutoCAD or LANDCADD command.

The accelerator keys are defined in the menu files. There are really two menu files which contain the accelerator key information: the MNU file and the MNS file. Since the MNU file is the original template menu file, and the other files originate from it, the best strategy for customizing is to leave the MNU files alone. Instead, we will look at the MNS file. Like the MNU file, the MNS file is a plain ASCII (character based) file which can be edited by any text editor. When you use the "out of the box" version of AutoCAD, the file to edit would be ACAD.MNS, which acts as a *source* file for the *compiled* menu files ACAD.MNC and ACAD.MNR. In LANDCADD, the proper file to edit is EPAWSWAP.MNS. This file contains the basic LANDCADD functions, which are active when any of the other LANDCADD menus are used.

After creating a back-up copy of the original file, examine the **EPAWSWAP.MNS** file. Use the **Search** function in your text editor to find the section of the menu with the heading *****ACCELERATORS**. It resembles the listing below:

```
***ACCELERATORS
[CONTROL+"L"]^O
[CONTROL+"R"]^V
ID_Undo      [CONTROL+"Z"]
ID_Cut       [CONTROL+"X"]
ID_Copy      [CONTROL+"C"]
ID_Paste     [CONTROL+"V"]
ID_Open      [CONTROL+"O"]
ID_Print     [CONTROL+"P"]
```

```
ID_New      [CONTROL+"N"]
ID_Save     [CONTROL+"S"]
```

The keyboard macros here are somewhat confusing since there are two separate formats used. The first two macros follow the easiest format to understand. The syntax works this way:

```
[keysequence]command
```

In the first line **[CONTROL+"L"]** is the key sequence. You hold down the **CTRL** key then press **L**.

The command is **^O** which sends AutoCAD the command **CTRL-O**.

A good question: Why do we have a keyboard macro **CTRL-L** sending AutoCAD the command **CTRL-O**? This is done because the Windows operating system has reserved **CTRL-O** as the standard command for opening documents. The AutoCAD use of **CTRL-O** has always been to toggle **Ortho** mode on and off. In Release 13, AutoCAD has adopted the Windows standard and has abandoned its use of **CTRL-O** for **Ortho** mode. Instead, AutoCAD has provided the alternative **CTRL-L** for **Ortho**. Within AutoCAD, however, **^O** (**CTRL-O**) still activates **Ortho**. So If you type **CTRL-L**, the macro translates this to **^O** within AutoCAD and toggles **Ortho** mode.

The second line is a similar substitution. In this case, AutoCAD is translating the keyboard entry **[CONTROL+"R"]** into the AutoCAD command **^V** (**CTRL-V**). The Windows command **CTRL-V** has been reserved for pasting the contents of the Clipboard into a document. AutoCAD has traditionally used **CTRL-V** to toggle to another viewport. The keyboard macro translates **CTRL-R** to **^V** within AutoCAD and performs the viewport toggle.

The remaining eight lines define accelerator keys by using an **ID TAG**. This ID tag is defined and used in other places in the menu file. The accelerator keys simply use the function that was defined in another place in the menu. You can accept the entries here, or you could translate them into the format used in the first two lines. In the case of the line:

```
ID_Print [CONTROL+"P"] could become
[CONTROL+ÓPÓ]^c^c_PLOT
```

The key sequence is entered first (**[CONTROL+"P"]**) then the AutoCAD command follows. The **^c^c** sequence causes AutoCAD to cancel any existing command before trying to execute the following command. The underscore (_) tells foreign-language versions of AutoCAD that the following command will be in English. The **PLOT** command executes normally. If you want to execute a command *transparently* (without interrupting the active command) you can precede the command with a single quotation mark (') rather than the **^c^c** sequence. Some examples of new accelerator keys follow:

```
[CONTROL+"P"]^c^c_PLOT          ;;Plot command
[CONTROL+"N"]^c^c_NEW           ;;New command
[CONTROL+"S"]^c^c_QSAVE         ;;Qsave command
[CONTROL+"W"]'_ZW               ;;calls ZW which is defined in the ACAD.PGP file
["INSERT"]^c^c_DDINSERT         ;;Ddinsert command
["DELETE"]^c^c_ERASE            ;;Erase command
[CONTROL+"M"]_MID               ;;Midpoint Osnap
[CONTROL+"I"]_INT               ;;Intersection Osnap
[CONTROL+"Q"]_QUAD              ;;Quadrant Osnap
```

Note that you can use any function keys (F1 through F12), named keys (Insert, Home, Delete, etc.), and even number pad keys for accelerator keys. These can be used by themselves or in combination with letters or numbers.

Although you can use the accelerator keys to define single-step macros (as was done in the ACAD.PGP file) there is a difference between the two types of keyboard macros. In the ACAD.PGP file, the keyboard macros require you to press **<Enter>** after the command alias before AutoCAD executes the command. With accelerator keys, the **<Enter>** command is automatically included. In the case of the line **["DELETE"]^c^c_ERASE** the Erase command begins as soon as the **DELETE** key is pressed. The **<Enter>** key need not be used at all. This can be confusing to both new and experienced AutoCAD users, so place your single-step macros in the ACAD.PGP file to keep the format consistent.

With three keyboard customization options available, how do you decide which to use? Here are some guidelines:

The accelerator key definitions will override any other definitions created in the **ACAD.LSP** or **LC_KB.LSP** files. Accelerator keys are best used to work with **CTRL**, **SHIFT**, undefined function keys (**F3**, **F11** and **F12**) and the named (**HOME**, **INSERT**, etc.) keys. This is the only way to use these keys in AutoCAD Release 13.

The simple AutoLISP commands found in the **LC-KB.LSP** or **ACAD.LSP** files are best used to define brief multi-step commands (like ZOOM Window). These commands are slightly more complicated than the other forms of macros, but are not difficult to build. The definitions created in the **LC-KB.LSP** and **ACAD.LSP** will override the definitions created by the **ACAD.PGP** file.

The command aliases created in the **ACAD.PGP** file are best used for single-step commands. These are the easiest type of keyboard macro to create and modify. They can be added to the **ACAD.PGP** file at any time and consume very little system memory.

Be careful of conflicts when defining **any** keyboard macro. Always test the keyboard combination you want to use to make sure that the combination does not already perform a function in AutoCAD.

AutoCAD Toolbars

AutoCAD Release 13 for Windows contains another very useful feature in its standard and customizable toolbars. See your AutoCAD documentation or any of the AutoCAD customizing books to assist you in controlling the display and content of the standard AutoCAD toolbars. We will examine the process of building a custom toolbar using existing icons, then we will build a custom icon to load the Project Manager.

To begin, we will open the "Toolbars" dialogue box.

`Command:`
Type **TBCONFIG** or pick Customize Toolbars from the Tools pull-down menu. This opens the "Toolbars" dialogue box (Figure 22.2).

Figure 22.2

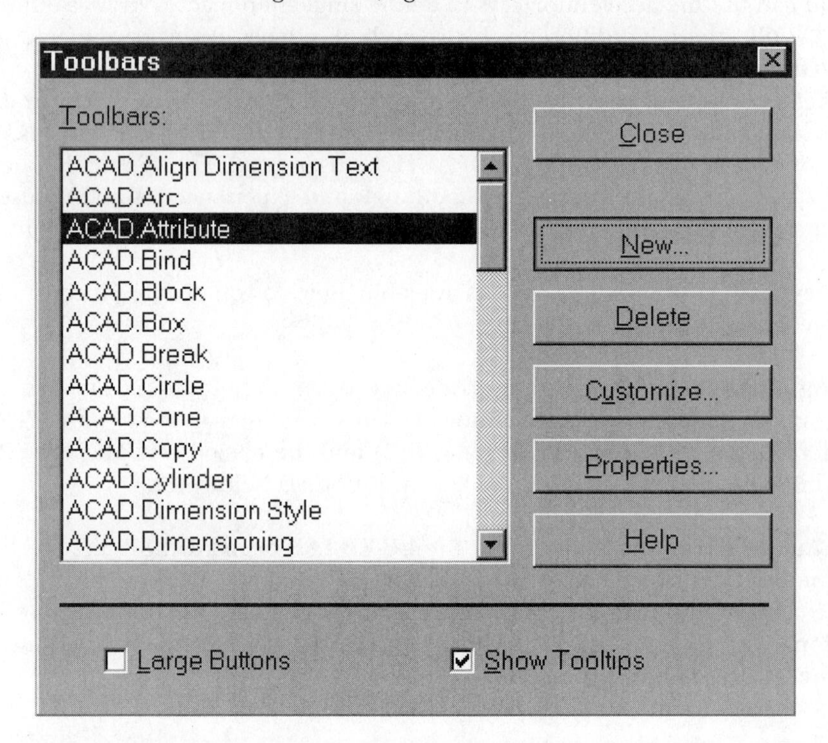

Pick **New** to build a new, customized toolbar. This opens the "New Toolbar" dialogue box (Figure 22.3).

Figure 22.3

New Toolbar

Toolbar Name:

LandCADD

Menu Group:

ACAD

OK

Cancel

Help

In this dialogue box you can enter the name of the new toolbar. To create a toolbar specifically for LANDCADD, enter the name **LANDCADD**. In the **Menu Group** pop-down box, leave the entry **ACAD**. The **Menu Group** refers to the menu with which the LANDCADD toolbar will be associated. Since the basic AutoCAD menu group is always accessible in LANDCADD, we will let our new toolbar remain associated with this menu.

To create a **Help** message that explains the purpose of the new toolbar, highlight the **ACAD.LANDCADD** toolbar title, then pick **Properties...**. This opens the "Toolbar Properties" dialogue box (Figure 22.4).

Figure 22.4

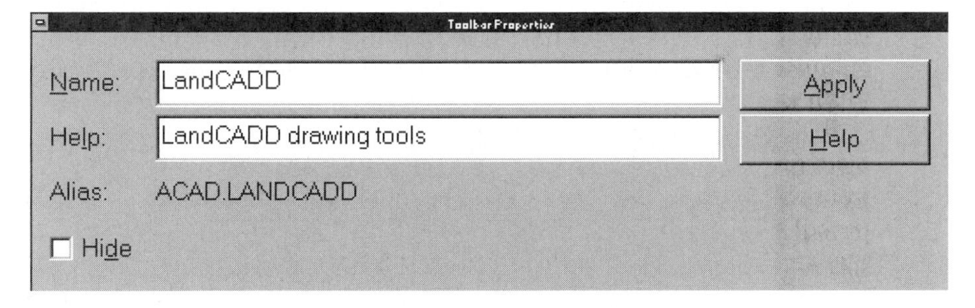

For the **Help:** message, use the line **LANDCADD drawing tools**, then pick **Apply**. Be sure to leave the **Hide** toggle box unmarked to make the toolbar display on the screen. Close the "Toolbar Properties" dialogue box with the Windows closing tool at the top left corner (there is no **Close** option) (Figure 22.5).

Figure 22.5

This returns us to the "Toolbars" dialogue box. We now have a new (and very small) toolbar near the top of the drawing area. This toolbar will hold copies of the icons we select from other menus.

Select **Customize** to open the "Customize Toolbars" dialogue box (Figure 22.6).

Figure 22.6

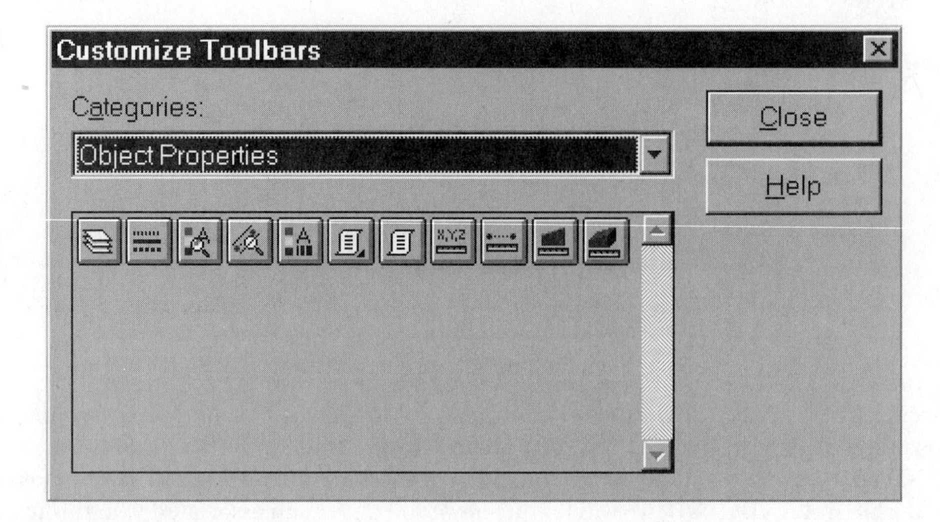

This dialogue box features a list of 13 different categories in the pop-down list. When an item on the list is highlighted, the icons are shown in the box below.

To add one of these icons to your LANDCADD toolbar, simply pick an icon, then drag it onto your LANDCADD toolbar. The icon will remain in the new toolbar, and is ready to use. Pick as many AutoCAD icons as you wish to add to your toolbar (Figure 22.7).

Figure 22.7

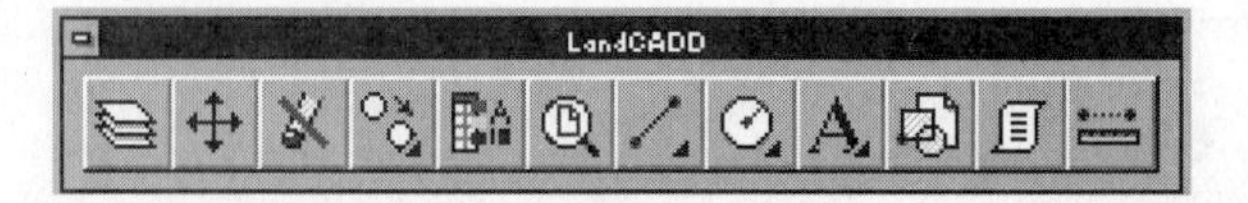

When you are finished adding icons to your toolbar, close the "Toolbars" dialogue box. This will force AutoCAD to rebuild the ACAD.MNS file, then recompile the ACAD.MNC and ACAD.MNR files.

To build an icon and command that does not exist, you can add a blank button to your toolbar, then define its function and create its button. We will create a simple Project Manager (**PM**) button to allow us to call the Project Manager with a toolbar button.

The simplest way to open the "Toolbars" dialogue box is to move your cursor to any icon on the LANDCADD toolbar, then click the right mouse button. This opens the desired dialogue box without any typing.

Pick **Customize**, then move to the **Custom** category on the pop-down list. Pick the blank button (on the left) and drag it to your LANDCADD toolbar (Figure 22.8.)

Figure 22.8

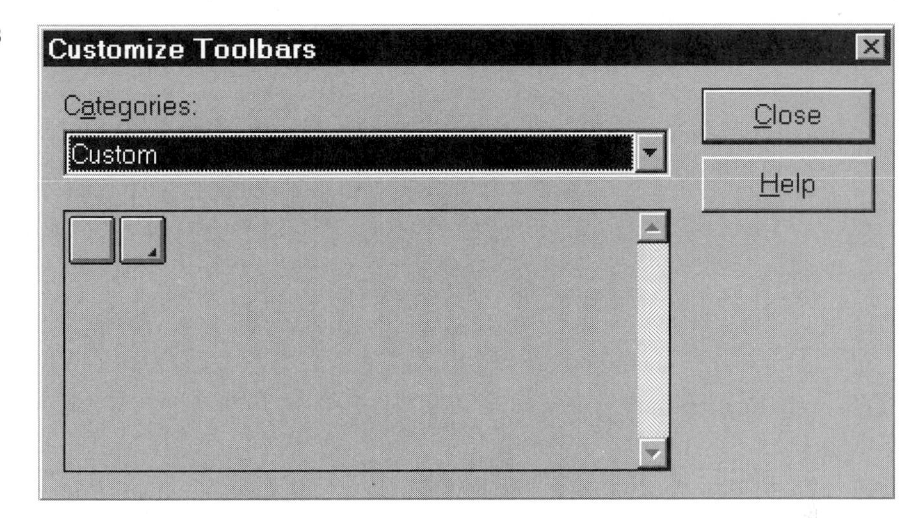

Once this blank icon is placed on your toolbar, pick it with your right mouse button. This opens the "Button Properties" menu toward the bottom of your screen (Figure 22.9).

In the **Name:** text box, enter **Project Manager.** In the **Help:** text box, enter **Eagle Point Project Manager**. The entry in the **Name:** box will appear as the ToolTip, while the **Help:** text appears on the status bar (below the Command: prompt). Both the ToolTip and the help text will appear if you hold your cursor over the icon for a few seconds.

In the **Macro** box below, place the cursor after the **^C^C** (which cancels the current command) and type **PROJMAN**. This will open the Project Manager any time we pick the button. We could also load and execute any AutoLISP command using a toolbar but-

Figure 22.9

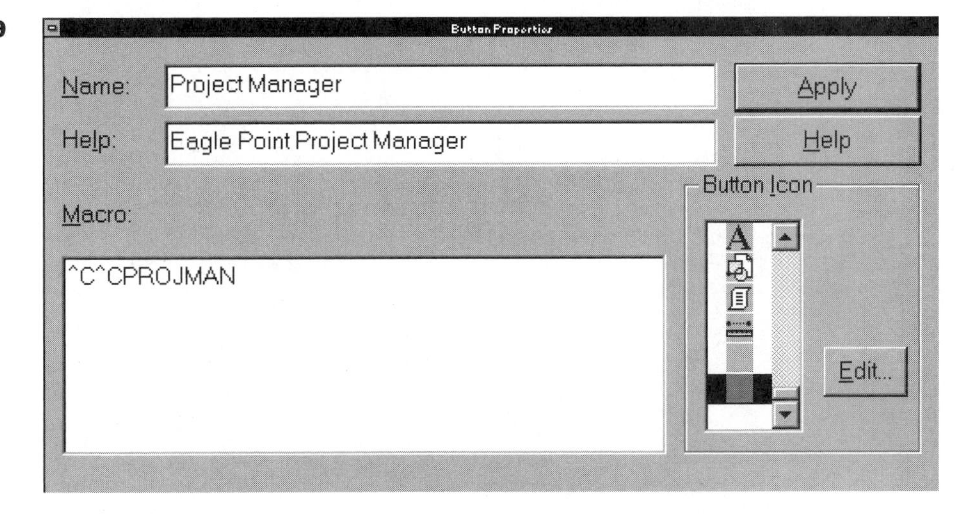

ton. The command syntax is the same as that found in the accelerator keys. In fact, any AutoLISP command or text string recognizable by AutoCAD can be entered here.

Move to the **Button Icon** area, then pick **Edit...**. This opens the "Button Editor" dialogue box (Figure 22.10).

Use the pens to create the simple block letters **PM**. Use the **Grid** toggle to place grids on the icon to make it easier to draw regular shapes. Because the icon display may be very small, it is best to keep the icons as simple as possible.

Pick **Save** to save the icon. AutoCAD will save this icon as a **Bitmap** file with the format **ICONnnnn.BMP** where **nnnn** is a number assigned by AutoCAD. Pick **Close** to return to the "Button Properties" dialogue box, then pick **Apply** to apply the changes to the button in the LANDCADD toolbar (Figure 22.11).

You can also build your icons in other, more versatile painting programs that will allow you to create bitmap files. As long as the output is 16 pixels by 16 pixels and uses the

Figure 22.10

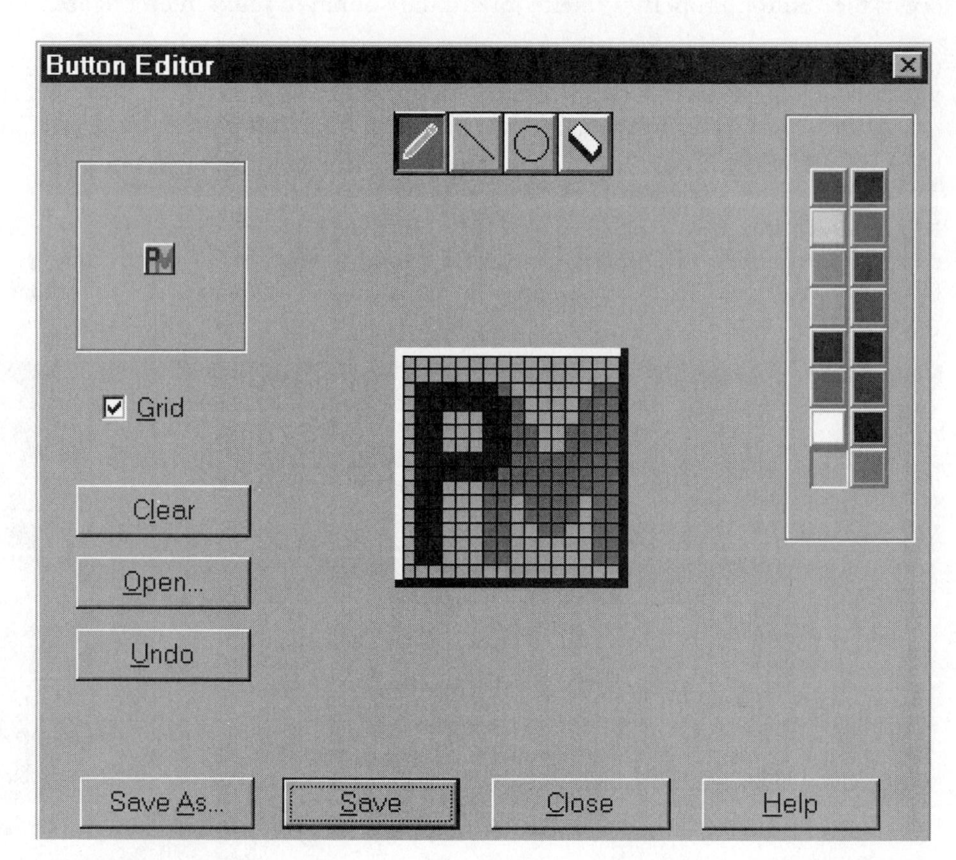

Figure 22.11

BMP format, the file can be used as an icon. To load an icon created from an outside program, be sure to move it into the same folder where the LANDCADD menus are stored (usually EPWIN\PROGRAM). After the file is in the correct folder, you can retrieve it using the **Open** option in the "Button Editor" dialogue box. You can also use one of AutoCAD's own icons as a basis for your new edited buttons. You can add symbols, change colors or make other modifications which give the button a different meaning than the original. When you save the icon, AutoCAD saves it to a new file name, so you won't overwrite one of the original bitmap files.

LANDCADD Commands

Advanced LANDCADD users with AutoLISP experience can build buttons for LANDCADD commands. To build a macro to run a LANDCADD command, you must know the command name and the names of any required files and menus which must be loaded before the command can run. Often, LANDCADD requires that certain AutoLISP routines load before others can run properly. In order to see if these routines have executed, you can force AutoLISP to search for the presence of a variable which is defined only if the "foundation" files have loaded. You can use the presence (or absence) of the variable to help you load the command or abort the macro. Once you devise a sequence, you can load the macro in the **Macro:** section of the "Button Properties" dialogue box.

As an example, if we wished to load the LANDCADD function PICKLAYER, we would first need to determine if the LANDCADD core functions had loaded. If the core functions have loaded, then the PICKLAYER command can execute. If they are not loaded, the command should abort. If we examine the **LC_CORE.LSP** file (one which automatically loads when we open the LANDCADD menu), we can see that one of the first functions to load involves the variable **EPLC_LF**. We will use this variable to determine whether or not the macro should continue or abort. See the macro (expanded to make it easier to read) below:

`^C^C`	clears any pending commands
`(if eplc_lf`	checks to see if the variable EPLC_LF exists

`(progn`	executes these next steps in order:
`(load "e:/epwin13/program/lisp/lc_pckly")`	loads LC_PCKLY.LSP
`(c:eplc_picklay)`	issues the PICKLAYER command
`)or` `(princ "^MCan't run Pick Layer function without the LANDCADD menu")`	warns that the function can't run and aborts
`)` `princ`	suppresses last line

This entire macro must be entered without extra spaces or lines. The text in the **Macro:** box is small and can be difficult to follow, so be patient and careful. The macro looks like this when entered in the **Macro:** box (Figure 22.12).

Figure 22.12

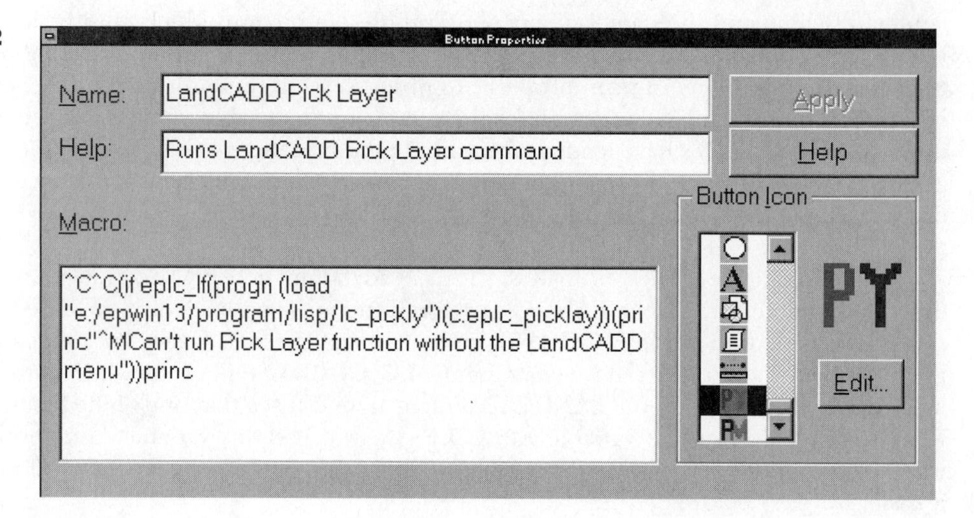

chapter

Establishing Office Standards

One of the key elements in using CAD effectively and efficiently is through the use of standardized routines and methods known as *office standards.* Office standards apply to three basic areas of CAD. First, standards should be established for basic **procedures**. Standardized procedures usually entail such items as file naming and storage, back-ups, virus scanning, handling outside consultant drawings, importing text, etc. These procedural standards are **external** to the CAD drawing process, but are applied to all drawings handled by the design office. Second, standards should also be applied to the various drawing **settings**. These include normal drawing settings such as units, limits, linetypes, layers, colors, dimension styles which must be included in a drawing, along with AutoCAD system settings such as blipmode, dimscale, and ltscale. These office standards are **internal** in that they are a part of the drawing process, and are stored with the drawings when they are saved. These internal settings are crucial to the efficient use of AutoCAD design tools, as well as determining the "look and feel" of the drawing itself. A third area where office standards are applied is **output**. Use of colors, pen widths, line weights and other plotting or printing conventions relate to the appearance of the work when it is produced as hard copy. These standards are both **internal** and **external** in that the settings are contained in the drawing itself, but are read by a hard-copy device which must be configured to correctly and uniformly interpret the information contained in the drawing file.

The benefits of standardization take several forms. First, you take advantage of the computer's suitability for handling routine or redundant tasks. If you consistently use a particular group of settings or drawing blocks, you should save them in a way that allows them to be used over and over. This frees you (and other staff members) from performing some of the tedious, repetitive, chores in drafting—thus permitting you to spend more of your time on creative design tasks. Any drafting process you need to perform more than once should be automated or standardized so that you (and others) won't need to start the same task from scratch next time. Let the computer store and retrieve this work so you never need to duplicate your efforts. When working in any CAD or computer-related task, always keep in mind the adage "DO IT ONCE." This type of efficiency saves both time and tedium.

In addition to the time- and tedium-saving benefits, standardization also helps you maintain a uniform "look and feel" in your drawings. You should set uniform styles in your drawings for text fonts, text size and proportion, dimensions, leaders, and other call-outs and tags. You should also use a core of standard symbols for such objects as plants, sprinklers, north arrows, scales, etc. The styles and settings you define and use can be placed in your drawings using one of two methods. First, the styles and settings can be stored in *prototype drawings* which are used as templates for new drawings. Some designers use prototypes which might appear to be blank drawings, but which contain predefined styles and settings that will be used as the drawing is built. Other designers build more elaborate prototype drawings which contain a title block, a corresponding text block and other commonly used blocks as well as the appropriate settings for the paper size and drawing scale. Generally speaking, these more complex prototypes take longer to build and are more complicated to store and retrieve properly, but they save a great deal of time in the long run.

A second method for inserting styles and settings into a drawing involves placing them in a drawing block which is then imported (inserted) into your drawing sometime during its construction. This method is complicated by the status of certain drawing components in AutoCAD. Some settings (limits, units, ltscale, etc.) are retained in the original drawing regardless of the settings of any block inserted into the drawing. The drawing insertion process will *not* overwrite the existing settings. Other drawing components (blocks, text and dimension styles, layers, linetypes) *can* be imported into a drawing along with the block which contains them. These will *add* to the existing list of blocks and styles in the drawing, but will not *overwrite* existing blocks or styles. The only way to overwrite existing block, style or layer definitions is to deliberately redefine them in the drawing. (Or—in the case of drawing blocks—insert a new drawing block with the same name as an existing one. This will only redefine the drawing block if you allow it.) Because of the conflicts between existing and imported styles, layers and block definitions, it is generally wise to avoid importing them in this way.

You can choose your standard drawing symbols from those supplied with the LANDCADD program, or you can build your own symbols for the job. The central issue of using customized drawing blocks relates to their display (and automatic retrieval) in existing LANDCADD icon menus. LANDCADD allows you to display your custom plant and sprinkler head symbols in the existing icon menus without any programming or menu modification. If the symbols are built and named properly, they will be displayed alongside LANDCADD's symbols in the appropriate places. Displaying other custom symbols in LANDCADD is more complicated. LANDCADD users with AutoLISP experience can modify LANDCADD's menu code to display both LANDCADD and custom symbols in the same icon menu. The task of building custom drawing blocks, creating slides and slide libraries, modifying LANDCADD's menus and then storing the drawing blocks in symbol libraries is time-consuming and very exacting. Do not attempt this type of customization without the guidance of a person with experience in AutoLISP and/or menu modification. If you lack the experience to modify menus but still wish to use custom blocks in your drawings, you can keep a set of standard symbols in a library, then insert them manually. Although this method does not require programming experience, it does require a disciplined procedure for file naming and storage, as well as reference documentation containing pictures of custom blocks and their drawing names.

Another way to standardize the "look and feel" of your drawings is through the use of standard title blocks, cover sheets, and detail drawing pages. These can be stored as drawing blocks, or as prototype drawings. In either case, these standard sheets can serve as templates for each paper size and type of drawing commonly used in your office. As with styles, settings and symbols, it takes time and effort to build the title blocks, cover sheets and detail pages, but their ongoing use can save hundreds of hours of work in the long run.

Another area which is greatly affected by standardization is coordination. If your procedures, styles and settings are defined and standardized, staff members can work on each other's drawings (either directly, or with Xrefs) with a minimum of difficulty. Different pieces of drawings can be added together into larger drawings if necessary, because all of the styles and settings will be consistent. Individual drawings in a project will all have the same "look and feel" if the basic styles and settings are consistent. Standardized procedures, styles and settings can be passed on to consultants and other firms whose drawings will be included in a set of project plans. All of the drawings can share the desired features to make them look unified and integrated.

The office standards you build do no good if they are not communicated to every member of the office team. Take the time to document all of your standards in a CAD Procedures Manual, and keep a copy at each CAD station. Include sections for styles, settings, layering, title blocks, standard drawing blocks, file naming conventions, back-up procedures, etc. Be as complete and precise as you can be in your documentation. The time and effort spent in the production of this document can leverage into hundreds of hours of saved time in the long run. A CAD Procedures Manual will greatly assist CAD staff members in learning the office system and adhering to it. In addition, a procedures manual is an invaluable training aid for new staff members, or contract employees who produce CAD drawings for your office. If you take the time to establish office standards, you should always record them and make them available to anyone who might need them.

Procedures

Building a set of standardized office procedures should begin with the initial configuration of the CAD station computers. It may be desirable to be sure that the boot-up procedure loads files and programs consistently from one computer to the next. With DOS machines, you can examine the AUTOEXEC.BAT file and the CONFIG.SYS file to be sure that the correct drivers, memory managers, disk cache and other programs are being loaded in the correct sequence and with the correct settings for maximum CAD performance. While these files often reflect differences in computer hardware from one system to the next, the same type of drivers and files are usually required for each system in your office. Windows users should examine their memory management settings to be sure that a permanent swap file is being used, and that it is large enough for AutoCAD to function properly.

Anti-virus software should be loaded at boot-up to both scan and clean viruses which can be introduced from infected disks. Since CAD businesses routinely do business with many other computer users, you should assume that every disk coming into your office is infected. This is cheap insurance compared to the time lost if a virus wrecks your hard drive file storage system. Anti-virus software is usually updated frequently to provide protection from newly developed viruses. Your anti-virus software company probably has updated files available on the World Wide Web to help you keep your virus protection up-to-date. When anti-virus software is installed, users often complain that the boot-up procedure is slower and that they resent the time wasted on the virus scan. It would be a serious mistake to bypass your office standards for a few minutes of time savings at the beginning of the day. The anti-virus scanning procedure should *not* be optional or debatable. The costs of virus mischief in your files and programs dwarf the inconvenience of a daily virus scan.

To start programs, be sure that your CAD stations have a consistent way to start AutoCAD, LANDCADD, word processing programs (including a simple ASCII text editor) and any other applications which most of your CAD staff need to use. On DOS machines, some companies provide a start-up menu to make opening programs fast and easy. These start-up menus are usually coupled with batch files which actually start the programs. Be sure that any menu program you use does not use too much memory, or cause conflicts with drivers or programs. Computers equipped with Windows are easy to configure to establish a desktop with short-cuts (Windows 95 and Windows NT 4.0) or program groups (Windows 3.1 and Windows NT 3.5) that make it simple and easy to start programs. While imposing office standards, remember that the primary purpose of standardization is productivity and ease of use, not rigid conformity. CAD designers, like everyone else, need to be able to express some individuality. The arrangement of the computer desktop, the "wallpaper" pattern in the background, the type of screen saver used and the presence or absence of a start-up menu does not usually have any influence on productivity.

Unlike the "window dressing" issues of screen appearance, back-up procedures *must* be very carefully and uniformly implemented and monitored. Establish a systematic approach for backing up both work stations and servers. Back-ups basically take these forms:

System Back-up

A system back-up is used to back up all of the software programs, drivers, memory management programs and hardware settings which are present on a work station. This type of back-up is performed as often as necessary to keep up with the installation of new software programs, hardware and peripheral devices. A full system back-up need only be performed when a new device or program is installed or upgraded. Once this type of back-up is performed, a copy can be stored on-site, while a second copy might be stored at a safe, off-site location (a safe-deposit box is sometimes used for this purpose). The primary purpose of this type of back-up is to restore a work station system after a hard-drive failure. Even if all of the data is lost from a hard disk, the system can be restored to full productivity once the hard disk is replaced and formatted. Simply insert the back-up media, perform a full restore, and the work station should be ready to use. The easiest way to implement a system back-up procedure is to regard it as part of the installation process for any new piece of hardware or software which goes into a work station. As soon as the physical installation is complete, the back-up finalizes the process.

Project File Back-up

Project files should be backed up on a separate schedule from system back-ups. Project files may change daily, so the back-up schedule you employ must take this into account. The system you use will depend on the type of back-up device you use, your back-up media and its storage capacity, and the number of active project files and their size. The project file back-up procedure should employ two types of back-ups. You should perform a daily back-up and a weekly back-up of the project files stored on your system.

The daily back-up might be either *complete* or *incremental.* A *complete* back-up would include all of the active project files stored in your system. These will include AutoCAD and LANDCADD drawing files, but might also include correspondence, spreadsheet files, databases and other related work files. Projects which become inactive should be archived onto back-up media (with two or more copies) then stored off-site. Use a systematic approach when deciding when projects become inactive. If several months pass without referring to a project, you might want to consider removing the project files from your system to leave more room for newer projects. An *incremental* back-up selectively copies only those files which have been changed since the last incremental back-up. If your back-up media has limited space, and you have many megabytes of files to backup, an incremental back-up may be desirable. Often, a complete back-up will require that someone be present to change tapes or cartridges as they fill up, so that the back-up can run through to completion. While some back-up programs operate in the background of the computer system (users are frequently not aware of their operation), a staff member must still attend to the tape or cartridge-changing chores. The incremental back-up may solve this problem if the number of files (and their required space) does not exceed the capacity of a single tape or cartridge. If the back-up can be *unattended,* it can be performed at the end of the work day, freeing your staff from this uninspiring (if essential) duty.

The weekly back-up should be complete, and probably should be removed from the premises when concluded. Weekly back-ups coordinate with nightly back-ups to form a full back-up strategy. Remember that your most recent incremental back-up will only save the files changed since the previous incremental back-up. If you want to keep a copy of the files which *haven't* been changed since the last incremental back-up, you need to store them in a complete back-up, performed weekly. Restoring all of your project files after a catastrophe (disk failure, theft, fire, earthquake, etc.) requires using the most recent *complete* back-up, followed by the most recent *incremental* back-up.

Here are some examples of good back-up strategy, incorporating a combination of complete and incremental back-ups. The exact implementation will depend on the type of back-up media used and the number and size of the files to be backed-up.

1. **Monday through Thursday**—nightly incremental back-ups; the same tape or cartridge is used each night. If the media storage capacity is large, or the number of files (and their file size) is small, each night's back-up can be kept separate from the previous night's back-up. If the media storage capacity is smaller (or if the number and size of the files is larger) each night's back-up can overwrite the previous night's back-up. The primary idea is to create a system which can be performed *unattended* at the end of each day.

   Friday—complete back-up; stored off-site when completed. If this back-up is stored off-site, even if your office burns down, you can restore your computer systems with a maximum loss of one week's work.

2. **Monday through Thursday**—nightly complete back-ups; a different tape or cartridge is used each night. Each Monday night back-up overwrites last Monday night's back-up.

   Friday—complete back-up; stored off-site when completed.

3. **Monday through Friday**—complete back-up nightly; tapes or cartridges taken off-site. After a back-up is completed, the tape or cartridge is taken to a safe, off-site location, and swapped with the previous tape or cartridge. Although this system only requires the use of two sets of tapes (or cartridges) you can use as many as you wish to provide security and minimize the wear on individual tapes or cartridges. While this system is the most secure, it also requires the most time and effort. Large firms with many employees will often assign this task to one (very responsible) person as an important part of their daily duties.

Note that these procedures can be implemented with stand-alone work stations or with a network. The only procedural difference lies in the fact that most networks use a central server to store project files, while individual work stations disperse project files across several computer systems.

As we saw earlier, the keys to creating successful back-up strategy lie in **pre-planning**, **standardization**, and **good record-keeping**. Each of these principles will enable you to select an appropriate system, communicate it to your coworkers and implement it across the CAD network or individual work stations.

File Management

While we discussed the particulars of AutoCAD file management earlier in this book, we will discuss some of the principles which underlie a sound approach to a standardized system for naming and storing CAD system and client project files.

As you begin to develop a CAD drawing system, remember that you will have two types of CAD files to store. First, you will accumulate a series of standardized drawings that are used by the office, but are not specific to any one project. We will call these **Library Files**. Library files fall into a few separate categories: SHEETS, DETAILS, BLOCKS, and TEXT.

- SHEETS typically include prototype title blocks, cover sheets, and detail drawing pages built for each paper size used by the office. If Paper Space is used extensively, one title block, cover sheet and detail drawing page will be required for each paper size. If Paper Space is not used extensively, you may want to keep a copy of each drawing sheet at each scale you routinely use in the office. Although this creates a large number of individual drawing files, each file takes up relatively little disk space, so there should not be a significant impact on hard drive storage.
- DETAILS will include standard details which enlarge and illustrate complex assemblies, installation techniques and building specifications which are best illustrated rather than explained in writing. Details are usually segregated by functional categories like Planting, Paving, Wood Construction, Sports Equipment, Irrigation, Walls and Planters, etc.
- BLOCKS would include objects which are routinely used in the annotation and symbolism of your drawings. You can make extensive use of the blocks provided by LANDCADD, but may find customized blocks more suitable for some purposes. Tick marks, arrowheads, call-outs, generalized planting and irrigation legends, customized plant symbols and irrigation heads would be examples of blocks which should be stored in a library.
- TEXT would be specifications and notes which are frequently used in drawings. These text files can be prepared in a word processing program, then stored as ASCII files. These files can be inserted into an AutoCAD (or LANDCADD) drawing where a text style can be assigned to the block. AutoCAD's MTEXT command makes handling paragraph text blocks relatively simple and easy.

Organizing Library Files is relatively straightforward. Keep each category separate in its own folder (or directory), and build sub-categories as required. Be sure to develop a uniform naming system for your Library Files. You can break the eight letter drawing name (using long file names should be avoided until they are more universally accepted) into separate two or three character modules to help you name your files. These modules can reflect the organization of the drawings within the category, a drawing number sequence, or some predefined code, or a combination of the three. There is no *right* or *wrong* way to organize and name files. Any system is better than no system. Be ready to modify your system if it becomes obvious that it no longer meets your needs. The key to organizing and naming files is to anticipate the type and number of files you will be working with. With this information in mind, develop a system then **document your system**. Place your documentation in your CAD Procedures Manual.

The documentation should include a chart which explains the overall organization of the system, the system used to name the drawing files and a list of drawings found in each category. Obviously, as the list of CAD files grows, this listing will require periodic updates. The Details section will require the most work to maintain. New drawings will accumulate rapidly in this area, so the file list will require the most frequent updates. In addition to maintaining an updated list of Detail drawings, hard copies of each Detail drawing should be stored in a Details Library binder set. As the number of CAD details grows, this binder set will ultimately replace the hand drawn details which nearly all offices currently have in place.

The organizational effort required to build and maintain a library is pointless unless everyone in your office understands it and can use it effectively. Building the system is a one-time effort, but maintaining it requires ongoing work. Creating and maintaining Detail drawings is an excellent way to train new CAD personnel, and to fill in slow times at a design office. The time and effort spent in building and maintaining a library pays off in making everyone's job easier.

Sample Library Organizational Chart:

Library Files (see Note below)

1. Sheets

    A. Title blocks

    B. Cover sheets

    C. Detail sheets

2. Details

    A. Planting

    B. Irrigation

    C. Paving

    D. Handicap details

    E. Walls and planters

F. Seating

G. Ad infinitum

3. Drawing blocks

A. Annotations

B. Dig Alert®

C. Plants

D. Sprinklers

E. Legends

4. Text blocks

A. General notes

B. Soil specifications

C. Irrigation notes

D. Planting notes

Note: it may be very desirable to specify all of your library drawings as **read-only**. This will prevent anyone from modifying the library files and replacing the existing ones with unapproved revisions. A single person should be in charge of making approved revisions to library files. If the library files are **read-only**, a CAD operator can open and edit the drawing, but cannot save it under the same name as the original drawing. This prevents an original drawing from being accidentally overwritten with a revised copy. See your operating system and AutoCAD documentation for information regarding **read-only** files.

The second major group of files to store are **Client Files** which (obviously) relate to clients and their projects. The method you use to name and store these drawings will vary considerably, depending on the number of clients you have, the number of projects you have for each client, the number of drawings you store in each project, and the number of plotted sheets per drawing file. Another complicating factor may be the presence of non-CAD files stored alongside CAD files. If you consider that all files can be organized by project number, a fairly clear organizational picture will appear:

Sample Client File Organizational Chart:

Client Files

1. Client name

A. Project Number

1) Drawings
   a) Site Analysis Plan
   b) Concept Plan
   c) Base Plan
   d) Construction Plan
   e) Demolition Plan
   f) Grading Plan
   g) Hardscape Plan
   h) Planting Plan
   i) Irrigation Plan
   j) Parking and Roadway Plan
   k) Ad infinitum
2) Correspondence
   a) Proposals
   b) Meeting minutes
   c) Billings
3) Spreadsheets
   a) Cost estimates
   b) Materials estimates
   c) Irrigation data
4) Database Files
   a) Planting lists
   b) Product price lists
   c) Equipment lists
   d) Vendor lists
5) Specially revised drawing blocks, details, text blocks
   a) Project- or client-specific details
   b) Project- or client-specific blocks
   c) Project- or client-specific text blocks
   d) Ad infinitum

While this system might not be appropriate for *all* situations, some elements can be used in almost any CAD filing system. First, each client ought to be kept separate from other clients. This arbitrary first level distinction helps break the drawings up into manageable groups. For each client, there will (we hope) be multiple projects, each bearing a separate number. This project number might contain a year module, followed by a sequential module (i.e., **97-127** for the 127th job in 1997) or some other category module. Once a job number is assigned to a project, it can be applied to all correspondence, billing, estimate files, CAD drawings, etc. Each job number would have a few separate directories for drawings, correspondence, spreadsheets, database files, and for client- or project-specific revised drawings or blocks. By placing all of the project files within a single project number folder (directory), you simplify the process of locating files, backing up files, and removing inactive projects from the system hard disk(s). You also employ a standardized organizational system for secretarial, bookkeeping and CAD staff. Since the distinction between these tasks is often blurred in the modern computer-oriented design business, it is very convenient to have all staff members be able to access all types of business files.

The most variable portion of this system relates to the naming of individual drawing files. The system you employ must take into consideration the number and variety of drawing files you will store, and the number of plotted sheets you will produce per drawing. A modular naming system will probably work best for most any file naming system, but its specific implementation will depend on your needs. Using common three- or four-letter abbreviations for the general category of drawing (IRR for irrigation, PLN for planting, HDS for hardscape, etc.) is often helpful. A second and third number module can be used to reference the sequence number of the drawing, and the number of plots in the drawing. Your specific filing system needs should dictate the naming system you use. As with the library filing system, there is no *right* or *wrong* way to name your files. Just be sure to build an organized system and **document it completely**. All of your organizational effort is wasted if you don't communicate it to all of the CAD and office staff.

Handling Outside Drawings

Just as naming and storing files requires a systematic approach, so does handling drawings produced by those outside your office. When an outside CAD consultant or firm provides drawings to your firm, a number of technical and practical questions must be resolved before the drawing can be used. As with establishing standards within an office, the key to working with outside firms is **communication**. You should have a clear understanding of the nature of the files you are receiving, and an established procedure for handling them so that they can be used quickly and efficiently.

When you discuss receiving a file from an outside consultant, a number of issues should be raised:

1. What software (and version) was used to produce the drawing file? Make sure that your version of AutoCAD will read the file that you are receiving. Version 13 AutoCAD should read drawings produced on older versions of AutoCAD, but if the drawing was produced in another program, there may be some difficulty translating the file into an AutoCAD-friendly format.

2. Is the file in a format which can be read by your version of AutoCAD (and LANDCADD)? AutoCAD can read both DWG and DXF format files (although Version 13 AutoCAD has some difficulty handling DXF files), so be sure that the file is in one of these formats. Translating files from one format to another almost always results in lost definitions of fonts, blocks, layers, attributes and colors. Be sure to check your translated result against the original drawing, or with the consultant him/herself to be sure that major discrepancies are resolved.

3. What special fonts and menus appear on the drawing? Do you have a copy of the new fonts being used, or is there an appropriate substitute? If the drawing was produced by a third-party AutoCAD add-on (like LANDCADD) are the menus requested when the drawing opens? (This setting can be changed easily—see your AutoCAD documentation for details.) If this is the case, the drawing will take longer to open and must be resaved with a currently available menu to prevent this problem from resurfacing each time the drawing is opened.

4. What units are used in the drawing? If the source file is in different units than you normally use, you will need to convert the drawing. This occurs frequently when engineering offices provide drawings to architecture offices. A drawing using engineering scale uses one foot as the base unit, while a drawing using architectural units uses one inch as the base unit. The net result is that an engineering must be scaled up by a factor of 12x to convert it to architectural units. LANDCADD has a routine in the **Tools** pull-down menu to convert drawings from one unit system to another.

5. What is the layering system used in the drawing? If the layers conform to AIA, CIS or some other system, be sure to get a copy of the standards used. If the layer system is peculiar to the client, be sure to determine the nature of the system and whether or not you must adhere to it when you do your drawing. If the system is radically different than yours, you will need to spend extra hours on the project. Be sure to consider this when establishing your fees.

6. Are there special blocks or special linetypes defined in the drawings? Determine if there are any blocks which can or cannot be modified in your drawing work. There may be some blocks which you can use repeatedly, but others which you might wish to modify for your use. Be sure to find out this information before spending too long working on your drawing.

7. Are there Xref drawings attached to the file? Are they included with the file? Some consultants make extensive use of Xrefs, so be ready to redefine the path to the Xrefs after they are copied onto your computer system.
8. Does the consultant use Paper Space or Model Space, or a combination of both? What is the scale of the Model Space viewport? Is there a title block included in the drawing?
9. What type of file compression is used on the drawing file? Does your office have the correct program version to decompress the file?
10. What kind of storage media is being used to bring the file into the office? Does your office have the correct drive type to read the file?

Under ideal circumstances, your consultant will have an abbreviated copy of his/her office standards to accompany the drawing file that is being provided to you that will answer these questions. Conversely, you should also have a set of standards in place if you produce drawings for other offices. Be ready to provide a written summary of the answers to the questions above when acting as a consultant for other CAD firms.

Once you receive a file, and the important information which should accompany it, you are ready to put it into your CAD system. These are some of the important steps to follow:

- Scan for viruses as soon as the file is read by a drive. If the file arrives on a tape, Zip cartridge or via modem, **always** scan it before you open it with AutoCAD or LANDCADD.
- Convert units as required. Work produced in engineering units must be translated to architectural units (and scaled by a factor of 12X) or vice versa.
- Examine the layers, colors, fonts, limits and other basic settings. Do they correspond with those on the original drawing? Convert and adjust these factors as necessary to bring the drawing in line with your office standards.
- Examine the block definitions. Are there blocks which should contain attributes? Are the attributes correctly translated?
- Resolve any Xrefs and reset the path to each Xref so that it can be found by the drawing. Bind each to the original drawing if appropriate.
- Examine the Model Space and Paper Space objects. Be sure you understand the settings used in the display of each Model Space viewport.
- After the drawing has been translated to the standards of your office, save it under a new name, in an appropriate location in your file system. Be sure to preserve the original file under another file name. You may need to refer back to it if you inadvertently erase an important feature.

- Erase, purge and **Wblock** the drawing to eliminate unused information which inflates the drawing size unnecessarily. If necessary, reorganize the layers so that large hatch patterns can be isolated and placed on frozen layers so that they do not slow down the handling of the drawing.

If you follow these steps in a systematic fashion, you should be able to use the drawing fairly quickly. If you know the characteristics of the file you will be receiving, you should be able to anticipate which drawings can be used almost immediately, and which will require more time to translate. Be sure to discuss your results with the consultant who provided the drawing. If your translation is badly distorted or contains unacceptable errors, be sure to communicate the nature of your problems with your consultant. The consultant may have a solution to the problem which can be easily performed at their end of the process. Effective communication with your consultant can save you time and prevent future problems.

Handling Hard Copies

Another issue that all designers confront is that of old hand-drawn hard copies with information which must be included as part of a CAD drawing set. When you have a hard copy page that must be converted to CAD format, you have a few basic choices at your disposal:

1. **Scan the image, then trace or convert it**

   This option requires the most expensive hardware and software, but takes the least amount of time to complete. To work effectively with D or E-size blue-line drawings you need a large format scanner (priced from $10,000 to $25,000) and software to convert the scanned image from dots (raster images) to lines, arcs, circles, etc. (vector images). Scanners vary in their speed, resolution capability and ease of use. If you decide to buy a scanner, you should consider how many drawings you will need to scan, what kind of condition the drawings are in, and how you will use the information in the scanned images. At the higher end, expensive scanners can handle E-size blue-line drawings in 15 seconds at 400 dots per inch (dpi) resolution. This type of performance might be appropriate if you need to scan hundreds of documents per month, and use the information to create working AutoCAD drawings. The high resolution of these plotters make them very accurate and capable of filtering out "background noise" found in most hard copy drawings. Paper wrinkles, speckling and red lines can all appear as drawing objects unless they are filtered out during scanning, or removed later with the conversion software. Less expensive scanners will take longer to scan the images, and may produce lower resolution output. These might be appropriate if you only scan a few dozen drawings per month, and the information from the scanner is mostly used in the background of your AutoCAD drawings. Lower resolution scanners may require more time spent filtering out unwanted objects from the scanned file.

Once a drawing is scanned, it is stored as a raster image consisting of hundreds of thousands (millions, in some cases) of dots. These dots can be edited in raster editing software packages (like Photo Shop or even Microsoft Paint), but are not directly useful in AutoCAD. You can import a raster image into AutoCAD, but it remains as a series of dots on the screen. None of the objects created by the dots can be edited with AutoCAD drawing tools. However, this raster file can be used as a tracing background, so that AutoCAD objects can be drawn over the top of the image of dots. Using this overlay method is time-consuming and can require large amounts of labor, but still may be faster than the manual methods described later.

To convert the scanned image into a useful form for AutoCAD, special software must be employed. Depending on the complexity and ability of the software, a raster file of dots can be converted into a vector file of lines, circles, arcs, editable text and other CAD objects (some packages even recognize and convert dimensions). Some types of this *vectorization* software can be operated automatically so that the conversion into CAD objects does not require any attention from the operator. The resulting image will still require editing and clean-up, since there are bound to be objects which cannot be translated accurately. Other vectorization software requires that a user assist with the conversion from dots to AutoCAD objects. This software trades the extra time spent converting objects (the software asks for input periodically during the conversion) for more accurate results, which require much less clean-up when completed. Vectorization software prices range from about $1000 up to $5000.

2 **Manually digitize the drawing**

This method requires the use of a large-format digitizer. A drawing is mounted on the digitizer, then traced in AutoCAD. This results in very accurate drawings, but requires a substantial amount of labor to complete. Some large format digitizers are mounted on pneumatic tables, and take up a large amount of floor space (not unlike old drafting tables). Other large-format digitizers are flexible mats which are made to roll out onto a desktop for occasional use. These represent an attractive alternative for those offices where space is at a premium. Large-format digitizers begin at around $2000.

3. **Manually scale the drawing**

The final method is the slowest and most labor-intensive: manually scale the drawing from the original, then draw it in AutoCAD. This requires only the use of a scale and AutoCAD, but may require dozens of hours of tedious work before a drawing is completed. While this method may require the least capital investment, it will be very expensive on a per-drawing basis.

4. **Send out the drawing to a scanning/digitizing service**

   Remember that there are services available that will scan your drawings and convert them to vector files for you. You can also find services that employ drafters who manually digitize drawings for a per-sheet fee. The expense of the scanning equipment and conversion software makes the use of these services very attractive if you have only a few drawings to convert, and a short time to complete your project. Be aware that current technology may allow some services to send their work to digitizing specialists in production houses in other countries where labor costs are lower. While this is simply a reflection of the international use of CAD, the drafting standards used may not meet your requirements. Be sure to investigate the quality of the work before you send mission-critical work to any conversion service.

Importing Text

While not central to the CAD drawing process, the importing of text files occurs frequently. It is wise to have a routine established for a standard way of producing correctly formatted ASCII files from the office word processing program, and a standard way of importing them into a drawing. The Windows platform offers the best tools for this type of cross-program file handling. In Windows 95 and Windows NT 4.0, you can use the Windows Explorer to simply "drag and drop" ASCII text files into AutoCAD/LANDCADD drawings. You can also use the Windows Clipboard to "copy and paste" ASCII text files from one application to another. Both Windows and DOS users of AutoCAD Release 13 can take advantage of the Import Text option contained in the Mtext command dialogue box.

Be sure that the entire office staff understands how to create ASCII files, and that the CAD designers know how to import the text into their CAD drawings. If you use one of the Windows platforms, be sure to let your staff members know about the Windows Notepad program, which creates pure ASCII text files. You can open and edit ASCII files in this program before you place them in your AutoCAD/LANDCADD drawing.

Settings and Styles

While the office procedures are important to provide the environment and organizational scheme for working with drawings, they do not relate to the actual content of the drawings. The Settings and Styles relate to the portions of the drawing which can be preset and standardized for drawing efficiency. We will examine those standardized components relating to drawing content in this section.

Start-up Drawing

When you open LANDCADD, you automatically start the Eagle Point Project Manager program, with a default start-up drawing in the background. When you install LANDCADD Version 13 "out of the box," the start-up drawing is EPA_SEED.DWG in Windows; EPSTRT13.DWG in DOS. This blank drawing is basically a clone of the original ACAD.DWG, the default drawing provided with AutoCAD.

Both EPA_SEED.DWG and EPSTRT13.DWG have a number of default settings that are not really suitable for architectural drawings, and should be modified. While it is easy to modify this drawing and tailor it to your needs, a better approach is to change the default setting to a drawing you create specifically for this purpose. In Windows versions of LANDCADD, the default start-up drawing selection is contained in the file EP.INI. The example below shows the file modified to use the drawing MYSTART.DWG. As you can see, simply change the drawing name and path shown after the environment variable STARTUPDWG. The drawing you specify will then load as the default drawing. Note that this setting will override the default drawing set in the AutoCAD configuration file ACAD.CFG.

EP.INI

```
acad
C:\epwin13\PROGRAM\;C:\epwin13\DWG\;C:\R13\COM\SUPPORT;C:\R13\
COM\
FONTS;C:\R13\WIN\SUPPORT
acadexe C:\R13\WIN\ACAD.EXE
STARTUPDWG C:\13dwgs\MYSTART.DWG
ep C:\epwin13\
epdwg C:\epwin13\DWG\
epsupport C:\epwin13\SUPPORT\
```

DOS versions of LANDCADD store the start-up drawing name in the LANDCADD.BAT file found in the EP directory. The edited version below shows the file modified to load the default start-up drawing as MYSTART.DWG. Notice that the default start-up drawing is listed on the same line as, and immediately after the ACAD.EXE command. Changing the drawing name and path listed here changes the default start-up drawing.

LANDCADD.BAT

```
@echo off
C:\EP\PROGRAM\EPADLOGO
set acad=C:\EP\PROGRAM;C:\EP\DWG;C:\R13\DOS;C:\R13\DOS\SUP-
PORT;C:\R13\COM\SUPPORT;C:\R13\COM\FONTS
set acadcfg=C:\R13\DOS
set acaddrv=C:\R13\DOS\DRV
set EP=C:\EP\
set EPDWG=C:\EP\DWG\
```

```
set EPSUPPORT=C:\EP\SUPPORT\
set dos4g=quiet
set dos4gvm=@C:\EP\support\dcsdmsed.vmc
set $path=%path%
set path=C:\EP\PROGRAM;%path%
type C:\EP\SUPPORT\SWTAB.TXT
C:\R13\DOS\ACAD.EXE C:\13DWGS\MYSTART.DWG
type C:\EP\SUPPORT\RETTAB.TXT
set acad=
set acadcfg=
set acaddrv=
set EP=
set EPDWG=
set EPSUPPORT=
set dos4g=quiet
set dos4g=
set dos4gvm=
set path=%$path%
set $path=
```

When you build your default start-up drawing, consider the settings you would want to have in effect if you left the Project Manager to begin a new drawing. You may not do this often if you always build new drawings with existing prototypes. Nonetheless, it might be easier to have your most commonly used settings in effect if you leave the Project Manager without specifying a prototype. As an example, if you most often worked in architectural units at 20 scale (1" = 20'-0"), with C-size paper, you could set the drawing up so that the default drawing had settings appropriate for those choices. With the architectural units, C-size, and 20 scale parameters in mind, the following settings could be saved with the drawing:

AutoCAD settings

Units	Architectural
Limits	0,0 to 480', 360'
Ltscale	120 (to start)
Dimscale	240
Grid	20'
Snap	1'

Blipmode	Off
UCSicon	Off
Viewres	3000
Standard text style	Romans.shx

Eagle Point Project Manager settings

Scale (Horizontal)	240
Text Height	.125"

Layers

Whether you are building a default set-up drawing or developing a prototype title block for later use, one of the most important areas of standardization to address is that of **layers**. Layers are the most important tool available for grouping, displaying and plotting information in your drawing. Using layers properly allows you to display and plot several drawings from a single CAD file, and assists you and your staff in organizing your drawing and maintaining a uniform "look and feel" in your drawings. To this end, the layers you use, including the layer names, colors and linetypes assigned to them, should be standardized for your office.

There are several benefits of using standardized layering. First, using layers permits groups of objects (drawn on the same layer) to be displayed at the same time with the same color which visually (on screen, at least) implies their relationship to each other. When objects have the same color and the same linetype on screen, it is also reflected in the plotted version of the drawing. Objects drawn on the same layer will have the same line thickness and linetype on the plotted copy, thus communicating a relationship even on a blue-line print. Another benefit from standardized layering comes when drawing are merged. If standardized layers are used, new, imported objects are placed on their respective layers in a predictable way. They will occupy layers which are already identified, and control of those objects is no more complicated than control of existing objects on the drawing. If individual layering schemes are used, the complication factor escalates dramatically. The number of layers may double after two drawings are merged, and one of the designers will need to learn the layer names used in the other drawing. Colors, linetypes and objects must be changed or relocated to bring the drawing under control. The time spent in this type of endeavor is unproductive and wasteful. A final advantage of standardized layering relates to communication with consultants and other members of a project team. If standard layer names are communicated to each team member and used throughout a project, each party can complete their task in less time and with less complication. Transferring information between drawings and between designers is simplified because the underlying organization of the files is identical.

A layering system should be implemented with the goal of having enough layers to provide for all of the categories of information placed on a drawing, but with no more than are justified. Unused layers increase drawing file size unnecessarily, and complicate the organization of existing drawing information. (If there are many unused layers when the drawing is completed, you can simply purge the extra layers as a final step before the drawing is saved.) The system should take into consideration both the objects that are included in your drawings and those which are likely to be added later by other designers or consultants. There are several standard layering systems that are promoted by different professional organizations involved in CAD management. We will examine these standards later in this section.

Once you (and your staff) have decided on a layering system, document it thoroughly in your CAD Procedures Manual. After documenting the layering standards, you can place the layers in all of your prototype drawings and your default start-up drawing. By building the standard layers into your prototype drawings, you avoid the time-consuming chore of creating each layer individually. If you work with several clients who each use different layering standards, you might choose not to place the layers in your prototype or default drawings. There are several different approaches to automating layer construction. Some designers employ AutoLISP routines or automated scripts to build layers, while still others will use blocks which contain different layer definitions. In either case, different layer schemes can be imported into your drawing during the set-up phase of drawing construction. All drawings produced by your office should contain some scheme of standard layers, with the drawing information arranged according to the layers used. Anyone receiving your drawings should be provided with a list of layers, their assigned colors, linetypes and suggested plotted line widths. This will assist other designers in integrating their drawing system with yours, or (at least) illustrate your adherence to your clients' layering standards.

Standard CAD Layering Systems

There are several standard systems for CAD layering being used in the construction and design industries. While it is rare to see any of these systems implemented in a rigid, formal way, they form a basis for many of the office layering systems used today. Most of these systems employ a modular approach to naming layers. Each layer name might consist of three modules, each separated with a dash. Some of these standards use named modules, while others use numerical modules. Proponents of each system make arguments that theirs is the superior one, but that decision will be up to CAD users like you and your clients.

AIA—The American Institute of Architects has developed a very widely-used standard spelled out in its 1990 publication *CAD Layer Guidelines: Recommended Designations for Architecture, Engineering, and Facility Management Computer-Aided Design.* Although a newly updated version is scheduled for release in March 1997, the AIA standards are currently among the most widely used by architectural designers. In the AIA **Long Format** system, the first module divides the layers into eight *major group* designations which correspond to the traditional disciplines separating the various document sheet numbers:

A—Architecture, Interiors and Facilities Management

S—Structural

M—Mechanical

P—Plumbing

F—Fire Protection

E—Electrical

C—Civil Engineering and Site Work

L—Landscape Architecture

The second module consists of four characters, and is called the *minor group.* The minor group is usually a category of work within the specific discipline. In landscape architecture, we see minor groups with such names as:

PLNT—Plants and Landsape Materials

IRRIG—Irrigation System

WALK—Walks and Steps

SITE—Site Improvements

ELEV— Elevations

SECT—Sections

DETL—Details

SHBD—Sheet Border and Title

SCHD—Schedules and Title

The third module (if used) is another four character group, called the *modifier.* The modifier creates subdivisions of the minor group. In a planting plan, layer names might show modifiers to the minor group PLNT:

L-PLNT-TREE—New Trees

L-PLNT-TXST—Existing Trees to Remain

L-PLNT-TDMO—Existing Trees to Be Removed

L-PLNT-GRND—Ground Covers and Vines

L-PLNT-BEDS—Rock, Bark and Other Landscaping Beds

L-PLNT-TURF—Lawn Areas

L-PLNT-PLAN—Schematic Planting Plans

A fourth module could be applied to the layer names if there were specific needs in the project for even more specific designations or divisions. Although AutoCAD will permit the use of layer names as long as 31 letters, the size of dialogue box text boxes (and common sense) make it impractical to use layer names longer than about 16 characters.

The AIA standard also provides for **Short Format** layer names, which contain the major group letter, followed by two-letter abbreviations of the minor group and modifier. Using the Short Format layer names, we would see the layers above named this way:

L-PLNT-TREE—LPLTR

L-PLNT-TXST—LPLTX

L-PLNT-TDMO—LPLTD

L-PLNT-GRND—LPLGR

L-PLNT-BEDS—LPLBE

L-PLNT-TURF—LPLTU

L-PLNT-PLAN—LPLPL

There are several advantages to using modular layer names of the type advocated by AIA. The modular approach makes it easy to use the **Filters** function in AutoCAD's "Layer Control" dialogue box. With this feature, you can display only those layers which begin with **L-PL**. This is important since the AIA system can create a large number of layers in a drawing.

The AIA system was intended as a *guideline* for layer naming, not as a dogmatic system which must be rigidly adhered to. If you base your own layering system on the AIA guidelines, feel free to add different minor groups or modifiers to suit your own organizational system.

CSI—The layering system developed by the Construction Specifications Institute (CSI) is based on their divisions of the construction sequence and the specification numbering system that accompanies it. These sixteen divisions are designated in the first two digits of a five-digit layer number.

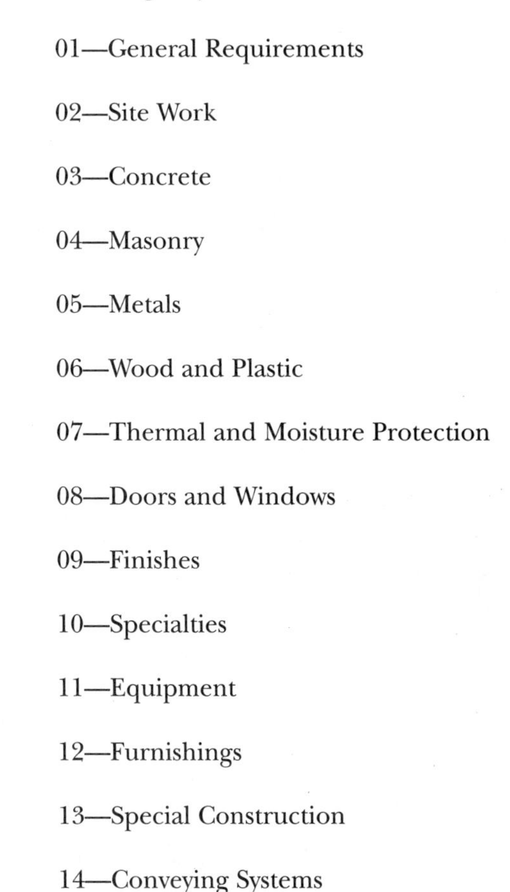

01—General Requirements

02—Site Work

03—Concrete

04—Masonry

05—Metals

06—Wood and Plastic

07—Thermal and Moisture Protection

08—Doors and Windows

09—Finishes

10—Specialties

11—Equipment

12—Furnishings

13—Special Construction

14—Conveying Systems

15—Mechanical

16—Electrical

The remaining three digits are used as more specific designations within the sixteen areas. Although this layering system is somewhat cryptic (and usually requires the use of a detailed CSI index), it does coordinate with CSI's extensive array of construction document preparation tools. Like AIA, CSI is revising its layering system in a new Uniform Drawing System (UDS), which is also scheduled for a March 1997 release date. While the coordination of these two standards remains to be seen, there has been an attempt to standardize layering into some sort of national standard. This National Computer Aided Drafting and Design Standard has been advocated by the National Institute of Building Sciences (NIBS).

LANDCADD uses the CSI numbering system in combination with AIA descriptive terminology to form a sort of hybrid layering system. The LANDCADD automatic layering system (if used) builds layers that reference the first four digits of the CSI numerical system, followed by one or two other modules, similar to the AIA minor group and modifier. LANDCADD refers to these modules as **Sublayers** or **Nodes**. We will examine the LANDCADD tools for layering later in this section.

ISO—The International Standards Organization has developed a series of world-wide standards for all types of electronic documentation. ISO standards are used for drawing sheets, hatch patterns, drawing symbols and other objects. The intention of the ISO layering standards is to promote a system that can not only be used in a consistent form from country to country, but also from CAD program to CAD program. (That it might be easier to get two countries to cooperate more easily than two CAD software companies will pass without comment.) It remains to be seen what type of standards will be adopted by ISO, but those firms doing international business (particularly in Europe, where ISO is an emerging standard) should watch these developments closely.

LANDCADD Layering

LANDCADD has several tools available to make these layering standards more accessible and easier to implement. You can define, edit and build a complete set of layers using the **Layers ▸** option found under the **Tools** pull-down menu. When you select this option, the bottom selections of the menu relate to layer creation (Figure 23.1).

When you select the **Set Sublay File** option, you open the "Enter Sublayer file name" dialogue box. From this dialogue box, you can select the file which corresponds to the layering standard that you want to employ. The file you select can be one supplied by LANDCADD, or one you build yourself (Figure 23.2).

The AIA.NOD file will load the AIA standard layers associated with Civil Engineering and Landscape Architecture. While most of the important minor groups and modifiers are loaded, you can always build more layers if necessary.

Figure 23.1

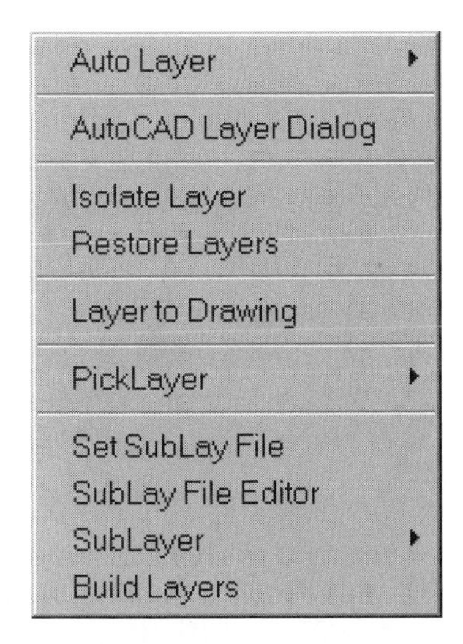

Figure 23.2

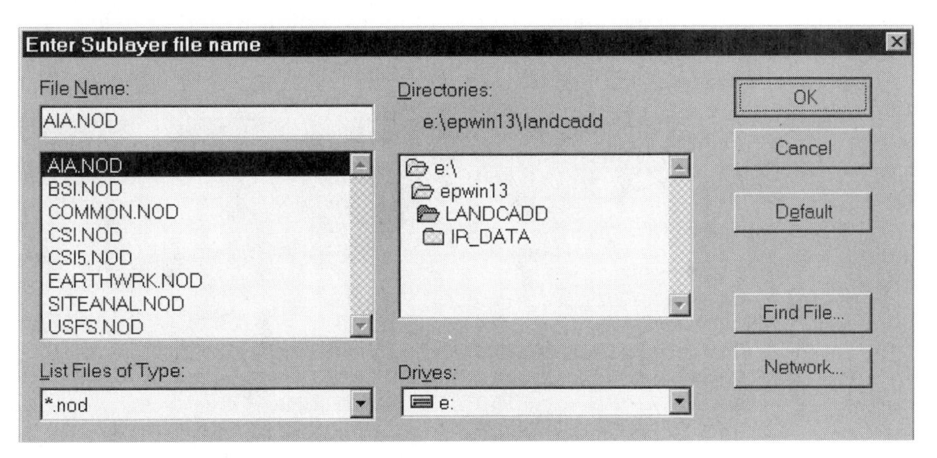

The following files are associated with these organizational layering standards:

AIA.NOD	American Institute of Architects
BSI.NOD	National Institute of Building Sciences
COMMON.NOD	LANDCADD's common names for layers
CSI.NOD	LANDCADD's modified version of Construction Specifications Institute layers

CSI5.NOD	Original 5-digit Construction Specifications Institute layers
EARTHWRK.NOD	Eagle Point layers built with Civil Engineering software
SITEANAL.NOD	LANDCADD layers built with Site Analysis software
USFS.NOD	U.S. Forest Service layers

Once a layer file has been selected, the layer names can be edited and their properties modified before they are inserted into the drawing. This is accomplished through the use of the **SubLay File Editor**. This editor contains complete tools for modifying the layers (Figure 23.3).

Each layer name is listed along with a box to illustrate its color and linetype. It is very easy to edit any (and all) of these properties in this dialogue box, but it is not possible to add or delete layers. Since the NOD files are simply ASCII text files, it is a relatively simple matter to edit them in a common text editor to add or delete layers. The format

Figure 23.3

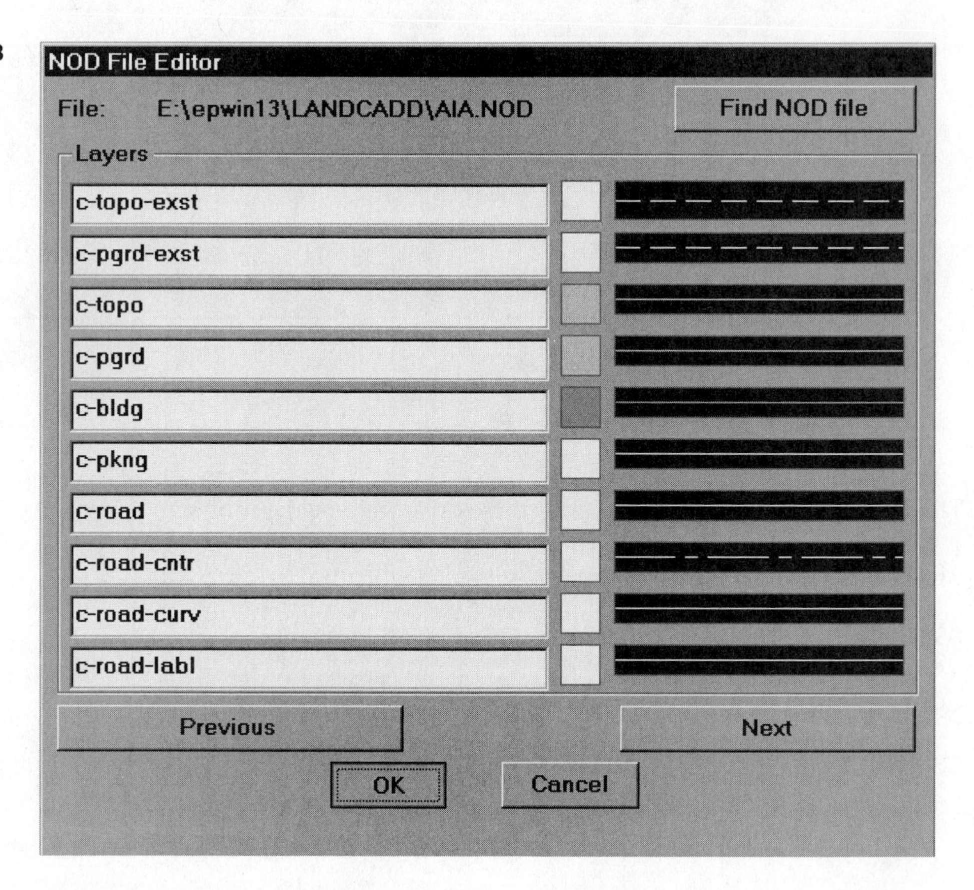

of these files is simple and easy to follow. The following list illustrates the first five layers from the AIA.NOD file:

1-4-4	the character lengths for each module—begins every NOD File
1	☐ layer number—begins the layer definition
c-shbd-nrth	layer name
7	layer color
continuous	layer linetype
2	☐ layer number
c-shbd-scal	layer name
7	layer color
continuous	layer linetype
3	☐ layer number
c-shbd	layer name
7	layer color
continuous	layer linetype
4	☐ layer number
l-site-mtch	layer name
7	layer color
dashed	layer linetype
5	☐ layer number
l-elev	layer name
7	layer color
continuous	layer linetype

If you want to add a layer at the end of the list, simply enter the next number in the sequence, assign a layer name, layer color and layer linetype to complete the description. You can re-enter the SubLay File Editor to check your work and make last minute modifications.

It is a simple matter to build a NOD file for each major client with individual layering requirements. These layer definitions are stored in small, simple files which can be retrieved and used whenever a new client drawing is created. This may be an attractive alternative to storing more complex prototype drawings or layer blocks which exist solely to add layers into your drawings. These prototypes and layering blocks consume more disk space and may be more awkward to store.

The layer definitions in the NOD file (whether provided by LANDCADD or built from scratch) do not take effect until the layers are actually added to the drawing. The layers are added to the drawing when you select the **Build Layers** option. After you select this option, a few moments pass as the layers are built. The new layers will then appear in the "Layer Control" dialogue box (Figure 23.4).

When LANDCADD is installed, you can select which layering standard you want to have as the default. You can select from AIA, CSI and Common layering names. The **Auto Layer** switch is intended to control whether LANDCADD will automatically add layers when new objects are placed in the drawings using LANDCADD routines. If this feature is turned off, no new layers are built, and objects are simply placed on the current layer.

As we saw before, the LANDCADD CSI layering definition is really a hybrid between the five-digit CSI system and the major group, minor group and modifier system employed

Figure 23.4

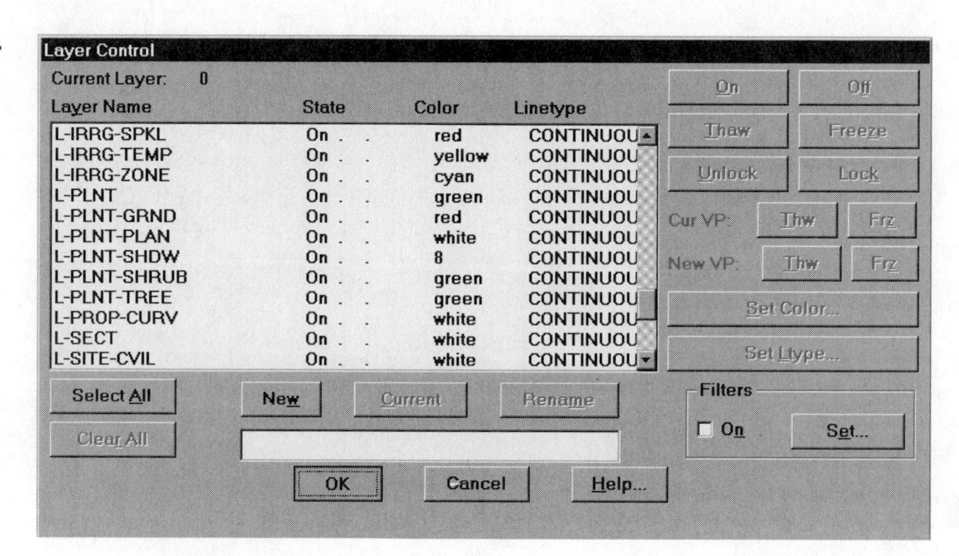

by AIA. When you use the LANDCADD CSI system, the first module of the layer name (called a **Node** in LANDCADD) is a four-digit number which is taken from the CSI layer list. The second and third modules are alphabetical, and are descriptive like the AIA minor groups and modifiers. This allows the cryptic number scheme found in CSI layering to be more specific and descriptive. As an example, the CSI category **0214** relating to roadways is broken into five layers:

0214-roads-cl

0214-roads-curve

0214-roads-label

0214-roads-line

0214-roads-point

The module divisions allow you to use the **Filters** option in the "Layer Control" dialogue box, so that you can display only those layers which begin with the desired number **0214**. LANDCADD layer tools also allow you to control the display of layers by selecting them from the module (node) number. This means that you could also shut off all layers which have the second module named **roads**. The LANDCADD tools allow more variety and precision than the AutoCAD **Filters** option, in that you can control the layers by individual modules, rather than by wild card characters in a dialogue box.

Another valuable LANDCADD tool which allows you to define layers in drawings is the Layers ▸ option Layer to Drawing. This option allows you to build a new drawing while including layers from objects you choose in the current drawing. When you open the new drawing, the layers corresponding to the objects you selected will be predefined in your new drawing.

Blocks

The LANDCADD program contains over 1000 (one thousand) drawing blocks which can be inserted into your drawings. Although the retrieval of these blocks is relatively simple in most cases, there may be certain blocks that you may wish to embed directly in your drawings, rather than needing to retrieve them through a menu or dialogue box. The blocks you decide to store in your drawing might be large, complex objects like title blocks, text blocks (placed inside the title block or in the drawing itself), or library blocks, which might contain a number of other, nested blocks within them. On the other hand, the embedded blocks might be relatively simple, like north arrows, scales, callouts, arrowheads, or tick marks. If you decide that you wish to embed the block definition in one of your prototype drawings, simply insert the block into the drawing, but instead of specifying an insertion point, press the **<Esc>** key. This will insert the block definition without displaying the block anywhere in the drawing.

Remember that block definitions are very efficient in terms of drawing size and disk space usage. A block definition of a group of objects consumes considerably less space than the objects drawn individually. Nonetheless, be sure that the blocks you embed in your prototypes are really required. If you find they are not, you should remove them from your prototype drawings.

Text and Dimension Styles

One of the easiest ways to standardize the "look and feel" of your drawings is to standardize the text style used in your drawings. When the fonts, proportions and text sizes are consistent, the drawings will look more unified. In addition, it is much easier to enter text in a drawing quickly if the styles have been predefined. If the styles are predefined, you simply need to select which style to use, rather than defining it from scratch. You can use fonts from any source you choose, but if you use fonts not included in AutoCAD or LANDCADD, you need to be sure that AutoCAD and LANDCADD can find the fonts you are using. There are two ways to accomplish this. First, you can copy the fonts into the normal location for AutoCAD fonts, the R13/COM/FONTS folder (or directory). If there are too many fonts involved, or if you want to keep the fonts separate from AutoCAD's native fonts, you can store them in any directory you choose, so long as it is defined in the AutoCAD "search path". The AutoCAD search path is found in the file which defines the operating environment for AutoCAD or LANDCADD.

In the EP.INI file, the information following the **acad** statement is the search path that AutoCAD and LANDCADD will follow when looking for files (including fonts). DOS LANDCADD users can find the search path in the LANDCADD.BAT file. In DOS, the search path is found immediately following the **acad=** statement. If these statements do not exceed the 127 character limit, you can add another directory (path included) to the end. This will enable AutoCAD and LANDCADD to find the font files you are using.

Another important ingredient in the "look and feel" of drawings relates to dimension styles. Because dimension styles can be complex and time-consuming to create, it is especially valuable to predefine them in your drawings. As with every facet of standardization, document every aspect of the styles included in your prototype drawings. Show how the styles are defined, named and stored. Be sure that every CAD designer knows the office standard styles and uses them. Using standardized styles will prevent conflicts from arising when identically named, dissimilar styles are used when two drawings are merged. The original drawing settings for all styles will always take precedence over the inserted styles if they both have the same names. This can lead to considerable confusion unless the styles are identical in the first place. If you settle on specific styles and style names early in the organizational scheme, you will avoid problems later.

Linetypes

Pre-loading linetypes can also save time and avoid confusion in the drawing process. As long as the number of linetypes does not inflate the file size too greatly. If there are several unused linetypes in the drawing, you can always purge them when the drawing is complete. Be sure to document which linetypes are predefined in your prototype drawings.

Custom Aliases, Macros and Toolbars

Once the general CAD filing and layering system is defined and the prototy drawings are created and organized, you should turn your attention to standardizing the customized routines and modifications that you want to make available to all of the LANDCADD and AutoCAD designers in your office. These customized routines are stored in only a few places:

1. The ACAD.PGP file which stores simple keyboard aliases for single-step commands
2. The ACAD.LSP and LC_KB.LSP files, which can load more complex macros
3. The ACAD.MNS or EPAWSWAP.MNS file, which contains the accelerator key definitions and the toolbar definitions.

As you learn the capabilities and limitations of AutoCAD and LANDCADD, and as you learn your own working habits and the types of tasks you perform most often, you will find that there are some functions which can be automated or accelerated by the use of some special abbreviation or macro. Do not hurry to develop these additional tools until you have a solid understanding of LANDCADD and AutoCAD, so that you don't spend time to develop a tool which already exists in some location you didn't know about. As you begin to build some new command aliases, macros and accelerator keys, be extremely thorough in refining and testing them. If you ultimately want to make the tools available to others, this debugging process is especially important.

Once these files are refined and thoroughly tested, you can copy them to all of the work stations to make the same tools available to all CAD designers. Again, there are no *right* or *wrong* aliases, macros and toolbars to use, so be as creative as you wish. But be sure to **document your tools**. Be sure that the alias, macro and toolbar definitions are clearly documented and that the documentation is available to all users. Be sure to communicate your intentions to all of your staff before you overwrite their files with yours. If users have their own toolbars, AutoLISP routines and command aliases already defined, you may be able to combine your work and theirs into appended files which contain both types of customization. If there are conflicts between your definitions and those of other staff members, try to keep standardization as a high priority. If different designers use the same machines, keeping consistent macro definitions will be required to maintain everyone's sanity.

Be especially careful to back up all of the original files before you replace them with newer versions. These older versions may be very helpful if your new "improved" version has a bug or an unexpected conflict that causes problems. Be sure to provide a way to restore the old versions of these customized files if you run into trouble. After the dust settles and your customized tools are in place, be sure to create a new system back-up to reflect the new files. Restoring a crashed system to its customized configuration requires a current back-up.

Plotting Standards

The logical connection of the standardized layers, colors and linetypes found in your office standards and the final appearance of your drawing is caused by the relationship between screen drawing colors and plotter pen numbers, and the pen numbers to the plotted line widths. The standards defined here will appear in the output which emerges from your plotter or printer. As you define your layers and screen colors, you should always consider your final drawing, and the line-weight hierarchy you want to use.

If you will be using a pen plotter, you have the choice of (usually) four different pen widths. Use .70mm for very heavy lines, .50mm for medium-heavy lines, .35mm for medium lines and .25mm for light lines. Although there are several colors of ink available, most designers use black ink for reproducible plots on vellum or mylar. Since most pen plotters feature six or eight pens, the most important feature of the line-weight scheme is which colors will correspond to which pen numbers. Once these assignments have been made, creating a standard plot simply depends on placing the correct pens in the correct locations in the plotter. While there are many drawbacks associated with pen plotters, complication is not one of them.

If you are using an ink-jet plotter, there are many more choices at your disposal. First, plotter pen widths are adjustable from .005" (approximately .13mm) to .080" (over 2.00 mm). Second, many ink-jet plotters feature color plotting. There are normally 256 discrete colors available in most AutoCAD applications. Although color plots are valuable for rendering and presentation drawings, they are seldom suitable for reproducible plots. However, the gray colors assigned in AutoCAD (AutoCAD Color Index colors 8, 9, 250 through 255) can be used as half-tone colors in plots. This gives you the option of placing some objects in the "background" of your drawing while placing others in the "foreground" by use of plotting colors. You could use this to diminish the prominence of a house and surrounding hardscape in a planting plan or irrigation plan. Another choice relates to the plotting quality for the individual plot. Many plotters offer Draft Mode, Final Mode and even Enhanced Mode. The higher plot quality settings routinely use more plotting ink, take more time for a plot, and often consume more plotter memory. Obviously, there is more flexibility built into ink-jet plotters than you will use for reproducible plots. If you plan to produce rendered drawings, presentation drawings and other full-color applications, you will use many more of these advanced features.

When creating a standard plotting scheme for an ink-jet plotter, you will probably only use four or five line weights for your plots. If this is the case, you may decide to use pen widths relatively close to the pen widths provided for pen plotters. But because ink-jet plotters are less likely to clog and skip at narrow line widths, you can probably use finer lines than would be possible with pens. This makes it possible to read smaller text and numbers than would be possible with pen plots or with hand-drawn work. The chart below indicates approximate line widths in millimeters and inches:

Drafting Pen Size	Width in Inches	Width in Millimeters
000000	.005	.13
0000	.007	.18
000	.010	.25
00	.012	.30
0	.014	.35
1	.020	.50
2	.024	.60
2-1/2	.028	.70
3	.031	.80
3-1/2	.039	1.00

If you are using a laser printer for some of your drawing output, you may have the line width flexibility of an ink-jet plotter, but not the color assignments. Since the smaller plots generated by a laser printer will almost certainly require a different line-weight hierarchy for legibility of small details, you might wish to create a different standard for laser prints than for plots. If you want to create a full-scale test plot of a portion of a large drawing, you may want to plot with standard plotter line-weight assignments in order to correctly assess the quality of the large drawing. If this is the case, you would need two separate configurations for the same device.

After you have experimented with your output devices and have settled on a standard set of plotting conventions (layers/colors/pen numbers/pen widths), you are ready to: a) commit these standards to a table in your CAD Procedures Manual, and b) create configuration files for your devices so that these standard conventions are accessible to all of the CAD staff. The configuration file which saves these settings is the PCP (Plotter

Configuration Parameters) file. You can save these plotting settings to a named PCP file in the "Plot Configuration" dialogue box. You will need to create a PCP file for each configuration you use for each device. If you need to plot drawings which make use of your clients' plotting standards, you should create a PCP file for each of those configurations as well. The PCP file is another simple ASCII file which can be edited in Windows Notepad or in the DOS Edit function. An example of a PCP is illustrated below:

```
;Created by AutoCAD on 10/23/1996 at 15:27
;From AutoCAD Drawing e:\13dwgs\base
;For the driver: Hewlett-Packard HP-GL/2 devices, ADI 4.2 - for Autodesk by
HP
;For the device: HP DesignJet 650C

VERSION = 1.0                This header information is placed by the driver file in effect
                             when the PCP file was created

UNITS = _I
ORIGIN = 0.00,0.00
SIZE = E
ROTATE = 0
HIDE = _N
PEN_WIDTH = 0.010000
SCALE = _F
PLOT_FILE = _NONE
FILL_ADJUST = _N
OPTIMIZE_LEVEL = 1

BEGIN_COLOR = 1              AutoCAD Color Index number (color 1 = red)
 PEN_NUMBER = 1              Plotter "pen" number (ink-jet plotters have "virtual pens")
 HW_LINETYPE = 0             *Plotter linetype for this color—normally set to 0 (continuous)
 PEN_WEIGHT = 0.007000       Line width assigned to this color (.007")
END_COLOR
BEGIN_COLOR = 2              AutoCAD color                     (color 2 = yellow)
 PEN_NUMBER = 2              Plotter pen number                (2)
 HW_LINETYPE = 0             Plotter linetype                  (0)
 PEN_WEIGHT = 0.024000       Line width assigned to this color (.024")
END_COLOR

BEGIN_COLOR = 3              AutoCAD color                     (color 3 = green)
 PEN_NUMBER = 3              Plotter pen number                (3)
 HW_LINETYPE = 0             Plotter linetype                  (0)
 PEN_WEIGHT = 0.007000       Line width assigned to this color (.007")
```

END_COLOR		
BEGIN_COLOR = 4	AutoCAD color	(color 4 = cyan)
PEN_NUMBER = 4	Plotter pen number	(4)
HW_LINETYPE = 0	Plotter linetype	(0)
PEN_WEIGHT = 0.012000	Line width assigned to this color (0.12")	
END_COLOR		
BEGIN_COLOR = 5	AutoCAD color	(color 5 = blue)
PEN_NUMBER = 5	Plotter pen number	(5)
HW_LINETYPE = 0	Plotter linetype	(0)
PEN_WEIGHT = 0.007000	Line width assigned to this color (.007")	
END_COLOR		

*linetypes normally controlled in the drawing, not at the plotter

PGP files are created every time you build a set of plotting pen, color and line width assignments and then saved to a file in the "Plot Configuration" dialogue box. AutoCAD does not provide sophisticated text editing tools to make modifying the PGP file quick and easy. By using the "search and replace" functions of an ASCII text editor, you can edit PCP files more conveniently. Obviously the more colors and pens you use, the larger the chore of building a PCP file. Remember that the parameters indicated for the unused colors and pens are not important, so do not spend time editing settings you will never use.

Once the PCP files are completed, you can retrieve the one you want to use for plotting when you select the plotting device and the plotter defaults. Using this system helps you adhere to the "DO IT ONCE" adage that we saw at the beginning of this section.

Preparing a set of CAD standards for your Procedures, Settings and Styles, and Plotting can be a daunting task. But the time spent developing and documenting these standards can pay huge dividends. You can become much more productive with LANDCADD much more quickly if you implement office standards *before* you accumulate a large number of non-standard projects. *Any* standard is better than no standard at all. If you anticipate your needs, you can develop a workable system of standards that can be used by your CAD staff almost immediately.

chapter

Using CAD on a Network

Introduction

One of the most important trends in business computer use is Networking. If there are more than two computers in your office, someone will probably engage you in a discussion on how (and why) to build a network. If you work with other CAD users for any period of time, the odds are good that your computers will (sooner or later) be connected by a CAD network.

With apologies to Jeff Foxworthy:

If your Windows desktop looks different every day—you may be on a network.

If your Simpsons background wallpaper now sports the company logo—you may be on a network.

If your Dilbert screen saver mysteriously became Flying Toasters—you may be on a network.

If your customized AutoCAD toolbars and macros just disappeared this morning—you may be on a network.

If you just started getting e-mail from people in your company that you never met (or wanted to)—you may be on a network.

If you don't recognize any of the files shown in your Windows Explorer—you might be on a network.

There are several advantages to running computers in a network. Networks allow you to share data and resources with other users, and use equipment and time more efficiently. Shared data storage allows all users in an office access to all of the drawing, correspondence, and project management files for any active project. This sharing allows for extensive use of powerful AutoCAD tools like Xrefs, prototype drawings and block libraries, and makes using office standards much easier. If all work files are shared and stored centrally, making back-ups is greatly simplified. On more extensive networks, users in remote sites can share project information and data files. Different members of a project team in different offices (or different cities) can share drawing information and improve communication. Networks also allow more efficient use of computer resources. Printers, plotters, CD-ROM drives or other storage devices can be shared by all users connected to the network. Each user can have access to each device as though it were connected to that user's personal work station. This improves office efficiency and eliminates the need for the "sneaker net" system of transferring files by floppy disk. Networks can even allow for shared communication links with distant networks and (of course) the Internet.

Network Components

Networks have a number of components which work together to form a cohesive system:

Network Operating System (NOS)—an operating system which allows the sharing of devices, communication between computers and sharing of files. Network operating systems are produced by "big name" software companies such as Microsoft (Windows NT, Windows 95, Windows for Workgroups), Novell (NetWare), Artisoft (LANtastic), and Banyan (VINES).

Network Interface Card (NIC)—a peripheral card which plugs into a PC motherboard to allow connection to the network. This device should attach to the fastest bus connection on your computer. The bus connection between the card and your computer's motherboard determines the speed and bandwidth of the communication between the network card and your computer. Pentium computers usually have ISA (Industry Standard Architecture) 16-bit slots and PCI (Peripheral Component Interface) 32-bit slots available. Usually 486-based computers only have ISA slots available. Use a PCI network card if your computer supports it.

Network Cabling—although there are several types of cabling available, you will most likely see either coaxial cable (with its characteristic black outer color and large metal connectors) or unshielded twisted pair (UTP) wire in a network. UTP wiring is becoming more popular than coaxial cable. Traditional UTP wiring is called 10BaseT cable. The name 10BaseT indicates the cable carries **10** megabits per second, direct current (also called **Base**band) signaling and is **T**wisted pair wire. Newer UTP wire may be manufactured to conform to Category 5 standards. Category 5 wiring is the highest quality available, and can be used for 100BaseT connections. 100BaseT network components can carry information 10 times as fast but are (obviously) more expensive. Under most any circumstances, if you are running network cables for a new network, you should use CAT5 network cabling. Since cabling is the most difficult component of a network to upgrade after it is installed, it is best to plan for growing network use and buy the best cable you can afford.

Server—a computer used to perform some specific task in the network. There are *file servers, printer (or plotter) servers, communication servers* and *application servers.* A file server is used to store files for the users in the network to access. Printer (or plotter) servers are used to allow access to an output device from any machine on the network. A communication server is used to provide access to outside networks or the Internet. An application server would be used to house centralized versions of software programs that all users would access to do their work. The type of server used depends on the basic philosophy of the network. A *server-based* network depends on a powerful centralized computer to handle these tasks for dozens of users. Such a server might feature hundreds of megabytes of RAM, many disk drives, and the fastest processor(s) on the market. In a server-based network, all computers communicate with the server, but not necessarily with each other. A *peer-to-peer* network spreads the server tasks over several computers. With a peer-to-peer network, no single central computer is in charge of the network, and several work station computers might have server responsibilities. One computer might act as a printer server, while another functioned as a plotter server, while a third performed as a file server. Peer-to-peer networks generally do not require expensive equipment and highly-trained personnel to install. In a peer-to-peer network, computers can all communicate with each other, not just with a central server.

Client—a computer which requires access to a server computer (or computers) to function fully. In a *server-based* network, a client computer may depend on the server to access important programs, to print files, and to communicate with the outside world. In this type of network, client computers are often equipped with smaller hard drives, less RAM and slower processors. In a *peer-to-peer* network, the distinction between a client and a server becomes less significant. Any computer with a peripheral device connected to it might, in a peer-to-peer network, act as both a client and a server.

Hub—a connecting device which accepts network cables from multiple computers. If a file server were used, all of the client computers would be connected to a hub, then the hub would be connected to the file server. A hub allows several computers to be connected in parallel, rather than in series. Computers connected in series perform like Christmas lights: if one piece of cable is damaged, a large segment of the network goes down with it. Computers connected in parallel are individually connected to the hub, so if one cable is damaged, only the affected computer goes down. Hubs are relatively inexpensive and are considered very basic devices for network construction.

Network Topology—this term relates to the organizational structure used in creating the physical network of computers. The type of cabling used, the network adapter cards and the network operating system all interact to create the topology of the network.

If one network is to connect with another network in a remote location (across town or across the world), another group of devices and connections are required.

Bridge—a device which allows two networks with different communication systems (perhaps caused by different cabling or different network topologies) to connect with each other. A bridge could be a software program or an actual hardware device. Bridges don't actually "direct traffic" by sending information to specific destinations. This type of function is performed by a router.

ISDN—an Integrated Services Digital Network line which provides inexpensive, high-speed connection from one network site to another. Companies and individuals making extensive use of Internet services often purchase ISDN service, which can be provided by a local telephone company. Like traditional telephone service, ISDN service is charged based on connection time. The more it is used, the higher the bill every month.

Router—a device which regulates and controls data which is being sent from one Local Area Network (LAN) to another. Typically a router is used to sense outgoing data being sent to a remote LAN. It then uses the ISDN line to connect to the remote side, then transmits the outgoing data. The remote LAN which receives the data also has a router which then sends the data to the correct computer work station. After the LAN transaction is complete, the router senses the lack of traffic, then disconnects the line. Routers typically are controlled by the network operating system.

Firewall—a special type of router which examines all incoming data and prevents unauthorized access from remote sites. Data is typically sent in pieces called *packets.* Each packet can be sent with an authorizing address attached to it. If the firewall does not detect the presence of this authorized address, it does not accept the data. Firewalls are an important part of network security.

Gateway—a special translation program that allows one network operating system to communicate with another. A gateway will allow mainframe computers to communicate with PCs, PCs to Macs, or Macs to minicomputers.

CAD Network Issues

Should we run a client-server or a peer-to-peer network?

For the majority of cases, CAD departments should be set up with modified peer-to-peer networks. Individual work stations should have enough hard disk space and RAM available to run the required software, while a central *file server* (for storing files, not running programs) would have a large hard disk (or disks) which can store data files for the entire office. In this modified peer-to- peer network, the work stations are more capable than traditional "client" computers, while the file server is not as powerful as a traditional "application server." This modified peer-to-peer approach makes sense for CAD users for several reasons:

1. Programs run slower from a network application server. CAD programs are usually very large, and require large amounts of data to be accessed, loaded into memory, and rewritten continuously during a session. This is especially true with AutoCAD 13. Trying to run AutoCAD through a network server can be slow for one user, but can become absolutely agonizing if several users log on together. The most efficient way to run AutoCAD is to load it from a fast hard disk connected to your own computer.

2 Architectural CAD networks are seldom large enough to justify the expense of setting up a true server-based network. With small networks, it may be difficult to justify the purchase of a dedicated server which will perform no other function than to run the network. True client-server networks are technically more challenging to build, more expensive to purchase and more difficult to maintain than peer-to-peer networks.

3. Peer-to-peer networking is built into Windows for Workgroups, Windows 95 and Windows NT. If these operating systems are used, no extra network software may be required. The lower expense may allow the purchase of other peripheral devices such as extra printers or plotters.

4. Peer-to-peer networks are easy to set up and run. Setting up a small peer-to-peer network with Windows 95 or Windows NT Workstation can often be accomplished in less than a day.

What about software licenses?

CAD professionals should always be sticklers for software licensing. You have two basic choices for licensing AutoCAD. First, you can license each copy of AutoCAD independently of the rest, so that you will have a separate license for each CAD computer in your network. This is the correct way to install AutoCAD if you want to load it from each work station. Your second option is to buy a single network license for all of your AutoCAD stations. You purchase a network version of AutoCAD, then install it on your server. The network version of AutoCAD comes with a *dongle.* A dongle is a security device which plugs into the printer port of the server, and is programmed for a set number of nodes (user stations). Although you can run AutoCAD from as many work stations as you wish, there is a set limit as to how many users can access AutoCAD at any one time. This means that you could install a five-node license on a ten-user network, but only five users could log on to the AutoCAD program at any one time. While the network version of AutoCAD allows central control of the program and licensing, it sometimes causes problems during the log-on procedure, and adding more nodes onto an existing license can be time-consuming and difficult.

LANDCADD is generally purchased as single-station software. Like AutoCAD, LANDCADD generally works best when run from a local hard disk. Several files and settings are saved from the previous session of LANDCADD, which can create real chaos if the program is run from a central application server. One user may log on to find their settings overwritten by another user who logged on later. This can make it difficult to access files and retrieve certain settings and databases.

How do most CAD users set up a network?

The most widely used way to run CAD on a network is to use a file server to store working files and office standard files. In the CAD world, the file server is used to store title blocks, detail drawings, specialized drawing blocks and text blocks. If the same file server is used to store other office files, it may also be used to store correspondence, proposals, specifications, and other word processing files and templates. The file server should be equipped with a very large, very fast hard drive. If your file server will be expected to serve several users and store several gigabytes of files, consider using SCSI (Small Computer Systems Interface) drives. These drives offer faster file access and greater "throughput" than other types of hard disks. There are several "flavors" of SCSI drives (each with its own speed and data throughput rate) from which to choose. As with most choices relating to computers, buy a larger, faster drive than you think you will ever need, and it should serve you well—for two years or less.

While working and office files are typically stored on the server, it makes sense to store a copy of your customized files to the server as well. If you use a customized set of keyboard macros, AutoLISP programs, accelerator keys or other tools, you should store them in a central location for loading onto new systems or to aid in recovery from a local hard disk failure. Since the file server is the "repository" for your working files, it should be the computer whose files are backed up daily. While it is possible (even desirable) to back up a server from a tape drive connected directly to it, be careful. If you choose to place the back-up device in your server, it is best to use a separate controller card for your tape drive. That way the controller data channel will be fully available to the back-up device, while leaving the disk drive controller channel unaffected. If it is not possible to place two controller cards in your computer, consider placing the tape (or other) back-up device in a work station computer and perform the back-up over the network. This will result in slower back-ups, but if back-ups are scheduled for non-work hours, the slower process will not be important.

Output devices (mostly laser printers and plotters) are either connected to work stations throughout the network (which act as print or plot servers), to the file server, or directly to the network. The type of connection between the output device and the network is very important if speed is a consideration.

Traditional output devices must be connected to a computer to function properly. The computer could be a work station (acting as a plot or print server) or the file server. These devices can be connected through a computer's communication (serial) port or through its printer (parallel) port. Although it is possible to connect a plotter with a serial (RS-232) connection, this type of connection is limited to about 2400 baud (bytes/second). While mechanical pen plotters are connected as serial devices, this type of connection is considered unacceptably slow for laser printers and modern (inkjet and electrostatic) plotters. A more common connection is through a parallel printer port, which operates at a more respectable 9600 baud rate. Most laser printers and modern plotters have parallel port models with a characteristic Centronics connector in the back. Centronics connectors have large sockets, connect with wire clips and have a bar-type connector device, rather than individual connector pins normally associated with RS-232 connectors.

Newer output devices can act as "stand alone" machines without a direct connection to any single computer. These devices use the fastest type of connection available: a direct network connection. Many new printers and plotters are available with built-in network adapters. These network printers and plotters can receive data at nearly 500,000 baud, about 50 times faster than a parallel connection. These devices do not require the use of a plot or print server computer for their connection to the network. Instead, they are simply connected to the rest of the network with normal network cabling. Although this type of connection is very fast, it requires that the output device contain a substantial amount of memory to receive large files as it prints (or plots) them. This built-in memory is required to take the place of the print or plot server which can store plot requests in its memory. Since memory is relatively inexpensive (about $2.00 per megabyte at this writing), adding ample memory to a plotter or printer should not add too much to its cost. If you expect your output device to receive heavy use, consider buying one with a built-in network connection.

As previously stated, the preferred method for running AutoCAD and LANDCADD on a network is from local disk drives. This approach provides for best performance and ease of use. But running the software from the local disk does require more work when installing programs, upgrading and customizing them. The key to doing any network operation efficiently is to use the network capability to work to your advantage. Once files are available on the file server, they can be accessed by all of the work stations. This suggests an important strategy when installing software: use your file server as an "installation server."

If your file server has plenty of room on the hard disk (it should), copy the entire installation CD (or floppy disk set) to the server. When you want to install the program on a work station, simply perform the installation from the file server drive. The file server then acts as the original CD-ROM drive. You still must meet the licensing and security requirements at the local hard disk, but the installation does not require CD-ROM drives at every work station, nor does it require toting disks all over the office when installing a patch or software upgrade. Depending on which version of AutoCAD you use, you may be required to supply an individual personalization disk for each installation or upgrade, or have a dongle present before AutoCAD will run. LANDCADD installations may require a specific key code number for its software-lock versions, or a dongle for its hardware-lock versions. After the installations are completed, you can delete the original program files from the server. Office standard menus, AutoLISP routines and keyboard macros can also be sent to individual work stations without difficulty if they are stored on the file server.

Although peer-to-peer networks are relatively easy to install, and relatively inexpensive (at least compared to three years ago), they should be planned and installed with great care. The adage "an ounce of prevention is worth a pound of cure" is especially applicable to networks in general and CAD networks in particular. Plan your overall network scheme with thoughts toward heavy usage, large drawing files and frequent printer and plotter use. If in doubt about how to set up your CAD network, hire a consultant to plan it with you. Whatever expense is incurred in the planning stages will be more than offset by the savings in the long run. Be very careful about your cable and connections. The problems caused by poor cabling and connections can be very difficult to trace and may persist for months before you find their cause. Finally, make sure that there is a clear-cut assignment of network administrative duties. Someone in the office should be responsible for backing up files, installing software and generally looking after the network. Ideally, that person should be hired with those duties in mind. If this is not the case, be sure that the person performing these essential duties has the proper training and experience to do them correctly. While many CAD networks are "home grown," administering them should not be taken lightly. Network administration on even small networks can grow to be a substantial job. Be sure that enough time and expertise are dedicated to administering your network properly. The health of your business may depend on it.

appendix

LANDCADD Command Synopsis

Landscape Design Module

The Landscape Design module of LANDCADD provides some of the most useful tools for developing construction and presentation planting plans. Using Landscape Design, you can insert plan view, elevation view and 3-D plant symbols which, in turn, will have AutoCAD attributes assigned to them. Plant symbols have two attributes: plant name and planting size. The symbols range from the most rudimentary circles to very complex presentation trees and shrubs. You can also create, store and retrieve your own customized plant symbols. Once the symbols are inserted, they can be labeled with several styles of plant labels. Groundcover areas can be drawn using several different edging patterns, while groundcover areas and quantities can be calculated and placed in the drawing as attributes. Both plant and groundcover attributes can be tallied and compiled to generate a bill of materials report.

In addition to plant and groundcover symbol utilities, the Landscape Design module features an extensive plant database which, in its original form, contains plants representing all U.S.D.A. climate zones. You can search this large database for plants which meet certain criteria, and develop customized lists of plant names for use in later projects. You can create your own plant records, adding species not already included in the original file of over 1000 plants. To more fully customize the LANDCADD program, you can develop smaller database files for plants which meet your local requirements. To assist in compiling estimates, the module also gives you the tools to assign a cost, category and comment to each plant. This helps to organize and simplify the estimate report, which can be printed, inserted into the drawing or saved to a file.

Planting pull-down menu

Label Symbol used to label existing plant symbols in a drawing

Veg Line... used to create specialized polylines for plant and groundcover borders

Plant Mix used to create groundcover labels and plant species mixes for enclosed polyline areas

Figure A1.1

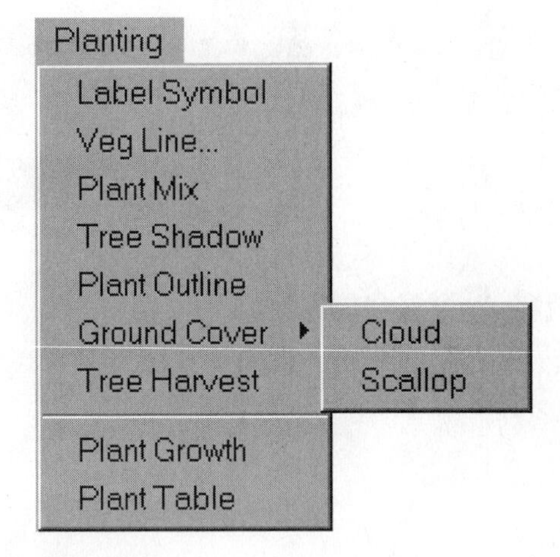

Tree Shadow used to create shadows on existing plants and trees in a drawing

Plant Outline surrounds a group of plants with a heavy polyline made of arcs

Groundcover ▸ creates polyline "clouds" or "scallops" around desired area

Tree Harvest removes selected trees by size or percentage

Plant Growth scales plant symbols up or down to simulate plant growth over time; correlated to plant database growth rate

Plant Table creates a schematic legend of plant symbols and plant names

Symbols pull-down menu

The options in the **Symbols** pull-down menu are intended to assist you in placing plant symbols into your drawing, along with a system for building your own 2-D and 3-D symbols. LANDCADD's selection of 2-D and 3-D plant symbols is available though a series of options within four major icon menus: **Broadleafs**, **Conifers**, **Palms_Cactus** and **Shrubs** (Figure A1.2).

Broadleafs contains an icon menu of Simple Plan, Detailed Plan, Simple 3D, Detailed 3D and Section View broadleaf tree symbols

Conifers contains an icon menu of Simple Plan, Detailed Plan, Simple 3D, Detailed 3D and Section View coniferous tree symbols

Figure A1.2

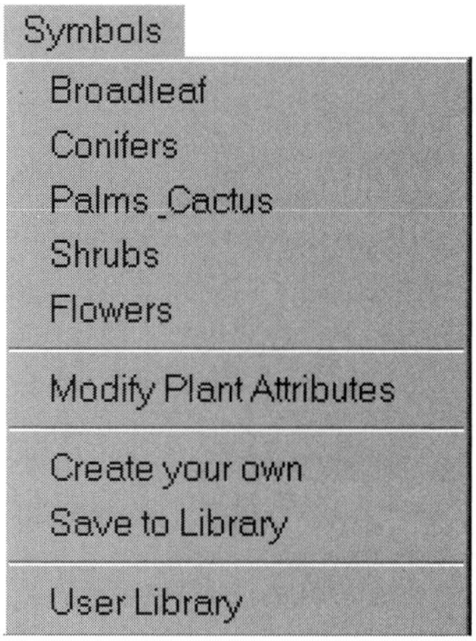

Symbols
Broadleaf
Conifers
Palms_Cactus
Shrubs
Flowers
Modify Plant Attributes
Create your own
Save to Library
User Library

Palms–Cactus contains an icon menu of Simple Plan, Detailed Plan, Simple 3D, Detailed 3D and Section View palm trees and a few tropical plants and cacti symbols

Shrubs contains an icon menu of Simple Plan, Detailed Plan, Simple 3D, Detailed 3D and Section View shrub symbols

Flowers contains Detailed 3D and Section View flower symbols

Modify Plant Attributes allows the redefinition of plant symbol attributes; new names and sizes can be selected from a plant list

Create Your Own allows you to create your own 2-D and 3-D symbols for inclusion into the User Library of plant symbols

Save To Library allows you to save the symbol on the screen into the User Library; assigns block name to the symbol and creates a slide file

User Library displays the symbols defined in the User Library

Database pull-down menu

The options found in the **Database** pull-down menu are intended to help you search through LANDCADD's extensive plant database, add your own plant records and compile plant lists for your projects (Figure A1.3).

Configuration allows you to select the database file for plant searches; sets the name category (scientific, common or code) used to display plant names

Figure A1.3

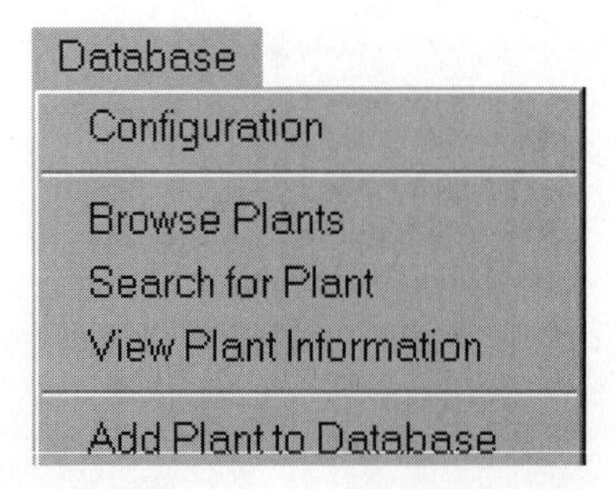

Browse Plants allows you to read through all of the plant names in a database

Search for Plant allows you to sort through a plant database, searching for plants by sorting criteria that you choose

View Plant Information allows you to examine the database record of any plant you have inserted into the drawing; select the plant from the screen

Add Plant to Database allows you to add a plant to the active database; displays all screens required to fully describe a new plant

Common Menus

The following **Tools** and **Presentation** pull-down menus are common to the Landscape Design, Irrigation Design and Site Planning modules. These menus provide access to miscellaneous tools which are useful in many drawing situations.

Tools pull-down menu

The selections listed under **Tools** are intended to make common design tasks easier and more intuitive. Although some of these functions have been made moot by the advances in AutoCAD Release 13 and the Windows graphical user interface, others remain very relevant and convenient (Figure A1.4).

Blocks ▸ opens the LANDCADD block options (Figure A1.5).

North Arrows... displays an icon menu of LANDCADD's north arrow icon menu; insert, scale and rotate arrow symbol

Scales... displays an icon menu of LANDCADD's scale symbols; insert and set scale factor

Figure A1.4

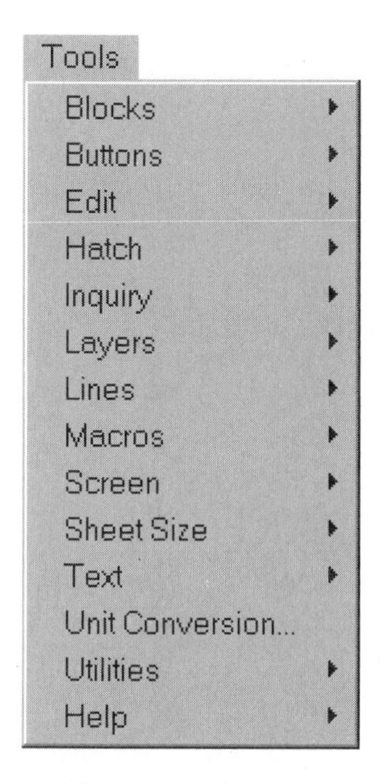

Figure A1.5

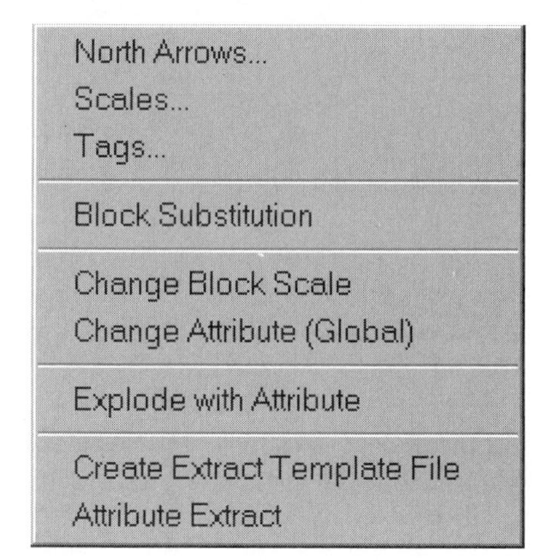

Tags... displays an icon menu of LANDCADD's tags, balloons, leaders and other labeling symbols

Block Substitution opens the block substitution commands; allows the swapping of one block for another—substitutions can be saved to a file

Change Block Scale allows specific blocks to be scaled and rotated; scaling is independent in X, Y and Z axes

Change Attribute (Global) allows all of the occurrences of a block to be redefined with a new attribute

Explode with Attribute explodes symbol and turns attributes into text on screen

Create Extract Template File allows you to create a template file for extracting attributes in the drawing; used in AutoCAD functions to create attribute extract files

Attribute Extract counts occurrences of different attributes and places them in a file; file can be used by database or spreadsheet programs

Buttons ▸ opens the LANDCADD buttons options[1] (Figure A1.6).

Figure A1.6

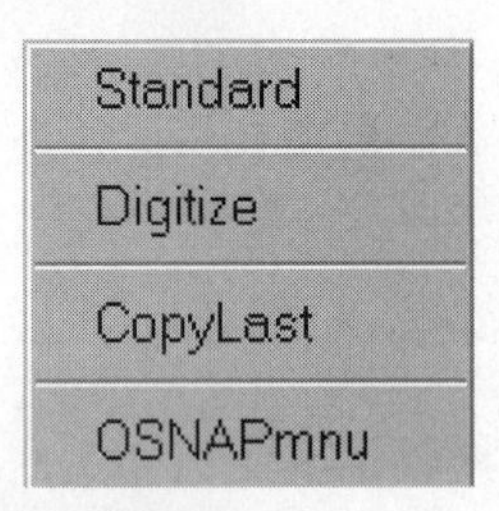

Standard reassigns buttons on pointing devices to LANDCADD's standard button configuration (uses Redraw, Cancel, Snap, Ortho, Grid, Coordinates, Isoplane and Tablet)

Digitize reassigns buttons on pointing devices to LANDCADD's Digitize option (uses Line, Arc, Polyline, Circle, Dtext, Erase, Break, Change, Copy and Move)

CopyLast reassigns buttons on pointing devices to LANDCADD's CopyLast option (uses Copy/Last, Move/Last, Redraw, Cancel, Snap, Ortho, Grid, Coordinates, Isoplane, and Tablet)

OSNAPmnu reassigns buttons on pointing device to LANDCADD's OSNAPmnu option (uses Center, Endpoint, Insert, Nearest, Intersection, Midpoint, Node, Perpendicular, Quadrant, Quick and None)

[1]LANDCADD button options will not function in Windows applications where the active driver causes the digitizer to function in mouse mode.

Command	Description
Edit	opens the LANDCADD edit options (Figure A1.7).
Copy and Rotate	copies and rotates objects that you select
Move and Rotate	moves and rotates objects that you select
MultiCopy	makes multiple copies of objects you select
MultiScale	scales multiple objects relative to their center points or midpoints rather than from a common point
Polyline Trim	trims objects to a polyline cutting edge
Reverse Polyline	reverses the sequence of points placed in a polyline; does not affect the shape of the polyline
Polyline Selection	creates a selection set bordered by a selected polyline; the polyline must exist before the selection set can be created
Hatch ▸	opens the LANDCADD hatch options (Figure A1.8).
Acad Bhatch	opens the AutoCAD Bhatch command
Bomanite	opens the Bomanite hatch pattern icon menu, then follows LANDCADD's hatching command
Brick	opens the brick hatch pattern icon menu, then follows LANDCADD's hatching command

Figure A1.7

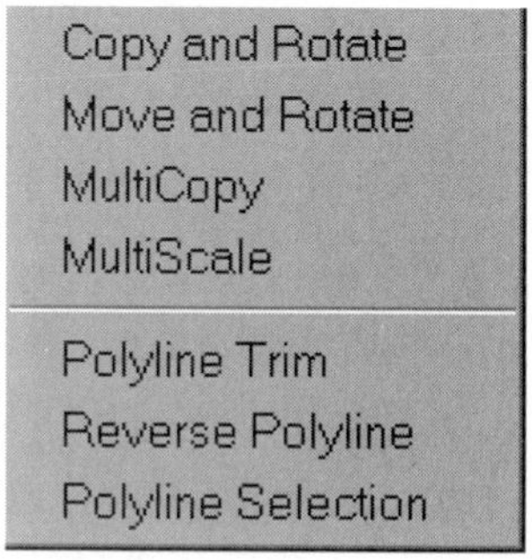

Copy and Rotate
Move and Rotate
MultiCopy
MultiScale

Polyline Trim
Reverse Polyline
Polyline Selection

Figure A1.8

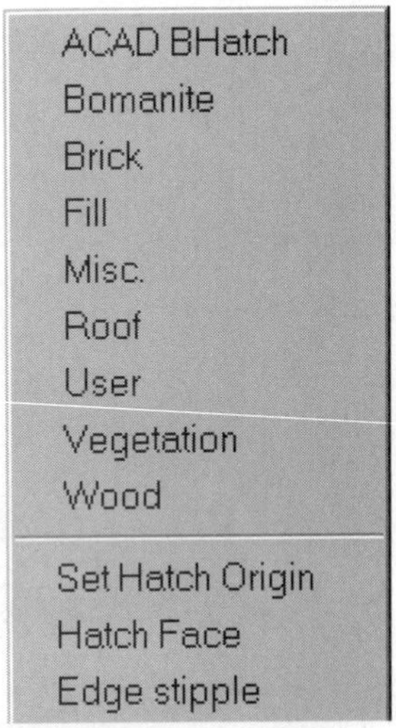

ACAD BHatch
Bomanite
Brick
Fill
Misc.
Roof
User
Vegetation
Wood
Set Hatch Origin
Hatch Face
Edge stipple

Fill opens the fill pattern icon menu, then follows LANDCADD's hatching command

Misc. opens the miscellaneous hatch pattern icon menu, then follows LANDCADD's hatching command

Roof opens the roofing hatch pattern icon menu, then follows LANDCADD's hatching command

User creates a manually controlled hatch pattern at the command line

Vegetation opens the vegetation hatch pattern icon menu, then follows LANDCADD's hatching command

Wood opens the wood hatch pattern icon menu, then follows LANDCADD's hatching command

Set Hatch Origin allows you to specify a starting location for a given hatch pattern

Hatch Face places a hatch on a three-dimensional face such as a wall or roof plane

Edge stipple creates a graduated hatch pattern of stipples; dense toward the outside, sparse toward the center

Inquiry ▸ opens the LANDCADD inquiry options (Figure A1.9).

Figure A1.9

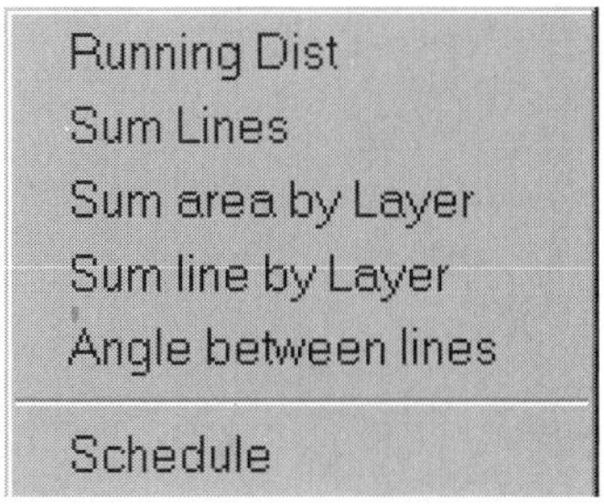

Running Dist
Sum Lines
Sum area by Layer
Sum line by Layer
Angle between lines
Schedule

Running Dist allows you to measure the distance traveled through a series of points selected on a drawing; creates and measures temporary line segments

Sum Lines measures and totals the lengths of a selection set of lines and polylines; filters out other objects

Sum area by Layer reports the area defined within closed polylines on a given layer; the layer of the selected object is used

Sum line by Layer measures and totals the lengths of lines and polylines on a given layer; the layer of the selected object is used

Angle between reports the angle between two lines you select lines

Schedule creates a full inventory of blocks used in a drawing; attributes are listed for each block; *the resulting inventory can be viewed or sent to a file*

Layers ▸ opens the LANDCADD layers options (Figure A1.10).

Auto Layer ▸ controls the Auto Layer function

Auto Layer ON turns automatic layer creation on or off; LANDCADD commands will build new layers when this feature is turned on

Auto Layer OFF

AutoCAD Layer Dialog opens the AutoCAD "Layer Control" dialogue box

solate Layer Iisolates layer of selected object by freezing all other layers

Restore Layers use after "Isolate Layer"; thaws all layers frozen by that command

Figure A1.10

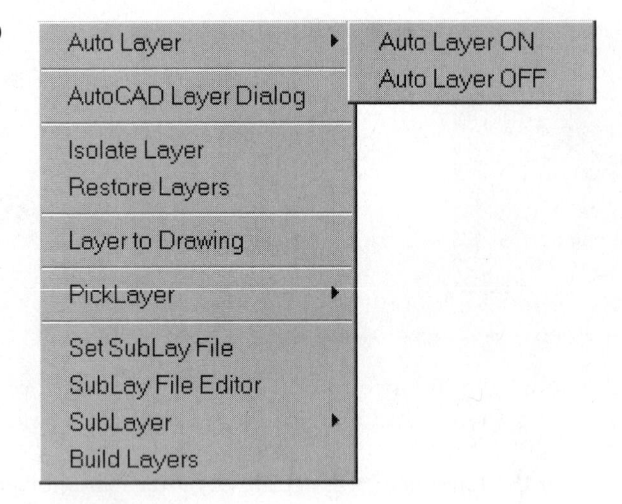

Layer to Drawing selects all objects on layers represented by objects selected, then creates a new drawing file with the objects; variation on WBLOCK command

Figure A1.11

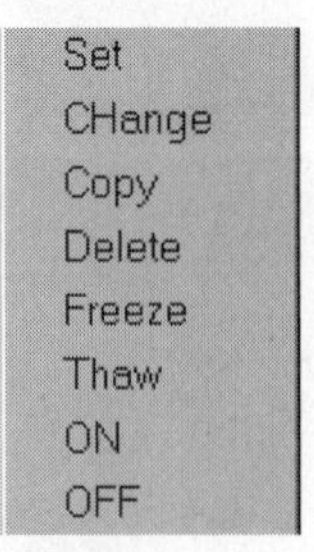

PickLayer ▶ (Figure A1.11).

Set sets current layer to that of the object selected

Change changes a selection set layer to that of another object selected

Copy copies a selection set to the layer entered at the command line

Delete erases all objects on a layer entered at the command line

Freeze freezes the layer(s) of the selection set chosen

Thaw thaws layers affected by "Freeze" command above

ON turns on a layer affected by "OFF" command below

OFF turns off the layer(s) of the selection set chosen

Set Sublay File selects the layer names built in the drawing from a file you select; sublayer files use the extension **.NOD** — AIA, CSI, USFS and other choices are available

Sublay File Editor opens a sublayer file editor which can alter the name, color and linetype assigned to each layer in the sublayer file

SubLayer ▸ tools to control the Sublayer functions; many of these tools are superceded by the layer control functions in AutoCAD Release 13 (Figure A1.12).

Figure A1.12

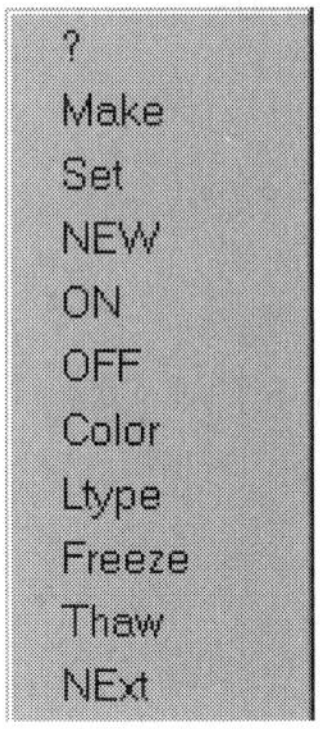

?
Make
Set
NEW
ON
OFF
Color
Ltype
Freeze
Thaw
NExt

? allows you to search for a layer name by node (a portion of the layer name) at the command line

Make allows you to create a new layer name at the command line; makes the new layer the current layer

Set allows you to set a new layer by searching the layer names by node (a portion of a layer name)

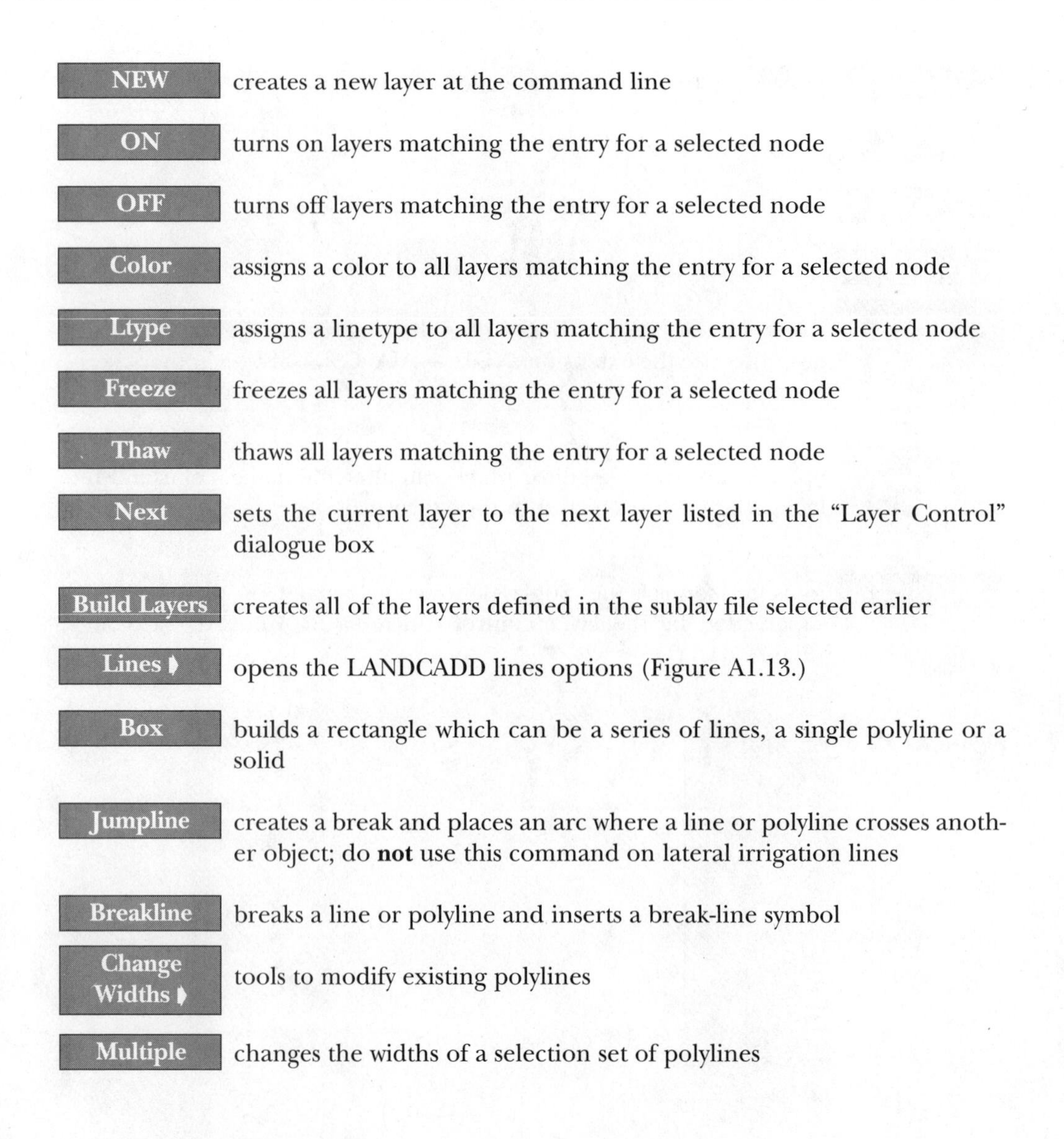

NEW creates a new layer at the command line

ON turns on layers matching the entry for a selected node

OFF turns off layers matching the entry for a selected node

Color assigns a color to all layers matching the entry for a selected node

Ltype assigns a linetype to all layers matching the entry for a selected node

Freeze freezes all layers matching the entry for a selected node

Thaw thaws all layers matching the entry for a selected node

Next sets the current layer to the next layer listed in the "Layer Control" dialogue box

Build Layers creates all of the layers defined in the sublay file selected earlier

Lines ▸ opens the LANDCADD lines options (Figure A1.13.)

Box builds a rectangle which can be a series of lines, a single polyline or a solid

Jumpline creates a break and places an arc where a line or polyline crosses another object; do **not** use this command on lateral irrigation lines

Breakline breaks a line or polyline and inserts a break-line symbol

Change Widths ▸ tools to modify existing polylines

Multiple changes the widths of a selection set of polylines

Figure A1.13

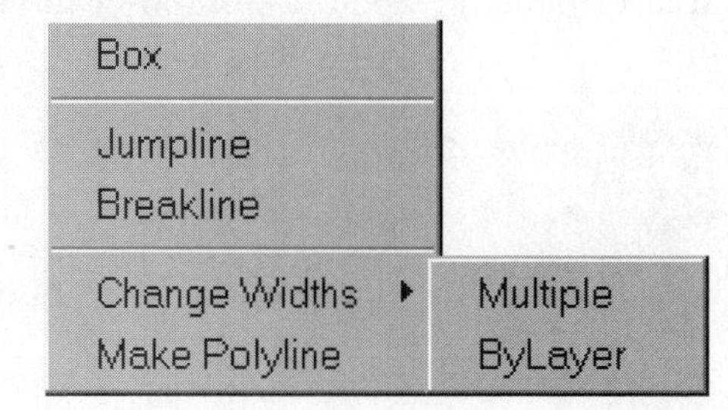

Bylayer changes the width of all polylines on the layer of a selected object

Make Polyline creates a single polyline from a series of continuous lines arcs and/or polylines

Macros ▸ opens the LANDCADD macros options (Figure A1.14).

Create allows you to create macros from keyboard entries; records keystrokes in AutoLISP files; no menu or icon selections can be entered in a macro

Save saves macros to a LISP file

Append allows additional macros to be added to a LISP file

Load loads an existing LISP file

Screen ▸ opens the LANDCADD screen options (Figure A1.15).

Figure A1.14

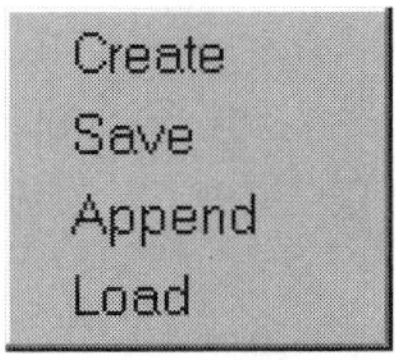

Create
Save
Append
Load

Figure A1.15

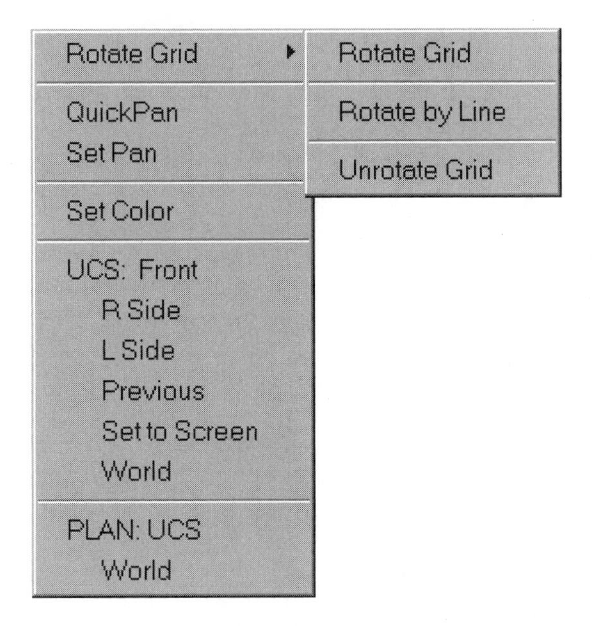

Rotate Grid ▸	tools to modify the display of grids on the screen
Rotate Grid	rotates grid and cursor to an angle you select
Rotate by Line	rotates grid and cursor to the angle of a line you select
Unrotate Grid	returns grid and cursor to **0°** rotation
Quick Pan	establishes temporary keyboard entries for panning around on the screen
Set Pan	sets scaling of pans used above
Set Color	sets drawing color for new objects
UCS: Front	rotates UCS 90° about the X axis, then resets plan view to the new UCS; view is elevation from bottom of World plan view
R Side	rotates UCS 90° about the X axis, moves origin point, then resets plan view to new UCS; view is elevation from right side of World plan view
L Side	rotates UCS 90° about the X axis, moves origin point, then resets plan view to new UCS; view is elevation from left side of World plan view
Previous	restores the previous UCS
Set to Screen	sets the current UCS to align with the view currently displayed
World	resets current UCS back to align with AutoCAD's default World coordinate system
Plan: UCS	sets current view to align with the current UCS
World	restores current UCS back to align with AutoCAD's default World coordinate system and resets the current view to align with the World coordinate system
Sheet Size ▸	opens the LANDCADD sheet size options (Figure A1.16).
Locate	locates a template rectangle on your drawing to represent the sheet edges at the paper size and plotting scale you select; does not reset limits
Move	moves template to new location

Figure A1.16

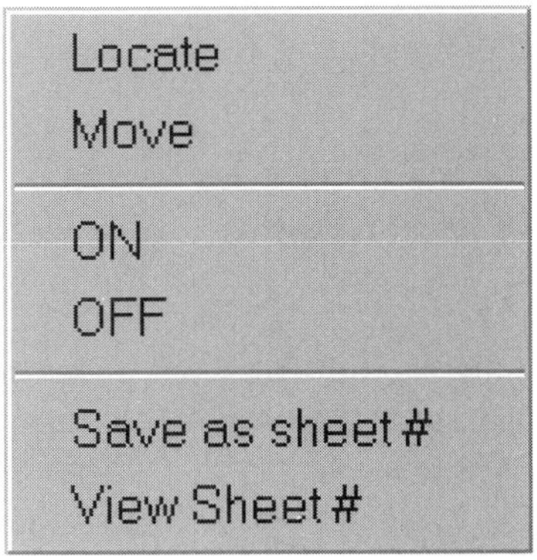

Locate
Move
ON
OFF
Save as sheet #
View Sheet #

ON turns on the template layer

OFF turns off the template layer

Save as sheet # creates a view that you define; saves the view under a name you enter

View Sheet # restores a view that you name at the command line

Text ▸ opens LANDCADD's text options (Figure A1.17).

Insert Text File inserts a selected text file at a point you specify, and at a size and style that you enter

Figure A1.17

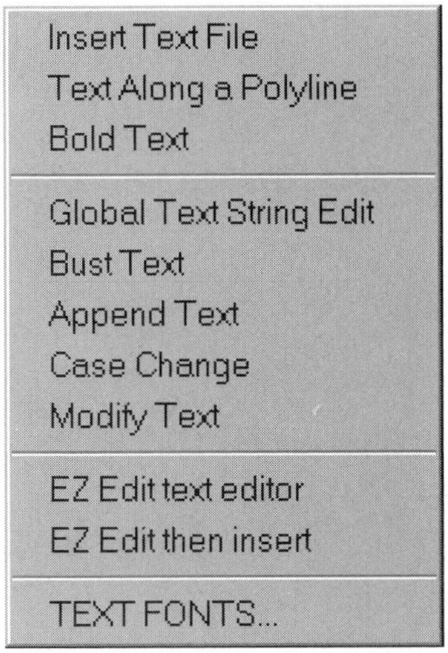

Insert Text File
Text Along a Polyline
Bold Text
Global Text String Edit
Bust Text
Append Text
Case Change
Modify Text
EZ Edit text editor
EZ Edit then insert
TEXT FONTS...

Button	Description
Text Along a Polyline	inserts text you enter at the command line along an existing polyline you choose
Bold Text	creates multiple copies of new or existing text to create a bold effect; number of copies is controlled by a value you enter
Global Text String Edit	allows case-sensitive editing of multiple lines of text; will replace all occurrences of an old string of text with a new string that you enter at the command line
Bust Text	breaks a line of text you select into two separate lines at the end of a case sensitive text string you enter at the command line; places the two lines of text on top of each other
Append Text	combines two lines of text into a single string on the screen
Case Change	changes line of text to all upper or all lower case letters
Modify Text	allows you to modify the height, width, obliquing angle, rotation or style of a selected line of text
EZ Edit	edits text using an existing DOS text editor used in an AutoCAD shell text editor
EZ Edit then insert	allows you to create text in the DOS text editor before inserting the text; you can specify the center, middle or right point of the text insertion and its style, height and rotation
TEXT FONTS...	displays an icon menu of LANDCADD text fonts; *build a new text style*
Unit Conversion	opens a dialogue box allowing conversions between architectural, decimal and metric units
Utilities ▸	opens LANDCADD's utilities options (Figure A1.18).
Configure Tablet	allows you to configure your digitizer tablet to establish menu areas; Windows versions may not support this option
Calibrate New	allows you to calibrate your digitizer to a drawing you want to trace; used after drawing is mounted on tablet

Figure A1.18

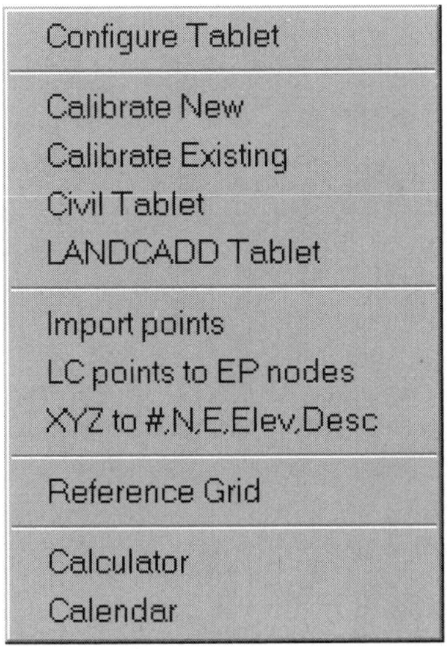

Configure Tablet
Calibrate New
Calibrate Existing
Civil Tablet
LANDCADD Tablet
Import points
LC points to EP nodes
XYZ to #,N,E,Elev,Desc
Reference Grid
Calculator
Calendar

Calibrate Existing allows you to relocate the drawing mounted on the digitizer; used after Calibrate New

Civil Tablet switches the command definitions on the digitizer to an Eagle Point civil engineering module that you select

LANDCADD Tablet switches the command definitions on the digitizer back to LAND-CAAD module commands

Import points allows a file of spot elevations or other surveying data points into LAND-CADD

LC points EP nodes converts LANDCADD point objects into Eagle Point node objects to which can be read by Eagle Point civil engineering modules

XYZ to #, N, E, Elev, Desc converts spot elevation files expressed in X,Y,Z format to N,E, Elev format; converts Quadrangle files to Surface Modeling files

Reference Grid places a pattern of major and minor grid lines on the drawing at an angle and base point you specify in a dialogue box; grid lines are placed on layers you select

Calculator displays an interactive calculator which can use distances, perimeters, areas, and angles from the drawing along with numbers you enter

Calendar displays a monthly calendar with a screen for entering notes for a particular day

Presentation pull-down menu

The selections listed under the **Presentations** pull-down menu are provided to allow you to create illustrative views and sequences to help you and your clients better visualize your projects. This menu also provides a direct link to Virtual Simulator (in Windows) or Virtual Illustrator (in DOS) to allow rendering and animated walk-throughs for your design projects. A stereo viewer also builds primitive 3-D wire-frame models which can be rotated and examined from different viewpoints (Figure A1.19).

Perspective: Create creates a perspective view of a drawing featuring 3-D objects; you set a lens focal length, an observation point and a target viewing point

Off converts a perspective view into an isometric view for editing

Isometric to Persp. changes an isometric view to a perspective view for realistic viewing

Elev. View creates a perspective elevation view of a drawing; requires parallel polylines at front and back of view

Make Slideshow creates a series of slides in a progression through the drawing; the progression can follow a radial path or an existing polyline; slides can use **hide** and **shade** options

Figure A1.19

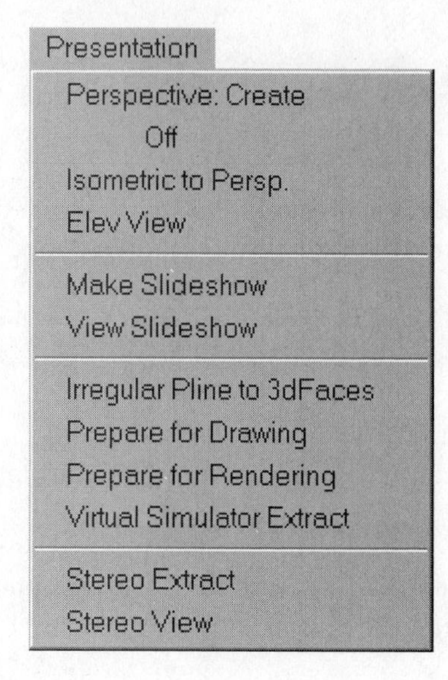

View Slideshow views a slideshow created above or in AutoCAD

Irregular Pline to 3dFaces converts closed polylines into 3-D faces which can be hatched, shaded or rendered

Prepare for Drawing returns drawings to original condition after "Prepare for Rendering" is run; surfaces are returned to 3-D objects. This function is used for Virtual Image, a DOS program not available in Windows versions of LANDCADD.

Prepare for Rendering converts 3-D objects to surfaces which can be rendered; allows you to create another file to save the results. This function is used for Virtual Image, a DOS program not available in Windows versions of AutoCAD.

Virtual Simulator Extract creates 3-D faces from 3-D objects at actual size, then opens Virtual Simulator program if installed *and configured correctly*

Stereo Extract creates a file which is used by the Stereo program which runs in a DOS window; will not run in Windows 95 and Windows NT platforms

Stereo View stereoscopic viewing program for AutoCAD 3-D objects; functions in a DOS shell; will not function in Windows 95 and Windows NT platforms

Estimating pull-down menu

The options shown in the **Estimating** pull-down menu are used to develop cost estimates for planting, irrigation and construction plans. You can create costs, categories and comments for items which are used in your drawings, then allow LANDCADD to tally the items and generate a report. The report can be formatted to function as a materials list for contractors, or as a detailed cost estimate for clients (Figure A1.20).

Configure sets up Estimating program with dialogue boxes repeating some functions in Report Definition and Select Blocks

Label w/attribute inserts a two-line visible attribute which you can label as desired

IncCALCS allows you to create attributes for non-block objects — can include measured areas, perimeters, volumes, etc.; attributes are invisible but are included in estimate counts

Figure A1.20

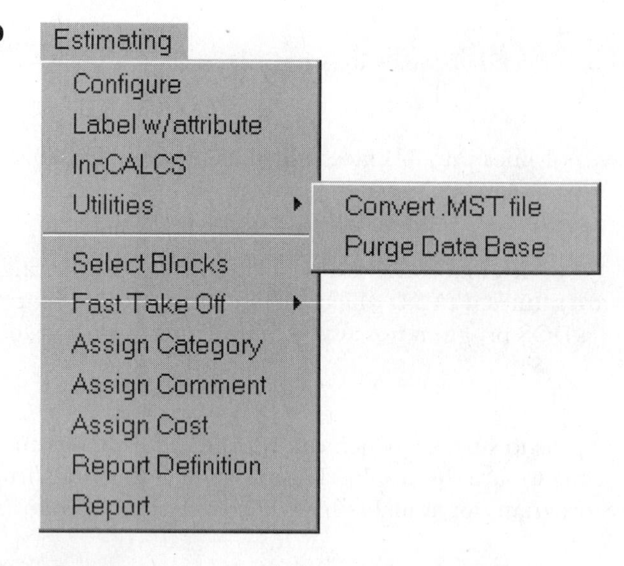

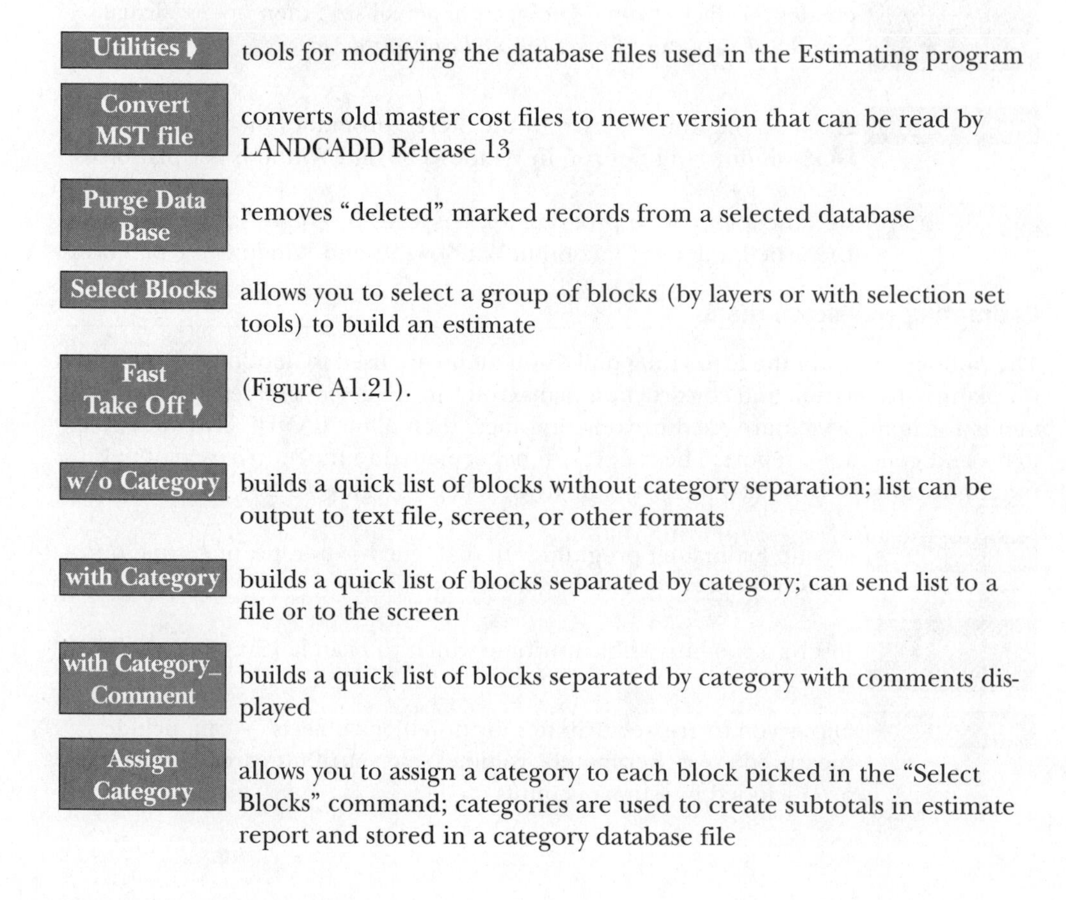

Utilities tools for modifying the database files used in the Estimating program

Convert MST file converts old master cost files to newer version that can be read by LANDCADD Release 13

Purge Data Base removes "deleted" marked records from a selected database

Select Blocks allows you to select a group of blocks (by layers or with selection set tools) to build an estimate

Fast Take Off (Figure A1.21).

w/o Category builds a quick list of blocks without category separation; list can be output to text file, screen, or other formats

with Category builds a quick list of blocks separated by category; can send list to a file or to the screen

with Category_ Comment builds a quick list of blocks separated by category with comments displayed

Assign Category allows you to assign a category to each block picked in the "Select Blocks" command; categories are used to create subtotals in estimate report and stored in a category database file

Figure A1.21

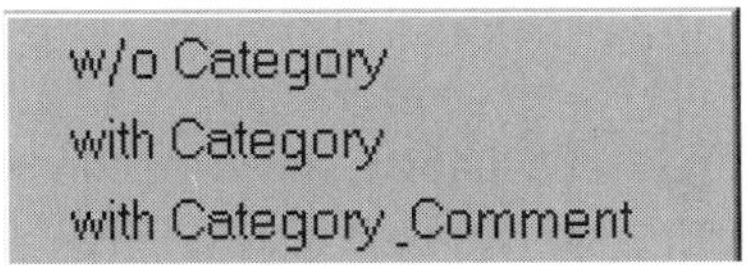

w/o Category
with Category
with Category_Comment

Assign Comment allows you to assign comments to any blocks picked in the "Select Blocks" command; comments are stored in a comments database file and may or may not be displayed in the final estimate report

Assign Cost allows you to assign a cost to each block picked in the "Select Blocks" command; costs are used in final estimate and stored in a price database file

Report Definition allows you to define the format of the report created; sets attributes used, categories, costs, cost modifiers, subtotal prices and final price of the blocks picked in the "Select Blocks" command

Report writes the report and offers several output file options, and can send the report directly to the printer (if configured)

Irrigation Design Module

The LANDCADD Release 13 Irrigation Design module equips the landscape design professional with tools to create accurate irrigation designs more quickly and easily than had previously been possible in LANDCADD. Irrigation head symbols can be inserted manually or automatically and, like plant symbols, have attributes assigned to them. The coverage arcs attached to the head symbols aid in alignment and quick detection of weakly covered areas. More detailed distribution analysis templates can also be used to evaluate system performance. Once sprinklers are organized into separate zones, the total flow for the area can be determined, along with the precipitation rate and station run time. The attributes assigned to the sprinkler head symbols assist in maintaining an accurate count of each type of sprinkler used, while the flow characteristics of each head are also used in the automatic pipe sizing routine.

The automatic pipe sizing routine is the most important time-saving feature in the Irrigation Design module. Once lateral lines are drawn to connect a group of heads into a zone, you can use the automatic sizing command to convert each pipe length into a block with attributes for size, material, length, and friction loss. Pipes are automatically labeled with a selected style, and the friction loss information can be used to determine the critical head (with the lowest available pressure) for each zone, and the total friction loss accumulated. In addition to head and pipe insertion, the Irrigation Design module can be used to lay out drip irrigation systems, place various types of irrigation equipment symbols, and assemble component symbols into irrigation detail drawings.

Heads pull-down menu

The **Heads** options are set up to allow you to insert and edit sprinkler head symbols, and to analyze and edit your head layout after it is established. You can also enter and edit data files for the sprinkler head manufacturers and models that you use most often in your drawings (Figure A1.22).

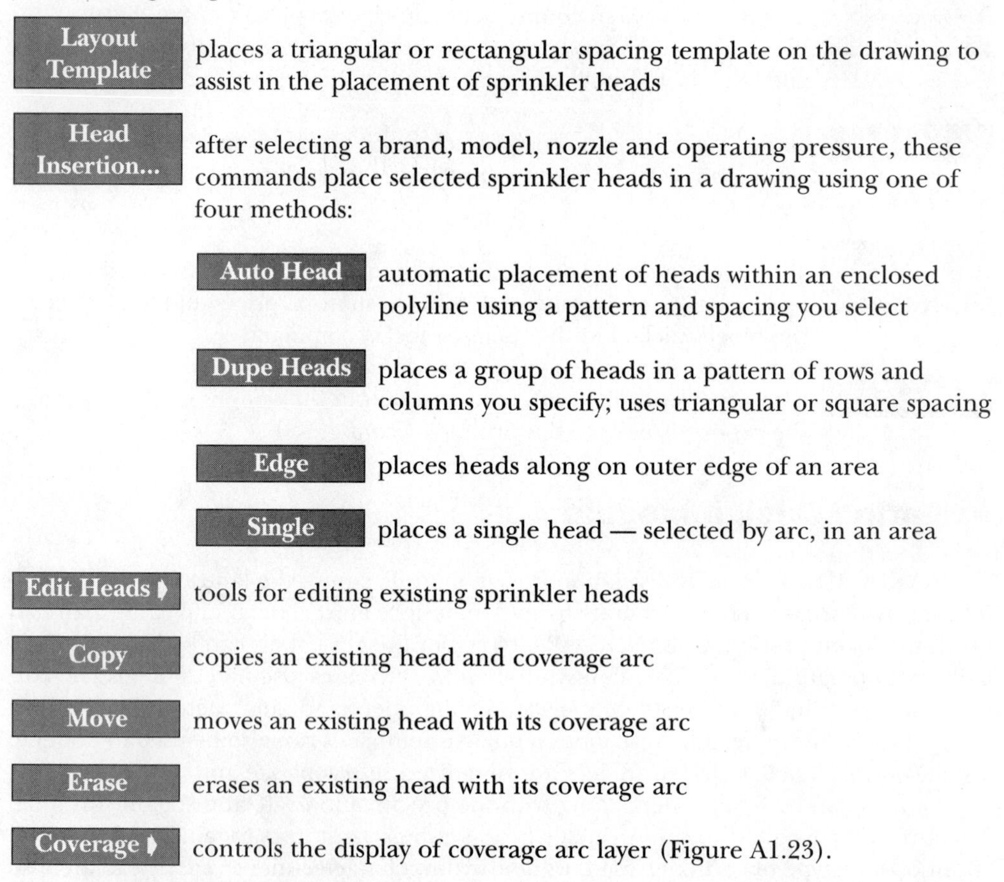

Layout Template places a triangular or rectangular spacing template on the drawing to assist in the placement of sprinkler heads

Head Insertion... after selecting a brand, model, nozzle and operating pressure, these commands place selected sprinkler heads in a drawing using one of four methods:

- **Auto Head** automatic placement of heads within an enclosed polyline using a pattern and spacing you select
- **Dupe Heads** places a group of heads in a pattern of rows and columns you specify; uses triangular or square spacing
- **Edge** places heads along on outer edge of an area
- **Single** places a single head — selected by arc, in an area

Edit Heads ▸ tools for editing existing sprinkler heads

Copy copies an existing head and coverage arc

Move moves an existing head with its coverage arc

Erase erases an existing head with its coverage arc

Coverage ▸ controls the display of coverage arc layer (Figure A1.23).

Figure A1.22

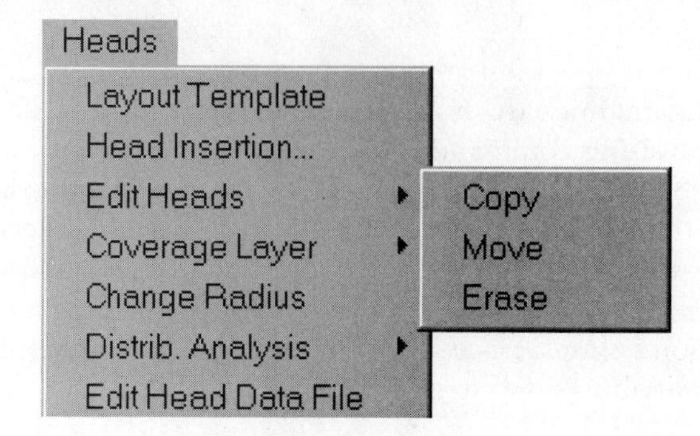

Figure A1.23 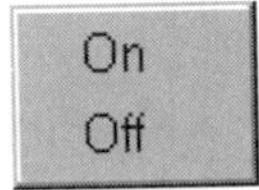

ON turns on layer containing coverage arcs

OFF turns off layer containing coverage arcs

Change Radius scales the radius of coverage arcs, but does not relocate heads or change head attributes

Distrib. Analysis ▸ (Figure A1.24).

Do Analysis places coverage templates for each sprinkler head to assist in analyzing sprinkler distribution uniformity; templates should be erased before another analysis is performed

Analysis Layer ▸

ON turns on layer containing distribution analysis templates

OFF turns off layer containing distribution analysis templates; does not remove old templates

Edit Head Data File allows you to edit head information and add new irrigation head information into existing data files; allows insertion of brand, model, pressure, arc and radius performance data

Zones pull-down menu

The **Zones** pull-down menu is set up to allow you to define and analyze irrigation zones or stations which contain the heads you have already inserted into your drawing. You can determine the total flow and gross precipitation rate of a station, and label the zone with a predefined tag with attributes to show station number, gallons per minute and valve size (Figure A1.25).

Define allows you to build a closed polyline around a group of heads that can be defined as a sprinkler zone

Figure A1.24 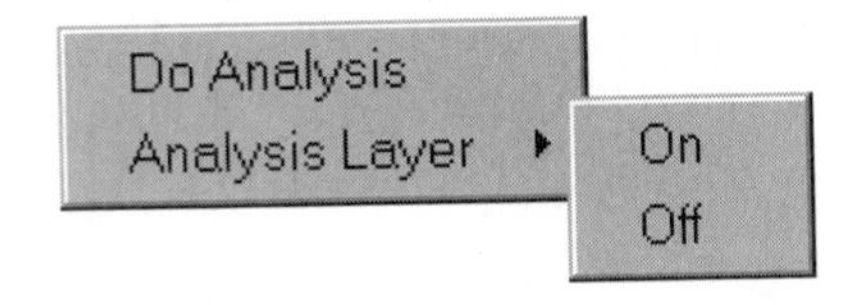

Figure A1.25

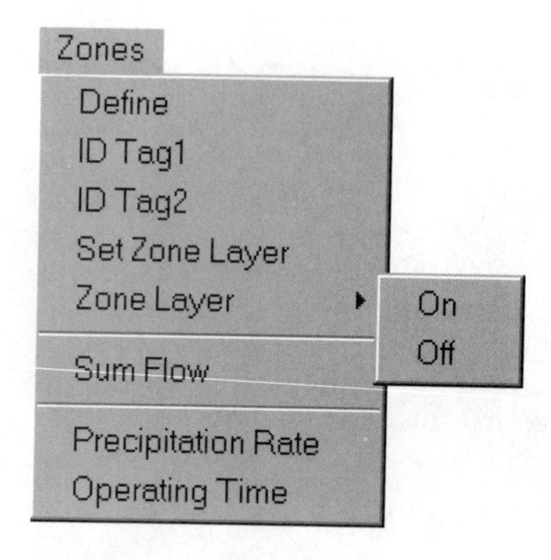

IDTag1	allows you to place a tag in a zone which has already been equipped with a control valve; zone tag information is taken from attributes attached to control valve
IDTag2	places a different style of ID Tag than above, but functions the sam way
Set Zone Layer	makes the **Zones** layer the current layer in the drawing; may not function in older versions of LANDCADD 13
Zone Layer ▸	controls display of irrigation zones layer
ON	turns on **Zones** layer; may not function in older versions of LANDCADD 13
OFF	turns off **Zones** layer; may not function in older versions of LANDCADD 13
Sum Flow	totals the flow from a selected group of sprinkler heads; does not require that a polyline be built around the zone before use
Precipitation Rate	calculates the average precipitation rate for a group of sprinkler heads; the area covered (including oversprayed areas) is included in this calculation
Operating Time	calculates watering time based upon the precipitation rate (entered or calculated as above), the number of watering days available and the number of inches of water required per week

Pipe pull-down menu

The **Pipe** pull-down menu contains the tools needed to lay out piping for the heads and zones laid out in the drawing. You can draw lateral lines, main lines and drip irrigation piping, then assign sizes to each pipe used. The pipe sizes can be labeled using several different styles, and can be edited. Once pipe has been sized, the pressure losses can be examined, as well as the lowest-pressure head in a zone (Figure A1.26).

Draw Mainline creates polyline main lines with a width you assign; no attributes are assigned as these polylines are built

Draw Lateral creates lateral pipe lines as lines; no attributes are assigned as these lateral lines are built — lines should pass through sprinkler symbols to be used in pipe sizing commands

Drip... allows the insertion of drip emitters (adjacent to existing plants or placed manually) and drip tubing; drip area can also be shown as a hatched area with a polyline border

Pipe Size ▸ creates pipe size labels and converts lateral pipe lines and polylines into blocks

U-Size converts existing pipe lines or polylines to blocks (or draws new) with attributes for size and type of pipe that you select; you can select from several different label styles and adjust label size and layer

Auto Size converts existing pipe lines (not polylines) to blocks as above, but sizes and labels automatically based on sprinkler flows and locations; cleans up overlapping pipe intersections and pipes overhanging sprinklers

Figure A1.26

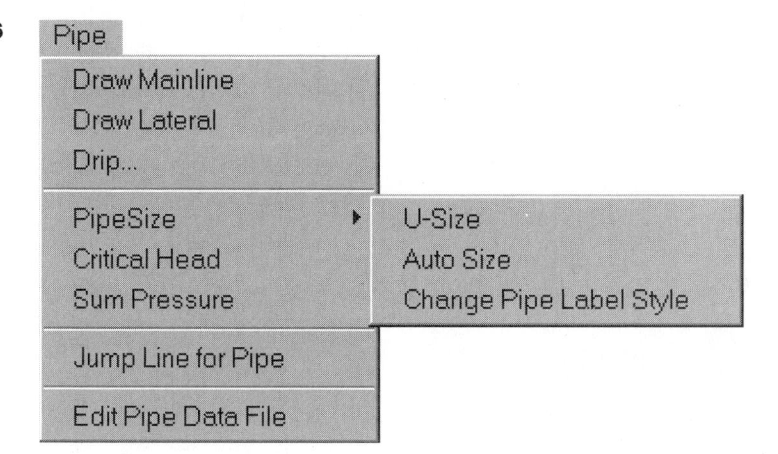

Change Pipe Label Style allows you to modify an existing pipe label by changing its style or orientation

Critical Head identifies the head on a labeled station which has the highest accumulated friction loss; valve and lateral piping must be built prior to use of this command

Sum Pressure totals the friction losses in a group of pipes you select

Jump Line for Pipe used to build a break in a pipe line, then supply an arc to indicate pipes which cross but do not connect; will work effectively on pipes which have been converted to blocks

Edit Pipe Data File allows you to modify, delete or create pipe data files

Symbols pull-down menu

This pull-down menu allows you to insert symbols for additional equipment required in an irrigation system. Selected section and isometric details are included to simplify building your own detail assemblies. The **Equipment Table** function allows you to generate a bill of materials when the job has been completed (Figure A1.27).

Section Details displays an icon menu of LANDCADD's irrigation section details; insert, scale and rotate symbols—can be used together to create more complex assemblies

Isometric Details... displays an icon menu of LANDCADD's irrigation isometric details; insert, scale and rotate symbols

Equipment... displays an icon menu of LANDCADD's irrigation equipment symbols; symbols are comparable to those automatically inserted in previous routines and comparable attributes can be assigned to them

Figure A1.27

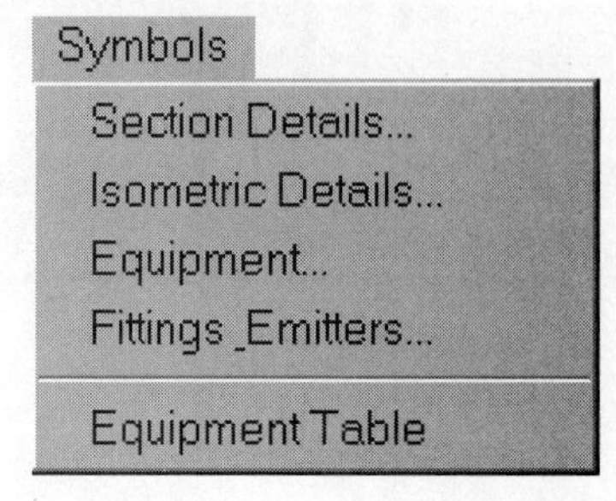

Fittings_ Emitters.. displays an icon menu of LANDCADD's drip emitters and a two irrigation fittings

Equipment Table builds an equipment table listing irrigation components (heads, valves piping, etc.) used in the drawing; size of the table can be controlled by setting text size; size of symbols is that used in the drawing itself

Site Planning Module

The Site Planning Module contains important routines and options for land planning and development. The tools here are usually best suited for larger scale and commercial projects. Property lines can be constructed using legal description or surveying data, using either a traverse or side shot mode. Existing and proposed spot elevations can be placed and labeled at the appropriate X,Y and Z coordinates. 3-D buildings and retaining walls can be constructed very efficiently using 3-D polylines or 3-D faces, or even selected from a library of LANDCADD symbols. You can also take advantage of a variety of parking layout features to insert parking stalls along curves or straight edges, specifying stall angle, width and length. The Site Planning module also has a large selection of 2-D and 3-D symbols for large site layout. You can choose from sports fields, recreational shelters, site furnishings and signs. You can even use a golf fairway routing routine to lay out a rudimentary golf course.

Some of the most powerful tools are available in the Topography section of the Site Planning module. You can draw (or trace) contours at specified intervals, and label contours in different ways. There are tools available to copy, edit and hatch contours, "heavy up" existing contours, or draw interpolated contours between two existing ones. You can even calculate the slope between two points across the face of a slope. Grid cell tools are useful for editing terrain maps with an overlain grid system. Objects can be shifted to the elevation of a grid cell or to the elevation of a specified face of a TIN (Triangulated Irregular Network). 3-D roadways can be generated from a cross-section profile, and placed at a specific elevation on the drawing. To examine the profile of a terrain model, you can place a polyline over the area, superimpose it onto the terrain map, then generate a plan view profile. The Site Planning module can work with Eagle Point's Surface Modeling module or other comparable program, which generates the TIN and grid networks used in the Topography section.

Layout pull-down menu

This pull-down menu allows you to lay out site property lines, utility lines and place retaining walls on a site plan (Figure A1.28).

Figure A1.28

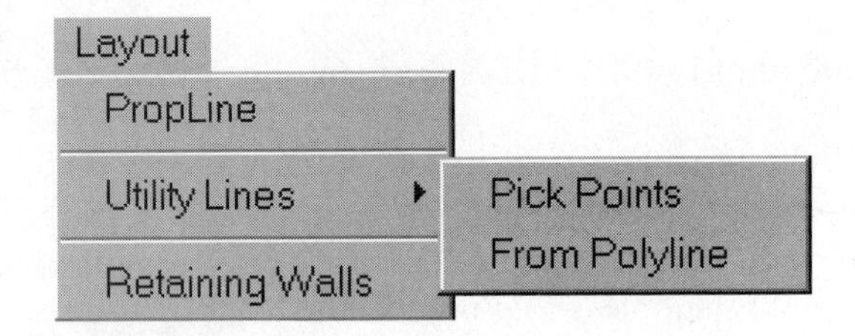

PropLine allows you to input property line information from surveying data, legal descriptions or civil engineering information; uses several different angle and distance measuring schemes to place property lines and label them

Utility Lines ▸ allows you to place utility lines broken with an abbreviation label

Pick Points draws utility lines by picking points on the drawing

From Polyline draws a property line to replace an existing polyline on the drawing

Retaining Walls places 3-D retaining walls on a drawing built of 3-D blocks; faces of blocks may have blank 3-D faces or patterns attached (this command ignores architectural units — multiply architectural values by 12)

Topography pull-down menu

The **Topography** pull-down menu allows you to create and manipulate contour lines, place spot elevations, and manipulate portions of a grid surface model map (Figure A1.29).

Set Elevation allows you to set the current elevation to that of a selected object on the drawing

Spot Elev ▸ allows you to insert spot elevation labels into your drawing

Existing places spot elevation at location and elevation you specify; displays number, elevation and name of the point as attributes

Figure A1.29

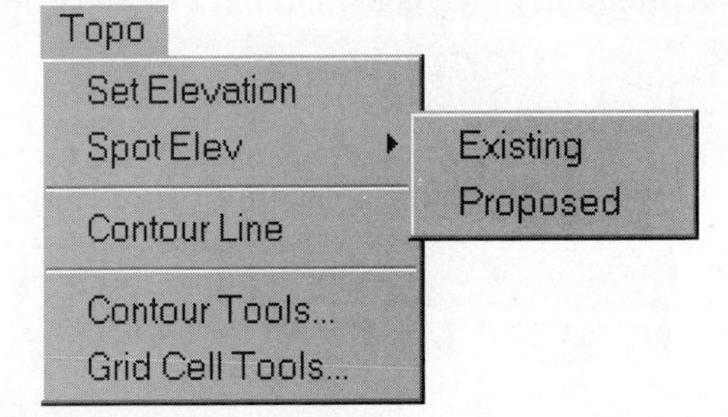

Proposed works as above; blocks are placed on different layer

Contour Line allows you to draw existing and proposed contours as polylines or as freehand sketch lines; contours will begin at an elevation you specify, and will be built at a selected interval as additional lines are drawn

Contour Tools... provides tools for copying, editing, labeling and trimming contour lines; extra vertices can be removed from polylines while others can be widened

Grid Cell Tools... provides tools for editing the 3-D faces created in terrain map; polylines can be superimposed over 3-D terrain, then used to create proximity or viewshed models.

Buildings pull-down menu

The **Buildings** pull-down menu contains tools to allow you to place building footprints on a site, build 3-D buildings with roofs, windows and doors, and to insert pre-drawn building symbols (Figure A1.30).

Footprint allows you to create a building footprint made of 3-D faces or a 3-D polyline; height and wall thickness can be adjusted

2D Door (recently added) allows you to insert single or double door swings into existing footprint; breaks existing wall without exploding

2D Window (recently added) allows you to insert double line for window openings; breaks existing wall without exploding

Figure A1.30

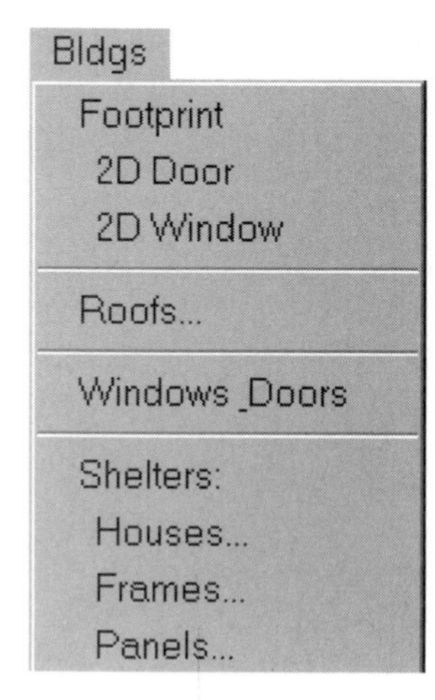

Roofs... displays an icon menu of LANDCADD's different options for building roofs; roofs are created from 3-D faces with defined peak and gutter heights or roof pitch

Windows & Doors displays an icon menu of LANDCADD's window and door symbols; can be inserted in plan or elevation view at elevation you select — symbols are pre-sized and not scaled at insertion

Shelters:

Houses... displays an icon menu of LANDCADD's 3-D house symbols

Frames... displays an icon menu of LANDCADD's 3-D gazebo and garden shelter symbols

Panels... displays an icon menu of LANDCADD's 2-D building panels; these elevation symbols can be placed on different planes and assembled into 3-D structures

Circulation pull-down menu

The **Circulation** pull-down menu relates to the placement of roads and parking lots. The **Parking** routines allow you to build parking stalls in linear, curved or large open areas. You can also access the portion of LANDCADD's extensive block library relating to placing people and vehicle symbols on a plan (Figure A1.31).

Roads places a 2-D roadway on the screen with adjustable width; road is built as center line is created

Parking ▸ tools for creating parking stalls along lines and curves, in rows or along islands

Fill Linear creates linear array of parking spaces with adjustable stall sizes and angles; allows inclusion of islands, stripes and handicap parking stalls and symbols

Curved creates parking stalls along an existing polyline arc; size and angles can be adjusted — functions as above

Figure A1.31

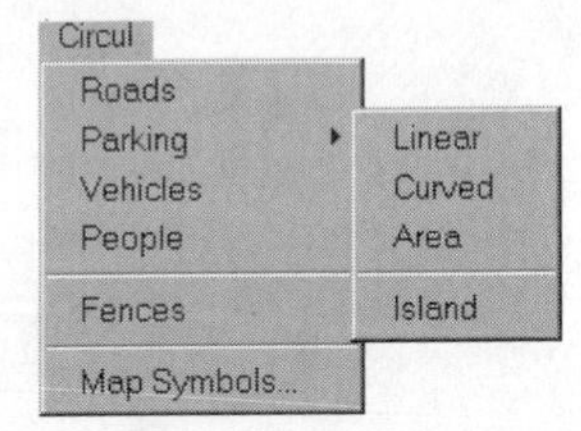

Area creates parking stalls in rows within a given area; size of stalls, number of rows or distance between rows can be used to vary number of stalls created

Island creates center or side islands with rounded ends for use in parking lot layout

Vehicles displays an icon menu of LANDCADD's plan view and 3-D vehicle symbols; includes cars, trucks, boats, bicycles, etc.

People displays an icon menu of LANDCADD's plan view and 3-D people symbols

Fences creates fence panels and posts from samples you select; the fence line can be entered or will follow an existing polyline (this command ignores architectural units — multiply architectural values by 12)

Map Symbols... displays an icon menu of LANDCADD's mapping symbols; dialogue box allows symbols to be scaled, layer assigned and attributes attached before insertion

Recreation pull-down menu

The **Recreation** pull-down menu contains tools and blocks relating to recreational site improvements. There are routines for building schematic golf course layouts along with inserting different types of sports fields and swimming pools (Figure A1.32).

Camping... displays an icon menu of LANDCADD's camping symbols, including tents, barbeques and RVs; symbols can have elevation specified and attributes attached before insertion

Golf ▸ tools for laying out rudimentary golf fairways, greens and tees, and a driving range

Routing Fairway after specifying the diameter of the green and tee, along with distance(s) to landing area(s) (specified in yards); allows you to lay out fairways and place a label for each hole

Figure A1.32

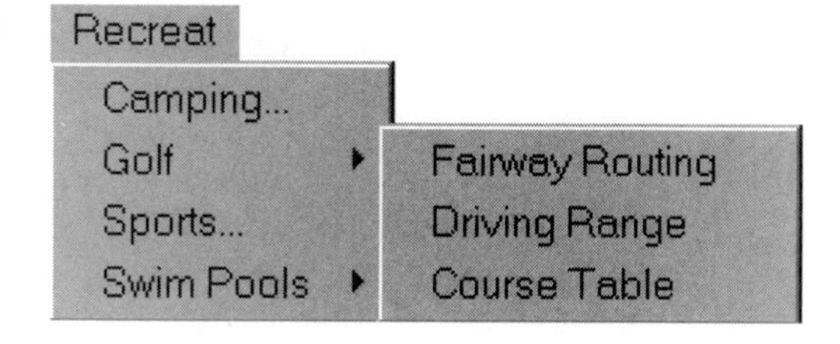

Driving Range inserts a driving range symbol at actual size

Course Table creates a small table which lists each hole with its distances and the total number of holes and yards

Sports... displays an icon menu of LANDCADD's sport field symbols; symbols can have attributes attached, be inserted at a specified elevation, and are inserted at actual size

Swim Pools ▸ places swimming pools of various sizes in the drawing (Figure A1.33).

Free Form... displays an icon menu of LANDCADD's free form pool symbols; each pool symbol can be scaled in X, Y and Z axes, assigned an attribute and layer before insertion

25 Meters x 5 Lanes inserts a 25 meter, 5 lane competition swimming pool at a specified elevation and layer at actual size; an attribute can be attached before insertion

Figure A1.33

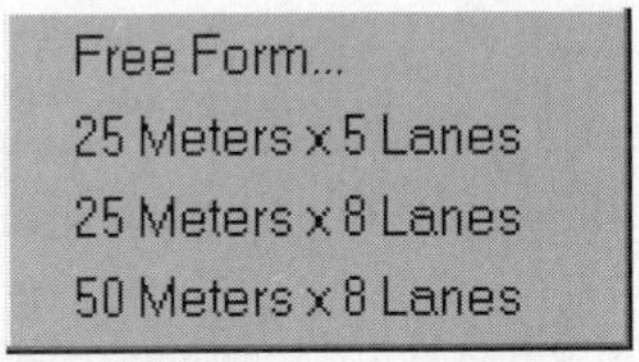

Free Form...
25 Meters x 5 Lanes
25 Meters x 8 Lanes
50 Meters x 8 Lanes

25 Meters x x 8 Lanes inserts a 25 meter, 8 lane competition swimming pool at a specified elevation and layer at actual size; an attribute can be attached before insertion

50 Meters x 8 Lanes inserts a 50 meter, 8 lane competition swimming pool at a specified elevation and layer at actual size; an attribute can be attached before insertion

Hardscape pull-down menu

The **Hardscape** pull-down menu allows you to place various types of site improvements on a plan. This area provides access to more of LANDCADD's extensive symbol libraries (Figure A1.34).

Site Furn... displays an icon menu of LANDCADD's plan view and 3-D site furnishing symbols; includes chairs, tables, grates, bollards, etc.

Figure A1.34

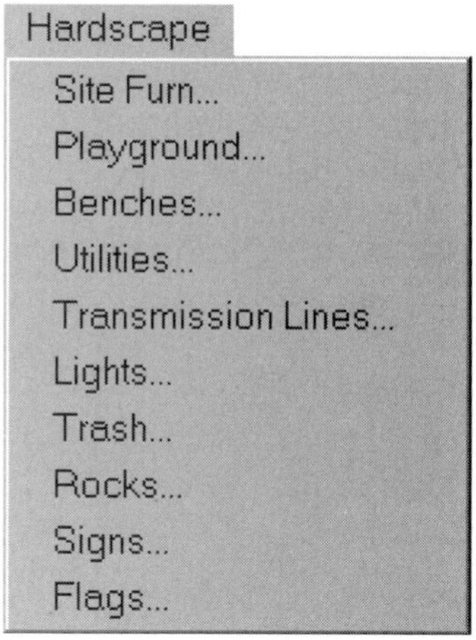

Hardscape
- Site Furn...
- Playground...
- Benches...
- Utilities...
- Transmission Lines...
- Lights...
- Trash...
- Rocks...
- Signs...
- Flags...

Playground displays an icon menu of LANDCADD's 3-D playground symbols; includes climbers, swings, bars, slides, etc.

Benches... displays an icon menu of LANDCADD's 3-D bench symbols

Utilities... displays an icon menu of LANDCADD's plan view and 3-D electrical utility symbols

Transmission Lines... allows you to insert transmission towers at a specified height and distance apart; symbols can be placed along a line or will follow an existing polyline (this command ignores architectural units — multiply architectural values by 12)

Lights... displays an icon menu of LANDCADD's plan view and 3-D light fixtures

Trash... displays an icon menu of LANDCADD's plan view and 3-D trash receptacles

Rocks... displays an icon menu of LANDCADD's section view and 3-D rocks

Signs... displays an icon menu of LANDCADD's 3-D street signs, private and public signage

Flags... displays an icon menu of LANDCADD's flag symbols which are inserted on a flagpole; specify direction flag points when inserting

Site Analysis Module

The Site Analysis module of LANDCADD is an evaluation tool used to examine terrain models and grid systems. The different grid cells can be categorized based on their slope percentage, directional aspect and elevation. In this way, a site can be systematically studied and appraised, with data reported visually and numerically. Using the mapping tools, you can establish search criteria for slope, aspect and elevation, then obtain a graphic representation of which grids meet the criteria. The Site Analysis module also includes a link to ArcCAD, a GIS (Geographical Information System) program which also works within AutoCAD.

Using the tools in the Site Analysis module, you can inspect the proximity projection from a point or polyline placed across the terrain. You can also examine the effects of obstructions on visibility from any viewpoint on the terrain. Shadows can be placed on the model based on latitude, date and time of day. Finally, various mapping symbols are available for annotation and clarification.

ArcCAD Link pull-down menu

This pull-down menu provides tools to link with ESRI's ArcCAD program. This allows you to coordinate landscape design work with existing GIS (Geographic Information System) data created with ArcCAD (Figure A1.35).

Figure A1.35

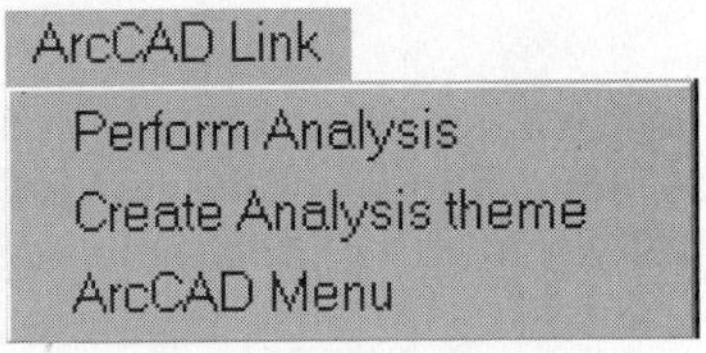

ArcCAD Link
- Perform Analysis
- Create Analysis theme
- ArcCAD Menu

Perform Analysis analyzes an existing terrain model by reading grid cell characteristics for slope, aspect, elevation and surface area; attributes are assigned to a block in each grid cell, and are read in the "Create Analysis theme" command

Create Analysis Theme reads attributes created in "Perform Analysis" command, then stores them in an ArcCAD database file; database "theme" file is stored under a name you specify

ArcCAD menu opens the ArcCAD menu if this program is installed; ArcCAD is a GIS (Geographic Information Systems) software program which works within AutoCAD and has extensive mapping capabilities

Surface pull-down menu

The **Surface** pull-down menu allows you to use LANDCADD's surface modeling tools to create and analyze terrain maps from spot elevation files. The analytical tools allow you to color-code and evaluate various physical characteristics of the site and generate reports on your findings (Figure A1.36).

Figure A1.36

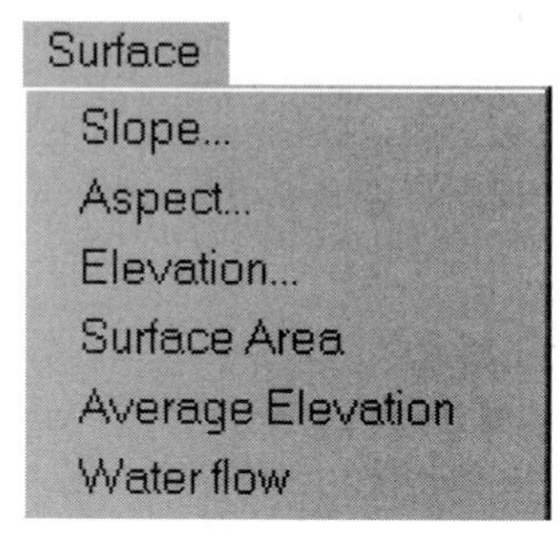

Slope examines existing terrain model, then assigns each grid cell to a slope percent classification whose upper and lower limits you define; results can be displayed as colored grids, hatched grids, and/or a draped series of 3-D faces — a legend of slope classes can also be generated

Aspect examines existing terrain model, then assigns each grid cell to a directional aspect classification; as with slopes, results can be displayed as colored grids, hatched grids, and/or a draped series of 3-D faces — a legend of aspect classes can be generated

Elevation examines existing terrain model, then assigns each grid cell to an elevation class whose upper and lower limits you define; results can be displayed as colored grids, hatched grids, and/or a draped series of 3-D faces — a legend of elevation classes can also be generated

Surface measures and reports the surface area of a selected group of 3-D faces

Average Elevation calculates and reports the average elevation from a series of selected 3-D faces

Water flow analyzes displayed TIN (Triangulated Irregular Network), then traces runoff path of water falling at a point or throughout the terrain model — places a 3-D polyline to mark water path

Proximity pull-down menu

The **Proximity** pull-down menu lets you run the **Proximity** routine to analyze the proximity of the grids in the drawing to a 3-D point (Figure A1.37).

Figure A1.37

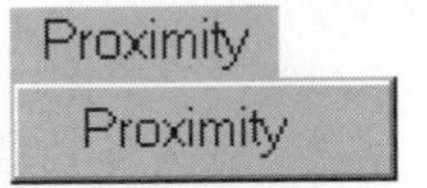

Proximity marks the measured 3-D or plan distance from a point or polyline on the drawing; the region within the measured proximity may be hatched with a scale and angle you specify

Shadow pull-down menu

The **Shadow** pull-down menu lets you observe the shadows cast by different objects at different times of the day or year (Figure A1.38).

Figure A1.38

Shade generates shadow patterns from 3-D objects based on their height and spread, along with the defined latitude, date and time — shadow pattern is created as a hatch whose layer, direction and density you specify

Views pull-down menu

This pull-down menu allows you to evaluate the views of a site from a specific spot or from a specific path. In addition, you can analyze the views after a defined obstruction is placed on the site (Figure A1.39).

Figure A1.39

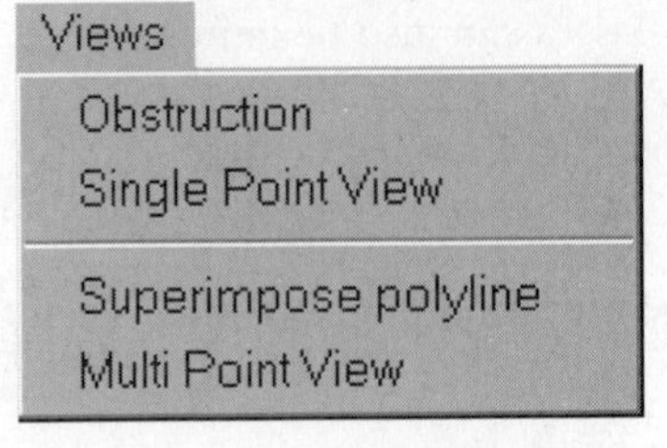

Obstruction builds vertical faces around the edges of selected grids to create a view obstruction; height and layer of obstruction must be specified

Single Point View examines terrain map from a specific viewpoint and elevation to determine the "viewshed" based on obstructions, surrounding terrain and desired viewing distance; "viewshed" can be displayed as a hatched area or a surrounding polyline

Superimpose Polyline uses an existing line or polyline to create a 3-D polyline that is draped over the terrain map, conforming to its various elevations; original line or polyline can be deleted

Multi Point View creates extended "viewshed" based on the vertices of a polyline; grid cells are classified according to how frequently they are viewed along the viewing path and a legend can be created to display these results

Maps pull-down menu

This pull-down menu allows you to display the results of the various analyses performed in other routines. Searches and sorts allow you to graphically search for specific portions of the site with characteristics you choose (Figure A1.40).

Figure A1.40

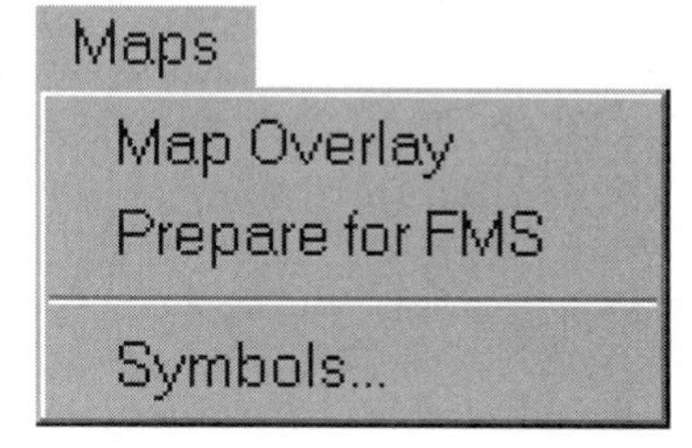

Map Overlay displays selected grids by sorting, adding or subtracting areas from results of slope, elevation, aspect, and view commands; can be used to search and display for multiple sort criteria — only functions with enclosed polylines created in prior commands

Prepare for FMS labels grid cells with attributes from Site Analysis commands for use in ArcCAD; ArcCAD must be installed for this command to function

Symbols... displays an icon menu of LANDCADD's site mapping symbols which are inserted in the drawing at a size, elevation and layer you specify; attributes can be assigned to each symbol

appendix

Plotting From Windows 95 and Windows NT

Plotting from Windows 95 and Windows NT presents special problems for AutoCAD users. To begin with, the plotter devices used by AutoCAD may not be directly supported by Windows 95 and NT. This means that you may not be able to find a matching output device when trying to configure the plotter in the Windows 95 (or NT) Control Panel. Even if you do find the plotter listed as a possible Windows system printer (or have a Windows 95 driver disk from the manufacturer), the printer driver required is not necessarily optimized for AutoCAD performance. In addition, these system printer drivers generally do not allow access to the various plotter options supported by AutoCAD. It is far better (and easier) to use the plotter drivers supplied by AutoCAD or by the manufacturer of the plotter device. These ADI (AutoCAD Device Interface) drivers are extensively tested with AutoCAD, and allow access to all of the plotter options supported by AutoCAD.

The problem arises when one of these drivers attempts to send an output file (custom tailored to the plotter device) through the Windows 95 interface. Simply put, AutoCAD tries to send information directly to the output device without consulting the Windows 95 (or NT) operating system. Since Windows 95 and Windows NT control access to all hardware devices (including modems, printers, plotters, hard disks, etc.) the plotter information is intercepted by the operating system and cannot get through. To behave properly, the ADI drivers would need to send their raw data directly to the Windows 95 (or NT) *spooler* function. The spooler function accepts raw information from a program, then interprets it and sends it out to the output device. Unfortunately, ADI drivers do not have the ability to write to the Windows 95 (or NT) spooler. As a result, you must develop a system for writing the output file to a disk, then send it through the Windows 95 (or NT) interface out to the plotter device. You can use AutoCAD's AUTOSPOOL feature to write these plot files to the hard disk, then use a batch file to send the file to the plotter device.

The following steps will guide you through the process of building the system for correctly writing these plot files, then sending them to the plotter.

1. **Build a PLOTS directory.**

   You will need a new, dedicated directory specifically for the batch file and the plot files which will be written to the hard disk. This is important because it keeps the plot files away from your program files and simplifies troubleshooting later.

2. **Configure AUTOSPOOL.**

   You can set up AutoCAD to create spool files automatically. By setting the default plot file name to AUTOSPOOL, then specifying a few other parameters, you can establish the placement and file name extensions for the spool files that AutoCAD writes.

   While running AutoCAD, follow these steps:

   `Command:`
   Type **CONFIG**.

```
Configuration menu
 0. Exit to drawing editor
 1. Show current configuration
 2. Allow detailed configuration
 3. Configure video display
 4. Configure digitizer
 5. Configure plotter
 6. Configure system console
 7. Configure operating parameters

Enter selection <0>:
```

   Type **7** to configure the operating parameters.

```
Configure operating parameters

 0. Exit to configuration menu
 1. Alarm on error
 2. Initial drawing setup
 3. Default plot file name
 4. Plot spooler directory
 5. Placement of temporary files
 6. Network node name
 7. Automatic-save feature
 8. Speller dialect
 9. Full-time CRC validation
10. Automatic Audit after DXFIN, or DXBIN
11. Login name
```

```
12. File locking
13. Authorization
14. Use long file names

Enter selection <0>:
```
Type **3** to change the Default plot file name.

```
Enter default plot file name (for plot to file)
or . for none <.>:
```
Type **AUTOSPOOL** for the default plot file name. You are then returned to the "Configure operating parameters" menu.

```
Enter selection <0>:
```
Type **4** to set the Plot spooler directory.

```
When plotting to a file, AutoCAD writes the output to the
plot spooler directory if the special name AUTOSPOOL is
given as the plot file name. The plot spooler directory
is ignored for normal plotting.

Enter plot spooler directory name
<.>:
```
Enter the directory **PLOTS**. Be sure to specify the full path to this directory (i.e., C:\PLOTS) You will then be returned to the "Configure operating parameters" menu.

```
Enter selection <0>:
```
Type **6** to set the Network node name.

On network and multi-user systems, a unique "network node name" should be specified for each user. This name is addedto plot spooler output files to ensure uniqueness.

```
Enter network node name (1 to 3 characters) <ac$>:
```
At each work station running AutoCAD, enter a unique three-character string to label the output from each station. The plot files created will use this three-character string as an extension to the file name. This allows you to identify the source of any plot files which might be written across a network. The string can be initials, computer numbers, etc. As you finish the string, you will be returned to the "Configure operating parameters" menu again.

```
Enter selection <0>: 0
```
Type **0**.

If you answer N to the following question, all configuration changes you have just made will be discarded.
Keep configuration changes? <Y>
Type **Y** to keep your configuration changes.

3. **Configure AutoCAD**.

Once AUTOSPOOL is configured, you can set up AutoCAD to perform a specific command after it creates the plot file. The command is stored in an AutoCAD system variable called **acadplcmd**. This can be set easily in AutoCAD 13.

Pick **Options** from the **AutoCAD and EP** menu. Select **Preferences...**, then pick the tab marked **Misc** (Figure A2.1.)

In the **Plot Spooling** text box, type **C:\PLOTS\ASPOOL.BAT %S %C** . This command line is not case-sensitive, so use either upper case or lower case letters as you wish. Be sure to include the correct path to the plot spooler directory if yours is different from this.

The **%S** identifies the name of the plot file just created, including the file name and unique extension specified in the "Network node name" selection.

The **%C** identifies the name of the plotter to which the plot file is to be sent. When you have multiple plotters available, this name is used to direct the file to the appropriate destination. The plotter name is created when you configure the plotter in the AutoCAD **CONFIGURE** command.

Figure A2.1

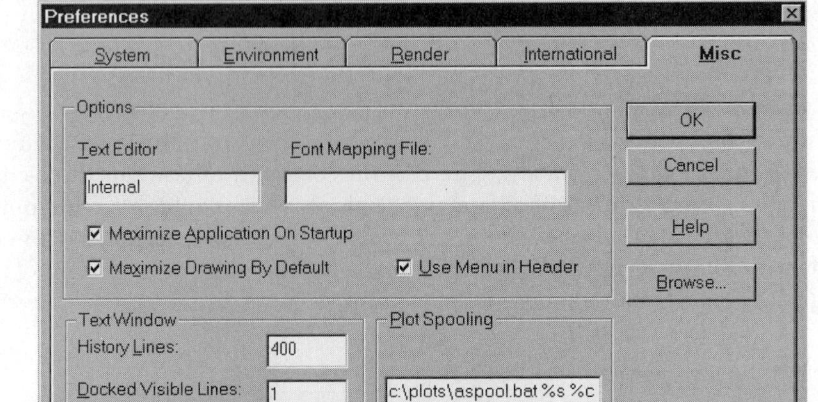

4. Check the names of your AutoCAD plotter devices, network node names and network plotter devices.

A vexing problem when building a plotting system relates to the precise names of the principal devices in the network and in AutoCAD. In order for the batch file (which we will create later) to work properly, all of the device names must agree with and adhere to the standards set out in batch file language. This means that all names should be **eight letters or less without punctuation or spaces**.

For example, if you had a Hewlett Packard laser printer connected to AutoCAD, its name could be LASERJET, but not LASERJET IV, or even LASERJET_IV. This is because the batch file language does not acknowledge long file names or spaces within file names. Even though the printer name is not a file, the batch file uses this name (identified by **%C**) as a value in its execution. Check all of your AutoCAD plotter device definitions (in the Configuration menu) to be sure that each name conforms to the eight-character limit. If the names do not conform to the eight character limit, change them so that they do. Note that you can change a plotter name without changing any other definition parameters.

You should also examine the names of each of the network work station computers. Each computer in a network must have a unique name to function properly. Be sure that each name is less than eight characters, without spaces or punctuation. While the name USER1 is acceptable, USER 1 is not. Since the name of any computer connected to a plotter is used in the batch file, be sure to check each plot server name (and all network plotters and printers) to make sure that each one adheres to the eight-character limit.

Finally, check the name of each of the Windows system printer devices to make sure that their names also adhere to the eight-character naming standard. As before, you can rename a printer without changing any other definition parameters. By adjusting all of the device names to the eight-character limit, you assure that the names will be passed correctly through the batch file and that the plots will be routed to their proper destinations.

5. Build the ASPOOL.BAT batch file.

The most difficult portion of this process to predict is the batch file itself. The batch file will depend on the plotter devices which are used in your system. The more plotter choices you have, the more complicated your batch file.

To build a batch file, use Notepad or the DOS Edit function. Be sure to place the batch file in the PLOTS directory.

The following illustration pertains to a network with the following characteristics:

1. A Hewlett Packard laser printer is connected to a server called SERVER2. Its AutoCAD plotter name (and Windows 95 network printer name) is LASERJET.
2. A Hewlett Packard inkjet plotter is connected to a work station called USER3. Its AutoCAD plotter name (and Windows 95 network printer name) is DESIGNJT.
3. A Summagraphics network plotter named as SUMMAGRP. Its AutoCAD plotter name is also SUMMAGRP.
4. A Hewlett Packard inkjet color printer is connected to this work station through the printer port LPT1. This printer is not connected to the network. Its AutoCAD plotter name is DESKJET.

The batch file below is numbered for the comments which appear below (do not use line numbers in your batch file):

```
    ASPOOL.BAT
 1.  @ECHO OFF
 2.  IF %2 == LASERJET GOTO LASERJET
 3.  IF %2 == DESIGNJT GOTO DESIGNJT
 4.  IF %2 == SUMMAGRP GOTO SUMMAGRP
 5.  IF %2 == DESKJET GOTO DESKJET
 6.  GOTO ERROR
 7.  :LASERJET
 8.  ECHO SPOOLING TEMPORARY FILE TO LASER JET PRINTER...
 9.  COPY/B %1 \\SERVER2\LASERJET
10.  GOTO END
11.  :DESIGNJT
12.  ECHO SPOOLING TEMPORARY FILE TO DESIGN JET
     PLOTTER...
13.  COPY/B %1 \\USER2\DESIGNJT
14.  GOTO END
15.  :SUMMAGRP
16.  ECHO SPOOLING TEMPORARY FILE TO SUMMAGRAPHICS PLOTTER
17.  COPY/B %1 \\SUMMAGRP
18.  GOTO END
19.  :DESKJET
20.  ECHO SPOOLING TEMPORARY FILE TO HP DESKJET PRINTER
21.  COPY/B %1 LPT1
22.  GOTO END
23.  :END
24.  ECHO DELETING TEMPORARY FILE...
```

```
25.   PAUSE
26.   DEL %1
27.   EXIT
28.   :ERROR
29.   ECHO ERROR SPOOLING TEMPORARY FILE
```

This lengthy batch file is really a simple pattern repeated for each output device, with a beginning and ending section to take care of housekeeping duties.

Line 1 turns off the visible display of the command lines.

Lines 2 through 5 direct the output of the batch file based on the **%2** variable — which is derived from the **%C** defined in the AutoCAD **ACADPLTCMD** system variable. Depending on the value, the batch file jumps to one of the modules named below.

Line 6 directs the batch file to jump to the ending section (lines 28 and 29) if the **%2** variable does not match any of the names listed.

Lines 7 through 10 are the actions required if the name of %2 matches LASERJET. The line "SPOOLING TEMPORARY FILE TO LASER JET PRINTER" is displayed first. Then the plot file — named by the variable **%1** — is copied (the **/B** tells the system that the file will be **binary**) to the network address **SERVER2\LASERJET**. This printer name follows the Universal Naming Convention (UNC) for network devices. Both Windows 95 and Windows NT support this system, as do Novell and other network system vendors. This address indicates that the device is named LASERJET and that it is connected to the network computer SERVER2. The ending line of this module sends the program to the **END** portion.

Lines 11 through 14 repeat these same instructions for the DESIGNJT plotter. The output is sent to a different network device, but the structure of the module is identical.

Lines 15 through 18 repeat the instructions for the SUMMAGRP plotter, but its network address does not show a server connection. This is because the device is a network plotter connected directly to the network. Its built-in network card allows it to have a direct connection without the need for a server,

Lines 19 through 22 give the same instructions to the DESKJET printer, but its address is the local printer port LPT1. Since this is not a network printer, it is not accessible to other users.

Lines 23 through 27 contain the END section. This section sends the message that the temporary plot file will be deleted, then pauses for the user to press any key. This step assures that you will be able to read the message in the DOS window that opens. Without this PAUSE line, the information scrolls off the screen before you can read it. This line can be omitted after the batch file has been debugged. After you press any key, the temporary plot file is deleted, and the routine is closed.

Lines 28 and 29 provide an error message if the **%2** variable does not match any printer/plotter named in the top section. This will occur if you do not name the plotters or printers correctly, or if you plot to any device not included in the definitions of the batch file.

6. **Test your results.**

Test your batch file with all of the output devices. Open a simple drawing, then use the "Device and Default Selection" dialogue box to select one of the output devices. Be sure that the names listed here match the names you used in the batch file. Plot the drawing to each output device to make sure it works properly. Once the batch file is completed and all of the devices work correctly, you can copy the batch file (with modifications for each computer) to all of the other work stations on the network (Figure A2.2).

When you plot, be sure that the toggle **Plot to a file** is active (Figure A 2.3.)

This option activates the AUTOSPOOL feature that creates the temporary spool file and drives the ASPOOL batch file.

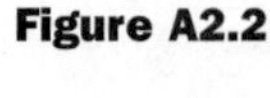
Figure A2.2

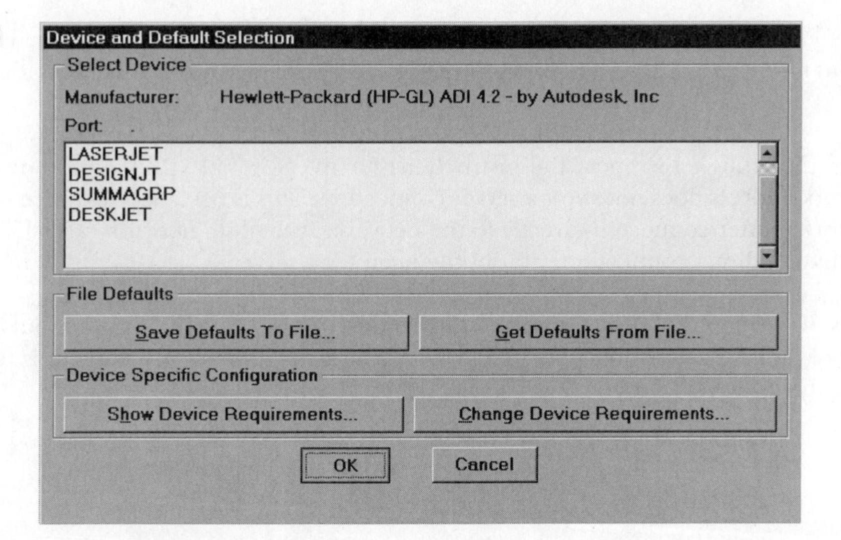

Figure A2.3

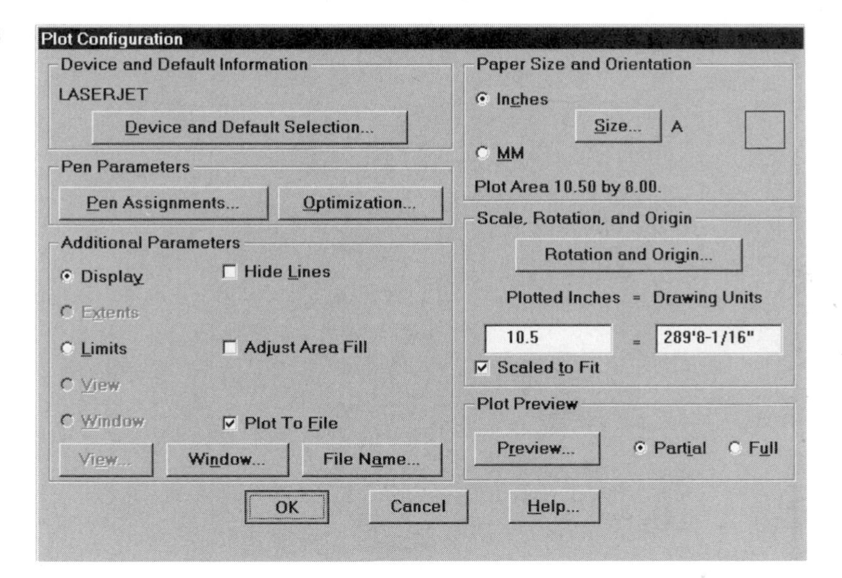
Plot Configuration
Device and Default Information
LASERJET
Device and Default Selection...
Pen Parameters
Pen Assignments...
Optimization...
Additional Parameters
Display
Extents
Limits
View
Window
Hide Lines
Adjust Area Fill
Plot To File
View...
Window...
File Name...
Paper Size and Orientation
Inches
MM
Size...
A
Plot Area 10.50 by 8.00.
Scale, Rotation, and Origin
Rotation and Origin...
Plotted Inches = Drawing Units
10.5 = 289'8-1/16"
Scaled to Fit
Plot Preview
Preview...
Partial
Full
OK
Cancel
Help...

appendix

Resources and References

AutoCAD 13 Instructor
James Leach
Irwin Graphics Series, Times Mirror Publishing, 1996

The Illustrated AutoCAD Quick Reference
Ralph Grabowski
Delmar Publishers, 1996

AutoCAD for Architects
Alan Jefferis and Michael Jones
Delmar Publishers, 1994

AutoCAD13 Instant Reference
George Omura and Paul W. Richardson
Sybex Publishing, 1995

Cadence Magazine
"AutoCAD Tips and Tricks"
October 1996

Inside AutoCAD Release 13 for Windows and Windows NT
F. Soen, D. Pitzer, H. Fulmer, J. Boyce, M. Peterson, R. Gesner, J. Beck, K. Coleman, A. Morris, T. Boersma, J. Fitzgerald
New Riders Publishing, 1995

AutoCAD Performance Tuning Toolkit
Michael Todd Peterson
New Riders Publishing, 1996

Managing AutoCAD in the Design Firm
Karen A. Vagts
Addison-Wesley Publishing Company, 1996

Cadence Magazine
"Windows 95 or Windows NT?"
February 1997

Maximizing AutoCAD R13
Rusty Gesner, Mark Middlebrook, Tony Tanzillo
Autodesk Press, 1996

How Software Works
Ron White
Ziff Davis Press, 1993

How Networks Work
Frank J. Derfler, Jr. and Les Freed
Ziff Davis Press, 1993

PC Magazine
"Your Next Operating System"
September 1995

Running Microsoft Windows NT Workstation Version 4.0
Craig Stinson and Carl Siechert
Microsoft Press, 1996

CAD Layer Guidelines: Recommended Designations for Architecture, Engineering and Facility Management Computer-Aided Design
American Institute of Architects Press, Washington D.C. 1990

Absolute Beginner's Guide to Networking
Mark Gibbs
Sams Publishing, 1995

Autodesk University Sourcebook 1995
"Maximum Network and Hardware Performance"
Robert Green
Miller Freeman Inc., 1995

AutoCAD Tech Journal—Summer 1996
"Plotting in Windows 95 or Windows NT"
Dale Fugier
Miller Freeman Inc., 1996

appendix

AutoCAD Release 14

During the late spring of 1997, Autodesk introduced the newest version of AutoCAD, Release 14. This long-anticipated release is worthy of plenty of attention. If you have not yet tried it, you should. There are numerous improvements that make the program run faster than ever before, and others that make it work more efficiently and easily.

Here are some of the most important improvements that will be of interest to LAND-CADD users:

1. **Demand Loading**—this feature loads the 3D portions of AutoCAD code only when 3D objects are used. This is one of the reasons that Release 14 runs so much faster than Release 13. In addition, Xref files can also be demand-loaded so that frozen layers and objects outside the current viewing window are not loaded unless they are made visible. Although you do not see this feature directly, it can result in a dramatic increase in speed.
2. **AutoSnap and Osnap markers**—this feature indicates object snap points with a colored marker (and even a tool-tip if you choose) as your cursor moves over objects on the screen. This can greatly speed your ability to snap to specific points on objects as you draw in AutoCAD (Figure A4.1).
3. **Real-time Pan and Zoom**—these are now easily accessible options of the Pan and Zoom commands. Not only can you see the results of your pans and zooms while you perform them, you can also switch between them with a cursor menu (Figures A4.2, A4.3, and A4.4).
4. **Layer and Linetype Properties dialogue box**—these redesigned dialogue boxes now behave more like Windows 95 and Windows NT 4.0 file boxes. Layers can be renamed, sorted and even deleted (finally!) using Windows 95-like commands such as CTRL- and SHIFT- button combinations. In addition, layers can be sorted by clicking on the column headings above (Figure A4.5).
5. **Image Editing**—Release 14 now provides useful tools for inserting and editing raster images within an AutoCAD drawing. Raster images (such as those created within paint or drawing programs, or clip-art images supplied by others)

Figure A4.1

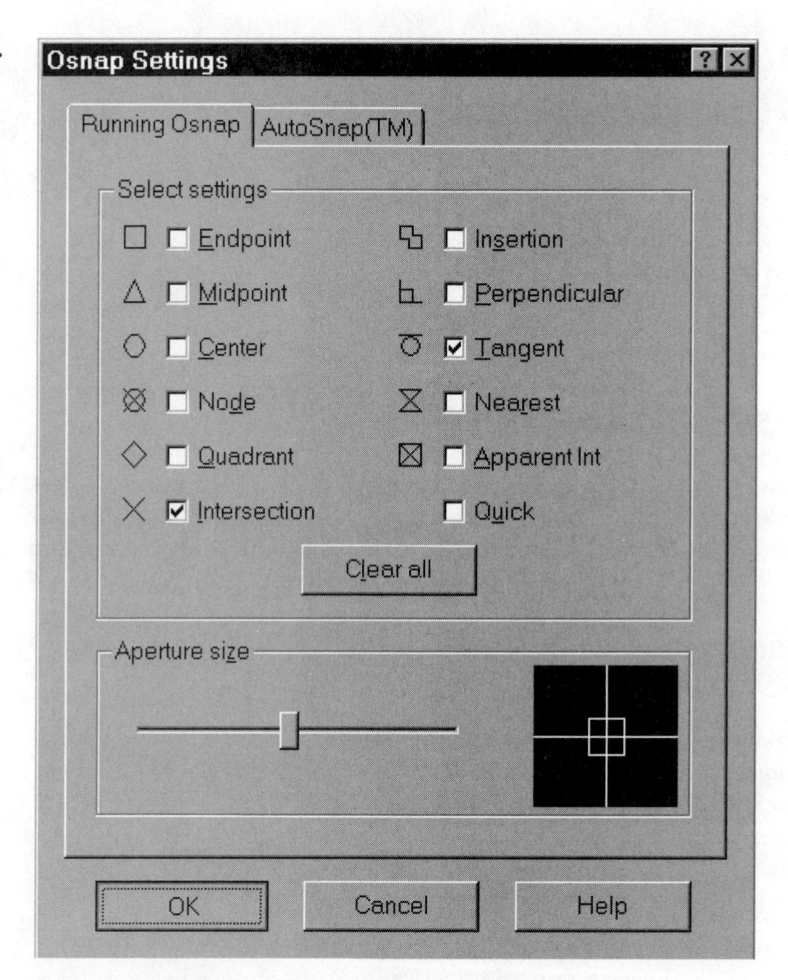

Figure A4.2

Figure A4.3

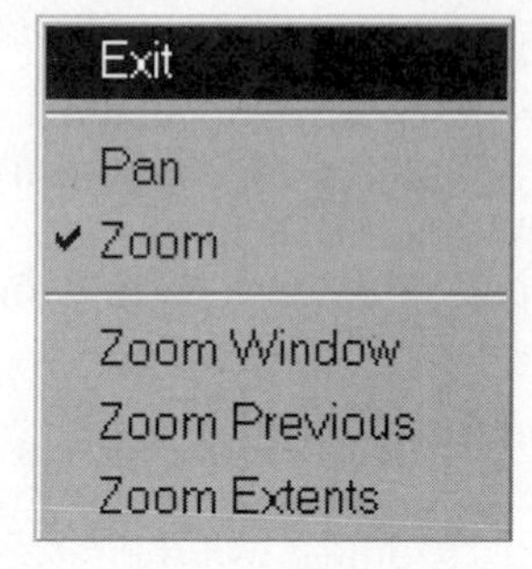

Figure A4.4

can be easily inserted into your AutoCAD drawings. Once there, they can then scaled, rotated, copied and can be used as cutting edges for trimming or extending other AutoCAD objects. You can also crop images, control their brightness and color, and control the display of a surrounding frame. These tools can be very helpful when you wish to include a photo, logo, clip-art or scanned image in an AutoCAD drawing (Figure A4.6).

6. **Xref Drawing Editor**—this dialogue box simplifies and clarifies the process of attaching and detaching externally referenced drawings. The improved dialogue box makes it much easier to understand where Xref drawings are stored and allows much greater control over how each Xref is loaded. If you choose, you can use the "demand-loading" (accessible in the Preferences dialogue box) feature to control how much of an Xref drawing is loaded into memory. Frozen layers and objects outside the current view can be excluded from memory (Figure A4.7).
7. **Light Weight Polylines and Hatches**—both polylines and hatches are now stored as single objects, which helps reduce drawing size, reduce memory usage as well as speed loading and regenerating the drawing.
8. **Solid Fill Hatch**—you can now use a solid color for a hatch pattern. Solid fill hatches have minimal impact on drawing sizes (Figure A4.8).
9. **Plot Preview**—the **Preview...** option in the Plot command proves a Microsoft-standard print preview screen. You can use the real-time zoom and pan within the screen to inspect specific areas in the plot. The **Preview** command yields the same screen, using the current settings in the **Plot** command.

Figure A4.5

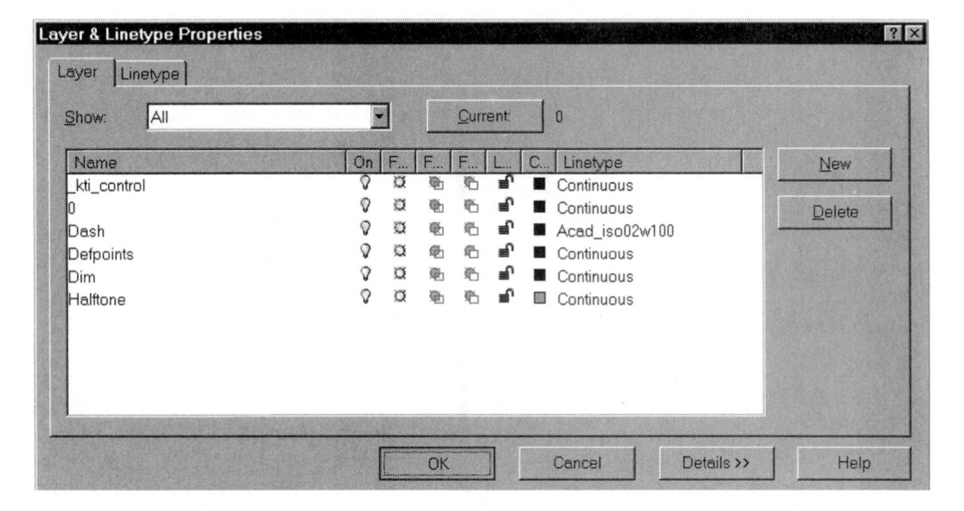

Figure A4.7

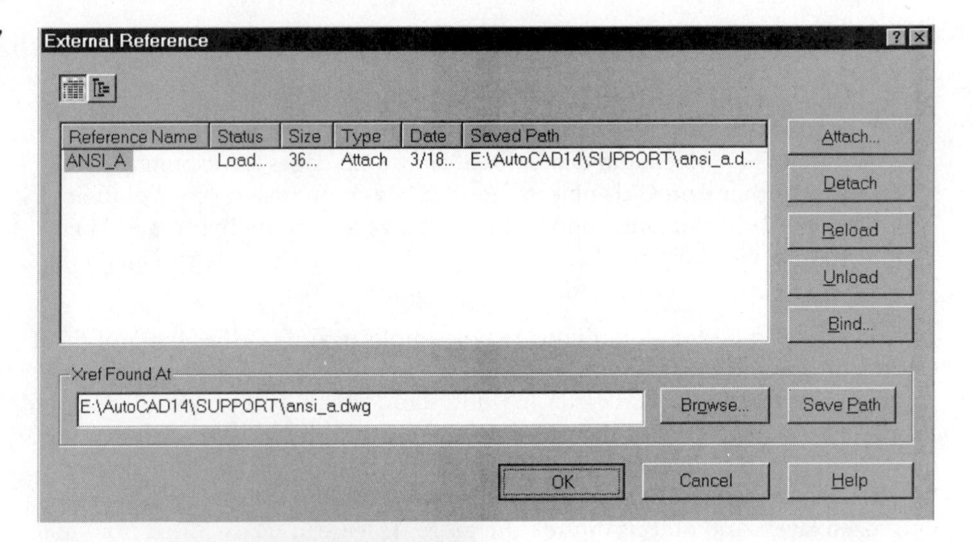

Figure A4.8

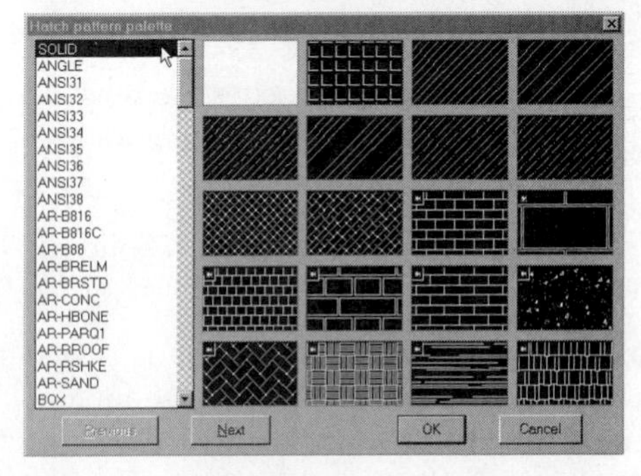

If you need to change the plot settings, you still must use the **Plot** command (Figure A4.9).

Figure A4.9

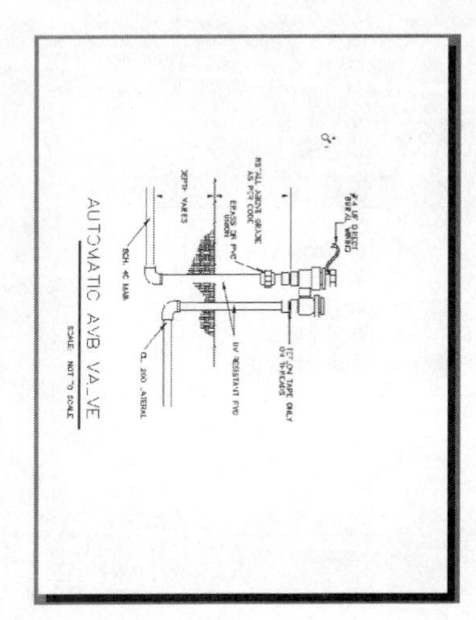

10. **Command Line Editing**—this long-awaited improvement allows you to use the Up Arrow and Down Arrow to scroll through previous entries on the command line (similar to the old DOSKEY command). You can also use the clipboard to paste entries onto the command line. These features can be very helpful to those who make use of AutoLISP commands and those who reload commands frequently during the debugging process.

11. **Object Properties Editing**—two improved tools make editing object properties easier than ever. First, the Object Properties toolbar is now interactive, and responds to any objects selected on the screen. Once objects are selected, you can change the layer, color, or linetype of the objects in the toolbar, and they are instantly updated. Second, the Make Object's Layer Current tool finds the layer of the selected object and makes it the current layer. This is another example of a tool that simply speeds the drawing process (Figures A4.10 and A4.11) .

Figure A4.10 **Figure A4.11**

12. **Match Properties**—this convenient tool allows you to select a group of objects, then change their properties to match that of a selected target object. You can use this tool to change regular object properties such as linetype, color, or layer, but you can also use it to change text style and dimension style as well as hatch pattern, linetype scale and thickness. This tool is another real time saver (Figure A4.12).

Figure A4.12

13. **Plotting**—AutoCAD 14 now prints directly to any output device defined in Windows 95 or Windows NT 4.0. This eliminates the cumbersome batch file preparation required for AutoCAD 13 to plot correctly in these environments. AutoCAD 14 can also be configured to plot directly to any network output device from within the Preferences dialogue box.

AutoCAD 14 and LANDCADD

As of this writing, the current version of LANDCADD operates identically whether used with AutoCAD Release 13 or 14. AutoCAD Release 14 simply provides greater speed and convenience. According to promotional literature and trade show information, Eagle Point has some new upgrades and revisions planned for late 1997 and early 1998. These include reorganized modules and the ability to have multiple modules open simultaneously. In addition, Eagle Point plans to offer all of its programs in a platform-independent mode. This means that you could run LANDCADD from AutoCAD (Release 13 or

14), or other CAD programs such as MicroStation, Visio, or TurboCAD. Up to this point, all Eagle Point products have been primarily tied to AutoCAD, so this represents a new direction for the company. For further developments on the subject, investigate Eagle Point's Web site at **www.eaglepoint.com**.

index

B

C

D

E

F

G

H

I

J

L

M

N

O

P

Q

R

S

T

U

V

LI[illegible] ESS

[illegible] y

Educational Software/Data

You the customer, and Autodesk Press incur certain benefits, rights, and obligations to each other when you open this package and use the software/data it contains. **Be sure you read the license agreement carefully, since by using the software/data you indicate you have read, understood, and accepted the terms of this agreement.**

Your rights:
1. You enjoy a non-exclusive license to use the enclosed software/data on a single microcomputer that is not part of a network or multi-machine system in consideration for payment of the required license fee, (which may be included in the purchase price of an accompanying print component), or receipt of this software/data, and your acceptance of the terms and conditions of this agreement.
2. You own the media on which the software/data is recorded, but you acknowledge that you do not own the software/data recorded on them. You also acknowledge that the software/data is furnished as is, and contains copyrighted and/or proprietary and confidential information of Autodesk Press or its licensers.
3. If you do not accept the terms of this license agreement you may return the media within 30 days. However, you may not use the software during this period.

There are limitations on your rights:
1. You may **not** copy or print the software/data for any reason whatsoever, except to install it on a hard drive on a single microcomputer and to make one archival copy, unless copying or printing is expressly permitted in writing or statements recorded on the diskette(s).
2. You may **not** revise, translate, convert, disassemble or otherwise reverse engineer the software data except that you may add to or rearrange any data recorded on the media as part of the normal use of the software/data.
3. You may **not** sell, license, lease, rent, loan, or otherwise distribute or network the software/data except that you may give the software/data to a student or and instructor for use at school or, temporarily at home.

Should you fail to abide by the Copyright Law of the United States as it applies to this software/data your license to use it will become invalid. You agree to erase or otherwise destroy the software/data immediately after receiving note of Autodesk Press's termination of this agreement for violation of its provisions.

Autodesk Press gives you a **limited warranty** covering the enclosed software/data. The **limited warranty** can be found in this package and/or the instructor's manual that accompanies it. This license is the entire agreement between you and Autodesk Press interpreted and enforced under New York law.

Limited Warranty
Autodesk Press warrants to the original licensee/purchaser of this copy of microcomputer software/data and the media on which it is recorded that the media will be free from defects in material and workmanship for ninety (90) days from the date of original purchase. All implied warranties are limited in duration to this ninety (90) day period. THEREAFTER, ANY IMPLIED WARRANTIES, INCLUDING IMPLIED WARRANTIES OF MERCHANTABILITY AND FITNESS FOR A PARTICULAR PURPOSE ARE EXCLUDED. THIS WARRANTY IS IN LIEU OF ALL OTHER WARRANTIES, WHETHER ORAL OR WRITTEN, EXPRESSED OR IMPLIED. If you believe the media is defective, please return it during the ninety day period to the address shown below. A defective diskette will be replaced without charge provided that it has not been subjected to misuse or damage.

This warranty does not extend to the software or information recorded on the media. The software and information are provided "AS IS." Any statements made about the utility of the software or information are not to be considered as express or implied warranties. Autodesk Press will not be liable for incidental or consequential damages of any kind incurred by you, the consumer, or any other user.

Some states do not allow the exclusion or limitation of incidental or consequential damages, or limitations on the duration of implied warranties, so the above limitation or exclusion may not apply to you. This warranty gives you specific legal rights, and you may also have other rights which vary from state to state. Address all correspondence to:
Address all correspondence to:

Autodesk Press
ATTN: CADD Marketing Manager
3 Columbia Circle
P. O. Box 15015
Albany, NY 12212-5015

Minimum Hardware System Requirements

* **Note:** LANDCADD software is not required, and any system capable of running AutoCAD R14 Windows will work.

Operating System: Windows 3.x or later (Win 95 or NT4 recommended)
CPU Type : Pentium 90 or compatible processor,
Memory: Required: 24 MB for Windows 95, 32 MB for NT
Recommended: 64 MB
Graphics: Anything supporting AutoCAD
CD-ROM Speed: 2X
Hard Drive Space: 50 MB of free hard-disk space